TECHNIQUES OF CHEMISTRY

ARNOLD WEISSBERGER, *Editor*

VOLUME IV

ELUCIDATION OF ORGANIC STRUCTURES BY PHYSICAL AND CHEMICAL METHODS

Second Edition

TECHNIQUES OF CHEMISTRY

ARNOLD WEISSBERGER, *Editor*

VOLUME I
PHYSICAL METHODS OF CHEMISTRY, in Five Parts
(INCORPORATING FOURTH COMPLETELY REVISED AND AUGMENTED
EDITION OF PHYSICAL METHODS OF ORGANIC CHEMISTRY)
Edited by Arnold Weissberger and Bryant W. Rossiter

VOLUME II
ORGANIC SOLVENTS, Third Edition
John A. Riddick and William S. Bunger

VOLUME III
PHOTOCHROMISM
Edited by Glenn H. Brown

VOLUME IV
ELUCIDATION OF ORGANIC STRUCTURES BY PHYSICAL AND CHEMICAL METHODS, Second Edition, in Three Parts
Edited by K. W. Bentley and G. W. Kirby

TECHNIQUES OF CHEMISTRY

VOLUME IV

ELUCIDATION OF ORGANIC STRUCTURES BY PHYSICAL AND CHEMICAL METHODS

Second Edition

Edited by

K. W. BENTLEY

Reckitt & Colman
Hull, England

AND

G. W. KIRBY

Department of Chemistry
Loughborough University of Technology
Loughborough, England

PART II

WILEY-INTERSCIENCE

A DIVISION OF JOHN WILEY & SONS, INC.

New York · London · Sydney · Toronto

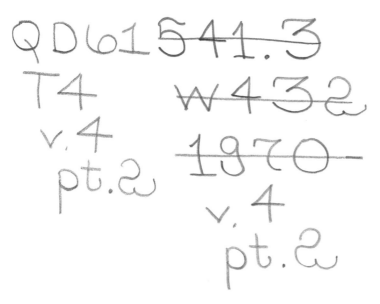
Library of Congress Cataloging in Publication Data:

Main entry under title:

Elucidation of organic structures by physical and chemical methods.

(Techniques of chemistry, v. 4)
1963 ed. published under title: Elucidation of structures by physical and chemical methods.
Includes bibliographies.
1. Chemistry, Physical organic—Addresses, essays, lectures. I. Bentley, Kenneth Walter, ed. II. Kirby, Gordon William, 1934–   ed.

QD61.T4 vol. 4 [QD476] 540′.28s [547′.1′22] 72–273

ISBN 0-471-92896-8

Printed in the United States of America.

10 9 8 7 6 5 4 3 2 1

AUTHORS OF PART II

G. O. ASPINALL
   York University, Toronto, Canada
K. W. BENTLEY
   Reckitt & Colman, Hull, England
J. P. CANDLIN
   Imperial Chemical Industries Ltd., Runcorn, England
ANGELO FONTANA
   Institute of Organic Chemistry, University of Padua, Padua, Italy
R. A. C. RENNIE
   Imperial Chemical Industries Ltd., Runcorn, England
ERNESTO SCOFFONE
   Institute of Organic Chemistry, University of Padua, Padua, Italy
Z. VALENTA
   University of New Brunswick, Fredericton, Canada
B. C. L. WEEDON
   Queen Mary College, London, England

INTRODUCTION TO THE SERIES

Techniques of Chemistry is the successor to the Technique of Organic Chemistry Series and its companion—Technique of Inorganic Chemistry. Because many of the methods are employed in all branches of chemical science, the division into techniques for organic and inorganic chemistry has become increasingly artificial. Accordingly, the new series reflects the wider application of techniques, and the component volumes for the most part provide complete treatments of the methods covered. Volumes in which limited areas of application are discussed can easily be recognized by their titles.

Like its predecessors, the series is devoted to a comprehensive presentation of the respective techniques. The authors give the theoretical background for an understanding of the various methods and operations and describe the techniques and tools, their modifications, their merits and limitations, and their handling. It is hoped that the series will contribute to a better understanding and a more rational and effective application of the respective techniques.

Authors and editors hope that readers will find the volumes in this series useful and will communicate to them any criticisms and suggestions for improvements.

Research Laboratories   ARNOLD WEISSBERGER
Eastman Kodak Company
Rochester, New York

PREFACE

The elucidation of the structure of unknown compounds and the synthesis of compounds of defined structures rank high among the aims of organic chemistry. The synthesis of complex structures resembles the composing of a jigsaw puzzle whose parts are formed according to certain rules as the composing proceeds. The establishment of the rules and techniques for the formation of bonds and for the transformation of groups and their application constitutes a major part of organic chemistry. The design of successful and elegant syntheses is the crowning accomplishment, but it seems very difficult to formulate general directions for this work. Apprenticeship with the masters—personal or through study of the original literature—remains the initiation.

Similar considerations apply to the elucidation of the structure of unknown compounds, but here the respective compound is at hand as the substrate of the work, rather than being a concept and aim. Some guidance is received from the study of the intact molecule, particularly by physical methods, and elucidation of structure lends itself, therefore, more readily to comprehensive presentation than multistep synthesis. Moreover, the degradative methods used in the elucidation of structure lack documentation in the reviewing literature. The present volumes are designed to fill this gap and to give guidance to the chemist who tries to unravel the structure of unknown organic synthetic or natural products.

In preparing this second edition of a book originally published more than eight years ago almost the whole work has been rewritten by a new team of contributors, thus ensuring properly structured up-to-date surveys of fields that have changed markedly in recent years. The emphasis is on results and interpretation rather than physical principles and manipulative techniques, wherever these are more fully dealt with in other volumes of this series, though no details essential to an understanding of the main text have been omitted. The subjects of some chapters in the first edition are no longer included, as they have received full and detailed treatment elsewhere (e.g., molecular rearrangements and biogenetic theory), and the subjects of other chapters have in some cases been combined to achieve a more rational and uniform treatment. New topics covered in this edition include X-ray crystallography, certain aspects of nuclear magnetic resonance spectroscopy, and stereoselective synthesis.

This rewriting has resulted in a much larger work, which is more conveniently divided into three volumes than two as in the first edition.

Hull, England   K. W. BENTLEY
Loughborough, England   G. W. KIRBY
February 1972

CONTENTS

Chapter **X**

DEHYDROGENATION AND ZINC DUST DISTILLATION

Z. Valenta

---

1   INTRODUCTION

Dehydrogenation is a valuable tool in the structure elucidation of natural products and preparation of aromatic substances. The carbon skeletons of a great number of sesqui-, di-, and triterpenes, steroids, and alkaloids were deduced primarily from the conversion of the compounds under investigation into recognizable aromatic substances. It is this "aromatization" reaction that will be the main topic of this chapter. The dehydrogenation of alcohols to ketones, used extensively in industrial processes and in the laboratory, formation of enones from saturated ketones by the action of quinones, catalytic oxidation with molecular oxygen and a metal catalyst [1], and similar processes involving only the oxidation of functional groups will not be included.

The dehydrogenation leading to an aromatization is very often the method of choice in structure elucidations, since the products contain few, if any, asymmetric carbon atoms and their carbon skeleton and often even the position of substituents can be recognized with the help of typical ultraviolet spectra. These aromatic dehydrogenation products can then be synthesized or compared with known compounds, and their identification provides an important clue for the structure elucidation of the more complex compound under investigation. It should be pointed out, however, that this powerful tool of degradative chemistry should be used with caution and regard for established principles of organic chemistry. A consideration of bond energies reveals that a dehydrogenation at elevated temperatures need not be (and in fact very seldom is) a simple removal of hydrogen. Carbon–carbon and carbon–heteroatom bond breakages leading to a loss of substituents and ring openings are often encountered. In addition, other transformations such as retropinacol rearrangement, ring expansion and contraction, formation of new rings, migration of substituents, and even reduction must be anticipated. While these "side reactions" cause the relatively low yields encountered in the dehydrogenation of complex molecules, they are not an unsurmountable obstacle to the correct interpretation of dehydrogenation results and they provide an additional challenge to the investigator's chemical reasoning. It will be one of the main purposes of this chapter to review not only the commonly encountered transformations, but also to point out the more unexpected and sometimes very surprising results obtained in dehydrogenation studies.

Structure elucidation is, of course, not the only field in which dehydrogenation has been applied. Its use in the synthesis of aromatic systems and in studies aimed at the elucidation of dehydrogenation mechanisms is frequently encountered in the chemical literature and will be briefly reviewed here.

The most commonly used methods for dehydrogenation, limited to the "aromatization" reaction as stated above, are (1) catalytic action of transition metals, (2) action of sulfur or selenium, and (3) "hydrogen transfer" to other organic compounds such as quinones without the use of a catalyst.

These processes will first be dealt with separately in a brief way, and some of their characteristics and possible modes of action will be described. Following the survey of the individual dehydrogenation agents, most of the

dehydrogenation results are arranged on the basis of the chemical transformation type rather than on the basis of different dehydrogenation agents
or reaction conditions. This is done mainly because *the products* of dehydrogenation with different agents are very often the same or at least similar
and analogous side reactions are encountered with both metal catalysts and
sulfur or selenium (although, of course, *the mechanism* of action may be quite
different).

The dehydrogenation with sulfur, selenium, and platinum metals has been
extensively reviewed by Plattner [2], his review containing references up to
1946. Earlier results were summarized by Ruzicka in 1928 [3] and Linstead
in 1937 [4]. Jackman [5] has reviewed hydrogen transfer reactions in which
both the donor and acceptor are organic molecules. An extensive review by
Beroza and Coad [6] describes the use of a microreactor connected with a gas
chromatograph. Recent articles on selenium dehydrogenation of diterpenes
by Carman [7] and of steroids by Shöntube and Janák [8] provide good
examples of the use of modern analytical methods in this field.

2   CATALYTIC DEHYDROGENATION

Theoretical Considerations

The pioneering work of Sabatier and Senderens [9] on catalytic dehydrogenation in the vapor phase with nickel catalysts and of Zelinsky [10] with
platinum metal catalysts initiated an extensive use of metals in the dehydrogenation of organic compounds in the last five decades.

Parallel with the application of the catalytic dehydrogenation as a preparative and degradative method was the thorough investigation of the
theoretical aspects of metal catalysis. Although a detailed description of these
theoretical studies is clearly outside the scope of this chapter, and although
"far too little is known about the mechanism at present" [11], it should be of
interest to summarize very briefly some of the results of these investigations.

In a review, Trapnell [12] discusses some of the factors responsible for
effective catalysis. First of all, a metal must contain the correct lattice
spacing in order to catalyze a particular reaction; i.e., the catalyst atoms
must be at certain favorable distances, and in a certain geometric pattern,
for the most effective adsorption and reaction [13]. Thus, a so-called *geometric* or *lattice spacing factor* is operative in catalysis. Second, an active metal
must have orbitals available for surface bonding; i.e., an *electronic factor* is
operative. The atomic *d*-orbitals of transition metals are considered responsible for high catalytic activity [14, 15].

Furthermore, in an effective catalysis, the adsorption of the substrate on
the metal surface should be rapid and weak; i.e., both the heat and activation
energy (in the case of an activated process) of adsorption should be small

[12]. From the geometric standpoint, weak adsorption can be achieved by strain introduced into the molecule when it is adsorbed at more than one point [16]. It is interesting that this reasoning is similar to the "incomplete fit" or "transition state fit" hypothesis of enzyme catalysis.

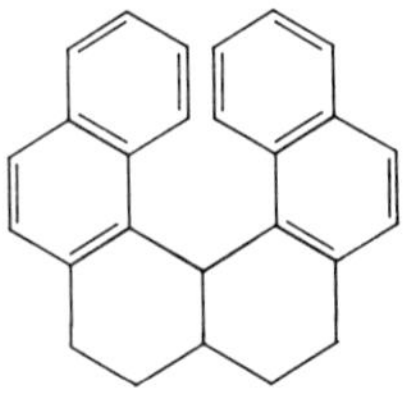

1 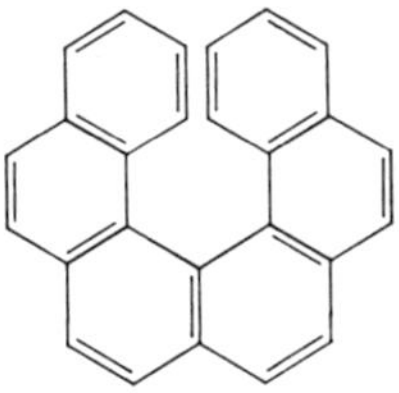   **2**

From the electronic standpoint, weak adsorption is the consequence of a limited number of *d*-orbitals available for bonding. In a study of adsorption of H_2 and ethylene on different transition metals, Beeck [17] found that the heat of chemisorption decreased with decreasing availability of atomic *d*-orbitals.

It is probably little more than a coincidence that the metals of the platinum and palladium triads have both the most favorable geometry and electronic factor for effective catalysis. While platinum and palladium are the two metals most commonly used in hydrogenations and dehydrogenations, it should be pointed out that rhodium seems to possess the best characteristics [12] and might well be tried when other metals fail. Newman found, for instance, that the dehydrogenation to helicene ($1 \rightarrow 2$) proceeds best with rhodium on Al_2O_3 in benzene (acting as solvent and H_2 acceptor) at 300°C [18].

While all the mechanisms involving hydrogen in a catalytic reaction are still controversial, detailed work on parahydrogen conversion, deuterium exchange, and similar exchange reactions has led to some clarification in the field [11, 12, 19]. It appears probable that reactions involving addition, exchange, or abstraction of hydrogen proceed through the adsorption of the reacting molecules on the catalytic surface followed by their transformation into a so-called "half-hydrogenated state" of the type:

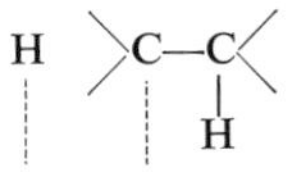

where each dotted line marks a bonding to the catalyst. Taylor [20] postulated the simultaneous removal of all six hydrogens on dehydrogenation of cyclohexane, and Braude, Linstead, and Mitchell [21] came to a similar conclusion in their study of the disproportionation of cyclohexene. The English authors postulated a termolecular reaction mainly because no detectable amount of cyclohexadiene was formed and because cyclopentene, cycloheptene, and cyclooctene did not disproportionate under similar conditions. Actually, these findings do not prove the simultaneous removal of *all* hydrogens [21], but only indicate that the reacting molecules probably do not desorb at an intermediate stage. A certain support for the theory that cyclohexane is attacked at several points simultaneously comes from the work of Balandin [22, 23] who studied the reaction of six-membered rings on active metal surfaces. He concluded that only metals possessing crystal faces with hexagonal symmetry (i.e., faces capable of forming six metal–carbon bonds) were active as catalysts.

It is likely that all metals act by the same basic mechanism and that differences in reactivity are caused by a different geometric and electronic factor [24].

Experimental Procedure

Catalysts

The most commonly used dehydrogenation catalysts are palladium and platinum, applied mostly with carriers. The preparation of platinized charcoal was described by Packendorff and Leder-Packendorff [25], palladized charcoal by Zelinsky and Turova-Pollak [26] and by Diels and Gädke [27], active palladium by Willstätter and Waldschmidt-Leitz [28], active platinum by Loew [29], platinized and palladized asbestos by Zelinsky and Borisoff [30], and palladium on barium sulfate by Mozingo [31]. Platinum on aluminum oxide has been used by Pines, Ipatieff, and their coworkers [32, 33], and Hernandez and Nord [34] described the preparation of rhodium and other metals on synthetic high polymers.

Zelinsky and his collaborators were the first to show that the method of preparation of the catalyst influences its activity and mode of action upon the substrate [26, 30, 35, 36]. Linstead and his collaborators [37, 38] investigated the effects of the various preparative methods on the activity of the catalysts and came to the following conclusions:

1. The catalysts prepared by the method of Willstätter and Waldschmidt-Leitz [28] are highly active; their activity can be improved by precipitation at higher dilution.

2. Platinum and palladium catalysts made under identical conditions have very similar activities. Palladium seems to have a greater tendency to produce side reactions.

3. The carrier has a considerable influence on activity, the order of activity being metal on charcoal > metal on asbestos > metal "black."

4. The course of the dehydrogenation of substances containing a quaternary cyclic carbon is influenced by the nature of the carrier, but not by the choice of metal.

Palladium on charcoal in combination with sulfur has been used for cyclodehydrogenation [39]. Catalysts and reaction conditions used in industrial dehydrogenation and aromatization processes have been described in summarizing articles [40–42].

Reaction Conditions

The temperature necessary for dehydrogenation varies widely and depends mainly upon the oxidation state of the substrate and the presence of quaternary carbon atoms. While tetralin can be dehydrogenated at 200°C or even lower, decalin requires 300°C [37] and the perhydronaphthalenes containing quaternary carbon atoms require a temperature of 325°C or higher [38, 43].

The duration of dehydrogenation also varies widely, but for workable yields the reaction should in general not be run for more than several hours.

The type of apparatus used for dehydrogenation will depend mainly upon the temperature and the amount of compound available. For degradative studies, it is often possible simply to heat the compound with the catalyst. It is recommended to perform the reaction in an inert atmosphere by sweeping with a slow stream of nitrogen or carbon dioxide. The reaction is thus driven to completion by removal of hydrogen, and side reactions involving hydrogenation are brought down to a minimum [37]. The reaction vessel is usually provided with a sealed-on air- or water-cooled reflux condenser and the low volatile products are collected in appropriate cold traps. This procedure should, of course, be applied only for substances boiling at a sufficiently high temperature. For lower boiling substances and for the dehydrogenation of a very small amount of substance a sealed tube can be used. In order to remove the hydrogen formed, it is often advantageous to use a hydrogen acceptor when the sealed tube technique is applied [18, 44–47].

Microdehydrogenations in a system connected to a gas chromatograph [6] and dehydrogenation of perhydroazulenes and perhydronaphthalenes for only 1–3 min [48] have recently been described.

Solvents

Various solvents have been used for catalytic dehydrogenation, both as hydrogen acceptors and as diluents. Benzene [18, 44–46], acetone [47], and maleic [49, 50], and fumaric and cinnamic acid [51] are the commonly used hydrogen acceptors, and naphthalene, quinoline, mesitylene, p-cymene, Dowtherm, sym-trichlorobenzene, xylene, triethylbenzene, and diphenyl ether have been used as solvents. The choice of an appropriate solvent makes it possible to perform the dehydrogenation at reflux at a chosen temperature.

Vapor Phase Technique

Günthard, Plattner, and coworkers described an apparatus for quantitative studies of dehydrogenation with palladized charcoal in the vapor phase [52–54]. Other apparatus for the vapor phase technique was previously described by Levitz and Bogert [55] and Ruzicka and Stoll [56].

Deuschel [57] used an apparatus with recycling of products, and Nunn and Rapson [58] described an apparatus for dehydrogenation of azulenes in vacuum.

3 DEHYDROGENATION WITH SULFUR AND SELENIUM

Theoretical Considerations

Sulfur was used for the dehydrogenation of rosin by Vesterberg [59] at the turn of the century and by others before him. Ruzicka [60, 61] applied this powerful degradative tool in a systematic way for his numerous structure elucidations in the terpene field. Diels [62] introduced the use of selenium in the chemical dehydrogenation; it is now used in preference to sulfur in most cases. Some of the advantages of the selenium modification are a simpler workup procedure and the fact that selenium does not enter organic molecules readily. In contrast to this, sulfur-containing products are sometimes formed [63–66]. The higher temperatures required for the selenium dehydrogenation often lead, of course, to more deep-seated changes in the reacting molecules.

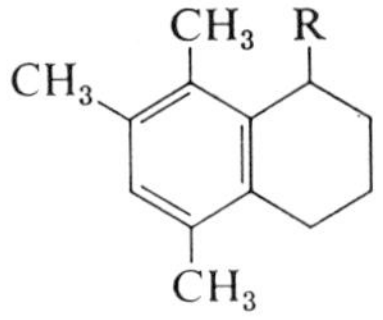

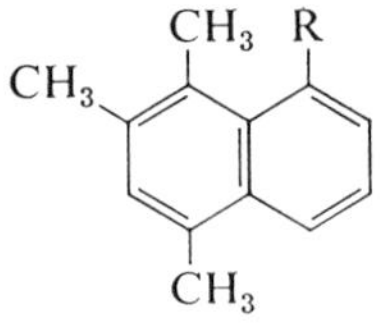

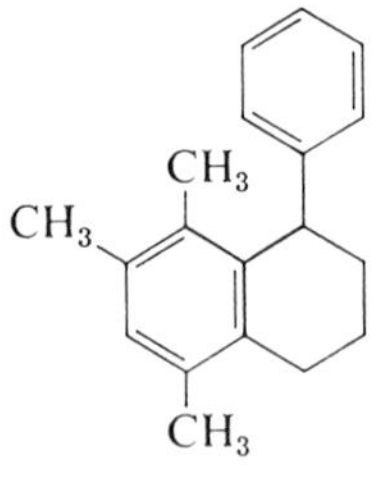

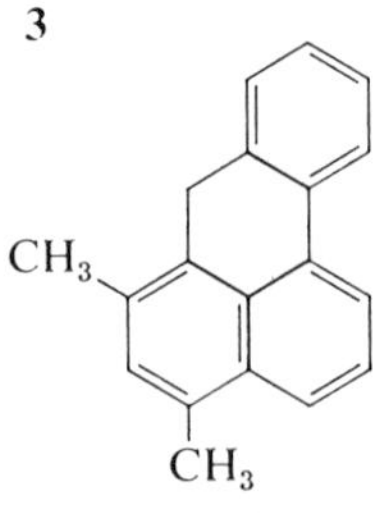

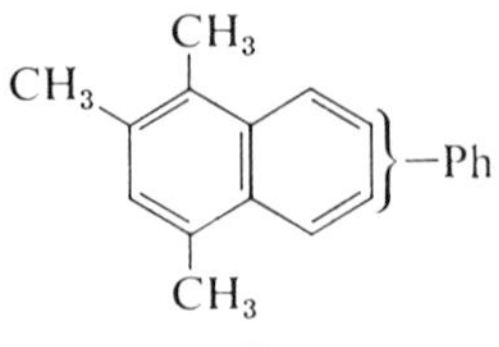

Very little is known about the exact mode of action of sulfur and selenium in the dehydrogenation reaction. There are, however, two interesting aspects of the reactivity of these two elements which should be considered at this point: (1) as Plattner points out [2], most perhydro compounds (containing no double bonds or oxygen functions) resist dehydrogenation with sulfur or selenium even at temperatures above 350°C; and (2) many, but not all, rearrangements and other side reactions observed in chemical dehydro-genation are similar to those encountered with acid-catalyzed reactions in solution, at least as far as the products are concerned. As an example, the interesting investigation of Cocker and his coworkers [67–69] of the de-hydrogenation of some substituted tetrahydronaphthalenes with selenium may be considered. Cocker observed that compound 3 (R = ethyl) lost the ethyl group during dehydrogenation to give 4 (R = H). Ethyl hydrogen selenide was proved as a product of the reaction. Similarly, compound 3 (R = cyclohexyl) yielded 4 (R = H), whereas the hydrocarbon 3 (R = methyl) was dehydrogenated without a loss of carbon to give 4 (R = methyl). The corresponding phenyl derivative 5 yielded two products; one was formed by cyclization (6) and the other by phenyl migration (7).

On the basis of this selected evidence, the following working hypothesis can be formed: the ease of dehydrogenation will depend upon the type of hydrogen which has to be abstracted first from the molecule. In a completely hydrogenated molecule (containing no oxygen functions), the ion (or free radical) remaining after the abstraction of hydrogen ion (or atom) receives no special stabilization; the abstraction of hydrogen is therefore difficult, and the dehydrogenation will only proceed sluggishly and at very high temper-atures. On the other hand, compounds containing an unsaturation or a potential unsaturation (e.g., a hydroxyl group) can lose hydrogen in an allylic or benzylic position, and this removal will be facilitated by the stabilization of the resulting ion or free radical. There seems to be no critical evidence available which would make it possible to decide between an ionic and a free radical mechanism, and it is actually probable that both homolytic and heterolytic cleavages are involved, their relative importance being determined by such factors as temperature and the type of bond to be broken.

Cocker's experiments [67–69] have been chosen purposely since they point out the difficulty of explaining the results of a chemical dehydrogen-ation by a unique mechanism. While most of his findings can be explained by an ionic mechanism, the loss rather than migration of the ethyl and cyclo-hexyl group in 3 (R = ethyl or cyclohexyl) seems to be indicative of a homolytic cleavage. The frequently observed loss of quaternary alkyl groups and several other transformations involving a carbon–carbon bond breakage (see Section 8) must similarly be assumed to be pyrolytic in nature.

From the several possible alternatives, the following pathway for the dehydrogenations described by Cocker may be examined: attack of selenium upon 3 leads to the abstraction of a hydride ion from one of the five benzylic positions, to give 3a, 3b, and three other ions. The loss of a proton converts 3a into 8, which on further attack with selenium can be converted into 8a. This ion can now aromatize by the loss of a proton, by migration of R and the loss of a proton, or by the loss of R$^+$. The formation of 4

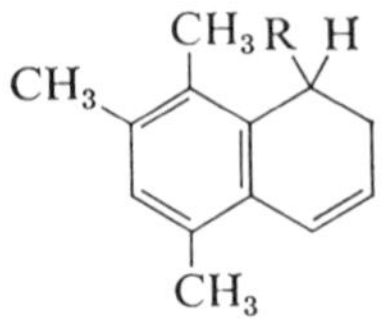

8

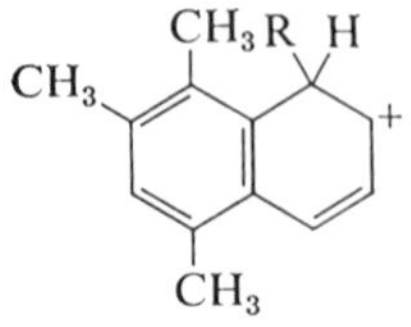

8a

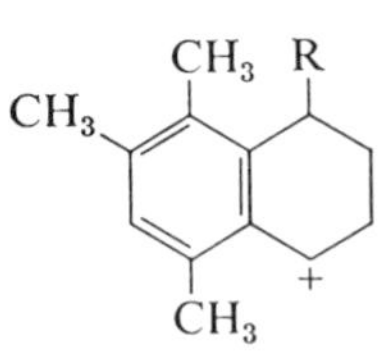

3a

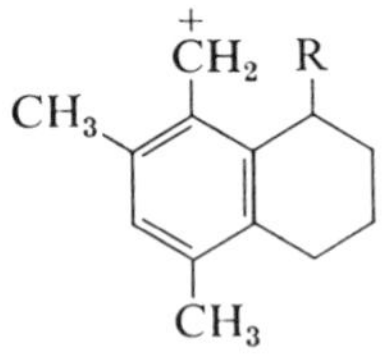

3b

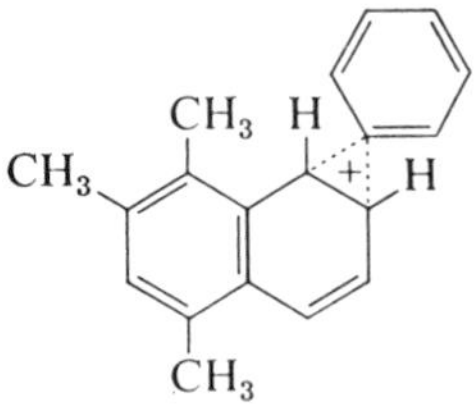

9

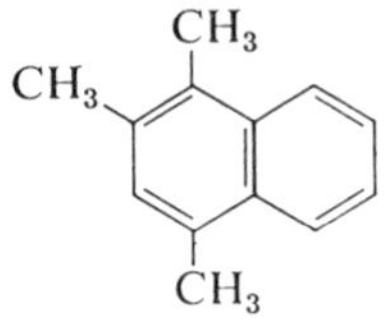

10

(R = methyl) from 8a (R = methyl) by the loss of a proton and of 7 from 8a (R = phenyl) through the ion 9 is well explained on the basis of this hypothesis. However, the formation of 10 from 8a (R = ethyl or cyclohexyl) by the loss of the substituent can hardly be reconciled with the known behavior of carbonium ions. A homolytic cleavage of the free radical equivalent of 8a or of 3 itself is more likely in this instance.

Since all the above transformations can also be explained by using the corresponding free radical intermediates, one is forced to the conclusion that all or at least some breakages involved in the chemical dehydrogenation are homolytic.

Similarly, the formation of the cyclic product 6 from compound 5 can proceed by way of the ion 11a or the free radical 11b.

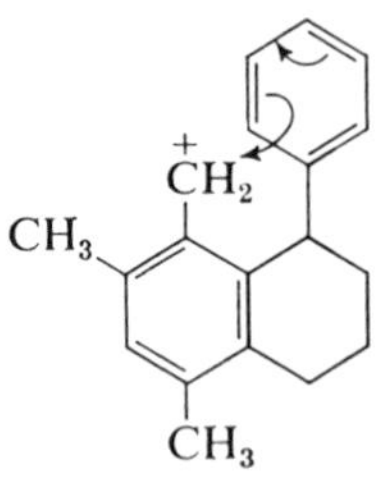

11a

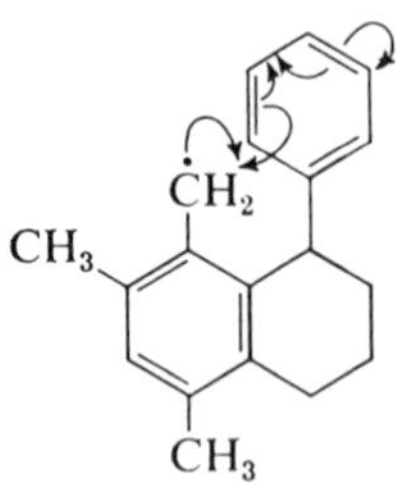

11b

An important study concerned with the detailed involvement of selenium in the dehydrogenation has been undertaken by Orchin [70] who concluded that a reasonable pathway involves a π-complex of selenium with a double bond or an aromatic ring probably first rearranging to a hydroperselenide. Thermal dissociation can then lead to a selenol which can then either lose hydrogen selenide to give an olefin, oxidize to a diselenide, or undergo a homolytic C-Se bond cleavage followed by further radical reactions.

Experimental Procedure

The technique used in dehydrogenation with sulfur and selenium has been described by Plattner [2].

Sulphur

The compound is heated with the calculated amount of sulfur to 200–220°C for several hours [71]. Excess of sulfur and prolonged heating at temperatures higher than 220°C in the presence of sulfur should be avoided since there is a danger of side reactions. The temperatures should be raised gradually at first to avoid excessive frothing. The evolution of hydrogen sulfide usually begins at about 180°C.

Selenium

The compound is heated with powdered selenium in a flask provided with a sealed-on reflux condenser. Low volatile products are condensed in attached cold traps. An excess of selenium may be used, since there is less danger of side reactions producing selenoorganic compounds. The reaction is usually conducted at temperatures ranging between 280 and 350°C, the best working range being about 300–340°C. More rearrangements and destruction of compounds can be expected with increasing temperature. The evolution of hydrogen selenide usually begins just under 280°C. The reaction with selenium usually requires longer heating than that with sulfur. An effective dehydrogenation of a compound containing quaternary carbon atoms and relatively few unsaturations usually requires at least 6–8 hr and in some cases the reaction was run up to 4 days. Compounds requiring relatively few changes for complete aromatization, on the other hand, can be successfully dehydrogenated with selenium in an hour or less. Bartlett, Dickel, and Taylor [72] found, for instance, that ibogaine (12) yielded the products 13 and 14 on heating with selenium to 180–300°C for 12 min and to 300–317°C for an additional 18 min.

Small amounts of material and relatively volatile compounds can be dehydrogenated in a sealed tube [73] and microquantities (1 mg or less) have been reacted in glass capillaries [8].

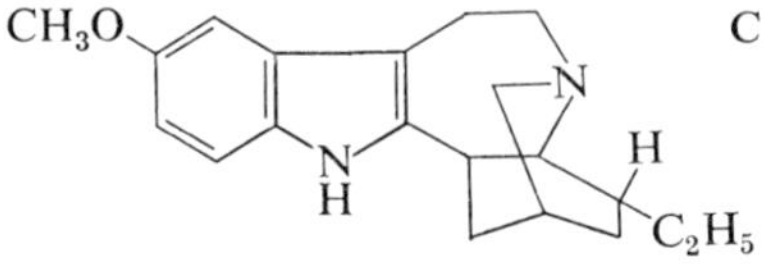

12

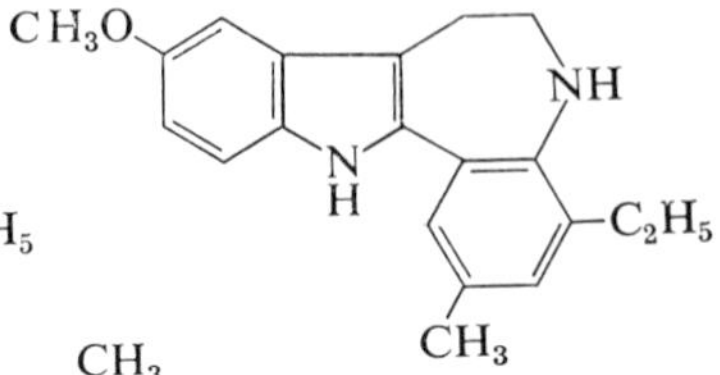

13

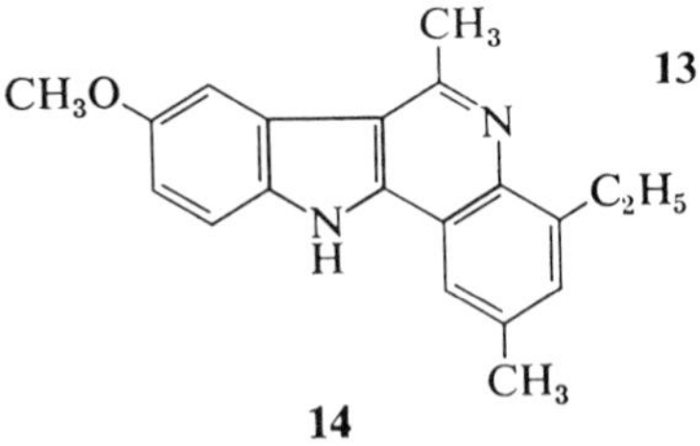

14

Solvents

The use of solvents is usually restricted to the catalytic dehydrogenations. There are, however, a few instances in which a solvent has been used in order to achieve controlled conditions with sulfur and selenium. These are almost invariably dehydrogenations involving only minor changes in the molecule. Naphthalene, quinoline [74], and dimethylformamide [75] are examples of the solvents used.

4   ISOLATION OF PRODUCTS

In the case of dehydrogenation with sulfur, all volatile products are usually distilled under vacuum; traces of sulfur in the distillate are relatively easily removed in subsequent operations. The crushed selenium dehydrogenation mixture is extracted in a Soxhlet apparatus with ether, benzene, or chloroform. A similar extraction procedure is recommended for catalytic dehydrogenations in which adsorbing carriers (e.g., charcoal) were used.

The mixture of products can then be separated into neutral, basic, and acidic fractions. The neutral and basic fractions can be developed by careful chromatography on alumina and the acidic fraction by chromatography or partition chromatography on silica [76]. Very similar compounds with basic or acidic groups can be separated by a countercurrent distribution using ether or chloroform and an appropriate buffer solution.

The purification can then be completed by recrystallization or, in the case of oily products, by preparation of crystalline derivatives. Picrates and trinitrobenzolates are often the derivatives of choice for aromatic hydrocarbons. The free hydrocarbons can be regenerated from these derivatives by chromatography on alumina. A highly active grade of alumina should be used for the decomposition of trinitrobenzolates.

Highly volatile products condensed in the course of dehydrogenation in attached cold traps can be worked up separately or combined with the residue.

A very useful modern method of separation and identification of volatile substances is the vapor phase chromatographic technique [77, 78]. Several extensive applications of this method for the analysis of dehydrogenation mixtures have been reported during the past decade [6–8, 79, 80]. These make it clear that, as an alternative to product separation, the analysis of chromatographic patterns can provide important structural information under favorable circumstances. The identification of aromatic products is facilitated by the development and improvement of new physical techniques in the last few years. Available methods are discussed elsewhere in this volume.

5 OTHER METHODS OF DEHYDROGENATION

Virtually all dehydrogenations used as a degradative tool are performed either catalytically or with sulfur and selenium. Other dehydrogenation methods, such as the bromination of von Baeyer and Villiger [81], treatment with manganese dioxide and sulfuric acid [82], and the use of dialkyl sulfides [83], have found no general application. Zinc dust distillation, a method frequently used for the degradation of natural products [84], is discussed in Section 10.

On the other hand, many preparative dehydrogenations of unsaturated substances can be performed with the use of chloranil and related quinones. Chloranil was first introduced by Arnold and Collins [85] and has since been used by numerous workers. Dehydrogenations with quinones have been reviewed by Jackman [5]. Walker and Hiebert recently reviewed [86] the properties and reactivity of 2,3-dichloro-5,6-dicyano-1,4-quinone.

Braude, Jackman, Linstead, and their coworkers [87–91] have investigated the action of various quinones upon dihydrobenzenes, dihydronaphthalenes, and similar donors and came to the following conclusions:

1. The reactions are essentially bimolecular.
2. The reactions are faster in polar than in nonpolar solvents. Light or benzoyl peroxide has no influence, and no coupling products are obtained.
3. The reactivity of quinones is enhanced by electron-attracting, and reduced by electron-donating, substituents. Rate constants and activation energies can be correlated with the quinone reduction potentials.
4. Product catalysis, observed with quinones of low potential, is due to quinol formation.
5. The rate at which the reaction proceeds increases with the resonance energy gain of the donor resulting from the reaction.

They postulate a two-step heterolytic mechanism, involving a rate-determining hydride ion transfer followed by a fast proton transfer:

$$RH_2 + Q \xrightarrow{\text{slow}} R \cdot H^+ + QH^- \xrightarrow{\text{fast}} R + QH_2.$$

The existence of a pronounced stereoelectronic factor was made probable in a study reported by Mechoulam and his coworkers [92]. During dehydrogenations with chloranil in the cannabinol series, a ready dehydrogenation took place only when the departing hydrogen was essentially perpendicular to the plane of a double bond or an aromatic ring.

Gardner, Wulfman, and Osborn [93] converted compound 15 into 16 with chloranil in a 32% yield. Similarly, Treibs and coworkers [94] converted the heptindoles 17 and 19 into compounds 18 and 20 in 20 and 25%

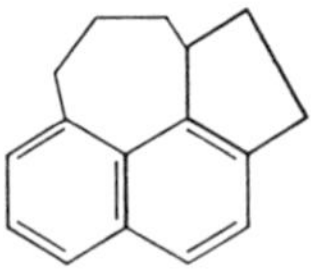

15

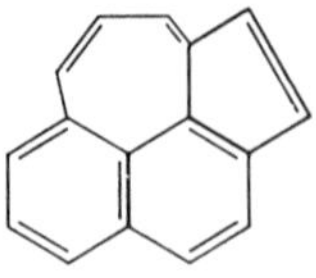

16

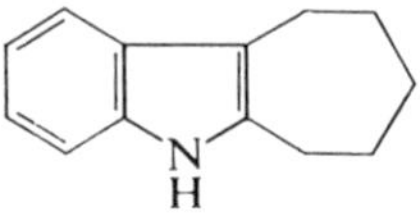

17

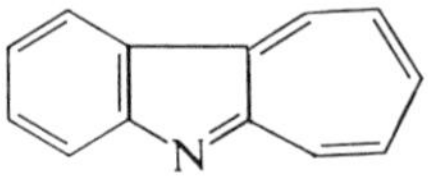

18

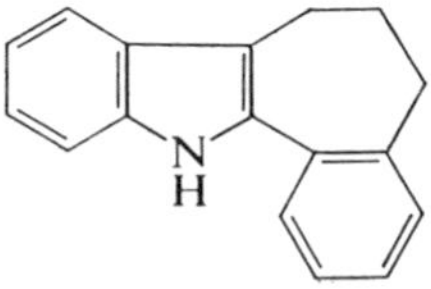

19

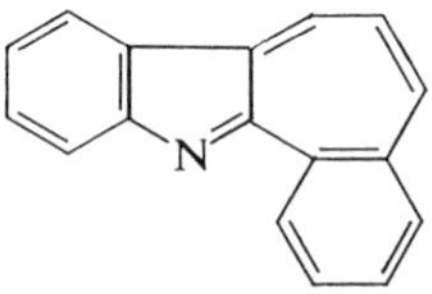

20

yields, respectively. Dehydrogenation to bufadienolides by chloranil and 2,3-dichloro-5,6-dicyano-1,4-quinone has been reported by Sarel and coworkers [95].

Linstead and coworkers [96] investigated the action of high potential quinones upon hydroaromatic systems blocked by quaternary methyl groups and found that in contrast to most catalytic and sulfur or selenium dehydrogenations the aromatization was achieved by a rearrangement of the Wagner-Meerwein type without a loss of the methyl group. Treatment of the diene 21 with tetrachloro-1,8-diphenoquinone at 150°C in a sealed tube gave isodurene (**23**), and the dehydrogenation of 1,1-dimethyltetralin (**24**) with 2,3-dichloro-5,6-dicyano-1,4-benzoquinone at 80°C gave 1,2-dimethyl-naphthalene (**27**) in an almost quantitative yield. These rearrangements, proceeding according to Linstead and his coworkers through **22** and through **25** and **26**, respectively, are well explained by the proposed ionic mechanism. Migratory aptitudes of methyl, phenyl and styryl groups observed

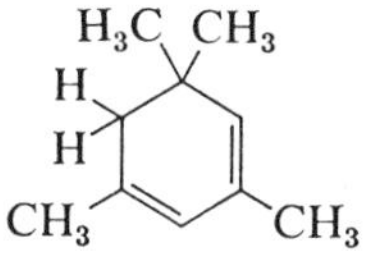

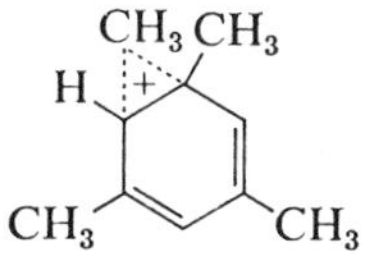

in quinone-induced dehydrogenation of 1,1-disubstituted dihydronaphthalenes by Herz and Caple [97] provide further supporting evidence. If these transformations are considered as a model of an ionic dehydrogenation, the conclusion becomes inevitable that the dehydrogenations of similar compounds containing quaternary alkyl groups with sulfur and selenium cannot proceed by a simple ionic mechanism (see page 8).

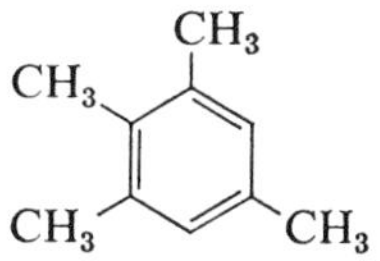

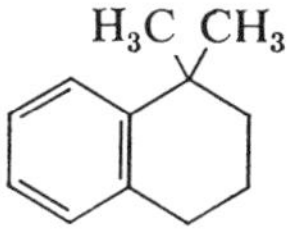

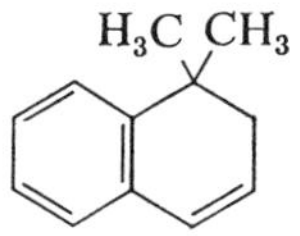

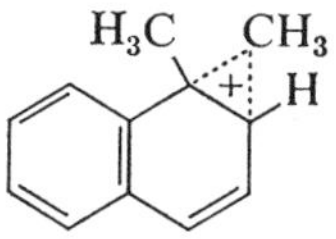

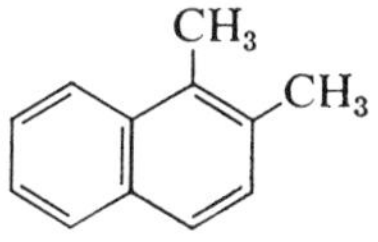

There are exceptions to the rule that no elimination of angular substituents takes place during dehydrogenations effected by quinones. Dannenberg [98, 99] reported that partial rearrangement and partial elimination of methyl groups takes place during high temperature dehydrogenation of steroids and 1,1-dimethyl dihydronaphthalenes.

The dehydrogenation with high potential quinones is mostly applied for preparative purposes and is only of limited use for degradation of complex molecules.

6 BEHAVIOR OF C–H BONDS

Normal Dehydrogenation

Dehydrogenation involving only the abstraction of hydrogen without skeletal changes usually proceeds very readily. With a few exceptions, only six-membered rings are dehydrogenated. The important exceptions are five- and seven-membered rings which can become aromatic due to special ring fusion, such as azulenes and similar compounds.

In general, the catalytic method is preferred for normal dehydrogenation. In fact, with the exception of perhydroazulenes, perhydro compounds resist dehydrogenation with sulfur and selenium even at higher temperatures. *trans*-Decalin is resistant to selenium at 350°C [100], 2-benzyl-*trans*-decalin is not dehydrogenated with selenium at 320–330°C [101] and 2-methyldecalin gave only a poor yield of 2-methylnaphthalene at 320–350°C [102]. Cyclohexane and its homologs are only slightly attacked on heating with sulfur to 300°C [103], and tetrahydrocadinene could not be dehydrogenated at 200–260°C [61]. On the other hand, unsaturated compounds usually dehydrogenate quite readily with sulfur or selenium. Tetralin [82, 104], cyclohexylbenzene [103], 2-cyclohexylnaphthalene [105], and 1-(1'-naphthyl)-cyclohexene [106] were dehydrogenated with sulfur to the corresponding fully aromatic substances and many normal dehydrogenations with selenium gave good yields of aromatic products.

Catalytic dehydrogenation of hydrocarbons not containing quaternary carbon atoms usually gives very good results. Among the compounds dehydrogenated by this method are menthane [107], cadinene [56], decalin [37, 52, 108, 109], 2-benzyldecalin [101], hexahydrofluorene [101], hydrindane [110], and octohydroanthracene [111]. Linstead and his coworkers [37, 38, 112, 113] studied the dehydrogenation of tetralin, decalin, octalin, octahydroanthracene, octahydrophenanthrene, and some monoterpenes with platinum and palladium catalysts. They found that tetralin dehydrogenates quantitatively at its boiling point, whereas decalin requires 300°C. Octalin disproportionates first to decalin and tetralin at a lower temperature, so that a good yield of naphthalene can only be obtained at 300°C or higher.

Several dehydrogenation studies with hydrogenated azulenes have been reported. A recent study [48] describes the effective treatment of perhydroazulenes and perhydronaphthalenes with palladium–charcoal for only 1–3 min. Anderson and Nelson [114] dehydrogenated hexahydroazulene with several catalysts and obtained best results with palladium–charcoal on alumina. The yield with sulfur was inferior. Günthard, Plattner, and coworkers [52–54] made a quantitative study of the vapor phase dehydrogenation of various hydrogenated azulenes with palladium on charcoal. They found that the yields increase with decreasing load on the catalyst and a shorter contact time. The catalytic surface was found to diminish during the dehydrogenation with resultant decrease of yield. Yields up to 30% were obtained with *cis*-bicyclo[5.3.0]decane and up to 60% with hydroazulenes containing a hydroxy group or a double bond in the cyclopentane ring. Further examples are the synthesis of *S*-guaiazulene and 4,5-benzazulene [115].

Complete aromatization of unsaturated substances can also be achieved with high potential quinones (see Section 5). Nakasaki [116] reported the dehydrogenation of hydroaromatic compounds (tetralin, phenylcyclohexane, tetrahydrocarbazol) with diphenyl disulfide in good yield.

Disproportionation and Hydrogenation

The reversible nature of the catalytic dehydrogenation and the thermodynamic properties of the reactants and products are responsible for the frequent occurrence of hydrogenation effects. These effects can manifest themselves in three ways: (*1*) "migration" of hydrogen in one molecule, (*2*) hydrogen transfer between like donors and acceptors (disproportionation), and (*3*) hydrogen transfer between unlike molecules (transfer hydrogenation). Many of the reported hydrogen migrations catalyzed by metals have been summarized by Braude and Linstead [90].

Knoevenagel and his collaborators [117] were the first to report a case of disproportionation. They found that $\Delta^{2,5}$-dihydroterephthalic ester yielded a mixture of terephthalic ester and the *cis*- and *trans*-hexahydro derivatives when heated with palladium black at 140°C. Wieland [118] observed the conversion of dihydronaphthalene into naphthalene and tetralin with evolution of heat on treatment with palladium black. Further examples of disproportionation are the studies with cyclohexene [119], methylcyclohexene [120], limonene [113, 121], α-pinene [113, 122], allyl-Δ^1-cyclohexene [123] and the octalins [37].

The exact nature of the disproportionation reaction has been discussed by Linstead, Braude, and their coworkers [21, 124]. They made a quantitative study with cyclohexene and cyclohexadiene and found that the disproportionation proceeded at high rates and at temperatures which were much lower than those necessary for the corresponding dehydrogenation. Since the reactions proceeded stoichiometrically without evolution of hydrogen and no spectroscopically detectable amount of cyclohexa-1,3-diene was formed during the disproportionation of cyclohexene, they suggested that the disproportionation is very likely a linked process in which the rate-determining step involves both the donor and acceptor molecules and that the reaction with cyclohexene is probably termolecular (see also page 4). The importance of thermodynamic factors can be demonstrated by the fact that 9,10-dihydroanthracene, the disproportionation of which would involve the loss of benzenoid character in one ring, only undergoes a small amount of dehydrogenation under these conditions [21].

The hydrogen transfer between unlike molecules is the logical extension of hydrogen migration and disproportionation and can be used both as a method for dehydrogenation and hydrogenation. Braude, Linstead, and Mitchell [21] list the references for the use of tetralin, cyclohexanol, and other alcohols as donors for the reduction in the presence of nickel or palladium and maleic acid, cinnamic acid, benzene, and acetone as acceptors for catalytic dehydrogenation. The palladium–maleic acid dehydrogenation method [49] was found very useful in certain cases [50]. Adkins and his coworkers [44–46] made a thorough study of dehydrogenation with benzene as acceptor in a sealed tube using palladium or nickel catalysts. Braude, Linstead, and their coworkers studied the palladium-catalyzed transfer hydrogenation of ethylenic and acetylenic bonds [21], nitro compounds [125] and other acceptors [126] by cyclohexene; guaiol and other sesquiterpenes have been catalytically dehydrogenated with fumaric or cinnamic acid acting as hydrogen acceptors [51].

Reduction of double bonds and disproportionation also happen frequently on dehydrogenation with sulfur and selenium. Examples are the production of the saturated hydrocarbon $C_{29}H_{50}$ from oleanolic acid [127], the reduction of cholesterilene to cholestane [128], reduction of allyl side chains [103, 123, 129, 130], conversion of indenes into indanes [131], and the disproportionation of 1-naphthylcyclohexene to a mixture of cyclohexyl- and phenylnaphthalenes with selenium [132]. By analogy, many transformations such as the dehydrogenation of khusilic acid [133], the conversion of limonene into p-cymene, and the dehydrogenation of selinene and eudesmol to 1-methyl-7-isopropylnaphthalene [61], among others, may be assumed to involve hydrogenation rather than double bond migration.

Although the mechanism of hydrogen transfer by sulfur and selenium is undoubtedly quite different from its catalytic counterpart, it must be assumed that the reaction of the hydrogen donor with both sulfur and selenium is reversible. It is interesting in this connection that hydrogen sulfide and selenide partially decompose to the elements at higher temperature [134].

As a direct consequence of the hydrogenating effects in catalytic and chemical dehydrogenation, double bonds in unsaturated side chains and in rings that cannot aromatize are invariably reduced. Furthermore, five- and seven-membered rings that cannot become part of an aromatic system remain in a fully saturated state. Further examples are the dehydrogenation of dextropimaric acid [135], of a tetrahydronaphthylacrylic ester [136], of hydrogenated cyclopentanonaphthalenes and phenanthrenes by Adkins and Hager [46], and of cevine by Jacobs and his coworkers [137]; the dehydrogenation of hexahydroindane to indane [110]; and the isolation of γ-methylcyclopentenophenanthrene from the dehydrogenation of sterols.

Functional groups (hydroxy groups, ketones, carboxy groups) situated in a side chain or in a ring are also frequently reduced (see Section 7). Homogeneous and catalytic hydrogen transfer reactions have been reviewed by Jackman [5].

Partial Dehydrogenation

It is usually desirable to use drastic dehydrogenation conditions in degradative studies. Incomplete dehydrogenation can thus be reduced to a minimum, and the workup procedure is simplified. Partial dehydrogenation is mostly due to a limited amount of sulfur or selenium in a chemical dehydrogenation, a low reaction temperature, a short reaction time, and the resistance of the substrate to dehydrogenation.

Octahydrochrysene was obtained from dodecahydrochrysene on treatment with two atoms of sulfur [138]. Cook and Hewett [139] obtained a mixture of 1,2-benzanthracene and a tetrahydro derivative by treatment of dodecahydrobenzanthracene with selenium. Other examples of incomplete dehydrogenation are the formation of tetrabyrine and ketoyobyrine [140] from yohimbine with selenium, the dehydrogenation of an amino acid derived from annotinine with palladium [141, 142], and the catalytic dehydrogenation of decahydroquinoline [143].

There are many cases reported in which six-membered rings containing quaternary carbon atoms resisted dehydrogenation under relatively mild dehydrogenation conditions. Adkins and England [45] found that the tetramethyldecalin (28) was converted into the tetralin (29) in a 90% yield and the octahydroanthracene (30) into 31 in a 68% yield on treatment with platinum–charcoal in benzene. Harris and Sanderson [135] obtained the naphthalene derivative 33 by the dehydrogenation of dextropimaric acid (32) with palladium–charcoal. 33 could be converted into 1,7-dimethylphenanthrene (pimanthrene) under more drastic conditions. Dreiding and Pummer [144] found that the methyl ether of 1-methyl-estrone (34) gave a 70% yield of the corresponding equilenin derivative (35), with epimeric C-14 with palladium–charcoal at 350°C in 5 min. Only phenanthrenes could be isolated after a 30 min treatment.

Further examples of partial dehydrogenation are the mild sulfur treatment of methyl vinhaticoate by King and King [145] and of totarol by Short and Wang [146].

7   BEHAVIOR OF C–O BONDS

Functional groups containing oxygen can be retained unchanged or can be reduced, oxidized, or eliminated. Their behavior will in general depend on their position, on the nature of the dehydrogenating agent, and on the reaction conditions. For easier reference, the different functional groups will be dealt with separately.

Alcohols

Primary Alcohols

Eschinazi [147] found that a catalytic amount of palladium on barium sulfate converts aldehydes and primary alcohols to hydrocarbons with one less carbon atom. The primary alcohols probably lose carbon monoxide after being dehydrogenated to the corresponding aldehydes. The compounds

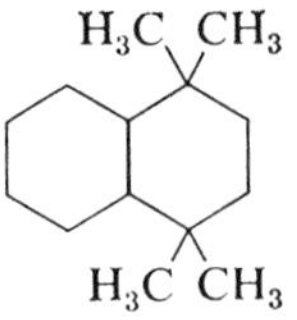

28

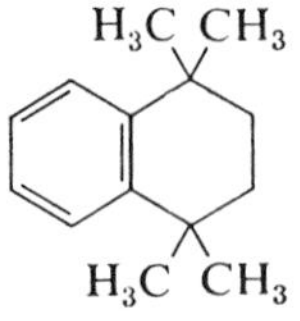

29

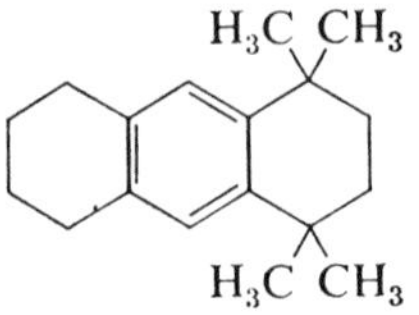

30

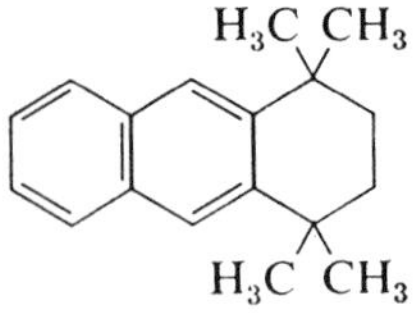

31

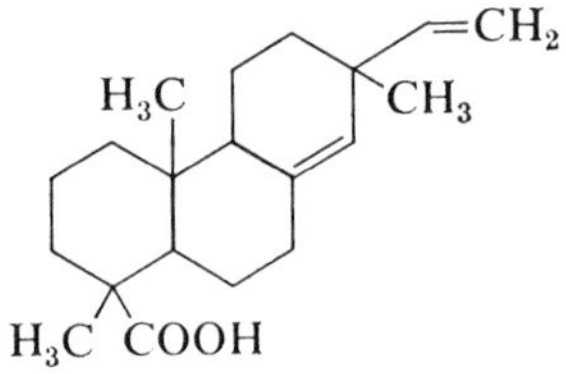

32

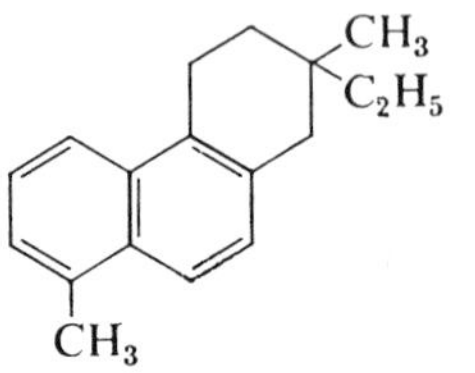

33

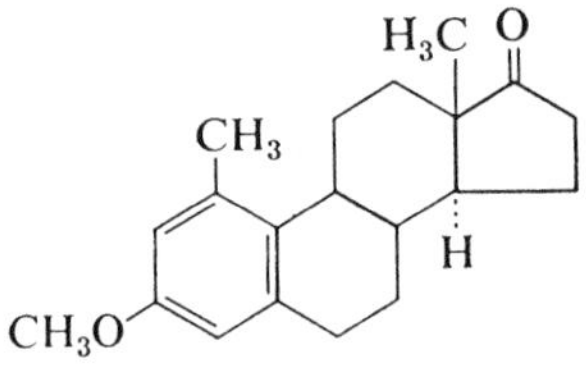

34

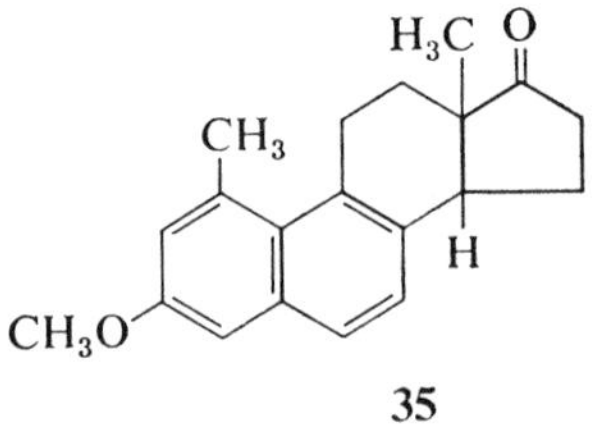

35

investigated by Eschinazi were citral, furfural, 2-phenylethyl alcohol, and benzyl alcohol. Secondary alcohols, ketones, and esters were unaffected under the conditions used. Newman and his coworkers [136, 148–150] investigated the behavior of primary hydroxyl groups in the tetralin series on dehydrogenation with palladium–charcoal. In the one case studied, the –CH$_2$–CH$_2$OH group lost carbon monoxide. The –CH$_2$OH group attached directly to a ring could not be retained to any extent either. They found that hydrogenolysis to a methyl group or decomposition to hydrogen and carbon monoxide or a combination of the two always occurred. It is not practical to protect the alcoholic function by acetylation, since hydrogenolysis still takes place. The one exception was the acetate **36**, which gave a 63% yield of β-naphthylethyl acetate. The lability of the primary acetoxyl group was also demonstrated by Büchi and Rosenthal [151] who obtained guaiazulene (**38**) in the dehydrogenation of the helenalin transformation product **37** with palladium.

Sulfur and selenium also remove the alcohol function in most cases. Arnold [152] obtained methylazulene by dehydrogenation of the azulene precursor **39** with sulfur or palladium. Tatevosyan and Vardanyan [153]

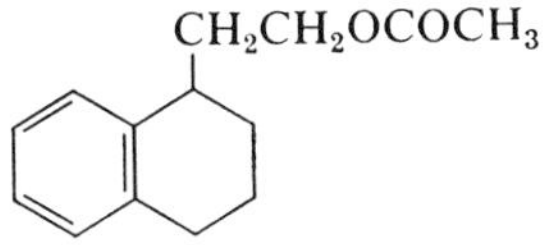

$$CH_2CH_2OCOCH_3$$

36

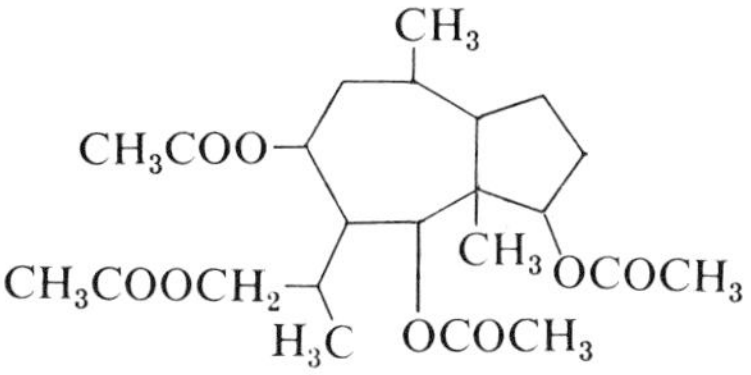

37

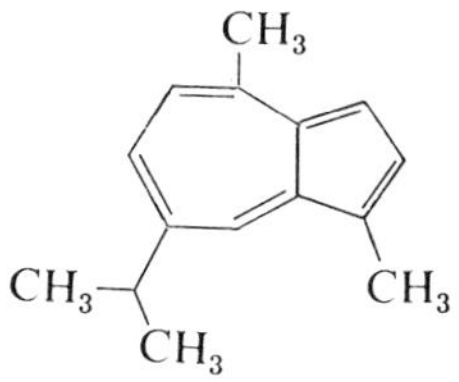

38

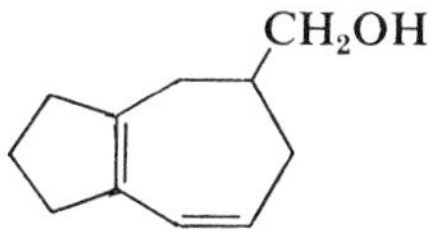

39

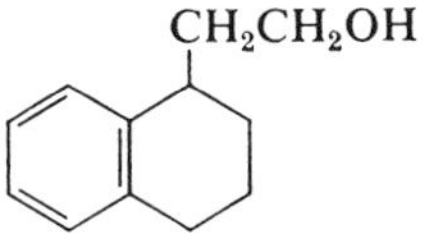

40

found that the tetralin **40** gives 1-ethylnaphthalene with selenium at 300°C, but that sulfur dehydrogenation at 190°C preserves the hydroxyl group to give β-1-naphthylethyl alcohol in a 66% yield.

Not surprisingly, –CH$_2$OH groups attached to a quaternary carbon atom are invariably eliminated. Some examples are the dehydrogenation of abietinol [71], hederagenin [154], and a triol derived from veatchine [155].

Secondary Alcohols

Secondary hydroxyl groups situated in a ring or in a side chain are usually eliminated. Examples in which this elimination leads to a change in the carbon skeleton are described in Section 8. The presence of ring-standing hydroxyl groups is very often advantageous, especially in chemical dehydrogenation. Water is often eliminated before dehydrogenation begins, and the molecule is labilized to further attack. Ruzicka and his coworkers [156] in their study of the selenium dehydrogenation of cholic acid recommend

complete removal of the water in the reaction vessel before the dehydrogenation starts to avoid frothing. Further examples of the elimination of
hydroxyl groups are the selenium dehydrogenation of cevine [137], yohimbine
[140], and lanostadienol [157]. Newman and Zahm [150] observed the
reduction of secondary hydroxyl groups in the side chain of substituted
tetralins during a palladium dehydrogenation.

The elimination of the oxygen function is, however, not always complete,
and it is often possible to isolate phenols, at least in small quantities. Adkins
and Davis [44] obtained a 20% yield of 2-naphthol in the dehydrogenation of
2-decalol with platinum in benzene and a 65% yield of thymol from menthol
with a nickel catalyst. Wessely and Grill [104] investigated the dehydrogenation of β-tetralols and β-decalols with sulfur at 170–290°C and found
that a good yield of 2-naphthol could be obtained at a sufficiently low
temperature. Linstead and Michaelis [112] found similarly that liquid
dehydrogenation of β-tetralols and β·decalols with palladium under sufficiently
mild conditions gave moderate yields of β-naphthol.

$$CH_3 \quad\quad OH$$
$$\text{—CH—CH}_3$$
$$N$$

41

Examples in which the dehydrogenation of natural products led to the
partial retention of the hydroxy group are quite numerous. 1,8-Dimethyl-2-
hydroxypicene and 1,2,5-trimethyl-6-hydroxynaphthalene (hydroxyagathal-
ene) were obtained from many triterpenes [158]. Short and Wang [146]
obtained 1-methyl-7-hydroxyphenanthrene in the dehydrogenation of totarol
with selenium. Honigmann [159] converted neoergosterol into dehydroneo-
ergosterol with platinum, and Cook and his coworkers [160] similarly
obtained 9-phenanthrol from octahydro-9-phenanthrol. Cocker and his
coworkers [161] obtained a substituted α-naphthol in the dehydrogenation of
desmotropo-ω-santonin with palladium. It is interesting that the compound
$C_8H_{11}NO$ obtained by the selenium dehydrogenation of cevine [162] has been
ascribed the structure **41** [163].

Tertiary Alcohols

Naturally enough, tertiary hydroxyl groups are readily eliminated under a
variety of conditions. Eudalene (1-methyl-7-isopropylnaphthalene) is formed
in the dehydrogenation of eudesmol with sulfur [61] and from 2-(hydroxyiso-
propyl)-8-methyl-1,2,3,4-tetrahydronaphthalene by dehydrogenation with
selenium [164]. Several tertiary hydroxyl groups are eliminated in the
dehydrogenation of cevine [137]. Both artabsin (**42**) [165] and matricin (**43**)
[166] dehydrogenate to chamazulene (**44**), whereas patchouly alcohol (**45**)
gives guaiazulene (**38**) [167].

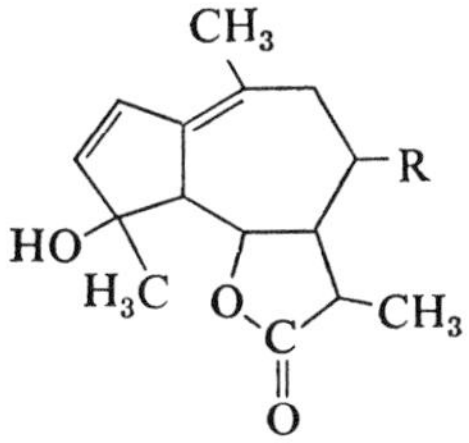

42: R = H
43: R = OAc

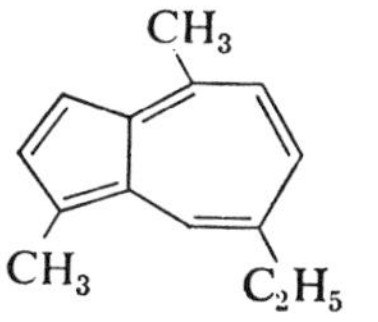

44

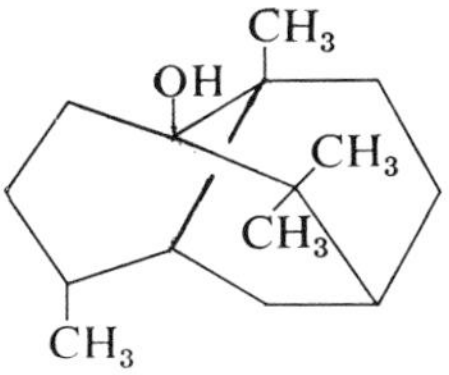

45

Methyl Ethers

Numerous cases are described in which an aromatic methoxyl group survived dehydrogenation, both catalytic [47, 144, 168, 169] and chemical [72, 145, 146, 170, 171]. However, Kon and Ruzicka [172] observed in one case that the methoxyl was only retained if the temperature of the selenium dehydrogenation was kept below 300°C, and Cocker and his coworkers [68, 161] reported two cases of elimination of an aromatic methoxyl group in the hydronaphthalene series with palladized charcoal.

Carbonyl Compounds

Aldehydes

The aldehyde group is invariably lost during dehydrogenation. As was already mentioned, Eschinazi [147] found that palladium converted aldehydes to hydrocarbons with one less carbon atom and carbon monoxide. In agreement with this finding, Newman and Bye [148] reported a quantitative loss of carbon monoxide from aldehydes in the tetralin series on treatment with palladium–charcoal.

Ketones

The behavior of ketones on dehydrogenation depends upon their position and the reaction conditions. Ketones in six-membered rings can be converted into the corresponding phenols under relatively mild conditions. Under more drastic conditions, complete elimination of oxygen usually takes place. Side-chain ketones can sometimes be retained with sulfur or a catalyst under mild conditions.

Darzens and Lévy [173] prepared phenol and α-naphthol from cyclohexanone and α-tetralone, respectively, by treatment with sulfur. Fieser, Hershberg and Newman [174] obtained the corresponding hydroxybenzpyrene by a short action of sulfur upon the ketone **46** at 220–230°C. After treatment with selenium at 320°C, 1,2-benzopyrene was isolated as the main product.

Ruzicka [73] obtained small amounts of the corresponding phenols in the

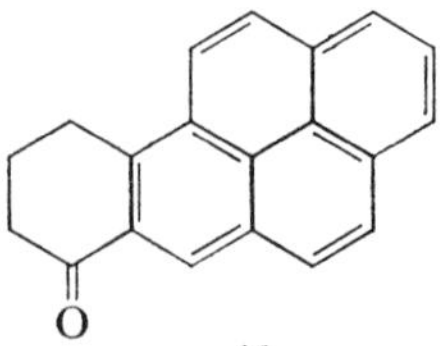

46

dehydrogenation of *trans*-β-decalone and 3-methyl-Δ^2-cyclohexenone with selenium at 260°C. Horning [175] dehydrogenated 3,5-dimethyl-Δ^2-cyclohexenone with sulfur to the corresponding phenol in a 26% yield. Charonnat and Girard [176] investigated the influence of added silicates upon the yield of phenol in the dehydrogenation of cyclohexanone with sulfur.

Catalytic dehydrogenation of cyclic ketones to phenols is also quite common. Ruzicka and Mörgeli [177] prepared 7-methyl-1-naphthol from 7-methyl-1-tetralone with palladium in a 60% yield; Adkins and Davis [44] obtained a 20% yield of 2-naphthol from 2-decalone with platinum in benzene; and Horning and coworkers [169, 178] obtained yields of phenols varying between 40 and 84% in the dehydrogenation of 5-substituted 3-methyl-Δ^2-cyclohexenones with palladium–charcoal in different solvents at reflux temperature. Mosettig and Duval [179] dehydrogenated 1- and 4-ketotetrahydrophenanthrenes to the corresponding phenanthrols with palladium black in boiling xylene or naphthalene in excellent yields. The dehydrogenation of α-tetralone, *trans*-α-decalone, and the *cis*- and *trans*-β-decalones with palladium was studied by Linstead and Michaelis [112]. They obtained 19–46% yields of the corresponding naphthols, together with some naphthalene. 2,2'-Binaphthyl was formed as a by-product in the dehydrogenation of *trans*-β-decalone. Compounds further removed from the aromatic state required more drastic conditions, and these led to an increase of oxygen elimination. The dehydrogenation of some terpene ketones was described by

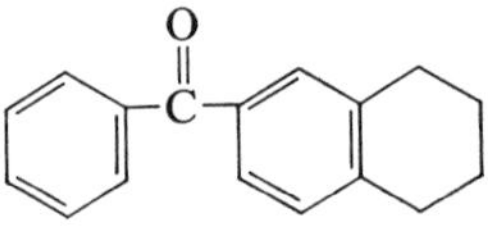

47

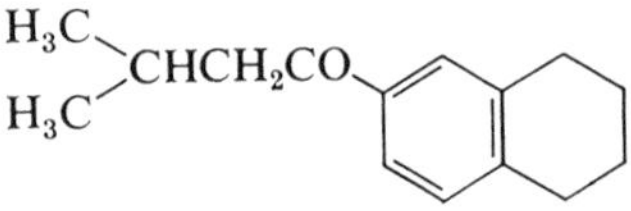

48

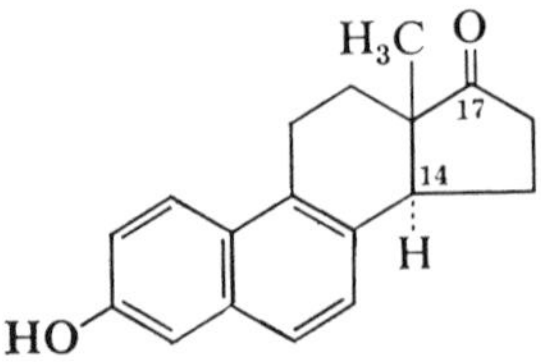

49

Linstead, Michaelis, and Thomas [113]. Cook and Somerville [180] dehydrogenated 3,4-benzocycloheptane-1,2-dione to 3,4-benzotropolone in a 10% yield with palladium–charcoal in boiling trichlorobenzene.

Keto groups situated in a side chain are usually reduced if sufficiently drastic conditions are used. However, mild treatment with sulfur and catalytic dehydrogenation sometimes leave the keto group intact. Barbot [181] obtained a 70% yield of methyl β-naphthyl ketone from β-acetyltetralin with sulfur, and Orchin and his coworkers [182] similarly obtained phenyl β-naphthyl ketone from the sulfur dehydrogenation of the ketone **47**. Catalytic dehydrogenation of **47** gave mostly the corresponding hydrocarbon. Fieser and Leffler [183] dehydrogenated the ketone **48** with palladium and obtained a mixture of β-isopentylnaphthalene and isobutyl β-naphthyl ketone.

Newman and his coworkers [136, 148–150] made a detailed study of the catalytic dehydrogenation of side-chain ketones in the tetralin series. A considerable reduction to ethyl was found to take place on dehydrogenation of acetyltetralins. However, small yields of methyl naphthyl ketones could be obtained, the best cases being 1,2,3,4-tetrahydro-5-acetylnaphthalene (51% yield) and 1,2,3,4-tetrahydro-2-acetylnaphthalene (55% yield). Only small yields of naphthylpropanones could be obtained from the dehydrogenation of tetralins substituted with the $-CH_2-CO-CH_3$ group. Furthermore, somewhat higher dehydrogenation temperatures were necessary in this series. Good yields (63–90%) of ketones were obtained from tetralins substituted with the $-CH_2-CH_2-CO-CH_3$ group.

Five-membered ring ketones are usually retained intact under mild conditions, whereas a drastic treatment eliminates both the oxygen and the carbon atom of the carbonyl group with a resultant split of the five-membered ring. Johnson and coworkers [184] obtained 4,5-benzindanone-1 by dehydrogenation of 3'-keto-3,4-dihydro-1,2-cyclopentenonaphthalene with sulfur for 1 hr at 220°C. Bachmann and Dreiding [168] investigated the dehydrogenation of equilenin (**49**) with palladium on charcoal. They found that treatment at 250°C yielded isoequilenin (epimerization at C-14), whereas heating of the methyl ether of **49** to 350°C resulted in the formation of 1-ethyl-2-methyl-7-methoxyphenanthrene. They were able to show that the missing carbon atom was eliminated as carbon dioxide and postulated a hydrolysis of the cyclopentanone ring by water present in the reaction vessel. It is interesting to note that the 17-hydroxy analog of **49** was oxidized and isomerized to isoequilenin with palladium at 250°C. Similar dehydrogenation results were obtained with 1-methyloestrone by Dreiding and Pummer [144] (see page 19). A split of a cyclopentanone ring similar to the one described by Bachmann and Dreiding [168] also takes place during the selenium dehydrogenation of some aconite alkaloids.

Some interesting ring closures leading to oxygen heterocycles have been observed by Orchin and his coworkers [185–187]. They reported the dehydrogenation of the ketone **50** to 1,9-benzoxanthene (**52**) with palladium on charcoal and found that the probable intermediate **51** is also converted into **52** on treatment with palladium.

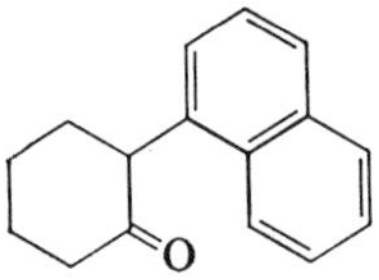

50 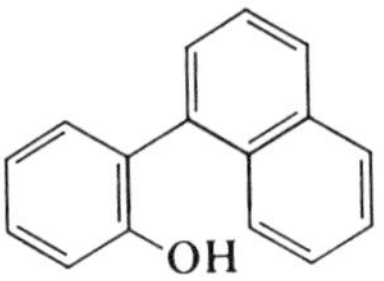   **51** 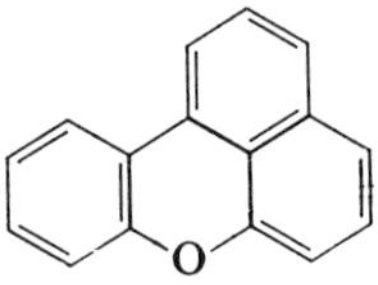   **52**

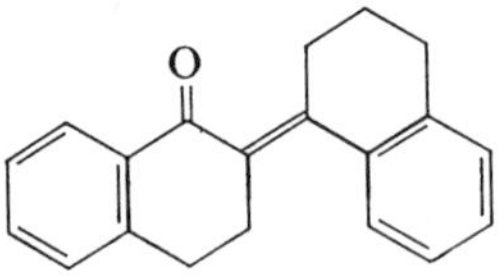

53 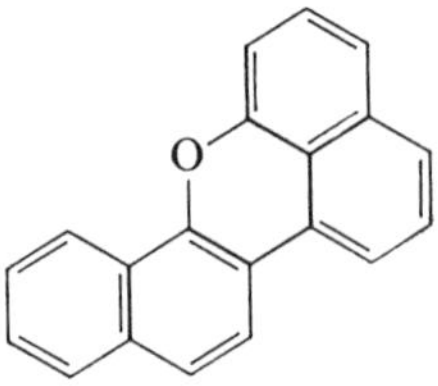   **54**

Similarly, *o*-phenylphenol yielded 8% of diphenylene oxide. Both the ketone **53** and its dihydro derivative were converted into dibenzo[*c,kl*]xanthene (**54**). Small amounts of 1,2′-dinaphthyl, naphthalene, and 2-(1′-naphthyl)-1-naphthol were also isolated from the dehydrogenation mixture. Cyclodehydrogenations of this type are interesting models for the biogenesis of natural products containing the aromatic ether moiety.

Two more aspects of the dehydrogenation of six-membered ring ketones deserve attention:

1. Ketonic groups can cause rearrangements of the Wagner-Meerwein type in analogy to secondary hydroxyl groups. These rearrangements are probably caused by the initial reduction of the ketone to a secondary alcohol and can be prevented by reducing the keto group to a $-CH_2-$ group by one of the standard reduction methods before the dehydrogenation is started.

2. If it is impossible to locate a keto group because of complete oxygen elimination during a direct dehydrogenation, the place of attachment of the oxygen function can be labeled by a Grignard reaction followed by the dehydrogenation of the tertiary alcohol to the homologous hydrocarbon [145].

Carboxylic Acids and Their Derivatives

Carboxyl Groups

Carboxyl groups attached to a quaternary carbon atom are invariably split off, both in a chemical and a catalytic dehydrogenation [47, 59, 135, 155,

188–193]. However, primary and secondary carboxyl groups can often be retained, especially in the dehydrogenation with sulfur or with a catalyst at a relatively low temperature. Darzens and Lévy [194–198] prepared substituted naphthoic acids and phenanthrenecarboxylic acids by the dehydrogenation of the appropriate tetrahydro derivatives with sulfur. The carboxyl group was usually unaffected, but the dehydrogenation of 1-methyltetralin-4-carboxylic acid was accompanied by a decarboxylation to 1-methylnaphthalene [199]. Further examples of the retention of carboxyl groups can be found in the investigations of Fieser and Hershberg [200, 201], of Cohen and his coworkers [202–204], and of Newman [205]. Dehydrogenation with selenium usually leads to a complete decarboxylation, although small amounts of aromatic acids can sometimes be isolated [73, 155, 206]. Similarly, catalytic dehydrogenation at 300°C or higher usually eliminates the carboxyl group [47, 190]. Ipatieff and his coworkers [207] investigated the decarboxylation of acid salts by the action of various catalysts.

Although in most cases the carboxyl group is either retained or completely eliminated, other transformations can take place under certain conditions. One of the more unexpected ones is the reduction to a methyl group in some dehydrogenations with selenium. Windaus and Thiele [208], for example, obtained 2,3-dimethylnaphthalene from the dihydro derivative of the maleic anhydride adduct of vitamin D_2, and Thiele and Trautmann [209] converted 2,3-naphthalic anhydride into 2,3-dimethylnaphthalene on treatment with selenium and p-cyclohexylphenol as a hydrogen donor. Similarly, 1,8-naphthalic anhydride was converted into 1-methylnaphthalene, one carboxyl group being eliminated in this case. Ruzicka [73] reported the conversion of the anhydride 55 into 1,6,7-trimethylnaphthalene with selenium at

55

350°C and Kulsi [133] observed this reduction during the dehydrogenation of khusilic acid. A similar reduction has been observed by Wiesner and his coworkers [155] in the selenium dehydrogenation of a dicarboxylic acid obtained from veatchine (see page 52).

It is clear that this reduction represents a real danger of erroneous interpretation of the dehydrogenation results. It should be pointed out, however, that it has been observed only in selenium dehydrogenations and, with few exceptions [133], only in cases in which the reducible carboxyl groups were present in the form of a cyclic anhydride or were so situated that they could form a monomeric anhydride during the reaction. This limitation makes it probable that the reduction takes place before the opening of the anhydride ring and proceeds presumably through the lactone and possibly cyclic ether intermediates. Unfortunately, no clear case of a lactone reduction with selenium necessary for the verification of this assumption seems to have been reported. One possible example is the identification of S-guaiazulene in the mixture of products from the dehydrogenation of hydrogenated lactucin by Šorm and his coworkers [210].

Another example of a possible reduction of a carboxyl group is the dehydrogenation of a transformation product of the diterpene cafestol. Djerassi and his coworkers [190, 211] proposed the structure 56 for cafestol on the basis of degradative studies, physical measurements of various transformation products, and the formation of the phenol 58 in the dehydrogenation of diacid 57 with palladium–charcoal.

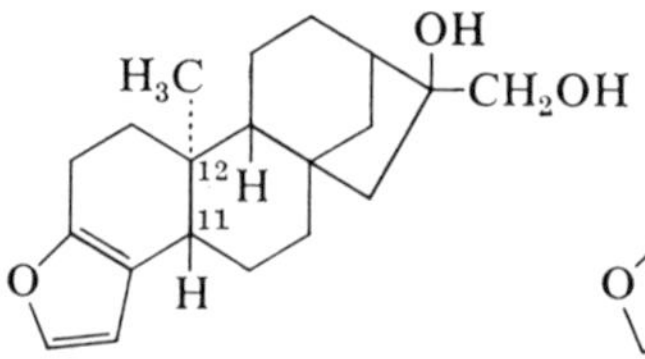

56 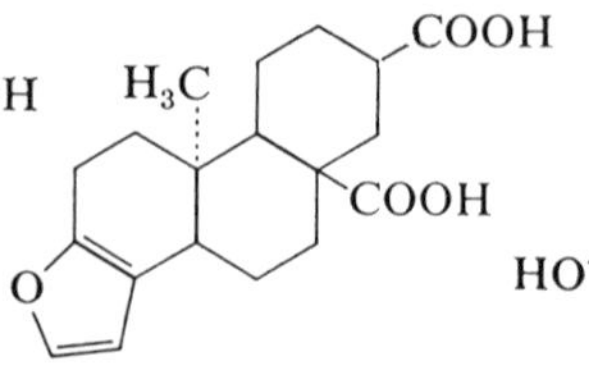   57 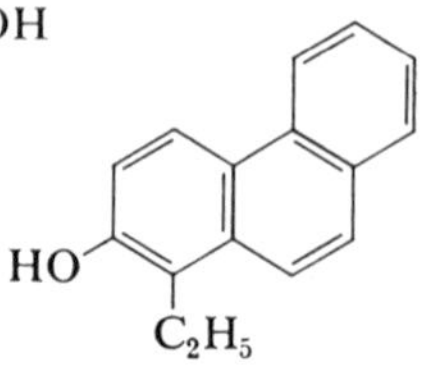   58

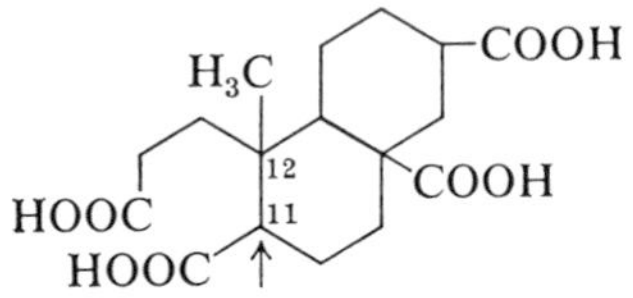

59 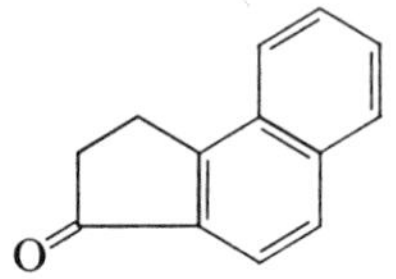   60

Haworth and Johnstone [191] converted cafestol into the tetracarboxylic acid 59, which yielded the ketone 60 and 1-ethyl-2-methylnaphthalene on dehydrogenation with selenium. On the basis of this dehydrogenation, they suggested that the angular methyl group in cafestol (and in 59) is attached at C-11. Djerassi and his coworkers [211] consider their evidence for structure 56 quite compelling and propose that Haworth's results are best explained by the reduction of the carboxyl group at C-11 (in structure 59) to a methyl group or a migration of the angular methyl group from C-12. Although a reduction of the C-11 carboxyl group to a methyl group is quite feasible in this case, it would be surprising that a methyl group rather than the propionic acid side chain (or an ethyl group after decarboxylation) would eliminate at C-12. A preferential elimination of the ethyl group has always been observed in similar cases [45, 135, 192]. A possible explanation could be the assumption that the most favorable conformation of the decalin system in **59** places the C-12 methyl group into an axial position and that axial substituents eliminate with greater ease [212]. However, no clear-cut evidence is available which would substantiate this assumption. The migration of the C-12 methyl group to C-11 during the dehydrogenation of **59** or of the possible intermediate **61** is a more likely explanation of the dehydrogenation

61

results. The cafestol dehydrogenation studies demonstrate the difficulties involved in the unambiguous interpretation of some dehydrogenation results and stress the importance of other degradative studies in the structure elucidation of complicated natural products.

Another reaction involving a carboxyl group has been observed with compounds containing nitrogen. An amino acid obtained as a degradation product of annotinine [141] was converted to a lactam on dehydrogenation with palladium (see page 54).

Ester Groups

Esterified carboxyl groups attached to a quaternary carbon atom are readily split off [145, 192, 213–215]. Other ester groups, however, quite often survive both catalytic and chemical dehydrogenation. Newman and his coworkers [136, 148, 149] studied the catalytic dehydrogenation of some esters in the tetralin series and found that the carbomethoxy group is quite stable in the temperature range 280–320°C. Similar dehydrogenations are described by Johnson and his coworkers [184] and by Zelinsky and his group [216, 217]. Fieser and Fieser [218] reported the retention of an ester group in sulfur dehydrogenation.

Cases are reported in which an ester group has been hydrolyzed during the dehydrogenation [73, 219] and in which two suitably located ester groups formed an anhydride [73]. Furthermore, suitably constituted amino esters can be transformed into lactams during the dehydrogenation (see the dehydrogenation of yohimbine and annotinine, page 53).

Lactones

Several examples of the dehydrogenation of five-membered lactones in sesquiterpenes have been reported. Almost invariably, the lactone ring was opened with the formation of a hydrocarbon with one less carbon atom. Examples are the dehydrogenation of artabsin **42** [165], matricin **43** [166], lactucin [210, 220], arborescine [221], and tenulin [222]. The one exception is the selenium dehydrogenation of lactucin by Šorm and his coworkers [210], who reported the formation of guaiazulene in addition to chamazulene. Tsuda and his coworkers [223] obtained 1,6-dimethyl-7-ethylnaphthalene in the dehydrogenation of the lactone **62** with selenium. Another example

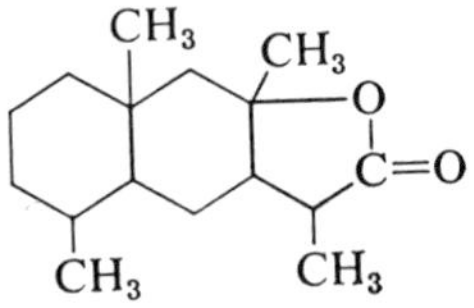

62

of a complete elimination of a lactone group is the dehydrogenation of annotinine with selenium [224] (see page 54).

Furan Rings

King and King [145] dehydrogenated vinhaticoic acid methyl ester (**63**) with selenium and obtained the phenol **64** together with 1,8-dimethyl-2-

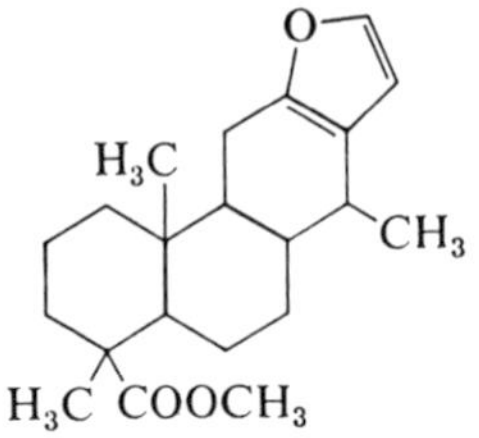

63 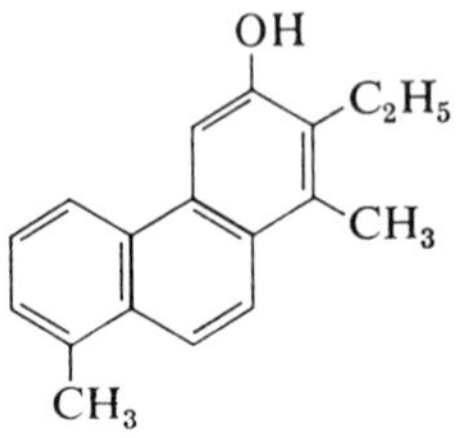   **64**

ethylphenanthrene. Mild dehydrogenation with sulfur yielded the partially dehydrogenated compound **65**, whereas a more vigorous sulfur treatment led to a compound, $C_{18}H_{14}O$, which was probably **66**. Similarly, the catalytic

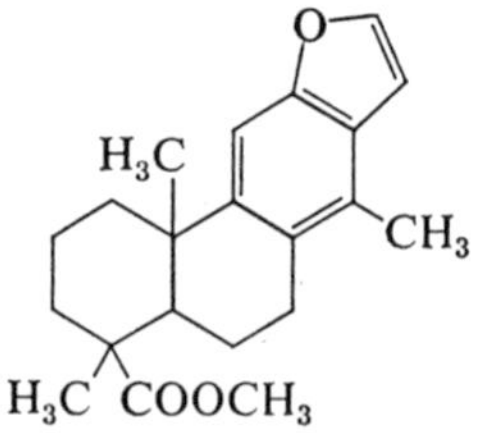

65 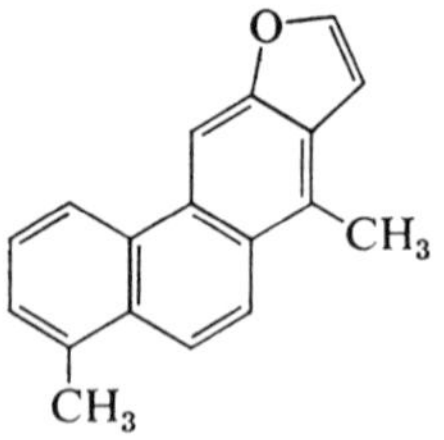   **66**

dehydrogenation of the diacid **57** derived from cafestol [190, 211] yielded the phenol **58** (see page 28). Thus it can be seen that the attachment of a furan ring can be established by dehydrogenation under a variety of conditions.

Wynberg [75] obtained β-phenylfuran by the dehydrogenation of the dihydrofuran **67** with sulfur in boiling dimethylformamide.

Artemazulene (**68**) [166, 210, 221] and linderazulene (**69**) [166, 210, 222] are formed by the dehydrogenation and, in some cases, cyclization of suitable derivatives of the guaianolides.

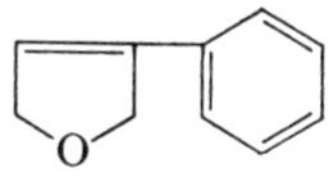

67 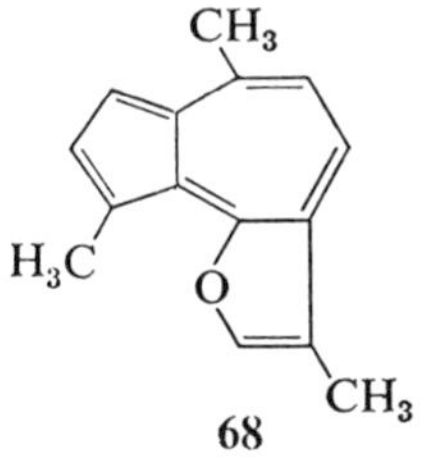   68 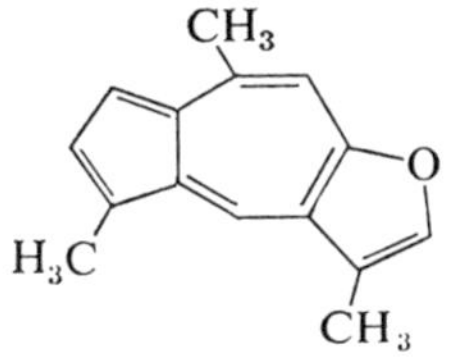   69

8   BEHAVIOR OF C–C BONDS

Dehydrogenations involving mostly or exclusively the cleavage of C–H and C–O bonds have been described so far. Actually, a great number of dehydrogenations result in a change of the carbon skeleton, caused by the elimination or migration of carbon-containing substituents; expansion, contraction, and cleavage of rings; and formation of new rings. Naturally enough, these skeletal changes are predominant under drastic conditions and should always be suspected in dehydrogenations with selenium and in catalytic reactions at higher temperatures.

Although it is not possible to formulate clear and consistent rules for all the skeletal changes encountered in the dehydrogenation, it will be attempted to rationalize as many of them as possible on the basis of the modern ideas about the reactivity of organic molecules.

Elimination of Alkyl Groups

Quaternary Alkyl Groups

The presence of an alkyl group attached to a quaternary carbon atom does not usually prevent a dehydrogenation to a completely aromatic molecule. The fact that quaternary alkyl groups are eliminated relatively easily is of importance in structure elucidations since a great number of natural products contain angular methyl groups. It is also important in this connection that these eliminations are only rarely accompanied by a 1,2 migration of the alkyl group. Exceptions to this rule and the structural prerequisites for alkyl migration are discussed on page 37.

The elimination of methyl groups was recognized for the first time by Ruzicka and his coworkers [61, 225, 226] in their dehydrogenation of selinene and eudesmol to eudalene with sulfur. Other eliminations by sulfur are the dehydrogenation of abietic acid to retene (1-methyl-7-isopropylphenanthrene)

[59, 227], the conversion of ionene into 1,6-dimethylnaphthalene [82], the dehydrogenation of fichtelite to retene [227], and the dehydrogenation of methyl vinhaticoate [145] and agathic acid [188]. In general, only slightly more vigorous conditions are necessary for sulfur dehydrogenations of compounds containing quaternary carbon atoms.

Zelinsky [228, 229] was unable to dehydrogenate 1,1-dimethylcyclohexane with a platinum catalyst, and similar failures with the catalytic dehydrogenation were reported by Cook and his coworkers [189] and by Clemo and Dickenson [230]. While the available evidence indicates that the catalytic elimination of angular alkyl groups is sometimes difficult, many examples are available in which sufficiently drastic conditions led to aromatization. Ruzicka and Waldmann obtained retene (85% yield) by dehydrogenating abietic acid [231] and fichtelite [232] with palladized charcoal. Methane was evolved together with hydrogen during the dehydrogenation. Palladized charcoal was also found effective in the dehydrogenation of dextropimaric acid [135].

Linstead and his coworkers [37, 38, 113] studied the behavior of quaternary methyl groups in catalytic dehydrogenation. They found that the disproportionation observed in the case of unsubstituted octalins (see page 17) was hindered by an angular methyl group. As a consequence, 9-methyloctalins required a higher dehydrogenation temperature (about 330°C). About the same temperature was also necessary for the dehydrogenation of unsubstituted decalins and 9-methyldecalins. In this case the addition of an angular substituent makes little difference. An interesting case is the catalytic dehydrogenation of selinene [113]. Eudalene and tetrahydroselinene are formed in a slow reaction at the unusually low temperature of 205°C.

The English authors [37, 38] also observed that a trace of 1-methylnaphthalene was formed during the dehydrogenation of 9-methyldecalin, and some 1,2-dimethylnaphthalene was formed as a by-product in the dehydrogenation of 1,1-dimethyltetralin. These anomalous migrations seem to take place especially when platinum or palladium catalysts precipitated on asbestos are used.

That catalyst metals precipitated on charcoal are relatively "safe" in this respect was further demonstrated by Adkins and England [45]. They found that the dehydrogenation at 350–375°C of 2,2,3-trimethyltetralin with platinized charcoal gave an 80% yield of 2,3-dimethylnaphthalene and only traces of 1,2,3-trimethylnaphthalene, whereas nickel on kieselguhr yielded about 40% of the trimethylnaphthalene. Since nickel on nickel chromite was intermediate in its action, it is likely that here again the percentage of migration is a function of the carrier rather than the metal. It is probable that active sites on certain carriers promote the rearrangement quite independently of the actual catalytic dehydrogenation. A similar variation in reactivity of the three catalysts was observed in the dehydrogenation of 1,1-dimethyl-1,2,3,4-tetrahydrophenanthrene and the corresponding ethyl derivative [45].

Adkins and England also dehydrogenated 2,2-dimethyltetralin and 2-methyl-2-ethyltetralin with platinized charcoal and obtained a 40% yield of 2-methylnaphthalene in both cases. Rather surprisingly, 2-methyl-2-ethyl-decalin was converted into 2-methylnaphthalene in a better yield (80%). It can be seen that an ethyl group was eliminated in preference in both cases. This preferential elimination is quite general, in both a catalytic and a chemical dehydrogenation. Similar reactions were observed with the corresponding 1,1-dialkyltetralins [44]. The tetramethyl compounds **28** and **30** (see page 20) resisted complete dehydrogenation even at higher temperatures [45]. Adkins and Hager [46] studied the dehydrogenation of various hydrogenated cyclopentanonaphthalenes, 1,2-cyclopentanophenanthrenes, and chrysenes containing angular substituents. Satisfactory alkyl eliminations were obtained, especially with platinum and palladium catalysts.

In analogy to the reaction with sulfur, selenium dehydrogenations do not appear to be hindered by quaternary carbon atoms. This can, perhaps, be explained partly by the fact that all selenium dehydrogenations are performed at relatively high temperatures. Actually, most quaternary alkyl groups are split off at approximately the same temperature (300–350°C), both with a catalyst and with selenium. There are a great number of examples of the selenium dehydrogenation of compounds containing angular substituents. Many diterpenes, triterpenes, steroids, alkaloids, and synthetic compounds containing quaternary carbon atoms have been successfully dehydrogenated by this method.

As already mentioned, ethyl groups eliminate in preference to methyl groups in most cases in the chemical dehydrogenation. For example, Barker and Clemo [233] obtained 2-methylphenanthrene from 2-methyl-2-ethyl-1,2,3,4-tetrahydrophenanthrene, and a similar elimination was observed in some diterpenes of the pimarane type (see, e.g., structure **32**, page 20) [135, 192].

Other Alkyl Groups

The retention or elimination of alkyl groups attached to carbon atoms other than quaternary seems to depend upon the nature of the alkyl group and the conditions of the reaction. While there is no unambiguous case reported in which a nonquaternary methyl group eliminated, longer side chains are lost quite often, presumably in the form of olefins in the catalytic reaction and as mercaptans and selenides or olefins in the chemical dehydrogenation. Elimination can be expected especially when selenium is used or in catalytic dehydrogenations at high temperature.

Fieser and Hershberg [171] observed the elimination of the butyric acid side chain from the tetralin **70** with selenium and Short and Wang [146] reported the elimination of an isopropyl group in a similar treatment of totarol. The maleic anhydride adduct of vitamin D_2 yields 2,3-dimethyl-naphthalene with selenium and β-naphthoic acid and naphthalene on treatment with palladized charcoal; the side chain bearing rings C and D is eliminated in both cases [208]. Cadinene is converted into 1,6-dimethyl-naphthalene on vigorous treatment with palladium [113], whereas a mild

CH_3O $CH_2CH_2CH_2COOH$

70

catalytic dehydrogenation [56] and even a prolonged treatment with selenium [234] yield 1-isopropyl-4,7-dimethylnaphthalene (cadalene). The dehydrogenation of 1,1,3-trimethyl-2-*n*-butylcyclohexane with selenium at 390–400°C yields a mixture of *m*-xylene and 1,3-dimethyl-2-*n*-butylbenzene [100]. The investigations of Cocker and his coworkers [67–69] of the dehydrogenation of tetralins **3** (see page 8) with selenium are interesting in this connection. As mentioned earlier, a methyl group in position 1 was retained, a phenyl group migrated, and ethyl and cyclohexyl groups were eliminated. Clearly, the complete aromatization of **3** is hindered in this case, since the R group and the methyl group in the *peri* position (together with an additional methyl group at C-2) are forced into one plane during aromatization. The retention of the methyl group in this case is therefore a good demonstration of its general resistance to elimination.

It can thus be seen that with the exception of methyl groups the behavior of nonquaternary alkyl groups is quite variable. The increase of the proportion of elimination with increasing temperature, however, seems to be quite general for all methods used. Another important factor is the position of the substituent. As would be expected, substituents in position 1 of a naphthalene nucleus and in positions 1 and especially 4 of a phenanthrene nucleus, for example, will eliminate with a greater ease than substituents in other positions. The retention of ethyl groups and longer side chains in relatively unhindered positions is quite common even under relatively drastic conditions.

All the reactions described so far involve a complete retention or a complete elimination of the substituent in question. There are many cases described, however, in which a methyl group appears in the product in a position occupied by a longer side chain or a ring in the starting material.

For example, the dehydrogenation of agathic acid (71) gave agathalene (72) [235], which was also obtained from the similarly constituted diterpene sclareol (73) [236, 237].

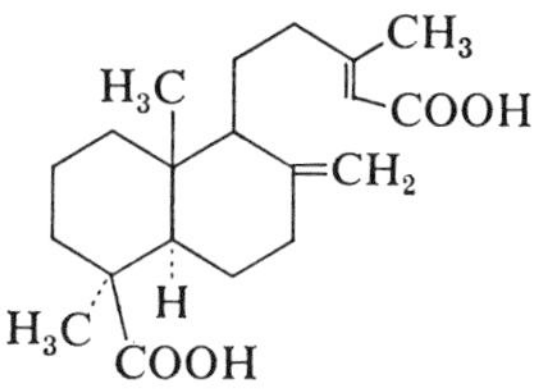

71 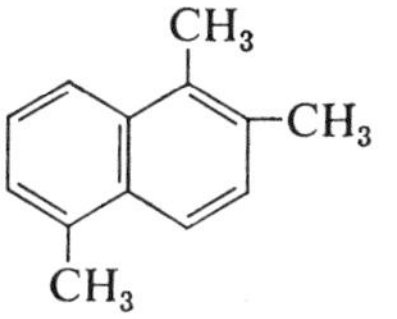   72 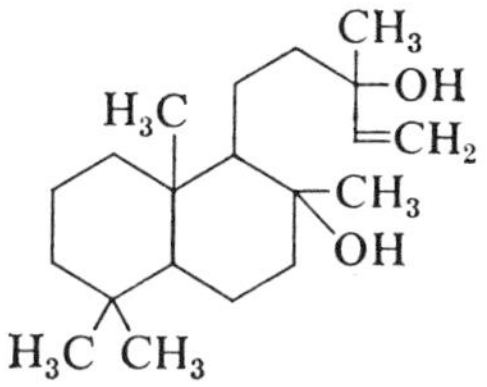   73

Agathalene was also formed in the dehydrogenation of many pentacyclic triterpenes [158] in which Ring C was broken instead of a side chain and in the dehydrogenation of tetracyclosqualene (77) and ambrein (78) [238]. It is difficult to decide whether the "new" methyl group was formed by a partial elimination of the side chain or ring or whether it migrated from the neighboring angular position. There are only very few instances recorded in which a cleavage was observed between two aliphatic carbon atoms. Matsumoto [239] dehydrogenated the hydrocarbon 74 with selenium at 350–370°C and obtained a 60% yield of the corresponding phenyl derivative, 75, together

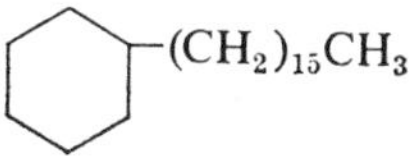

74 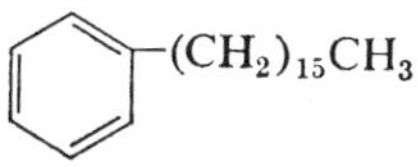   75

with 5% of pentadecane and small amounts of toluene and other hydrocarbons. This is clearly not too good a case, since the percentage of total elimination is quite small. Nevertheless, it is interesting that pentadecane and toluene were formed in preference to hexadecane and benzene. Buchta and Kallert [170] obtained 14% toluene, 17% 1-methyl-6-methoxynaphthalene, and 65% 1-(β-phenylethyl)-6-methoxynaphthalene in the dehydrogenation of compound 76 with selenium. In this case, of course, the aliphatic C–C bond is weakened by its dibenzylic position.

There is an obvious similarity between compound 76 and tetracyclosqualene (77) and ambrein (78). It appears reasonable to assume that a pyrolytic cleavage facilitated by the doubly allylic position takes place between C-11 and C-12. A similar split is also probable in the case of agathic acid and sclareol and was, in fact, proposed by Ruzicka [235]. The dehydrogenation of the triterpenes is somewhat more involved. It is interesting that triterpenes of the

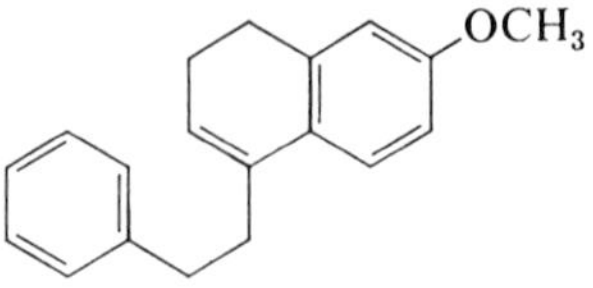

76

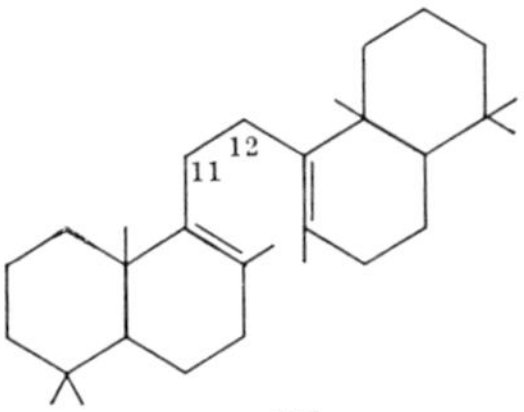

77

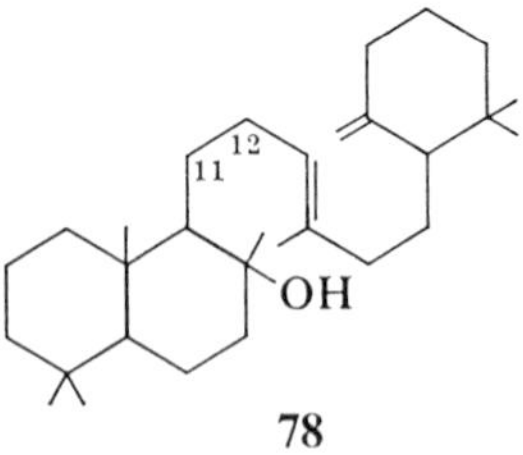

78

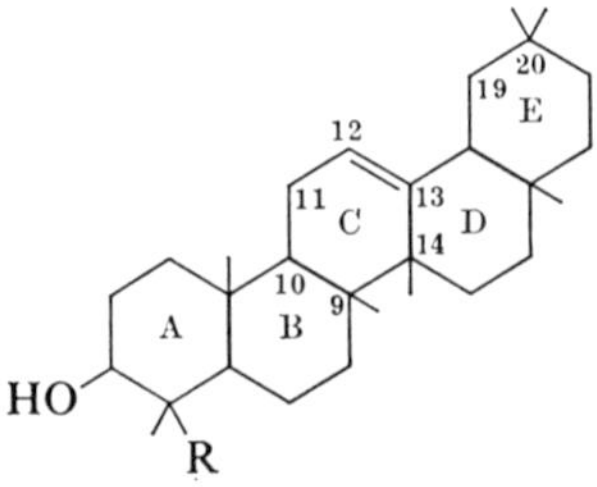

79

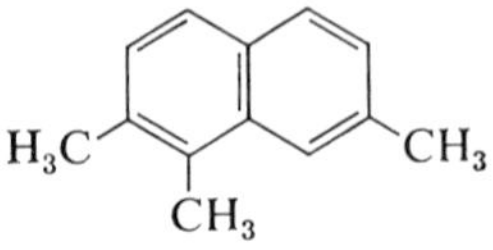

80

β-amyrin type (**79**) yield, among several other products, 1,2,7-trimethyl-naphthalene (**80**), 2,7-dimethylnaphthalene, and the already mentioned agathalene (**72**) [158]. The formation of the trimethylnaphthalene (**80**) can be explained either by a methyl migration from C-20 to C-19 and a split of the bonds between C-9 and C-14 and C-12 and C-13 or by a split of the bond between C-11 and C-12, and an elimination of the angular methyl at C-20. The second alternative is much more likely since the migration of a methyl group from a *gem*-dimethyl position by the action of selenium has only been observed in compounds containing an oxygen function in the α position (see page 37). Thus the dehydrogenation probably proceeds by the cleavage of the C-9 to C-14 bond (possibly preceded by the introduction of an additional labilizing double bond between C-10 and C-11), followed by changes analogous to the ones considered in the case of tetracyclosqualene and ambrein. 2,7-Dimethylnaphthalene would then be formed by an alternative route by the cleavage of the C-12 to C-13 bond.

A formally similar side-chain elimination is encountered in the well-known conversion of the sterols into γ-methylcyclopentenophenanthrene (**81**) on treatment with selenium. It is assumed in this case [240–243] that the steroid side chain eliminates completely and that this elimination causes the migration

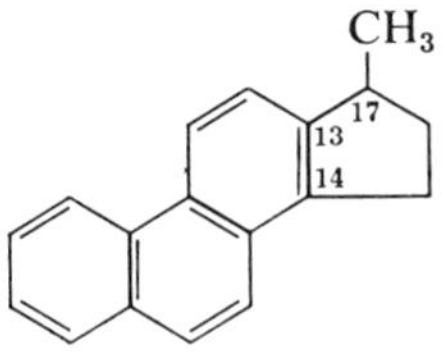

81

of the neighboring angular methyl group (see also the dehydrogenation of solanidine, page 50). The mechanism of this transformation is not clear, but probably involves the migration of the methyl group to a carbonium ion or free radical at C-17. Model compounds without the C-17 side chain, but containing an angular methyl group in the C-13 [244] or C-14 position [245], gave cyclopentenophenanthrene by methyl elimination. Similarly, 8-methylhydrindane was converted into indane on treatment with palladium–charcoal [246].

An interesting recent study [8] deals with the purity and content of **81** in steroid dehydrogenation mixtures.

Migration of Alkyl Groups

Three types of alkyl migration have been described already. They are the rearrangement of *gem*-dimethyl to 1,2-dimethyl compounds in the reaction with high potential quinones (see Section 5), a similar rearrangement encountered in the catalytic dehydrogenation (see page 32), and the possible migration of angular methyl groups during the elimination of α-standing side chains (see above).

An alkyl migration of common occurrence is the retropinacol rearrangement of compounds containing partial structure **82**. Triterpenes containing this moiety [158, 247–249] yield 1,2,5,6-tetramethylnaphthalene on dehydrogenation with selenium. The oxygen elimination probably proceeds through the ion **83**. It is interesting that terpenes containing the structural unit **84**

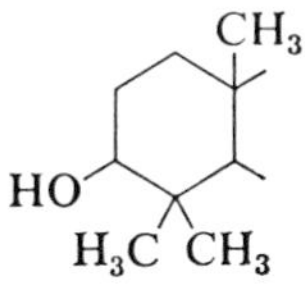

82 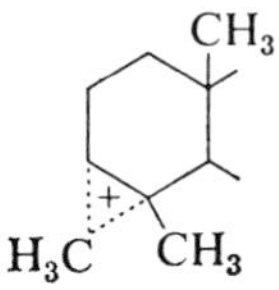   **83** 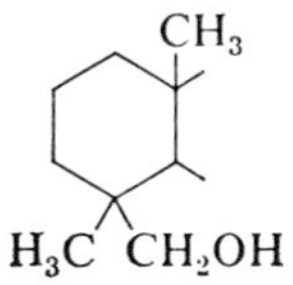   **84**

give 1-methyl- and *not* 1-ethyl-substituted aromatic hydrocarbons on dehydrogenation [71]. The elimination of the angular –CH$_2$OH group (see page 19) is clearly faster than the rearrangement. However, if water is eliminated before the dehydrogenation by a standard method, 1-ethyl compounds are obtained [71, 145]. Not surprisingly, acetylation of compounds with the partial structure **82** strongly suppresses the rearrangement [250, 251]. Acetic acid will eliminate through a cyclic mechanism indicated in **85** in analogy to the Tchugaev reaction and other pyrolytic *cis* eliminations

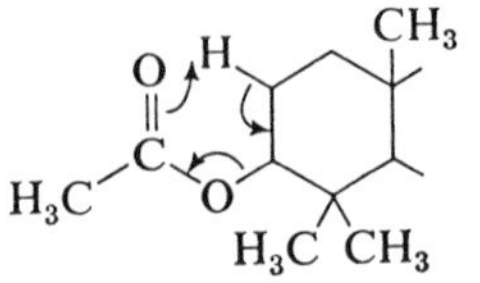

85

[252]. Rearrangements of the Wagner-Meerwein type can also be expected with suitably situated keto groups [206].

Migrations of nonquaternary alkyl groups are quite rare and are encountered mainly in the reaction with selenium at relatively high temperatures. It has been known for some time that α-alkylnaphthalenes are converted into β-alkylnaphthalenes on silica gel at about 420°C and even lower (350°C), in the case of α-phenylnaphthalene [253]. Cocker and his coworkers [69] showed that phenyl groups migrate to the naphthalene β position (see page 8) on treatment with selenium. In analogy to the elimination of alkyl groups, migrations will sometimes occur when substituents are located in relatively hindered positions of the various aromatic nuclei. These migrations or eliminations can often be prevented by using sulfur or a catalyst instead of selenium. For example, Fieser and his coworkers [254] reported the conversion of **86** into 6-methylchrysene (**87**) in a

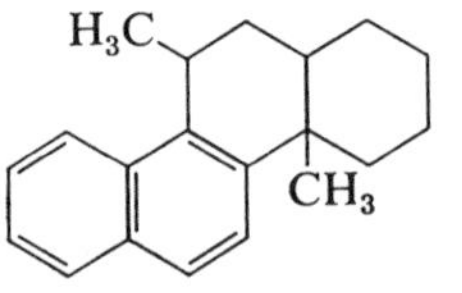

86 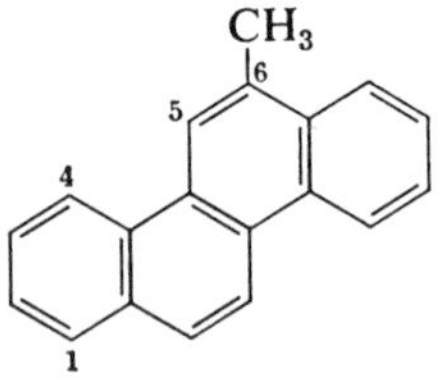   **87**

low yield on treatment with selenium. However, Newman [255] was able to prepare even 4,5-dimethylchrysene in a low yield by a mild sulfur dehydrogenation. Jones and Ramadge [256] reported the conversion of ketones **88**

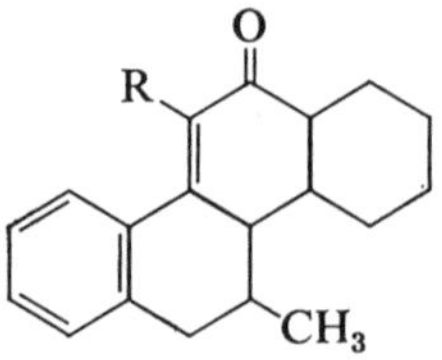

88: R = CH$_3$
89: R = H

and **89** into chrysene and a small amount of monomethylchrysene by a Clemmensen reduction followed by a selenium dehydrogenation of the unisolated intermediate.

If their structural assignments are correct and if, furthermore, the reaction does not involve a more deep-seated rearrangement of the molecule, this dehydrogenation would be one example of a loss of nonquaternary methyl groups. The methyl group elimination postulated by Ruzicka and his coworkers [236] in the dehydrogenation of a sclareol transformation product can be explained by an alternative cyclization of Ring C on treatment with formic acid.

An interesting methyl group migration was observed by Haworth and his coworkers [215]. The dehydrogenation of **90** with selenium yielded 1,8-dimethylphenanthrene (**92**) together with a small amount of the expected

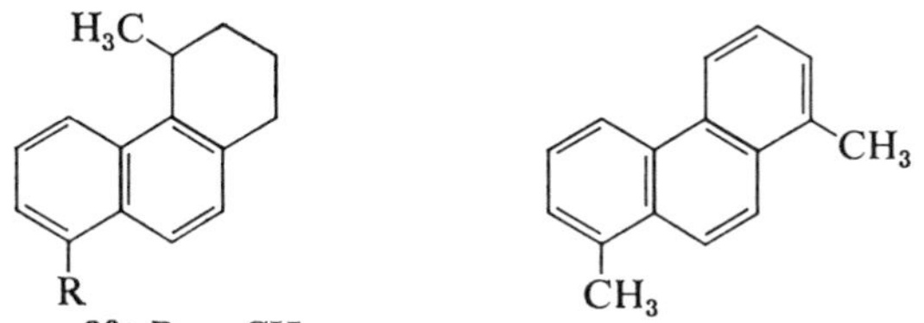

90: R = CH₃
91: R = H

92

1,5-dimethylphenanthrene. On the other hand, compound **93** was converted into 4-methylphenanthrene with selenium. It is attractive to assume that this "methyl migration" proceeds through the spiro intermediate **94** rather than a

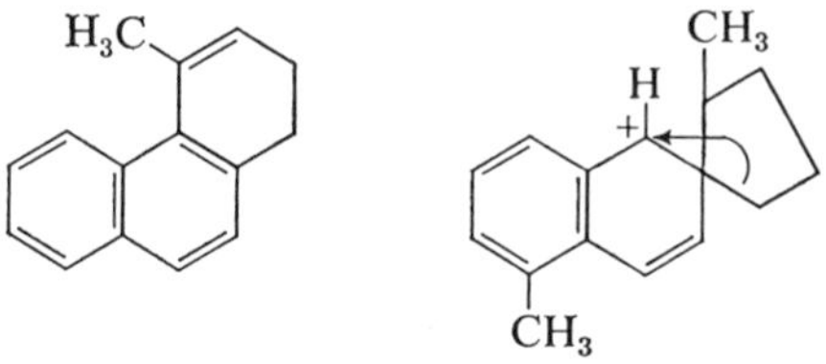

93   94

bicyclo[2.2.1] intermediate as originally considered by Haworth (see next section for a discussion of the expansion of spiro rings). Bachmann and Edgerton [257] obtained 4-methylphenanthrene by the dehydrogenation of **91** with palladium–charcoal.

The formation of the phenanthrene **96** on treatment of **95** with selenium [258] was explained satisfactorily by the well-understood dienone-phenol rearrangement [259].

The selenium dehydrogenation of some sesquiterpenes yields *Se*-guaiazulene (**97**) instead of *S*-guaiazulene (**38**) [260, 261]. In this case the methyl group migrates from C-1 to C-2 at high temperature.

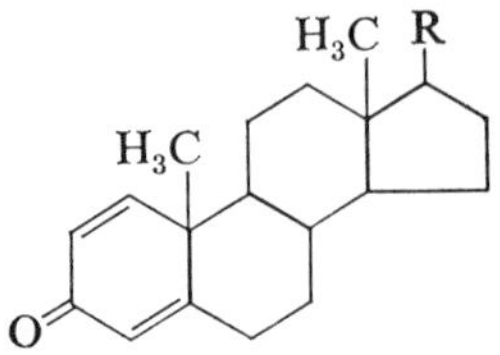

95

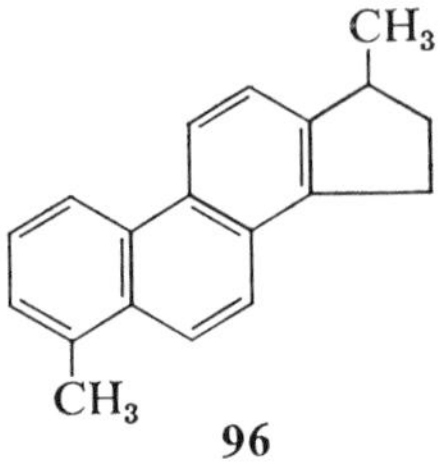

96

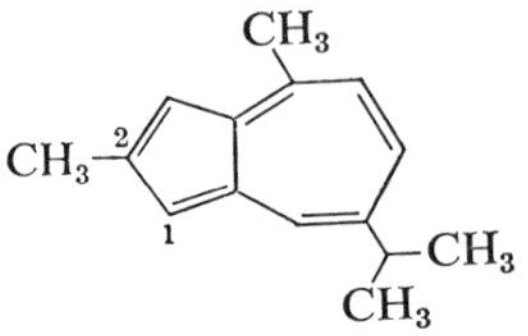

97

Expansion of Rings

Ring expansions are frequently encountered in selenium dehydrogenations and with sulfur and catalysts at relatively high temperatures. Spiro compounds are particularly susceptible to this type of skeletal change. Two rearrangements proceeding through spiro intermediates have already been mentioned (see above). Clemo and Ormston [262] obtained naphthalene in the dehydrogenation of cyclopentanespirocyclohexanone and the corresponding hydrocarbon with selenium. Cook and Hewett [139] reported the conversion of **98** into pyrene and of **99** into 1-methylpyrene (**100**) on treatment with selenium. It is interesting that the dehydrogenation of **98** involves the loss of two carbon atoms, whereas all carbons are retained in the case of **99**. The likely reason is a greater tendency of an ethyl group to eliminate.

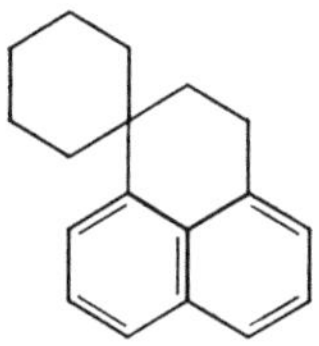

98 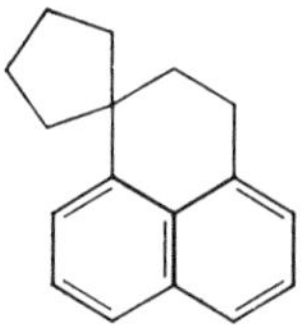   99 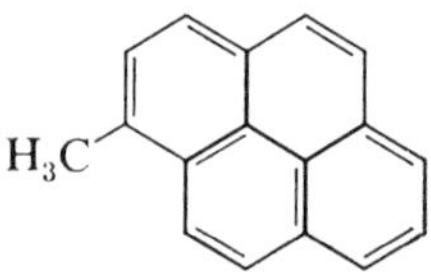   100

Šorm and his coworkers [263] obtained naphthalene derivatives in the dehydrogenation of some transformation products of the unusual sesquiterpenes acorone (**101**) and isoacorone. The product of a Grignard reaction yielded **102**, whereas the two derived dienes, acordiene and isoacordiene, gave the hydrocarbons **103** and **104**, respectively.

It is, of course, quite likely that many transformations of this type take place during the preparation of the starting materials rather than during the actual dehydrogenation, especially when acidic reagents are employed.

Sen Gupta and Chatterjee studied the behavior of many spiro compounds in the dehydrogenation with platinum–charcoal and selenium [264–269]. They found that the catalytic reaction gives better yields and less elimination. The catalytic dehydrogenation of compound **105** (R = H or CH$_3$) gave the corresponding 1-substituted phenanthrenes **106** with preferential migration

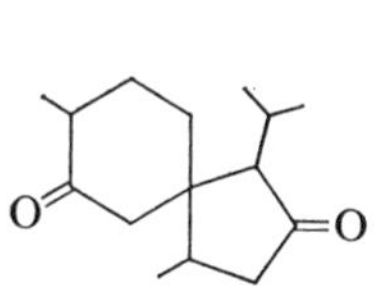

101 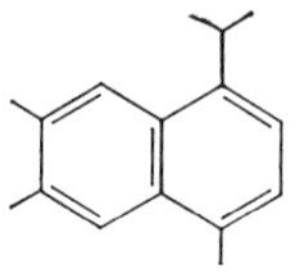   **102** 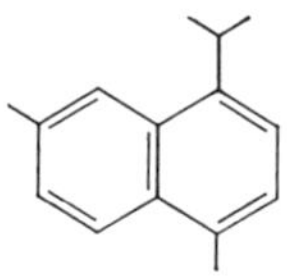   **103** 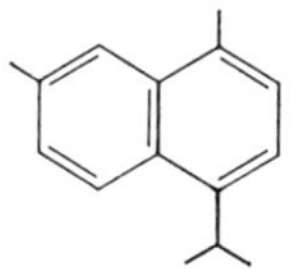   **104**

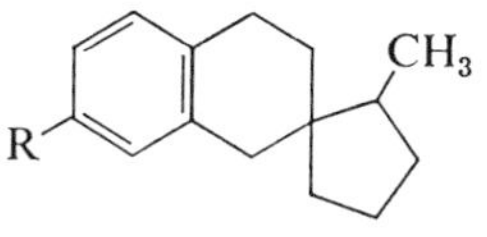

105 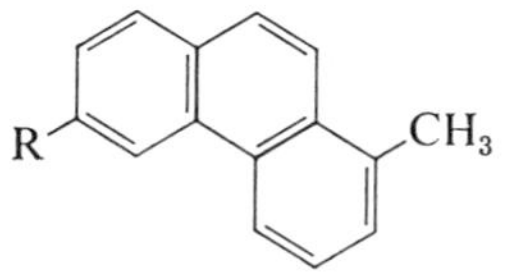   **106**

of the –CH$_2$– group. Similar results were obtained with analogous compounds containing additional rings. An unusual influence of substitution was observed in the catalytic dehydrogenation of **107** and **108**. While **107** yielded 2-methylphenanthrene, **108** gave 3,6-dimethylphenanthrene (**109**) and a mixture of 2,6- and 2,7-dimethylanthracenes. The selenium dehydrogenation of **105** (R = H) appeared to give small amounts of phenanthrene and a

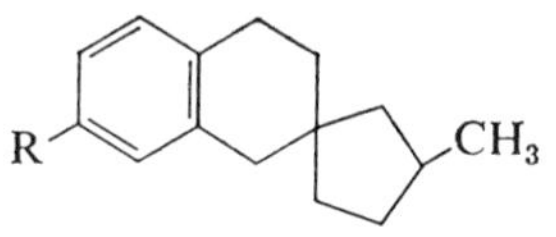

107: R = H
108: R = CH$_3$

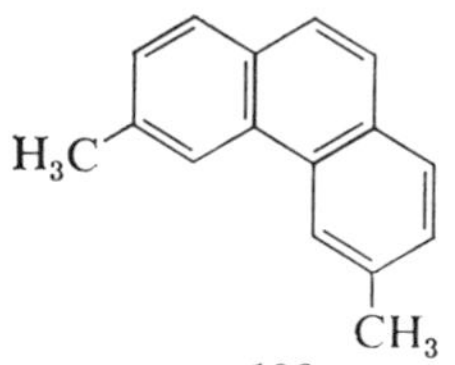

109

similar treatment of 1,2,3,4-tetrahydronaphthalene-2,1′-spirocyclohexane was reported to yield only phenanthrene and anthracene.

Levitz and Bogert [270] converted 4-methyl-1,2,3,4-tetrahydronaphthalene-1,1′-spirocyclopentane into 9-methylphenanthrene with palladium–charcoal and Marvel and Brooks [271] similarly prepared naphthalene and 1-methylnaphthalene from spirodecane and 3-methylspirodecane, respectively. The observed inertness of these spiranes to selenium can be explained by the general resistance of perhydro compounds to chemical dehydrogenation. Surprisingly, Zelinsky and Elaghina [272] obtained 1-methyl-2-ethylbenzene as the main product of a vapor phase dehydrogenation of spirodecane over platinum–charcoal. Adkins and his coworkers obtained a 25–40% yield of phenanthrene from 1,2,3,4-tetrahydronaphthalene-1,1'-spirocyclopentane with nickel on kieselguhr [44] and a 15–50% yield of chrysene from the corresponding tetrahydrophenanthrene derivative with various catalysts [45]. Ring expansions must also be involved in the complex catalytic dehydrogenation of isolongifolene [43].

Some instances are also reported in which spiranes did not rearrange during the dehydrogenation, particularly under relatively mild conditions [189, 213, 273].

A ring enlargement of a different type was first observed during the dehydrogenation of sterols. Diels and his coworkers obtained chrysene in the dehydrogenation of cholesterol with palladium–charcoal [27] and of cholic acid with selenium [274]. According to Ruzicka [275, 276] this ring enlargement requires temperatures above 400°C.

It is impossible to decide whether this rearrangement takes place directly on the steroid nucleus or on the initially formed γ-methylcyclopentenophenanthrene (81). Positive model experiments are available for both alternatives. Ruzicka and Peyer [131] obtained a good yield of naphthalene from α-methylindane with selenium or palladium–charcoal at 450°C. Naphthalene was also formed from β-methyl-, from α- and β-ethyl-, and from isooctylindane. Adkins and Mayer [46] converted the hydrocarbon 110 into chrysene with a platinum catalyst in a 10% yield.

The fact that a ring enlargement of this type can sometimes take place under relatively mild conditions is demonstrated by the experiments of Denisenko [277] who obtained 1,3-diphenylpropane from the hydrocarbon 111 and 1-methylphenanthrene from 112 on treatment with platinum–

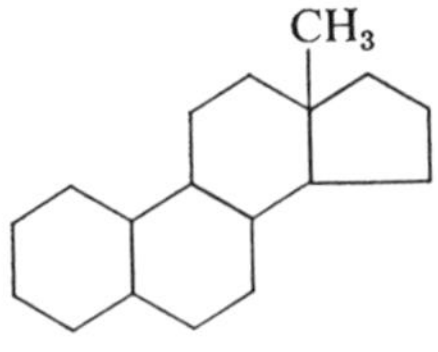

110

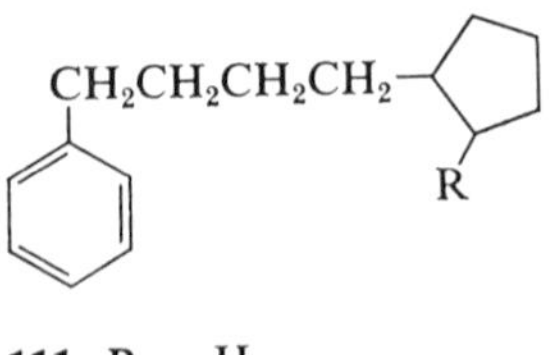

111: R = H
112: R = CH₃

charcoal at 310°C. Similarly, Nenitzescu and Cioränescu [278] reported the formation of a small amount of naphthalene in the vapor phase dehydrogenation of α-methylindane over platinized charcoal at 310–350°C. It is not certain whether these catalytic reactions are simple ring enlargements or involve a cyclization of an aliphatic side chain formed as an intermediate (see page 44). Robinson [279] obtained phenanthrene by a sulfur dehydrogenation of 1-cyclopentylindane.

The formation of naphthalene as a by-product in the catalytic dehydrogenation of hydroazulenes [54] represents another type of ring enlargement.

Contraction of Rings

Ruzicka and Seidel [100] studied the behavior of cycloheptanes and cyclooctanes in the dehydrogenation with selenium in the temperature range 370–440°C. Cycloheptane yielded toluene at 440°C, cyclooctane was converted into p-xylene at 390–410°C, and the corresponding alkyl derivatives yielded various mono- and polysubstituted benzenes, characterized by their conversion into benzenecarboxylic acids.

Pines and his coworkers [32] prepared fluorene in a reaction of cycloheptylbenzene with platinum on alumina at 380°C, Turova-Pollak and Rappoport [280] obtained p-xylene together with traces of the meta isomer in the vapor phase dehydrogenation of methylcycloheptane over platinum–charcoal at 310°C, and Horn and his coworkers [281, 282] obtained substituted naphthalenes and azulenes in the dehydrogenation of various bicylo[5.5.0] and bicyclo[5.4.0] compounds over palladium–charcoal. The contraction of dimethyldecahydrochrysenes to isopropylbenzofluorenes has recently been reported [283]. Other examples are the dehydrogenation of ibogaine (12) [72] and cevine (135) [137].

Cleavage of Rings

Three-Membered Rings

The cleavage of three-membered rings was often observed during the dehydrogenation of mono- and sesquiterpenes. Zelinsky and Levina [284] found that the catalytic treatment of carane produced p-cymene, whereas thujane cleaves by an alternative route to give a cyclopentene derivative which resists dehydrogenation. Obviously, a cleavage of this type is a pyrolytic reaction which is not necessarily governed by the subsequent aromatization.

Cadalene is formed in the sulfur dehydrogenation of the tricyclic sesquiterpene copaene [285] and the formation of an azulene together with cadalene in the dehydrogenation of ledol [286] was reported. The conversions of aromadendrene [287], globulol [288], and himbaccol [289] into guaiazulene (38) by dehydrogenation also involve the cleavage of three-membered rings.

Four-Membered Rings

Zelinsky [122] obtained pinane and some aromatic compounds on passing pinene vapor over palladized asbestos at 190°C. Linstead and his coworkers [113] observed the formation of approximately equimolecular amounts of

CH$_3$

103

p-cymene and pinane on boiling pinene (**103**) with platinized charcoal at 156°C. Pinane itself could be converted into *p*-cymene at 300°C in a slow reaction.

In contrast to this specific cleavage of the C-6 to C-7 bond, Pines and his coworkers [33] reported the formation of a mixture of *o*-cymene, *p*-cymene, and 1,2,3- and 1,2,4-trimethylbenzenes on treatment with platinized alumina.

A four-membered ring cleavage takes place during the dehydrogenation of annotinine and its transformation products (see page 54).

Five-Membered Rings

Nonbridged five-membered rings are relatively stable to dehydrogenation. For example, cyclopentenophenanthrenes are obtained from many steroid derivatives and several products containing cyclopentane rings were formed in the dehydrogenation of cevine (see page 50). Barrett and Linstead [290] reported the stability of *cis*-bicyclo[3.3.0]octane to platinum at 310°C. Further confirmatory evidence is provided by the work of Zelinsky and his coworkers [291, 292] and of Errington and Linstead [246].

However, instances have been reported in which cyclopentanes had been hydrogenated to aliphatic hydrocarbons under sufficiently drastic catalytic conditions [293–295] and in which substituted benzenes were also formed [296] (see also page 43). While it is not known with certainty whether this formation of aromatic substances proceeds by ring enlargement or by an opening and subsequent closing of the ring, it has been shown repeatedly that aromatics can be formed by a catalytic cyclodehydrogenation of aliphatic hydrocarbons. This reaction is one of the methods used for gasoline refining [41, 42]. This catalytic side reaction does not represent a great danger in degradative studies, since the yields are generally low and special catalysts are often required.

A cleavage of five-membered rings may be expected, however, if their presence hinders complete aromatization or if they are labilized by oxygen substituents. For example, 1,2,8-trimethylphenanthrene is the main product of the selenium dehydrogenation of tetracyclic triterpenes [158, 297]. The five-membered ring D is completely eliminated in this case. The formation of vetivazulene from tricyclovetivene [298], of retene from phyllocladene [299], and of 1,7-dimethylphenanthrene from rimuene [299] also involves the cleavage of a five-membered ring. The labilizing effect of an oxygen function is shown by the work of Dreiding and Pummer [144] with 1-methylestrone (24), of Bachmann and Dreiding [168] with equilenin (49), and by the results obtained in the dehydrogenation of some aconite alkaloids.

Six-Membered Rings

The formation of naphthalene derivatives in the dehydrogenation of pentacyclic triterpenes (see page 36) is an example of the cleavage of six-membered rings. This type of cleavage is, of course, not too common since other structural elements are usually eliminated in preference to six-membered rings.

Not surprisingly, bicyclo[2.2.2]octanes dehydrogenate readily [300, 301], the dehydrogenation being accompanied by a reverse Diels-Alder reaction. Elaghina and Zelinsky [302] obtained diphenylmethane in the dehydrogenation of 2,3-benzobicyclo[3.3.1]-2-nonene with platinized charcoal at 300°C. Fluorene was formed in addition when a more active catalyst was used.

An interesting border case was described by Wiesner and his coworkers [303, 304]. The dehydrogenation of 114 ($n = 10$ and 14) with palladium–charcoal in boiling naphthalene yielded the corresponding phthalic anhydrides

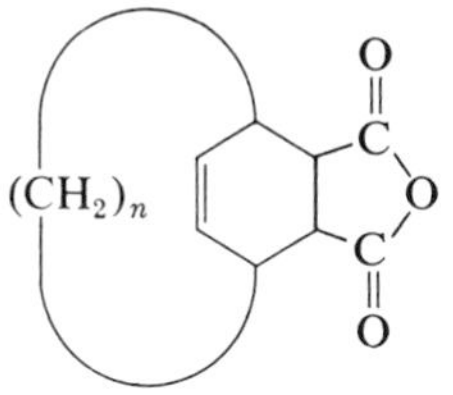

114 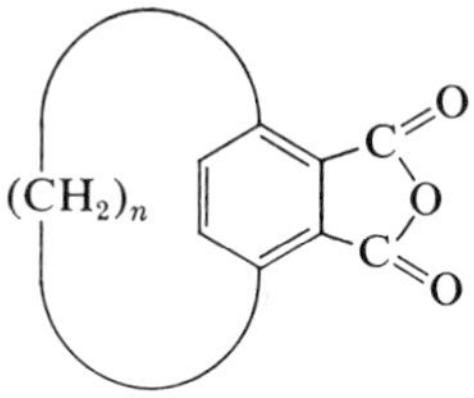   115

115 ($n = 10$ and 14). The solvent free treatment of 114 ($n = 9$) with palladium–charcoal gave mostly starting material while a dehydrogenation with selenium yielded the diacid corresponding to 115 ($n = 9$) together with some 1,4-nonamethylenebenzene. The catalytic dehydrogenation of 114 ($n = 8$) gave only 1,3-cyclododecadiene formed by a reverse Diels-Alder reaction.

A cleavage due to ring strain was also observed by Cram and Allinger [305] in the (2,2)paracyclophane series.

Formation of New Rings

Ring closures are often encountered both in a catalytic and a chemical dehydrogenation. One of the classical examples is the formation of cadalene in the sulfur dehydrogenation of the monocyclic sesquiterpene zingiberene (116) [61, 306].

Since the isomeric sesquiterpene bisabolene (117) yields only benzenoid hydrocarbons under similar conditions, it can be assumed that the cyclization of (116) precedes the actual dehydrogenation and is caused by a favorable

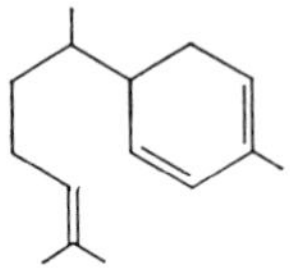

116 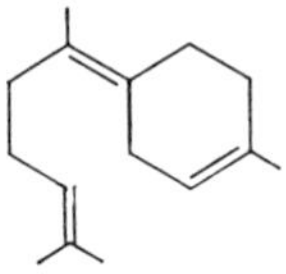   117

disposition of the reacting double bonds. A similar reaction has been observed in the dehydrogenation of agathic acid (71) [235] and of elemol [307]. 1,7-Dimethylphenanthrene was obtained in addition to the expected agathalene (72) from agathic acid. Büchi and Pappas [188] suggested that this cyclization was probably catalyzed by the hydrogen sulfide (or selenide) formed during the dehydrogenation and showed, furthermore, that the additional naphthalene derivative obtained in the dehydrogenation had the tricyclic structure 118, formed by an alternative cyclization.

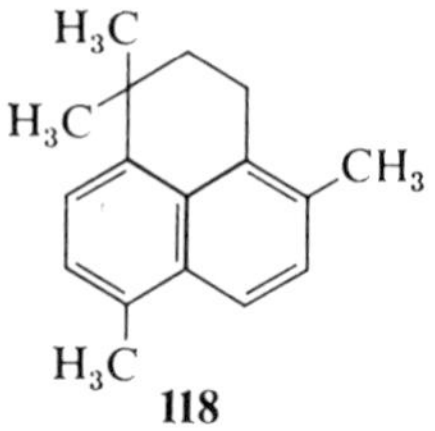

118

A great many cyclizations of synthetic compounds have been described. For example, Zelinsky and his colleagues [308] reported the formation of fluorene from diphenyl- or dicyclohexylmethane and of phenanthrene from sym-diphenylethane over platinum–charcoal at 300°C. Berger [309] obtained phenanthrene by heating o,o'-bitolyl with sulfur, while Orchin and Woolfolk [310] obtained 4-methylfluorene from the same compound with palladium–charcoal at 450°C. Orchin and his coworkers also reported the formation of fluorene from 2-methylbiphenyl [311], of a mixture of fluorene (as the major product), 9-methylfluorene, and phenanthrene from 2-ethylbiphenyl [312], of fluoranthene from 1-phenylnaphthalene [313], and of perylene from 1,1'-binaphthyl [314]. All these catalytic cyclodehydrogenations required temperatures of 450°C or higher. Crawford and Supanekar [39] used palladium in combination with sulfur to produce benzonaphthofluoranthenes.

Sen Gupta and Chatterjee [267] obtained an almost quantitative yield of pyrene from 4-ethylphenanthrene with platinum–charcoal at a surprisingly low temperature (300–320°C). They also reported the formation of pyrene and its derivatives from suitably substituted spiro compounds [267–269]. The formation of fluorene from cycloheptylbenzene [32] has already been mentioned. Deuschel [57] reported the double cyclization of 1,4-dicyclohexyl-2,5-dimethylbenzene at 400°C.

Some examples of ring formation during a selenium dehydrogenation are the formation of naphthalene from 1,2-diethyl-1-cyclohexene at 420°C [131], the dehydrogenation of vitamin A [315], the formation of small amounts of pyrene from 4,5-dimethylphenanthrene [316], and the cyclization to **6** described by Cocker and Jenkinson [69] (see page 9). The surprising formation of the cyclopentenone **60** in the selenium dehydrogenation of a cafestol transformation product [191] deserves a mention at this point (see page 28).

In addition to γ-methylcyclopentenophenanthrene (**81**) and chrysene, several pentacyclic hydrocarbons are formed during the dehydrogenation of sterols and bile acids. Bachmann et al. [317] showed by the synthesis of a comparison sample that the hydrocarbon, m.p. 274°C, formed in the selenium dehydrogenation of cholic acid (**119**) at various temperatures [156, 276, 318] is very probably the naphthofluorene **120**. The formation of **120** is simply explained by the elimination of the C-13 angular methyl group and cyclization of the side chain followed by elimination of the oxygen function.

The naphthofluorenes obtained on similar treatment of the sterols [275, 319] however, are not formed by an analogous cyclization since the synthetic compound **121** was found to be different from the pentacyclic hydrocarbon obtained from cholesterol [320]. Bergmann [321] suggested that these hydrocarbons have, in fact, the structure **123** (R = H or CH_3) and are formed by the cleavage of the five-membered ring followed by a double cyclization of the intermediate **122**. This proposal was based on the possibly analogous formation of 9-methyl-11*H*-naphtho[2,1-*a*]fluorene from strophanthidin. The close similarity of the ultraviolet spectra of **121** and the hydrocarbon obtained from cholesterol [320] and the probable identity of a synthetic specimen with the naphthafluorene from strophanthidine [321] seem to substantiate Bergmann's postulate. The main objection is clearly the unprecedented nature of the transformation.

A different type of ring closure is involved in the formation of picene (**125**) by the dehydrogenation of cholic acid (**119**) or cholatrienic acid with selenium at 360°C [156, 318]. Ruzicka and his coworkers [276] explain this reaction by the cleavage of the bond between C-13 and C-17 followed by a cyclization to a six-membered Ring D, aromatization, and a further ring closure to **124** under decarboxylation. Picene is then formed by the expansion (possibly by cleavage and reformation) of Ring E. The preferential formation of

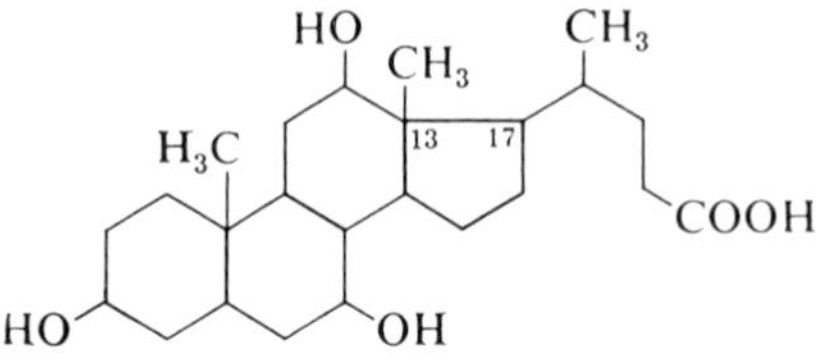

119

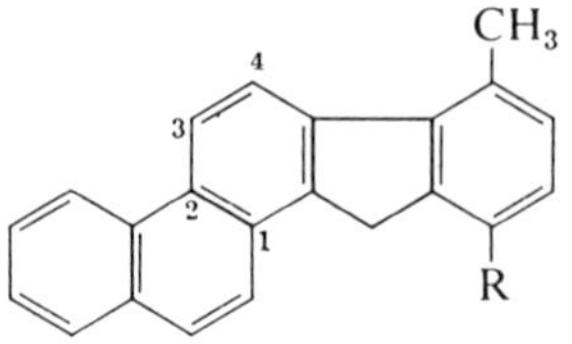

120: R = H
121: R = isopropyl

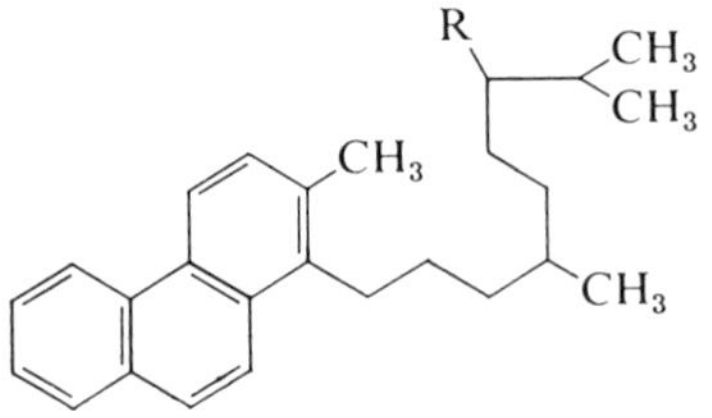

122

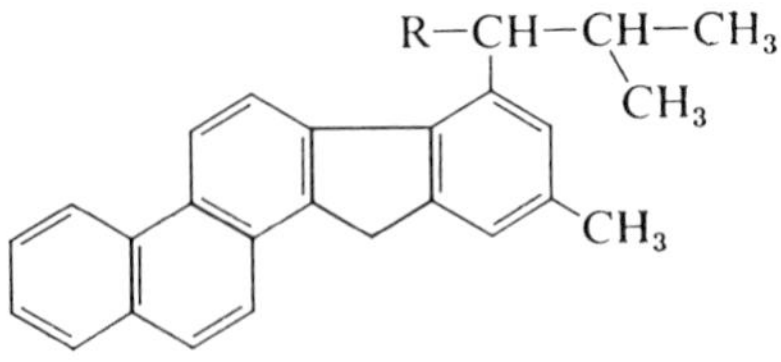

123

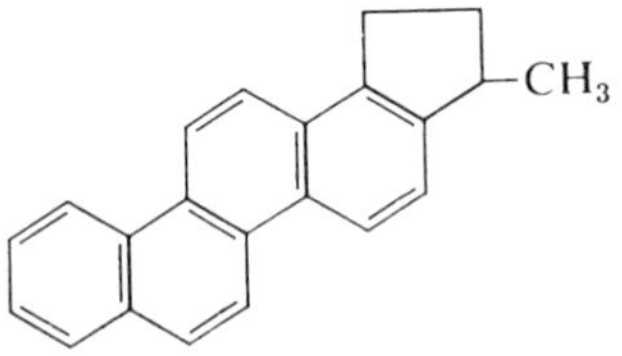

124

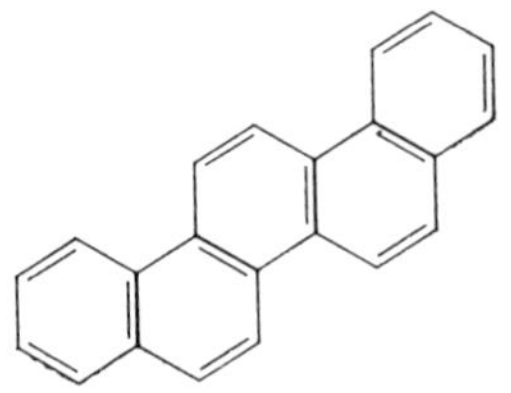

125

chrysene in the dehydrogenation of sterols at high temperatures (above 400°C) can probably be explained by a rapid elimination of the side chain under these conditions.

9   BEHAVIOR OF C–N BONDS

Simple Compounds

There is no fundamental difference between the behavior of cyclic C–C and C–N bonds in dehydrogenation. Since a tertiary nitrogen atom usually hinders a complete aromatization in a way analogous to a quaternary carbon atom, a cleavage of one of the three C–N bonds will take place under suitable conditions. The aromatization of tertiary bases of the type **126** is usually

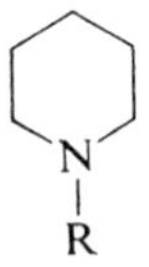

126

accompanied by the loss of the R (alkyl) group. The complete elimination of ring-standing tertiary nitrogen atoms has also been reported [72, 137].

In general, nitrogen-containing rings dehydrogenate more readily than the corresponding carbocycles. Janot and his coworkers [322] observed a rapid dehydrogenation of piperidine over palladized charcoal at 180–200°C. The same catalyst dehydrogenated cyclohexane rings more slowly at temperatures above 280°C. Decahydroquinoline can be converted into 5,6,7,8-tetrahydro-quinoline by a partial dehydrogenation [143]. Since partial hydrogenation of quinoline yields 1,2,3,4-tetrahydroquinoline, both tetrahydro derivatives are readily available.

The catalytic dehydrogenation of piperidines [30, 323, 324] and pyrrolidines [323, 325, 326] gives the corresponding aromatic hydrocarbons without difficulty. Similarly, decahydroquinoline [143] and decahydroisoquinoline [327] dehydrogenate quite readily. 5,6,7,8-Tetrahydroquinoline gives a good yield (85%) of quinoline with palladium–charcoal at 300°C [328]. Witkop [329] found that cis-decahydroisoquinoline dehydrogenated more readily than the corresponding trans isomer with palladium at 210°C in analogy to the results obtained in the decahydroquinoline series [143]. A similar difference in the dehydrogenation tendency was observed by Lettir, Goutarel, and Janot [330] in their study of yohimbane and alloyohimbane. Tetrahydrocarbazole is readily dehydrogenated by chloranil [331] or in a catalytic reaction [332, 333].

An interesting rearrangement was observed by Prelog and his coworkers [334] during the dehydrogenation of 3-acetyl-1,4-dimethyl-1,2,5,6-tetra-hydropyridine with selenium or palladium. The formation of 2,3,4-trimethyl-pyridine in this reaction was explained by a ring opening through β-elimi-nation of the nitrogen followed by a ring closure between the nitrogen atom and the carbonyl function.

In connection with their study of alkaloids, Späth and Galinovsky [335] investigated the catalytic dehydrogenation of α-piperidone, dihydrocarbo-styril, and their derivatives. Prelog and Balenović [336] studied the behavior of bicyclic alkaloid models in the dehydrogenation with palladium–charcoal and selenium and found no fundamental differences between the two reagents. 1-Azabicyclo[2.2.1]heptane yielded γ-picoline, and quinuclidine (1-azabi-cyclo[2.2.2]octane) was converted into 4-ethylpyridine. Surprisingly, quino-line was obtained in small quantity as the only isolable product from the dehydrogenation of 1-azabicyclo[4.4.0]decane.

Alkaloids

Dehydrogenation has been used effectively for the structure elucidation of alkaloids in the recent past. While a complete compilation of data on this subject is outside the scope of this review, several typical examples will be given in order to demonstrate the behavior of natural bases during the dehydrogenation.

The alkamines of the *Solanum* group yielded 2-ethyl-5-methylpyridine (**128**) and γ-methylcyclopentenophenanthrene (**81**) on selenium dehydrogenation, indicating a close structural relationship to the steroids [337]. In agreement with this, the structure of the alkamine solanidine was found to be **127**

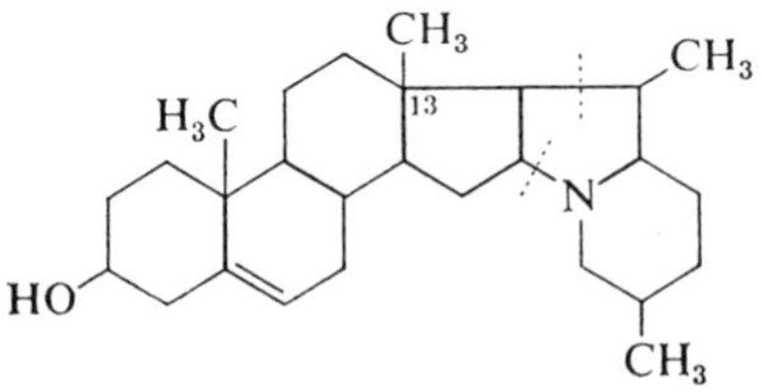

127 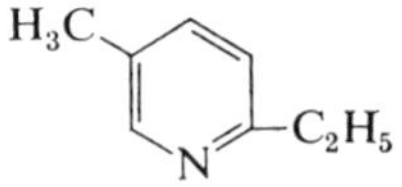   128

[338, 339]. The relationship to steroids and the formation of the two dehydrogenation products is quite clear. It is interesting that in this case the formation of γ-methylcyclopentenophenanthrene must proceed by a migration of the C-13 methyl group, provided that the two dehydrogenation products are not formed by alternate ring cleavages. The formation of other pyridine bases was, however, not reported. This result favors the pathway proposed for the similar transformation of steroids (see page 37).

Alkamines of the *Veratrum* group which have been subjected to selenium dehydrogenation yielded the typical basic product **128**, but no **81** was obtained [337]. For example, Jacobs and his coworkers [340, 341] reported the formation of **128** together with 2-ethyl-5-methyl-3-hydroxypyridine in the dehydrogenation of jervine. No phenanthrenes were formed, but several other products were isolated, some of which appeared to be homologs of 1,2-benzofluorene and others of β-phenylnaphthalene on the basis of ultraviolet spectra. Fried and coworkers [342] proposed the modified steroid structure **129** for jervine. This formulation explains the dehydrogenation results very satisfactorily.

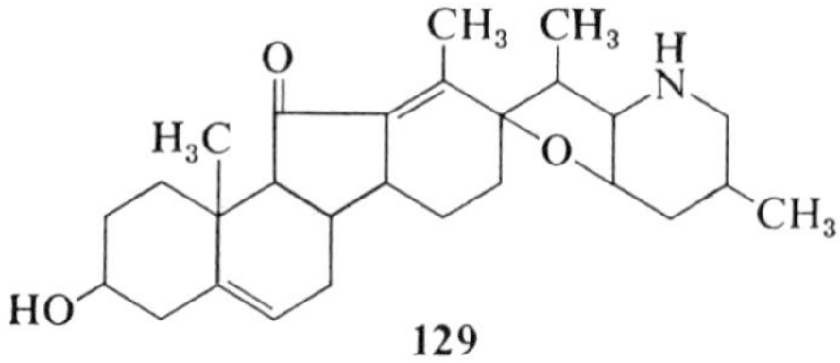

129

A large number of dehydrogenation products were obtained from the alkamine cevine [337]. In a summarizing article, Jacobs and Pelletier [137] rationalized the results of dehydrogenation and suggested that cevine contained the basic skeleton **130**, which differs from the skeleton of jervine only by one additional C–N bond. Among other compounds, the following dehydrogenation products were formed: β-picoline, the pyridine (128), cevanthridine (131), the base 132, 4,5-benzindane (133), and the hydrocarbons $C_{17}H_{16}$, $C_{18}H_{18}$, $C_{19}H_{20}$, and $C_{24}H_{30}$ which probably all contain the skeleton

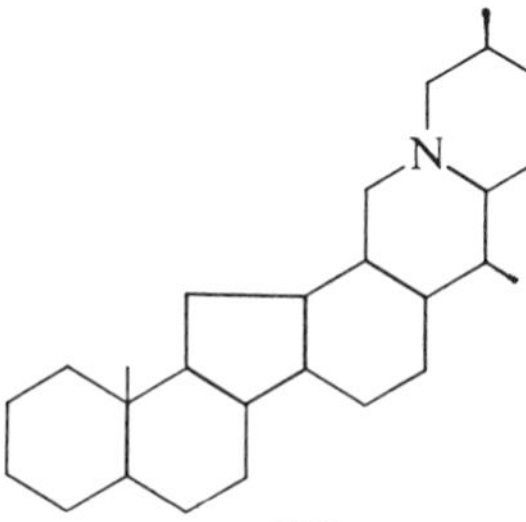

130

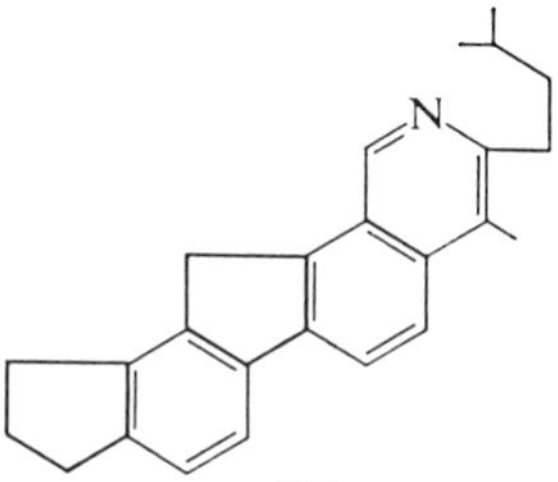

131

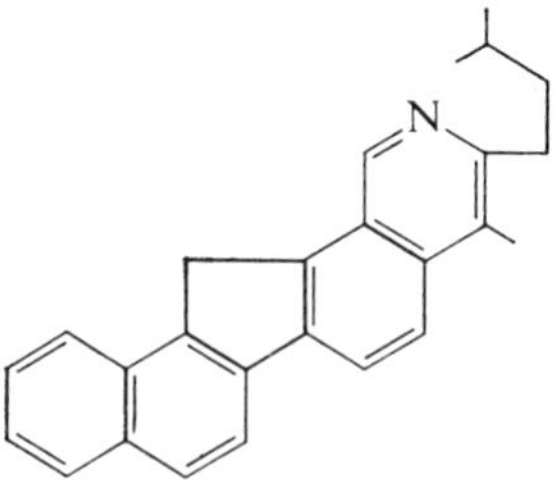

132 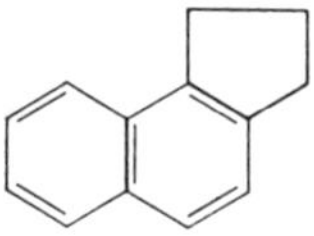   133 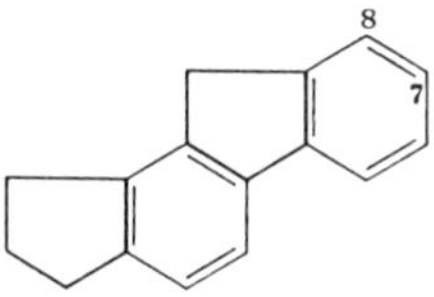   134

134 with additional substituents at C-7 and/or C-8. Barton, Jeger, Prelog, and Woodward [163] confirmed the deductions of Jacobs and Pelletier and proposed the complete structure 135 for cevine. They suggested that the base

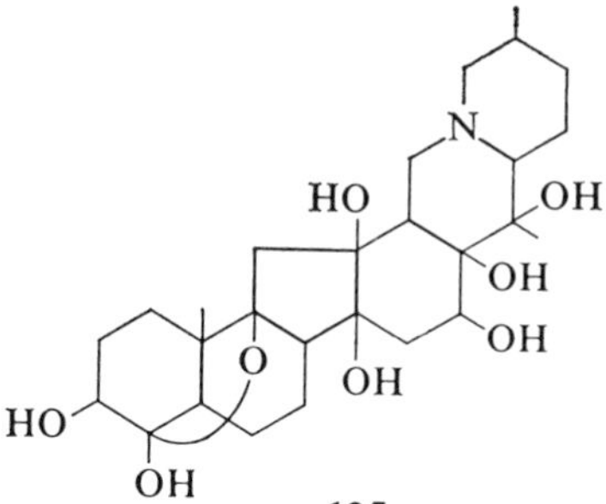

135

$C_8H_{11}NO$, which is also one of the dehydrogenation products, is 2-(1′-hy-droxyethyl)-5-methylpyridine (41). The cevine dehydrogenation is interesting from several standpoints: (1) it shows that polyoxygenated compounds can be successfully dehydrogenated if the oxygen functions are favorably situated; (2) the presence of oxygen functions labilizes some of the bonds in the molecule and partly accounts for the multiplicity of products—the contraction of Ring A is clearly caused by the presence of a pinacolic system; and (3) the cleavage of every one of the three C–N bonds has to be postulated in order to explain all dehydrogenation products.

Dehydrogenation proved to be of great importance in the structure elucidation of some diterpene alkaloids. The isomers veatchine and garryine yielded the phenanthrene 136 and the azaphenanthrene 137 on treatment

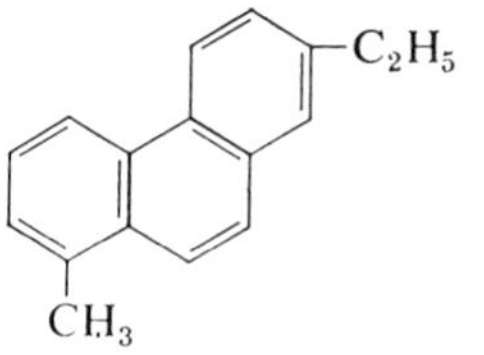

136

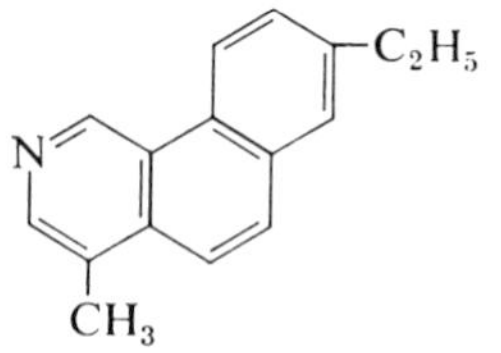

137

with selenium; and this remarkable result, together with other degradative work, led Wiesner and his coworkers [343] to propose structure 138 for dihydroveatchine. They also proposed structure 139 for dihydroatisine,

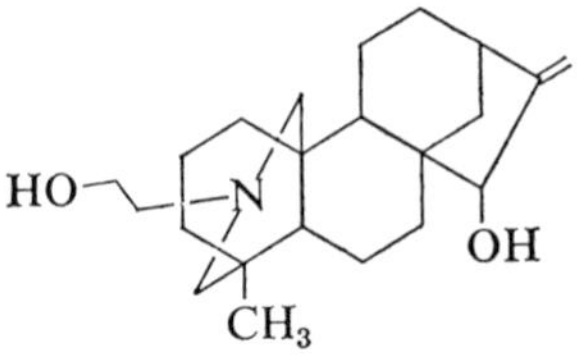

138

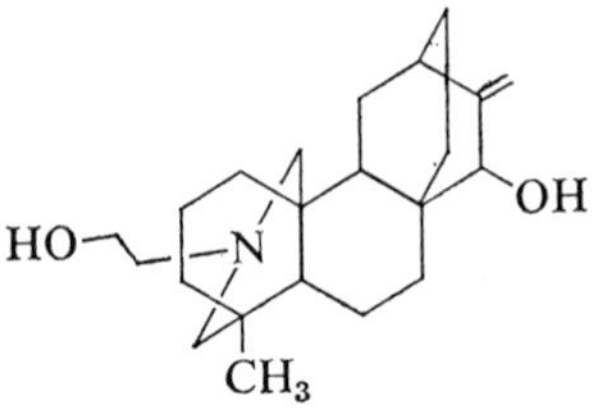

139

based mainly on the complete parallelism between the reactions of atisine and veatchine and the formation of 1-methyl-6-ethylphenanthrene in the dehydrogenation of atisine [344]. The formation of the azaphenanthrene 137 in addition to 136 indicates the comparable stability of C–C and C–N bonds in the dehydrogenation. In analogy to similar reductions of carboxyl groups reported previously (see page 27), the dicarboxylic acid 140 yielded 1,7-dimethylphenanthrene together with 1-methyl-7-carboxyphenanthrene in selenium dehydrogenation [155].

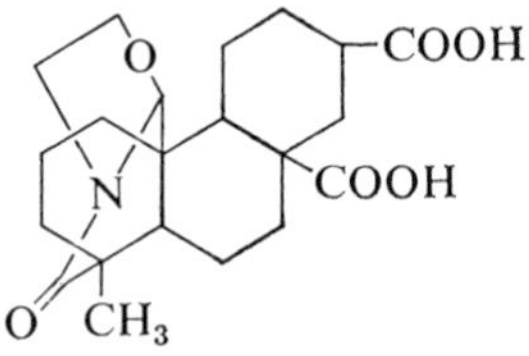

140

Many indole alkaloids give typical dehydrogenation products. Since the subject has been reviewed [140, 345], only a few examples will be given. Yohimbine (**141**) yielded three dehydrogenation products, yobyrine (**142**), tetrabyrine (**143**), and ketoyobyrine (**144**). The structures of all three products were proved by synthesis [140]. The formation of **142** and **143** requires no comment. The remarkable formation of ketoyobyrine was explained by the

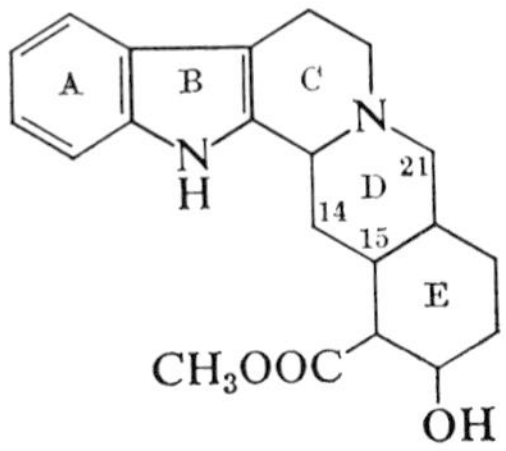

141

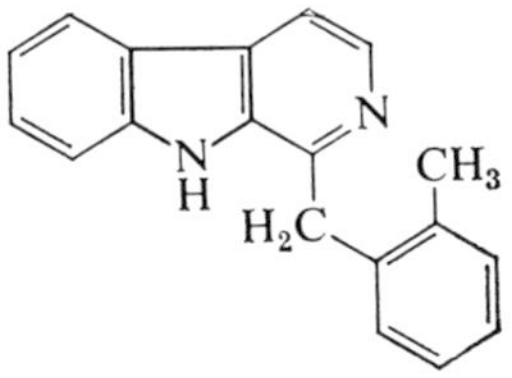

142

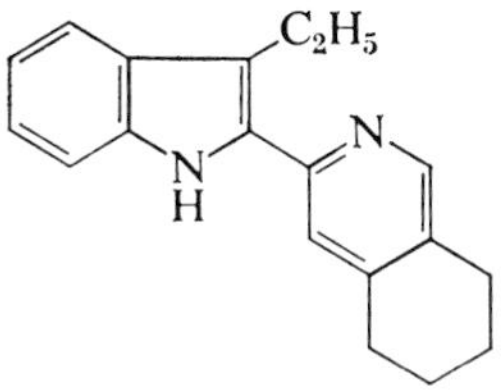

143

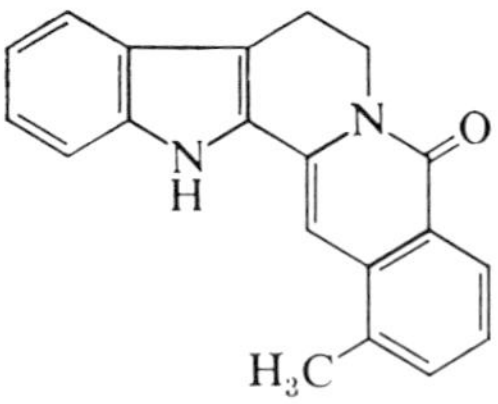

144

aromatization of Ring E, cleavage of the N–C-21 bond, rotation about the C-14 to C-15 bond, lactam closure, and final dehydrogenation [140].

The structure, chemistry, and dehydrogenation of ajmaline were reported by Woodward [346]. Deoxydihydroajmaline (**145**) yielded the three bases **146**, **147**, and **148**. It is interesting that in this case the secondary nitrogen atom

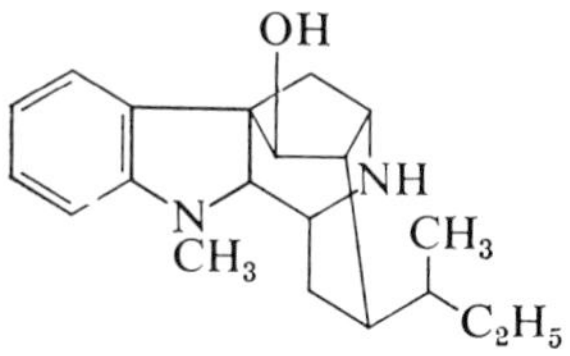

145

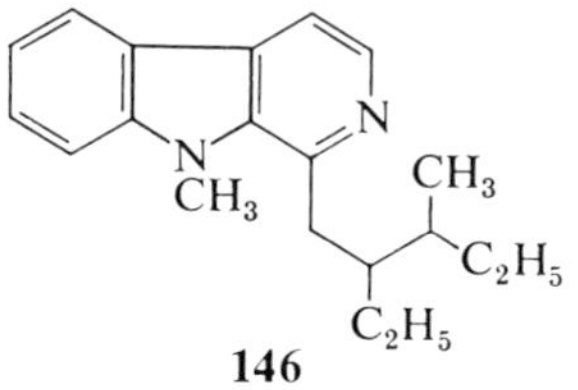

146

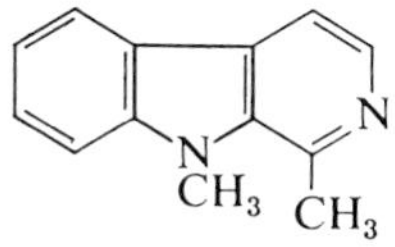

147

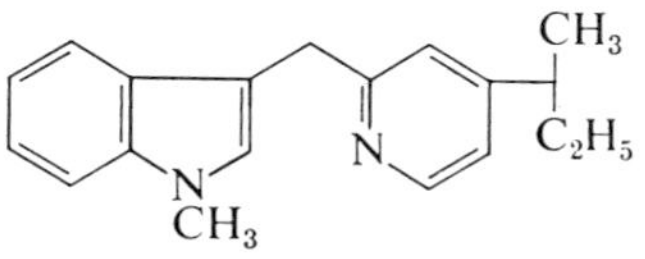

148

is a member of two six-membered rings, and one or the other ring is preserved in the dehydrogenation by the cleavage of different C–C bonds.

The interesting dehydrogenation of ibogaine (12) leading, together with
other degradative work, to structure elucidation by Bartlett, Dickel, and
Taylor [72] has already been mentioned (page 11).

Dehydrogenation was also of great importance in the study of annotinine
(149) [142]. 8-*n*-Propylquinoline (150) was obtained together with several
other products in the selenium dehydrogenation. It can be seen that 150 can
be formed from 149 in two different ways. Similarly, the degradation product
151 yielded the quinolone acid 152 on treatment with palladium charcoal
[142]. In analogy to the results obtained by Späth and Galinovsky [335]
with *N*-methyl-α-piperidones and dihydrocarbostyrils, no C–N bond was
cleaved in this case. The formation of an aromatic acid in the dehydrogenation
of the amino acid 153 was reported by Anet and Marion [347]. Valenta,
Wiesner, and their coworkers [141] were able to show that this acid had
the unexpected structure 154 and that the decarboxylated compound 155
was also formed. Since the structure of annotinine has been rigorously
proved by extensive degradative work [142, 348] and the structure of acid

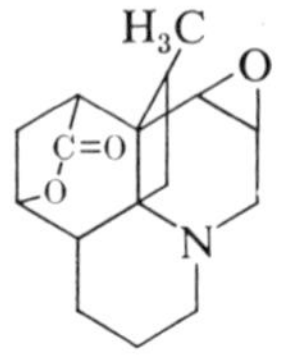

149

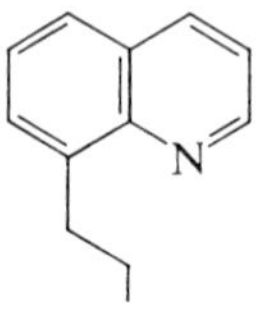

150

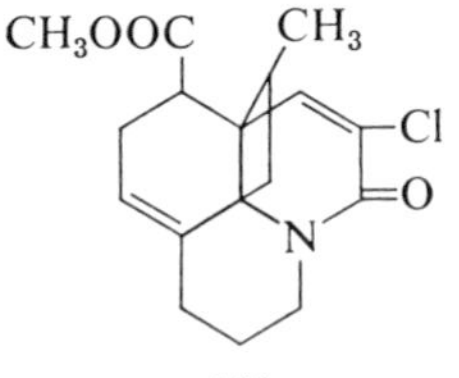

151

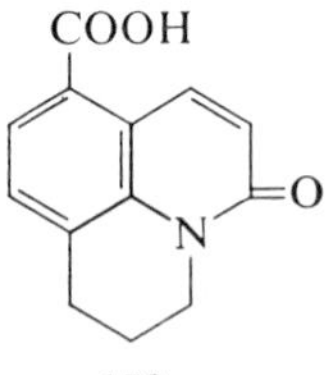

152

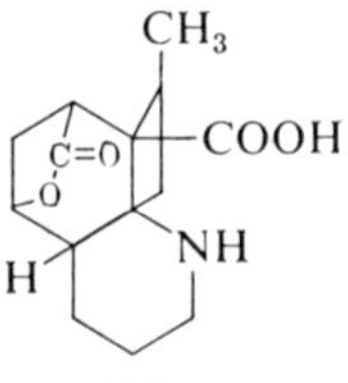

153

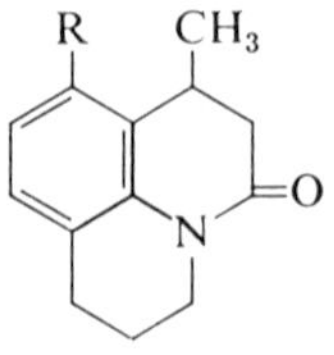

154: R = COOH
155: R = H

154 by synthesis [141], there is no doubt about the reality of the unusual rearrangement **153** → **154**. It is interesting that only one hydrogen molecule and one water molecule are abstracted in the reaction. The dehydrogenation of the methyl ester of **153** also gives **154**. The catalytic dehydrogenation of **153** at a higher temperature yielded mainly **155** and the corresponding quinolone, formed by the abstraction of an additional molecule of hydrogen.

10 ZINC DUST DISTILLATION

The distillation of natural products with zinc dust is a method for elucidating the basic skeleton of the molecules under investigation. Its advantages are the relatively simple experimental procedure and short reaction time and the fact that it yields, in most cases, easily identifiable oxygen-free aromatic degradation products. However, the yields are usually very low, and the high temperature required for the reaction often leads to excessive bond breaking and rearrangements.

In this brief section, no attempt has been made to cover the subject exhaustively. Representative examples of the application of this method to various classes of natural products, with special emphasis on most suitable substrates, have been recorded. In addition, some distillations leading to rearranged products have been summarized in order to point out the limitations of the method.

Experimental Procedure

The substance is thoroughly mixed with 10–100 parts of zinc dust and placed at the closed end of a Pyrex tube between cleaned asbestos plugs. Additional zinc dust is added to the other side of the plug, and the open end (usually slightly bent) is connected to an appropriate cold trap. The whole system is flushed thoroughly with hydrogen or nitrogen gas, and the zinc dust is then heated with a small open flame or electrically, starting at the open end. The reaction is usually finished in several minutes. In most cases, the products condense in the cooler part of the tube.

The actual reaction temperature has not been recorded in many cases, but it seems that temperatures close to 400°C are the most suitable.

The amount of substance in one tube varies widely, but it is recommended to use less than 100 mg. The reaction time is thus shortened, and charring and other side reactions are brought down to a minimum.

Zinc dust (Merck, *pro analysi*), zinc dust (Fisher), and a reagent prepared electrolytically in the laboratory [349] have been used. The advantages and disadvantages of zinc oxide present in commercial zinc dust and formed during distillation have been discussed [350].

The following selected references can be consulted for experimental details:

(a) Distillation of yohimbine in an electrically heated apparatus [351].

(b) Distillation of atromentin [352].

(c) Distillation of boletol [349].

(d) Distillation of gramine [353].

(e) Distillation of terramycin derivatives [354].

(f) Clar's "zinc dust fusion" of high-molecular quinones [355].

(g) Identification of polycyclic hydrocarbon derivatives (3–10 mg) by zinc dust fusion followed by paper chromatography [356].

(h) Distillation of o-quinones [350], (i) small scale fusion of alkaloids coupled with mass spectrometry and gas chromatography [357].

Application

The zinc dust distillation has been applied mostly for the investigation of three classes of substrates: (1) phenols (Table 10.1), (2) quinones (Table 10.2), and (3) alkaloids (Table 10.3).

Distillation of Phenols

The method was first applied by von Baeyer [358], who found that phenol and cresol could be reduced to benzene and toluene, respectively, by passing their vapors over heated zinc dust. It was soon found that this method can

Table 10.1 Distillation of Phenols

Substance	Products (yield)	Ref.
Phenol	Benzene	358
Cresol	Toluene	358
4,4-Dihydroxybiphenyl	Biphenyl	359
Hexahydroxybiphenyl	Biphenyl	360, 372
9-Hydroxyanthracene	Anthracene	361
m-Hydroxydiphenylamine	Diphenylamine	367
4-Hydroxyquinaldine	Quinaldine	370
Oxypyridine	Pyridine	368
4-Anilinocarbostyril	4-Anilinoquinoline	369
Estrone	Chrysene	373
1,2-Dimethyl-7-hydroxyphenanthrene	1,2-Dimethylphenanthrene	362
Dedimethylaminoterrarubein (158)	Naphthacene (2.5%)	354
Terranaphthoic acid	α- and β-Methylnaphthalenes, naphthalene	354
Rubrofusarin	Naphthalene, small amount of anthracene	374

be generally applied for the conversion of phenols to the corresponding aromatic hydrocarbons. 4,4′-Dihydroxybiphenyl [359] and a hexahydroxybiphenyl [360] yielded biphenyl, while 9-hydroxyanthracene could be reduced to anthracene [361]. 1,2-Dimethylphenanthrene was obtained by Butenandt et al. by the distillation of 1,2-dimethyl-7-hydroxyphenanthrene [362].

Graebe found that anisol was not reduced under similar conditions [363]. Further examples of the retention of ether groups are the conversion of dihydrolycorine anhydromethine (**156**) to the phenanthridine (**157**) [364] and the reduction of phenanthrene quinone to diphenanthro(9,10-*b*,9′,10′-*d*)furan [355]. Ether groups can, however, be removed under sufficiently drastic conditions. The distillation of *N*-demethylcodeine gave phenanthrene (20% yield) [365], which was also obtained by the distillation of methyl- and ethylmorphenols [366]. The distillation of **156** also produced phenanthridine and 1-methylphenanthridine in addition to **157**.

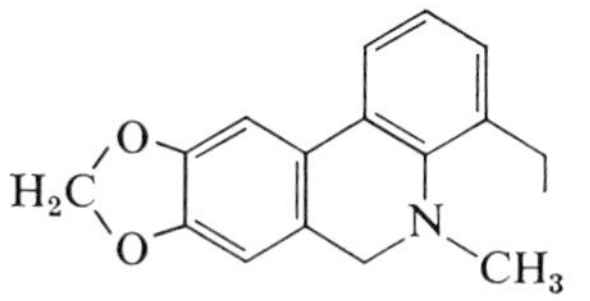

156 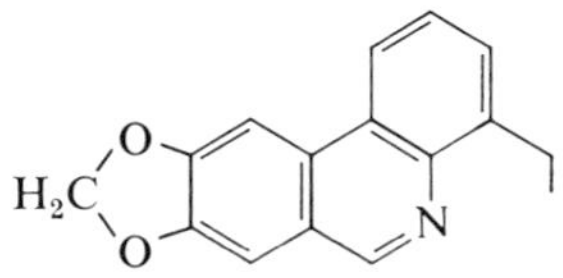   157

Several normal reductions of phenols containing nitrogen atoms have been reported. The distillation of 3-hydroxydiphenylamine yielded diphenylamine [367], oxypyridine could be converted to pyridine [368], 4-anilinocarbostyril was reduced to 4-anilinoquinoline [369], and 4-hydroxyquinaldine yielded quinaldine [370].

Zinc dust distillation played a significant part in the notable elucidation of the structure of the antibiotic terramycin [354]. The isolation of naphthacene (2.5% yield) from the distillation of a terramycin degradation product, dedimethylaminoterrarubein (**158**), revealed the basic skeleton of the molecule. Two other degradation products of terramycin were distilled with zinc dust. Terranaphthoic acid (1-methyl-2-carboxy-4,5-dihydroxynaphthalene) yielded α-methylnaphthalene together with small amounts of naphthalene and

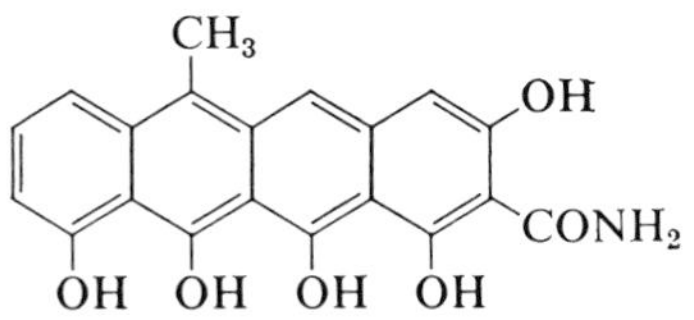

158 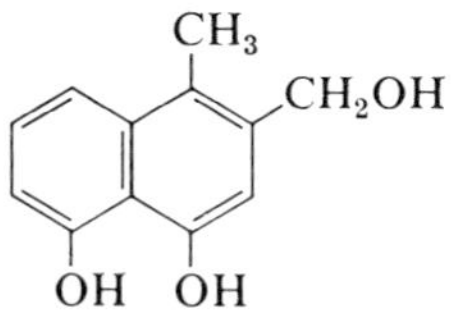   159

β-methylnaphthalene, while the alcohol **159** was reduced to 1,3-dimethyl-naphthalene. It is apparent from these examples that some alkyl groups migrate or are completely or partially eliminated under the usual reaction conditions (see page 64).

According to Kögl [349, 371], several phenol acetates gave a better yield of the aromatic hydrocarbon than the corresponding phenols. It is quite likely that phenol acetates in general can be used to advantage.

Distillation of Quinones

Using von Baeyer's method, nineteenth century chemists were able to reduce various quinones to the corresponding aromatic hydrocarbons. For example, anthraquinone was converted to anthracene [375] and naphthazarine to naphthalene [376].

The method was later successfully applied by Kögl and his school in a systematic investigation of fungal pigments [349, 352, 371, 377, 378]. The conversion of boletol [349] to anthracene and that of atromentin [352], polyporic acid [377], and muscarufin [378] to terphenyl played an important part in their structural elucidations. Gripenberg [379] obtained biphenyl, p-terphenyl, and benzobis[1,2-b,4,5-b']benzofuran in the distillation of thelephoric acid.

Table 10.2   Distillation of Quinones

Substance	Products (yield)	Ref.
Naphthazarine	Naphthalene	376
Anthraquinone	Anthracene	375, 381
Anthraquinone	Anthracene (80%) + bianthryl (10%)	355
Alizarin	Anthracene	375, 381
Indanthren	Anthracene	382
Triacetylboletol	Anthracene	349
Phenanthrene quinone	Phenanthrene (80%) + diphenanthro [9,10-b,9',10'-d]furan (10%)	355
Thelephoric acid	Biphenyl, p-terphenyl + benzobis [1,2-b,4,5-b,]benzofuran	379
Atromentin	Terphenyl (5%)	352
Muscarufin	Terphenyl	378
Polyporic acid	Terphenyl (10%)	377
Dioxynaphthacene quinone	Naphthacene	383
Dibenzpyrene quinone	Dibenzpyrene	384
Violanthrone	Violanthrene	355
Anthanthrone	Anthanthrene	355
Hypericine	Mesoanthrodianthrene	380

Clar [355] introduced a modification suitable for the reduction of high molecular weight quinones. This, the so-called "zinc dust fusion," is performed at a lower temperature (200–290°C) in a NaCl–ZnCl$_2$ melt. By this method, anthracene and phenanthrene are produced in an 80% yield from the corresponding quinones. An important application combining zinc dust fusion with paper chromatography for the identification of the resulting hydrocarbons was reported by Gasparic [356]. Substituent halogen atoms, as well as sulfo, nitro, sulfonyl chloride, carboxy, hydroxy, amino, and acetoamido groups, were eliminated to give the hydrocarbons. Davies and Ennis [350] have found that the best yield of unrearranged hydrocarbons during the reduction of polycyclic o-quinones is obtained when zinc oxide present in commercial zinc dust is first converted into metallic zinc by a treatment with hydrogen gas. It should be noted, however, that these authors have not compared their procedures with the zinc dust fusion method [355].

Brockmann [380] has reported the reduction of the photodynamic pigment hypericine to mesoanthrodianthrene.

Distillation of Alkaloids

The zinc dust distillation of a great number of alkaloids has resulted in the production of simple aromatic and heteroaromatic substances. The success of the method in this field can be attributed to three factors:

1. Zinc dust is a strong dehydrogenation agent under the drastic reaction conditions.

2. Cleavage of C–C and C–N bonds is facile at the reaction temperature.

3. Pyridines, pyrroles, indoles, and other heteroaromatic substances are relatively stable to zinc dust.

In many cases the reaction is purely of historical interest, since similar or better results can be obtained by a catalytic or chemical dehydrogenation. The zinc dust distillation can be used with advantage, however, if the removal of all oxygen functions is essential and in the degradation of molecules offering resistance to normal dehydrogenation. The two methods quite often yielded widely different products, and the combined results simplified a structural interpretation.

As a rule, the yields in the alkaloid field are extremely low (usually less than 1%) and the results must therefore be interpreted with caution.

PYRIDINES

The distillation of the hydrochloride of coniine (2-*n*-propylpiperidine) by A. W. Hofmann [385] yielded 2-*n*-propylpyridine, which was also obtained in a similar reaction from granatanine hydrochloride (**160**) [386]. Cincholoipon (3-ethyl-4-carboxymethylpiperidine), a degradation product of cinchona alkaloids, was converted to 3-ethylpyridine by Skraup [387]. Jacobs and

Table 10.3 Distillation of Alkaloids

Substance	Products (yield)	Ref.
Coniine	2-*n*-Propylpyridine	385
Granatanine	2-*n*-Propylpyridine	386
Cincholoipon	3-Ethylpyridine	387
Lupanine	2-Ethylpyridine	389
Sparteine	Pyrroles	390
Matrinic acid	1-Methylquinolizidine	391, 392
Cevine	2-Ethyl-5-methylpyridine + β-pipecoline	388
Gramine	Skatole + indole (5–10%)	353
Yohimbine	Skatole + 3-ethylindole + harman	
	+ isoquinoline + *p*-cresol	351, 393, 394
Ajmaline	Ind-*N*-methylharman + carbazole	395
Aspidospermine	Indoles + a pyridine	406
C-Dihydrotoxiferine I	Skatole + 3-ethylindole + isoquinoline	396
Gelsemine	Skatole + 4,7-dimethylisoquinoline	399, 400
Echitamine	2,1'-Dimethylpyrrolo[2'.3'-3.4]	
	quinoline + echitamyrine	401–404
Quinamine	2,3-Dimethylindole	397
Eseroline	2-Methylindole	398
Retronecine	Pyrrole	405
Morphine	Phenanthrene (3–4%)	407
N-Demethylcodeine	Phenanthrene (20%)	365
Methylmorphimethin	Phenanthrene (10%)	408
Sinomenine	Phenanthrene	409, 410
Colchinol methyl ether	9-Methylphenanthrene	411, 412
Dihydrolycorine anhydromethine	Phenanthridine + 1-methylphenan-thridine + 1-ethyl-6,7-methylene-dioxyphenanthridine	364

Craig [388] reported that cevine (**161**) was degraded to β-pipecoline, 2-ethyl-5-methylpyridine, and three other bases in a zinc dust distillation followed by a catalytic reduction.

The distillation of lupin alkaloids provides an instructive example of the sensitivity of the reaction to relatively minor structural variations. While lupanine (**162**) was degraded to 2-ethylpyridine [389], sparteine (C=O in **162**

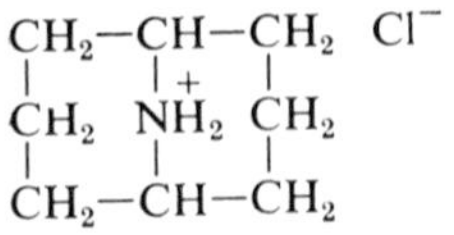

$$\begin{array}{c} CH_2-CH-CH_2 \quad Cl^- \\ | \quad \quad |_+ \quad | \\ CH_2 \quad NH_2 \quad CH_2 \\ | \quad \quad | \quad \quad | \\ CH_2-CH-CH_2 \end{array}$$

160

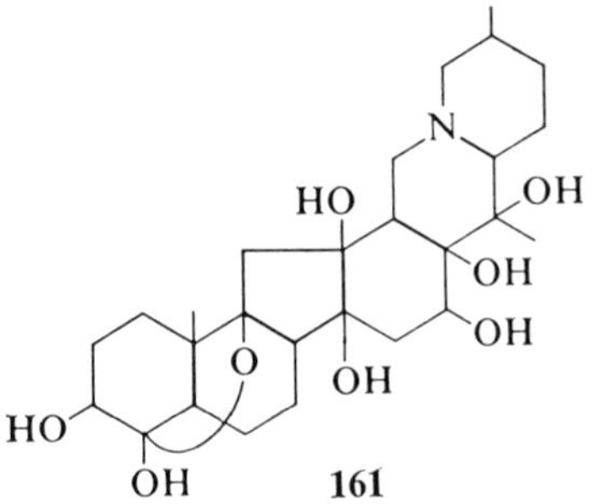

161

replaced by CH_2) yielded a mixture that gave typical pyrrole color reactions. Clemo and Raper [390] found that cytisine, a lupin alkaloid containing only six-membered rings, also yields pyrroles on distillation with zinc. Since other degradative evidence made the presence of a five-membered ring in sparteine unlikely, they postulated that the pyrroles are formed during the pyrolysis

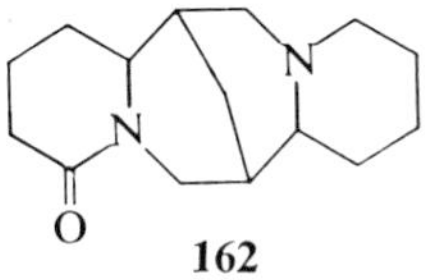

162

by a rearrangement and proposed the correct structure for sparteine. Surprisingly, the hydrochloride of matrinic acid (**163**) yielded 1-methylquinolizidine (**164**) on distillation [391, 392]. The survival of **164** is undoubtedly due to the fact that it vaporized before it could be degraded further. The conversion of **163** to **164** can proceed either by two independent reductive cleavages or by a reverse Diels-Alder reaction of a dehydromatrinic acid.

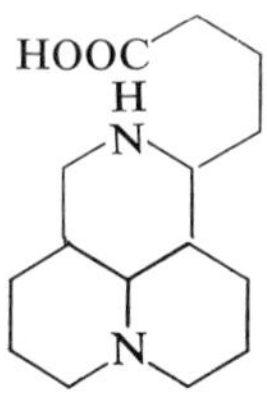

163 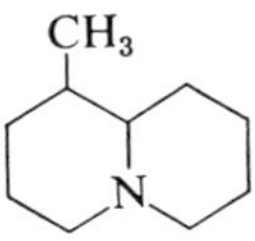   **164**

INDOLES AND PYRROLES

Indole and its homologs are obtained in the distillation of most indole alkaloids. Other frequently encountered degradation products are harman, substituted pyridines, isoquinoline and its homologs, and carbazole.

Winterstein and Walter [393] reported the isolation of indoles from the distillation of yohimbine (165). Witkop [394] showed that this mixture consisted of 3-ethylindole and skatole. In addition, he was able to isolate isoquinoline, harman, and a very small amount (0.1 %) of p-cresol [351]. It is interesting that all these degradation products were formed without a skeletal rearrangement. The diagnostic value of the method in this case is obvious.

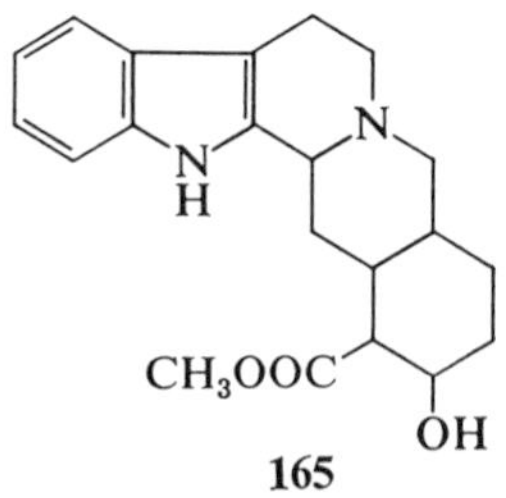

165

The distillation of ajmaline (166) [346] yielded carbazole in addition to ind-N-methylharman [395]. The misleading production of carbazole is a demonstration of the fact that new rings can be formed during the reaction.

C-Dihydrotoxiferine I yielded isoquinoline, 3-ethylindole, and skatole [396]; quinamine could be degraded to 2,3-dimethylindole [397]; and eseroline was converted to 2-methylindole [398]. The distillation of gelsemine

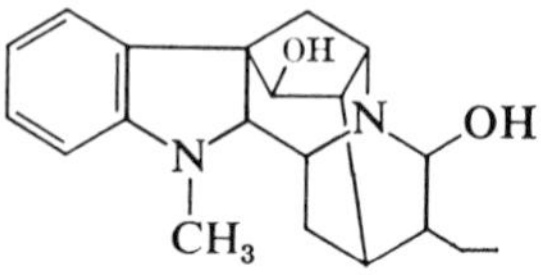

166

[399] yielded skatole and a base which is probably 4,7-dimethylisoquinoline [400]. Rearranged products were produced in the distillation of echitamine [401–404], (see page 64).

The distillation of the *Senecio* alkaloid retronecine [405] gave pyrroles as determined by the pine splinter test.

PHENANTHRENES

Distillations in the morphine series revealed the presence of a perhydro-phenanthrene skeleton in the molecule. Phenanthrene was obtained on treatment of morphine [3–4%] [407], N-demethylcodeine (20%) [365], methylmorphimethin (10%) [408], methylmorphenol [366], and sinomenine [409, 410]. Thus dehydrogenation, in addition to the elimination of the nitrogen bridge and the oxygen functions, took place in all cases.

Colchinol methyl ether yielded a small amount of 9-methylphenanthrene when subjected to a Hofmann degradation, followed by treatment with HI

Table 10.4 Distillation of Other Substances

Substance	Products	Ref.
Benzophenone	Diphenylmethane	355
Methylmorphenol	Phenanthrene	366
Columbin	1,2,5-Trimethylnaphthalene + *o*-cresol	206, 413
Hederagenin	Hydrocarbons	414

and a zinc dust distillation [412]. However, it was found [411] that the hydrocarbon (**167**) could also be converted to 9-methylphenanthrene by a similar treatment. Thus the presence of a seven-membered B-ring in colchicine could not be excluded on the basis of this degradation.

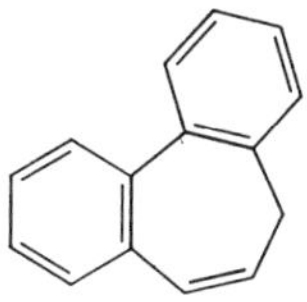

167

Distillation of Carbocyclic Compounds

The method has found virtually no application in the study of steroids and terpenes. Dehydrogenation with sulfur, selenium, or a catalyst, a method which leaves more endocyclic C–C bonds uneffected, has been used almost exclusively in this field.

However, suitably substituted molecules can be converted by zinc dust to useful degradation products (Table 10.4). The distillation of columbin (**168**) [206] yielded *o*-cresol and 1,2,5-trimethylnaphthalene [413]. The substitution of Ring A and the presence of a perhydronaphthalene unit in the molecule could be deduced from this result.

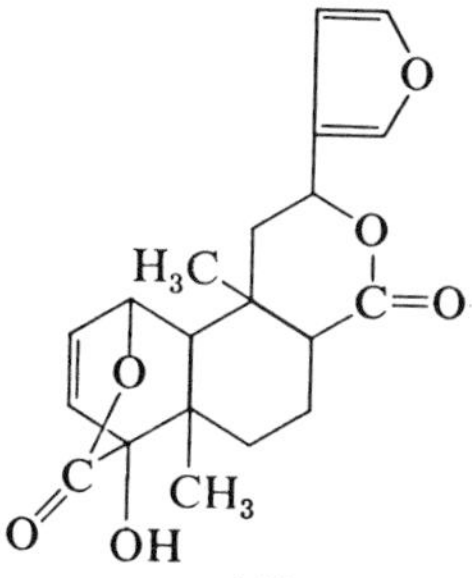

168

The production of aromatic hydrocarbons in the distillation of hederagenin has been reported [414].

Side Reactions

Because of the drastic reaction conditions, excessive cleavage of endocyclic and side-chain bonds is encountered. While normal dehydrogenation usually eliminates only quaternary alkyl substituents, zinc dust distillation frequently degrades aliphatic side chains. Examples are the production of 3-ethylindole and skatole in the indole alkaloid field (see page 62); the degradation of cincholoipon [387]; the loss of a methyl group in the distillation of dedimethyl-aminoterrarubein (158), terranaphthoic acid [354] and gramine [353]; and the degradation of an ethyl group in the distillation of dihydrolycorine anhydro-methine (156) [364].

Carboxyl groups and their derivatives are invariably eliminated. The distillations of yohimbine (165) [394], terranaphthoic acid, dedimethylamino-terrarubein (158) [354], muscarufin [378] and boletol [349] result in the complete elimination of the carboxyl carbon atom.

While hydroxyl groups and the corresponding acetates are always removed, ethers are sometimes preserved under sufficiently mild conditions (see page 56).

As would be expected, molecular rearrangements are frequently encountered. In most cases, however, they are of a predictable nature and do not necessarily impair the diagnostic value of the method. Migration of side chains is encountered in the distillation of eseroline [398] (indole methyl group from position 3 to 2), terranaphthoic acid, the phenol 159 [354] and columbin (168) [206, 413].

Several changes involving the ring system deserve mention at this point. Distillation of polycyclic o-quinones is accompanied by extensive ring con-traction to fluorene and fluorenone derivatives [350]. Enlargement of ring D during the distillation of estrone resulted in the production of chrysene [373]. Skeletal changes are involved in the production of carbazole from ajmaline [345, 395], pyrroles from sparteine and cytisine [390], and 9-methylphen-anthrene from colchinol methyl ether [411, 412]. The distillation of echitamine derivatives results in the production of 2,1'-dimethylpyrrolo[2'.3'-3.4]quino-line [401], and echitamyrine [403]. This complicated change involves, inter alia, the transformation of a dihydroindole to a quinoline. The for-mation of a new ring is involved in the transformation of hypericine to mesoanthrodianthrene [380].

Clar [355] reported the formation by dimerization of bianthryl (10%) and diphenanthro[9,10-b,9',10'-d]furan (10%) in the zinc dust fusion of anthra-quinone and phenanthrene quinone, respectively.

An extensive study of the behavior of hydrocarbons at high temperature has been undertaken by Badger [415].

References

1. R. P. A. Sneeden and R. B. Turner, *J. Amer. Chem. Soc.*, **77**, 190 (1955).
2. P. A. Plattner, in *Newer Methods of Preparative Organic Chemistry*, Interscience, New York, 1948, p. 21.
3. L. Ruzicka, *Fortschr. Chem. Phys. u. Phys. Chem.*, **A19**, No. 5 (1928).
4. R. P. Linstead, *Ann. Rept. Progr. Chem. (Chem. Soc. London)*, **33**, 294 (1937).
5. L. M. Jackman, *Adv. Org. Chem.*, **2**, 329 (1960).
6. M. Beroza and R. A. Coad, *J. Gas Chromatography*, **4**, 199 (1966).
7. R. M. Carman, *Aust. J. Chem.*, **16**, 225 (1963).
8. E. Schöntube and J. Janák, *Coll. Czech. Chem. Commun.*, **33**, 193 (1968).
9. P. Sabatier, *Catalytic Hydrogenation*, Akadem Verlagsgesellschaft, Leipzig, 1914.
10. N. Zelinsky, *Chem. Ber.*, **44**, 3121 (1911).
11. P. G. Ashmore, *Catalysis and Inhibition of Chemical Reactions*, Butterworths, London, 1963.
12. B. M. W. Trapnell, "Specificity in Catalysis by Metals," *Quart. Rev.* (London), **8**, 404 (1954).
13. I. Langmuir, *Trans. Faraday Soc.*, **17**, 617 (1921).
14. M. H. Dilke, E. B. Maxted, and D. D. Eley, *Nature*, **161**, 804 (1948).
15. D. A. Dowden, *J. Chem. Soc.*, **1950**, 242.
16. G. H. Twigg and E. K. Rideal, *Trans. Faraday Soc.*, **36**, 533 (1940).
17. O. Beeck, *Discussions Faraday Soc.*, **8**, 118 (1950).
18. M. S. Newman and D. Lednicer, *J. Amer. Chem. Soc.*, **78**, 4765 (1956).
19. D. D. Eley, *Quart. Rev.* (London), **3**, 209 (1949).
20. H. S. Taylor, *J. Amer. Chem. Soc.*, **60**, 627 (1938).
21. E. A. Braude, R. P. Linstead, and P. W. D. Mitchell, *J. Chem. Soc.*, **1954**, 3578.
22. A. A. Balandin, *Z. Phys. Chem.*, **B3**, 167 (1929).
23. A. A. Balandin, *Dokl. Akad. Nauk SSSR*, **97**, 449 (1954); through *Chem. Abstr.*, **49**, 10028 (1955).
24. H. S. Taylor, in *Frontiers in Chemistry*, Vol. 5, Interscience, New York, 1948.
25. K. Packendorff and L. Leder-Packendorff, *Chem. Ber.*, **67**, 1388 (1934).
26. N. D. Zelinsky and M. B. Turova-Pollak, *Chem. Ber.*, **58**, 1292 (1925).
27. O. Diels and N. Gädke, *Chem. Ber.*, **58**, 1231 (1925).
28. R. Willstätter and E. Waldschmidt-Leitz, *Chem. Ber.*, **54**, 113 (1921).
29. O. Loew, *Chem. Ber.*, **23**, 289 (1890).
30. N. Zelinsky and P. Borisoff, *Chem. Ber.*, **57**, 150 (1924).
31. R. Mozingo, *Org. Syntheses*, **26**, 77 (1946).
32. H. Pines, A. Edeleanu, and V. N. Ipatieff, *J. Amer. Chem. Soc.*, **67**, 2193 (1945).
33. H. Pines, R. C. Olberg, and V. N. Ipatieff, *J. Amer. Chem. Soc.*, **70**, 533 (1948).
34. L. Hernandez and F. F. Nord, *Experientia*, **3**, 489 (1947).

35. N. D. Zelinsky, *Chem. Ber.*, **59**, 156 (1926).
36. N. Zelinsky and N. Pavlov, *Chem. Ber.*, **56**, 1250 (1923).
37. R. P. Linstead, A. F. Millidge, S. L. S. Thomas, and A. L. Walpole, *J. Chem. Soc.*, **1937**, 1146.
38. R. P. Linstead and S. L. S. Thomas, *J. Chem. Soc.*, **1940**, 1127.
39. M. Crawford and V. R. Supanekar, *J. Chem. Soc. C*, **1966**, 2252.
40. B. C. Alsop and D. A. Dowden, *J. Chem. Phys.*, **51**, 678 (1954).
41. K. K. Kearby, in *The Chemistry of Petroleum Hydrocarbons*, Vol. 2 (B. T. Brooks, C. E. Boord, S. S. Kurtz, Jr., and L. Schmerling, eds.), Reinhold, New York, 1955, p. 221.
42. K. K. Kearby, in *Catalysis: Hydrogenation and Dehydrogenation*, Vol. 3 (P. H. Emmett, ed.), Reinhold, New York, 1955, p. 453.
43. J. R. Prahlad and Sukh Dev, *Tetrahedron*, **26**, 631 (1970).
44. H. Adkins and J. W. Davis, *J. Amer. Chem. Soc.*, **71**, 2955 (1949).
45. H. Adkins and D. C. England, *J. Amer. Chem. Soc.*, **71**, 2958 (1949).
46. H. Adkins and G. F. Hager, *J. Amer. Chem. Soc.*, **71**, 2962 (1949).
47. J. Heer and K. Miescher, *Helv. Chim. Acta*, **31**, 219 (1948).
48. H. Minato, M. Ishikawa, and T. Nagasaki, *Chem. Pharm. Bull.* (Tokyo), **13**, 717 (1965).
49. R. Majima and S. Murahashi, *Proc. Imp. Acad.* (Tokyo), **10**, 314 (1934).
50. E. Wenkert and D. K. Roychandhari, *J. Chem. Amer. Soc.*, **79**, 1519 (1957).
51. R. Pallaud and Huynh-An-Hoa, *Chim. Anal.* (Paris), **47**, 22 (1965).
52. H. H. Günthard, R. Süess, L. Marti, A. Fürst, and P. A. Plattner, *Helv. Chim. Acta*, **34**, 959 (1951).
53. E. Kováts, P. A. Plattner, and H. H. Günthard, *Helv. Chim. Acta*, **37**, 983 (1954).
54. E. Kováts, P. A. Plattner, and H. H. Günthard, *Helv. Chim. Acta*, **37**, 997 (1954).
55. M. Levitz and M. T. Bogert, *J. Amer. Chem. Soc.*, **64**, 1719 (1942).
56. L. Ruzicka and M. Stoll, *Helv. Chim. Acta*, **7**, 84 (1924).
57. W. Deuschel, *Helv. Chim. Acta*, **34**, 2403 (1951).
58. J. R. Nunn and W. S. Rapson, *J. Chem. Soc.*, **1949**, 825.
59. A. Vesterberg, *Chem. Ber.*, **36B**, 4200 (1903).
60. L. Ruzicka and J. Meyer, *Helv. Chim. Acta*, **4**, 505 (1921).
61. L. Ruzicka, J. Meyer, and M. Mingazzini, *Helv. Chim. Acta*, **5**, 345 (1922).
62. O. Diels, *Chem. Ber.*, **69A**, 195 (1936).
63. E. Buchta and W. Kallert, *Justus Liebigs Ann. Chem.*, **573**, 220 (1951).
64. L. F. Fieser, *J. Amer.Chem. Soc.*, **55**, 4977 (1933).
65. L. Ruzicka and A. G. van Veen, *Justus Liebigs Ann. Chem.*, **476**, 70 (1929).
66. H. I. Thayer and B. B. Corson, *J. Amer. Chem. Soc.*, **70**, 2330 (1948).
67. W. Cocker, B. E. Cross, J. T. Edward, D. S. Jenkinson, and J. McCormick, *J. Chem. Soc.*, **1953**, 2355.
68. W. Cocker, B. E. Cross, and J. McCormick, *J. Chem. Soc.*, **1952**, 72.
69. W. Cocker and D. S. Jenkinson, *J. Chem. Soc.*, **1954**, 2420.
70. H. A. Silverwood and M. Orchin, *J. Org. Chem.*, **27**, 3401 (1962); W. T. House and M. Orchin, *J. Amer. Chem. Soc.*, **82**, 639 (1960).

71. L. Ruzicka and J. Meyer, *Helv. Chim. Acta*, **5**, 581 (1922).
72. M. F. Bartlett, D. F. Dickel, and W. I. Taylor, *J. Amer. Chem. Soc.*, **80**, 126 (1958).
73. L. Ruzicka, *Helv. Chim. Acta*, **19**, 419 (1936).
74. L. F. Fieser, M. Fieser, and E. B. Hershberg, *J. Amer. Chem. Soc.*, **58**, 1463 (1936).
75. H. Wynberg, *J. Amer. Chem. Soc.*, **80**, 364 (1958).
76. E. Lederer and M. Lederer, *Chromatography: A Review of Principles and Applications*, Van Nostrand, Princeton, N.J., 1957.
77. D. H. Desty and C. L. A. Harbourn, *Vapour Phase Chromatography* (Proceedings of the Symposium Sponsored by the Hydrocarbon Research Group of the Institute of Petroleum Held at the Institute of Electrical Engineers), Academic Press, New York, 1957.
78. A. I. M. Keulemans, in *Gas Chromatography* (C. G. Verver, ed), Reinhold, New York, 1957.
79. I. C. Nigam and L. Levi, *J. Chromatography*, **17**, 466 (1965).
80. T. Okamoto and T. Onaka, *Chem. Pharm. Bull.* (Tokyo), **11**, 1086 (1963).
81. A. v. Baeyer and V. Villiger, *Chem. Ber.*, **32**, 2429 (1899).
82. L. Ruzicka and E. A. Rudolph, *Helv. Chim. Acta*, **10**, 915 (1927).
83. J. J. Ritter and E. D. Sharpe, *J. Amer. Chem. Soc.*, **59**, 2351 (1937).
84. B. Witkop, *J. Amer. Chem. Soc.*, **79**, 3193 (1957).
85. R. T. Arnold and C. J. Collins, *J. Amer. Chem. Soc.*, **61**, 1407 (1939).
86. D. Walker and J. D. Hiebert, *Chem. Rev.*, **67**, 153 (1967).
87. E. A. Braude, A. G. Brook, and R. P. Linstead, *J. Chem. Soc.*, **1954**, 3569.
88. E. A. Braude, L. M. Jackman, and R. P. Linstead, *J. Chem. Soc.*, **1954**, 3548.
89. E. A. Braude, L. M. Jackman, and R. P. Linstead, *J. Chem. Soc.*, **1954**, 3564.
90. E. A. Braude and R. P. Linstead, *J. Chem. Soc.*, **1954**, 3544.
91. E. A. Braude, L. M. Jackman, R. P. Linstead, and G. Lowe, *J. Chem. Soc.*, **1960**, 3133; R. F. Brown and L. M. Jackman, *J. Chem. Soc.*, **1960**, 3144.
92. R. Mechoulam, B. Yagnitinsky, and Y. Gaoni, *J. Amer. Chem. Soc.*, **90**, 2418 (1968).
93. P. D. Gardner, C. E. Wulfman, and C. L. Osborn, *J. Amer. Chem. Soc.*, **80**, 143 (1958).
94. W. Treibs, R. Steinert, and W. Kirchhof, *Justus Liebigs Ann. Chem.*, **581**, 54 (1953).
95. S. Sarel, Y. Shalon, and Y. Yanuka, *J. Chem. Soc. D*, **1970**, 80, 81.
96. R. P. Linstead, E. A. Braude, L. M. Jackman, and A. N. Beames, *Chem. Ind.* (London), **1954**, 1174.
97. W. Herz and G. Caple, *J. Org. Chem.*, **29**, 1691 (1964).
98. H. Dannenberg, *Synthesis*, **1970**, 74.
99. H. Dannenberg and H. H. Keller, *Justus Liebigs Ann. Chem.*, **702**, 149 (1967).
100. L. Ruzicka and C. F. Seidel, *Helv. Chim. Acta*, **19**, 424 (1936).
101. J. W. Cook and C. L. Hewett, *J. Chem. Soc.*, **1936**, 62.
102. J. W. Barrett, A. H. Cook, and R. P. Linstead, *J. Chem. Soc.*, **1935**, 1065.
103. A. S. Broun and B. V. Ioffe, *Nauch. Byul. Leningr. Gos. Univ.*, **20**, 11 (1948); through *Chem. Abstr.*, **43**, 5376 (1949).

104. F. Wessely and F. Grill, *Monatsh. Chem.*, **77**, 282 (1947).
105. M. D. Bodroux, *Ann. Chim.* (Paris), **11**, 511 (1929).
106. R. Weiss and K. Woidich, *Monatsh. Chem.*, **46**, 456 (1925).
107. N. Zelinsky, **56**, 787 (1923).
108. H. Kaffer, *Chem. Ber.*, **57**, 1261 (1924).
109. N. Zelinsky, **56**, 1723 (1923).
110. N. D. Zelinsky and I. N. Titz, *Chem. Ber.*, **62**, 2869 (1929).
111. A. Maillard, *C.R. Acad. Sci.*, *Paris*, **200**, 1856 (1935).
112. R. P. Linstead and K. O. A. Michaelis, *J. Chem. Soc.*, **1940**, 1134.
113. R. P. Linstead, K. O. A. Michaelis, and S. L. S. Thomas, *J. Chem. Soc.*, **1940**, 1139.
114. A. G. Anderson and J. A. Nelson, *J. Amer. Chem. Soc.*, **73**, 232 (1951).
115. T. M. Jacob and S. Dev, *Chem. Ind.* (London), **1956**, 576. E. Kloster-Jensen, E. Kováts, A. Eschenmoser, and E. Heilbronner, *Helv. Chim. Acta*, **39**, 1051 (1956).
116. M. Nakasaki, *J. Chem. Soc. Japan, Pure Chem. Sect.*, **74**, 403 (1953).
117. E. Knoevenagel and B. Bergdolt, *Chem. Ber.*, **36**, 2857 (1903).
118. H. Wieland, *Chem. Ber.*, **45**, 484 (1912).
119. N. D. Zelinsky and G. S. Pavlov, *Chem. Ber.*, **66**, 1420 (1933).
120. N. Zelinsky, *Chem. Ber.*, **57**, 2055 (1924).
121. N. Zelinsky, *Chem. Ber.*, **57**, 2058 (1924).
122. N. Zelinsky, *Chem. Ber.*, **58**, 864 (1925).
123. R. J. Levina and D. M. Trachtenberg, *J. Gen. Chem. Russia*, **6**, 764 (1936).
124. R. P. Linstead, E. A. Braude, P. W. D. Mitchell, K. R. H. Wooldridge, and L. M. Jackman, *Nature*, **169**, 100 (1952).
125. E. A. Braude, R. P. Linstead, and K. R. H. Wooldridge, *J. Chem. Soc.*, **1954**, 3586.
126. E. A. Braude, R. P. Linstead, P. W. D. Mitchell, and K. R. H. Wooldridge, *Chem. Soc.*, **1954**, 3595.
127. L. Ruzicka, H. Hösli, and L. Ehmann, *Helv. Chim. Acta*, **17**, 442 (1934).
128. C. Dorée and V. A. Petrov, *J. Chem. Soc.*, **1934**, 1129.
129. A. Cohen, J. W. Cook, and C. L. Hewett, *J. Chem. Soc.*, **1935**, 1633.
130. J. W. Cook and G. A. D. Haslewood, *J. Chem. Soc.*, **1935**, 767.
131. L. Ruzicka and E. Peyer, *Helv. Chim. Acta*, **18**, 676 (1935).
132. J. W. Cook and C. A. Lawrence, *J. Chem. Soc.*, **1936**, 1431.
133. P. S. Kalsi, *Chem. Ind.* (London), **1970**, 276.
134. J. W. Mellor, *A Comprehensive Treatise on Inorganic and Theoretical Chemistry*, Vol. 10, Longmans, Green, London, 1940, pp. 114, 757.
135. G. C. Harris and T. F. Sanderson, *J. Amer. Chem. Soc.*, **42**, 2081 (1948).
136. M. S. Newman and F. T. J. O'Leary, *J. Amer. Chem. Soc.*, **68**, 258 (1946).
137. W. A. Jacobs and S. W. Pelletier, *J. Org. Chem.*, **18**, 765 (1953).
138. J. v. Braun and G. Irmish, *Chem. Ber.*, **65**, 883 (1932).
139. J. W. Cook and C. L. Hewett, *J. Chem. Soc.*, **1934**, 365.
140. V. Boekelheide and V. V. Prelog in *Progress in Organic Chemistry*, Vol. 3 (J. W. Cook, ed.), Academic Press, New York, 1955, p. 218.

141. Z. Valenta, K. Wiesner, C. Bankiewicz, D. R. Henderson, and J. S. Little, *Chem. Ind.* (London), B. I. F. Review, R40 (1956).
142. K. Wiesner, Z. Valenta, W. A. Ayer, L. R. Fowler, and J. E. Francis, *Tetrahedron*, **4**, 87 (1958).
143. M. Ehrenstein and W. Bunge, *Chem. Ber.*, **67**, 1715 (1934).
144. A. S. Dreiding and W. J. Pummer, *J. Amer. Chem. Soc.*, **7**, 3162 (1953).
145. F. E. King and T. J. King, *J. Chem. Soc.*, **1953**, 4158.
146. W. F. Short and H. Wang, *J. Chem. Soc.*, **1951**, 2979.
147. H. E. Eschinazi, *Bull. Soc. Chim. Fr.*, **1952**, 967.
148. M. S. Newman and T. S. Bye, *J. Amer. Chem. Soc.*, **74**, 905 (1952).
149. M. S. Newman and J. R. Maugham, *J. Amer. Chem. Soc.*, **71**, 3342 (1949).
150. M. S. Newman and H. V. Zahm, *J. Amer. Chem. Soc.*, **65**, 1097 (1943).
151. G. Büchi and D. Rosenthal, *J. Amer. Chem. Soc.*, **78**, 3860 (1956); W. Herz, A. R. De Vivar, J. Romo, and N. Viswanathan, *J. Amer. Chem. Soc.*, **85**, 19 (1963).
152. H. Arnold, *Chem. Ber.*, **80**, 123 (1947).
153. G. T. Tatevosyan and A. G. Vardanyan, *Zh. Obschch. Khim.*, **19**, 332 (1949); through *Chem. Abstr.*, **43**, 6609 (1949).
154. L. Ruzicka and G. Giacomello, *Helv. Chim. Acta*, **20**, 299 (1937).
155. K. Wiesner, J. R. Armstrong, M. F. Bartlett, and J. A. Edwards, *J. Amer. Chem. Soc.*, **76**, 6068 (1954).
156. L. Ruzicka, M. W. Goldberg, and G. Thomann, *Helv. Chim. Acta*, **16**, 812 (1933).
157. H. Schulze, *Z. Physiol. Chem.*, **238**, 35 (1936).
158. O. Jeger, *Fortschr. Chem. Org. Naturstoffe*, **7**, 1 (1950).
159. H. Honigmann, *Justus Liebigs Ann. Chem.*, **551**, 292 (1934).
160. J. W. Cook, C. L. Hewett, and C. A. Lawrence, *J. Chem. Soc.*, **1936**, 71.
161. W. Cocker, B. E. Cross, A. K. Fateen, C. Lipman, E. R. Stuart, W. H. Thompson and D. R. A. Whyte, *J. Chem. Soc.*, **1950**, 1781.
162. L. C. Craig and W. A. Jacobs, *J. Biol. Chem.*, **139**, 263 (1941).
163. D. H. R. Barton, O. Jeger, V. Prelog, and R. B. Woodward, *Experientia*, **10**, 81 (1954).
164. R. N. Chakravarti, *J. Indian Chem. Soc.*, **20**, 393 (1943).
165. V. Herout, L. Dolejš, and F. Šorm, *Chem. Ind.* (London), **1956**, 1236.
166. Z. Čekan, V. Herout, and F. Šorm, *Chem. Ind.* (London), **1956**, 1234.
167. G. Büchi and R. E. Erickson, *J. Amer. Chem. Soc.*, **78**, 1262 (1956); M. Dobler, J. D. Dunitz, B. Gubler, H. P. Weber, G. Buchi, and J. Padilaa, *Proc. Chem. Soc.*, **1963**, 383.
168. W. E. Bachmann and A. S. Dreiding, *J. Amer. Chem. Soc.*, **72**, 1323 (1950).
169. E. C. Horning and M. G. Horning, *J. Amer. Chem. Soc.*, **69**, 1359 (1947).
170. E. Buchta and W. Kallert, *Justus Liebigs Ann. Chem.*, **576**, 1 (1952).
171. L. F. Fieser and E. B. Hershberg, *J. Amer. Chem. Soc.*, **58**, 2382 (1936).
172. G. A. R. Kon and F. C. J. Ruzicka, *J. Chem. Soc.*, **1936**, 187.
173. G. Darzens and A. Lévy, *C.R. Acad. Sci.*, Paris, **194**, 181 (1932).
174. L. F. Fieser, E. B. Hershberg, and M. S. Newman, *J. Amer. Chem. Soc.*, **57**, 1509 (1935).

175. E. C. Horning, *J. Amer. Chem. Soc.*, **67**, 1421 (1945).
176. R. Charonnat and M. Girard, *Bull. Soc. Chim. Fr.*, **1949**, 208.
177. L. Ruzicka and E. Mörgeli, *Helv. Chim. Acta*, **19**, 377 (1936).
178. E. C. Horning, M. G. Horning, and G. N. Walker, *J. Amer. Chem. Soc.*, **71**, 169 (1949).
179. E. Mosettig and H. M. Duval, *J. Amer. Chem. Soc.*, **59**, 367 (1937).
180. J. W. Cook and A. R. Somerville, *Nature*, **163**, 410 (1949).
181. A. Barbot, *Bull. Soc. Chim.*, **47**, 1314 (1930).
182. M. Orchin, E. O. Woolfolk, and L. Reggel, *J. Amer. Chem. Soc.*, **71**, 1126 (1949).
183. L. F. Fieser and M. T. Leffler, *J. Amer. Chem. Soc.*, **70**, 3212 (1948).
184. W. S. Johnson, H. C. E. Johnson, and J. W. Petersen, *J. Amer. Chem. Soc.*, **67**, 1360 (1945).
185. M. Orchin, *J. Amer. Chem. Soc.*, **70**, 495 (1948).
186. M. Orchin and L. Reggel, *J. Amer. Chem. Soc.*, **73**, 1877 (1951).
187. M. Orchin, L. Reggel, and R. A. Friedel, *J. Amer. Chem. Soc.*, **71**, 2743 (1949).
188. G. Büchi and J. J. Pappas, *J. Amer. Chem. Soc.*, **76**, 2963 (1954).
189. J. W. Cook, G. A. D. Haslewood, and A. M. Robinson, *J. Chem. Soc.*, **1935**, 667.
190. C. Djerassi, H. Bendas, and P. Sengupta, *J. Org. Chem.*, **20**, 1046 (1955).
191. R. D. Haworth and R. A. W. Johnstone, *J. Chem. Soc.*, **1957**, 1492.
192. L. Ruzicka and F. Balas, *Helv. Chim. Acta*, **6**, 677 (1923).
193. L. Ruzicka, L. Ehmann, M. W. Goldberg, and H. Hösli, *Helv. Chim. Acta*, **16**, 833 (1933).
194. G. Darzens and A. Lévy, *C.R. Acad. Sci., Paris*, **194**, 2056 (1932).
195. G. Darzens and A. Lévy, *C.R. Acad. Sci., Paris*, **200**, 2187 (1935).
196. G. Darzens and A. Lévy, *C.R. Acad. Sci., Paris*, **201**, 902 (1935).
197. G. Darzens and A. Lévy, *C.R. Acad. Sci., Paris*, **202**, 427 (1936).
198. G. Darzens and A. Lévy, *C.R. Acad. Sci., Paris*, **203**, 669 (1936).
199. G. Darzens and A. Lévy, *C.R. Acad. Sci., Paris*, **199**, 1131 (1934).
200. L. F. Fieser and E. B. Hershberg, *J. Amer. Chem. Soc.*, **57**, 1508 (1935).
201. L. F. Fieser and E. B. Hershberg, *J. Amer. Chem. Soc.*, **57**, 1851 (1935).
202. A. Cohen, *Nature*, **136**, 869 (1935).
203. A. Cohen, J. W. Cook, and C. L. Hewett, *J. Chem. Soc.*, **1936**, 52.
204. A. Cohen and F. L. Warren, *J. Chem. Soc.*, **1937**, 1315.
205. M. S. Newman, *J. Org. Chem.*, **9**, 518 (1944).
206. D. H. R. Barton and Dov Elad, *J. Chem. Soc.*, **1956**, 2090.
207. V. N. Ipatieff, J. E. Germain, and H. Pines, *Bull. Soc. Chim. Fr.*, **1951**, 259.
208. A. Windaus and W. Thiele, *Justus Liebigs Ann. Chem.* **521**, 160 (1935).
209. W. Thiele and G. Trautmann, *Chem. Ber.*, **68**, 2245 (1935).
210. L. Dolejš, M. Souček, M. Horák, V. Herout, and F. Šorm, *Chem. Ind.* (London), **1958**, 530.
211. C. Djerassi, M. Cais, and L. A. Mitscher, *J. Amer. Chem. Soc.*, **80**, 247 (1958).
212. E. Wenkert, *Chem. Ind.* (London), **1955**, 282.
213. J. C. Bardham and S. C. Sengupta, *J. Chem. Soc.*, **1932**, 2520.
214. L. F. Fieser and H. L. Holmes, *J. Amer. Chem. Soc.*, **58**, 2319 (1936).

215. R. D. Haworth, C. R. Mavin, and G. Sheldrick, *J. Chem. Soc.*, **1934**, 454.
216. N. Zelinsky and N. Glinka, *Chem. Ber.*, **44**, 2305 (1911).
217. N. Zelinsky and N. Uklonskaja, *Chem. Ber.*, **45**, 3677 (1912).
218. L. F. Fieser and M. Fieser, *J. Amer. Chem. Soc.*, **57**, 1679 (1935).
219. M. S. Newman and A. S. Hussey, *J. Amer. Chem. Soc.*, **69**, 3023 (1947).
220. D. H. R. Barton and C. R. Narayanan, *J. Chem. Soc.*, **1958**, 963.
221. Y. Mazur and A. Meisels, *Chem. Ind.* (London), **1956**, 492.
222. D. H. R. Barton and P. de Mayo, *J. Chem. Soc.*, **1956**, 142.
223. K. Tsuda, K. Tanabe, I. Iwai, and K. Funakoshi, *J. Amer. Chem. Soc.*, **79**, 5721 (1957).
224. C. Bankiewicz, D. R. Henderson, F. W. Stonner, Z. Valenta, and K. Wiesner, *Chem. Ind.* (London), **1954**, 544.
225. L. Ruzicka and C. F. Seidel, *Helv. Chim. Acta*, **5**, 369 (1922).
226. L. Ruzicka and M. Stoll, *Helv. Chim. Acta*, **5**, 923 (1922).
227. L. Ruzicka, F. Balas, and H. Shinz, *Helv. Chim. Acta*, **6**, 692 (1923).
228. N. Zelinsky, *Chem. Ber.*, **56**, 1716 (1923).
229. N. D. Zelinsky, K. Packendorff, and E. G. Chochlova, *Chem. Ber.*, **68**, 98 (1935).
230. G. R. Clemo and H. G. Dickenson, *J. Chem. Soc.*, **1935**, 735.
231. L. Ruzicka and H. Waldmann, *Helv. Chim. Acta*, **16**, 842 (1933).
232. L. Ruzicka and H. Waldmann, *Helv. Chim. Acta*, **18**, 611 (1935).
233. R. L. Barker and G. R. Clemo, *J. Chem. Soc.*, **1940**, 1277.
234. L. H. Briggs, N. S. Gill, F. Lions, and W. I. Taylor, *J. Chem. Soc.*, **1949**, 1098.
235. L. Ruzicka and J. R. Hosking, *Helv. Chim. Acta*, **14**, 203 (1931).
236. L. Ruzicka, L. L. Engel, and W. H. Fischer, *Helv. Chim. Acta*, **21**, 364 (1938).
237. L. Ruzicka and M. M. Janot, *Helv. Chim. Acta*, **14**, 645 (1931).
238. I. Heilbron, W. M. Owens, and I. A. Simpson, *J. Chem. Soc.*, **1929**, 873. L. Ruzicka and F. Lardon, *Helv. Chim. Acta*, **29**, 912 (1946).
239. T. Matsumoto, *J. Chem. Soc. Japan, Pure Chem. Sect.*, **72**, 68 (1951).
240. E. Bergmann, *Chem. Ind.* (London), **54**, 175 (1935).
241. A. Cohen, J. W. Cook, and C. L. Hewett, *J. Chem. Soc.*, **1935**, 445.
242. J. W. Cook, *Chem. Ind.* (London), **54**, 176 (1935).
243. O. Rosenheim and H. King, *Chem. Ind.* (London), **1933**, 299.
244. J. W. Cook and A. Girard, *Nature*, **133**, 377 (1934).
245. A. Butenandt, H. Dannenberg, and W. Steidle, *Z. Naturforsch*, **9**, 288 (1954).
246. K. D. Errington and R. P. Linstead, *J. Chem. Soc.*, **1938**, 666.
247. D. H. R. Barton and K. H. Overton, *J. Chem. Soc.*, **1955**, 2639.
248. L. Ruzicka, M. W. Goldberg, and K. Hofmann, *Helv. Chim. Acta*, **20**, 325 (1937).
249. L. Ruzicka, H. Schellenberg, and M. W. Goldberg, *Helv. Chim. Acta*, **20**, 791 (1937).
250. L. Ruzicka and A. H. Lamberton, *Helv. Chim. Acta*, **23**, 1338 (1940).
251. L. Ruzicka and G. Rosenkranz, *Helv. Chim. Acta*, **23**, 1311 (1940).
252. D. J. Cram, in *Steric Effects in Organic Chemistry* (M. S. Newman, ed.), Wiley, New York, 1956, p. 304.
253. F. Mayer and R. Schniffner, *Chem. Ber.*, **67**, 67 (1934).

254. L. F. Fieser, L. M. Joshel, and A. M. Seligman, *J. Amer. Chem. Soc.*, **61**, 2134 (1939).

255. M. S. Newman, *J. Amer. Chem. Soc.*, **62**, 2295 (1940).

256. W. E. Jones and G. R. Ramadge, *J. Chem. Soc.*, **1938**, 1853.

257. W. E. Bachmann and R. O. Edgerton, *J. Amer. Chem. Soc.*, **62**, 2219 (1940).

258. H. H. Inhoffen, G. Stoeck, and G. Kölling, *Chem. Ber.*, **82**, 263 (1949).

259. R. B. Woodward, H. H. Inhoffen, H. O. Larson, and K. H. Menzel, *Chem. Ber.*, **86**, 594 (1953).

260. A. J. Haagen-Smit, *Fortschr. Chem. Org. Naturstoffe*, **12**, 1 (1955).

261. H. Pommer, *Angew. Chem.*, **62** 281 (1950).

262. G. R. Clemo and J. Ormston, *J. Chem. Soc.*, **1933**, 352.

263. V. Sýkora, V. Herout, J. Pliva, and F. Šorm, *Chem. Ind.* (London), **1956**, 1231.

264. S. C. Sen Gupta and D. N. Chatterjee, *Sci. Cult.* (Calcutta), **17**, 93 (1951); through *Chem. Abstr.*, **46**, 11168 (1952).

265. S. C. Sen Gupta and D. N. Chatterjee, *J. Indian Chem. Soc.*, **29**, 438 (1952).

266. S. C. Sen Gupta and D. N. Chatterjee, *J. Indian Chem. Soc.*, **31**, 11 (1954).

267. S. C. Sen Gupta and D. N. Chatterjee, *J. Indian Chem. Soc.*, **31**, 285 (1954).

268. S. C. Sen Gupta and D. N. Chatterjee, *J. Indian Chem. Soc.*, **31**, 911 (1954).

269. S. C. Sen Gupta and D. N. Chatterjee, *J. Indian Chem. Soc.*, **32**, 13 (1955).

270. M. Levitz and M. T. Bogert, *J. Org. Chem.*, **8**, 253 (1943).

271. C. S. Marvel and L. A. Brooks, *J. Amer. Chem. Soc.*, **63**, 2630 (1941).

272. N. D. Zelinsky and N. V. Elaghina, *C.R. Acad. Sci. URSS*, **52**, 227 (1946).

273. N. D. Zelinsky and N. I. Schuikin, *Chem. Ber.*, **62**, 2180 (1929).

274. O. Diels and A. Karstens, *Justus Liebigs Ann. Chem.*, **478**, 129 (1930).

275. L. Ruzicka and M. W. Goldberg, *Helv. Chim. Acta*, **20**, 1245 (1937).

276. L. Ruzicka, G. Thomann, E. Brandenberger, M. Furter, and M. W. Goldberg, *Helv. Chim. Acta*, **17**, 200 (1934).

277. Y. I. Denisenko, *Bull. Acad. Sci.* (*URSS*), *Classe Sci. Chim.*, **1944**, 337.

278. C. D. Nenitzescu and E. Cioránescu, *Chem. Ber.*, **69**, 1040 (1936).

279. R. Robinson, *J. Chem. Soc.*, **1936**, 80.

280. M. B. Turova-Pollak and P. L. Rappoport, *J. Gen. Chem.* (*USSR*), **13**, 353 (1943).

281. D. H. S. Horn, J. R. Nunn, and W. S. Rapson, *Nature*, **160**, 829 (1947).

282. D. H. S. Horn and W. S. Rapson, *J. Chem. Soc.*, **1949**, 2421.

283. A. Reisse, D. Cagniant, and P. Cagniant, *Bull. Soc. Chim. Fr.*, **1969**, 2115.

284. N. D. Zelinsky and R. J. Levina, *Justus Liebigs Ann. Chem.*, **476**, 60 (1929).

285. G. G. Henderson, W. M'Nab, and J. M. Robertson, *J. Chem. Soc.*, **1926**, 3077.

286. G. Komppa and G. A. Nyman, *C.R. Trav. Lab. Carlsberg*, *Ser. Chim.*, **22**, 272 (1938); through *Chem. Abstr.*, **32**, 6234 (1938).

287. C. B. Radcliffe and W. F. Short, *J. Chem. Soc.*, **1938**, 1200.

288. A. Blumann, A. R. H. Cole, K. J. L. Thieberg, and D. E. White, *Chem. Ind.* (London), **1954**, 1426.

289. A. J. Birch and K. M. C. Mostyn, *Aust. J. Chem.*, **8**, 550 (1955).

290. J. W. Barrett and R. P. Linstead, *J. Chem. Soc.*, **1936**, 611.

291. N. D. Zelinsky, S. E. Michlina, and M. S. Eventova, *Chem. Ber.*, **66**, 1422 (1933).

292. N. D. Zelinsky and I. N. Titz, *Chem. Ber.*, **64**, 183 (1931).

293. Y. I. Denisenko, *Chem. Ber.*, **69**, 1353 (1936).
294. N. D. Zelinsky and B. A. Kazansky, *C.R. Acad. Sci. URSS*, **3**, 168 (1934).
295. N. D. Zelinsky, B A. Kazansky, and A. F. Plate, *Chem. Ber.*, **66**, 1415 (1933).
296. B. A. Kazansky and A. F. Plate, *Chem. Ber.*, **69**, 1862 (1936).
297. E. R. H. Jones and T. G. Halsall, *Fortschr. Chem. Org. Naturstoffe*, **12**, 44 (1955).
298. G. Chiurdoglu and P. Tullen, *Chem. Ind.* (London), **1956**, 1094.
299. C. W. Brandt, *N.Z. J. Sci. Technol.*, **B34**, 46 (1952).
300. F. Hofmann, *Chem.-Ztg.*, **57**, 5 (1933).
301. B. A. Kazansky and A. F. Plate, *Chem. Ber.*, **68**, 1259 (1935).
302. N. V. Elaghina and N. D. Zelinsky, *C.R. Acad. Sci.* (*URSS*), **30**, 728 (1941).
303. M. F. Bartlett, S. K. Figdor, and K. Wiesner, *Can. J. Chem.*, **30**, 291 (1952).
304. K. Wiesner, D. M. MacDonald, R. B. Ingraham, and R. B. Kelly, *Can. J. Research*, **B28**, 561 (1950).
305. D. J. Cram and N. L. Allinger, *J. Amer. Chem. Soc.*, **77**, 6289 (1955).
306. L. Ruzicka and A. G. van Veen, *Justus Liebigs Ann. Chem.*, **468**, 133 (1929).
307. N. A. Sorensen and F. Hougen, *Acta Chem. Scand.*, **2**, 447 (1948).
308. N. D. Zelinsky, I. N. Titz, and M. V. Gaverdovskaja, *Chem. Ber.*, **59**, 2590 (1926).
309. H. Berger, *J. Prakt. Chem.*, **133**, 331 (1932).
310. M. Orchin and E. O. Woolfolk, *J. Amer. Chem. Soc.*, **67**, 122 (1945).
311. M. Orchin, *J. Amer. Chem. Soc.*, **67**, 499 (1945).
312. M. Orchin, *J. Amer. Chem. Soc.*, **68**, 571 (1946).
313. M. Orchin and L. Reggel, *J. Amer. Chem. Soc.*, **69**, 505 (1947).
314. M. Orchin and R. A. Friedel, *J. Amer. Chem. Soc.*, **68**, 573 (1946).
315. I. M. Heilbron, R. A. Morton, and E. T. Webster, *Biochem. J.*, **26**, 1194 (1932).
316. M. S. Newman and H. S. Whitehouse, *J. Amer. Chem. Soc.*, **71**, 3664 (1949).
317. W. E. Bachmann, J. W. Cook, C. L. Hewett, and J. Iball, *J. Chem. Soc.*, **1936**, 54.
318. L. Ruzicka and G. Thomann, *Helv. Chim. Acta*, **16**, 216 (1933).
319. O. Diels, W. Gädke, and P. Körding, *Justus Liebigs Ann. Chem.*, **459**, 1 (1927).
320. J. W. Cook, C. L. Hewett, W. V. Mayneord, and E. Roe, *J. Chem. Soc.*, **1934**, 1727.
321. E. Bergmann, *J. Amer. Chem. Soc.*, **60**, 2306 (1938).
322. M. M. Janot, J. Keufer, and J. LeMen, *Bull. Soc. Chim. Fr.*, **1952**, 230.
323. H. Adkins and L. G. Lunsted, *J. Amer. Chem. Soc.*, **71**, 2964 (1949).
324. A. P. Terentjev and S. M. Gurvich, *Sb. Statei Obsch. Khim*, **2**, 1105 (1953); through *Chem. Abstr.*, **49**, 5469 (1955).
325. M. Ehrenstein, *Chem. Ber.*, **64**, 1137 (1931).
326. N. D. Zelinsky and J. K. Yurjev, *Chem. Ber.*, **64**, 101 (1931).
327. E. Späth, F. Berger, and W. Kuntara, *Chem. Ber.*, **63**, 134 (1930).
328. V. Prelog and S. Szpilfogel, *Helv. Chim. Acta*, **28**, 1684 (1945).
329. B. Witkop, *J. Amer. Chem. Soc.*, **70**, 2617 (1948).
330. A. Lettir, R. Goutarel, and M. M. Janot, *Bull. Soc. Chim. Fr.*, **1954**, 866.
331. B. M. Barclay and N. Campbell, *J. Chem. Soc.*, **1945**, 530.
332. W. E. Babcock and K. H. Pausacker, *J. Chem. Soc.*, **1951**, 1373.

333. E. C. Horning, M. G. Horning, and G. N. Walker, *J. Amer. Chem. Soc.*, **70**, 3935 (1948).
334. V. Prelog, A. Komzak, and E. Moor, *Helv. Chim. Acta*, **25**, 1654 (1942).
335. E. Späth and F. Galinovsky, *Chem. Ber.*, **69**, 2059 (1936).
336. V. Prelog and K. Balenovic, *Chem. Ber.*, **74**, 1508 (1941).
337. V. Prelog and O. Jeger, in *The Alkaloids, Chemistry and Physiology*, Vol. 3 (R. H. F. Manske and H. L. Holmes, eds.), Academic Press, New York, 1953, p. 247.
338. L. C. Craig and W. A. Jacobs, *J. Biol. Chem.*, **149**, 451 (1943).
339. V. Prelog and S. Szpilfogel, *Helv. Chim. Acta*, **25**, 1306 (1942).
340. W. A. Jacobs, L. C. Craig, and G. I. Lavin, *J. Biol. Chem.*, **141**, 51 (1941).
341. W. A. Jacobs and Y. Sato, *J. Biol. Chem.*, **181**, 55 (1949).
342. J. Fried, O. Wintersteiner, M. Moore, B. M. Iselin, and A. Klingsberg, *J. Amer. Chem. Soc.*, **73**, 2970 (1951).
343. K. Wiesner, J. R. Armstrong, M. F. Bartlett, and J. A. Edwards, *Chem. Ind.* (London), **1954**, 132.
344. W. A. Jacobs and L. C. Craig, *J. Biol. Chem.*, **143**, 589 (1942).
345. J. E. Saxton, *Quart. Rev.* (London), **10**, 108 (1956).
346. R. B. Woodward, *Angew. Chem.*, **68**, 13 (1956).
347. F. A. L. Anet and L. Marion, *Chem. Ind.* (London), **1954**, 1232.
348. K. Wiesner, W. A. Ayer, L. R. Fowler, and Z. Valenta, *Chem. Ind.* (London), **1957**, 564.
349. F. Kögl and W. B. Deijs, *Justus Liebigs Ann. Chem.*, **515**, 14 (1935).
350. W. Davies and B. C. Ennis, *J. Chem. Soc.*, **1960**, 1488.
351. B. Witkop, *Justus Liebigs Ann. Chem.*, **554**, 83 (1943).
352. F. Kögl and J. J. Postowsky, *Justus Liebigs Ann. Chem.*, **440**, 32 (1924).
353. H. v. Euler, H. Erdtman, and H. Hellström, *Chem. Ber.*, **69**, 743 (1936).
354. F. A. Hochstein, C. R. Stephens, L. H. Convor, P. P. Regna, R. Pasternack, P. N. Gordon, F. J. Pilgrim, K. J. Brunings, and R. B. Woodward, *J. Amer. Chem. Soc.*, **75**, 5455 (1953).
355. E. Clar, *Chem. Ber.*, **72**, 1645 (1939); *Polycyclic Hydrocarbons*, Vol. 1, Academic Press, New York, 1964.
356. J. Gasparic, *Microchim. Acta*, **1966**, 288.
357. K. Biemann and G. Spiteller, *J. Amer. Chem. Soc.*, **84**, 4578 (1962).
358. A. v. Baeyer, *Justus Liebigs Ann. Chem.* 140, 295 (1866).
359. O. Döbner, *Chem. Ber.*, **9**, 130 (1876).
360. C. Liebermann, *Justus Liebigs Ann. Chem.*, **169**, 221 (1873).
361. C. Liebermann and Topf, *Chem. Ber.*, **9**, 1201 (1876).
362. A. Butenandt, H. A. Weidlich, and H. Thompson, *Chem. Ber.*, **66**, 601 (1933).
363. C. Graebe, *Justus Liebigs Ann. Chem.*, **152**, 66 (1869).
364. H. Kondo and S. Uyeo, *Chem. Ber.*, **70**, 1087 (1937).
365. E. v. Gerichten, *Chem. Ber.*, **31**, 51 (1898).
366. E. v. Gerichten and H. Schrötter, *Chem. Ber.*, **15**, 1484 (1882).
367. V. Merz and W. Weith, *Chem. Ber.*, **14**, 2346 (1881).
368. O. Fischer and E. Renouf, *Chem. Ber.*, **17**, 755 (1884).
369. St. v. Niementowski, *Chem. Ber.*, **40**, 4285 (1907).

370. L. Knorr, *Chem. Ber.*, **16**, 2596 (1883).
371. F. Kögl, H. Erxleben, and L. Jänecke, *Justus Liebigs Ann. Chem.*, **482**, 105 (1930).
372. L. Barth and G. Goldschmidt, *Chem. Ber.*, **12**, 1237 (1879).
373. A. Butenandt and H. Thompson, *Chem. Ber.*, **67**, 140 (1934).
374. H. Tanaka and T. Tamura, *Tetrahedron Lett.*, **1961**, 151.
375. C. Graebe and C. Liebermann, *Justus Liebigs Ann. Chem.*, *Suppl.*, **7**, 287 (1870).
376. C. Liebermann, *Justus Liebigs Ann. Chem.*, **162**, 333 (1872).
377. F. Kögl, *Justus Liebigs Ann. Chem.*, **447**, 78 (1925).
378. F. Kögl and H. Erxleben, *Justus Liebigs Ann. Chem.*, **479**, 11 (1930).
379. J. Gripenberg, *Tetrahedron*, **10**, 135 (1960).
380. H. Brockmann, *Fortschr. Chem. Org. Naturstoffe*, **14**, 147 (1957).
381. H. Fritzsche, *Z. Chem.*, **1869**, 392.
382. R. Scholl and H. Berblinger, *Chem. Ber.*, **36**, 3427 (1903).
383. S. Gabriel and E. Leupold, *Chem. Ber.*, **31**, 1272 (1904).
384. R. Scholl and H. Neumann, *Chem. Ber.*, **55**, 118 (1922).
385. A. W. Hofmann, *Chem. Ber.*, **18**, 109 (1885).
386. G. Ciamician and P. Silver, *Chem. Ber.*, **27**, 2850 (1894).
387. Z. H. Skraup, *Monatsch. Chem.*, **9**, 783 (1888).
388. W. A. Jacobs and L. C. Craig, *J. Biol. Chem.*, **120**, 447 (1937).
389. H. Thoms and K. Bergerhoff, *Arch. Pharm.*, **263**, 3 (1925).
390. C. R. Clemo and R. Raper, *J. Chem. Soc.*, **1933**, 644.
391. H. Kondo, *Arch. Pharm.*, **266**, 1 (1928).
392. K. Winterfeld and A. Kneuer, *Chem. Ber.*, **64**, 150 (1931).
393. E. Winterstein and M. Walter, *Helv. Chim. Acta*, **10**, 577 (1927).
394. B. Witkop, *Justus Liebigs Ann. Chem.*, **556**, 103 (1944).
395. D. Mukherji, R. Robinson, and E. Schlittler, *Experientia*, **5**, 215 (1949).
396. H. Wieland, B. Witkop, and K. Bähr, *Justus Liebigs Ann. Chem.*, **558**, 144 (1947).
397. K. S. Kirby, *J. Chem. Soc.*, **1945**, 528.
398. A. H. Salway, *J. Chem. Soc.*, **103**, 351 (1913).
399. B. Witkop, *J. Amer. Chem. Soc.*, **70**, 1424 (1948).
400. H. Conroy and J. K. Chakrabarti, *Tetrahedron Lett.*, **4**, 6 (1959).
401. A. J. Birch, H. F. Hodson, and G. F. Smith, *Proc. Chem. Soc.*, **1959**, 224.
402. H. Conroy, R. Bernasconi, P. R. Brook, R. Ikan, R. Kurtz, and K. W. Robinson, *Tetrahedron Lett.*, **6**, 1 (1960).
403. T. R. Govindachari and S. Rajappa, *Chem. Ind.* (London), **1954**, 1159.
404. J. A. Hamilton, T. A. Hamor, J. M. Robertson, and G. A. Sim, *Proc. Chem. Soc.*, **1961**, 63.
405. G. Barger, T. R. Seshadri, H. E. Watt, and T. Yabuta, *J. Chem. Soc.*, **1935**, 11.
406. B. Witkop, *J. Amer. Chem. Soc.*, **70**, 3712 (1948).
407. E. v. Gerichten and H. Schrötter, *Justus Liebigs Ann. Chem.*, **210**, 396 (1881).
408. L. Knorr, *Chem. Ber.*, **27**, 1144 (1894).
409. H. Kondo and E. Ochiai, *Justus Liebigs Ann. Chem.*, **470**, 224 (1929).

410. E. Ochiai, *J. Pharm. Soc. Japan*, **503,** 8 (1924); *Chem. Abstr.*, **18,** 1667 (1924).
411. J. W. Cook, G. T. Dickson, and J. D. Loudon, *J. Chem. Soc.*, **1947,** 746.
412. A. Windaus, *Justus Liebigs Ann. Chem.*, **439,** 59 (1924).
413. K. Feist, E. Kuntz, and R. Brachvogel, *Justus Liebigs Ann. Chem.*, **519,** 124 (1935).
414. A. W. van der Haar, *Chem. Ber.*, **55,** 1054 (1922).
415. G. M. Badger and T. M. Spotswood, *J. Chem. Soc.*, **1959,** 1635.

Chapter **XI**

REDUCTION METHODS

J. P. Candlin and R. A. C. Rennie

Structural determination of organic compounds involves the transformation of the unknown compound into a known compound either by synthesis or degradation techniques. With the advent of high resolution mass spectrometers, nmr and spectroscopic techniques, and the range of chromatographic techniques, the problem is frequently to convert the unknown into a form amenable to analysis by these tools. Reduction reactions have a large part to play in the necessary transformations and simplifications. However, they cannot be treated in isolation from other techniques. We have decided that the best plan is to attempt to survey the scope of the principal reducing reagents, exemplified where appropriate by applications in structure elucidation studies. To emphasize the interdependence of the various techniques, we have included a number of cases which are discussed in more detail.

Because the reactivity of a functional group and the possibility of side reactions depends on the other functional groups present and on steric effects, it is not possible to cover modifications of reagents that have been introduced from time to time in the space available. Besides the general references included in our survey, attention is also drawn to the compilation of reduction methods made in the Annual Reports of the Chemical Society (London) and to the series *Synthetic Methods of Organic Chemistry* (Theilheimer), Wiley-Interscience.

I THE HYDRIDE REAGENTS

Aluminum Hydrides

Lithium aluminum hydride reduces a wide selection of functional groups with the notable exception of carbon-to-carbon multiple bonds and ether linkages [1–3]. The ethers, diethyl ether, tetrahydrofuran, and diglyme are, therefore, the most commonly used solvents. Modifications to the original

Table 11.1   Functional Groups Reduced by $LiAlH_4$

Functional Group	Product
$\diagup\!\!=\!\!O$	$-CH-OH$
$-CO_2R$	$-CH_2OH + ROH$
$-CO_2H$	$-CH_2OH$
$-CONHR$	$-CH_2NHR$
$-CONR_2$	$-CH_2-NR_2$ or $\left[\begin{array}{c}-CH-NR_2\\ \mid\\ OH\end{array}\right] \rightarrow -CHO + R_2NH$
$-C\equiv N$	$-CH_2-NH_2$ or $[-CH=NH] \rightarrow -CHO$
$-C=NOH$	$-CH-NH_2$
$-C-NO_2$ (aliphatic)	$-C-NH_2$
$-CH_2OSO_2C_6H_5$	$-CH_3$
$-CH_2Br$	$-CH_3$
$-\underset{\diagdown O \diagup}{C-\!\!-C}-$	$-CH_2-\underset{\mid}{\overset{\mid}{C}}-$ OH

reagent have been introduced to give greater selectivity, the most important of which are:

lithium aluminum hydride ———➤ aluminum hydride

| ROH
▼

lithium dialkoxy aluminum hydrides

| ROH
▼

lithium trialkoxy aluminum hydrides.

The common functional groups reduced with lithium aluminum hydride are shown in Table 11.1.

All four hydrogen atoms of lithium aluminum hydride can be used for reduction, but at successively slower rates. This forms the basis for the preparation of milder and more selective reagents by replacement of two or three of the hydrogen atoms by alkoxy groups. Lithium diethoxy aluminum hydride is prepared by the controlled addition of 2 moles of ethanol to 1 mole of LiAlH$_4$ [4]. It is especially recommended for reduction of N,N-dimethylamides to aldehydes, compare Table 11.1. This offers an alternative to the Rosenmund reduction:

$$CH_3CH_2CH_2CON(CH_3)_2 \rightarrow CH_3CH_2CH_2CHO \ (90\%).$$

Lithium triethoxy aluminum hydride is prepared from 3 moles of ethanol and 1 mole of LiAlH$_4$ [5]. It is especially effective in reducing aliphatic and aromatic nitriles to aldehydes in high yield. Lithium trimethoxy aluminum hydride is similar to LiAlH$_4$ in reactivity.

Lithium tri-t-butoxyaluminum hydride, formed from LiAlH$_4$ and t-butanol, is a milder reagent than LiAlH$_4$ since it will reduce aldehydes, ketones, and acid chlorides, but not esters or nitriles [6]. It enables acid chlorides to be reduced in good yield to aldehydes [7]. This reagent also shows enhanced stereoselectivity compared to lithium aluminum hydride, and this has been of considerable use in the chemistry of steroids [8].

Aluminum hydride is most readily prepared by adding the calculated amount of 100% sulfuric acid to a tetrahydrofuran solution of lithium

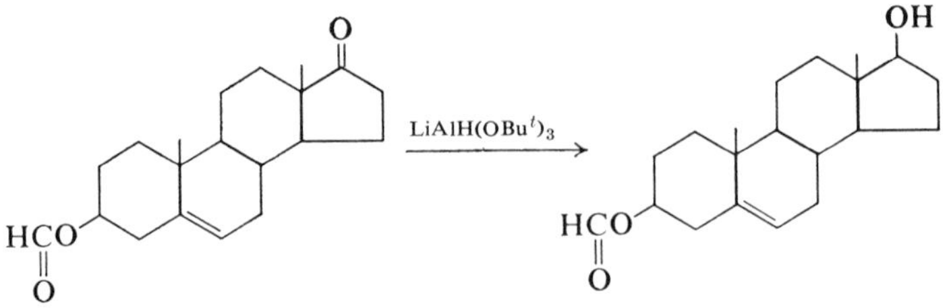

3β-Formyloxy-Δ^5-androsten-17-one   17β-ol

Table 11.2   Reduction of Oximes

PhCCH$_3$ $\longrightarrow$ PhCHCH$_3$ + PhNHEt

$\parallel$

NOH

NH$_2$

Reagent:

LiAlH$_4$ (ether)	80%	20%
LiAlH$_4$:AlCl$_3$ (1:4)	4%	96%
AlH$_3$	95%	5%

aluminum hydride [9]:

$$2LiAlH_4 + H_2SO_4 \rightarrow 2AlH_3 + 2H_2 + Li_2SO_4.$$

In many applications the heterogeneous mixture so formed can be used directly. Although it has a similar spectrum of reactivity to LiAlH$_4$, though possibly more reactive, it can be used with advantage in certain reductions. For example, in the reduction of carboxylic acids or esters in the presence of reactive halogen substituents:

$$ClCH_2CO_2H \xrightarrow{\text{LiAlH}_4} ClCH_2CH_2OH \ (5\text{–}13\%)$$

$$\xrightarrow{\text{AlH}_3} ClCH_2CH_2OH \ (70\%).$$

Since a nitro group is not reduced, it can be used for the reduction of other functional groups while leaving the nitro groups intact. In certain cases AlH$_3$ is to be preferred for the reduction of amide and nitriles (especially those with an active α-hydrogen atom) to amines, and in the reduction of enolizable keto esters.

Aluminum hydride shows differences from LiAlH$_4$ in the reduction of oximes (Table 11.2) and in the reduction of epoxides (Table 11.3).

Other useful modifications to the reactivity of LiAlH$_4$ have been made by the addition of various proportions of aluminum trichloride [10, 11].

Table 11.3   Reduction of Epoxides

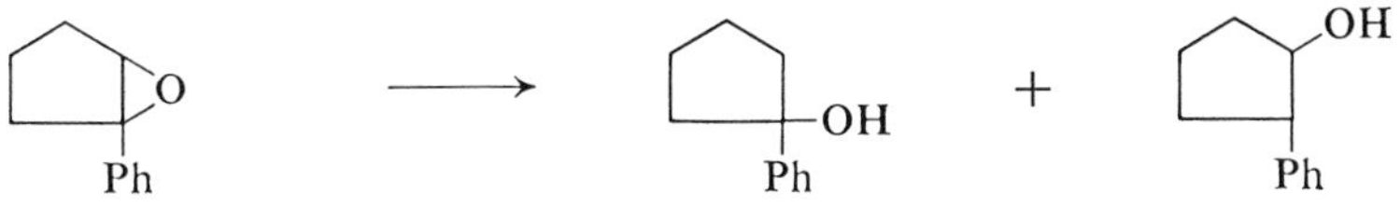

LiAlH$_4$	87%	13%
AlH$_3$	9%	91%

The most commonly used stoichiometric ratios correspond to Eqs. (11.1)–(11.4):

$$3LiAlH_4 + AlCl_3 \rightarrow 3LiCl + 4AlH_3 \tag{11.1}$$

$$LiAlH_4 + AlCl_3 \rightarrow LiCl + 2AlH_2Cl \tag{11.2}$$

$$LiAlH_4 + 3AlCl_3 \rightarrow LiCl + 4AlHCl_2 \tag{11.3}$$

$$LiAlH_4 + 4AlCl_3 \rightarrow LiCl + 4AlHCl_2 + AlCl_3. \tag{11.4}$$

One major difference between $LiAlH_4$ and the mixed hydrides is the ability of the latter to reduce some ethers. Tetrahydrofuran can be reduced to

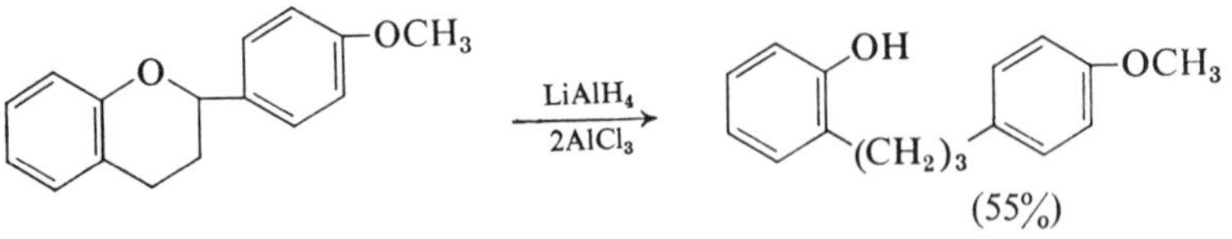

n-butanol at elevated temperatures and prolonged reaction times. The reagent can be useful for the cleavage of activated benzyl ethers. Thus 4′-methoxy flavan is cleaved, but not flavan [12].

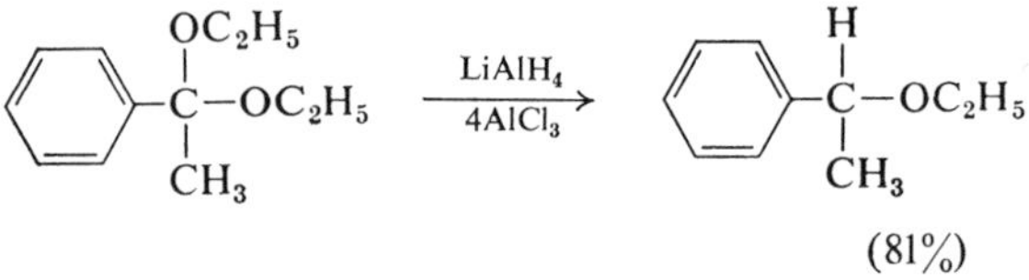

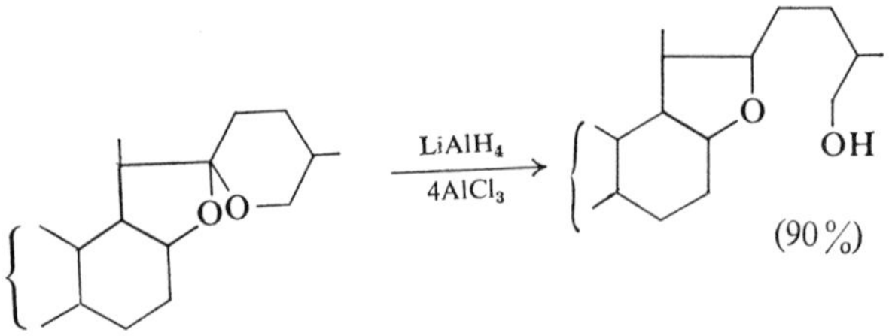

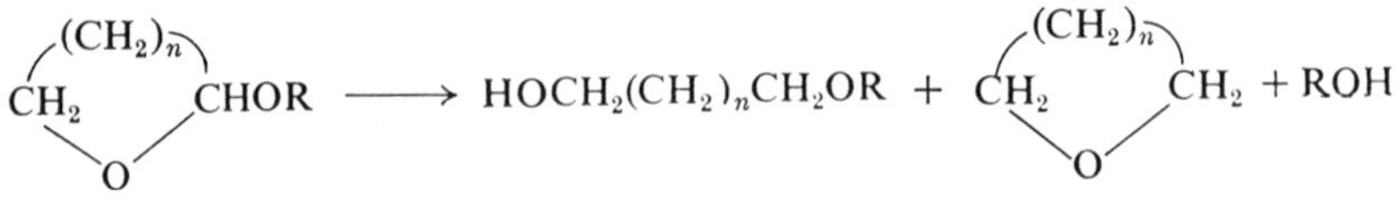

Acetals and ketals are reduced to ethers [13–15]. Hemithioketals and

hemithioacetals are reduced to the ω-hydroxyalkyl alkyl thioethers, with exclusive C–O bond cleavage [16, 17].

$$\text{[structure]}-\text{SCH}_2\text{Ph} \xrightarrow[4\text{AlCl}_3]{\text{LiAlH}_4} \text{PhCH}_2\text{S(CH}_2)_3\text{CH}_2\text{OH}$$

(63%)

The mixed hydrides differ from LiAlH$_4$ in the direction of ring opening of an unsymmetric epoxide. In general LiAlH$_4$ makes a nucleophilic attack at the least substituted carbon atoms whereas LiAlH$_4$–AlCl$_3$ attacks at the most substituted carbon atom [18].

The usual ether solvents are unsuitable for some LiAlH$_4$ reductions. In this case pyridine can be used and in general reductions proceed normally [19]. However, if a solution of LiAlH$_4$ in pyridine is allowed to stand for at least 24 hr a new, milder, reducing agent is formed. This is believed to be a tetrakis-(N-dihydropyridyl)-aluminate containing both 1,2- and 1,4-dihydro-pyridine groups [20].

This reagent reduces a ketone group but not an acid or ester group, and has been recommended for the selective reduction of a keto-acid to the corresponding hydroxy acid.

Sodium bis(2-methoxyethoxy) aluminum hydride has recently been introduced commercially. It is similar in reactivity to LiAlH$_4$ but is soluble in nonpolar solvents such as benzene [21]:

$$2\text{RCH}_2\text{CO}_2\text{R}^1 + 2\text{NaAlH}_2(\text{OOCH}_2\text{CH}_2\text{OCH}_3)_2 \longrightarrow$$

$$\text{NaAl(OCH}_2\text{R)}_2(\text{OCH}_2\text{CH}_2\text{OCH}_3)_2 + \text{NaAl(OR}^1)_2(\text{OCH}_2\text{CH}_2\text{OCH}_3)_2 \xrightarrow{\text{H}_2\text{O}}$$

$$2\text{RCH}_2\text{OH} + 2\text{R}^1\text{OH} + 2\text{NaAlO}_2 + 4\text{CH}_3\text{OCH}_2\text{CH}_2\text{OH}.$$

Although lithium aluminum hydride will not reduce an acetylene, it will selectively reduce a propargyl alcohol system (or a group reducible to such a system) to the corresponding olefin [22]. The reaction is believed to proceed through the aluminum alkoxide. As a consequence of this mechanism the olefin has the *trans* configuration [23]:

$$\text{CH}{\equiv}\text{C(CH}_2)_2\text{C}{\equiv}\text{CCO}_2\text{Et} \rightarrow \text{CH}{\equiv}\text{C(CH}_2)_2\text{CH}{\overset{t}{=}}\text{CHCH}_2\text{OH}.$$

A side reaction occasionally met with in the reduction of 1,4-butynediol systems is elimination of one hydroxyl group and formation of an allenic alcohol [24]. When one hydroxyl group is converted to a better leaving group, for example the tetrahydropyranyl ether, this elimination reaction becomes the predominant pathway [25].

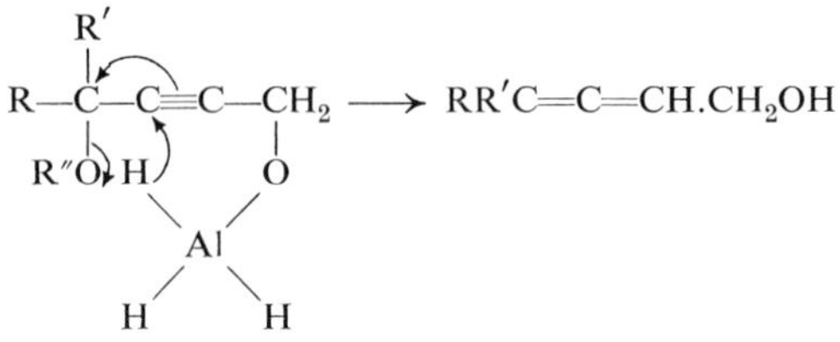

$$R\text{—}\underset{\underset{R''O\,H}{|}}{\overset{\overset{R'}{|}}{C}}\text{—}C\!\equiv\!C\text{—}CH_2 \longrightarrow RR'C\!=\!C\!=\!CH.CH_2OH$$

R″ = tetrahydropyranyl

A number of anomalous reductions involving hydride reagents are known and workers using these reagents for structural elucidation reactions have to be aware of the occasional possibility of rearrangements. As a further example, the rearrangement of a δ-enol lactone on reduction with lithium tributoxy aluminum hydride is quoted [26]. A similar rearrangement also takes place on reduction of γ-enol lactones [27]. The reaction is, in fact, an excellent synthetic procedure for these bicyclic systems.

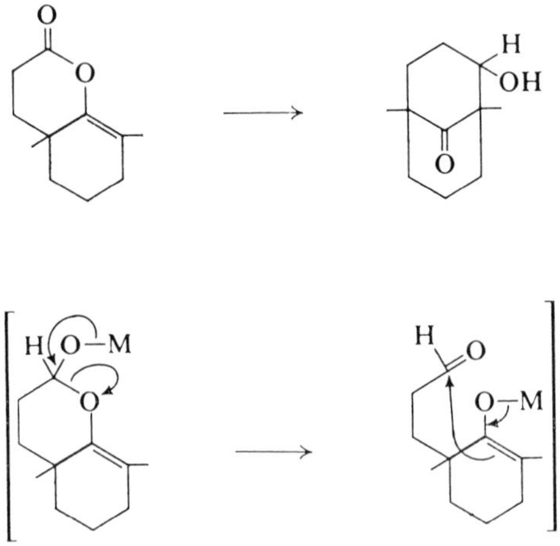

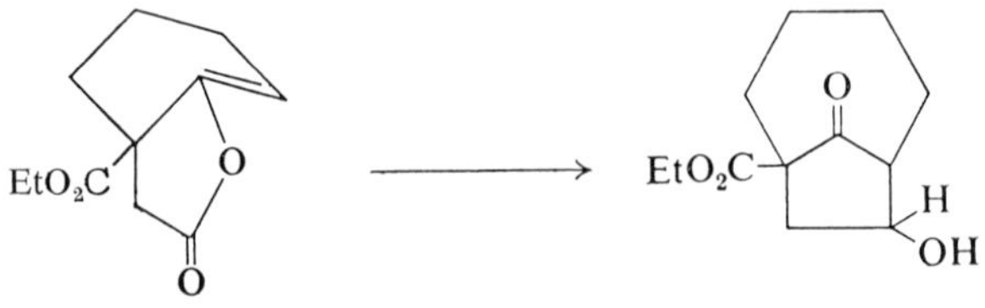

Aluminum Alkoxides

Aluminum alkoxides can reduce ketones to the corresponding alcohols. This reaction, known as the Meerwein-Ponndorf reduction, takes place by hydride transfer from the aluminum alkoxide [30]. It is conceptually the reverse of the Oppenaur oxidation [31]. Under forcing conditions diaryl ketones can be reduced to the hydrocarbon [28, 29]. The procedure is occasionally useful for the reduction of unsaturated aldehydes to the corresponding alcohols, the reduction of diaryl ketones, and the reduction of quinones to the parent hydrocarbon. For this latter reaction aluminum cyclohexoxide has been recommended [32] rather than the normally used aluminum isopropoxide.

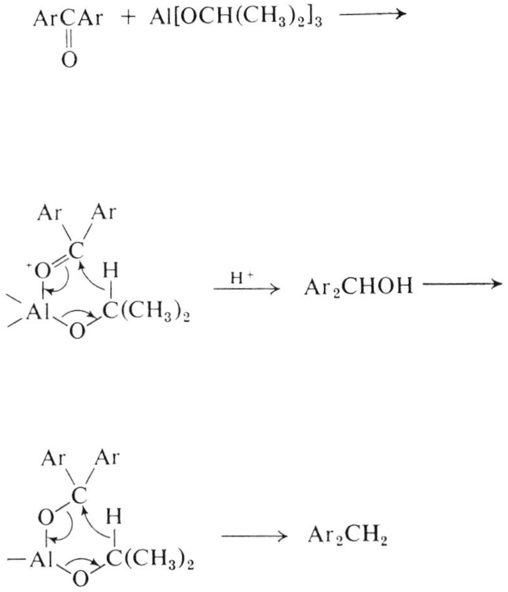

$$ArCAr + Al[OCH(CH_3)_2]_3 \longrightarrow$$
$$\underset{O}{\overset{\|}{}}$$

$$\xrightarrow{H^+} Ar_2CHOH \longrightarrow$$

$$\longrightarrow Ar_2CH_2$$

Boron Hydrides

Sodium borohydride is a considerably milder reducing agent than lithium aluminum hydride. Normally it is used in aqueous solution in which it reduces aldehydes and ketones to the corresponding alcohol, but is unreactive toward such compounds as epoxides, esters, lactones, carboxylic acids, nitriles, and nitro groups [33]. Aldehydes and ketones are reduced at markedly different rates, and selective reduction is possible. Although $NaBH_4$ is insoluble in diethyl ether, it is sufficiently soluble in diglyme and triglyme to be of use. In these solvents the rate of reduction of ketones compared to aldehyde is very slow. Acid chlorides are rapidly reduced in diglyme or dioxane solutions.

Table 11.4   Functional Groups Reduced by B_2H_6

Functional Group	Product
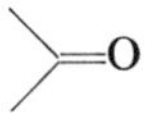 $\diagdown C{=}O$	 $-\overset{\displaystyle \mid}{C}H-OH$
$-CO_2R$	$-CH_2OH + ROH$ (slow)
$-CO_2H$	$-CH_2OH$ (fast)
$-C{\equiv}N$	$-CH_2NH_2$
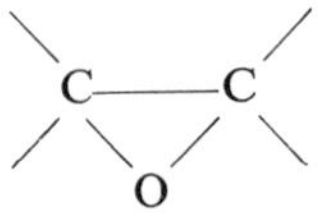 epoxide	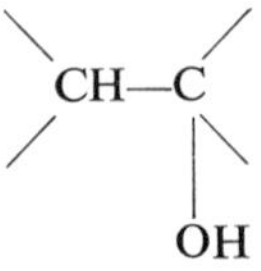 $CH-\overset{\displaystyle \mid}{\underset{\displaystyle OH}{C}}$
$-CH{=}CH-$	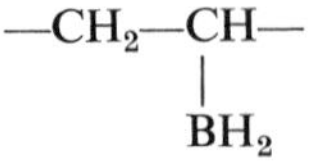 $-CH_2-\overset{\displaystyle \mid}{\underset{\displaystyle BH_2}{CH}}-$
$-C{\equiv}C-$	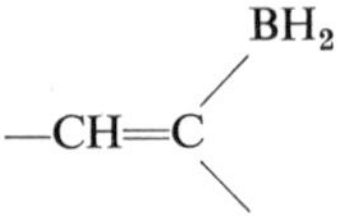 $-CH{=}C\diagup^{BH_2}$

Lithium borohydride, which can be formed from $NaBH_4$ and $LiCl$, is a more powerful reducing agent than $NaBH_4$ and will reduce an ester to an alcohol. It is more soluble in ether solvents than $NaBH_4$ but is otherwise of limited utility [34].

Diborane, B_2H_6, is a gas and is usually used in tetrahydrofuran or diglyme solution. Various methods of preparation have been proposed [34, 35], the most convenient of which is the addition of boron trifluoride diethyl etherate to a solution of $NaBH_4$ in diglyme:

$$3NaBH_4 + 4BF_3 \rightarrow 2B_2H_6 + 3NaBF_4.$$

The gaseous B_2H_6 can then be swept from the generating flask into the reaction vessel. Alternatively, the diborane can be generated *in situ*. However, in a number of cases product differences have been noted between internally and externally generated diborane [36]. This is probably due to unconsumed BF_3 acting as a Lewis acid. The principal functional groups reduced by B_2H_6 are shown in Table 11.4.

Nitro groups and aryl and alkyl halides do not react, neither do carboxylic acid salts nor acid chlorides. Primary amides evolve hydrogen but are not reduced. The reactivity sequence of diborane is acids > olefins > ketones > nitriles > epoxides > esters > acid chlorides.

This compares with the alkali metal borohydrides where the sequence is acid chlorides > ketones > epoxides > nitriles > acids. By the appropriate choice of reagent a number of selective reductions can be carried out. For example:

$$\text{Reduced} \xleftarrow{\quad B_2H_6 \quad}\!\!\!\!\times\!\!\!- \text{ Acid chloride} \xrightarrow{\quad NaBH_4 \quad} \text{Reduced}$$

$$\text{Reduced} \longleftarrow \text{Ester} \qquad \longrightarrow\!\!\!\!\times\!\!\!\longrightarrow$$

$$\text{Reduced} \xleftarrow{\quad B_2H_6 \quad} \text{Nitrile} \xrightarrow{\quad LiBH_4 \quad}\!\!\!\!\times\!\!\!\longrightarrow$$

$$\longleftarrow\!\!\!\!\times\!\!\!- \text{ Ester} \qquad \longrightarrow \text{Reduced.}$$

The notable difference between diborane and the other hydride reagents surveyed is in its reactions with olefins and acetylenes. Since the organoboranes so formed undergo protonolysis by carboxylic acids, the over-all process is reduction of an olefin to hydrocarbon or acetylene to olefin [37]:

$$3RCH{=}CH_2 \xrightarrow{\frac{1}{2}B_2H_6} (RCH_2CH_2)_3B \xrightarrow{3CH_3CO_2H} 3RCH_2CH_3 + (CH_3COO)_3B.$$

By using sodium borodeuteride to generate diborane and then decomposing with CH_3CO_2D, deuterium atoms can be introduced in specific locations. This is particularly useful in the reduction of acetylenes [38].

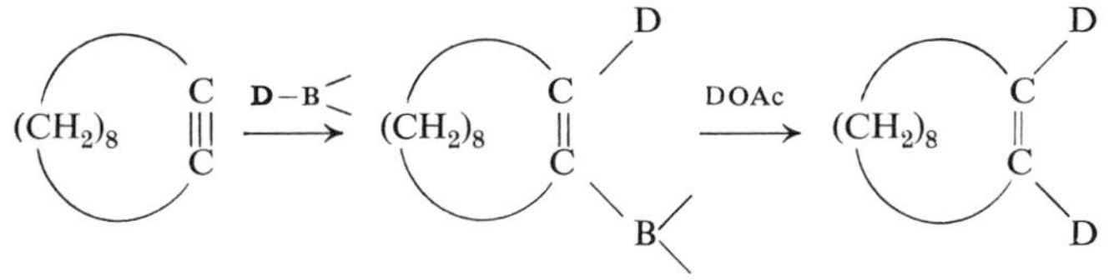

Diborane adds to unsymmetrical olefins to give, as the predominant product, that which has the boron atom bonded to the less highly substituted carbon atom. Increasing the steric hindrance in the olefin allows the reaction to be arrested at the dialkyl borane or monoalkyl borane stage. Thus reagents such as bis-2-methyl-2-butyl borane [39], diisocamphenyl borane, and 2,3-dimethyl-2-butyl borane can be used for selective hydroboration of less hindered olefins and acetylenes [34]. The organoboranes are especially useful for effecting clean monohydroboration of acetylenes, thus giving a good route to cis-olefines (Table 11.5).

Table 11.5

Acetylene	Reagent	Olefin	Yield (%)
1-Hexyne	B_2H_6	1-Hexene	7
1-Hexyne	Diisoamyl borane	1-Hexene	92
3-Hexyne	B_2H_6	cis-3-Hexene	80
3-Hexyne	Diisoamyl borane	cis-3-Hexene	90

Table 11.6 Comparison of Reactivity of Hydride Reagents[a]

	LiAlH₄ (Ether)	LiAlH(OBuᵗ)₃ (THF)	LiAlH₄ + AlCl₃ (Diglyme)	NaBH₄ (EtOH)	NaBH₄ + LiCl (Diglyme)	B₂H₆ (THF)	Bis-2-methyl 2-butyl borane
Olefine $C=C$						√	√
Acetylene $C\equiv C$						√	√
Aromatic 							
ArOR							
ArCH₂X	(√)		(√)				
Aldehyde R—CHO	√	√	√	√	√	√	√
Ketone R—CO—R'	√	√	√	√	√	√	√
Ketal R'₂C(OR)₂			√				
Acid RCO₂H	√		√			√	
Acid salt RCO₂M	√		√				
Acid chloride RCOCl	√	√	√	√	√	(√)	
Ester RCO₂R	√		√		√	(√)	
Lactone RCO—OR	√	(√)	√		√	√	√
Amide RCONH₂	√		√				
Nitrile RCN	√		√			√	
Nitro R—NO₂	√		√				
Oxide 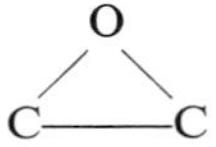	√		√		√		
Chloride R—Cl	√		√		√		
Sulfonate R—SO₂OR'	√		√		√		

[a] Adapted from H. C. Brown, *Hydroboration*, Benjamin, New York, 1962.

9-Borabicyclo[3.3.1]nonane, formed from cycloocta-1,5-diene and B_2H_6, has recently been recommended as a thermally and air stable hydroborating agent [40]. This reagent hydroborates rapidly and quantitatively with a selectivity comparable to diisoamyl borane.

As an aid to the selection of the most appropriate hydride reducing agent, a comparison of the reactivity of hydride reagents is given in Table 11.6.

2 DISSOLVING METAL REAGENTS

The most common reagents of the dissolving metal type are those involving the dissolution of sodium, lithium, and zinc, together with the lesser used metals calcium, aluminum, and magnesium. The reactions involve the transfer of electrons from the metal to the substrate to form a metal ion and an anion radical (or possibly a dianion) which then can disproportionate or react with protons either from the solvent or an added acid. Solvents which form useful reagents are liquid ammonia, certain alkylamines, alcohols, and hydrochloric, acetic, and formic acids.

Metal–Ammonia Reducing Agents

Probably the most important use of this type of reagent is in the reduction of aromatic systems, the Birch reduction [41–43]. Lithium and sodium have been used, with t-BuOH being the preferred added proton source. Lithium has been the favored metal, particularly in the standard Wilds-Nelson procedure where ether is used as a cosolvent [44]. However, if distilled liquid ammonia, which has thereby been freed from traces of iron salts, is used, sodium is as effective as lithium in the majority of cases. Standard procedures used in the steroid field have been described [45]. The mechanism of the reduction of benzene is thought to be [46]:

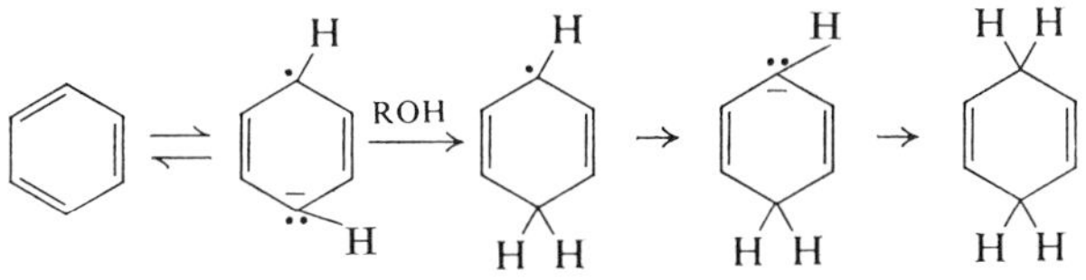

A proton source (i.e., an acid stronger than ammonia) is required to displace the initial equilibrium step to the right. Electron-withdrawing substituents should promote reduction whereas electron-donating substituents should retard reaction. The electron density of the radical ion can be considered to be greatest at positions *ortho* and *para* to electron-withdrawing substituents and preferential protonation at these sites is expected. Electron-donating substituents would be expected to protonate at the *meta*-position.

These predictions are approximately true, but other factors, including steric effects, have an influence on the products obtained [47].

In the absence of a proton source the initially formed nonconjugated diene can be isomerized into the conjugated diene which is further reduced. A wide range of Birch reductions have been reviewed [48].

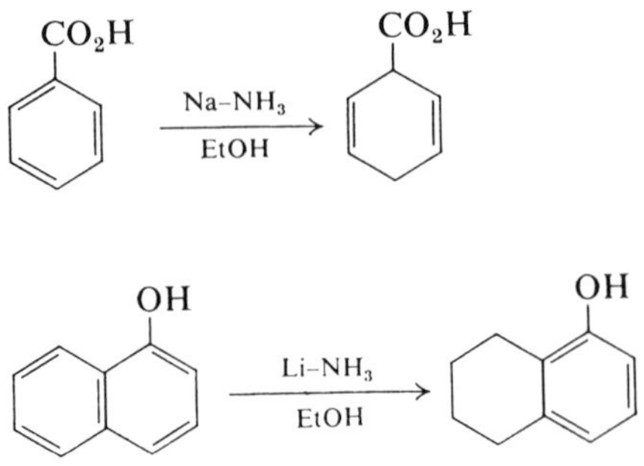

Nonconjugated olefins are not reduced by metal ammonia systems, but conjugated systems are reduced readily in the absence of proton donors. Terminal double bonds are reduced preferentially, with dimerization being a common side reaction [49]. 1,4-Dienes can be isomerized into the conjugated diene and be reduced, but 1,5-and higher dienes are not normally isomerized and therefore are not reduced [50].

Disubstituted acetylenes are readily reduced to *trans*-olefins by alkali metals in liquid ammonia. The reaction is believed to proceed through a radical ion which is protonated; then electron transfer forms an anion which is again protonated. The intermediate radical anion can react with another substituent as shown [51].

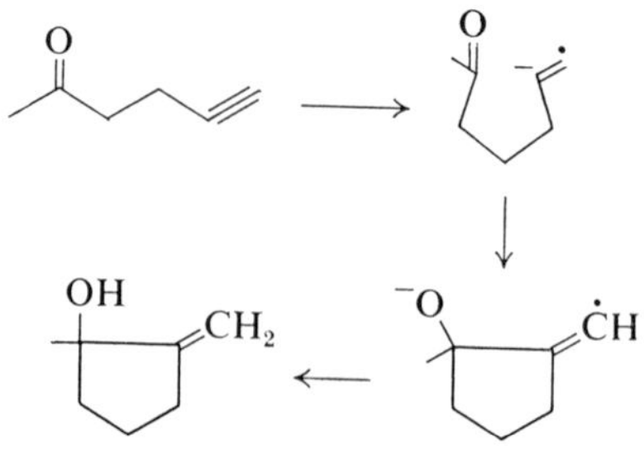

Terminal acetylenes can be reduced to terminal olefins. Part of the acetylene is utilized as the proton donor in the absence of an added proton source. Conversion of a terminal acetylene into its salt with sodamide enables selective reductions to be made because the salts are not reduced [42]:

$$CH_3(CH_2)_2C\!\!\equiv\!\!C\!\!-\!\!(CH_2)_4\!\!-\!\!C\!\!\equiv\!\!CH \xrightarrow[NH_3]{NaNH_2} CH_3(CH_2)_2C\!\!\equiv\!\!C\!\!-\!\!(CH_3)_4C\!\!\equiv\!\!C\!\!-\!\!Na$$

$$\downarrow Na, NH_3, NH_4Cl$$

$$CH_3(CH_2)_2\!\!-\!\!CH\overset{t}{\!=\!}CH(CH_2)_4C\!\!\equiv\!\!CH.$$

Allenes are also reduced to olefins. The reduction of certain cyclic acetylenes is thought to proceed by isomerization to the allene followed by reduction [52a]:

$$(CH_3)_2C\!\!=\!\!C\!\!=\!\!CH\!\!-\!\!CH_3 \xrightarrow{Na/NH_3} (CH_3)_2CHCH\!\!=\!\!CHCH_3 + (CH_3)_2\!\!=\!\!CHCH_2CH_3.$$

$$(34\% \; cis, 48\% \; trans) \qquad\qquad (18\%)$$

Conjugation of an olefin with a carbonyl group is sufficient to promote reduction [53].

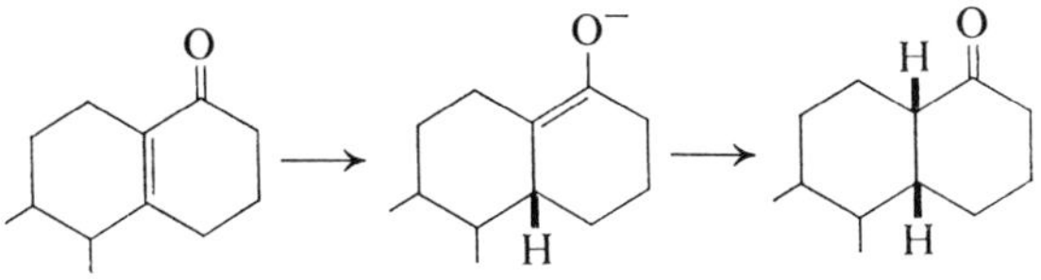

The initial enolate anion is sufficiently basic to abstract a proton from ammonia, but the second intermediate resists further electron addition because of the negative charge on the oxygen. However, if a proton source is added, this intermediate is protonated and can be reduced to the alcohol. The stereochemistry of this reduction is frequently different from that obtained by catalytic hydrogenation. Protonation of the β-carbanion is controlled by stereoelectronic factors which require the hydrogen to enter axially into a cyclohexanone ring. If the β-position is located at a ring junction, this frequently leads to a *trans* stereochemistry. A more detailed exposition is available [53, 54].

The less substituted of the two olefinic bonds of a dienone can be reduced selectively. The initial enolate anion can, however, react as a nucleophile either in a substitution or fragmentation reaction [55].

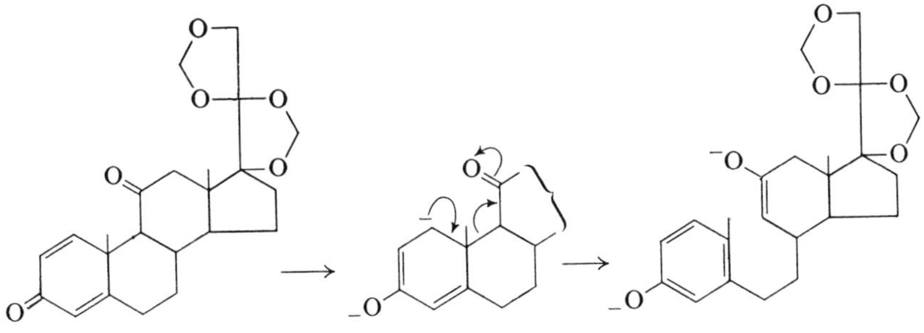

Control of the course of a reaction through the supply of a proton source makes it possible to reduce α,β-unsaturated ketones in the presence of an aromatic ring. The example given also compares the stereochemical course of catalytic hydrogenation and dissolving metal reduction [56].

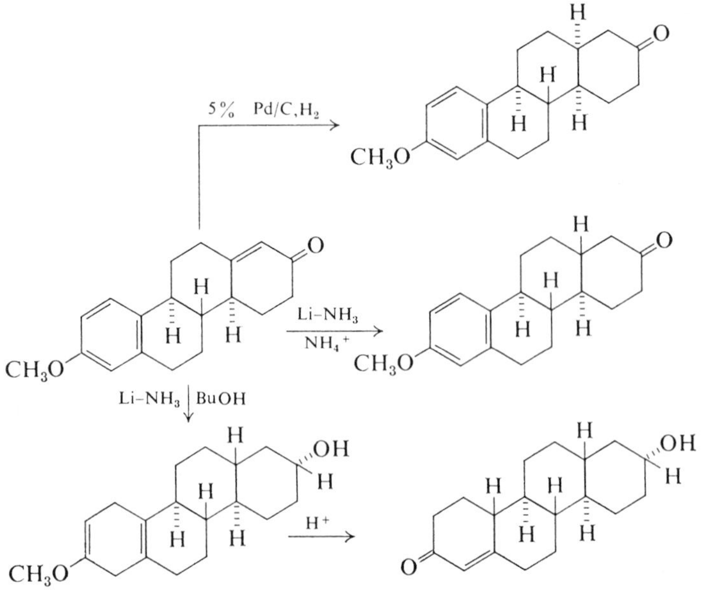

The use of alkylamines as solvents rather than liquid ammonia has been recommended [57–59]. Lithium is the preferred metal. The advantages over liquid ammonia lie in greater solvent power for the substrate and a somewhat more powerful reagent.

Metal–Acid Reducing Agents

Amalgamated zinc in strong aqueous acid reduces ketones to the corresponding hydrocarbon in what is generally known as the Clemmensen reduction [60, 61]. The same transformation can also be achieved by the Wolff-Kishner reduction (p. 96) or by desulfurization of dithioketals by Raney nickel [62].

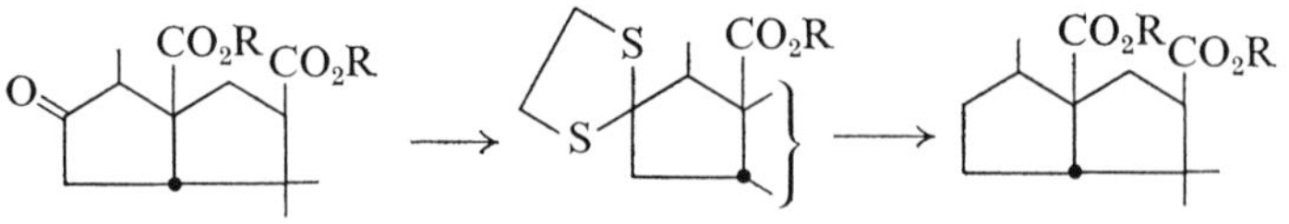

The choice of method will depend on the other substituent groups, and a comparison has been made [63]. However, for diaryl and alkyl aryl ketones, probably the best choice is a mixed $LiAlH_4$–$AlCl_3$ reduction [64] (see also Section 1).

A number of selective reductions by zinc and acids are described in the section on fission reactions (Section 3).

3 FISSION REACTIONS

Addition of an electron from a dissolving metal to a substrate generates an anionic center. If the carbon atom adjacent to the developing anion carries a good leaving group, then elimination can take place. The resulting radical species can then acquire a further electron to give an anion which may then be protonated by the reaction mixture or during the isolation procedure.

Stereoelectronic considerations require that the developing p-orbital at the α-carbon atom overlaps with the π-orbital system of the ion radical.

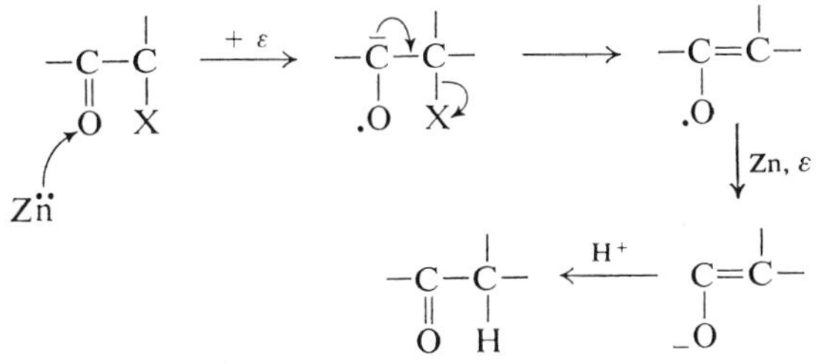

$$X = Cl, Br, I, \text{ tertiary nitrogen, OH, OAc}$$

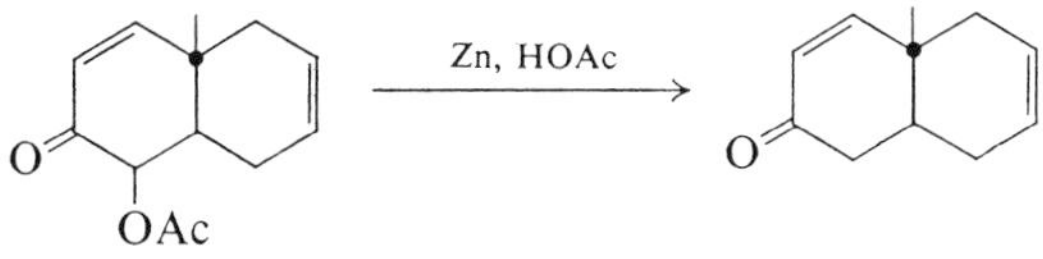

The cleavage of cyclohexanones with axial α-substituents therefore proceeds more readily than the corresponding equatorially substituted molecule [65].

A number of useful transformations of this type involve zinc and acetic acid or zinc–acetic acid and hydrochloric acid [66].

The reaction also takes place in vinylogous systems and has proved useful in investigations on steroids [67].

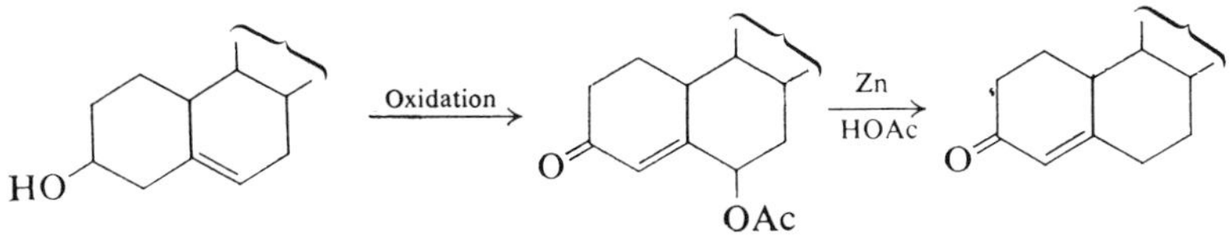

As a further example, this type of reaction provided a useful degradation in the investigation of the structure of the antibiotic tetracycline [68].

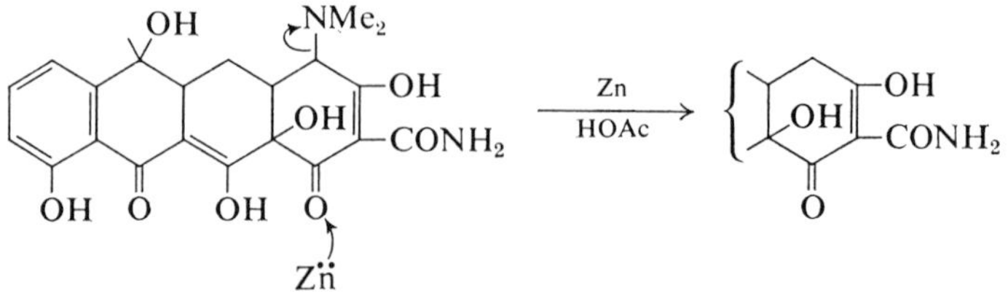

Fissions of this type brought about by lithium and sodium in liquid ammonia include cleavage of the carbon–nitrogen bond in N-benzyl groups. This is a valuable method for removing protective N-benzyl groups, for example, in peptide synthesis [43, 69, 81]. Also cleaved are quaternary nitrogen salts by the Birch-Emde fission. This reaction is of value in natural product investigations [70–73].

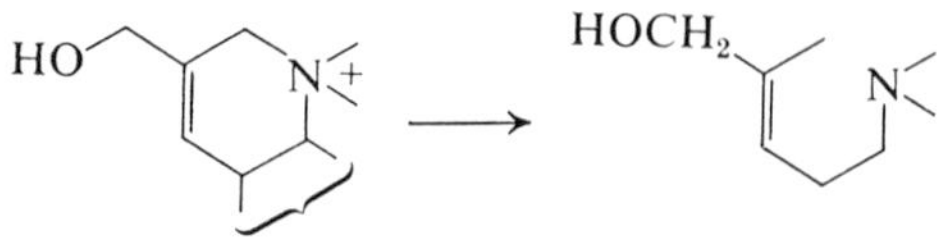

The carbon–oxygen bond of allyl, benzyl, and diaryl ethers are also cleaved, but not those of acetals or ketals [74, 75].

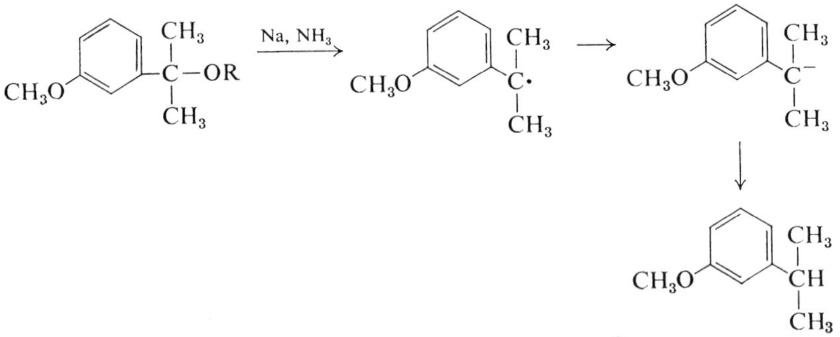

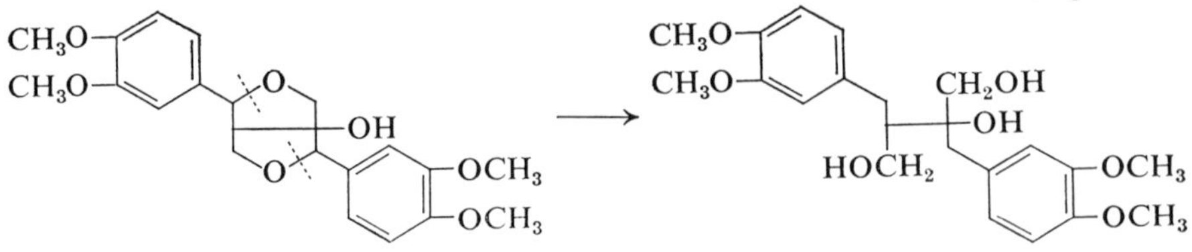

Fissions of this type are valuable degradation tools, the relative ease of fission of ethers with various substituents have been investigated empirically [76] and by molecular orbital calculations [77].

Although aliphatic alcohols are not reduced by $Na/Li-NH_3$, allyl and benzyl alcohols can be. The alcohol can be protected by prior formation of an alkoxide or tetrahydropyranyl ether.

The cleavage by Na–liquid ammonia reduction of diethyl phosphate esters of phenols [78] and enol esters [79] proceeds in a rather different manner. These esters can be prepared from diethyl-phosphonate.

$$Ar{-}O{-}PO(OC_2H_5)_2 \rightarrow Ar{-}H$$

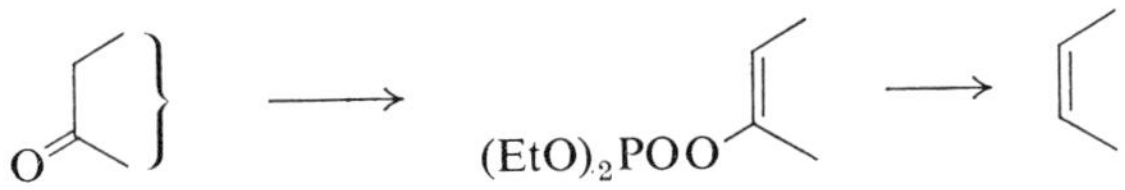

Procedures for the characterization of phenolic degradation products by this fission have been described [80].

Activated alkyl benzyl ethers are readily cleaved. Thioethers, including aliphatic ethers, are also cleaved by sodium–liquid ammonia. The removal of *S*-benzyl protecting group by $Na-NH_3$ is of considerable utility [81]. Thioketals are also cleaved by $Na-NH_3$ [82].

Carbon–halogen bonds can be cleaved readily with sodium in liquid ammonia [83]. Frequently the organic product is a mixture with formation of the dimer, the amine, and the olefin derived from the original carbon portion. 1,2-Dichloroalkanes give the olefin [84] while vinylic chlorides give either the corresponding olefin, with retention of configuration, or the acetylene [85]. Carbon–halogen bonds can also be reduced with zinc and acid [86] or sodium in alcohol [87]. Many of these reactions are thought to proceed through an organometallic intermediate [88].

4 REDUCTIONS INVOLVING HYDRAZINE

Diimide

Diimide, $HN{=}NH$, is generated as a transient intermediate by the oxidation of hydrazine or the decarboxylation of dipotassium azodicarboxylate [89, 90]. The reagent reduces olefins with high stereospecificity ($>98\%$) by *cis* addition and is thought to proceed through a cyclic transition state [91].

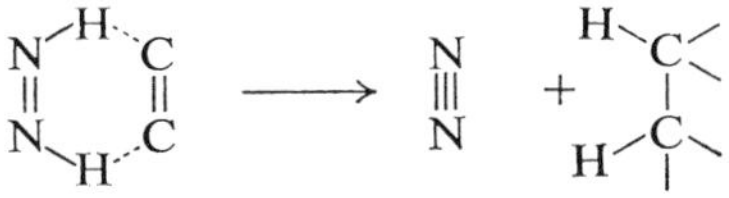

In general, symmetrical multiple bonds are reduced (e.g., $-C{\equiv}C$, $-C{=}C-$, $-N{=}N-$, $-O{=}O-$), whereas unsymmetrical multiple bonds (e.g., $-C{\equiv}N$, NO_2, $-N{=}C$, $S{=}O$) are not. The reagent does not hydrogenolyze Ar–I,

Ar–Br, R–Br, R–SH, R–S–S–R, and R–S–C–SR′. The lack of reactivity of this reagent with the C–S bond can be useful in selective reduction. The high stereoselectivity and the steric relationships enforced by the cyclic transition state always leads to attack from the least hindered side of a substrate, even if this leads to the least thermodynamically favorable product; for example, the reduction of α-pinene proceeds as shown [92].

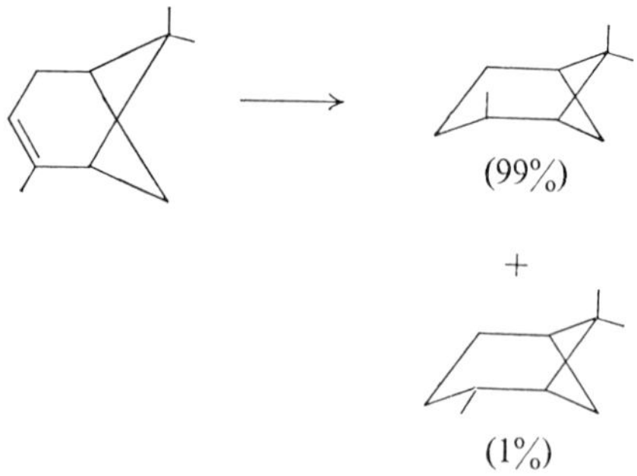

(99%)

+

(1%)

The effect of conformation of cyclic olefins is shown by the reduction of *cis,trans,trans*-$\Delta^{1,5,9}$-cyclododecatriene to *cis*-cyclododecene by diimide [93].

Diimide has been used in the total synthesis of mycophenolic acids which confirmed the gross structure and established the configuration of the side-chain double bond. A selective catalytic hydrogenation of **1** → **2** could not be achieved. Diimide brought about the required reduction smoothly [94].

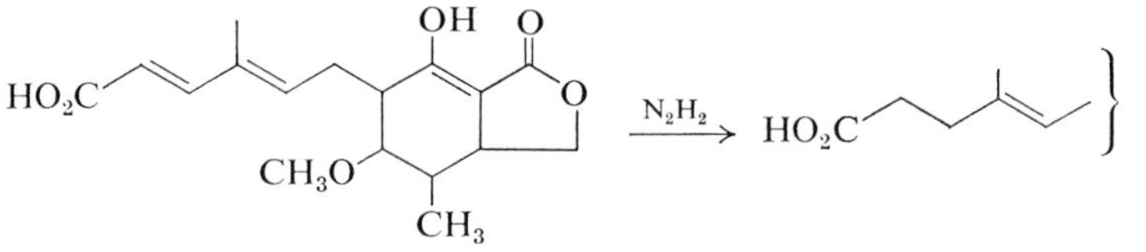

1   **2**

Hydrazine

Hydrazine alone or together with metal catalysts can effect a number of reductions, but these are of limited utility [89, 95].

Wolf-Kishner Reduction

This technique for the reduction of a carbonyl compound to the corresponding hydrocarbon involves formation of the hydrazone with hydrazine, preferably *in situ*, and base-catalyzed decomposition of the hydrazone [96]. Various modifications of the technique have been proposed, the most general useful one is that of Huang-Minlon [97]. Others have been proposed by Gates and Tschudi [98], Lock [99], Barton [100], Nagata-Itazaki [101,

102], and Gardner et al. [103]. The reaction can be carried out at room temperature in dimethyl sulfoxide [104]. Unfortunately, this procedure requires the preformed hydrazone. The Nagata-Itazaki procedure appears to be the most effective for forcing the reduction of highly bonded carbonyl groups [101, 102].

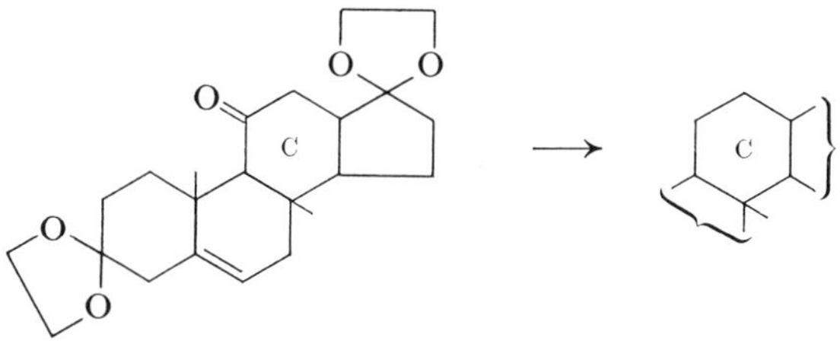

5   CATALYTIC HYDROGENATION

The catalyzed reduction of an organic substrate by hydrogen is a reaction of wide applicability and is capable of giving high specificity. The ease of reduction of a particular functional group is effected not only by the steric environment within the molecule but also by the choice of catalyst, solvent, reaction temperature, and hydrogen pressure. In the more difficult cases of reduction the determination of the best reaction conditions is a fine art with, on occasions, the added complication of variable results.

Table 11.7 gives a guide to the selection of reaction condition for reductions of functional groups.

Heterogeneous Catalysis

Most heterogeneous catalysts consist of the pure metal either unsupported or supported [105–108]. There are a few exceptions, for example, metal sulfides [109–113] and copper chromite, but these catalysts are used only for special applications. The unsupported catalysts are usually in the form of finely divided metal [114] and are used where the products of the hydrogenation tend to inhibit the reaction [115].

Supported catalysts are physically supported, finely divided metals on the surface of a material with a large surface area [116]. They are usually prepared by impregnation [117, 118] or ion-exchange [119, 120] from a solution of the metal complex followed by chemical reduction. The effective surface area of the metal is thereby substantially increased. Recent improvements in activity and selectivity in supported catalysts have been directed towards increasing the dispersion of catalyst on high surface area supports, for example, molecular sieves [121].

Table 11.7 Recommended Procedures for Catalytic Reduction of Functional Groups[a]

Functional Group	Catalyst and Conditions	Refs.
Acetylene → olefin (*cis*)	5% Pd/BaSO$_4$, addition of quinoline or base recommended, methanol solvent, 25°, and 1 atm H$_2$	155, 165, 166, 221
Olefin → alkane	5% Pd/C in alcohol or ethyl acetate, or PtO$_2$ in acetic acid solvent, 25–75°, and 1–5 atm H$_2$	222–225
Aromatic → cycloalkane	5% Rh/C or 5% Ru/C, alcohol/water solvent, 25–75°, and 1–50 atm H$_2$	178, 193, 226, 227
Phenol, aniline → substituted cycloalkanes	5% Rh/C or 5% Ru/C, alcohol/water solvent, 75–125°, and 50–100 atm H$_2$	169, 186, 193, 230
N-Heterocyclic → saturated	5% Rh/C or 5% Ru/C, methanol/water solvent, 100°, and 50–100 atm H$_2$; PtO$_2$ in glacial HOAc, 25–50°, and 1–5 atm H$_2$	111, 228, 231
Carbonyl → alcohol	5% Ru/C, methanol/water solvent, 25–50°, and 1–5 atm H$_2$; 5% Rh/C, acetic acid solvent, 25°, and 4 atm H$_2$	193, 229
Ester → alcohol	Copper barium chromite, dioxane solvent, 175–200°, and 300 atm H$_2$	216–219, 232
Carboxylic acid → alcohol	Re$_2$O$_7$ or Rh$_2$O$_3$, water solvent, 150–200°, and 200–500 atm H$_2$	200–202, 233
Nitro → amine	5% Pd/C or 5% Pt/C or PtO$_2$, alcohol or glacial acetic acid solvent, 25°, and 1 atm H$_2$	234–236,
Nitro → oxime	5% Pd/C deactivated by Pb salts or pyridine, water solvent, 150°, and 30–50 atm H$_2$	237, 238

Table 11.7   (Continued)

Functional Group	Catalyst and Conditions	Refs.
Nitrile → amine	PtO$_2$, acetic anhydride solvent, 25°, and 5 atm H$_2$; Raney Ni, alcohol saturated with NH$_3$ solvent, 25°, and 10 atm H$_2$	239, 240
	Hydrogenolysis	
Aryl chloride	5% Pd/C, alcohol solvent + KOH or MgO, 25°, and 1 atm H$_2$	241
Benzyl chloride	10% Pd/C, alcohol or ethyl acetate solvent, 25°, and 1 atm H$_2$	242
Benzyl alcohol or ether	5% Pd/C, alcohol solvent, 25°, and 1 atm H$_2$	243, 244
Acid chloride (Rosenmund reduction)	5% Pd/BaSO$_4$, aromatic solvent, 25°, and 1 atm H$_2$	245

[a] A comprehensive list of the products obtained by the heterogeneous catalytic hydrogenation of functional groups can be obtained from Refs. 105 and 106.

Commonly used supports include carbon, alumina, silica, clays, asbestos, alkaline metal carbonates and sulfates, and molecular sieves. Organic supports have been used occasionally. A palladium/silk hydrogenation catalyst has been described which gives optically active products [122, 123], and platinum/nylon catalysts are claimed to have certain advantages over normal supports [124].

The choice of the best support for a catalyst is complex and the factors involved are not fully understood [106, 125]. Activated charcoal is usually the most suitable support since it is inert to most solvent systems and the products are rarely adsorbed strongly enough to inhibit the reaction. Alumina is a good second choice whereas supports like alkaline earth carbonates or sulfates can give rise to increased selectivity [126].

The physical properties of a support also effect the activity and selectivity. Such properties as surface area [127], mean pore diameter, pore size distribution, particle size [125], and agglomerate size, all determine the effectiveness of the catalyst system.

The amount of catalyst used is often greater than would be thought necessary, but is advantageous in overcoming deactivation of the catalyst by the product or adventitious poisons. In order to prevent side reactions it is often better to increase the rate of reaction by increasing the catalyst concentration.

Higher temperatures usually increases the rate of reduction, but often the selectivity is decreased. The yield of *cis* isomers in the reduction of an olefin is usually lower at higher temperatures [128, 129], moreover hydrogenolysis reactions can become more important at elevated temperatures [130, 131].

Commonly used solvents are water, methanol, or acetic acid, but most solvents which are themselves not hydrogenated are suitable. Complete solubility of the substrate in the solvent is not essential and many liquid substrates have been reduced in water as a suspension when the substrate was immiscible with water. The acidity of the media is critical to the success of some reductions (see Table 11.7).

The impurities in hydrogen gas may influence the activity of the catalyst. Thus the rate of hydrogenation of aldehydes to alcohols using ruthenium catalysts is increased by the addition of oxygen [132].

Platinum

Most functional groups with the exception of esters, carboxylic acids, and amides are hydrogenated by this catalyst under mild conditions (i.e., less than 70°C and pressures below 60 psi hydrogen).

Commonly used unsupported catalysts are reduced platinum oxide (Adam's catalyst) [133–136] and platinum black [137–139]. Chemical reduction of chloroplatinic acid by sodium borohydride [140, 141] or trialkyl silanes [142, 143] give more active catalysts than platinum oxide for the reductions of olefins and acetylenes.

Supported platinum catalysts are prepared by reduction, either using hydrogen [137] or hydride reagents [142, 143], of chloroplatinic acid in the presence of a carrier, usually activated charcoal or alumina. Alternatively, the impregnated support can be dried and then reduced in a stream of hydrogen [144]. The activity can be controlled to some extent by the method of preparation. Differences between batches of prepared catalysts and commercial catalysts can be expected.

Catalysts with modified reactivity can be obtained by the addition of acids [145, 146] or bases [147] or by chloride salts [148], for example Sn^{II}, Mn^{II}, Fe^{III}, and Ce^{IV}. Deactivation by the addition of electron donors [149, 150], for example, phosphines, amines, or metal ions [151] such as Hg^{II} or Zn^{II}, can be of use in controlling selectivity.

Platinum oxide on silica is a very active catalyst and can be used quantitatively for the determination of olefin unsaturation, the advantages being the short reaction time and the sharp end point [152]. This catalyst has also been used for the hydrogenation of a methyl ester without concurrent transesterification. Modified platinum catalysts can be used to selectively reduce unsaturated aldehydes to unsaturated alcohols [153, 154]:

$$C_6H_5CH{=}CHCHO \xrightarrow[\text{FeCl}_2/\text{Zn(OAc)}_2]{\text{Pt/PtO}_2/\text{H}_2} C_6H_5CH{=}CHCH_2OH.$$

Palladium

Most palladium catalysts are supported on activated charcoal, strontium carbonate, barium sulfate [155, 156], alumina, zeolites [157], or organic polymers [158, 159]. The use of palladium hydroxide either alone [160, 161] or supported on carbon [162] has been recommended for the reduction of nitro, keto, and aromatic groups.

Palladium catalysts appear to be very resistant to inhibitors or poisons [thus hydrogenolysis reactions (page 108) are only slightly affected by the presence of halide ions], although modifiers, for example, sulfur (especially in the gas phase [163]) and amines [164], do alter the specificity. Lindlar's catalyst, Pd/CaCO$_3$, modified by the additions of lead acetate and quinoline [165, 166], has been used for the selective hydrogenation of acetylenes to cis-olefins.

Rhodium

This catalyst is usually supported on alumina [167, 168] or charcoal [129], and is particularly effective for the hydrogenation of aromatic [169–173] and heterocyclic [168, 174–177] systems under mild conditions. Very little hydrogenolysis takes place [178, 179].

Inhibition by electron donors, such as sulfides, amines, occurs; for the reduction of heterocyclic systems, for example, pyridine and quinoline, it is necessary to perform the reduction under acid conditions [180].

Ruthenium

Supported ruthenium on charcoal or alumina [181] is the usual form of this catalyst, and the activity is similar to supported rhodium catalysts, being particularly effective for the reduction of aromatic [182–184] and heterocyclic systems, for example, pyridine [185], without concurrent hydrogenolysis [186–192]. Aliphatic aldehydes and ketones [193], including monosaccharides [194], are also reduced under mild conditions. However, it is possible to reduce α,β-unsaturated aldehydes to the saturated aldehyde quantitatively [195]. Ruthenium catalysts can be used under high pressure to hydrogenate carboxylic acids and anhydrides [196].

Most reductions carried out using ruthenium catalysts are promoted by traces of water but inhibited by acids.

Osmium

This metal has not been used extensively, but osmium on carbon or alumina has been recommended [197] for the hydrogenation of the carbonyl group in α,β-unsaturated aldehydes (but not ketones) and the nitro group in aromatic halonitro compounds. Heterocyclic nitrogen compounds have also been hydrogenated using osmium catalysts [198, 199].

Rhenium

Rhenium oxide or the finely divided metal is the most effective catalyst for the reduction of unsubstituted carboxylic acids and amides [200–202].

Nickel

Nickel catalysts have been used as finely divided metal [203, 204], supported [205] on Kieselguhr, charcoal, or as Raney nickel [105, 206, 207]. There are several forms of Raney nickel, each varying in activity, but all prepared by the general technique of digesting a nickel/aluminum alloy with alkali. The resulting finely divided nickel contains traces of aluminum.

The main use of nickel catalysts is for the hydrogenation of aromatic systems [208, 209] and the hydrogenolysis of organic sulfur compounds [210], although this latter reaction requires large amounts of catalysts to avoid inhibition by the sulfur compound. Catalytic hydrogenations are inhibited by organic compounds of sulfur, phosphorus, arsenic, but promoted by traces of platinum compounds [211].

Nickel Boride

This catalyst is obtained by borohydride reduction of an appropriate nickel salt [212] and it can be preformed in the presence of a carrier such as carbon [213, 214]. The selective hydrogenation of olefinic double bonds is efficiently carried out by nickel boride. Cobalt boride is also reported to catalyze the reduction of organic nitrogen–nitrogen multiple bonds using hydrazine as the reductant [215].

Copper Chromium Oxide

This catalyst is prepared by the thermal decomposition of copper ammonium chromite [216–219], and is primarily used for the reduction of esters, amides, and acyloins [220]. The conditions for hydrogenation are rather severe (200°C and 3000 psi hydrogen).

Model Mechanisms for Heterogeneous Hydrogenation

Hydrogenation of Olefins [223, 224, 246]

The most generally accepted mechanism for the hydrogenation of simple olefins is that of Horiuti and Polanyi [247, 248]. Under high hydrogen availability (or alternatively, a high catalyst/olefin ratio), the stereochemistry of the product is usually determined by the most favorable orientation for the adsorption of the olefin, whereas under conditions of low hydrogen availability (less than 1 atm hydrogen pressure) equilibration between the half hydrogenated states occurs, which may result in a mixture of products.

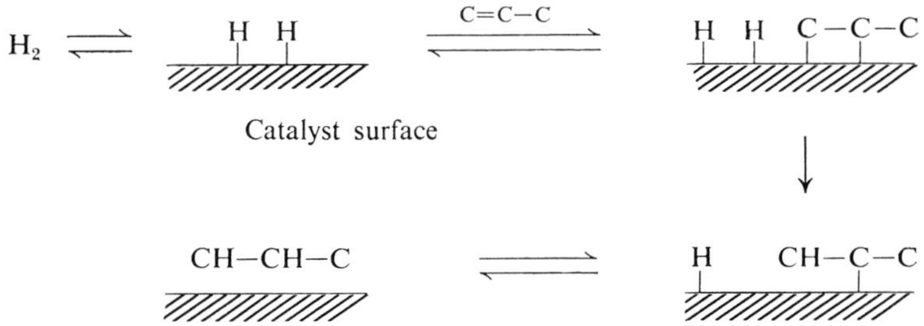

$$H_2 \rightleftarrows \quad \begin{array}{c} H \ \ H \\ | \ \ | \\ \text{//////////} \end{array} \quad \xrightarrow{C=C-C} \quad \begin{array}{c} H \ \ H \ \ C-C-C \\ | \ \ | \ \ | \ \ | \\ \text{//////////} \end{array}$$

Catalyst surface

$$\downarrow$$

$$\begin{array}{c} CH-CH-C \\ \text{//////////} \end{array} \quad \rightleftarrows \quad \begin{array}{c} H \quad CH-C-C \\ | \quad | \\ \text{//////////} \end{array}$$

Thus, for nonconjugated double bonds, most hydrogenation studies (especially those conducted at pressures greater than 1 atm) result in the *cis* addition of two hydrogen atoms from the less hindered side of the double bond [249].

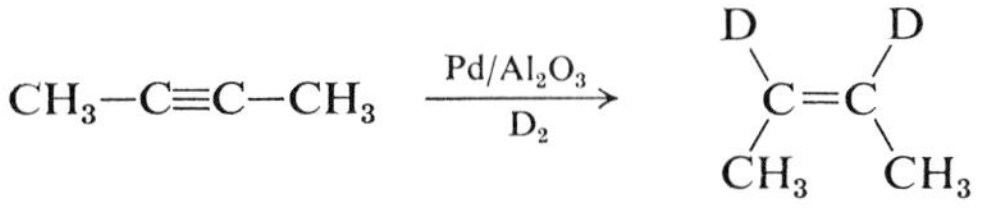

$$CH_3-C{\equiv}C-CH_3 \xrightarrow[D_2]{Pd/Al_2O_3} \begin{array}{c} D \qquad D \\ \diagdown \quad \diagup \\ C=C \\ \diagup \quad \diagdown \\ CH_3 \quad CH_3 \end{array}$$

cis

In some cases, however, anomalies appear with individual catalysts. Thus palladium catalysts tend to favor isomerization of olefins even in the presence of hydrogen, whereas platinum catalysts usually do not promote isomerization. In general the order of isomerization activity is $Pd > Ru > Rh > Pt > Ir$ [223, 224]. Also, exclusive *cis* addition does not always occur with palladium catalysts [250].

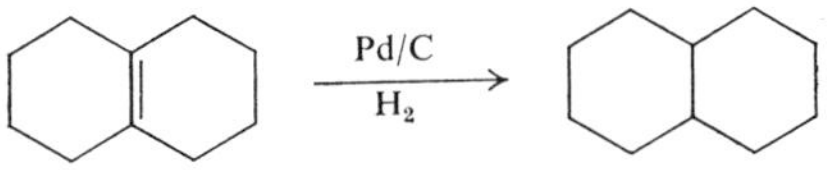

$$\text{(bicyclic alkene)} \xrightarrow[H_2]{Pd/C} \text{(decalin)}$$

(90%, *trans*)

In order to explain some of these features, an intermediate π-allyl species has been postulated [223, 224, 246, 251].

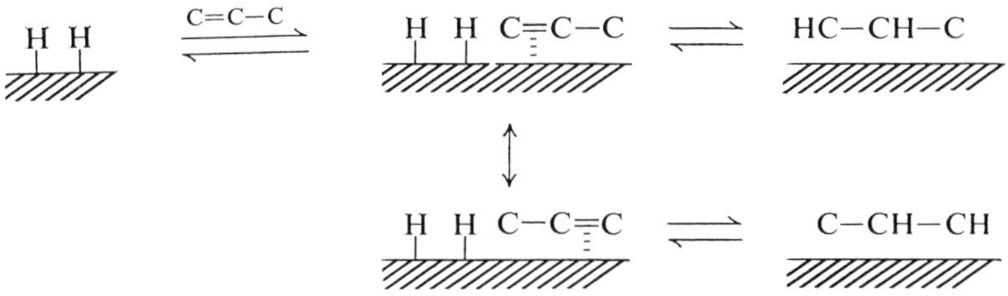

$$\begin{array}{c} H \ \ H \\ | \ \ | \\ \text{/////} \end{array} \xrightarrow{C=C-C} \begin{array}{c} H \ \ H \ \ C{=}C-C \\ | \ \ | \\ \text{//////////} \end{array} \rightleftarrows \begin{array}{c} HC-CH-C \\ \text{//////////} \end{array}$$

$$\updownarrow$$

$$\begin{array}{c} H \ \ H \ \ C-C{=}C \\ | \ \ | \\ \text{//////////} \end{array} \rightleftarrows \begin{array}{c} C-CH-CH \\ \text{//////////} \end{array}$$

Hydrogenation of α,β-Unsaturated Carbonyls (Olefin Group)

Hydrogenation of α,β-unsaturated carbonyls is influenced by several factors such as solvent composition, presence of acids and bases, catalyst/substrate ratio, and the order of introduction of hydrogen and substrate to the reaction mixture [248–252].

The Horiuti-Polanyi mechanism can be invoked to explain some of these effects [248]. Thus a 1,4 adsorbed species (favored by solvents of high dielectric constants) and 1,2 species (occurring in nonpolar solvents) may be formed.

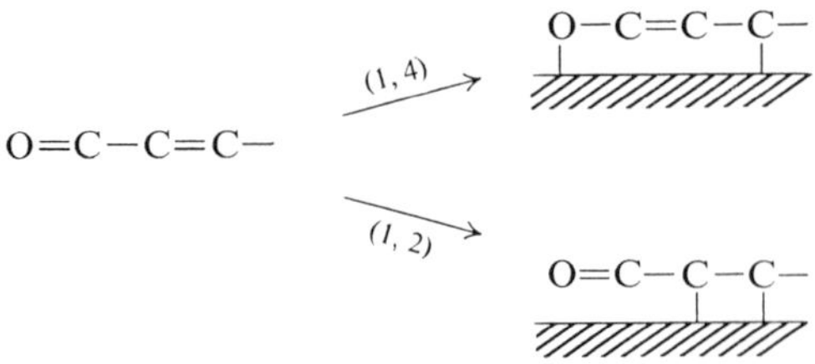

From models it can be predicted that 1,4 adsorption of certain unsaturated ketones, for example, $\Delta^{1,9}$-octalone-2, would result in a higher percentage of the *cis* isomer, whereas 1,2 adsorption would yield nearly equal amounts of *cis* and *trans* product. In the presence of acids, the protonated adsorbed

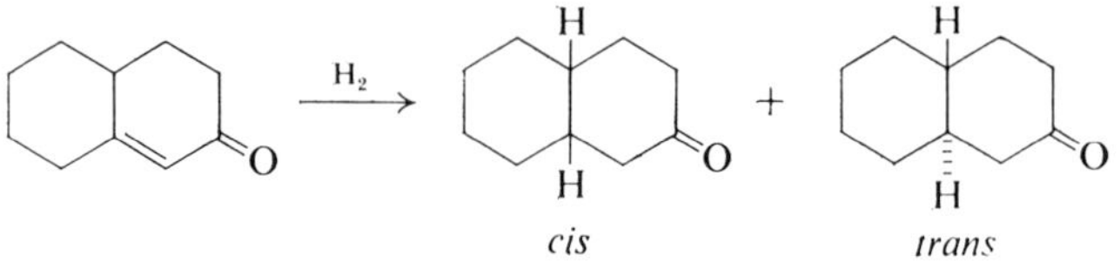

α,β-unsaturated species resembles the 1,4 adsorbed state, resulting in a predominantly *cis* product. In basic solutions the strong adsorption of the enolate ion governs the products obtained from hydrogenation, often resulting in a mixture of *cis* and *trans* products.

Hydrogenation of Carbonyl Groups

In general, axial alcohols are formed in acid solutions and equatorial alcohols in basic solutions when cyclic ketones are reduced. A proposed model is shown.

The prediction of the stereochemical course of any hydrogenation can never be absolutely certain. However, the principle that catalytic hydrogenation takes place by *cis* addition of hydrogen to that side of the molecule which is most easily accessible to the catalyst is of value in these cases where it is possible with confidence to predict which is the least hindered side of a molecule. The predictions are most likely to be met under conditions of mild and rapid hydrogenation. Conformational changes brought about by changes in the acidity of the reaction medium, or the presence of other groups distant from the reaction site, can effect the approach of the catalytically active site and can, therefore, change the stereochemical course. The possibility of using specific absorption effects, either through the choice or quantity of the catalyst, or by manipulation of functional groups, can be fruitful, particularly when steric approach effects are finely balanced.

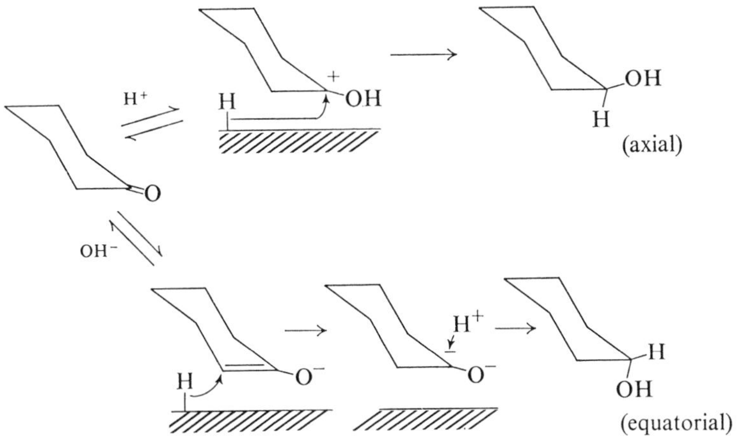

Selectivity and Stereochemistry

Olefins

The rate at which various olefins are hydrogenated by a given catalyst depends on its degree of substitution, unless complicated by steric factors. The rates are in general in the order mono > di > tri > tetra substituted olefin. Differences between the metals in the spread of the relative rates have

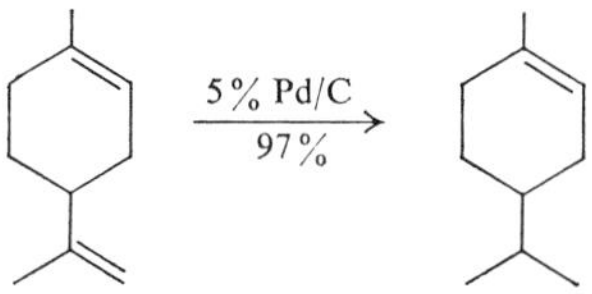

been employed to create selective catalysts. Thus reduction of limonene gave Δ^1-p-menthane in high yield [253]. Further examples are seen in the reduction of manoöl [254] and caryophyllene [255] where hydrogenation takes place first at the bonds indicated.

The primary product of reduction is that which arises by *cis* addition of hydrogen to the least hindered side of the molecule. It is, however, not always easy to determine which is the least hindered side. Moreover, the possibility of double bond migration before reduction must be borne in mind. For this reason, platinum is the preferred catalyst. In rigid molecules it is more evident which is the least hindered side, and the rule has been applied with good effect to steroids and triterpenes [256–258].

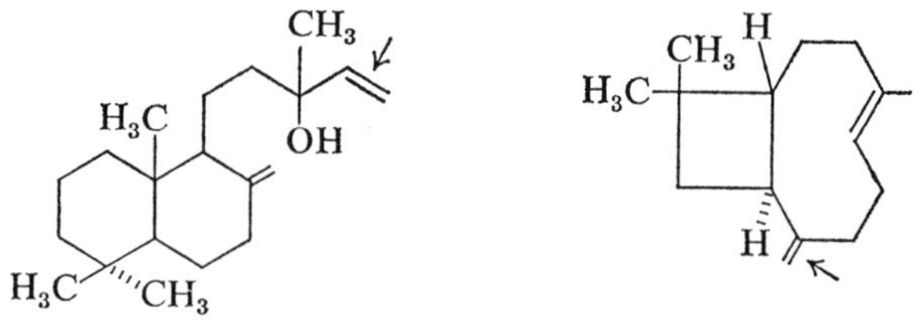

The effect on the stereochemical control of a hydrogenation of the presence of other functional groups and the choice of the metal has already been referred to. The point is illustrated by studies on the reduction of derivatives of 3α,19-dihydroxychlolest-5-ene and derivatives [259] (Table 11.8). It is evident that the 19-hydroxyl exercises a greater effect than the 3α substituent.

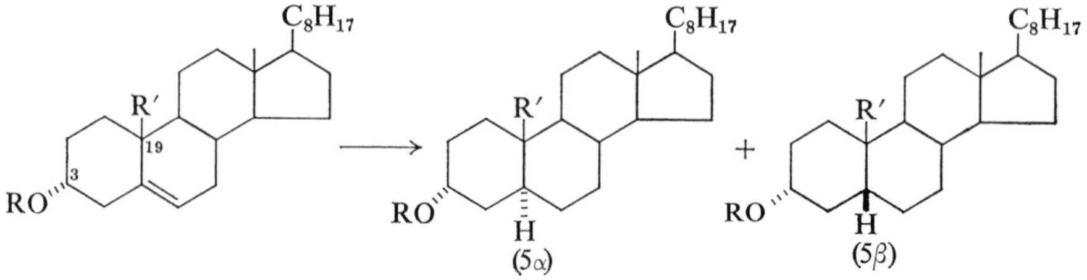

The catalytic hydrogenation of 7-acetoxy norbornadiene and some of its derivatives clearly shows the influence of catalyst and steric requirements [260]. Particularly striking is the effect of pre-reduction of the catalyst on product distribution. This was rationalized on the basis that prereduction

Table 11.8 Percentage of 5α- or 5β-dihydro Compounds Formed for Various Hydrogenation Catalysts

Substituents		Catalysts					
R (3α)	R′ (19)	Pt/AcOH		Rh/AcOH		Rh/iPrOH	
H	CH$_3$	27	63	38	62	54	46
Ac	CH$_3$	31	69	18	82	19	81
H	CH$_2$OH	10	90	2	90	10	90
Ac	CH$_2$OH	3	97		>99		>99
		(5α)	(5β)	(5α)	(5β)	(5α)	(5β)

Table 11.9 Reduction of 7-Acetoxynorbornadiene

Catalyst	syn (%)	anti (%)	Saturated (%)	Nortricyclene (%)	Diene[b] (%)
Pd/C	46.1	9.0	21.6	5.0	18.3
Pd/C[a]	28.8	26.6	14.1	21.7	8.9
Pt/C	35.9	3.0	29.3	6.3	25.6
Pt/C[a]	44.2	4.0	21.3	7.2	23.3

[a] Pre-reduced catalyst.
[b] Unreacted.

forms a species more ready to take part in coordination with the diene (Table 11.9).

Because of steric factors certain double bonds can be virtually inaccessible to the reagent, and the failure to hydrogenate C=C bonds in certain steroids

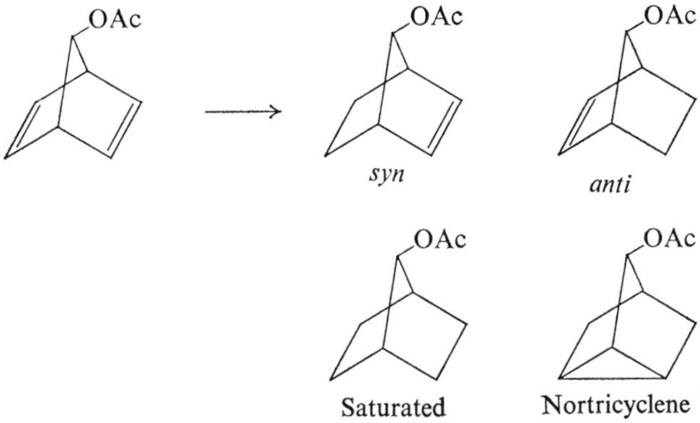

and triterpenes has been used for diagnostic purposes. Thus the unsaturated bond in α-amyrin and β-amyrin [261] is resistant to reduction, as is the $\Delta^{8,9}$ bond in the lanostene series [262].

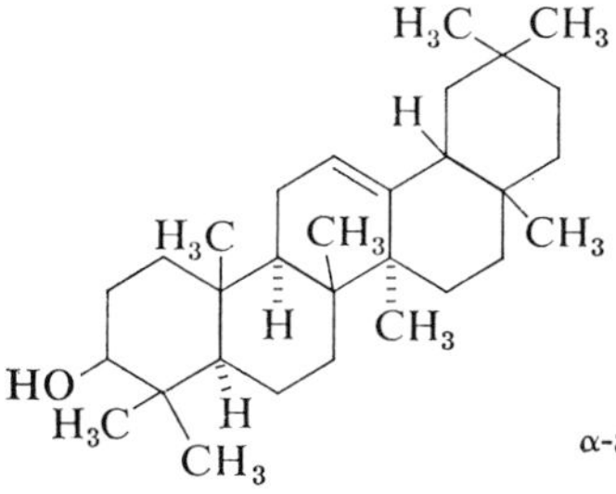

α-amyrin

Carbonyl Compounds

The stereochemistry of the alcohol formed on reduction of a ketone depends on the steric accessibility of the ketone and on the catalyst. Empirical studies led to the Auwers-Skita rules [263] which predict that reduction in neutral or basic solutions gives the more stable equatorial alcohol as the principal product, while reduction in acidic solution gives the axial alcohol as the principal product. The application of these rules to steroids [264] and the generality of the rule [265] have been discussed.

It is clear that it must be used with circumspection, for the presence of polar groups can lead to specific absorption effects which will effect the steric situation at the catalyst surface.

Hydrogenolysis Reactions (see also Hydrogenolysis Chromatography)

A number of different types of bonds undergo cleavage during conventional mild hydrogenation procedures.

Carbon–Halogen Bonds

The ease of fission of carbon–halogen bonds is in the order $RF \ll RCl < RBr < RI$. Benzyl, aryl, allyl, or vinyl halides are more readily cleaved than are alkyl halides. However, by the appropriate choice of catalyst double bonds may be reduced without concurrent hydrogenolysis [266].

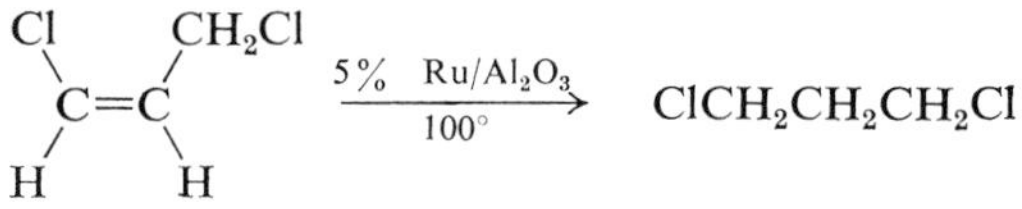

$$\underset{H}{\overset{Cl}{\underset{|}{C}}}=\underset{H}{\overset{CH_2Cl}{\underset{|}{C}}} \xrightarrow[100°]{5\% \ Ru/Al_2O_3} ClCH_2CH_2CH_2Cl$$

Allylic Oxygen Bonds $C=C-C{+}O-$

The allylic oxygen function is subject to hydrogenolysis besides normal saturation of the C=C bond [267, 268]. The appropriate choice of catalyst and control of solvent is important in maximizing the fission reaction. The presence of acid is beneficial.

Nemotin contains an allylic ester group which when reduced with platinum in acid media undergoes hydrogenolysis to give undecanoic acid [269].

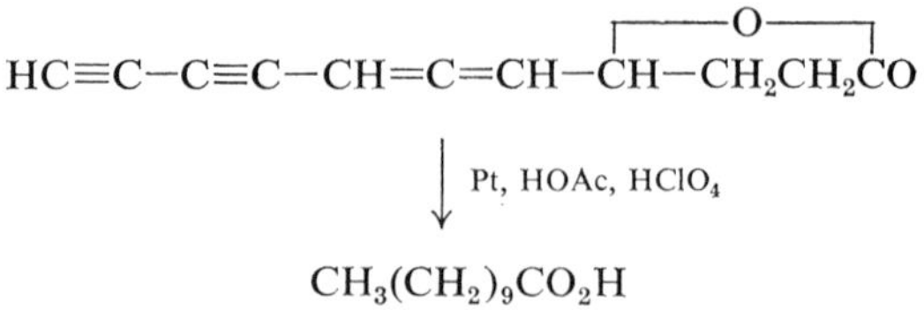

$$HC{\equiv}C-C{\equiv}C-CH{=}C{=}CH-\overset{\lceil \quad\quad O \quad\quad \rceil}{CH}-CH_2CH_2CO$$

$$\downarrow \text{Pt, HOAc, HClO}_4$$

$$CH_3(CH_2)_9CO_2H$$

It is to be noted that fission can be accompanied by bond migration, as can be seen in the case of ψ-santonin. The allylic ester is cleaved and the less thermodynamically favorable tri-substituted olefin is formed [270].

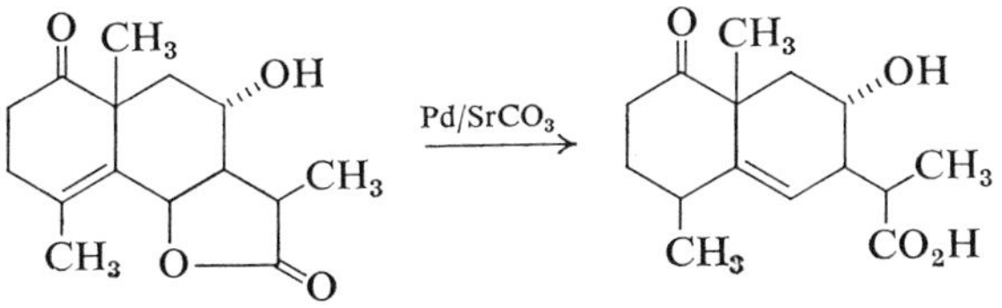

Allyl ethers and halides are also cleaved and, in general, the more polar the group the more ready the fission. If a molecule contains more than one allyl group, both can be cleaved. β-Chlorocodide gives as one reduction product the molecule resulting from fission of the allyl oxygen and chloride [271].

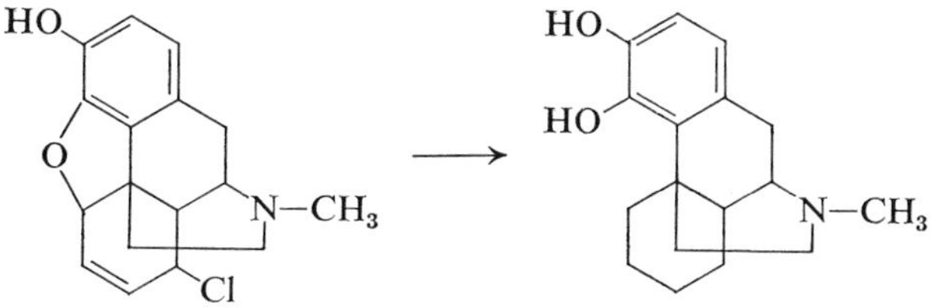

Vinyl Esters

Vinyl esters but not vinyl ethers undergo hydrogenolysis. The effect of the position of the double bond is nicely illustrated by the reduction of the butenolides shown [272]. Vinylogous systems are also cleaved as indicated by

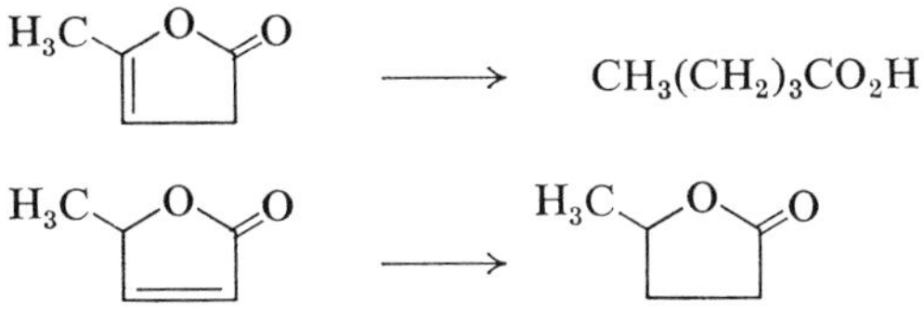

$$CH_3(CH_2)_3CO_2H$$

the cleavage of scillaren and schilliroside [273].

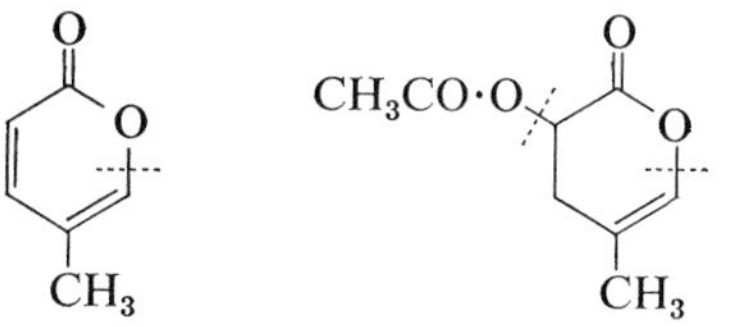

Benzyl Systems

$$ArCH_2\!\!-\!\!X \rightarrow ArCH_3 + HX$$

where $X =$ halogen, NH_2, NHR, NR_2, OH, OR, or OCOR. The ready cleavage of benzyl systems of this type has been utilized extensively for the removal of protecting groups used in synthesis. For this purpose it is essential to avoid ring saturation before hydrogenolysis can take place. The techniques have been reviewed [243].

Palladium is in general the preferred catalyst. Important uses include cleavage of carbohydrate benzyl ethers, benzyl esters, particularly in organophosphate synthesis, and the carbobenzyloxy group in peptide synthesis [243].

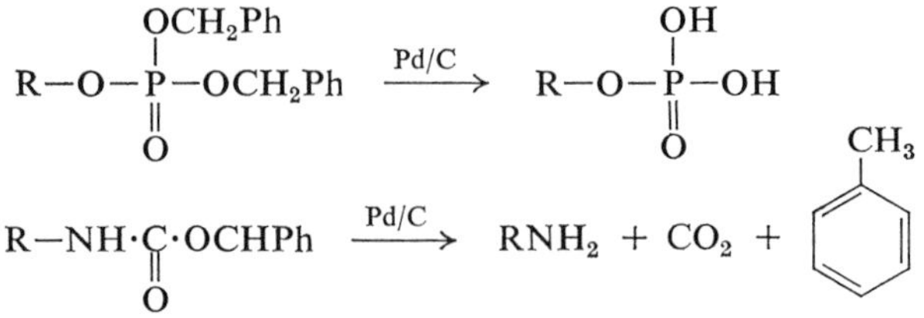

$$R-O-\underset{\underset{O}{\|}}{\overset{\overset{OCH_2Ph}{|}}{P}}-OCH_2Ph \xrightarrow{\text{Pd/C}} R-O-\underset{\underset{O}{\|}}{\overset{\overset{OH}{|}}{P}}-OH$$

$$R-NH\cdot\underset{\underset{O}{\|}}{C}\cdot OCHPh \xrightarrow{\text{Pd/C}} RNH_2 + CO_2 +$$

The removal of a benzyl alcohol function has on occasion proved useful in degradative studies, as in the case of the tetracycline shown [274]. It is

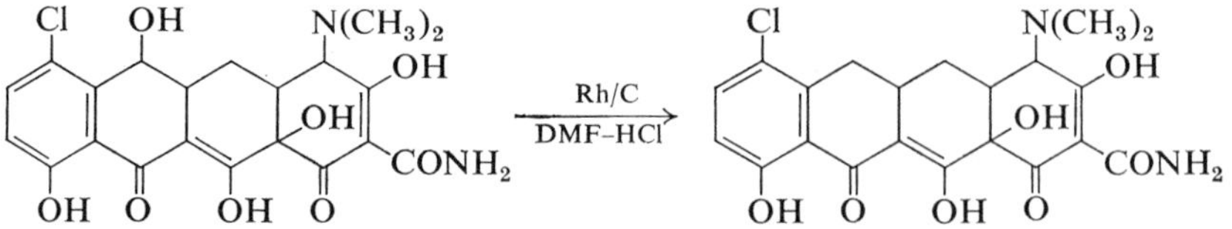

noteworthy that this catalyst avoided dehalogenation. An alternative mechanism, though not applicable to the case quoted, involves dehydration followed by reduction.

Three-Membered Rings

Except for specially strained rings or severe conditions [275] the cyclopropane ring is the only one to show C–C cleavage [276]. In allyl cyclopropanes the bond between the two least substituted carbon atoms is broken. The more substituted the ring the more difficult it is to hydrogenolyze. However, a phenyl or vinyl substituent increases the ease of cleavage, and fission takes place adjacent to these bonds [277].

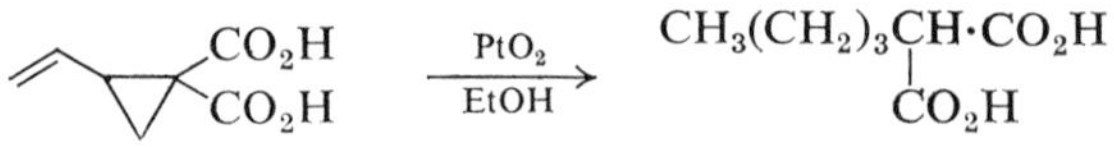

A further example is provided by the cleavage of the unusual amino acid hypoglycin, which occur to the extent of 20% at a and b [278, 279].

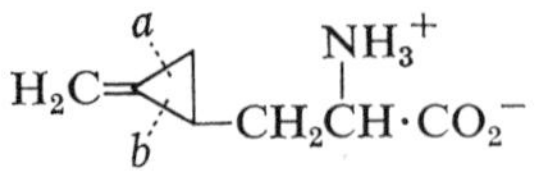

However, by proper choice of conditions it is possible to saturate an olefin without hydrogenolysis of a cyclopropane ring [280, 281].

The epoxide ring is quite readily cleaved, particularly by Pt or Raney nickel catalysts [282, 283].

In acid media the carbon–oxygen bond opens to give the more stable carbonium ion, leading to the least substituted alcohol as the predominant product. In neutral solution the epoxide generally opens to give the most highly substituted alcohol. However, it is not easy to predict with confidence. Differential reduction of diepoxides can be achieved, depending on steric factors [284].

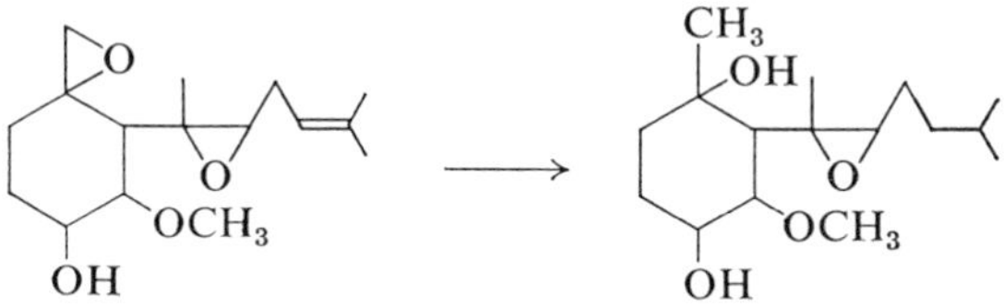

Hydrogenolysis Chromatography

Gas chromatography has been combined with hydrogenolysis to provide a valuable technique for the identification of organic and metal organic molecules [285–289]. In this technique a small tube packed with a hydrogenation catalyst is attached to the injection port of a gas–liquid chromatography machine, hydrogen is passed over the heated catalyst, and then the sample is injected. The products of hydrogenation are then swept directly into the chromatography column, separated, detected and, if desired, trapped for further identification by infrared or mass-spectrometric techniques [288, 290, 291]. By a suitable choice of conditions multiple bonds are saturated, and halogen, sulfur, nitrogen, oxygen, and certain metal atoms are hydrogenolyzed. This treatment results in considerable simplification and frequently results in the carbon skeleton of the original compound being identified. Since a wide range of functionally substituted molecules may have the same retention time, stripping down to the carbon skeleton can reveal structural relationships otherwise obscured. Moreover, authentic samples of identified materials are often not available and at times determination of the carbon skeleton may settle the question of identity. Hydrocarbons containing up to approximately twenty carbon atoms can be handled by this method.

The technique is particularly valuable in that only small quantities of material are required; in suitable cases microgram quantities are sufficient [292]. Efficient capillary tube collection devices have been developed [291].

Table 11.10 Examples of Hydrogenolysis

Compound	Hydrogenation Product

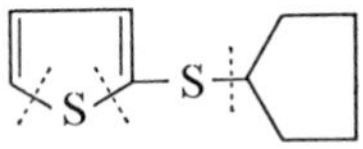

n-Butane,

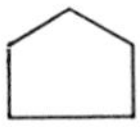

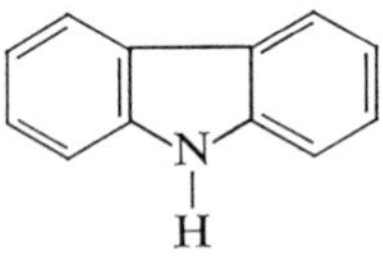

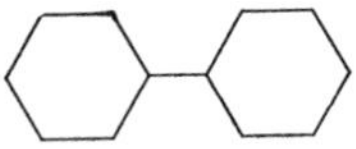

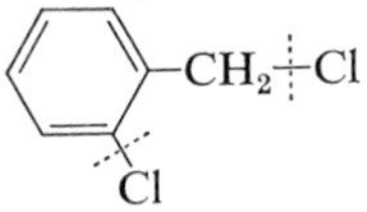

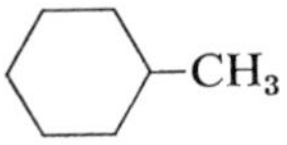

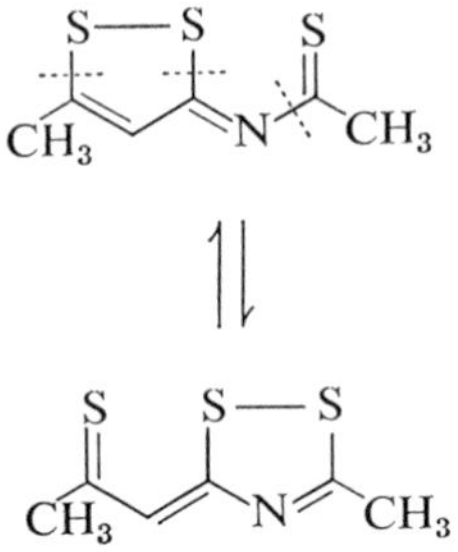

—CH$_2$—Cl

Cl

—CH$_3$

S—S   S

CH$_3$ N   CH$_3$

n-Butane, ethane

S   S—S

CH$_3$ N   CH$_3$

CH$_3$

CH$_3$—C—OH   Isobutane

CH$_3$

CH$_3$

CH$_3$CH$_2$CH—OH   *n*-Butane

CH$_3$

CH$_3$—CH—CH$_2$—OH   *n*-Propane, Isobutane

CH$_3$CH$_2$CH$_2$—CH$_2$—OH   *n*-Propane, *n*-butane

The major pathways of hydrogenolysis have been established for the main classes of compounds, and the skeleton present can now be deduced with reasonable confidence. The ease of hydrogenolysis of various classes of compounds is partly dependent on the catalyst. Palladium (1%) on Gas-Chrom. P or alumina is satisfactory for hydrogenation, desulfurization, dehalogenation, and deoxygenation. Platinum (5%) on glass beads is said to be preferable for hydrogenolysis of nitrogen [288].

The halides and sulfides cleave very readily to form the parent hydrocarbon. Primary aldehydes and amines give, in addition, products arising from cleavage of the adjacent bond. Specific examples are given in Table 11.10 [288].

In general a given compound subjected to hydrogenolysis gives the parent hydrocarbon and/or the next lower homolog. This is illustrated in Table 11.11.

In a complementary technique for handling small quantities of material *in situ*, reduction prior to paper chromatography can be of use. Reduction of a palladium chloride solution with formaldehyde, followed by application of the sample and exposure to a hydrogen atmosphere, has been described [293, 294].

Homogeneous Catalytic Hydrogenation

Although homogeneous catalytic hydrogenation by transition metal complexes has been known since 1938–1939 [295, 296], it is only in the last several years that catalysts of sufficient reactivity and selectivity have been discovered. There is, as yet, comparatively little literature on the utilization of these catalysts, but the indications are that they will prove to be a useful complement to heterogeneous systems.

The mechanisms of hydrogenation for several of these catalysts have been reviewed [297–304] and it is the interplay of the metal, its valency state, the nature and steric effect of the ligand, and the over-all stereochemistry of the transition state which gives hope of increased selectivity. It is possible that in the future specific catalysts will be designed for a particular reduction. Most catalysts only appear to reduce olefinic and acetylenic groups, but some homogeneous systems have been developed for the reduction of functional groups such as aldehyde and ketone, and for carbon–nitrogen linkages.

Some unique aspects of homogeneous catalysts are the ability to bring about specific deuteration by avoiding the scrambling usually found with heterogeneous catalysts, a technique which has been used as an alternative to ozonolysis for the location of olefinic bonds, and the remarkable selectivity for the hydrogenation of substituted olefinic bonds.

The activity of many homogeneous catalysts for hydrogenation of olefinic

Table 11.11 Generalized Products of Hydrogenolysis

Type of Compound		Main Product
Paraffin		Unchanged
Unsaturates	$RCH{=}CHR'$	RCH_2CH_2R'
	$RC{\equiv}CR'$	
Halides	$R{+}X$	RH
	$X{+}R{+}X$	
Sulfides	$R{+}SH$	RH
	$R'{+}SS{+}R$	$R'H, RH$
	$R{+}S$ (ring)	RH_2
Aldehydes	$R{+}CH{\mp}O$	RH, RCH_3
Acids	$R{+}CO_2H$	RH
Anhydrides	$R{+}C{-}O{-}C{+}R$ (both $C{=}O$)	RH
Alcohols: tertiary	$\underset{R\quad\; R'}{\overset{R''\quad OH}{\diagdown C \diagup}}$	$RR'R''CH$
secondary	$RR'HC{+}OH$	$RR'CH_2$
primary	$R{+}CH_2{+}OH$	RH, RCH_3
Ethers	$R{+}CH_2{+}O{+}CH_2{+}R'$	$RH, RCH_3, R'H, R'CH_3$
Ketones	$R{+}C{+}R'$ ($C{=}O$)	RCH_2R' (main), $RH, R'H$
Esters	$R{+}CH_2{+}O{-}C{+}R'$ ($C{=}O$)	$RH, RCH_3, R'H$
Amines: tertiary	$\underset{R{-}C{-}R''}{\overset{R'\quad NH_2}{\diagdown \;\diagup}}$	$RR'R''CH$
secondary	$RR'CH{-}NH_2$	$RR'CH_2$
primary	$R{+}CH_2{+}NH_2$	RH, RCH_3
Amides	$R{+}C{+}N{+}R'R''$ ($C{=}O$)	$RH, RCH_3, R'H, R''H$

bonds is equal to that of the most active heterogeneous catalysts. More reproducibility in reductions is also claimed, although in most cases either poisoning or acceleration of the reaction occurs with traces of oxygen. Inhibition of reduction by sulfur compounds is not as serious for certain homogeneous catalysts as for their heterogeneous counterparts.

Hydridochlorotris(triphenyl phosphine)ruthenium(II),RuHCl(PPh$_3$)$_3$ [305–307], has been claimed to be the most active catalyst for the hydrogenation of olefins. The reduction of acetylenes and dienes proceeds only to the monoene stage [306, 308], while internal, cyclic [309], and substituted olefins are hydrogenated very slowly [305, 308]. Selective hydrogenations are possible, as in the reduction of 1,4-androstadiene-3,17-dione to 4-andro-stene-3,17-dione [310].

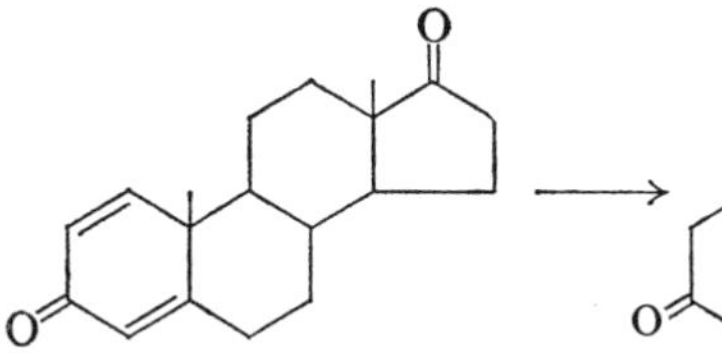

Other homogeneous ruthenium catalysts have been described, for example, ruthenium carboxylates/PPh$_3$/HBF$_4$, which are very active for the hydrogenation of olefins in aqueous solutions [311].

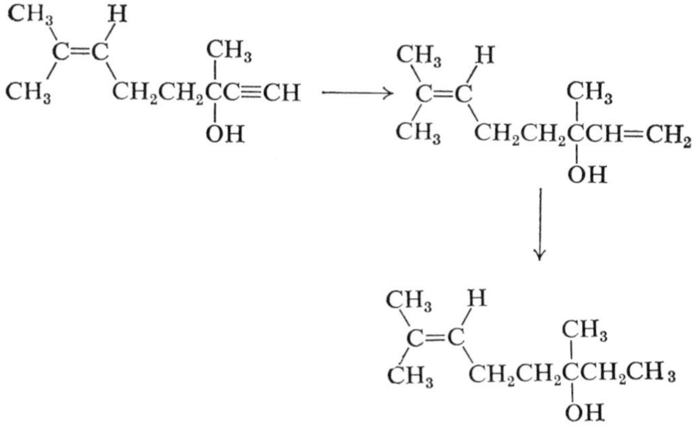

Chlorotris(triphenyl phosphine)rhodium(I), RhCl(PPh$_3$)$_3$ [312, 313], is perhaps the most intensively studied homogeneous catalyst, and several accounts of the mechanism of reduction are available [314–326]. This catalyst has been recommended for specific labeling by D$_2$ [321, 322, 327–329] and T$_2$ [330, 331] because of the absence of scrambling and isomerization [326]. Substituted olefins [332] and unsaturated compounds having nonreducible functional groups [332–337], such as acid, ester, cyano, aryl, and nitro groups, have been successfully reduced. This catalyst does not hydrogenolyze benzyl ether or amines and is not poisoned by thiols [338, 339]. The acetylenic group is more readily reduced than the olefinic, and the process proceeds by *cis* addition of hydrogen [338].

Further examples of the selectivity possible using RhCl(PPh$_3$)$_3$ are the

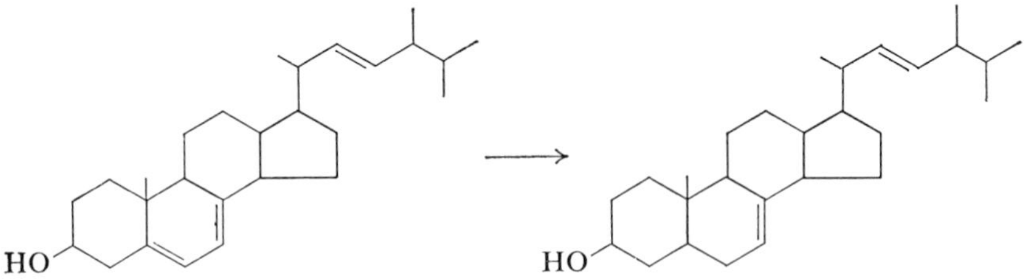

reduction of ergosterol to 5α,6-dihydroergosterol [340, 341] and pregna-5,16-diene-3β-ol-20-one to pregnenolone acetate [340] (in 88% yield).

Other hydrogenation reactions or hydrogen exchange reactions catalyzed by RhCl(PPh$_3$)$_3$ include the selective reduction of the vinyl group in linaloöl [340], 8,14-dihydrothebaine from thebaine [340], 3α-hydroxy-5β-androstane-17-one from 5β-androstane-3,17-dione [342], and Δ^4-α,β-cholestane-3-one to 5α- and 5β-cholestane-3-one [321, 343, 344].

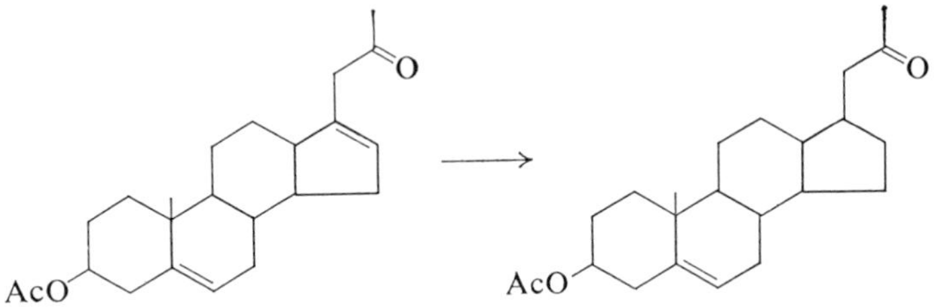

The solvent used for homogeneous hydrogenation reactions is important in determining both the rate of hydrogenation, the selectivity obtained [323], and the extent of scrambling during hydrogenation [326]. This may be the result of solvent molecules acting as coordinating ligands, thus altering the reactivity of the metal complex towards the substrate and hydrogen. By using optically active amide solvents with the homogeneous rhodium complex, Rhpy$_2$Cl$_2$(BH$_4$)(amide), hydrogenated products from substituted olefins can be obtained which contain up to 50% optical purity [345]. The same effect can be obtained using optically active phosphine ligands in the complex RhCl$_3$(PR$_3$)$_3$, from which hydratropic acid is formed by reduction of α-phenylacrylic acid in ~15% optical yields [346]. α-Substituted styrene derivatives are also reduced to hydrocarbons in $\sim$5% optical yields [347]. By using deuterium and the rhodium complex $[RhH_2(PPh_3)_2(solvent)_2]^+$ it is possible to reduce ketones to the corresponding alcohol in which deuterium is introduced specifically at the carbonyl carbon [348] in contrast to many heterogeneous catalysts which also promote β-deuteration [328]. The

$$\begin{array}{c} R \\ \diagdown \\ \diagup \\ R' \end{array} C = O \longrightarrow \begin{array}{c} R \\ \diagdown \\ \diagup \\ R' \end{array} CD - OD$$

explanation offered is that the keto form (and not the enol form) of the ketone is complexed to the rhodium during hydrogenation.

Aqueous or methanolic solutions of cobaltous cyanide, $[Co(CN)_5]^{3-}$, are efficient catalysts for the reduction of the olefinic group in a variety of activated unsaturated compounds, including conjugated dienes, styrenes, α,β-unsaturated aldehydes, ketones, acids, and so on [349, 350]. Selective hydrogenation of dienes to monoenes is possible since nonconjugated olefins are neither reduced nor isomerized [351]. Nickel cyanide solutions, $[Ni_2(CN)_6]^{4-}$, will also hydrogenate acetylenes [352], olefins [353], conjugated dienes [354], and olefins conjugated with other functional groups [355].

Palladium and platinum complexes containing stabilizing ligands, for example, phosphine and $SnCl_3^-$, can be used to hydrogenate multiple unsaturated fatty acid esters selectively [356–359], for example soya bean oil, and certain polyolefinic substrates to the monoene compounds [357, 360, 361].

In general, homogeneous hydrogenation catalysts can bring about selective hydrogenation, specific deuteration and tritation, and asymmetric synthesis without the problems often encountered with heterogeneous catalysts of isomerization, disproportionation, and hydrogenolysis. Almost all the results up to the present have been obtained with olefinic and acetylenic groups, while the more resistant groups, such as aryl, have been hydrogenated only with specific catalysts, usually Ziegler-type systems [362]. Other functional groups can be reduced, but at this stage in time these reactions are perhaps better performed using heterogeneous catalysts.

6 SELECTED EXAMPLES OF STRUCTURAL DETERMINATION BY REDUCTION TECHNIQUES

Structural Determination of an Organometallic Complex

LiAlD$_4$, Mass Spectroscopy, nmr [A. C. Cope and R. W. Siekman, *J. Amer. Chem. Soc.*, **87**, 3272 (1965)]. Azobenzene reacts with potassium tetrachloroplatinite(II) or palladium(II) dichloride to give colored complexes.

The ir and uv spectral evidence together with the fact that azobenzene was not liberated by the action of cyanide, triphenylphosphine, or amines led to its formulation as **3**.

The presence of a bond formed between the metal atom and the *ortho*-carbon atoms of the azobenzene were established by reduction with lithium aluminum deuteride followed by hydrolysis with deuterium oxide. This gave a deuterated hydrazobenzene, which for convenience of analysis was oxidized to the corresponding azobenzene. Mass spectrometric analysis of this azobenzene gave 4% d_0, 93% d_1, and 3% d_2 species. Azobenzene-2-d_1, azobenzene-3-d_1, and azobenzene-4-d_1, were synthesized by standard techniques. The nmr of these species differs sufficiently from the azobenzene liberated from the complex to be identified as azobenzene-2-d_1.

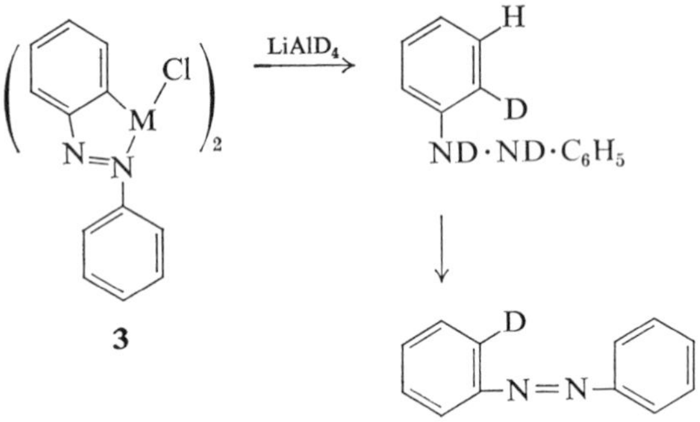

This method was subsequently used to determine the structure of complexes formed by Pt^{II} and Pd^{II} with dialkyl benzyl amines [A. C. Cope and E. C. Friedrich, *J. Amer. Chem. Soc.*, **90**, 909 (1968)] and vinyl pyridine [A. Kashura, K. Tanoka, and T. Izumi, *Bull. Chem. Soc. Japan*, **42**, 1702 (1969)].

Stereochemical Determination of Natural Product

Na, NH₃—Optical Activity Determination. ("The Relative and Absolute Configurations of Catechins and *epi*-Catechins," A. J. Birch, J. W. Clark-Lewis, and A. V. Robertson, *J. Chem. Soc.*, **1957**, 3586).

Catechin and *epi*-catechin had been shown by the classical structural studies of Freudenberg to be geometrical isomers, which are now interpreted in terms of the structures **4** (R = H) for *epi*-catechin (*cis* isomer) and **5** (R = H) for catechin (*trans* isomer).

The relative and absolute configuration had not been settled. The work of Birch and his coworkers settled this point, and it also confirmed that the previously known epimerizations and racemization of catechins in hot aqueous solutions occur by inversion of the 2-aryl group. Reductive fission of the benzylic oxygen of the tetramethyl ethers (**5**, R = CH$_3$) by sodium in liquid ammonia removed the center of asymmetry at the 2-aryl group. (−)-*epi*-Catechin tetramethyl ether gave excess (+)-1-(3,4-dimethoxyphenyl)-3-(2,4,6-trimethoxyphenyl)propane-2-ol. (+)-Catechin tetramethyl ether gave an excess of the (−)-enantiomorph. Thus (+)-catechin and (−)-*epi*-catechin have opposite configurations at position 3 and, therefore, (+)-catechin and (+)-*epi*-catechin have the same configurations at position 3. The absolute configuration was then established by Prelog's atrolactic acid method.

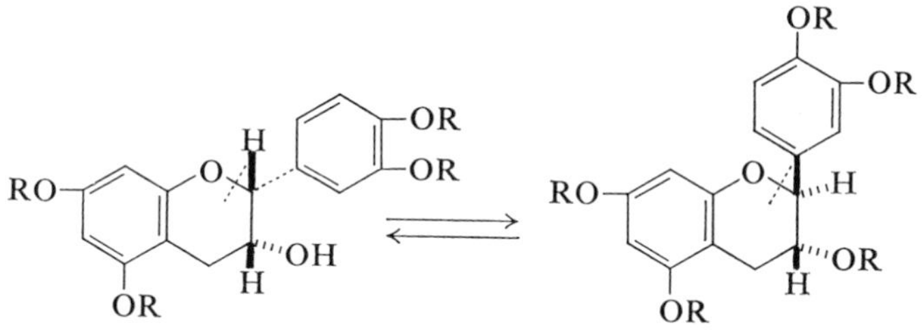

Of more general interest is the observation that the reduction of the optically active propanols required careful attention to conditions to avoid racemization. Moreover, further reduction to related substituted propanes also occurred. Birch put forward possible mechanisms for this and suggested that variation in results may be related to details of experimental technique such as concentration of ethanol, speed of reduction, and efficiency of stirring. The stereochemistry of the reaction of the tetramethyl catechins with AlCl$_3$–LiAlH$_4$ has also been investigated [K. Weinges, F. Toribio, and E. Paulus, *Justus Liebigs Ann. Chem.*, **688**, 127 (1965)].

Macrolide Antibiotic Structure

HI–P, LiAlH$_4$, H$_2$–Pt, Gas–Liquid Chromatography, Mass Spectroscopy [A. C. Cope, R. K. Bly, E. P. Burrows, O. J. Ceder, E. Ciganek, B. T. Gillis, R. F. Porter, and H. E. Johnson, *J. Amer. Chem. Soc.*, **84**, 2170 (1962). A. C. Cope, E. P. Burrows, M. E. Dereig, S. Moon, and W. D. Wirth, *J. Amer. Chem. Soc.*, **87**, 5452 (1965)].

Cope and his coworkers worked out a general method of determining the carbon skeleton of the larger macrolide antibiotics by reductive removal of oxygen atoms (first reference above). They found it to be amenable to variations and consider it indispensable in structural investigations of this class of compounds (second reference above). It is interesting to reflect that the method is a variation on the venerable hydriodic acid–red phosphorous method which was used by E. Fischer to prove, by reduction to *n*-hexane, that glucose contains a straight chain of six carbon atoms. The antibiotic

Fungichromin is hydrogenated by Pt–H$_2$ in methanol to give a crystalline decahydro derivative. The uptake of 5 moles of hydrogen based on a molecular weight of 671 and microanalysis supported the molecular formula C$_{35}$H$_{68-70}$O$_{12}$. The determination of the exact number of carbon atoms in a molecule of this size presents a problem since conventional elemental analysis and molecular weight determination are not sufficiently accurate. The removal of the oxygen atoms to give a derivative containing all the carbon atoms in their original arrangement was achieved by the sequence shown.

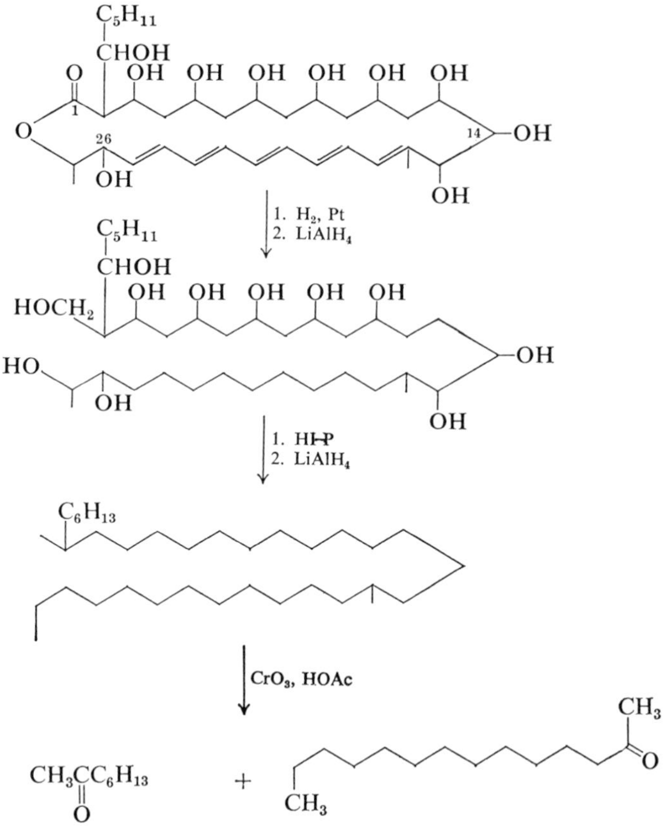

Hydriodic acid–red phosphorous removed all the oxygen atoms but the product contained iodine, which was reduced by treatment with LiAlH$_4$. Gas chromatography and mass spectrometry identification gave the molecular formula as C$_{35}$H$_{22}$ and confirmed that Fungichromin is a macrocyclic lactone rather than an ester. The C$_{35}$ hydrocarbon was oxidized to 2-octanone and 2-tetradecanone, whose identity was established by gas chromatography and mass spectrometry.

Periodate oxidative cleavage of the *vic*-glycol system gave two polyols which, by the same reduction sequence as just described, gave identifiable fragments. This enabled the oxygen functions to be located.

A similar sequence applied to the aglycone of the macrolide Rimocidin gave 3-methyluntriacontane in 34% yield from perhydro-Rimocidin. The identity was determined by gas chromatography, mass spectrometry, and synthesis, and therefore establishes the carbon skeleton of this antibiotic.

Exhaustive reduction of the antibiotic Spiramycin and comparison of the mass spectrum of the skeletal hydrocarbon with that derived from Magnamycin led to a revision of the structure of Spiramycin to include a 16-member lactone ring. The structural relationships of Spiramycins and Magnamycins were also clarified. [S. Omura, A. Nakagawa, M. Otani, T. Hata, H. Ogura, and K. Furahata, *J. Amer. Chem. Soc.*, **91**, 3401 (1969).]

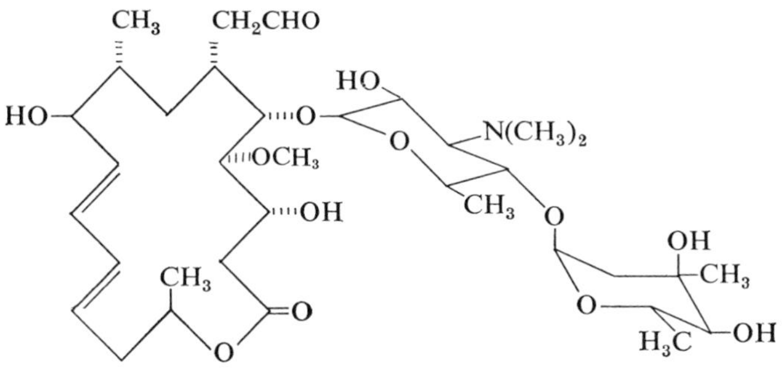

Biosynthesis of Ergot Alkaloids

Birch-Emde Fission, Kuhn-Roth Oxidation, Radioactivity Determination [H. G. Floss, W. Hornmann, N. Schilling, K. Kelly, D. Groeger, and D. Erge, *J. Amer. Chem. Soc.*, **90**, 6500 (1968)].

This example illustrates the use of the Emde-Birch fission during an investigation of the biosynthesis of ergot alkaloids. Ergot alkaloids originate from the combination of L-tryptophan, mevalonic acid, and methionine. The partial sequence shown in the scheme has been established.

Chanoclavine-1 could give rise to elymoclavine in two ways: by oxidative cyclization between the methyl groups (C-7) and the nitrogen, or by reaction between the hydroxymethyl group and the nitrogen, followed by hydroxylation at the methyl group. The latter reaction would involve an isomerization at the Δ^8 double bond. (*R,S*)-Mevalonic-2-^{14}C acid was fed to cultures of *Claviceps paspali* and the chanoclavine-1 isolated. A sample was then fed to *Claviceps* strain SD-5 which produces mainly elmyoclavine. The latter was submitted to an Emde-Birch fission, followed by Kuhn-Roth oxidation to give acetic acid (C-7 + C-8).

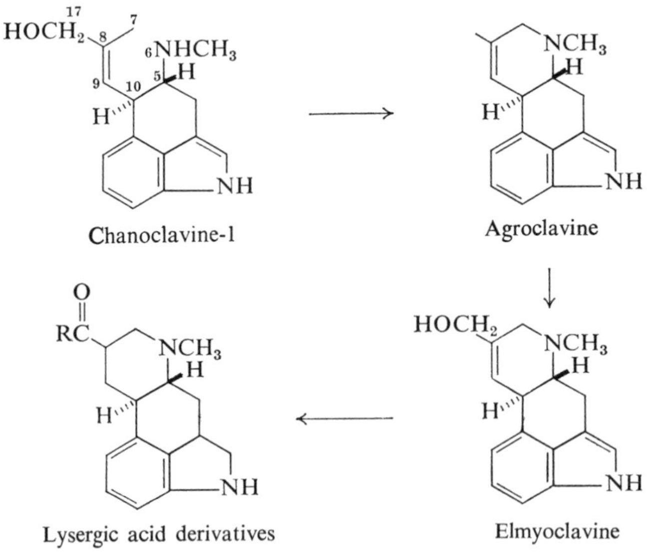

Chanoclavine-1

Agroclavine

Lysergic acid derivatives

Elmyoclavine

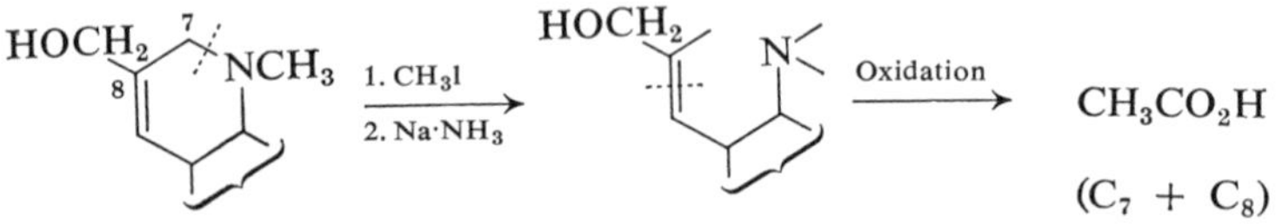

$$\text{HOCH}_2 \quad \text{NCH}_3 \xrightarrow[\text{2. Na·NH}_3]{\text{1. CH}_3\text{I}} \text{HOCH}_2 \quad N \xrightarrow{\text{Oxidation}} \text{CH}_3\text{CO}_2\text{H}$$

$$(C_7 + C_8)$$

Measurement of the radioactivity established that chanoclavine-1 labeled at C-7 gives rise to elymoclavine labeled at C-17. Thus the cyclization involves an olefin *cis–trans* isomerization.

Pheromone Identification

Microscale, Hydrogenolysis–Chromatography, Mass Spectroscopy, Infrared Analysis ["Isolation and Identification of Chief Component of Male Tassel Scent in Black-Tailed Deer," R. G. Brownlee, R. M. Silverstein, D. M. Schwarze, and A. G. Singer, *Nature*, **221**, 284 (1969)].

This example, which utilizes hydrogenolysis chromatography, illustrates what can be done with small quantities by means of modern techniques. The total amount of sample available for the various treatments was 30 μg.

The ir spectrum of the sample in carbon tetrachloride showed absorption at 5.63 μ (C=O) and 8.53 μ (C–O), ascribed to a five-membered lactone. The sample was recovered by gas chromatography. A 5-μg sample was analyzed by mass spectrometry and gave a parent ion at 196, equivalent to $C_{12}H_{20}O_2$ and a base peak at 85. This led to the partial structure **6** which contains a double bond or a ring in a straight or branched side chain

A 3-μg sample submitted to hydrogenolysis chromatography gave two peaks, one of which was not trapped. The other was identified as *n*-undecane,

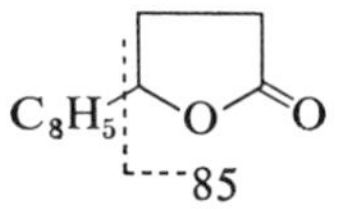

C_8H_5 —— 85

6

$CH_3(CH_2)_4CH=CHCH_2$

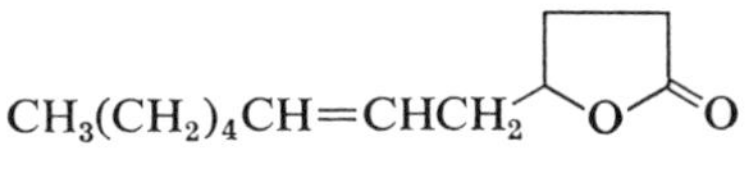

7

and the missing component was assumed to be *n*-dodecane. Therefore, the side chain was established as straight. A complementary technique to hydrogenation is that of micro-ozonolysis [M. Beroza and B. A. Bierl, *Anal. Chem.*, **39**, 1131 (1967)]. A 2-μg sample submitted to this technique gave hexanal, leading to the structure 7 for the pheromone. The olefin was presumed to be *cis* since no ir band at 10.4 μ was found. The structure was confirmed by comparison of ir, mass spectrometry, and gas–liquid chromatography data with a synthetic sample.

Structure-Activity Relationships in *t*-RNA

NaBT$_4$, Radioactivity Measurement, Chromatography, Biochemical Techniques [P. Cerutti, J. W. Holt, and N. Miller, *J. Mol. Biol.*, **34**, 505 (1968). T. Igo-Kemenes and H. G. Zachau, *Eur. J. Biochem.*, **10**, 544 (1969)].

Cerutti, Holt, and Mills have shown how reduction by sodium borotritiide can be used for the detection and determination of the unusual minor

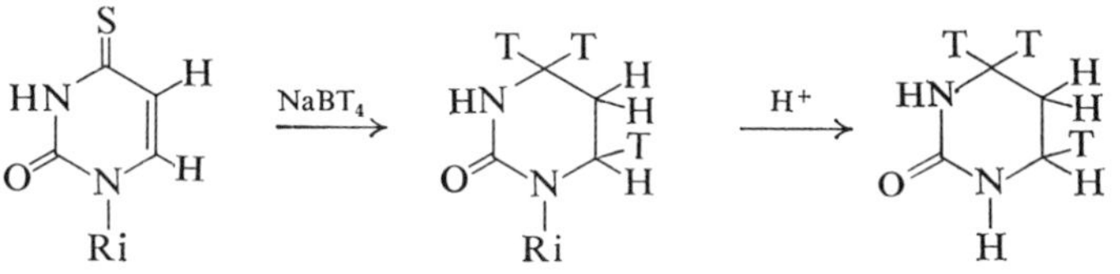

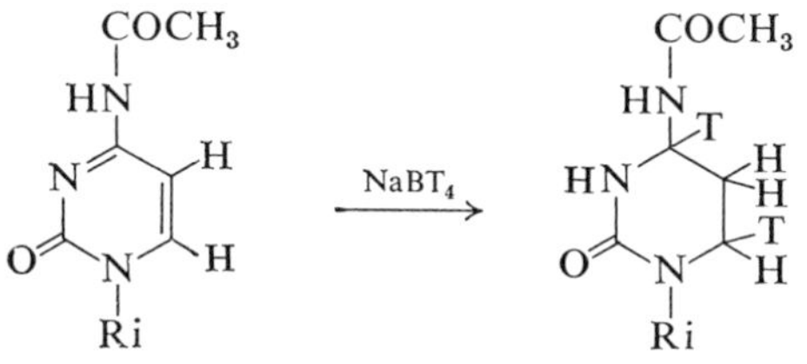

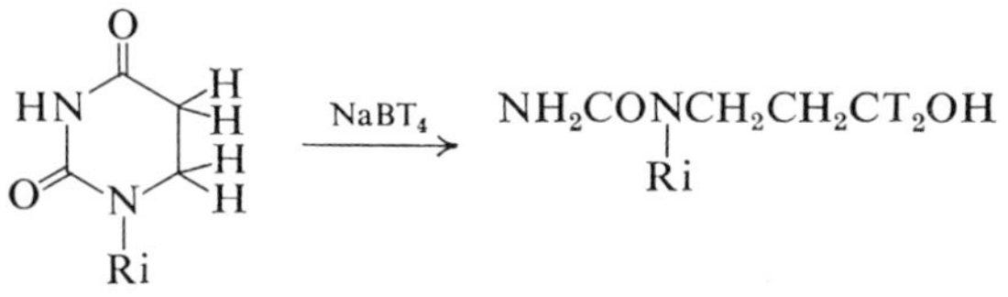

$NH_2CONCH_2CH_2CT_2OH$

Ri = ribose

nucleosides 4-thiouridine, N^4-acetylcytidine, and 5,6-dihydrouridine. The reduction proceeds as shown.

These three minor nucleosides and 3-methyl cytidine are the only constituents of t-RNA known to be susceptible to reduction by $NaBH_4$ in the dark. Isolation by ion exchange chromatography combined with thin layer chromatography and determination of radioactivity gives a quantitative analysis. Cerutti and coworkers suggest that since only t-RNA appears to contain dihydrouridine and 4-thiouridine, this technique allows the different types of cellular RNA to be distinguished. They found that yeast t-RNA contains an average of 3 residues of dihydrouridine per chain of 77 nucleotides, while *Escherichia coli* B t-RNA contains 2.5 residues, and *Bacillus subtilis* t-RNA contains 0.8 residues.

Igo-Kemenes and Zachau have shown how this technique may be used to probe structure–activity relationships in t-RNA. Dihydrouridine is reduced but N^4-acetylcytidine in position 12 of t-RNA is not reduced, although easily reduced as part of a trinucleotide. It is possible that pairing protects the group from reduction, and this agrees with a clover-leaf model. In addition, biological activity does not appear to be impaired in the reduced product.

However, it was shown that reduction of phenylalanine t-RNA by $NaBH_4$ caused a complete loss of aminoacyl accepting activity. Similarly treated valine t-RNA and formylatable methionine t-RNA lose only 20% of their activity. By using $NaBT_4$ it was shown that incorporation of tritium occurs mostly into 5,6-dihydrouridine and 4-thiouridine. Since two 5,6-dihydrouridines and one 4-thiouridine occur in the loop proximal to the 5^1 terminus of *Escherichia coli* phenylalanine t-RNA, it was concluded that this area is essential for enzyme recognition [L. Shugart and M. P. Stulberg, *J. Biol. Chem.*, **244**, 2806 (1969)].

References

1. (a) W. G. Brown, *Org. Reactions*, **6**, 469 (1951). (b) E. Schenker, *Newer Methods of Preparative Organic Chemistry*, Vol. 4 (W. Foerst, ed.), Academic Press, New York, 1968, p. 197.
2. (a) N. G. Gaylord, *Reduction with Complex Metal Hydrides*, Interscience, New York, 1956. (b) R. E. Lyle and P. S. Anderson, *Advances in Heterocyclic Chemistry*, Vol. 6 (A. R. Katritzky and A. J. Boulton, eds.), Academic Press, New York, 1966, p. 46.
3. H. Hörmann, *Newer Methods in Preparative Organic Chemistry*, Vol. 2 (W. Foerst, ed.), Academic Press, New York, 1963, p. 213.
4. H. C. Brown and A. Tsukamoto, *J. Amer. Chem. Soc.*, **81**, 502 (1959).

5. H. C. Brown and C. P. Garg, *J. Amer. Chem. Soc.*, **86**, 1085 (1964).
6. (a) H. C. Brown and C. J. Shoaf, *J. Amer. Chem. Soc.*, **86**, 1079 (1964). (b) H. C. Brown and R. F. McFarlin, *J. Amer. Chem. Soc.*, **80**, 5372 (1958).
7. H. C. Brown and B. C. Subba Rao, *J. Amer. Chem. Soc.*, **80**, 5377 (1958).
8. J. Fajkŏs, *Collect. Czech. Chem. Commun.*, **24**, 2284 (1959).
9. N. M. Yoon and H. C. Brown, *J. Amer. Chem. Soc.*, **90**, 2927 (1968).
10. M. N. Rerick, *The Mixed Hydrides*, Metal Hydrides Inc., Beverley, Mass., 1959.
11. E. L. Eliel, *Rec. Chem. Progr.*, **22**, 129 (1961).
12. B. R. Brown and G. A. Somerfield, *Proc. Chem. Soc.*, **1958**, 7.
13. E. L. Eliel and M. Rerick, *J. Org. Chem.*, **23**, 1088 (1958).
14. E. L. Eliel, V. G. Badding, and M. N. Rerick, *J. Amer. Chem. Soc.*, **84**, 2371 (1962).
15. G. R. Pettit and W. J. Bowyer, *J. Org. Chem.*, **25**, 84 (1960).
16. E. L. Eliel, L. A. Pilato, and V. G. Badding, *J. Amer. Chem. Soc.*, **84**, 2377 (1962).
17. E. L. Eliel, B. E. Nowak, R. A. Daignault, and V. G. Badding, *J. Org. Chem.*, **30**, 2440 (1965).
18. E. L. Eliel and M. N. Rerick, *J. Amer. Chem. Soc.*, **82**, 1362 (1960).
19. P. T. Lansbury, *J. Amer. Chem. Soc.*, **83**, 429 (1961).
20. P. T. Lansbury and J. O. Peterson, *J. Amer. Chem. Soc.*, **85**, 2236 (1963).
21. J. F. Corbett, *Chem. Commun.*, **1968**, 1257; see also *Chem. Eng. News*, 80 (October 13, 1969).
22. R. A. Raphael and F. Sondheimer, *J. Chem. Soc.*, **1950**, 3185.
23. B. L. Shaw and M. C. Whiting, *Chem. Ind.* (London), **1953**, 409.
24. E. B. Bates, E. R. H. Jones, and M. C. Whiting, *J. Chem. Soc.*, **1954**, 1854.
25. J. S. Cowie, P. D. Landor, and S. R. Landor, *Chem. Commun.*, **1969**, 541.
26. J. Martin, W. Parker, B. Shroot, and T. Stewart, *J. Chem. Soc.*, C, **1967**, 101.
27. W. Carruthers and M. I. Qureshi, *Chem. Commun.*, **1969**, 832.
28. A. L. Wilds, *Org. Reactions*, **2**, 178 (1944).
29. T. Bersin, *Newer Methods of Preparative Organic Chemistry*, Interscience, New York, 1948, p. 127.
30. R. B. Woodward, N. L. Wendler, and F. J. Brutschy, *J. Amer. Chem. Soc.*, **67**, 1425 (1945).
31. C. Djerassi, *Org. Reactions*, **6**, 207 (1951).
32. V. Bruckner, A. Karczag, K. Körmendy, M. Meszaros, and J. Tomasz, *Tetrahedron Lett.*, **1960**, 5.
33. H. C. Brown, *J. Chem. Educ.*, **38**, 173 (1961).
34. H. C. Brown, *Hydroboration*, Benjamin, New York, 1962.
35. G. Zweifel and H. C. Brown, *Org. Reactions*, **13**, 1 (1963).
36. (a) E. Breuer, *Tetrahedron Lett.*, **1967**, 1849. (b) K. M. Biswas, L. E. Houghton, and A. H. Jackson, *Tetrahedron, Suppl.*, No. 7, 261 (1966).
37. H. C. Brown and K. Murray, *J. Amer. Chem. Soc.*, **81**, 4108 (1959).
38. A. C. Cope, G. A. Berchtold, P. E. Peterson, and S. H. Sharman, *J. Amer. Chem. Soc.*, **82**, 6370 (1960).
39. H. C. Brown and G. Zweifel, *J. Amer. Chem. Soc.*, **81**, 1512 (1959).

40. E. F. Knights and H. C. Brown, *J. Amer. Chem. Soc.*, **90**, 5280, 5281 (1968).
41. A. J. Birch, *Quart. Rev.*, **4**, 69 (1950).
42. A. J. Birch and H. Smith, *Quart. Rev.* **12**,17 (1958).
43. G. W. Watt, *Chem. Rev.*, **46**, 317 (1950).
44. A. L. Wilds and N. A. Nelson, *J. Amer. Chem. Soc.*, **75**, 5366 (1953).
45. C. Djerassi, ed., *Steroid Reactions*, Holden-Day, San Francisco, 1963.
46. A. P. Krapcho and A. A. Bothner-By, *J. Amer. Chem. Soc.*, **81**, 3658 (1959).
47. A. P. Krapcho and M. E. Nadel, *J. Amer. Chem. Soc.*, **86**, 1096 (1964).
48. H. Smith, *Organic Reaction in Liquid Ammonia*, Vol. 1, Part 2, Wiley, New York, 1963.
49. W. Hückel and H. Bretschneider, *Justus Liebigs Ann. Chem.*, **540**, 184 (1939).
50. A. J. Birch. *J. Chem. Soc.*, **1947**, 1642.
51. G. Stork, S. Malhotra, H. Thompson, and M. Uchibayashi, *J. Amer. Chem. Soc.*, **87**, 1148 (1965).
52. (a) D. Devaprabhakara and P. D. Gardner, *J. Amer. Chem. Soc.*, **85**, 648 (1963). (b) N. A. Dobson and R. A. Raphael, *J. Chem. Soc.*, **1955**, 3558.
53. G. Stork and S. D. Darling, *J. Amer. Chem. Soc.*, **82**, 1512 (1960); **86**, 1761 (1964).
54. L. Velluz, J. Valls, and G. Nominé, *Angew. Chem. Inter. Ed.*, **4**, 181 (1965).
55. M. Tanabe, J. W. Chamberlin, and P. Y. Nishiura, *Tetrahedron Lett.*, **1961**, 601.
56. W. F. Johns, *J. Org. Chem.*, **28**, 1856 (1963).
57. R. A. Benkeser, R. E. Robinson, D. M. Sauve, and O. H. Thomas, *J. Amer. Chem. Soc.*, **77**, 3230 (1955).
58. R. A. Benkeser and E. M. Kaiser, *J. Org. Chem.*, **29**, 955 (1964).
59. R. L. Augustine, *Reduction*, Arnold, London, 1968, p. 126.
60. E. L. Martin, *Org. Reactions*, **1**, 155 (1942).
61. D. Staschewski, *Angew. Chem.*, **71**, 726 (1959).
62. G. Stork and F. Clarke, *J. Amer. Chem. Soc.*, **83**, 3114 (1961).
63. W. Reusch, *Reduction* (R. L. Augustine, ed.), Arnold, London, 1968, p. 200.
64. J. Blackwell and W. J. Higginbottom, *J. Chem. Soc.*, **1961**, 1405.
65. R. S. Rosenfeld and T. F. Gallagher, *J. Amer. Chem. Soc.*, **77**, 4367 (1955).
66. (a) A. C. Cope, J. W. Barthel, and R. D. Smith, *Org. Synthesis, Coll. Vol.* **4**, 218 (1963). (b) V. Prelog, K. Schenker, and H. H. Günthard, *Helv. Chim. Acta*, **35**, 1598 (1952).
67. L. F. Fieser, *J. Amer. Chem. Soc.*, **75**, 4377 (1953).
68. F. A. Hochstein, C. R. Stephens, L. H. Conover, P. P. Regna, R. Pasternack, P. N. Gordon, F. J. Pilgrim, K. J. Brunings, and R. B. Woodward, *J. Amer. Chem. Soc.*, **75**, 5455 (1953).
69. J. M. Stewart and J. D. Young, *Solid State Peptide Synthesis*, Freeman, San Francisco, 1969, p. 46.
70. S. Bhattachavji, A. J. Birch, A. Brack, A. Hofmann, H. Kobel, D. C. C. Smith, H. Smith, and J. Winter, *J. Chem. Soc.*, **1962**, 421.
71. G. Childs and E. J. Forbes, *J. Chem. Soc.*, **1959**, 2024.
72. E. Grovenstein, Jr., and L. C. Rogers, *J. Amer. Chem. Soc.*, **86**, 854 (1964).

73. H. G. Floss, U. Hornemann, N. Schilling, K. Kelley, D. Groeger, and D. Erge, *J. Amer. Chem. Soc.*, **90**, 6500 (1968).
74. A. J. Birch, *J. Chem. Soc.*, **1945**, 809.
75. A. J. Birch, G. K. Hughes, and E. Smith, *Aust. J. Chem.*, **7**, 83 (1954).
76. A. J. Birch, *J. Chem. Soc.*, **1947**, 102.
77. H. E. Zimmerman, *Tetrahedron*, **16**, 169 (1961).
78. G. W. Kenner and N. R. Williams, *J. Chem. Soc.*, **1955**, 522.
79. M. Fetizon, M. Jurion, and N. T. Anh, *Chem. Commun.*, **1969**, 112.
80. S. W. Pelletier and D. M. Locke, *J. Org. Chem.*, **23**, 131 (1958).
81. V. du Vigneaud, *Symposium on Peptide Chemistry*, Chem. Soc. Spec. Publ. No. 2, p. 49 (1955).
82. A. R. Pinder and H. Smith, *J. Chem. Soc.*, **1954**, 113.
83. A. Beverloo, M. C. Dieleman, P. E. Verkade, K. S. de Vries, and B. M. Webster, *Rec. Trav. Chim. Pays-Bas.*, **81**, 1033 (1962).
84. W. M. Schubert, B. S. Rabinovitch, N. R. Larson, and V. A. Sims, *J. Amer. Chem. Soc.*, **74**, 4590 (1952).
85. M. C. Hoff, K. W. Greenlee, and C. E. Boord, *J. Amer. Chem. Soc.*, **73**, 3329 (1951).
86. P. A. Levens, *Org. Syn. Coll. Vol.* **2**, 320 (1943).
87. S. B. Soloway, A. M. Damiana, J. W. Sims, H. Bluestone, and R. E. Lidov, *J. Amer. Chem. Soc.*, **82**, 5377 (1960).
88. P. E. Verkade, K. S. de Vries, and B. M. Webster, *Rec. Trav. Chim. Pays-Bas.*, **83**, 367 (1964).
89. S. Hünig, H. R. Müller, and W. Thier, *Angew. Chem. Int. Ed.*, **4**, 271 (1965)
90. C. E. Miller, *J. Chem. Educ.*, **42**, 254 (1965).
91. E. J. Corey, D. J. Pasto, and W. L. Mock, *J. Amer. Chem. Soc.*, **83**, 2957 (1961).
92. E. E. van Tamelen and R. J. Timmons, *J. Amer. Chem. Soc.*, **84**, 1067 (1962).
93. M. Ohno and M. Okamoto, *Tetrahedron Lett.*, **1964**, 2423.
94. A. J. Birch and J. J. Wright, *Chem. Commun.*, **1969**, 788.
95. A. Furst, R. C. Berlo, and S. Hooton, *Chem. Rev.*, **65**, 51 (1965).
96. Ref. 63, p. 171.
97. Huang-Minlon, *J. Amer. Chem. Soc.*, **68**, 2487 (1966); **71**, 3301 (1949).
98. M. Gates and G. Tschudi, *J. Amer. Chem. Soc.*, **78**, 1380 (1956).
99. G. Lock, *Monatsh. Chem.*, **85**, 802 (1954).
100. D. H. R. Barton, D. A. J. Ives, and B. R. Thomas, *J. Chem. Soc.*, **1955**, 2056.
101. W. Nagata and H. Itazaki, *Chem. Ind.* (London), **1964**, 1194.
102. French Patent, 1,424,400; *Chem. Abstr.*, **65**, 15454 (1966).
103. P. D. Gardner, L. Rand, and G. R. Haynes, *J. Amer. Chem. Soc.*, **78**, 3425 (1956).
104. D. J. Cram, M. R. V. Shahyun, and G. R. Knox, *J. Amer. Chem. Soc.*, **84**, 1734 (1962).
105. R. L. Augustine, *Catalytic Hydrogenation*, Arnold, London, 1965.
106. P. N. Rylander, *Catalytic Hydrogenation over Platinum Metals*, Academic Press, London, 1967.
107. D. V. Sokol'skii, *Hydrogenation in Solution*, Israel Program for Scientific Translations, Jerusalem, 1964.

108. "Catalytic Hydrogenation and Analogous Pressure Reactions," *Anal. N. Y. Acad. Sci.*, **145**(1), 1 (1967).
109. H. Greenfield and F. S. Dovell, *J. Org. Chem.*, **32**, 3670 (1967).
110. Netherland Patent 66,11772; *Chem. Abstr.*, **67**, 53614 (1967).
111. F. S. Dovell and H. Greenfield, *J. Amer. Chem. Soc.*, **87**, 2767 (1965).
112. U.S. Patent 3,350,450; *Chem. Abstr.*, **68**, 95486 (1968).
113. H. S. Broadbent and C. W. Whittle, *J. Amer. Chem. Soc.*, **75**, 975 (1953).
114. A. N. Mal'tsev, N. I. Kobozev, and L. V. Voronova, *Russ. J. Phys. Chem.*, **43**, 1581 (1969).
115. R. A. Barnes and H. M. Fales, *J. Amer. Chem. Soc.*, **75**, 975 (1953).
116. E. B. Maxted and J. S. Elkins, *J. Chem. Soc.*, **1961**, 5086.
117. F. G. Ciapetta and C. J. Plank, *Catalysis*, Vol. 1 (P. H. Emmett, ed.), Reinhold, New York, 1954, p. 315.
118. W. B. Innes, *Catalysis*, Vol. 1 (P. H. Emmett, ed.), Reinhold, New York, 1954, p. 245.
119. H. A. Benesi, R. M. Curtis, and H. P. Studer, *J. Catal.*, **10**, 328 (1968).
120. British Patent 1,146,437.
121. U.S. Patent, 3,013,990; *Chem. Abstr.*, **56**, 5442 (1962).
122. S. Akabori, S. Sakurai, Y. Izumi, and Y. Fujii, *Nature*, **178**, 323 (1956).
123. S. Akabori, S. Sakurai, Y. Izumi, and Y. Fujii, *Biokhimiya*, **22**, 154 (1956); *Chem. Abstr.*, **51**, 11356 (1957).
124. D. P. Harrison and H. F. Rase, *Ind. Eng. Chem.*, *Fundam.*, **6**, 161 (1967).
125. P. N. Rylander, M. Kilroy, and V. Coven, *Engelhard Ind. Tech. Bull.*, **6**, 11 (1965).
126. F. J. McQuillin, W. O. Ord, and P. L. Simpson, *J. Chem. Soc.*, **1963**, 5996.
127. R. L. Augustine, *J. Org. Chem.*, **28**, 152 (1963).
128. W. Cocker, P. V. R. Shannon, and P. A. Staniland, *J. Chem. Soc.*, *C*, **1966**, 41.
129. P. N. Rylander and D. R. Steele, *Engelhard Ind. Tech. Bull.*, **3**, 91 (1962).
130. H. A. Smith and B. L. Stump, *J. Amer. Chem. Soc.*, **83**, 2739 (1961).
131. G. N. Walker, *J. Org. Chem.*, **23**, 133 (1958).
132. British Patent 949,255; *Chem. Abstr.*, **60**, 15664 (1964).
133. R. Adams, V. Voorhees, and R. L. Shriner, *Org. Syn. Coll. Vol.* **1**, 463 (1941).
134. V. L. Frampton, J. D. Edwards, and H. R. Henze, *J. Amer. Chem. Soc.*, **73**, 4432 (1951).
135. W. F. Bruce, *J. Amer. Chem. Soc.*, **58**, 687 (1936).
136. C. W. Keenan, B. W. Gresemann, and H. A. Smith, *J. Amer. Chem. Soc.*, **76**, 229 (1954).
137. R. Baltzly, *J. Amer. Chem. Soc.*, **74**, 4586 (1952).
138. W. Theilaker and H. G. Drössler, *Chem. Ber.*, **87**, 1676 (1954).
139. A. N. Mal'tsev, N. I. Kobozev, and L. V. Vorondova, *Russ. J. Phys. Chem.*, **42**, 1034 (1968).
140. C. A. Brown, *J. Amer. Chem. Soc.*, **91**, 5901 (1969).
141. H. C. Brown and C. A. Brown, *J. Amer. Chem. Soc.*, **84**, 1494 (1962).
142. C. Eaborn, B. C. Pant, E. R. A. Peeling, and S. C. Taylor, *J. Chem. Soc.*, *C*, **1969**, 2823.
143. French Patent 1,376,744; *Chem. Abstr.*, **62**, 7632 (1965).

144. U.S. Patent 3,210,296; *Chem. Abstr.*, **63**, 17763 (1965).
145. E. B. Hershberg, E. Oliveto, M. Rubin, H. Staeudle, and L. Kuhlen, *J. Amer. Chem. Soc.*, **73**, 1144 (1951).
146. K. Kindler, H. G. Helling, and E. Sussner, *Justus Liebigs Ann. Chem.*, **605**, 200 (1957).
147. British Patent 1,123,837; *Chem. Abstr.*, **67**, 53890 (1967).
148. E. B. Maxted and S. Akhter, *J. Chem. Soc.*, **1959**, 3130.
149. E. B. Maxted and R. W. D. Morrish, *J. Chem. Soc.*, **1940**, 252.
150. E. B. Maxted and H. C. Evans, *J. Chem. Soc.*, **1937**, 603.
151. E. B. Maxted and A. Marsden, *J. Chem. Soc.*, **1940**, 469.
152. F. A. Vandenheuvel, *Anal. Chem.*, **28**, 362 (1956).
153. W. F. Tuley and R. Adams, *J. Amer. Chem. Soc.*, **47**, 3061 (1925).
154. P. N. Rylander, N. Himelstein, and M. Kilroy, *Engelhard Ind. Tech. Bull.*, **4**, 49 (1963).
155. D. J. Cram and N. L. Allinger, *J. Amer. Chem. Soc.*, **78**, 2518 (1956).
156. K. W. Rosenmund and E. Karg, *Chem. Ber.*, **75**, 1850 (1942).
157. British Patent 1,173,469.
158. O. A. Tyurenkova and E. I. Seliverstova, *Kinet. Katal.* **10**, 359 (1969).
159. O. A. Tyurenkova, *Russ. J. Phys. Chem.*, **43**, 1587 (1969).
160. S. Nishimura, M. Shimahara, and M. Shiota, *J. Org. Chem.*, **31**, 2394 (1964).
161. S. Nishimura, T. Itaya, and M. Shiota, *Chem. Commun.*, **1967**, 442.
162. W. M. Pearlman, *Tetrahedron Lett.*, **1967**, 1663.
163. R. L. Burwell, *Chem. Eng. News*, p. 56 (August 22, 1966).
164. British Patent 1,159,967; *Chem. Abstr.*, **71**, 112607 (1969).
165. H. Lindlar, *Helv. Chim. Acta*, **35**, 446 (1952).
166. H. Lindlar and R. Dubois, *Org. Synthesis*, **46**, 89 (1966).
167. G. E. Ham and W. P. Coker, *J. Org. Chem.*, **29**, 194 (1964).
168. M. Freifelder, R. M. Robinson, and G. R. Stone, *J. Org. Chem.*, **27**, 284 (1962).
169. P. N. Rylander and D. R. Steele, *Engelhard Ind. Tech. Bull.*, **3**, 19 (1962).
170. A. I. Meyers, W. N. Bevering, and G. Garcia-Munoz, *J. Org. Chem.*, **29**, 3427 (1964).
171. M. Freifelder, D. A. Dunnigan, and E. J. Baker, *J. Org. Chem.*, **31**, 3438 (1966).
172. M. Freifelder, Y. H. Ng, and P. F. Helgren, *J. Org. Chem.*, **30**, 2485 (1965).
173. U.S. Patent 3,349,139; *Chem. Abstr.*, **68**, 104616 (1968).
174. M. Freifelder, *J. Org. Chem.*, **27**, 4046 (1962); **28**, 602, 1135 (1963).
175. R. T. Rapla, E. R. Lavagnino, E. R. Shepard, and E. Farkas, *J. Amer. Chem. Soc.*, **79**, 3770 (1950).
176. M. Freifelder, *Advanc. Catal.*, **14**, 203 (1963).
177. U.S. Patent 3,159,639; *Chem. Abstr.*, **62**, 7732 (1965).
178. J. H. Stocker, *J. Org. Chem.*, **27**, 2288 (1962).
179. I. A. Kaye and R. S. Matthews, *J. Org. Chem.*, **28**, 325 (1963).
180. M. Freifelder, *J. Org. Chem.*, **26**, 1835 (1961).
181. U.S. Patent 3,244,644; *Chem. Abstr.*, **65**, 82 (1966).

182. N. S. Smirnova, A. S. Chegolya, and A. A. Ponomarev, *J. Org. Chem. USSR*, **1**, 1441 (1965).
183. R. E. Ireland and P. W. Schiess, *J. Org. Chem.*, **28**, 6 (1963).
184. R. T. Rapla and E. Farkas, *J. Org. Chem.*, **23**, 1404 (1958).
185. M. Freifelder and G. R. Stone, *J. Org. Chem.*, **26**, 3805 (1961).
186. A. E. Barkdoll, D. C. England, H. W. Gray, W. Kirk, Jr., and G. M. Whitman, *J. Amer. Chem. Soc.*, **75**, 1156 (1953).
187. British Patents 1,176,336 and 1,176,337.
188. A. A. Ponomarev, L. M. Ryzhenko, and N. S. Smirnova, *J. Org. Chem. USSR*, **5**, 73 (1969).
189. G. Ferrari and A. Andreetta, *Chim. Ind.* (Milan), **51**, 38 (1969).
190. British Patent 1,122,609; *Chem. Abstr.*, **69**, 66981 (1968).
191. French Patent 1,320,373; *Chem. Abstr.*, **60**, 2835 (1964).
192. Y. Takagi, T. Naito, and S. Nishimira, *Bull. Chem. Soc. Japan*, **38**, 2119 (1965).
193. G. Gilman and G. Cohn, *Advan. Catal.*, **9**, 733 (1957).
194. N. A. Vasyunina, G. S. Barysheva, and A. A. Balandin, *Bull. Acad. Sci., USSR*, **1969**, 772.
195. P. N. Rylander, N. Rakoneza, D. Steele, and M. Bollinger, *Engelhard Ind. Tech. Bull.*, **4**, 49 (1965).
196. J. E. Carnahan, T. A. Ford, W. F. Gresham, W. E. Grigsby, and G. F. Hager, *J. Amer. Chem. Soc.*, **77**, 3766 (1955).
197. P. N. Rylander and D. R. Steele, *Engelhard Ind. Tech. Bull.*, **10**, 17 (1969).
198. V. S. Sadikov and A. K. Akhailov, *J. Chem. Soc.*, **1928**, 438.
199. V. S. Sadikov and A. K. Akhailov, *Chem. Ber.*, **61**, 1797, 1801 (1928).
200. H. S. Broadbent and T. G. Selin, *J. Org. Chem.*, **28**, 2343 (1963).
201. H. S. Broadbent and W. J. Bartley, *J. Org. Chem.* **28**, 2345 (1963).
202. (a) H. S. Broadbent, G. C. Campbell, W. J. Bartley, and J. H. Johnson, *J. Org. Chem.*, **24**, 1847 (1959). (b) W. H. Davenport, V. Kollonitsch, and C. H. Kline, *Ind. Eng. Chem.*, **60**(11), 10 (1968).
203. W. Wenner, *J. Org. Chem.*, **15**, 301 (1950).
204. J. J. Lapporte and W. R. Schuett, *J. Org. Chem.*, **28**, 1947 (1963).
205. L. W. Covet, R. Connor, and H. Adkins, *J. Amer. Chem. Soc.*, **54**, 1651 (1932).
206. R. Schröter, *Newer Methods of Preparative Organic Chemistry*, Vol. 1, Interscience, New York, 1948, p. 61.
207. E. Lieber and F. L. Morritz, *Advan. Catal.*, **5**, 417 (1953).
208. A. B. Mekler, S. Ramachandran, S. Swaminathan, and M. S. Newman, *Org. Synthesis*, **41**, 56 (1961).
209. X. A. Dominguez, I. C. Lopez, and R. Franco, *J. Org. Chem.*, **26**, 1625 (1961).
210. G. R. Pettit and E. E. van Tamelen, *Org. Reactions*, **12**, 356 (1962).
211. D. R. Levering and E. Lieber, *J. Amer. Chem. Soc.*, **71**, 1515 (1949).
212. R. Paul, P. Buisson, and N. Joseph, *Ind. Eng. Chem.*, **44**, 1006 (1952).
213. C. A. Brown, *Chem. Commun.*, **1969**, 952.
214. C. A. Brown, *Chem. Commun.*, **1970**, 139.
215. J. M. Pratt and G. Swinden, *Chem. Commun.*, **1969**, 1321.
216. H. Adkins and R. Connor, *J. Amer. Chem. Soc.*, **53**, 1091 (1931).

217. W. A. Lazier and H. R. Arnold, *Org. Synthesis Coll. Vol.* **2**, 142 (1943).
218. H. Adkins, E. E. Burgoyne, and H. J. Scheider, *J. Amer. Chem. Soc.*, **72**, 2626 (1951).
219. G. Grundmann, Newer Methods of Preparative Organic Chemistry, Vol. 1, Interscience, New York, 1948, p. 103.
220. A. T. Blomquist and A. Goldstein, *Org. Synthesis Coll. Vol.* **4**, 216 (1963).
221. R. J. Tedeschi and G. Clark, Jr., *J. Org. Chem.*, **27**, 4323 (1962).
222. R. Mozingo, *Org. Synthesis Coll. Vol.* **3**, 685 (1955).
223. G. C. Bond and P. B. Wells, *Advan. Catal.*, **15**, 91 (1964).
224. S. Siegel, *Advan. Catal.*, **16**, 123 (1966).
225. R. L. Burwell, *Chem. Rev.*, **57**, 895 (1957).
226. P. N. Rylander, N. Rakoncza, D. Steele, and M. Bollinger, *Engelhard Ind. Tech. Bull.*, **4**, 95 (1963).
227. H. A. Smith, *Catalysis*, Vol. 5 (P. H. Emmett, ed.), Reinhold, New York, 1957, p. 175.
228. M. E. Derieg, B. Brust, and R. I. Fryer, *J. Heterocycl. Chem.*, **3**, 165 (1966).
229. E. Breitner, E. Roginski, and P. N. Rylander, *J. Org. Chem.*, **24**, 1855 (1959).
230. A. W. Burgstahler and Z. J. Bithos, *Org. Synthesis*, **42**, 62 (1962).
231. E. E. Schweizer and K. K. Light, *J. Amer. Chem. Soc.*, **86**, 2963 (1964).
232. L. I. Peterson, R. B. Hager, A. F. Vellturo, and G. W. Griffin, *J. Org. Chem.*, **33**, 1018 (1968).
233. R. A. Grimm, A. J. Stirton, and J. K. Weil, *J. Amer. Oil. Chem. Soc.*, **46**, 118 (1969).
234. H. C. Yao and P. H. Emmett, *J. Amer. Chem. Soc.*, **83**, 796, 799 (1961).
235. L. H. Klemm, J. W. Sprague, and E. Y. K. Mak, *J. Org. Chem.*, **22**, 161 (1957).
236. R. Adams and F. L. Cohen, *Org. Synthesis Coll. Vol.* **1**, 240 (1941).
237. W. K. Seifert and P. C. Condit, *J. Org. Chem.*, **28**, 265 (1963).
238. A. Lindenmann, *Helv. Chim. Acta*, **32**, 69 (1949).
239. W. H. Carothers and G. A. Jones, *J. Amer. Chem. Soc.*, **47**, 3051 (1925).
240. H. Greenfield, *Ind. Eng. Chem., Prod. Res. Develop.*, **6**, 142 (1967).
241. R. Baltzly and A. P. Phillips, *J. Amer. Chem. Soc.*, **68**, 261 (1946).
242. P. N. Rylander and D. R. Steele, *Engelhard Ind. Tech. Bull.*, **6**, 41 (1965).
243. W. H. Hartung and R. Simonoff, *Org. Reactions*, **7**, 263 (1953).
244. E. I. Klabunovskii, *Russ. Chem. Rev.*, **35**, 546 (1965).
245. E. Mosettig and R. Mozingo, *Org. Reactions*, **4**, 362 (1948).
246. E. K. Rideal, *Concepts in Catalysis*, Academic Press, London, 1968.
247. I. Horiuti and M. Polanyi, *Trans. Faraday Soc.*, **30**, 1164 (1934).
248. R. L. Augustine, D. C. Migliorini, R. E. Foscante, C. S. Sodano, and M. J. Sisbarro, *J. Org. Chem.*, **34**, 1075 (1969).
249. E. F. Meyer and R. L. Burwell, *J. Amer. Chem. Soc.*, **85**, 2877 (1963).
250. J. F. Sauvage, R. H. Baker, and A. S. Hussey, *J. Amer. Chem. Soc.*, **83**, 3874 (1961).
251. J. J. Rooney and G. Webb, *J. Catal.*, **3**, 488 (1964).
252. I. Jardine, R. W. Howsam, and F. J. McQuillin, *J. Chem. Soc.*, C, **1969**, 260.
253. W. F. Newhall, *J. Org. Chem.*, **23**, 1274 (1958).
254. J. R. Hosking and C. W. Brandt, *Chem. Ber.*, **68**, 1311 (1935).

255. L. Ruzicka, K. Huber, P. A. Plattner, S. S. Deshapande, and S. Studer, *Helv. Chim. Acta*, **22**, 716 (1939).
256. L. F. Feiser, *Experientia*, **6**, 312 (1950).
257. R. P. Linstead, W. E. Doering, S. B. Davis, P. Levine, and R. B. Whetsone, *J. Amer. Chem. Soc.*, **64**, 1985 (1942).
258. J. R. Lewis and C. W. Shoppee, *J. Chem. Soc.*, **1955**, 1365.
259. Y. Watanabe, Y. Mizuhara, and M. Shiota, *J. Org. Chem.*, **33**, 468 (1968).
260. W. C. Baird, Jr., B. Franzus, and J. H. Surridge, *J. Org. Chem.*, **34**, 2944 (1969).
261. J. Simonsen and W. C. J. Ross, *The Terpenes*, Vol. 4, Cambridge Univ. Press, Cambridge, 1957, pp. 118, 126, 128, 177.
262. Ref. 261, p. 76.
263. J. H. Brewster, *J. Amer. Chem. Soc.*, **76**, 6361 (1954).
264. D. H. R. Barton, *J. Chem. Soc.*, **1953**, 1027.
265. S. P. Findlay, *J. Org. Chem.*, **24**, 1540 (1959).
266. G. E. Ham and W. P. Coker, *J. Org. Chem.*, **29**, 194 (1964).
267. S. Nishimura, T. Onoda, and A. Nakamura, *Bull. Chem. Soc. Japan*, **33**, 1356 (1960).
268. D. A. Denton, F. J. McQuillin, and P. L. Simpson, *J. Chem. Soc.*, **1964**, 5535.
269. J. D. Bu'lock, E. R. H. Jones, and P. R. Leeming, *J. Chem. Soc.*, **1955**, 4270.
270. W. G. Dauben, W. K. Hayes, J. S. P. Schwarz, and J. W. McFarland, *J. Amer. Chem. Soc.*, **82**, 2232 (1960).
271. K. W. Bentley, *The Chemistry of the Morphine Alkaloids*, Oxford Univ. Press, Oxford, 1954, p. 136.
272. W. Cocker and S. Hornsby, *J. Chem. Soc.*, **1947**, 1157.
273. A. Stoll, *Experientia*, **10**, 282 (1954).
274. T. L. Fields, A. S. Kende and J. H. Boothe, *J. Amer. Chem. Soc.*, **83**, 4612 (1961).
275. J. H. Sinfelt, *Catal. Rev.*, **3**, 175 (1969).
276. J. Newham, *Chem. Rev.*, **63**, 123 (1963).
277. R. W. Kierstead, R. P. Linstead, and B. C. L. Weedon, *J. Chem. Soc.*, **1952**, 3610.
278. R. S. de Ropp, J. C. van Meter, E. C. de Renzo, K. W. McKerns, C. Pidacks, P. H. Bell, E. F. Ullman, S. R. Safir, W. J. Fanshawe, and S. B. Davis, *J. Amer. Chem. Soc.*, **80**, 1004 (1958).
279. D. K. Black and S. R. Landor, *J. Chem. Soc.*, C, **1968**, 288.
280. S. R. Poulter and C. H. Heathcock, *Tetrahedron Lett.*, **1968**, 5339.
281. M. T. Wuesthoff and B. Rickborn, *J. Org. Chem.*, **33**, 1311 (1968).
282. A. Fürst and Pl. A. Plattner, *Helv. Chim. Acta*, **32**, 275 (1949).
283. H. C. Brown and A. Suzuki, *J. Amer. Chem. Soc.*, **89**, 1933 (1968).
284. J. M. Ross, D. S. Tarbell, W. E. Lovett, and A. D. Cross, *J. Amer. Chem. Soc.*, **78**, 4675 (1949).
285. M. Beroza, *Anal. Chem.*, **34**, 1801 (1962).
286. M. Beroza and R. Sarmiento, *Anal. Chem.*, **35**, 1353 (1963).
287. M. Beroza and R. Sarmiento, *Anal. Chem.*, **36**, 1745 (1964).

288. C. J. Thompson, H. J. Coleman, R. L. Hopkins, and H. T. Rall, *J. Gas Chromatogr.*, **5**, 1 (1967).
289. T. A. Gough and E. A. Walker, *J. Gas Chromatogr.*, **8**, 134 (1970).
290. M. Beroza and F. Acree, *J. Assoc. Offic. Agr. Chem.*, **47**, 1 (1964).
291. R. G. Brownlee and R. M. Silverstein, *Anal. Chem.*, **40**, 2077 (1968).
292. U.S. Patent 3,230,046; *Chem. Abstr.*, **64**, 11034 (1966).
293. H. P. Kaufmann and D. K. Chowdhury, *Chem. Ber.*, **91**, 2117 (1958).
294. H. M. A. Elgamal and M. B. E. Fayez, *Z. Anal. Chem.*, **226**, 408 (1967).
295. M. Calvin, *Trans. Faraday Soc.*, **34**, 1181 (1938).
296. M. Iguchi, *J. Chem. Soc. Japan*, **60**, 1287 (1939).
297. J. Halpern, *Ann. Rev. Phys. Chem.*, **16**, 103 (1965).
298. J. Halpern, *Discussion Faraday Soc.*, **46**, 7 (1968).
299. J. Halpern, *Advan. Chem. Series*, **70**, 1 (1968).
300. J. P. Candlin, K. A. Taylor, and D. T. Thompson, *Reactions of Transition Metal Complexes*, Elsevier, Amsterdam, 1968.
301. C. W. Bird, *Transition Metal Intermediates in Organic Synthesis*, Academic Press, New York, 1967.
302. M. L. H. Green, *Organometallic Compounds, Vol. II, The Transition Elements* Methuen, London, 1968.
303. M. E. Vol'pin and I. S. Kolominikov, *Russ. Chem. Rev.*, **38**, 273 (1969).
304. J. E. Lyons, L. E. Rennick and J. L. Burmeister, *Ind. Eng. Chem., Prod. Res. Develop.*, **9**, 2 (1970).
305. P. S. Hallman, B. R. McGarvey, and G. Wilkinson, *J. Chem. Soc., A*, **1968**, 3143.
306. I. Jardine and F. J. McQuillin, *Tetrahedron Lett.*, **1966**, 4871.
307. P. S. Hallman, D. Evans, J. A. Osborn, and G. Wilkinson, *Chem. Commun.*, **1967**, 305.
308. British Patent 1,141,847; *Chem. Abstr.*, **71**, 23510 (1969).
309. I. Jardine and F. J. McQuillin, *Tetrahedron Lett.*, **1968**, 5189.
310. S. Nishimura and K. Tsuneda, *Bull. Chem. Soc. Japan*, **42**, 852 (1969).
311. P. Legzdins, G. L. Rempel, and G. Wilkinson, *Chem. Commun.*, **1969**, 825.
312. J. F. Young, J. A. Osborn, F. H. Jardine, and G. Wilkinson, *Chem. Commun.*, **1965**, 131.
313. Netherlands Patent 66,02062; *Chem. Abstr.*, **66**, 10556 (1967).
314. J. A. Osborn, F. H. Jardine, J. F. Young, and G. Wilkinson, *J. Chem. Soc., A*, **1966**, 1711.
315. F. H. Jardine, J. A. Osborn, and G. Wilkinson, *J. Chem. Soc., A*, **1967**, 1574.
316. S. Montelatici, A. van der Ent, J. A. Osborn, and G. Wilkinson, *J. Chem. Soc., A*, **1968**, 1054.
317. H. van Bekkum, F. van Rantwijk, and T. van der Putte, *Tetrahedron Lett.*, **1969**, 1.
318. C. O'Connor and G. Wilkinson, *Tetrahedron Lett.*, **1969**, 1375.
319. T. J. Katz and S. Cerefice, *J. Amer. Chem. Soc.*, **91**, 2405 (1969).
320. J. F. Biellmann and M. J. Jung, *J. Amer. Chem. Soc.*, **90**, 1674 (1968).

321. W. Voelter and C. Djerassi, *Chem. Ber.*, **101**, 58 (1968).
322. A. Hussey and Y. Takeuchi, *J. Amer. Chem. Soc.*, **91**, 672 (1969).
323. J. P. Candlin and A. R. Oldham, *Discussions Faraday Soc.*, **46**, 60 (1968).
324. G. C. Bond and R. A. Hillyard, *Discussions Faraday Soc.*, **46**, 20 (1968).
325. Y. Chevallier, R. Stern, and L. Sajus, *Tetrahedron Lett.*, **1969**, 1197.
326. R. L. Augustine and J. F. van Peppen, *Chem. Commun.*, **1970**, 495, 497, 571.
327. E. W. Garbisch, Jr., and M. G. Griffith, *J. Amer. Chem. Soc.*, **90**, 6543 (1968).
328. M. Fetizon and J. C. Gramain, *Bull. Chem. Soc. Fr.*, **1969**, 651.
329. D. S. Weinberg and M. W. Scoggins, *Org. Mass. Spectrometry*, **2**, 553 (1969).
330. H. Simon, O. Berngruber, and S. K. Erickson, *Tetrahedron Lett.*, **1968**, 707.
331. A. L. Odell, J. B. Richardson, and W. R. Roper, *J. Catal.*, **8**, 393 (1967).
332. J. J. Sims, V. K. Honwad, and L. H. Selman, *Tetrahedron Lett.*, **1969**, 87.
333. E. J. Smutny, H. Chung, K. C. Dewhirst, W. Keim, T. M. Shryne, and H. E. Thyret, *Amer. Chem. Soc., Div. Petrol. Chem. Preprints*, **14**(2), B100 (1969).
334. S. Takahashi, Y. Yamazaki, and N. Hagihara, *Bull. Chem. Soc. Japan*, **41**, 254 (1968).
335. A. J. Birch and K. A. M. Walker, *Tetrahedron Lett.*, **1967**, 3457.
336. F. H. Jardine and G. Wilkinson, *J. Chem. Soc., C*, **1967**, 270.
337. R. E. Harmon, J. L. Parsons, D. W. Cooke, S. K. Gupta, and J. Schoolenberg, *J. Org. Chem.*, **34**, 3684 (1969).
338. A. J. Birch and K. A. M. Walker, *Tetrahedron Lett.*, **1967**, 1935.
339. A. B. Hornfeldt, J. S. Gronowitz, and S. Gronowitz, *Acta Chem. Scand.*, **22**, 2725 (1968).
340. A. J. Birch and K. A. M. Walker, *J. Chem. Soc., C*, **1966**, 1894.
341. A. J. Birch and K. A. M. Walker, *Tetrahedron Lett.*, **1966**, 4939.
342. J. C. Orr, M. Mersereau, and A. Stanford, *Chem. Commun.*, **1970**, 162.
343. W. Voelter and C. Djerassi, *Chem. Ber.*, **101**, 1154 (1968).
344. I. Jardine and F. J. McQuillin, *Chem. Commun.*, **1969**, 503.
345. P. Abley and F. J. McQuillin, *Chem. Commun.*, **1969**, 477.
346. W. S. Knowles and M. J. Sebacky, *Chem. Commun.*, **1968**, 1445.
347. L. Horner, H. Siegel, and H. Buthe, *Angew. Chem. Int. Ed.*, **7**, 942 (1968).
348. R. R. Schrock and J. A. Osborn, *Chem. Commun.*, **1970**, 567.
349. J. Kwiatek, *Catal. Rev.*, **1**, 37 (1967).
350. J. Kwiatek, *Advan. Chem. Series*, **70**, 207 (1968).
351. M. G. Burnett, P. J. Connolly, and C. Kemball, *J. Chem. Soc., A*, **1968**, 991.
352. U.S. Patent 2,966,534; *Chem. Abstr.*, **55**, 8288 (1961).
353. T. Mizuta, H. Samajima, and T. Kwan, *Bull. Chem. Soc. Japan*, **41**, 727 (1968).
354. M. G. Burnett, *Chem. Commun.*, **1965**, 507.
355. W. M. Dennis, D. M. Rosenblatt, R. R. Richmond, G. A. Finseth, and G. T. Davis, *Tetrahedron Lett.*, **1968**, 1821.
356. H. Itani and J. C. Bailar, Jr., *J. Amer. Oil Chem. Soc.*, **44**, 147 (1967).
357. J. C. Bailar, Jr., and H. Itani, *J. Amer. Chem. Soc.*, **89**, 1592 (1967).
358. E. N. Frankel, E. A. Emken, H. Itani, and J. C. Bailar, Jr., *J. Org. Chem.*, **32**, 1447 (1967).

359. R. W. Adams, G. E. Bately, and J. C. Bailar, Jr., *J. Amer. Chem. Soc.*, **90,** 6051 (1968).
360. R. W. Adams, G. E. Bately, and J. C. Bailar, Jr., *Inorg. Nucl. Chem. Lett.*, **4,** 455 (1968).
361. H. A. Tayim and J. C. Bailar, Jr., *J. Amer. Chem. Soc.*, **89,** 4330 (1967).
362. S. J. Lapporte and W. R. Schuett, *J. Org. Chem.*, **28, 1963,** 1947.

Chapter **XII**

THE FISSION OF CARBON–CARBON BONDS

K. W. Bentley

I INTRODUCTION

Oxidative procedures have been widely used by chemists in the elucidation of the structures of complex natural products for many decades. One of the earliest examples of the use of such processes was Baeyer's oxidation of indigo to oxindole [1], and the reactions played important parts in the elucidation of the structures of brazilin [2–4], berberine [5, 6], strychnine [7–9], and the steroids [10]. In many cases deep-seated degradations occurred under the drastic conditions employed, but the simple fragments that emerged could generally be fitted together in such a way as to allow the proposal of rational complete structures. More detailed studies of oxidation processes and the development of more specific reagents has provided modern workers with a wide range of procedures specially suited for particular purposes. In particular the fission of carbon–carbon double bonds in natural products is a very important part of the processes of degradation, since the double bond provides a reactive site in the molecule and thus a point of relatively easy attack. The products of the fission may frequently be identified fairly readily, and reassembly of the fragments of the molecule often leads to a speedy solution of the structural problem including, naturally, the position of the double bond itself.

The processes of fission are of particularly wide application since double bonds may be introduced into naturally occurring substances by means of the Hofmann degradation procedure (see Chapter XIII), by elimination of water from alcoholic compounds, and by the elimination of hydrogen halides from substances containing halogens. Determination of the position of a double bond so introduced naturally affords valuable information regarding the position of the functional group in the original molecule.

More recently hydrolytic procedures for the degradation of compounds under milder conditions have become much more widely used, though the large fragments so obtained are frequently then subjected to oxidation in order to elucidate their structures further. Examples of such hydrolytic fissions will be found in Chapters XIV and XVI, and additional examples are given in Section 3 of this chapter.

2 OXIDATIVE METHODS

The methods available for the oxidation of organic compounds may be classified under the reagents concerned as each has particular applications and limitations. Cross references to the systems suffering oxidation are given where this is thought to be of value.

Potassium Permanganate

Reviews are available covering the use of potassium permanganate [11–13]. It has been widely used in organic chemistry because of its solubility in water, acetone, acetic acid, pyridine, and *t*-butanol, which it attacks only very slowly. It is an extremely powerful oxidizing agent which may be used either in the presence of an excess of nonoxidizable, usually sulfuric, acid, when it is reduced to manganous salts giving five equivalents of available oxygen per molecule,

$$MnO_4^- + 8H^+ + 5\varepsilon \rightarrow Mn^{2+} + 4H_2O$$
$$2KMnO_4 + 3H_2SO_4 \rightarrow K_2SO_4 + 2MnSO_4 + 3H_2O + 5O$$

or in weakly acid, neutral, or alkaline solution, when it is reduced to manganese dioxide, giving three equivalents of available oxygen,

$$MnO_4^- + 4H^+ + 3\varepsilon \rightarrow MnO_2 + 2H_2O$$
$$2KMnO_4 + H_2O \rightarrow 2KOH + 2MnO_2 + 3O.$$

In strongly alkaline solution permanganate ion may only be reduced to manganate, when only one equivalent of oxygen per molecule is available,

$$MnO_4^- + \varepsilon \rightarrow MnO_4^{2-}$$
$$2KMnO_4 + 2KOH \rightarrow 2K_2MnO_4 + H_2O + O.$$

In all cases the reactions involved in oxidation with permanganate involve changes in pH and, since the rate of oxidation of most substances is dependent on the pH, it is frequently desirable, especially in oxidations in neutral solution or in acetone, to minimize changes in pH by buffering the solution, bubbling carbon dioxide through it, or adding magnesium or zinc sulfate, with which hydroxyl ions form insoluble hydroxides. In many cases, however, either acid or alkali is present in such a concentration that changes in pH during oxidation are insignificant. In unbuffered neutral solution the oxidation of aldehydes to acids causes a drop in pH, whereas the oxidation of a secondary alcohol to a ketone results in a rise in pH,

$$3RCHO + 2MnO_4^- \rightarrow 3RCOO^- + H^+ + 2MnO_2 + H_2O$$
$$3R_2CHOH + 2MnO_4^- \rightarrow 3R_2CO + 2OH^- + 2MnO_2 + 2H_2O.$$

Potassium permanganate will effect the oxidation of certain alkanes, particularly at tertiary C–H bonds [13–16], for example, $(1) \rightarrow (2)$.

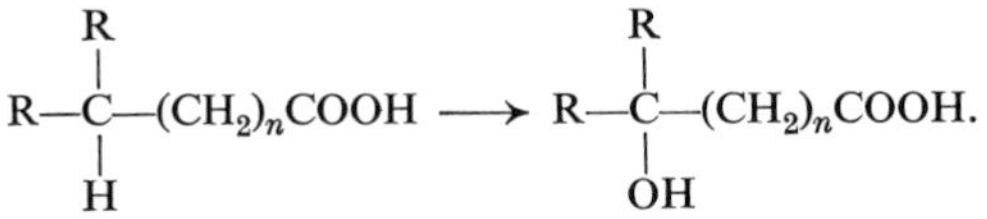

$$R\text{--}\underset{\underset{H}{|}}{\overset{\overset{R}{|}}{C}}\text{--}(CH_2)_n COOH \longrightarrow R\text{--}\underset{\underset{OH}{|}}{\overset{\overset{R}{|}}{C}}\text{--}(CH_2)_n COOH.$$

1   2

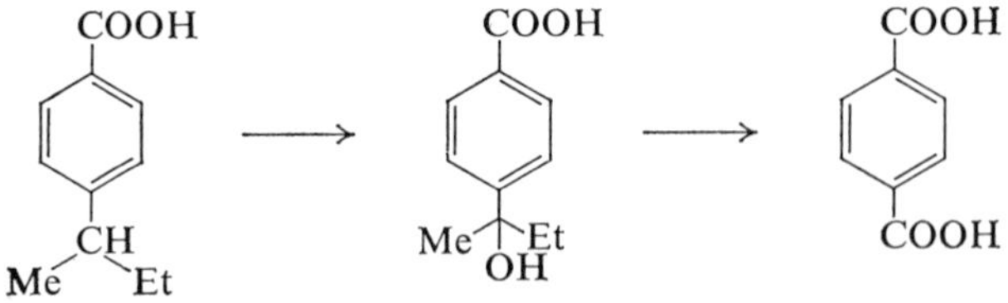

3   4   5

Benzylic tertiary –CH– is more readily oxidized and the acid (3) can be converted into the hydroxy acid (4) in 79% yield by permanganate. Allyl groups attached to aromatic nuclei are, however, generally oxidized to –COOH, (4 and 5), especially by excess of the reagent under vigorous conditions, and this process has frequently been used to indicate points of attachment of chains to aromatic rings.

Alkenes are very readily attacked by permanganate. The first stage in oxidation is the production of a 1,2-glycol (6 and 7), but further oxidation readily occurs and aldehydes (8), ketones (9), acids (11), and/or α-hydroxy ketones (10) may be produced if the substitution pattern of the original olefin (6) permits, and the pH conditions (neutral or slightly basic) are favorable. In

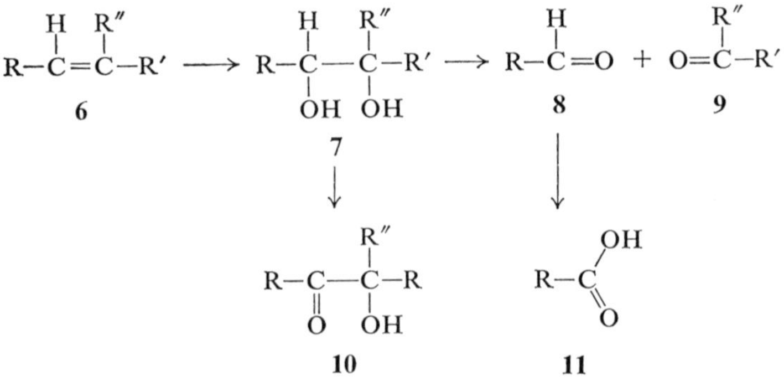

6   7   8   9   10   11

general, basic conditions are preferred for the preparation of 1,2-glycols in good yield.

The mechanism of the reaction has been studied and shown to involve a *cis*-cyclic manganese(V) ester, since experiments using ^{18}O labeled permanganate involve transfer of ^{18}O to the glycol. Where the concentration of OH⁻ is low, substantial amounts of α-hydroxy ketone can be formed, and this has been attributed to competition between attack of the initial cyclic ester by OH⁻ and MnO_4^- as shown in Formulas **12–18** [13, 17].

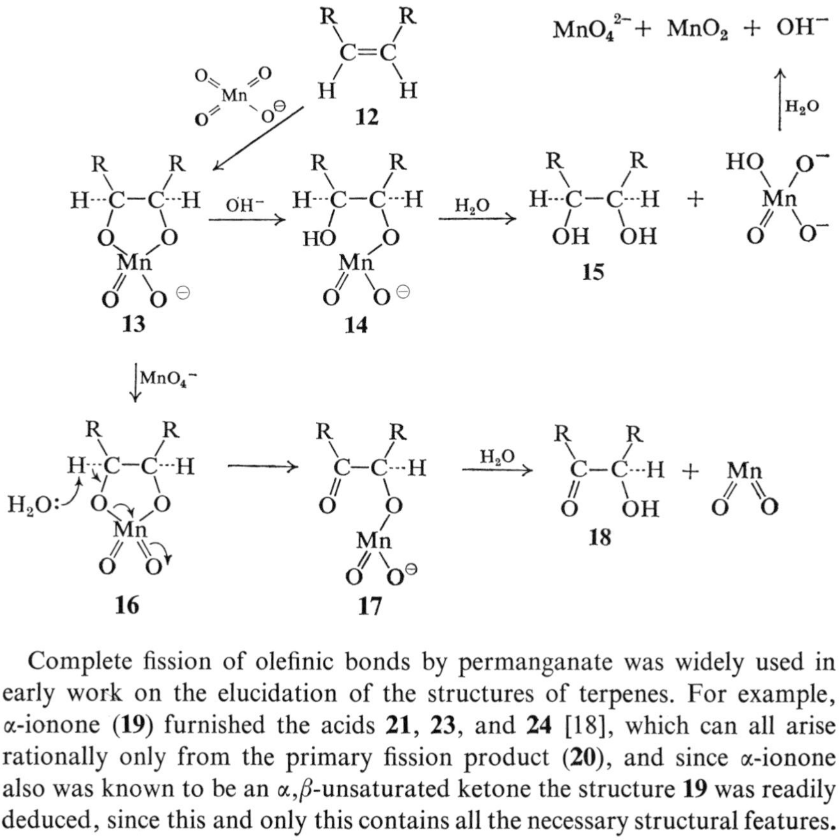

Complete fission of olefinic bonds by permanganate was widely used in early work on the elucidation of the structures of terpenes. For example, α-ionone (**19**) furnished the acids **21**, **23**, and **24** [18], which can all arise rationally only from the primary fission product (**20**), and since α-ionone also was known to be an α,β-unsaturated ketone the structure **19** was readily deduced, since this and only this contains all the necessary structural features.

In some cases the results of oxidation must be interpreted with caution. For example, camphene (**25**) with cold dilute permanganate affords the glycol (**26**) [19] which on further oxidation yields camphenylic acid (**30**) and camphenic acid (**31**) [29]. Camphenylic acid was identified as the α-hydroxy acid (**30**) by its oxidation to the ketone (**32**) by lead peroxide. The production of camphenic acid (**31**), however, caused great confusion in the early work on compounds related to camphor since its formation was not readily explicable

19 **20** **21** **22** **23** **24**

on the basis of the structure **25** for camphene. The acid doubtless arises by acyloin transformation of the hydroxy aldehyde (**27**) to the α-hydroxy ketone (**29**) and oxidation of this via the α-diketone (**28**).

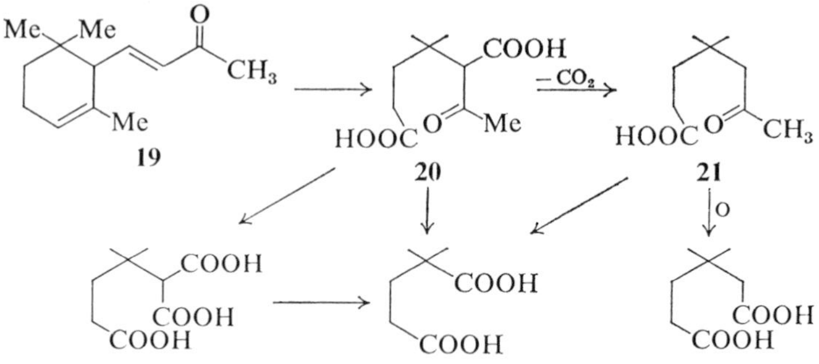

For other methods of fission of carbon–carbon double bonds, see pages 143, 150, 169, 177, and Section 3.

Aromatic nuclei bearing electron-donating groups, for example, phenols and aromatic amines, are very readily oxidized by permanganate. Unsubstituted nuclei are much more stable, and additional stability is conferred by electron-withdrawing substituents. The difference in behavior is illustrated by 1-amino (**33**) and 1-nitronaphthalene (**35**).

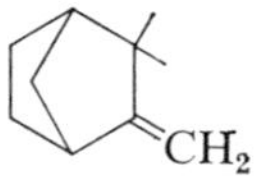

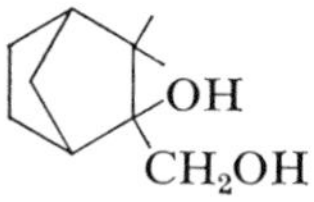

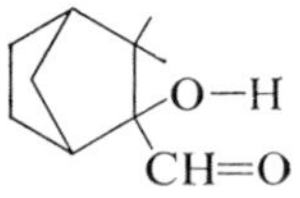

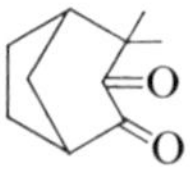

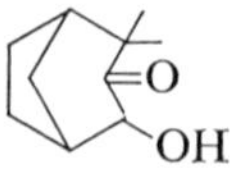

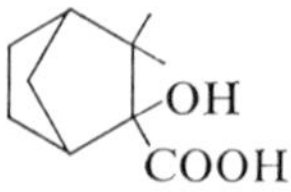

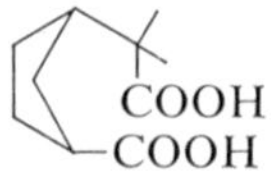

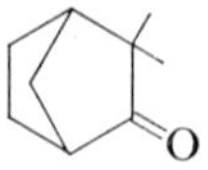

Quinoline (**36**) is oxidized to pyridine-2,3-dicarboxylic acid (**37**), but the heterocyclic ring in isoquinoline is more susceptible to oxidation, for example, the base **39**, derived from papaverine (**41**), is oxidized to the acid **40**,

25

26

27

28

29

30

31

32

33 **34**

35 → **38**

36 **37**

39 **40**

41

42 **43**

though papaverine itself is oxidized to the acid **40** and to the acid **43**, in which the carboxyl group stabilizes the heterocyclic ring, and further oxidation furnishes the tricarboxylic acid (**42**) [19–22]. As will be seen from the above, phenol ethers are much more stable than phenols towards oxidation. Phenyldihydrothebaine (**44**) is oxidized by permanganate to 4-methoxyphthalic acid (**45**) and benzoic acid (**46**) [23].

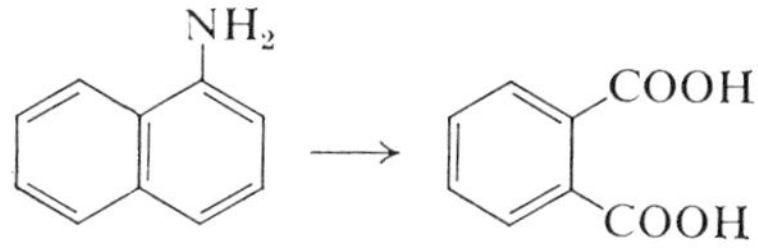

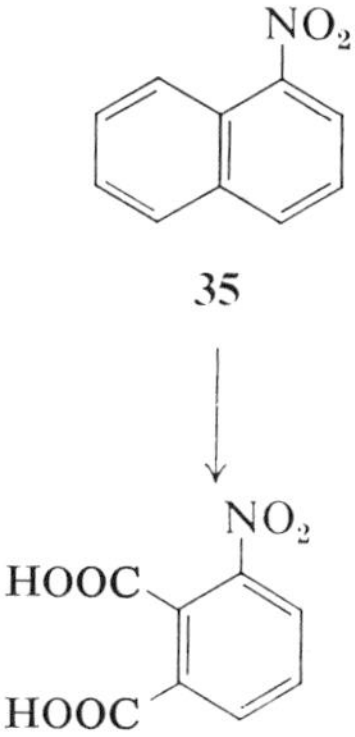

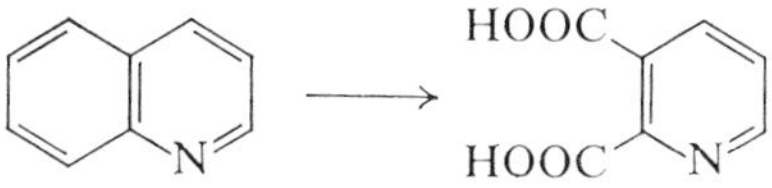

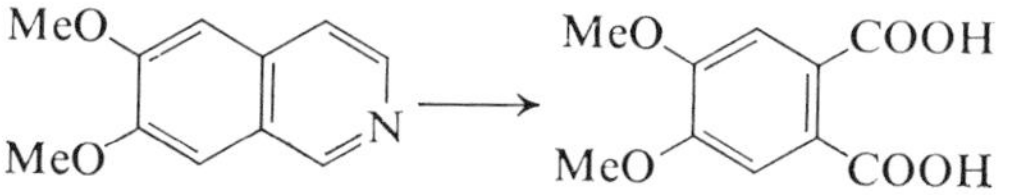

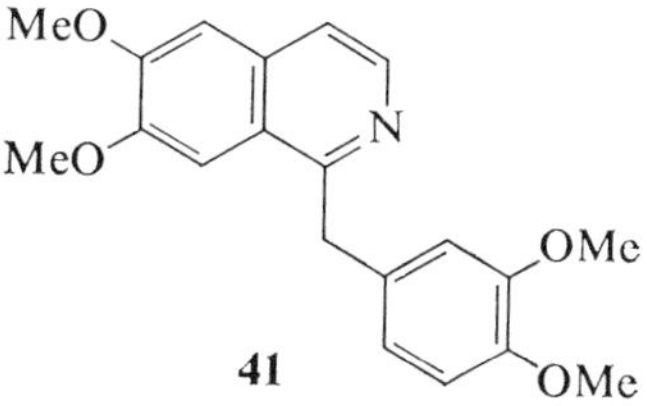

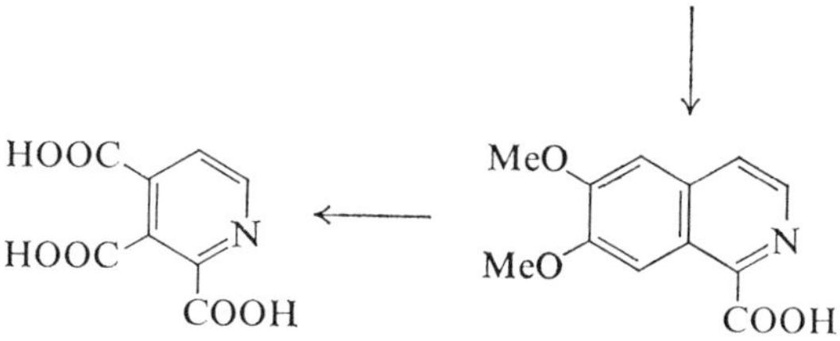

Chromic Acid and Related Compounds

By virtue of its powerful oxidizing properties and its solubility in water, acetic acid, acetic anhydride, and certain other organic solvents, chromic acid is probably the most widely used of all oxidizing agents in organic chemistry. Carbon–carbon single and double bonds may be broken directly, or the reagent may be used for oxidizing –CH$_2$– and –CHOH– to –CO– prior to further degradation.

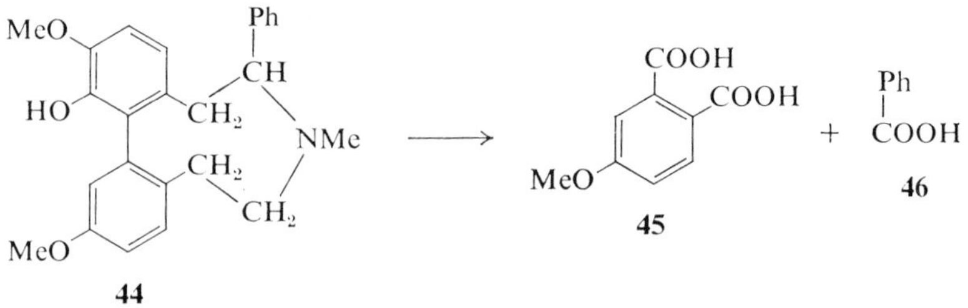

Hydrocarbon chains may be degraded to acetic acid, and this is the basis of the Kuhu-Roth determination of C-methyl groups [24], though chains linked to aromatic nuclei are usually oxidized to the aromatic carboxylic acid.

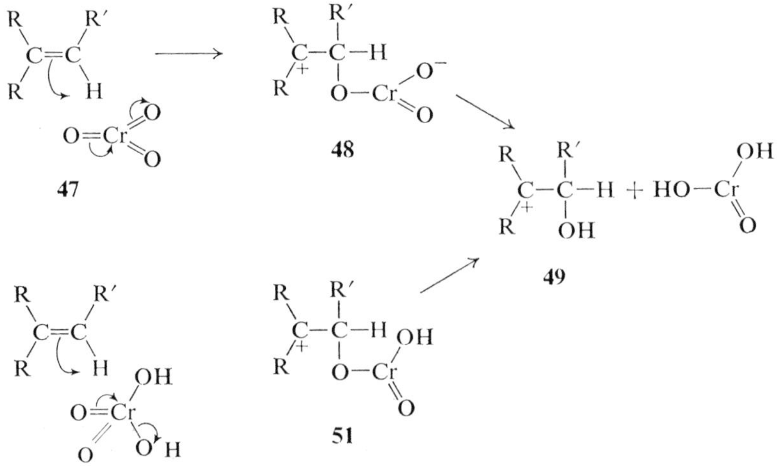

Chromic acid is widely used for the oxidation of carbon–carbon double bonds with results generally similar to those of permanganate oxidation, that is, the production of ketones and/or acids. The initial reaction is the donation of an electron pair from the olefinic double bond to chromium trioxide (47) or chromic acid (50), the product, 48 or 51, being rapidly hydrolyzed by water to the conjugate acid of an epoxide (49) [25]. The final product of the reaction is dependent upon the solvent in which oxidation takes place. Epoxides are generally obtained in acetic acid and acetic anhydride diluted with carbon disulfide. In the presence of water, hydration of the epoxide may occur to give the 1,2-glycols, which can be further oxidized by the chromic acid to ketols or by carbon–carbon bond fission to ketones and/or acids [26]. Ketols, when formed, may suffer dehydration to α,β unsaturated ketones. In aqueous sulfuric acid the reaction frequently involves rearrangement of the pinacol–pinacolone type, especially when the concentration of acid is high, for example, 2,3-dimethylbut-2-ene (52) is oxidized mainly to acetone (55) in dilute sulfuric acid, but mainly to pinacolone (53) if the acid concentration exceeds 50% [27].

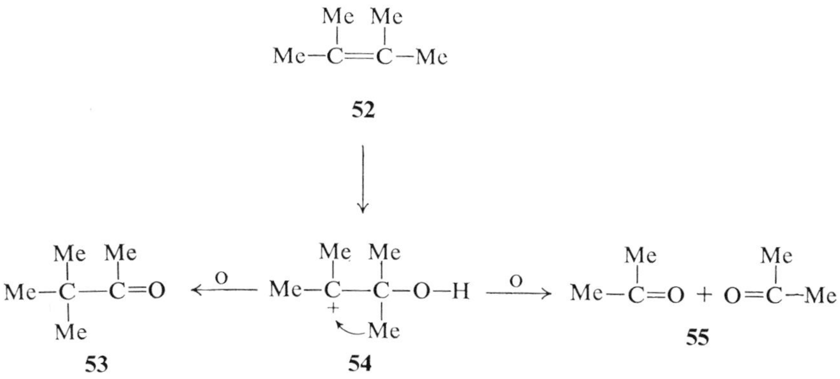

The oxidation of camphene (62) by chromic acid is illustrative of the reactions that can occur. Oxidation with chromic acid and acetic anhydride diluted with carbon disulfide yields camphene oxide (61) [28], whereas in acetic anhydride/carbon tetrachloride the products are principally the ketone camphenilone (58) and camphenilanaldehyde (56) [29], and all of these can arise from the carbonium ions (59) and (60), as shown. The isolation of epoxides in such oxidations is dependent upon a number of factors, principally their stability to ring opening.

The oxidation of camphene (62) to camphor (65), which occurs in approximately 80% yield in aqueous acids [30], cannot be explained in terms of initial attack of the double bond by chromic acid, since the methylene group of the former becomes the bridgehead methyl group of the latter, and in this case the initial reaction is the acid-catalyzed hydration and rearrangement of camphene, through the carbonium ions 63 and 64≡67, to isoborneol (66), which suffers normal oxidation of the secondary alcoholic group. In this connection it is of interest to note that the oxidation of 1-methyl-α-fenchene (72) gives mainly camphor (65), with only a small amount (6%) of fenchone (73). The last-named ketone presumably arises through the intermediates 69 and 70 [cf. oxidation of camphene (62) to camphor (65)] and this led to a suggestion that the carbonium ion 68, which would be expected to suffer rearrangement with ease, is not an intermediate in the oxidation to camphor, but rather that the intermediate is the cyclic compound 71 [31].

With chromyl acetate the oxidation of double bonds takes a similar course [26], and chromyl chloride affords good yields of chlorohydrins rather than epoxides in many cases, and cyclohexene (74) has been shown to give two chlorohydrins (76), cyclohexanone (77) and cyclohexenone (78), all of which could readily arise from the carbonium ion 75 [32, 33].

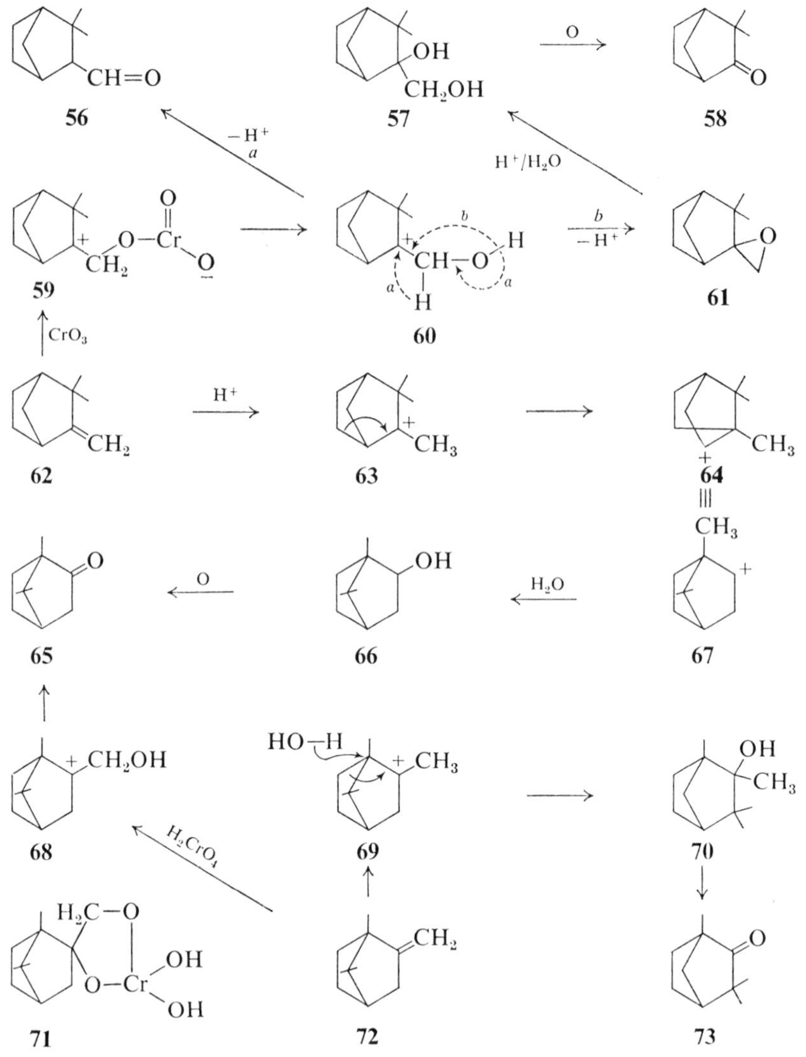

For other methods of fission of carbon–carbon double bonds see pages 139, 150, 169, 177, and Section 3.

The chromic acid oxidation of aromatic nuclei generally proceeds with nuclear oxidation rather than nuclear fission, for example, naphthalene gives 1,4-naphthaquinone and some phthalic acid [34, 35], 1-methylnaphthalene gives 5-methyl-1,4-naphthaquinone [36], anthracene and phenanthrene give anthraquinone [37] and phenanthraquinone [38], and the last-named quinone also results from the oxidation of 9- or 10-substituted phenanthrenes [39]. 3,4,6-Trimethoxyphenanthrene surprisingly gives 3,6-dimethoxy-phenanthra-1,4-quinone [23], whereas the corresponding 4-acetoxy-3,6-dimethoxy compound gives the 9,10-quinone [40].

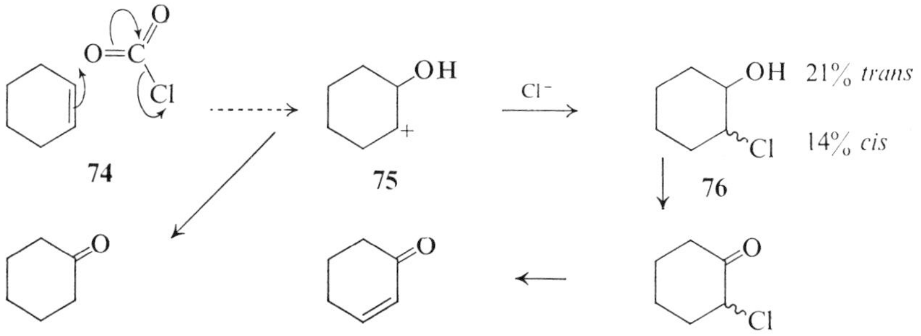

The oxidation of ketones with chromic acid involves fission of a carbon–carbon bond and the production of two carboxyl (**80**) groups or a carboxyl and carbonyl group (**81**). Considerable information of structural value can be obtained from a study of the nature of the products. Cyclic ketones degraded in this way give dicarboxylic acids containing the same number of carbon

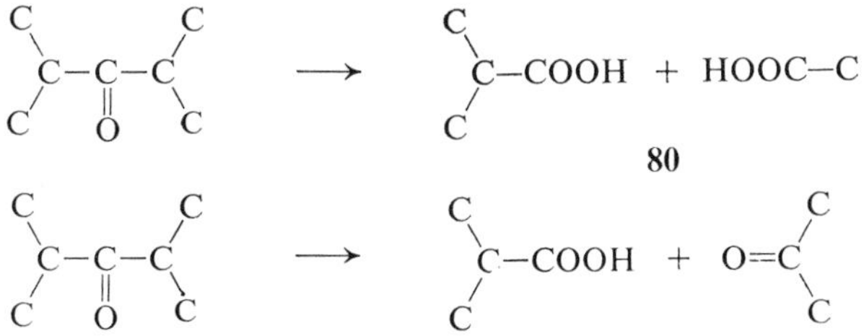

atoms, and providing these are derived from rings of six or more members (**82**), the acids (**83**) can be cyclized to ketones (**84**) containing one carbon atom less than the parent, which in turn can be oxidized to dicarboxylic acids (**85**). This sequence of reactions can be repeated until a derivative of cyclopentanone (**86**) is reached, when oxidation gives an acid (**87**) that forms an anhydride (**88**) rather than a ketone on heating. In this way the size of ring systems can frequently be determined.

The "rule" that permits the prediction of the size of a cyclic system from the behavior on heating of the dicarboxylic acid obtained on oxidation is known as the "Blanc rule." This rule states that the formation from the dibasic acid of a cyclic ketone indicates that the parent ring system was six-membered or larger, and the formation of an anhydride indicates that the parent ring system was five-membered or smaller. For failure of this rule, see page 200.

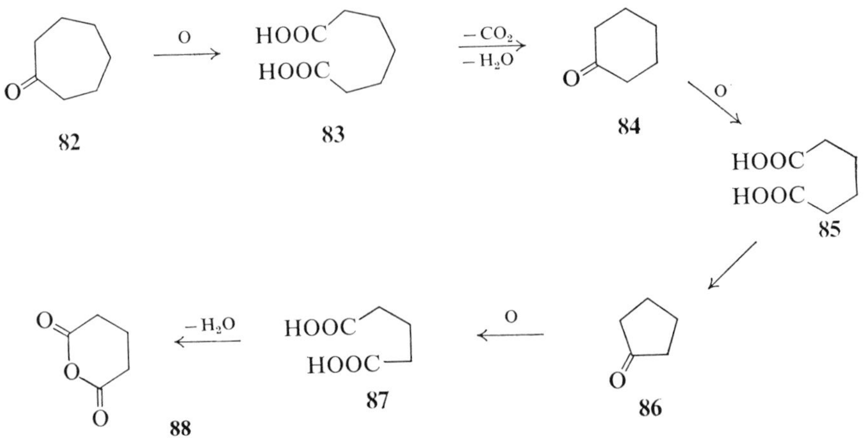

Carbonyl groups which facilitate further oxidation of this type may be generated in molecules by chromic acid oxidation of either –CHOH– or –CH$_2$– groups. Oxidation of alcohols can be achieved with chromic acid in aqueous acetic acid or water though other oxidizable groups, particularly olefinic bonds, may be affected simultaneously or preferentially in these solvents [26], and cleavage of phenyl alkyl carbinols may occur [41, 42]. The reaction may be carried out in acetone/dilute sulfuric acid [43] when, in the oxidation of secondary alcohols, the resulting ketone is effectively protected from further oxidation by the large excess of acetone. Further protection of the product from oxidation may be achieved by adjusting the concentrations of reactants so that a phase separation occurs, with the organic compound in an upper acetone layer and the oxidant in a lower aqueous layer [26].

The reagents of choice for the oxidation of alcohols when other sensitive groups are present are the chromium trioxide–pyridine complex [26, 44, 45] and *t*-butyl chromate [26, 46–49], though the latter causes fission of 1,2-diols [50, 51].

The oxidation of –CH$_2$– to –CO– by chromic acid is readily effected when the methylene group is activated by a double bond, though double bond oxidation is generally a competing reaction. The reaction is in general of minor importance in open-chain compounds, but has been widely used in the steroid and terpene fields. In general where two allylic methylene groups are present, competing oxidation occurs. For example 1-methylcyclohexene (**90**) gives both possible ketones **89** and **91** [52], and the olefin **93** gives the two ketones **92** and **94** [53].

Di-*t*-butyl chromate has been used as an oxidizing agent for allylic methylene groups, reaction being effected in a nonpolar solvent in the presence of acetic anhydride [46], though the nature of the active component in such a mixture is not clear [26]. As with chromic acid, competing oxidation

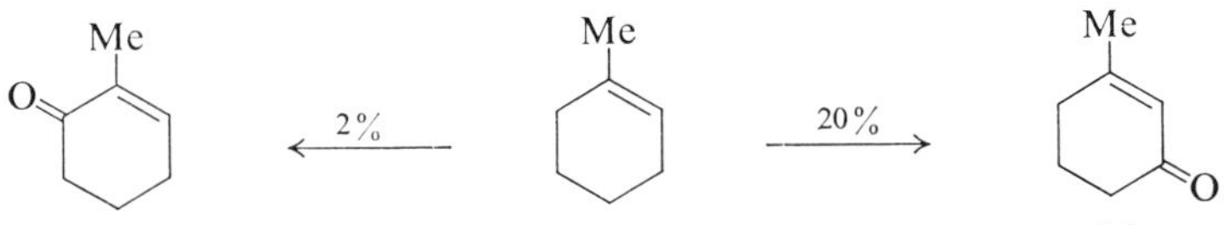

occurs at alternative methylene groups, but with this reagent double bonds are rarely attacked.

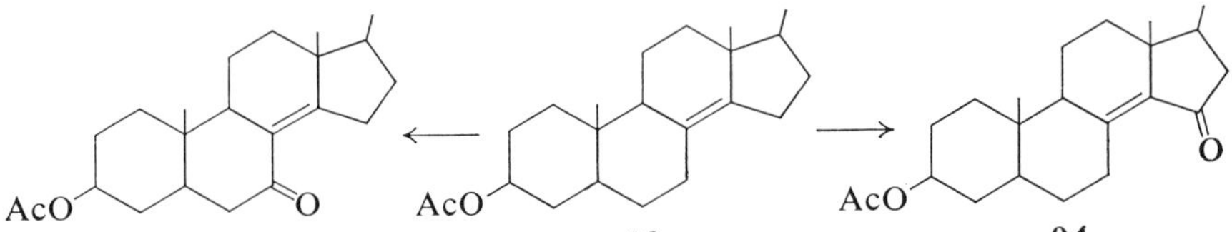

Saturated hydrocarbons may be oxidized by chromic acid under vigorous conditions. In the absence of activating groups, such as carbonyl or aromatic nuclei, tertiary C–H groups are generally attacked first. The sequence of reactions involves rupture of the C–H bond as the rate determining step, with initial formation of the alcohol C–OH, which can be further converted by dehydration into the olefin, which is readily oxidized [54]. The initial reaction has been represented as a concerted electrophilic substitution **95 → 97** [11].

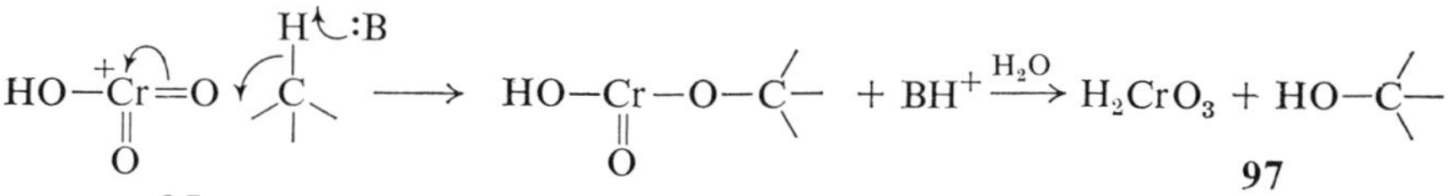

Oxidation of –CH_2– in this way frequently gives high yields of ketones, and further oxidation may be inhibited by carrying out the reaction in acetic anhydride, when diacetyl derivatives of the ketones are produced.

An improved procedure for the Kuhn-Roth oxidation using *t*-butyl chromate in pyridine in place of chromic acid [55], has been developed recently.

Ozone

The reaction of ozone with double bonds, followed by the decomposition of the resulting products, is one of the most reliable procedures for oxidative fission of unsaturated molecules and for determining the precise location of the unsaturation. Oxidation with potassium permanganate, though most effective, frequently gives misleading results, particularly in the case of unsaturated acids, as a result of the migration of the double bond, before fission, under the frequently alkaline conditions. Oxidation of the unsaturated compounds with ozone, however, almost invariably gives reliable results; seemingly anomalous products sometimes appear, but the processes whereby they arise are generally well understood and their production generally helps, rather than complicates, the solution of the structural puzzle.

The reaction of ozone with double bonds has been extensively studied by Harries [56], by Staudinger [57], and more recently by Criegee [58–64]; the work of these and others has been effectively reviewed by Long [65] and by Bailey [66].

Mechanism

The process is by no means a simple one, and the main steps in the mechanism have been elucidated by Criegee. The ozone reacts with the olefin **98** to give an initial product formulated as **99**, although the structure is uncertain; this primary product is unstable and is rapidly transformed through **100** into the zwitterion **102** and an aldehyde or ketone (**103**). Clearly, with an unsymmetrical olefin this fission may occur in either of two ways (see below).

The zwitterion **102** is subsequently stabilized (*1*) by reaction with a hydroxylic solvent, HOX, if such is present, to give the alkoxy or acyloxy hydroperoxide **105**; (*2*) by reaction with the aldehyde or ketone **103** to give the ozonide **101** and/or polymeric ozonides; (*3*) by polymerization to the peroxide **104**; or (*4*) by rearrangement. If the solvent is one that can react with the zwitterion (e.g., water, alcohols, organic acids), then formation of the hydroperoxide **105** predominates over the alternative processes since the solvent is present in large excess. If, however, the solvent is inert (paraffins, carbon tetrachloride, chloroform, methylene chloride, ethyl acetate, acetone, formamide, ether, tetrahydrofuran, nitromethane, acetic anhydride), the principal transformation product of the zwitterion is the monomeric ozonide **101**, when compound **103** is an aldehyde, or is a ketone in which the carbonyl group is activated in some way, for example, by an α-COOEt group. When compound **103** is a ketone *without* an activated carbonyl group, dimerization of the zwitterion **102** to the peroxide **104** generally occurs, as the ketonic carbonyl group is less susceptible to nucleophilic attack than is the carbonyl group of an aldehyde. This dimerization of the zwitterion may

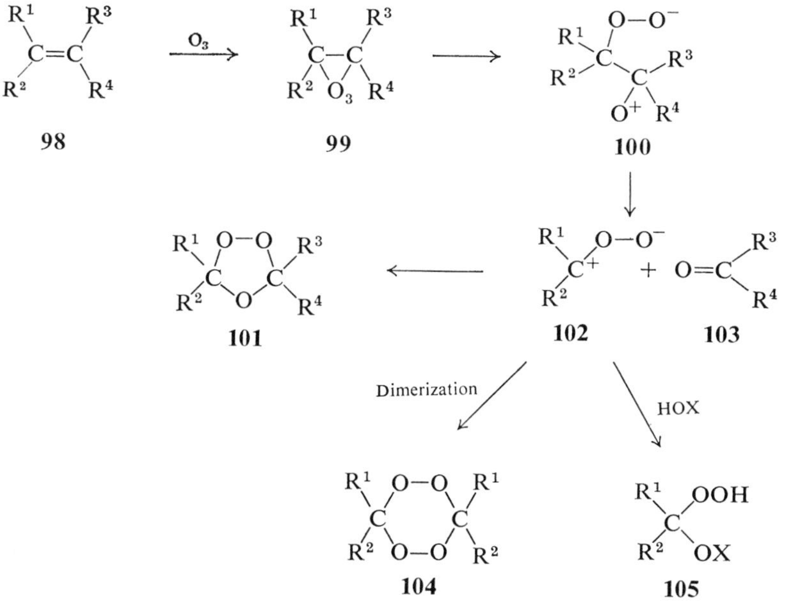

98   99   100

101   102   103

Dimerization   HOX

104   105

occur to a significant extent even when compound **103** is an aldehyde [67, 68] as is shown by the isolation of considerable amounts of benzaldehyde, formaldehyde, and acetaldehyde directly, before reduction or hydrolysis of the products, during the ozonolysis of stilbene, styrene, anethole, and isoeugenol methyl ether [69, 70, 71].

Steric hindrance can be a major factor in preventing interaction of the zwitterion (**102**) and an aldehyde, and is doubtless operative, whichever way fission of the intermediate of type **100** occurs, during the ozonolysis of stigmastadien-3-one (**106**), which gives 1 mole of aldehyde and 1 mole of peroxide per mole of olefin directly [72].

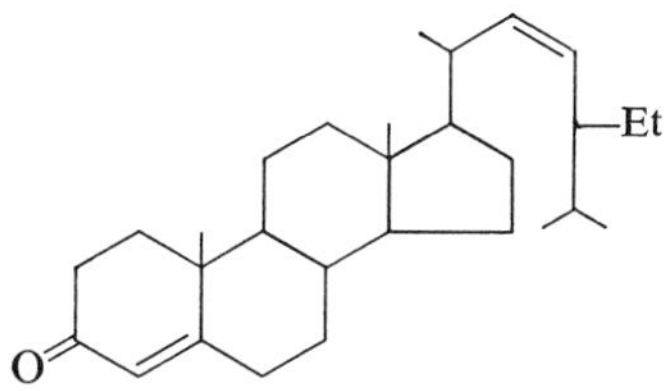

106

Ozonolysis of substances containing double bonds sterically hindered on one side frequently affords epoxides or their transformation products. For example, fenchene (**107**) gives a certain amount of the acid **111**, presumably via **108** and **109**, as well as the expected ketone **110** [73]. Similarly, the ozonolysis of β-isoquinine (**112**) may proceed through the epoxide **113**, since the product has been assigned structure **114** [74], although this has

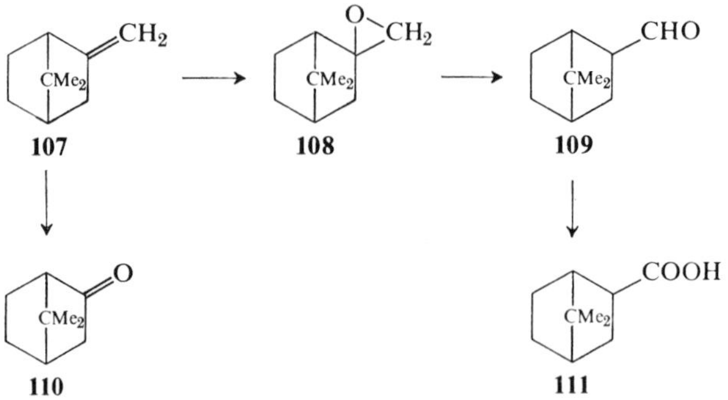

never been rigorously established and the possibility that it is in fact the epoxide **113** cannot be ruled out.

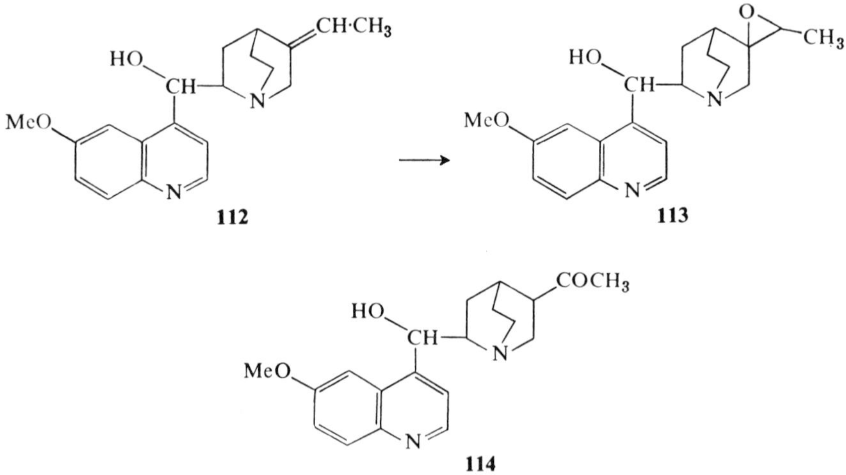

Many other examples of epoxidation have been observed during ozonolysis and the problem has been recently reexamined by Bailey and Lane [75] who have concluded that there is a competition between normal ozonide formation and partial cleavage, the competition being strongly dependent upon the bulk of the substituents at the double bond. These workers believe that as the bulk of the substituents increases, the ease of 1,3-dipolar addition of ozone to the double bond with the formation of the primary ozonide (**118**) decreases and the reaction proceeds with the preferential formation of π- or σ-complexes. The complex **116** could rearrange to the carbonium ion (**117** $\equiv$ **121**) which could lose oxygen to give the epoxide **120**, the aldehyde **119**, or the vinyl alcohol **122**; the epoxide could also rearrange **120** or **123** to give the same final products [75].

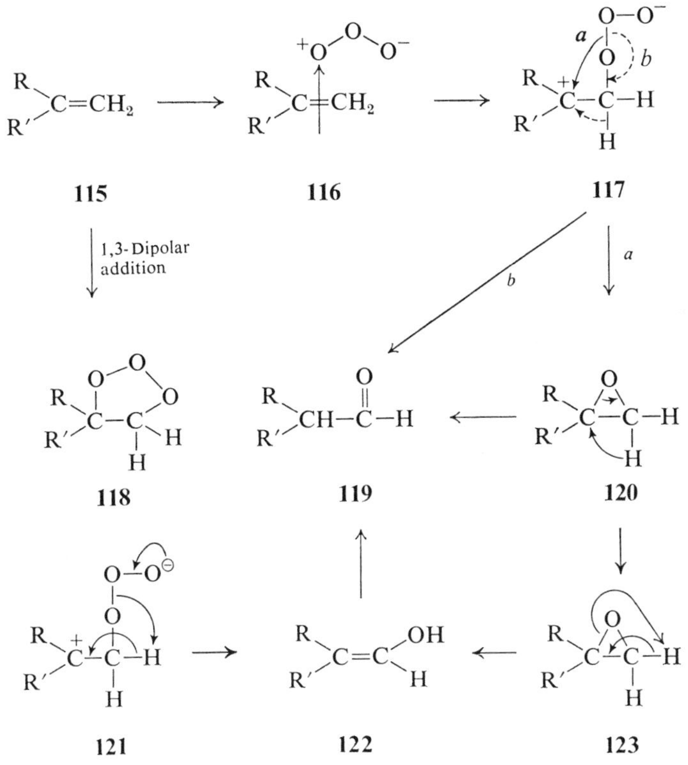

Examination of the Criegee mechanism as set out in formulas **98–105** shows that the initial ozonide (**99** $\equiv$ **124**) could split in two different ways to give zwitterions **126** and **130** and carbonyl compounds **127** and **129**. Interaction of zwitterions and carbonyl compounds could then give "cross ozonides" **131** and **132** as well as the ozonide **133**. Symmetrical cross ozonides can exist in *cis* (**132**) or *trans* (**131**) forms. Such cross ozonide have have been obtained and it has recently been shown [76, 77] that the *cis:trans* ratio in the product obtained in certain cases depends on the stereochemistry of the original olefin, which is incompatible with the mechanism set out above. It has accordingly been suggested that the initial reaction is the formation of a π-complex of ozone with the olefin, and that this is transformed through the primary ozonide (molozonide) in most *trans-* and unhindered *cis*-olefins into the primary zwitterion **125**, some of which is cleaved to the Criegee ion **126** and carbonyl component **127**. The carbonyl

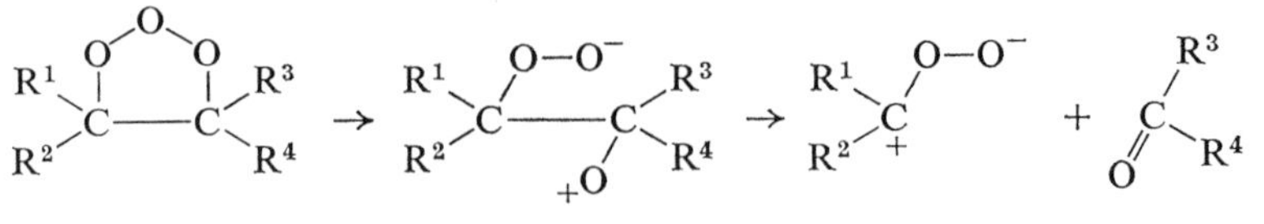

$$\underset{124}{\overset{R^1\ R^3}{\underset{R^2\ R^4}{C—C}}} \rightarrow \underset{125}{\overset{R^1\ R^3}{\underset{R^2\ R^4}{C—C}}} \rightarrow \underset{126}{\overset{R^1}{\underset{R^2}{C}}} + \underset{127}{\overset{R^3}{\underset{R^4}{C}}}$$

124   125   126   127

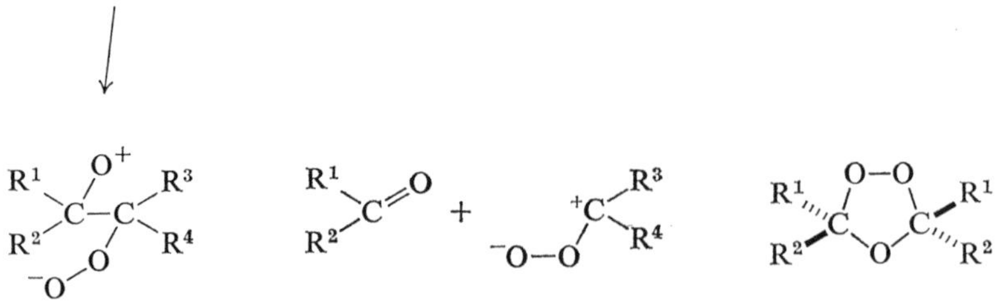

128   129   130   131

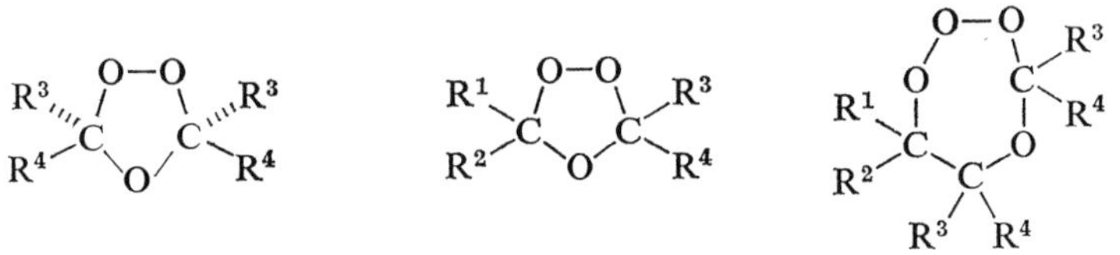

132   133   134

component and primary zwitterion can then react to give an intermediate of type **134**.

It is possible to explain the experimental findings by applying these ideas to the ozonolysis of *cis-* and *trans*-2-methylpen-3-ene. With the *trans*-olefin **135**, combination of the primary zwitterion **139** with the aldehyde **140**, derived by fission of the zwitterion, gives the compounds **141** and **143**, which decompose **142** to acetaldehyde **145** and the *cis* (**146**) and *trans* (**144**) cross-ozonides, in almost equal amounts, since the intermediates **141** and **143** are of comparable stability. With the *cis*-olefin **147**, however, the primary zwitterion **148** forms the intermediate **149** preferentially, since in this form alone there are only H:H nonbonded repulsions, and decomposition of this gives the *cis*-cross-ozonide **146**, which therefore predominates in the product [76, 77].

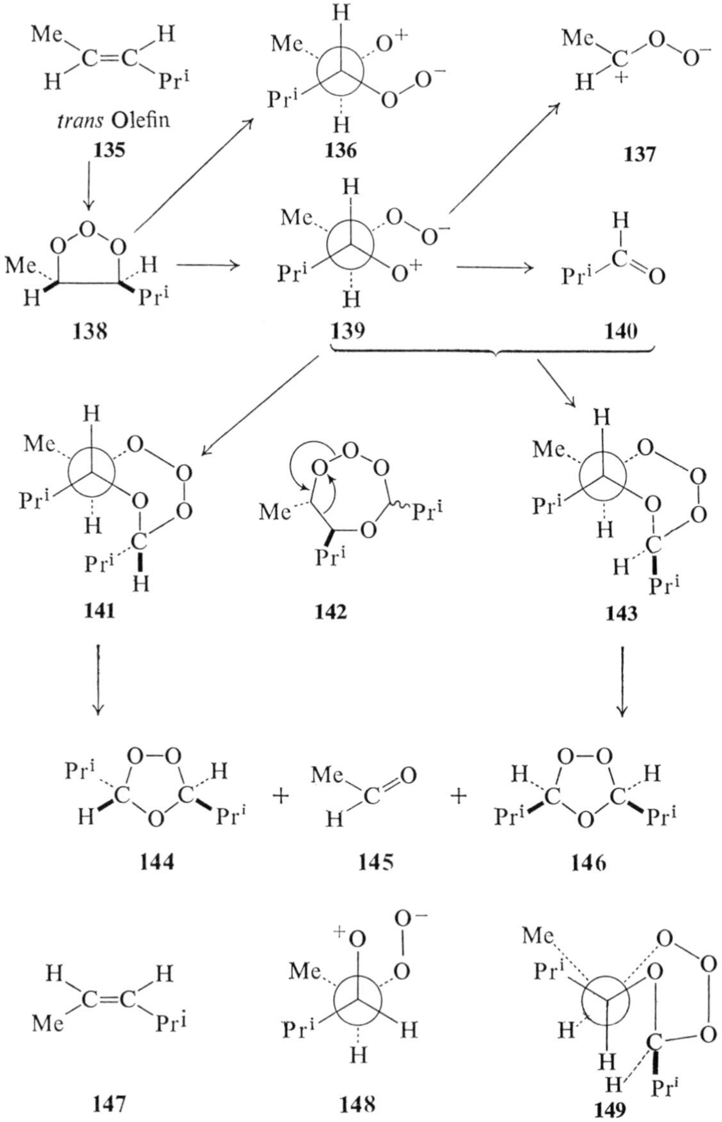

In recent years positive proof of the existence of primary ozonides of general structure **124** has been obtained by chemical and spectroscopic means [78, 79, 81]. Such primary ozonides are stable only below about −90° and are in general formed only from relatively unhindered olefins.

As can be seen from the foregoing, the mixtures obtained by the interaction of ozone with olefinic compounds are complex, and the viscous and explosive nature of the initial products may be attributed to the peroxide content. The ozonides of structure **101** are generally well-defined substances; those derived from pent-1-ene, hex-1-ene, hex-2-ene, oct-2-ene, oct-4-ene, 3,3-dimethylbut-1-ene, styrene, and stilbene have sharp melting and boiling points [63], and the ozonide **151** derived from 1,2-dimethylcyclopent-1-ene (**150**) has been synthesized from the diketone **152** through the peroxide **153** [82].

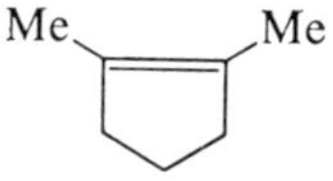

Compounds such as **150**, which have a fully substituted double bond in a five-membered ring, or in certain types of four-membered rings, are exceptions to the general statement that ozonides are not formed when the carbonyl compound **103** is an unactivated ketone. Steric factors are doubtless operative in these cases, as formation of the ozonide (**151**), which contains rings of five, six, and seven members, would be more easily accomplished than would formation of the corresponding ozonide during the oxidation of a derivative of cyclohexene. The latter process would require the construction of rings of five, seven, and eight members, and in such cases large polymeric ozonides or peroxides are formed instead [62, 63, 82].

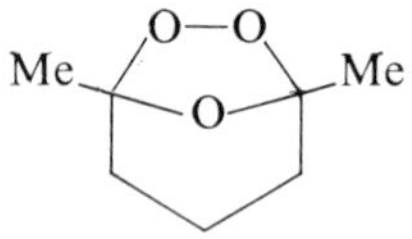

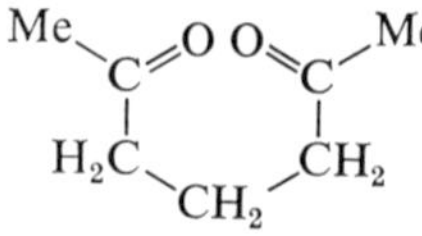

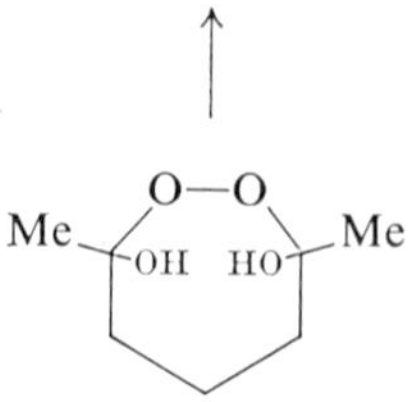

The structures of polymeric ozonides have not been extensively studied. Only in one case, the polymeric ozonide of phenanthrene, has a structure been assigned to a compound of this class with reasonable confidence (see below).

Reaction of the zwitterion **102** with a hydroxylic solvent affords a hydroperoxide of type **105**. Many of these substances are stable, and that obtained from tetramethylethylene **154** in the presence of isopropanol (i.e., **156**) has also been synthesized by the autoxidation of diisopropyl ether (**157**) [64, 83]. In the ozonolysis of cyclic olefins, when the other end of the chain obtained by fission is aldehydic, the hydroperoxide appears as a cyclic hydroxyperoxide where sterically possible; if steric factors prevent cyclization, the product is a polymeric hydroxyperoxide. This is in keeping with the known ability of hydroperoxides to give peracetals and hemiperacetals with aldehydes and ketones.

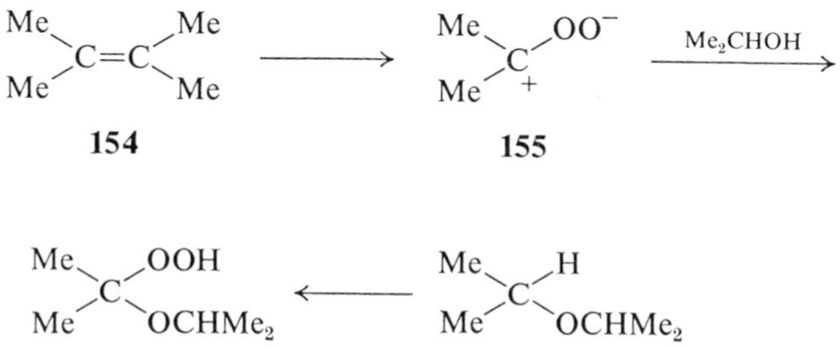

Experimental Procedures

The most common procedure for the ozonolysis of olefinic substances is the passage of oxygen containing 2–15% of ozone through a solution of the unsaturated compound in a suitable solvent. The ozonized oxygen is most often produced by the passage of a so-called "silent" electric discharge through dry oxygen, and a number of commercial ozonizers are available; a workable apparatus may be constructed in the laboratory quite simply [84–90].

Ozone may also be obtained in concentrations as high as 58% by electrolysis of sulfuric or perchloric acid solutions [91], and liquid and solid ozone have been prepared for use in the ozonolysis of substances that react readily with oxygen [66].

The ozonized oxygen is usually delivered into a cooled solution of the olefin by a descending tube terminating in a porous disk. The effluent gases may be led through turpentine or potassium iodide solution to absorb any unreacted ozone. If the unsaturated compound reacts readily with ozone, the termination of reaction is indicated by the appearance of ozone in the effluent gases, which will then liberate iodine from potassium iodide. If reaction with ozone is not rapid, or if overozonization of compounds containing aromatic nuclei is to be avoided, the reaction may be terminated when a test portion of the solution no longer immediately decolorizes a solution of bromine in chloroform [92]. Alternatively, the reaction may be followed by determining the concentration of ozone in the gas stream and effluent gases and measuring the flow rate, from which the rate of absorption of ozone may be calculated; the reaction may then be stopped when the theoretical amount of ozone has been absorbed [93].

Generally the best yields of desired materials are obtained by carrying out the reaction at low temperatures and in dilute solution.

The solvent used may be one that is inert to ozone in the presence of a reactive olefin (alkanes, carbon tetrachloride, chloroform, methylene chloride, ethyl acetate, acetone, acetic anhydride, formamide, dimethyl-formamide, tetrahydrofuran, nitromethane) or one that reacts readily with the zwitterion intermediate (water, alcohols, organic acids). Although methanol reacts readily with ozone when pure, it does not do so appreciably below $-20°$ if a reactive unsaturated compound is present. Hydroxylic solvents that react with the zwitterion intermediate are reported often to give better yields of normal nonperoxidic products after decomposition of the intermediate ozonolysis product than do the more frequently used inert solvents [68], a point not usually emphasized in general accounts of the reaction.

Decomposition of Peroxide Ozonolysis Products

The value of ozonolysis in the study of natural products depends on the identification of the products of the decomposition of the mixtures of peroxidic compounds resulting from the interaction of olefinic substances with ozone. Four different methods of decomposition—reduction, oxidation, hydrolysis, and thermal decomposition—have been employed, and the value of the first three will be separately discussed; thermal decomposition is dangerous and not recommended.

REDUCTION

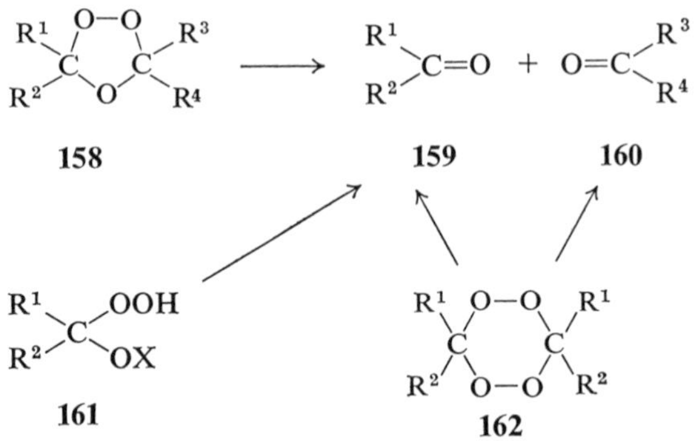

All the peroxidic products may be reduced as in formulas **158–162** to mixtures of aldehydes and/or ketones from the structures of which the position of the original unsaturation may be deduced and, in some cases, the entire structure of the parent olefinic compound inferred. It is clear, then, that reliable methods of reduction of ozonolysis products are of primary importance. Those that have been used include reduction with zinc and acetic acid [94], sodium bisulfite [95], sulfurous acid [96], sodium and liquid ammonia, sodium iodide [68, 82, 97], stannous chloride [72], and catalytic reduction.

Catalytic reduction at low temperatures is generally accepted as the best reductive process [93, 98–101], although it has been reported that the reduction often only goes 70% to completion and may be accompanied by the production of acids and rearranged materials, presumably from the ozonide (163 → 164 + 165), but the production of these materials may be minimized by reduction at still lower temperatures or by reduction with sodium and liquid ammonia or sodium iodide, also at very low temperatures.

$$\text{RCH} \underset{O}{\overset{O-O}{\diagup \diagdown}} \text{CHR} \quad \longrightarrow \quad \text{RCOOH} + \text{OHC}-\text{R}'$$

$$\textbf{163} \qquad\qquad\qquad \textbf{164} \qquad \textbf{165}$$

Treatment of the mixture resulting from ozonolysis with the reducing agent as soon as possible is essential if good yields of aldehydic and ketonic materials are to be obtained. Even sensitive dialdehydes such as glutaraldehyde, adipaldehyde, and pimelaldehyde may be obtained in yields of 50–75% by rapid working.

Highly polymerized ozonides are not generally reduced catalytically at room temperatures, and as higher temperatures favor side reactions, it is best to avoid production of such polymers during the ozonolysis, if possible by using a solvent (e.g., methanol) that readily converts the zwitterion intermediate into a hydroperoxide.

For convenience, hydrogenation of the peroxidic materials is carried out in the same solvent as the ozonolysis, wherever possible. Clearly, if the ozonolysis solvent is halogenated, it must be removed before reduction is attempted.

Reduction of the peroxidic materials has also been accomplished with lithium aluminum hydride and sodium borohydride [100, 102–107] and by Grignard reagents [79, 80, 100]. These processes naturally afford alcohols instead of the aldehydes and ketones obtained by the other methods of reduction referred to above.

OXIDATION

The oxidation of peroxidic ozonolysis products, as shown in formulas 166 → 167 and 168, affords acids. The process has been little used in the study of natural products, although it has recently become of considerable importance in the commercial synthesis of carboxylic acids (Ref. 7 and references therein).

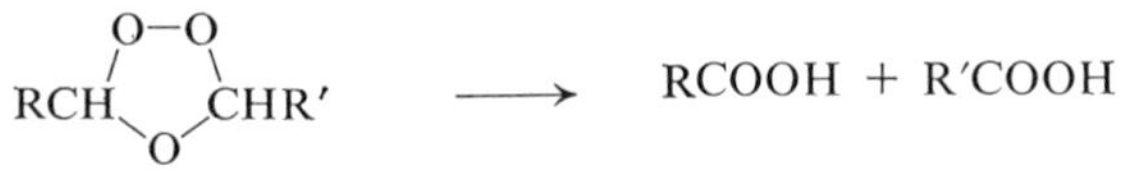

$$RCH\underset{O}{\overset{O-O}{\diagup\diagdown}}CHR' \longrightarrow RCOOH + R'COOH$$

166   167   168

HYDROLYSIS

Hydrolysis of ozonides and hydroperoxides will also yield aldehydes and ketones, as shown in formulas 169–173. However, this process is generally unsatisfactory because the hydrogen peroxide that is produced may effect oxidation of sensitive materials, particularly aldehydes, during the hydrolysis with the result that yields of desired products may be low.

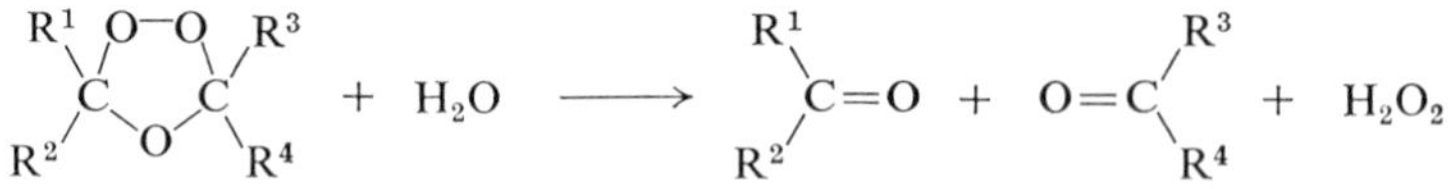

$$\underset{R^2}{\overset{R^1}{\diagdown}}\underset{O}{\overset{O-O}{C}}\underset{R^4}{\overset{R^3}{\diagup}} + H_2O \longrightarrow \underset{R^2}{\overset{R^1}{\diagdown}}C=O + O=C\underset{R^4}{\overset{R^3}{\diagup}} + H_2O_2$$

169   170   171

$$\underset{R^2}{\overset{R^1}{\diagdown}}\underset{OX}{\overset{OOH}{C}} + H_2O \longrightarrow \underset{R^2}{\overset{R^1}{\diagdown}}C=O + H_2O_2 + HOX$$

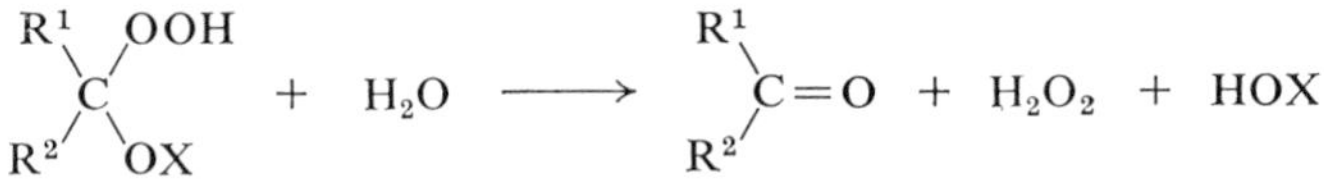

172   173

Apparently Anomalous Products of Ozonolysis

A number of reports are to be found in the literature of the production of substances during the ozonolysis of natural products that do not appear to arise by the normal process of formation of peroxidic substances followed by reduction or hydrolysis as outlined above. The processes by which these arise are, however, generally well understood. The formation of epoxides and their transformation products has been discussed above (p. 150) with the mechanism of ozonolysis.

The production of compounds 178 and 179 from santonide (174) when the primary ozonization product is heated with methanolic hydrogen chloride presumably proceeds through the epoxide 175, hydrolysis of which would yield the keto acid (176 ↔ 177), and esterification of this would lead to the normal ester 179 and the pseudoester 178 [107].

Abnormal ozonolysis products may be obtained in varying amounts if the olefin is of type 180 (X = N, O, or S) or type 181 with N, O, or S as a substituent in the aryl group. For example, in addition to formic acid, the

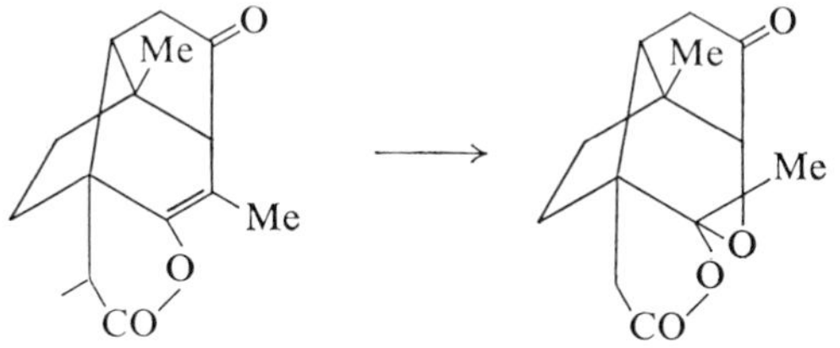

174 → 175

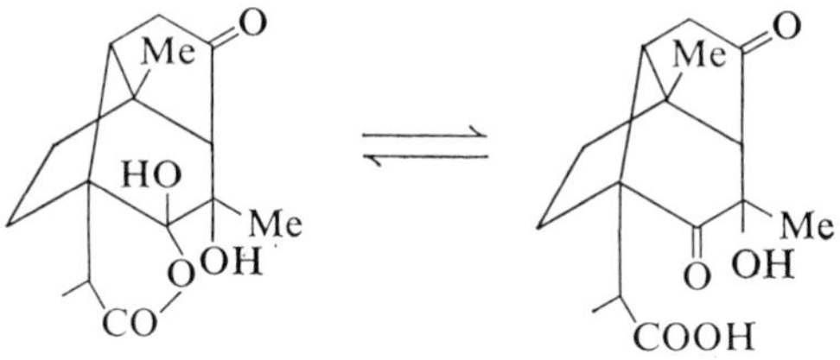

176 ⇌ 177

| MeOH/HCl (under 176)
| MeOH/HCl (under 177)

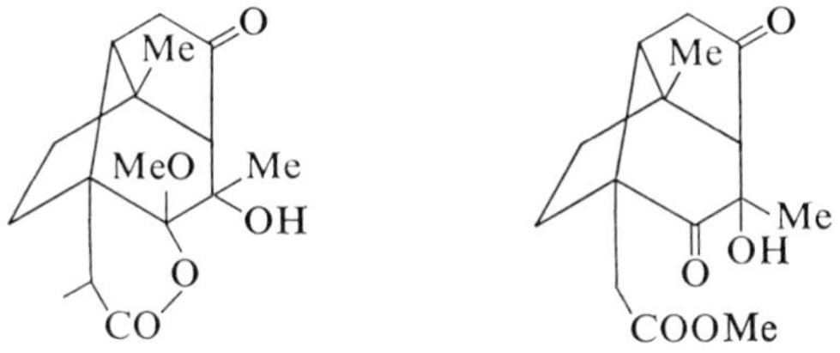

178   179

olefin **182** gives acetic acid instead of the expected lactic acid, and the amine **183** gives formic acid instead of diethylglycine [108].

The process in such cases may be represented as involving a rearrangement of the zwitterion intermediate (**188 → 192**). The extent to which this type of abnormal reaction occurs depends on the extent to which the zwitterion **188** is formed in preference to **185** during the ozonolysis of the olefin **186**, and

$$C=C-C-X \qquad C=C-Ar \qquad CH_2=CH-\underset{\underset{OH}{|}}{CH}CH_3 \qquad CH_2=CHCH_2NEt_2$$

180   181   182   183

this in turn depends on the electron-attracting power of the group X, increasing as this power decreases. The ease of rearrangement of the zwitterion **188** when formed also depends on the nature of the substituent X, increasing again as the electron-attracting power of X decreases or the electron-releasing power increases. Following rearrangement of the zwitterion, the loss of carbon monoxide and HX followed by oxidation (192 → 189) can account for the observed products.

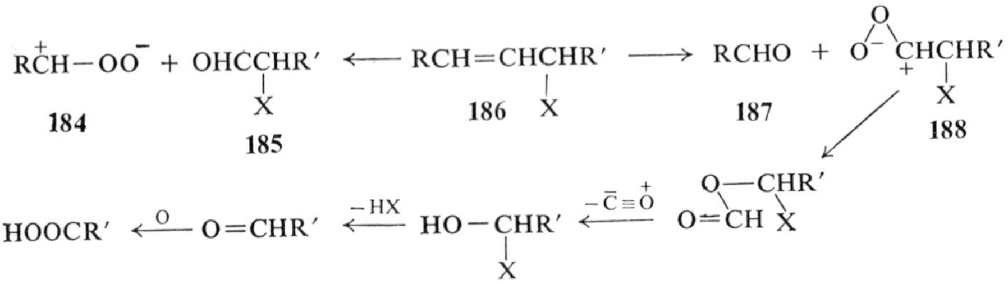

$$\overset{+}{R\overset{|}{C}H}-OO^- + \underset{\underset{185}{X}}{OHCCHR'} \longleftarrow \underset{\underset{186}{X}}{RCH=CHCHR'} \longrightarrow \underset{187}{RCHO} + \underset{\underset{188}{X}}{O^-\overset{+}{C}HCHR'}$$

184

$$\underset{189}{HOOCR'} \overset{O}{\longleftarrow} \underset{190}{O=CHR'} \overset{-HX}{\longleftarrow} \underset{\underset{X}{191}}{HO-\overset{|}{C}HR'} \overset{-\overset{+}{C}\equiv O}{\longleftarrow} \underset{192}{O=CH\ X}$$

α,β-Unsaturated acids, aldehydes, and ketones also contain the system 180 (X = O) and frequently give abnormal products during ozonolysis. For unsaturated acids the mechanism has been represented [109] by rearrangement of the ozonide (193 → 194 + 195); for unsaturated aldehydes and

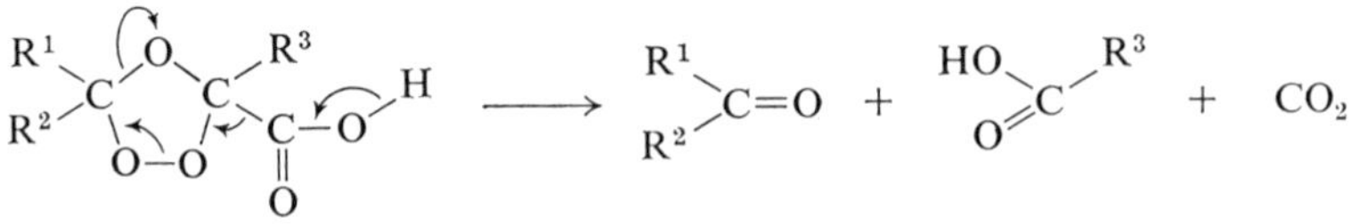

193   194   195

ketones the rearrangement is depicted as proceeding through the hydrate of the ozonide (196 → 197 + 198 + 199).

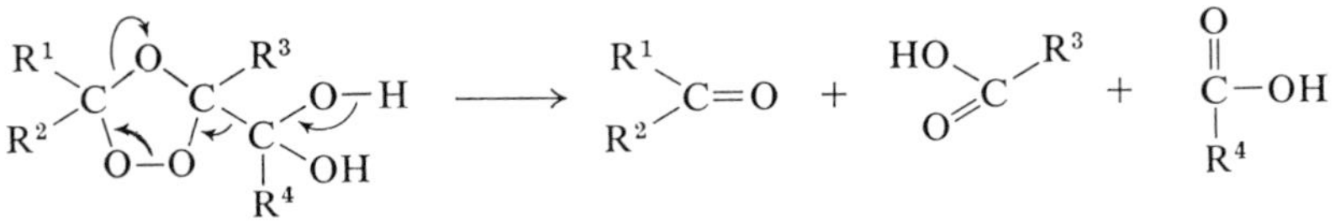

196   197   198   199

A similar process has been postulated during the ozonolysis of santonide (174). Hydrolysis of the initial product with methanolic hydrogen chloride yields, in addition to compounds 178 and 179 previously mentioned, the keto lactone 202, presumably as shown in 200 → 202 [107].

A process similar to that depicted in formulas 193 → 194 + 195 could be operative during the ozonolysis of allylic alcohols, but not of allylic ethers or allylic tertiary amines, which also give anomalous products of the same type. The mechanism has been criticized by Bailey [66] on the grounds that it fails to account for the abnormal products from these ethers and amines or for the production of carbon monoxide, which is frequently observed, and that it requires prior formation of the ozonide. The alternative mechanism of

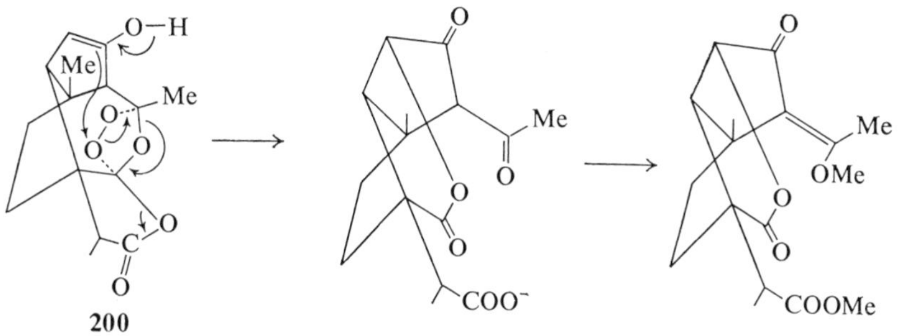

200 → **201** → **202**

zwitterion rearrangement, however, would require the formation of zwitterions that would on electronic grounds not be expected to be produced to any great extent, and in which carboxyl, formyl, or acyl groups would not be expected to migrate at all readily.

The zwitterion rearrangement mechanism does, however, satisfactorily account for the production of phenols during the ozonolysis of methoxylated methyl cinnamates [110, 111]. For example, in the ozonization of methyl-*p*-methoxy-cinnamate (**206**), the two zwitterions **203** and **208** would be formed in approximately equal amounts, and migration of the aryl group in **203** should occur readily to give, first, **204** and finally the phenol **205** [66].

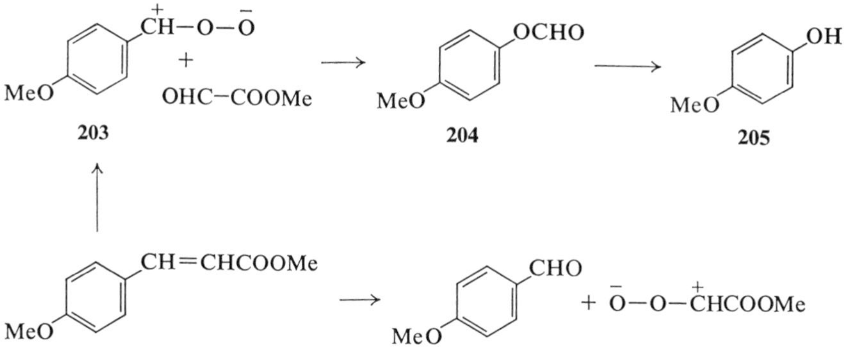

203 → **204** → **205**

206 → **207** + **208**

Abnormal products are also obtained in very small amounts from certain substances that do not belong to the general class **180** or **181**; for example, α-pinene (**209**) gives pinic acid (**210**) and pinonic acid (**211**) [112, 113], and vinylcyclohexene (**212**) gives the acids **213**, **214**, and **215** [114, 115]. In these

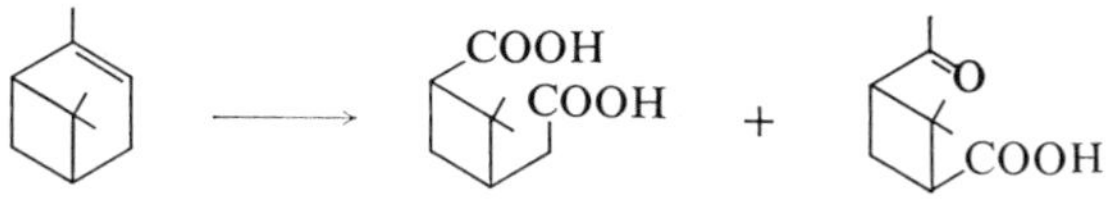

209 → **210** + **211**

and similar cases the abnormal products could have arisen from the intermediate zwitterion by rearrangement, or from the normal ozonolysis product by oxidation.

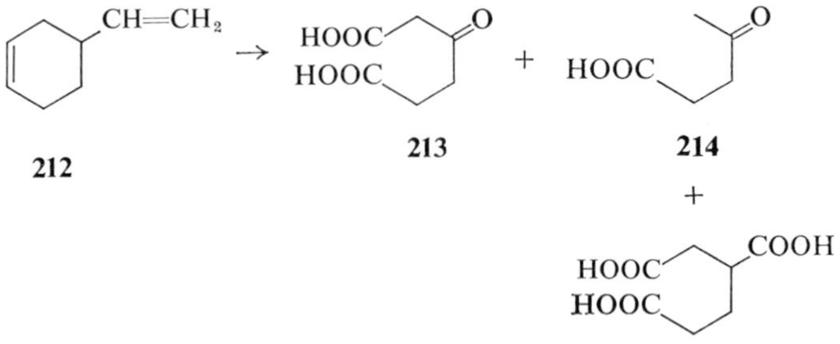

212   213   214

+

215

Other abnormal results of the ozonolysis of natural products, for which no rational mechanism of production can be advanced, have been reported. For example, carbon dioxide, methane, and hydrogen are obtained from oleic acid (216) [116] and acetaldehyde from humulone (217, R = isopropyl) and cohumulone (217, R = isobutyl) [117].

$$CH_3(CH_2)_7CH=CH(CH_2)_7COOH$$

216

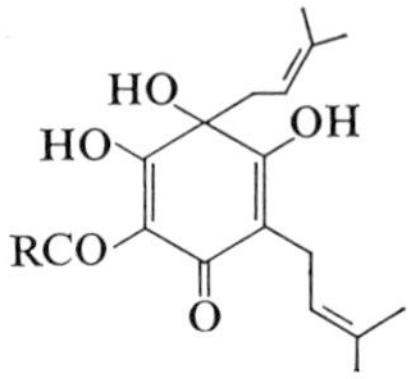

217

Milas and Nolan [118] have shown that abnormal reactions occur principally during ozonolysis carried out in inert solutions and may be avoided or minimized by ozonolysis in methanol, in which solvent the zwitterion is converted to the methoxy hydroperoxide more quickly than it rearranges. However, even alkoxy hydroperoxides sometimes rearrange easily. Camphene (218), for example, gives very little camphenilone (219) under normal conditions; in inert solvents the lactone 220 is obtained, presumably by the rearrangement of the zwitterion 221, together with an acid, probably 222, while in methanolic solution the ester 225 is obtained,

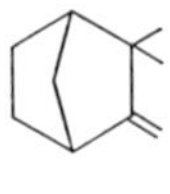

218 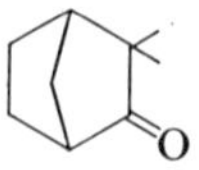   219 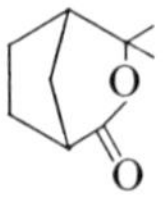   220

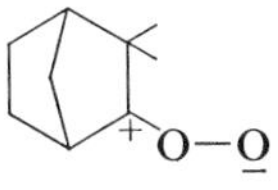

221

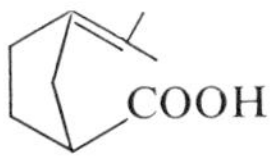

222

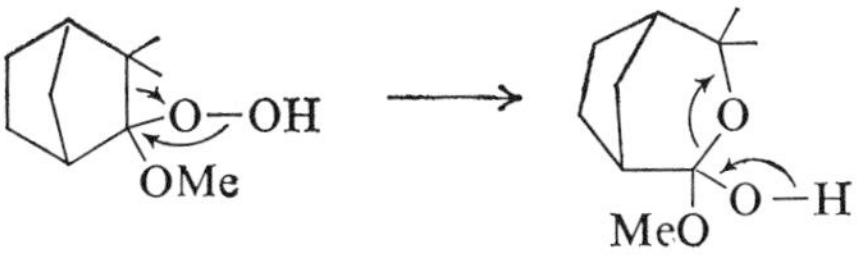

223 → **224**

225

presumably by the rearrangement of the methoxy hydroperoxide **223** through the intermediate **224** [67].

Ozonolysis of Aromatic Nuclei

Aromatic nuclei will react with ozone, though less rapidly than olefinic substances, and the reaction between ozone and unsaturated aromatic compounds can easily be stopped before the aromatic nuclei are disrupted. Benzene is split at all three "double bonds" if sufficient ozone is used. o-Xylene in this way gives glyoxal, methyl glyoxal, and diacetyl in the ratio 3:2:1, indicating that ozone reacts equally with the two Kekulé forms of the substance. This may be more accurately expressed by saying that the attack of ozone on a hybrid molecule gives two monoaddition products in equal amounts [66, 119, 120], and the mode of addition of the first molecule of ozone (i.e., **230** → **226** or **227** → **228**) naturally determines the mode of subsequent addition (i.e., **226** → **229** or **228** → **231**).

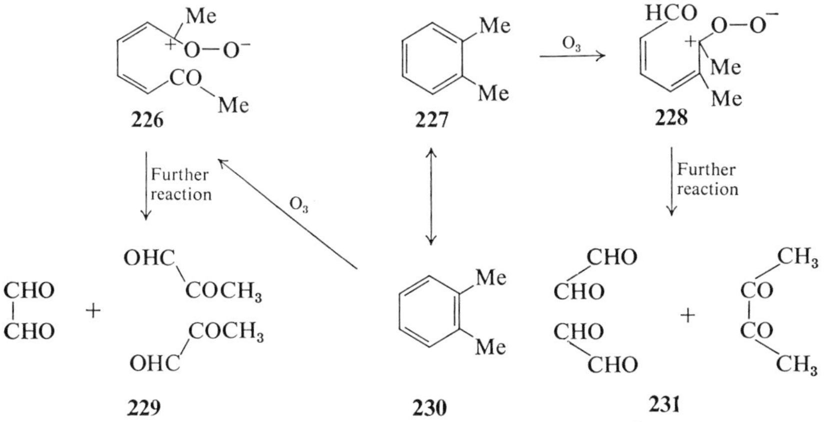

226

227

228

229

230

231

With ethers of catechol, initial fission of the aromatic ring appears to occur only between the two oxygen functions, although it must be admitted that only a few such compounds have been studied. For example, dihydrocodeine

(232, R = Me) and ethyldihydromorphine (232, R = Et) yield α-ozodi-hydrocodeine (233, R = Me) and α-ozodihydroethylmorphine (233, R = Et), respectively [121, 122], and further ozonolysis of these two compounds gives as end product codinal (234) [123]. More complex bases in this series behave similarly [124]. Thebaine (235), however, which contains a nonaromatic unsaturated ring, affords under similar conditions α-thebaizone (236) with fission of the enol ether double bond [125, 126].

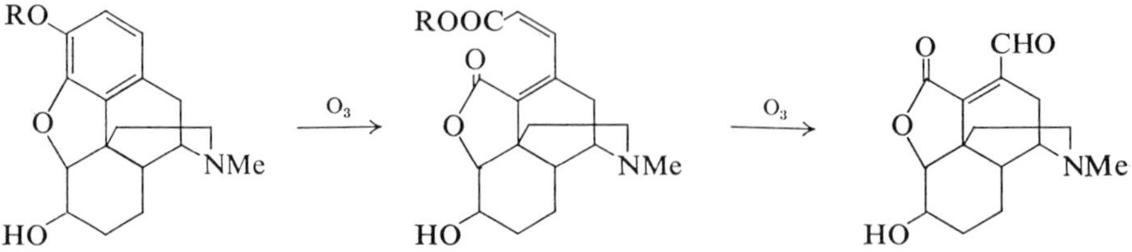

232   233   234

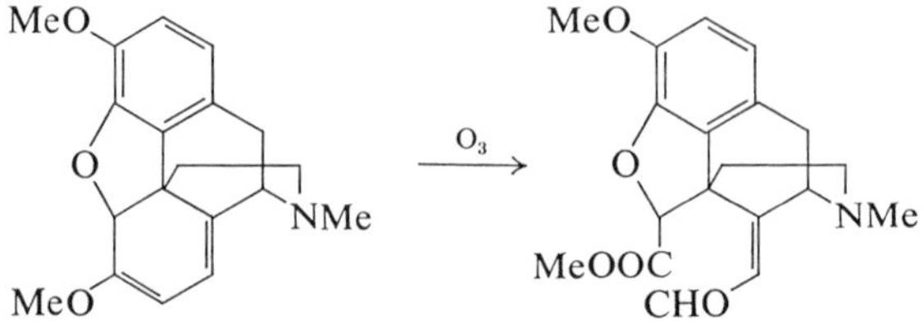

235   236

Use has been made of this unique fission of catechol nuclei by ozone in the total synthesis of strychnine when compound 237 was converted by ozone into the diester 238.

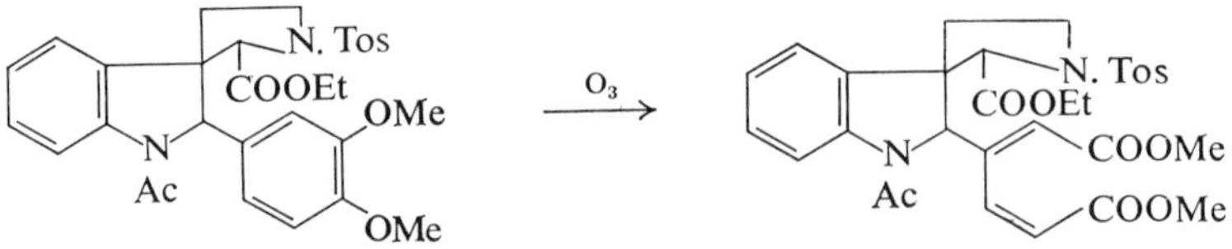

237   238

Naphthalene reacts readily with 2 moles of ozone, presumably both reacting with the same ring, and thereafter the reaction is slow. Methylated naphthalenes react preferentially with ozone at the 1,2 and 3,4 bonds in the ring bearing the methyl group.

Phenanthrene (239) reacts with ozone exclusively at the 9,10 bond, that is, the bond of lowest bond localization energy, and only 1 mole of ozone is

readily absorbed. In methanol solution the reaction proceeds as shown in
239 → **241**; compound **241** (R = H) can be precipitated, while **241** (R = Me)
is formed on further reaction with the solvent. The compounds **241** (R = H)
and **241** (R = Me) will both react with base to give the aldehydic acid **242**
(R = H) and the ester **243** (R = Me), respectively, in good yield [68]. In
inert solvents, however, the product is a polyozonide of probable structure
243.

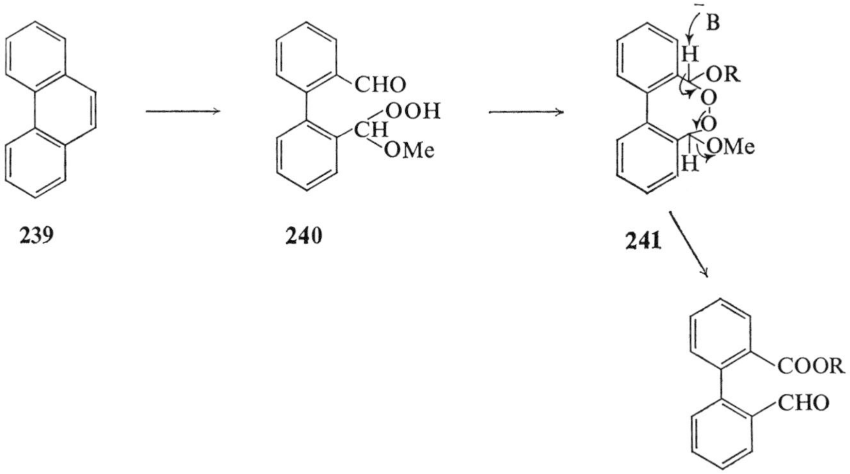

239   **240**   **241**

242

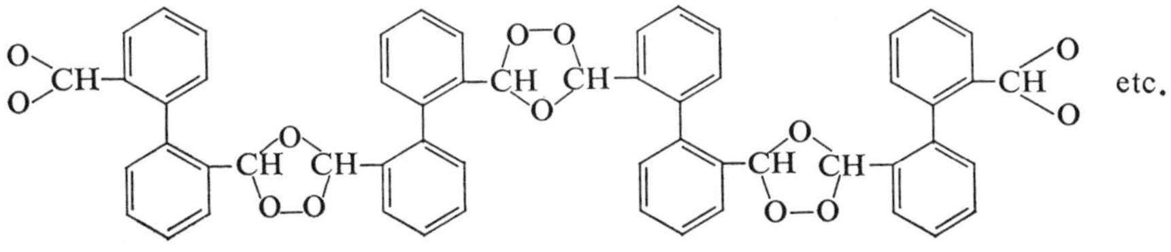

243

Ozone attacks pyrene (**244**), as does osmium tetroxide, at the 4,5 bond
[127]; but anthracene (**245**), which is attacked at the 1,2 bond by ozmium

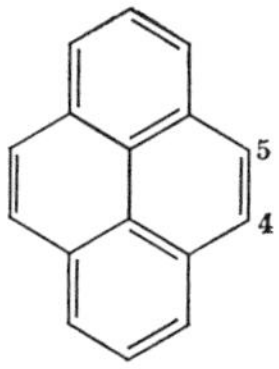

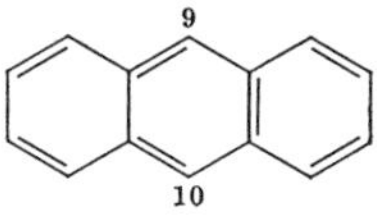

244   **245**

tetroxide, is attacked at the 9,10 bond by ozone, giving anthraquinone and anthrahydroquinone after reduction of the peroxidic initial product [128].

Ozonolysis of Heterocyclic Compounds

Pyridine will react with ozone less readily than benzene, but substituted pyridines will give products formally derived from both Kekulé forms, as is the case with benzene derivatives. With quinoline the initial major attack is at positions 5,6 and 7,8, that is, in the benzene ring, after which the pyridine ring is attacked slowly; there is, however, minor attack initially at position 3,4 in the pyridine ring [66]. The two rings are of much nearer equal reactivity in the case of isoquinoline in which the initial attack is 60% by 1 mole of ozone at position 3,4 (pyridine ring) and 40% by 2 moles of ozone at positions 5,6 and 7,8 (benzene ring); the final products are phthalic acid and cinchomeronic acid [129, 130].

Benzopyrylium (flavylium) salts likewise afford products of two alternative modes of fission; for example, the salt **246** ↔ **249** yields salicylaldehyde (**247**)

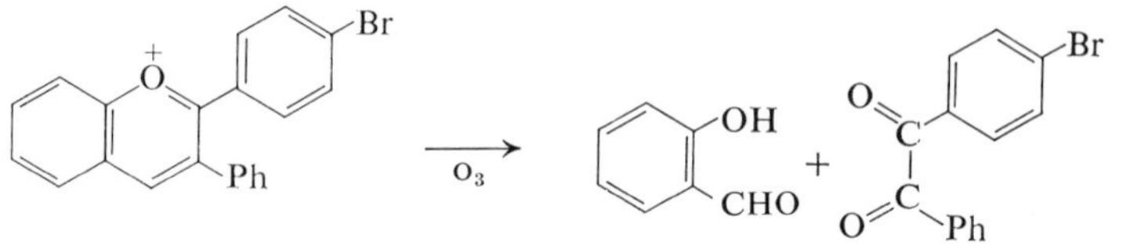

| 246 | 247 | 248 |

| 249 | 250 |

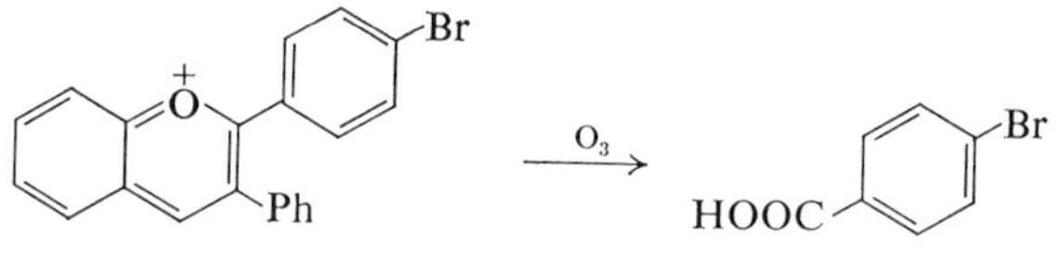

and *p*-bromobenzil (**248**) which must arise from fission of the molecule at positions 1,2 and 3,4 together with *p*-bromobenzoic acid (**250**), which must arise from fission at position 2,3 [131].

Indole (**251**) on ozonolysis affords *o*-aminobenzaldehyde (**252**), which gives a trimeric condensation product [132], and indole derivatives in general

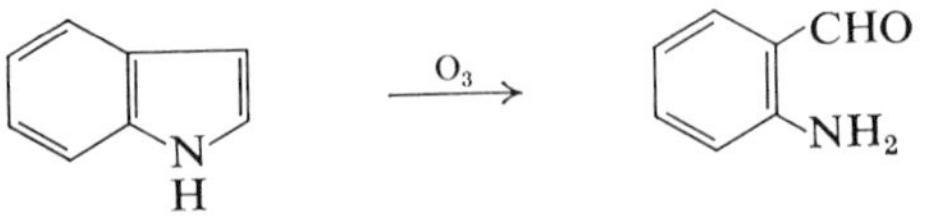

| 251 | 252 |

always undergo fission of the carbon–carbon double bond in the heterocyclic ring [66]. Acylamino aldehydes, acids, or ketones are obtainable in this way from 2-, 3-, and 2,3-substituted indoles [132–134]. Even such derivatives of indole as yohimbine (**253**) behave in the same way; this alkaloid on ozonolysis affords the amide **254**, which readily cyclizes to the base **255** under alkaline conditions [96].

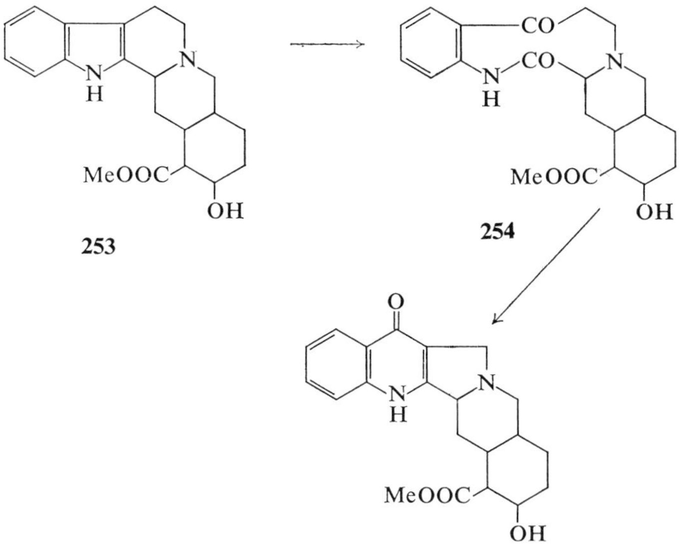

For other methods of fission of carbon–carbon double bonds see pages 139, 143, 169, and 177, and Section 3.

Lead Tetraacetate

Lead tetraacetate is a very widely used reagent for the oxidative fission of vicinal glycols, α-hydroxy ketones, α-hydroxy acids, α-diketones, and α-keto-acids. Since there are several methods of conversion of olefins into vicinal glycols (see p. 204) the process can be used as part of a method of the fission of carbon–carbon double bonds that is an alternative to and often more satisfactory than ozonolysis. Other methods of fission of vicinal glycols and similar systems are oxidation with periodic acid, ceric salts, and sodium bismuthate, which are separately described on pages 177, 185, and 187. The reaction has found considerable use in the elucidation of structures, in preparative organic chemistry, and for the estimation of vicinal glycol systems. Several reviews of the use of this reagent are available [135–139].

The fission of vicinal glycols may be represented most generally by the cyclic reaction mechanism **256** → **259** [137, 140, 141]. In support of this mechanism the following evidence may be cited.

(*a*) Stable cyclic esters analogous to those of type **258** derived from osmium tetroxide have been isolated [142, 143].

(*b*) With isomeric glycols the *threo* isomer (the racemic form with symmetrical diols) is more readily cleaved than the *erythro* (*meso* in symmetrical

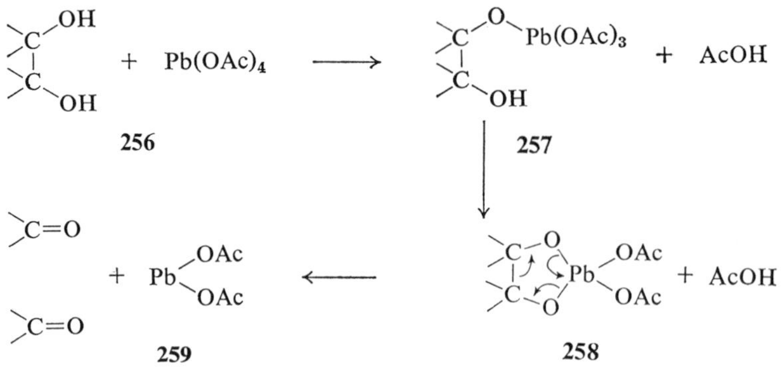

256   **257**

259   **258**

compounds) form, and this difference can be used to identify the diols when both are available [139].

With cyclic diols the *cis* form is oxidized more rapidly than the *trans* isomer when the rings have up to seven members; for nine- and larger-membered rings the *trans* form is the more easily cleaved [141].

The cyclic mechanism cannot always be involved, however, since certain *trans*-glycols in which the formation of such an intermediate is impossible, or would involve considerable distortion and strain in the molecule, are oxidized by lead tetraacetate, for example, *trans*-decalin-9,10-diol (**260**) and *trans*-1,2-dimethylcyclopentane-1,2-diol (**261**) [141, 144]. In such cases there are, however, very great differences in the rate of oxidation of the *cis* and *trans* isomers, for example, the *cis* isomers of the diols **260** and **261** are

260   **261**

oxidized 4000 and 3000 times, respectively, more rapidly than the *trans* forms [139]. Certain *trans*-diols of this type are oxidized only with great difficulty, for example the *trans*-camphane-2,3-diol (**262**) is very unreactive towards both lead tetraacetate and periodic acid in comparison with the *cis* isomer **263** [145], and the diols 1,6-anhydro-β-D-glucofuranose (**264**) and 1,6-anhydro-α-D-galactofuranose (**265**) are unaffected by these oxidizing reagents even over periods of several days [146, 147]. Such resistance to cleavage has been attributed to steric effects preventing distortion of the

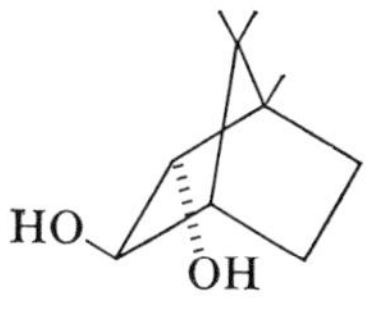

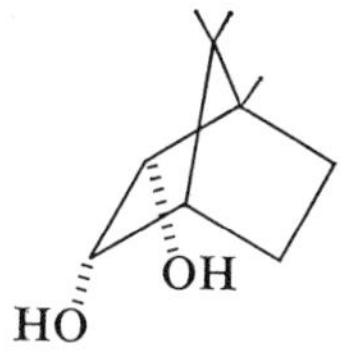

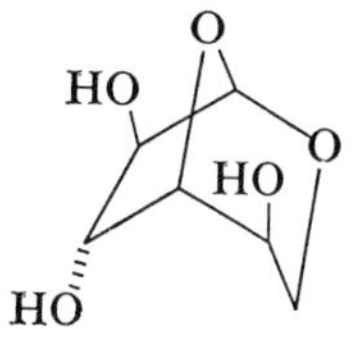

molecule in such a way as to permit formation of the cyclic intermediate [148]. Most resistant diols can be cleaved by lead tetraacetate in pyridine [149].

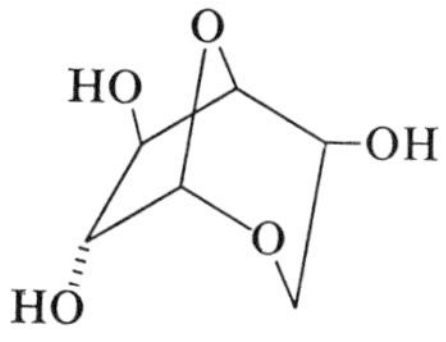

A reaction mechanism that could be operative in the cleavage of *trans*-glycols where formation of a cyclic intermediate is inhibited involves proton transfer to an external base (**266**). It may be noted that bases such as OAc⁻ will catalyze the oxidation, and the increased rate of reaction in pyridine may be a consequence of the high basicity of the solvent.

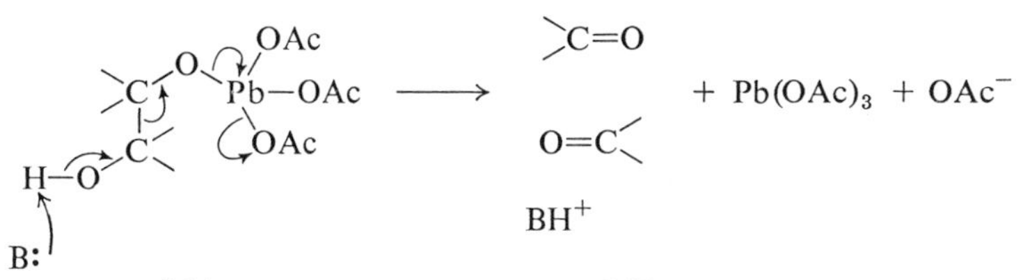

In general as the substitution of the glycol with alkyl groups increases, the rate of oxidation increases rapidly. This is particularly true of cyclic systems, and the oxidation of rigid *trans*-cyclic diols occurs most readily in ditertiary systems [139, 141]. This effect appears to be steric rather than electronic, since the more bulky the substituents are the more reactive is the diol. The decomposition of the intermediate **258** to ketones **259** involves the transformation of tetrahedral carbon to trigonal, and bulky groups may be expected to facilitate the formation of a transition state for such a process.

Lead tetraacetate oxidations may be effected in organic solvents such as benzene, chloroform, or nitrobenzene, but are most frequently carried out in glacial acetic acid, which should be purified by refluxing with chromium trioxide and fractional distillation before use. Mixtures of acetic acid with inert solvents have also been used. The reaction rate may be increased by the addition of aprotic solvents such as pyridine [149], or by the addition of water or alcohols [150, 151, 152], but water in high concentration retards the reaction and also destroys the reagent by hydrolysis [152]. Some oxidations are catalyzed by acids, for example, the oxidation of pinacol and of *cis-* and *trans*-cyclohexane-1,2-diol the best catalysts being strong acids. By protonation of the acetate, radical acids could facilitate either of the reactions **256 → 257 → 258**:

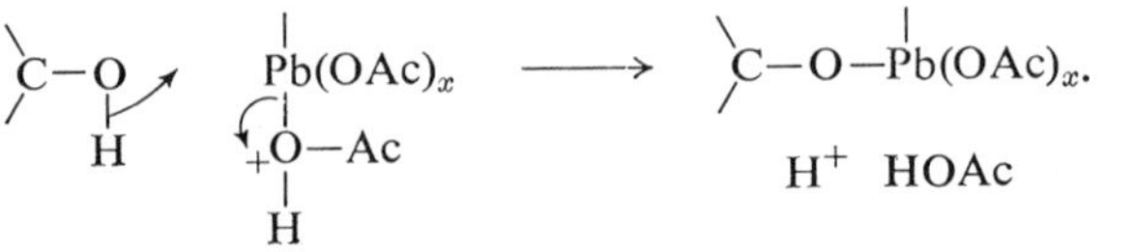

$$\ce{>C-O\bond{...}H + Pb(OAc)_x + O-Ac/H -> >C-O-Pb(OAc)_x. + H^+ HOAc}$$

Lead diacetate is formed during the reaction and is precipitated from benzene or chloroform and can easily be removed. It is soluble in acetic acid, but dilution with water will precipitate water-insoluble products which may be removed, and the lead may be removed from the solution of water-soluble products by hydrogen sulfide.

α-Hydroxy acids are very readily oxidized by lead tetraacetate with the formation of carbon dioxide and aldehydes or ketones, and the reaction may be represented as in formulas **268–270**. Since carboxylic acids can be

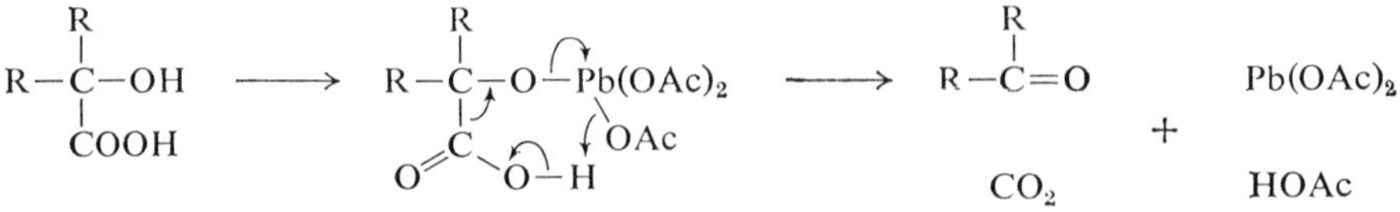

$$\ce{R-C(OH)(COOH) -> R-C-O-Pb(OAc)_2 (C=O, O-H, OAc) -> R-C=O + CO_2 + Pb(OAc)_2 + HOAc}$$

268   **269**   **270**

brominated and hydrolyzed to α-hydroxy acids, and since aldehydes obtained after oxidation by lead tetraacetate can be further oxidized to acids, a stepwise degradation of acids until chain branching is encountered can be achieved. Such a process is an alternative to the Barbier-Wieland degradation (Section 5).

The oxidation of an α-hydroxy acid was used in this way in the structural elucidation of methymycin. Methynolide (**271**), the sugar-free portion of methymycin, on oxidation with potassium permanganate afforded three products, of which the largest was assigned the structure **273** since it could be oxidized by lead tetraacetate to the ketone **274**, of known structure, which was also obtained in the permanganate oxidation [153].

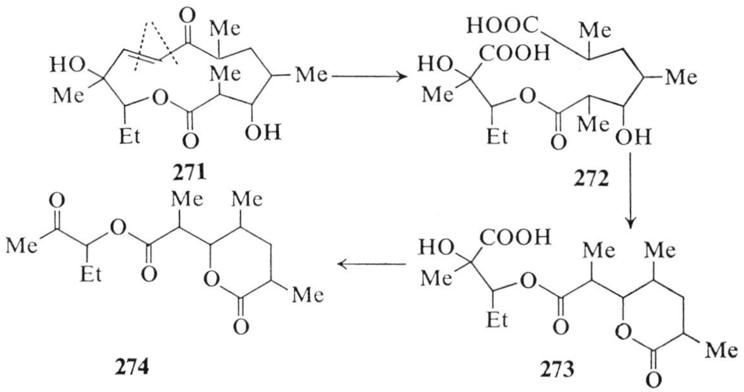

Lead tetraacetate will effect the oxidative decarboxylation of α-keto acids in quantitative yield in moist acetic acid solution, and the reaction may be represented as in formulas **275** → **278** [154]. The first stage of this process

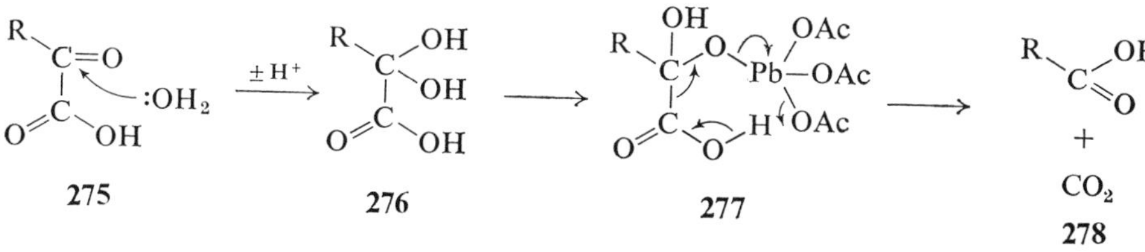

requires protic solvents for good yields, the reactions in aprotic solvents being slow and incomplete; esters or nitriles are obtained in the presence of alcohols or hydrogen cyanide.

Other fragmentations have also been observed in structural studies, as in the oxidations of jaconecic acid (**279**) and isojaconecic acid (**280**). Normal oxidation of the latter to the ketone **281** is also observed [155].

Lead tetraacetate can attack other systems in organic compounds as well as vicinal glycols, and a summary of the processes involved follows.

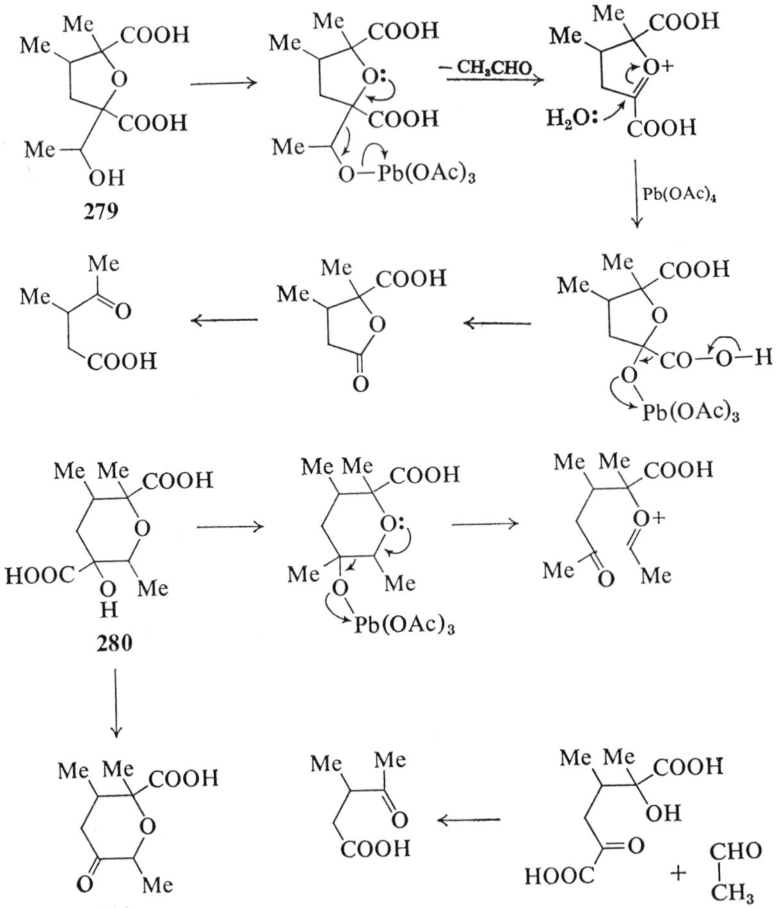

Alcohols (282) can be oxidized to aldehydes or ketones (283). In aqueous acetic acid the initial equilibrium is almost entirely on the side of 282 and oxidation only occurs at elevated temperatures. Oxidation in good yield can occur in benzene at the boiling point [140] and in pyridine at room temperature [156, 157]. Fragmentation has been observed with β,γ-unsaturated alcohols and β-hydroxy aldehydes and ketones [158], for example, 1-allylcyclohexanol (284) gives cyclohexanone (286) [159], and diacetone

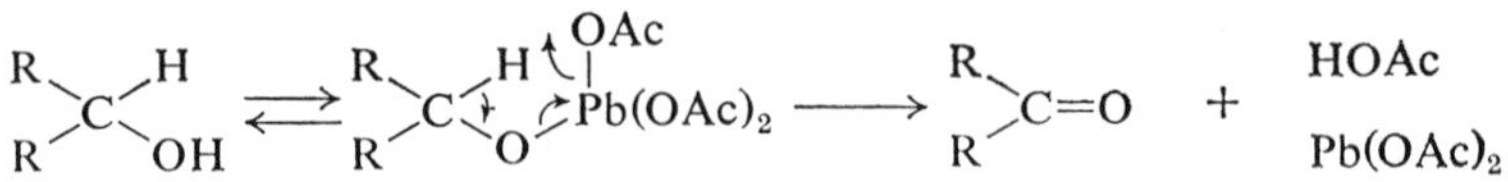

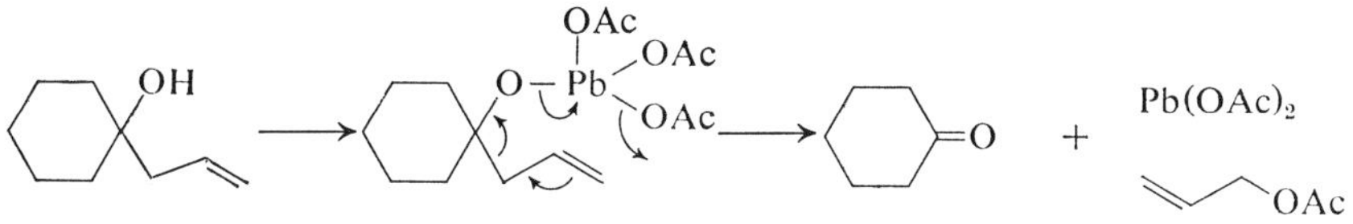

284   285   286

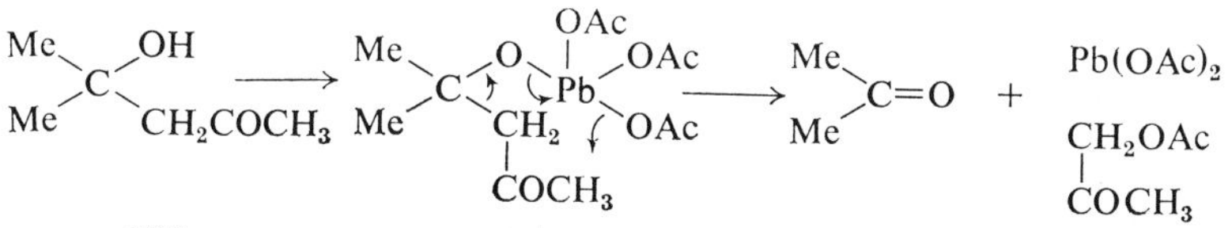

287   288   289

alcohol (**287**) gives acetone and acetoxyacetone (**289**) [158]. By analogous processes fragmentations of the steroidal β,γ-unsaturated alcohol (**290**) to formaldehyde and the acetoxy olefin (**291**), and the 5-hydroxy-α,β-unsaturated ketone (**292**) to the acetoxy ketone (**295**) occur [160]. Other carbon–carbon bond fissions observed during the oxidation of alcohols with

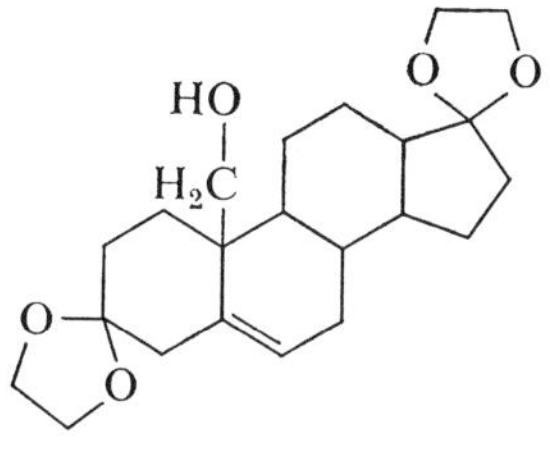

290 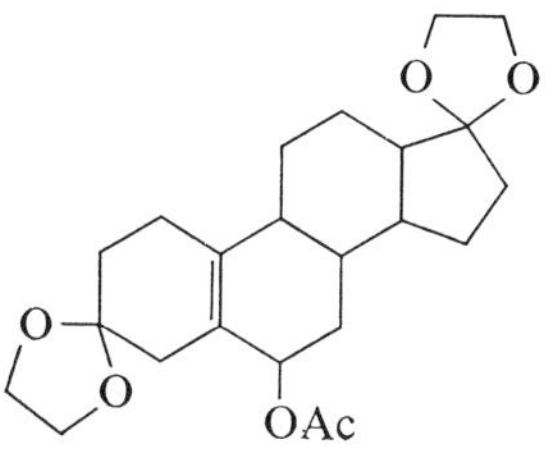   291 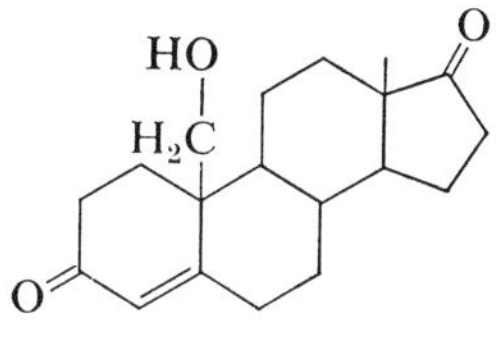   292

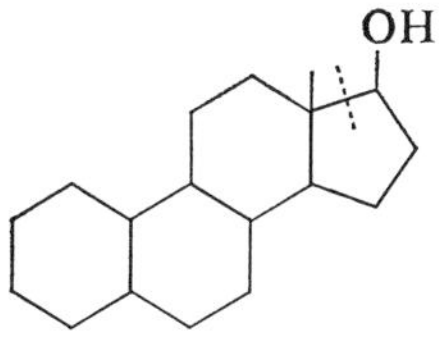

293 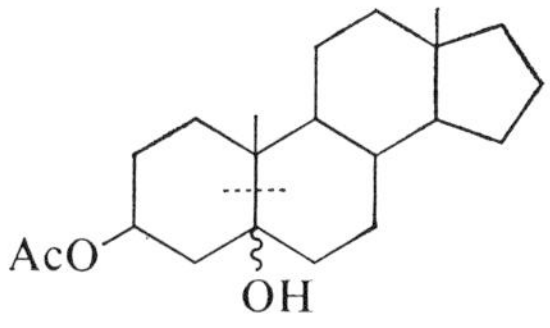   294 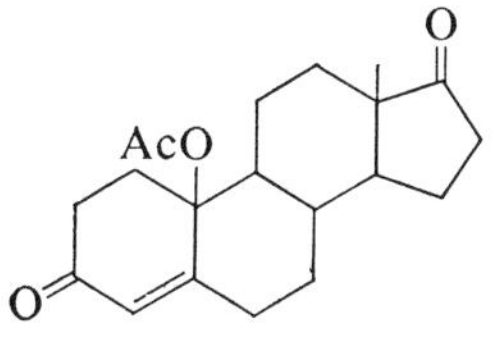   295

lead tetraacetate include those indicated in the skeleton formulas **293** and **294** [160–164], the fission of certain alcohols of general structure **296** [165], and the oxidation of the dihemiacetal (**297**) to the acid anhydride (**298**) [166].

Ketones containing reactive methylene groups are oxidized to α-acetoxy or α,α′-diacetoxy ketones, and the same reaction occurs with esters and

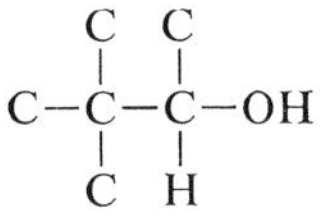

296

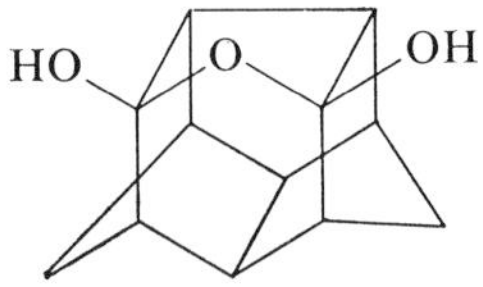

297

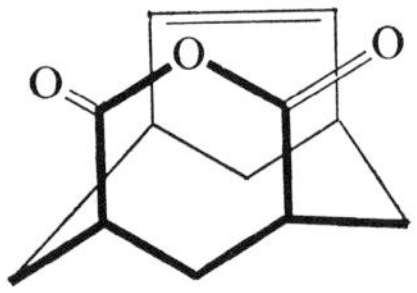

298

anhydrides though with less ease. The reaction is discussed in detail, with a list of examples, in a recent review by Criegee [138]. Similarly methylene groups adjacent to and activated by aromatic nuclei are oxidized to –CH(OAc)– [138].

Monobasic saturated acids are generally stable to lead tetraacetate, being oxidized only at elevated temperatures, with the exception of formic acid which is readily oxidized to carbon dioxide. As this acid is not oxidized by periodic acid, 1,2,3-triols, such as many carbohydrates, differ in their reactions with the two reagents, the central carbon atom being oxidized to carbon dioxide by lead tetraacetate but only to formic acid by periodic acid. β,γ-Unsaturated acids are oxidized readily with the loss of carbon dioxide (**299**)–(**301**) [167].

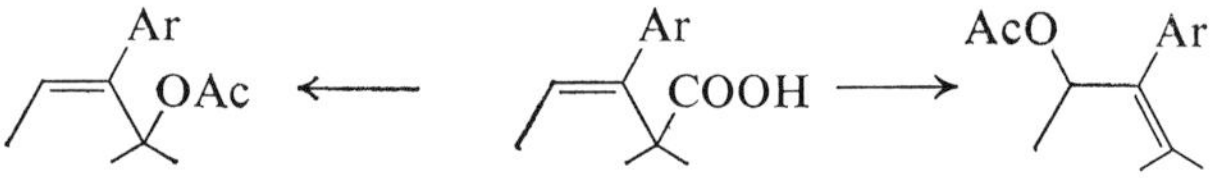

299

300

301

α,β-Dicarboxylic acids related to succinic acid are oxidized generally to carbon dioxide and an olefin (**302**)–(**304**) by lead tetraacetate in the presence

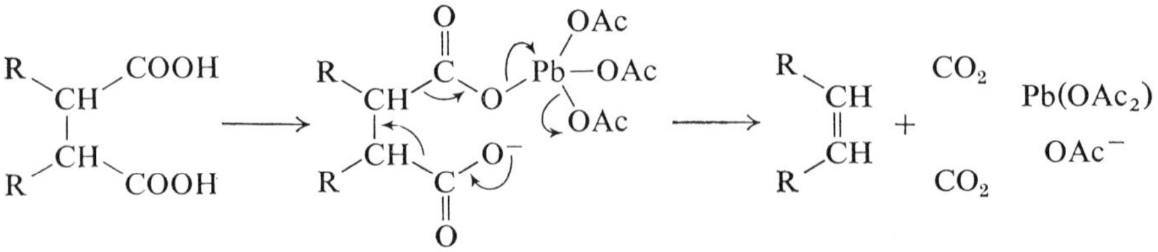

302

303

304

of pyridine, usually in benzene solution [168–172]. A summary of successful reactions is given by Criegee [138]. Other reactions sometimes take precedence over this degradation, for example in bridged ring systems where both carboxyl groups are suitably situated *endo* to a double bond anhydride formation may occur (**305**)–(**306**) [172].

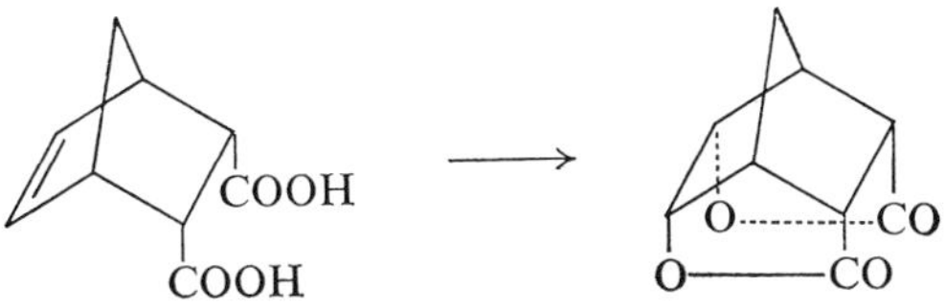

305 → **306**

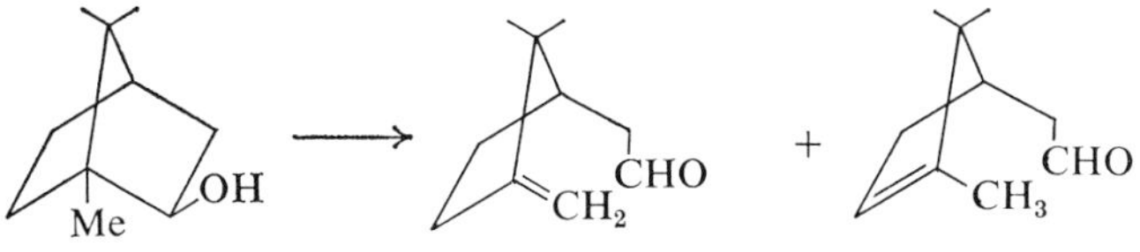

307 → **308** + **309**

Lead tetraacetate has been shown to oxidize isoborneol (**307**) to a mixture of the two aldehydes **308** and **309** [173].

Olefins can suffer the addition of two acetoxy groups to the double bond and can undergo allylic substitution [138]. Both of these reactions give poor

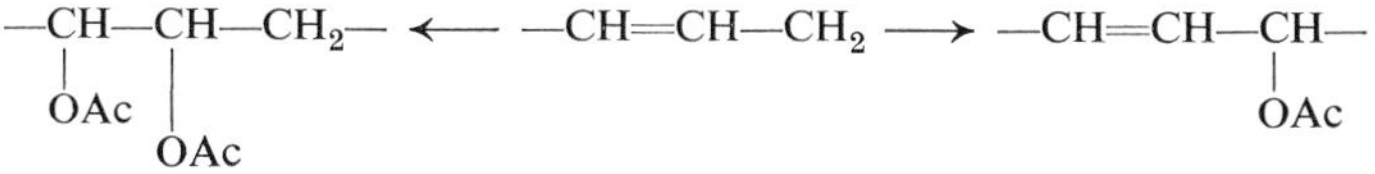

$$-CH-CH-CH_2-\ \longleftarrow\ -CH=CH-CH_2\ \longrightarrow\ -CH=CH-CH-$$
$$\quad|\qquad|\qquad\qquad\qquad\qquad\qquad\qquad\qquad\qquad\qquad|$$
$$\ OAc\quad|\qquad\qquad\qquad\qquad\qquad\qquad\qquad\qquad\qquad\ OAc$$
$$\qquad\quad OAc$$

yields and do not seriously compete with glycol fission.

Phenols are oxidized by lead tetraacetate to quinones and acetylation products, and reactive aromatic nuclei such as anthracene may suffer oxidation and/or substitution. The reactions have been summarized by Criegee [138].

Periodic Acid

Periodic acid is a specific oxidizing agent for vicinal glycols [174], converting them to carbonyl compounds. In this respect it is an alternative to lead tetraacetate and is of significant value in the controlled fission of carbon–carbon double bonds, which may be readily converted into 1,2-diol systems (see p. 204). The two processes are in general complementary since periodic acid and periodates are most commonly used in aqueous solutions for the oxidation of carbohydrates and water-soluble materials of low molecular weight, whereas lead tetraacetate, which is used in acetic acid or aprotic solvents, is invaluable for the oxidation of compounds insoluble in water. Oxidations by periodic acid of substances insoluble in water can, however, be accomplished in aqueous acetic acid, dioxan, or methanol with water contents as low as 10%. The use of periodic acid has been reviewed [139, 175].

Oxidations are effected with periodic acid itself or with sodium periodate in the presence of sulfuric acid. The periodate solutions can be standardized and the compound to be oxidized can be treated with a known volume of standard oxidant, and the progress of the reaction can be determined by removing aliquots of the mixture and estimating the consumed oxidant by the liberation of iodine from potassium iodide in neutral solution, in which iodates do not react with iodides.

The preparation of tetraethylammonium periodate has recently been reported [176]. This salt is freely soluble in many solvents and may well find a wide use in oxidation procedures.

In aqueous solution periodic acid, HIO_4, is present as H_5IO_6, and a typical oxidation of a 1,2-glycol may be represented by

$$\underset{\substack{| \quad | \\ OH \quad OH}}{R-CH-CH-R'} + H_5IO_6 \longrightarrow RCHO + R'CHO + HIO_3 + 2H_2O.$$

The reagent may also be used for the fission of α-hydroxy acids, ketones and aldehydes, α-keto acids, α-diketones, α-amino alcohols and ketones, for example:

$$PhCO-CH(OH)Ph\cdot + H_5IO_6 \longrightarrow PhCOOH + PhCHO + HIO_3 + 2H_2O$$

$$\underset{\substack{| \quad | \\ OH \quad NH_2}}{CH_3CH-CH-COOH} + H_5IO_6 \longrightarrow$$

$$CH_3CHO + OHC-COOH + NH_3 + HIO_3 + 2H_2O.$$

For the reaction a cyclic mechanism, similar to that proposed for oxidations with lead tetraacetate, has been suggested (**310 → 313**) [140, 173, 174, 177].

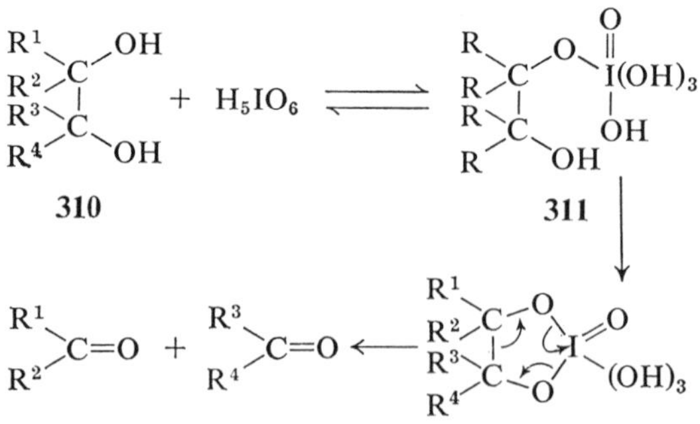

The kinetics of the oxidation have been studied and shown to be compatible with such a mechanism [178].

There are close similarities between oxidations with the two reagents. As with lead tetraacetate, *cis*-cyclic diols are oxidized more rapidly than their *trans* isomers. Certain diols, such as *trans*-decalin-9,10-diol (**260**) [141], *trans*-1,2-dihydroxy-1,2-dimethylcyclopentane (**261**) [144], and the carbohydrate derivatives **264** [146] and **265** [147] are greatly resistant to oxidation with periodic acid as well as with lead tetraacetate and, as with the latter reagent, in such cases formation of a cyclic ester is either impossible or subject to great steric hindrance.

Unlike lead tetraacetate, periodic acid oxidizes ditertiary glycols only very slowly and, under the acid conditions of the reaction, pinacolone rearrangement may occur more rapidly than oxidation.

Polyhydric alcohols can be completely degraded by periodic acid, for example, *myo*-inositol (**314**) is converted into formic acid [179, 180, 181]. Protection of one or more of the hydroxyl groups by esterification or etherification effectively prevents complete disruption, for example, the glycoside

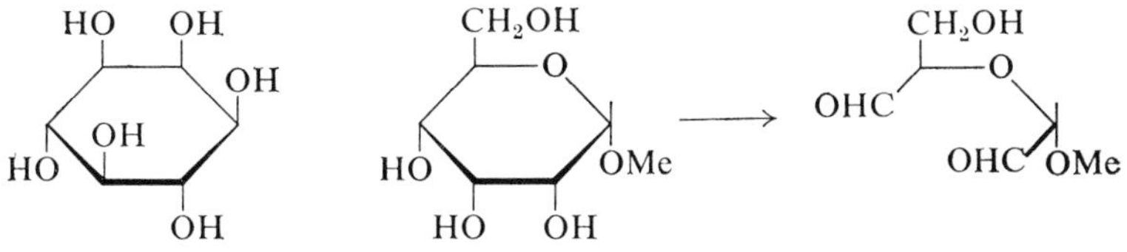

314   315   316

315 loses only one carbon atom, as formic acid, in being oxidized to the stable dialdehyde **316** [182]. The hemiacetal linkage in free sugars does not survive the oxidation.

α-Hydroxy acids react very slowly with periodic acid in contrast to their behavior with lead tetraacetate. Tartaric acid (**318**), which is at once a vicinal diol and an α-hydroxy acid, reacts rapidly with periodic acid only as a vicinal diol, giving pyruvic acid (**319**) which is then very slowly oxidized to carbon dioxide and formic acid [183, 184], whereas with lead tetraacetate it reacts as a vicinal diol (**318 → 319**) and as an α-hydroxy acid (**320**).

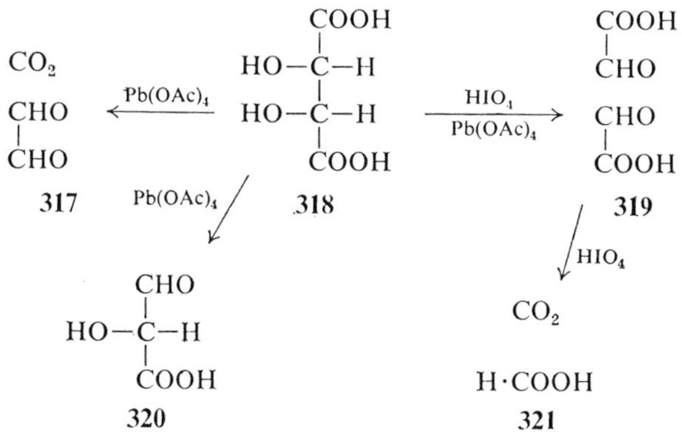

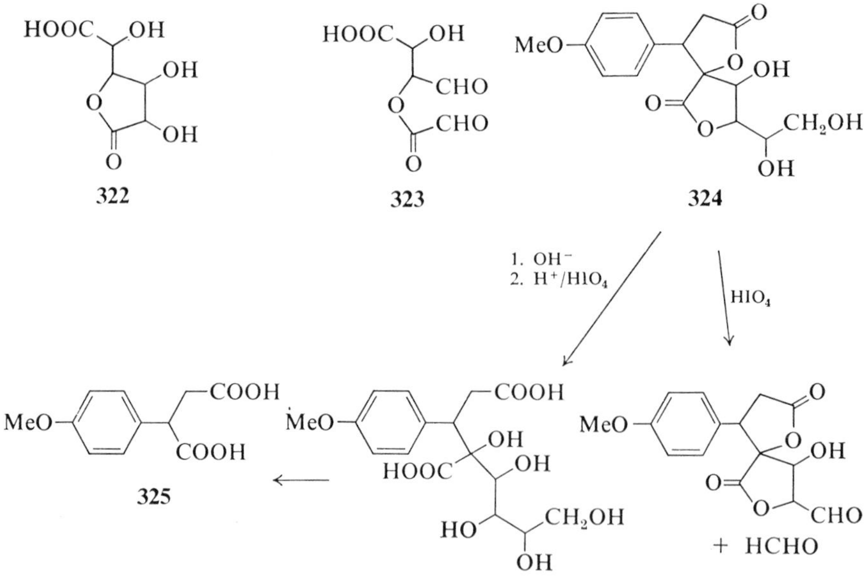

322

323

324

1. OH⁻
2. H⁺/HIO₄

HIO₄

325

326

327

+ HCHO

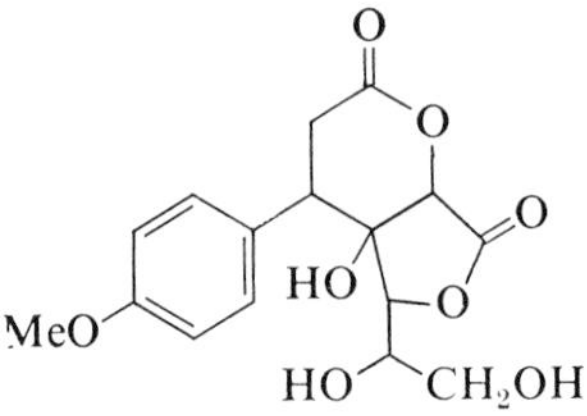

328

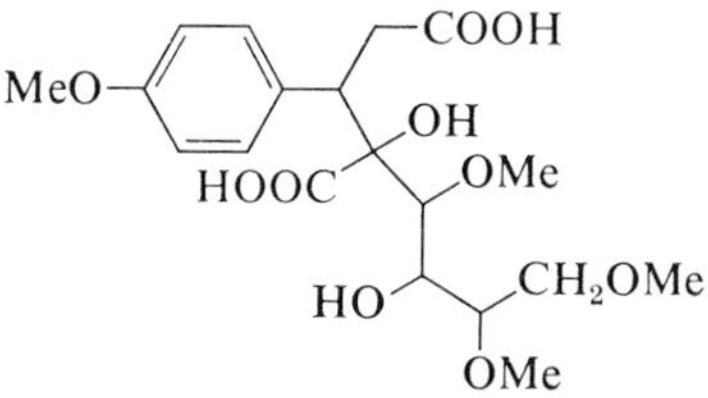

329

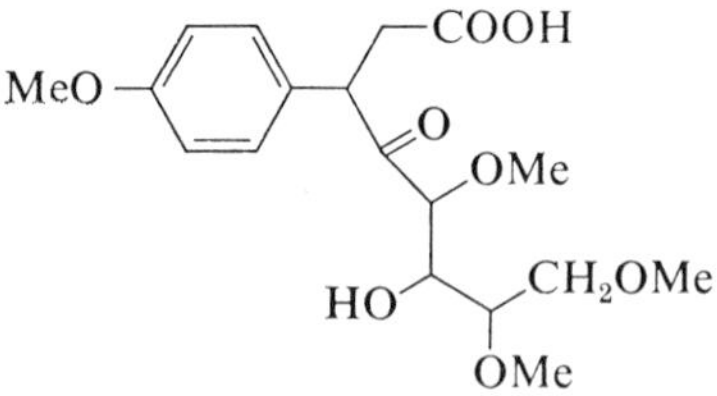

330

Lactones are unaffected by periodic acid provided they are stable to hydrolysis under the acid conditions of the reaction. For example the lactone **322** from D-saccharic acid is oxidized only to the ester **323** [185, 186], the α-hydroxy acid system being unattacked.

The dilactone leucodrin methyl ether (**324**) [187, 188] in acid solution reacts with periodic acid with the absorption of two equivalents of oxygen only, to give 1 mole of formaldehyde by fission of the vicinal diol (**327**).

Opening of the lactone rings in alkalies followed by acidification results in the diacid **326**, in which recyclization is slow, and this solution reacts with periodic acid with the absorption of eight equivalents of oxygen to give formaldehyde, three equivalents of acid (presumably formic acid), and *p*-anisylsuccinic acid (**325**). On the basis of the oxidant consumed, 1-anisyl-2-ketoglutaric acid would have been expected, but since this must presumably be oxidized further, it must be concluded that complete oxidation was not achieved [189]. The alternative structure **328**, subsequently disproved [187, 188], would explain these facts, and in particular the production of the acid **325**, but would require the production of glycollic acid, which was not detected [189]. The diacid **329** from leucodrin tetramethyl ether resisted attack by periodic acid, but was oxidized to the ketone **330** by lead tetraacetate [189].

Periodic acid has been widely used in the study of carbohydrates, especially the sugars and amino sugars derived from antibiotics, which are of low molecular weight and relatively easily dissolved in aqueous media. The antibiotic novobiocin (**331**) on hydrolysis with methanolic hydrogen chloride affords a mixture of three methyl glycosides. The major product (**332**) was shown to contain two methoxyl groups, one hydroxyl group, one *gem*-dimethyl group, and a carbamoyl group, and to give a reducing sugar on hydrolysis. It was stable to periodate, but after acid hydrolysis (**333**) it consumed one equivalent of this oxidant with the generation of 1 mole of formic acid (**334**), showing that the free hydroxyl group is adjacent to the glycosidic link. Alkaline hydrolysis of the glycoside afforded urea and a glycosidic vicinal diol (**335**), which was oxidized by one equivalent of periodate to a dialdehyde (**336**) that gave glyoxal (**337**) on acid hydrolysis. Acid hydrolysis of the *N*-free glycoside (**335**) gave a sugar (**338**) that consumed two equivalents of periodate, with the production of 2 moles of formic acid. The aldehydic sugar **333** was converted via the thioacetal into the reduced compound **339**, which when oxidized by periodate gave acetaldehyde and a nonvolatile product, converted by bromine into $(-)$-α-methoxy-β-hydroxyisovaleric acid (**340**). Only the formula **332** for the glycoside accounts satisfactorily for these findings. The other two glycosides obtained in the hydrolysis of novobiocin were the α-glycoside isomer of **332** and a positional isomer of **332** in which the –OH and –OCONH$_2$ groups were interchanged; of these the latter failed to react with periodate after hydrolysis of the glycosidic link, but gave the glycoside **335** on alkaline hydrolysis, and can therefore only have the indicated structure [190–193].

The antibiotic streptomycin (**341**) may be degraded into a diguanidino tetrahydroxycyclohexane and a sugar residue. This sugar, streptobiosamine, can be hydrolyzed to an amino sugar, *N*-methyl-L-glucosamine, but the other component, streptose, could not be isolated. The structure of streptose was

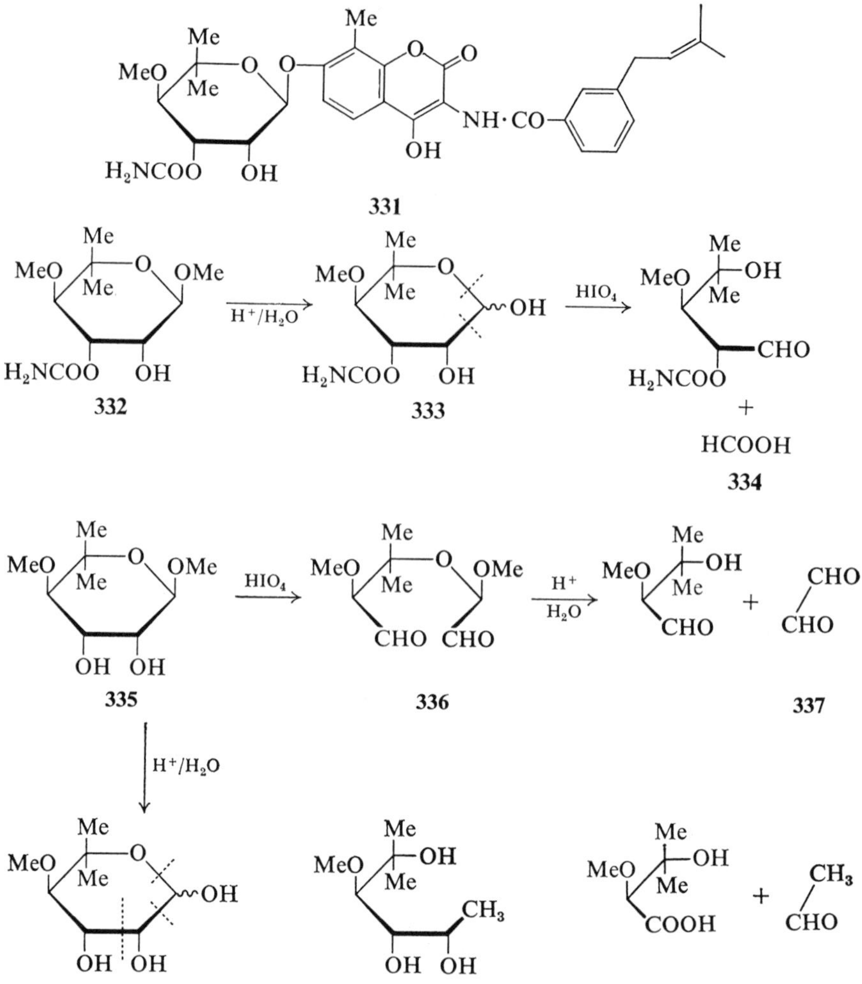

331

332    $\xrightarrow{\text{H}^+/\text{H}_2\text{O}}$ 333 $\xrightarrow{\text{HIO}_4}$ 334

+

HCOOH

334

335    $\xrightarrow{\text{HIO}_4}$ 336 $\xrightarrow[\text{H}_2\text{O}]{\text{H}^+}$ 337

$\downarrow \text{H}^+/\text{H}_2\text{O}$

338   339   340

determined in the following manner. Treatment of streptomycin (341) with ethyl mercaptan and hydrogen chloride gave streptidine (342) and a dithio-acetal, which was acetylated to the compound 343. This was desulfurized with Raney nickel and hydrolyzed to the deoxy-sugar bisdesoxystreptose (344), which on oxidation with periodate followed by hydrolysis gave hydroxyacetaldehyde and acetoin (345). Hydrolysis of the dithioacetal 343 with aqueous mercuric chloride gave the aldehyde 346, oxidative hydrolysis of which yielded the acid 347, and this on oxidation with periodate absorbed 2 moles of the reagent and gave oxalic acid, glyoxalic acid, and acetaldehyde (348) [194–195].

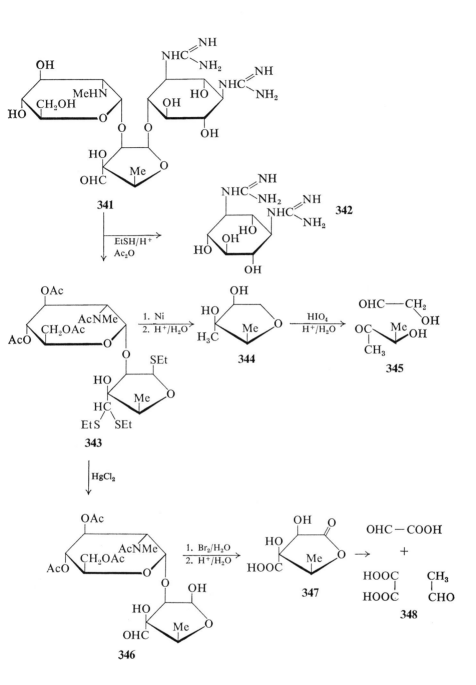

341

342

EtSH/H⁺
Ac₂O

343

1. Ni
2. H⁺/H₂O

344

HIO₄
H⁺/H₂O

345

HgCl₂

346

1. Br₂/H₂O
2. H⁺/H₂O

347

348

α-Amino alcohols are oxidized by periodate providing a primary or secondary amino group is present. For example the diamine actinamine (**349**) is oxidized by periodate with the consumption of 6 moles of the oxidant and complete fission of the molecule to formic acid, whereas the diacetyl derivative **350** consumes only 2 moles of oxidant and gives 1 mole of formic acid (**351**) [196].

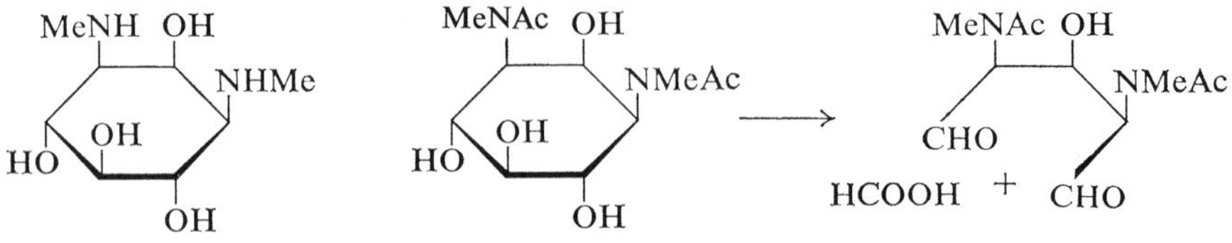

349   350   351

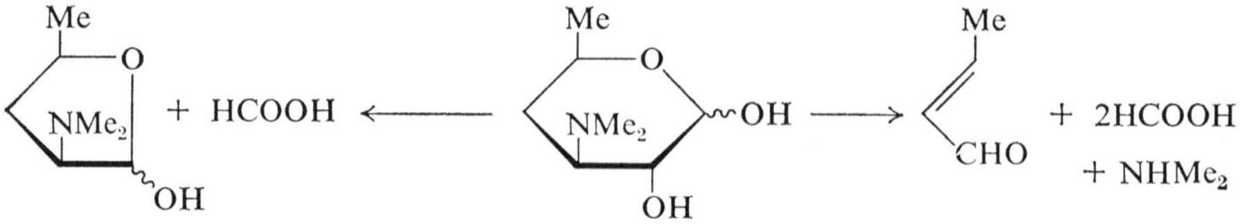

352   353   354

α-Tertiary amino alcohols, however, are not immune to fission, for example desosamine (**353**) readily reacts with 1 mole of periodate to give the masked aldehyde **352** and formic acid and also reacts less easily with a second mole of oxidant to give crotonaldehyde, dimethylamine, and 2 moles of formic acid (**354**) [197, 198]. Periodic acid has been used as a method of estimation of certain vicinal hydroxyaminoacids, by measurement of the amount of ammonia evolved during oxidation [175].

The oxidation of suitably substituted steroid side-chains can be effected in various ways according to the nature of the substituent. For example the triol system present in the ketone **355** is completely oxidized to the diketone **356** [199], whereas the ketol system of Reichstein's substance S (**357**) affords the acid **358** [200].

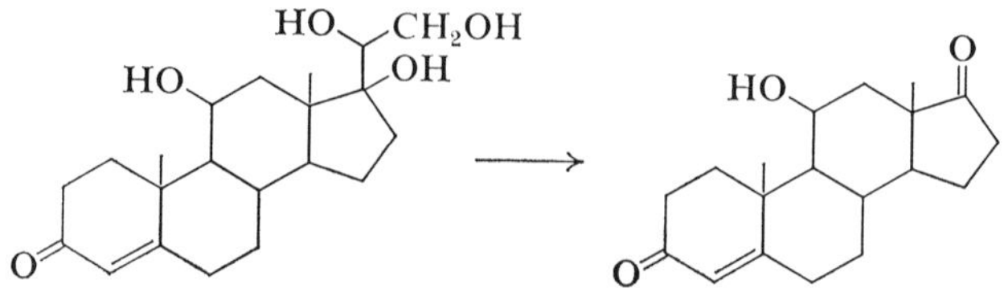

355   356

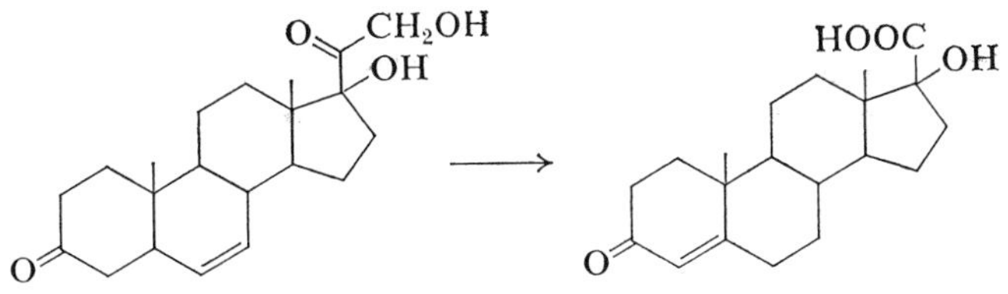

357 → 358

As previously mentioned, the use of periodic acid allied to reactions for the dihydroxylation of double bonds gives a very useful method for fission of olefinic linkages. It is possible to effect fission of double bonds directly by the use of sodium periodate and catalytic amounts of either potassium permanganate [201–206] or osmium tetroxide [207, 208]. In the former case permanganage is the active oxidizing agent, being itself reduced to manganate; the double bond is oxidized to an α-hydroxy ketone, which is cleaved by the periodate. The manganate ion is subsequently reoxidized to permanganate which is thus continuously available for oxidation. When the products include aldehydes, these are usually oxidized to the corresponding acid, but formaldehyde is more resistant to oxidation if the pH is kept at 7–8, and terminal methylene groups may be detected, though the yield of formaldehyde is not quantitative. Quantitative yields of acetone may be obtained in this way from compounds containing isopropylidene groups.

With periodate and osmium tetroxide, hydroxylation of the double bond first occurs (see p. 209) and the periodate then cleaves the resulting glycol and oxidizes the osmic acid to oxmium tetroxide, which is made continuously available for oxidation. Aldehydes are not oxidized by osmium tetroxide, and quantitative yields are frequently obtained. The method is particularly suitable for the estimation of terminal methylene groups.

Side reactions with periodate as an oxidizing agent are not extensive. Reactive C–H bonds may be oxidized and the product of such a reaction may then be cleaved as a reactive glycol or hydroxy ketone. Such processes result in overconsumption of the reagent in relation to the number of cleavable bonds present in the starting material and limit the applicability of the reagent in quantitative analytical processes [139, 209, 210].

Phenols are oxidized by periodic acid to quinones [139, 211, 212].

Sodium Bismuthate, Iodoso Compounds

Sodium bismuthate in acid media [213] and aryl iodosoacetates and benzoates [152, 214–216] will also in general cleave 1,2-diols. These reagents are less specific than periodic acid, and like lead tetraacetate they will effect the rapid oxidation of α-hydroxy acids to ketones and carbon dioxide.

The reactions may be represented as proceeding by the mechanisms illustrated by structures **359–363**.

The oxidation of α-hydroxy ketones to α-diketones by bismuth oxide in acetic acid is probably a related process [217].

Sodium bismuthate is quite stable in acid media and the reactions may be carried out in water, aqueous alcohols, dioxane, or acetic acid. There is no

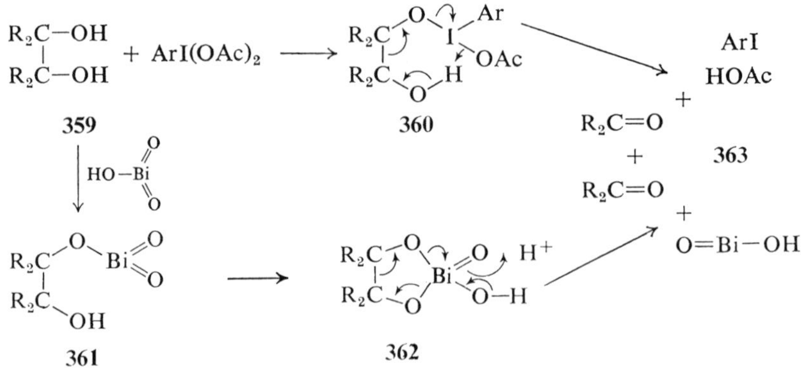

difference in the rate of oxidation of *cis-* and *trans-*cyclohexane-1,2-diols with this reagent, but the reaction probably occurs at the surface of the particles of sparingly soluble sodium bismuthate, where the structure of the diol may be of no importance.

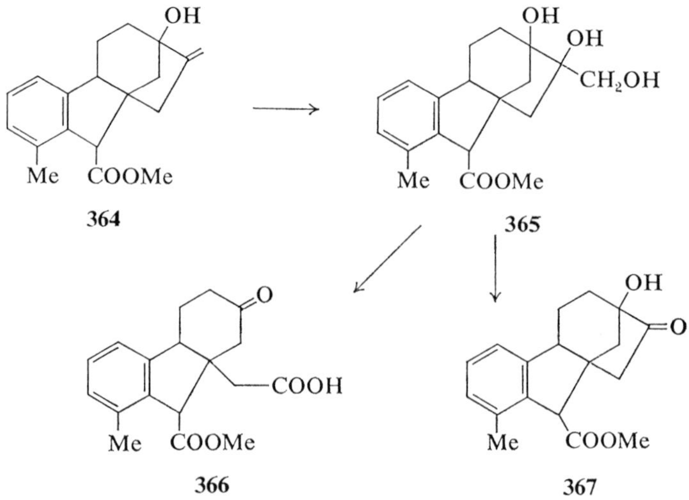

Aryl iodoso acetates may be prepared most conveniently by the oxidation of aryl iodides by hydrogen peroxide in acetic anhydride. They are used as oxidizing agents most commonly in acetic acid. The reaction may be followed by the removal of aliquots of the solution and measuring the iodine liberated on addition of potassium iodide.

These two reagents have no obvious advantages over lead tetraacetate and periodic acid, and have not been widely used. Sodium bismuthate was used in studies leading to the elucidation of the structure of allogibberic acid, a product of hydrolysis of gibberellic acid. The methyl ester of allogibberic acid (364) was oxidized to the triol 365, from which sodium bismuthate gave the ketol 367 and the keto acid 366. The ketol 367 was also obtained from allogibberic acid by ozonolysis and esterification; it was oxidized to the acid 366 by sodium bismuthate and must therefore be an α-hydroxy ketone and since it is not oxidized by bismuth oxide to an α-diketone, the hydroxyl group must be tertiary [218].

Ceric Salts, Vanadium(V)

Vicinal glycols can be oxidized with carbon–carbon double bond fission by ceric salts [219] and pentavalent vanadium compounds, such as ammonium vanadate, in acid solution [220]. The oxidation processes are essentially different from those involved with lead tetraacetate and periodic acid since they are one-electron transfer rather than two-electron transfer reagents and free radical mechanisms are involved. The reactions may be represented generally by the following mechanisms: $368 \rightarrow 370 \rightarrow 372 \rightarrow 373$ and $368 \rightarrow 371 \rightarrow 372 \rightarrow 373$. The oxidation with ceric salts is applicable to all glycols, but the vanadium(V) oxidation is applicable only to ditertiary glycols, since with secondary glycols the C–H bond of the CH–OH group is broken in preference to the C–C bond, and the oxidation process is then the same as for secondary alcohols $374 \rightarrow 375$. In a similar manner oxidation of acetoin (376) with vanadium(V) gives biacetyl (377) whereas oxidation of methyl acetoin (378) gives acetone and acetaldehyde (379).

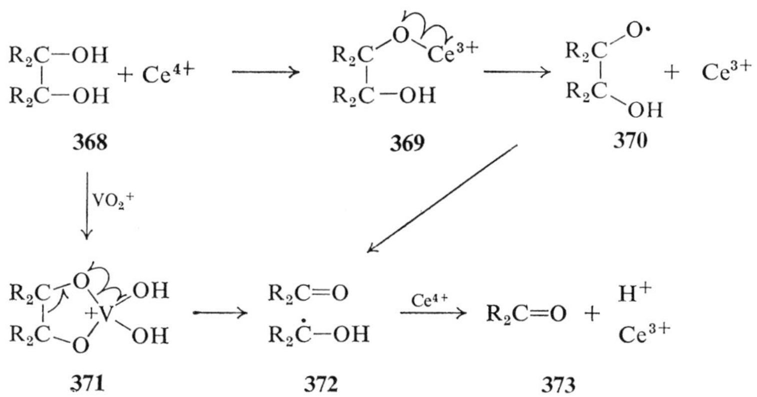

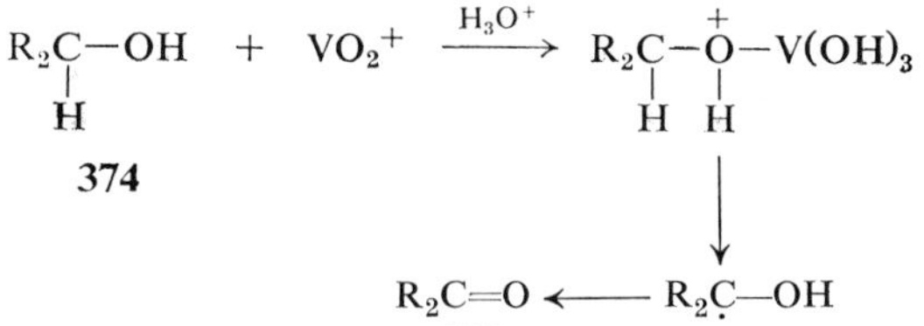

$$R_2C-OH + VO_2^+ \xrightarrow{H_3O^+} R_2C-\overset{+}{O}-V(OH)_3$$

$$\underset{H}{|} \qquad\qquad \underset{H\ H}{|\ |}$$

374

$$R_2C=O \longleftarrow R_2\overset{.}{C}-OH$$

375

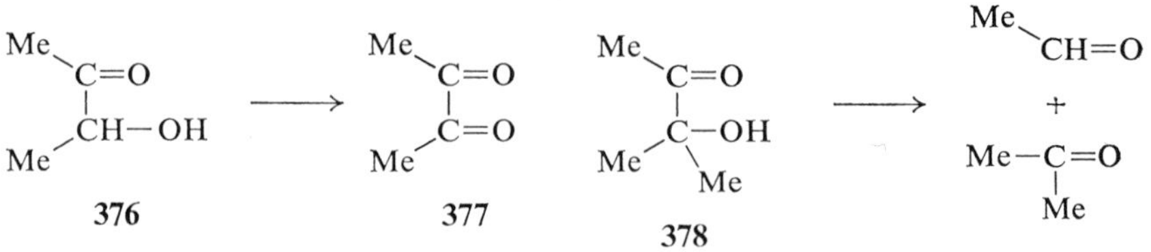

376 → **377** → **378** → **379**

With α-hydroxy acids, vanadium(V) results in oxidation to an aldehyde and carbon dioxide, whereas ceric salts can oxidize such acids to α-keto acids and to carbon dioxide and a lower acid [221]:

$$RCHO + CO_2 \xleftarrow{V^V} \underset{\underset{OH}{|}}{RCHCOOH} \xrightarrow{Ce^{vi}} RCOCOOH \xrightarrow{Ce^{iii}} RCOOH + CO_2.$$

Ketones are oxidized by both reagents first to α-hydroxy ketones and subsequently to acids with carbon–carbon bond fission [219, 220].

The reactions have been used only occasionally for the elucidation of structures, since a much more extensive background to the two-electron transfer oxidations with lead tetraacetate and periodic acid is available.

Hydrogen Peroxide, Peracids

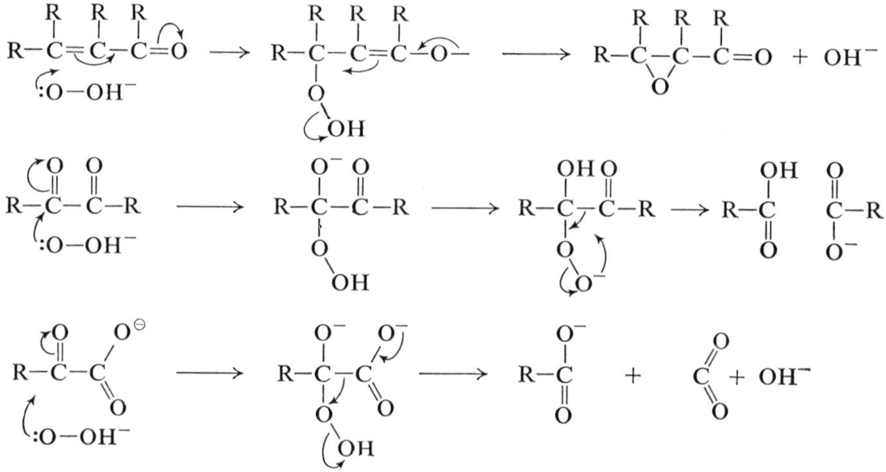

Hydrogen peroxide in alkaline solution will oxidize α,β-unsaturated ketones to the corresponding epoxides, α-diketones to carboxylic acids, and α-keto acids to carbon dioxide and lower acids. The salts of α,β-unsaturated acids are unaffected. The mechanisms of these processes are similar and involve attack of the electron-deficient carbon atom by a hydroperoxide anion [222–224]. Oxidoketones bearing an enolizable hydrogen atom may in many cases react with alkali to give the corresponding α-diketone, which may be further oxidized:

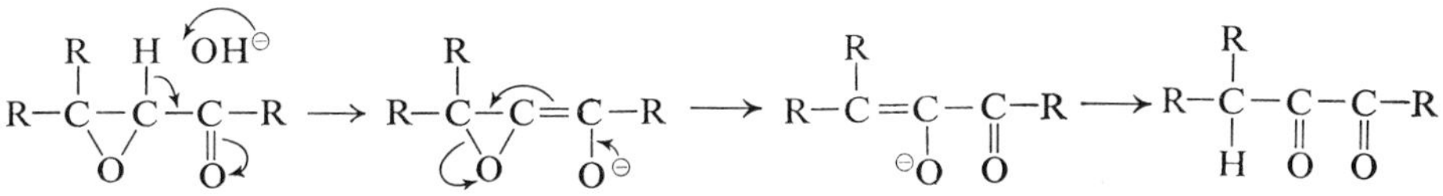

This process does not occur with oxido cyclic ketones because the intermediate enolate ion would be highly strained, but in such compounds the oxide ring may be opened by a different route, involving ring contraction **380 → 382**, followed by cleavage of the three-membered ring.

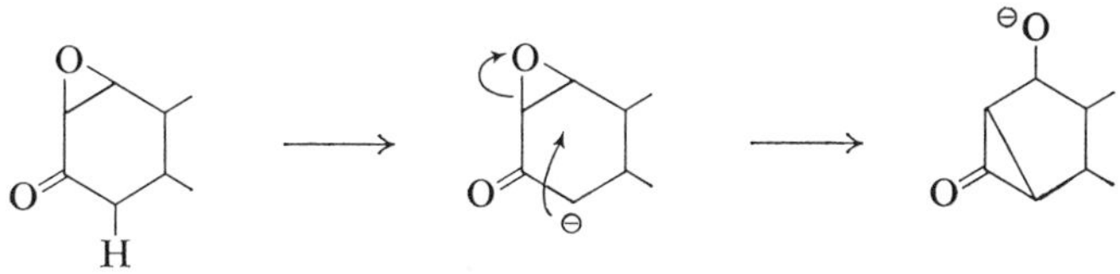

380   381   382

This reagent has been used for the degradation of tropolones, which may be oxidized in the α-diketone form. For example, puberulic acid (**383**) affords a good yield of aconitic acid (**385**) when oxidized by hydrogen peroxide in $1N$ sodium hydroxide at low temperatures [225], whereas more vigorous oxidation affords no products of structural significance [226]. The reagent

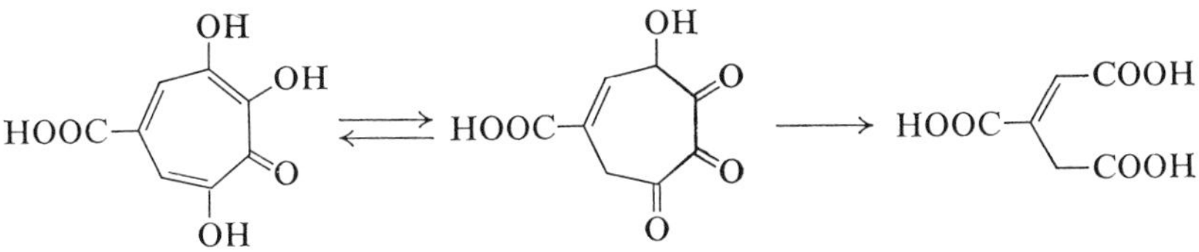

383   384   385

can, however, cause hydroxylation of tropolones [227], and this can affect the mode of fission, for example, hinokitiol (**387**) yields the acids **390** and **391**, presumably by way of the hydroxylated tropolones **386** and **388** [228–230].

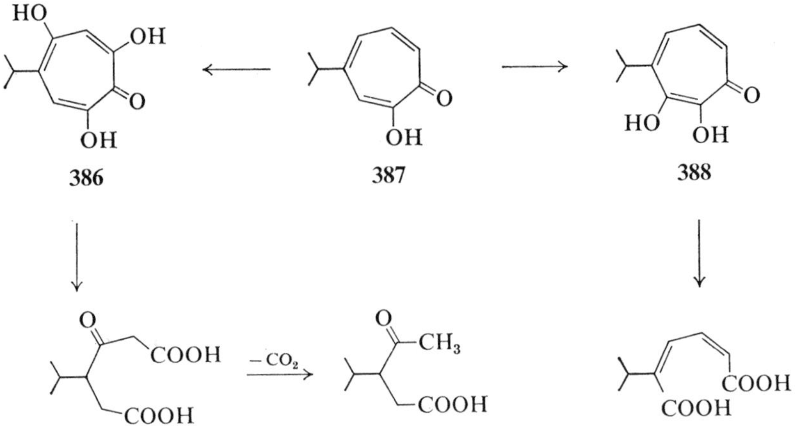

386   387   388

389   390   391

The reagent was used to oxidize a hydroxy-α-diketone during work on the determination of the point of attachment of the side chain in lanosterol. The diketone 392, derived from lanosterol, on oxidation with alkaline hydrogen peroxide gave a monobasic acid formulated as the keto acid 393 by the original workers [231, 232], but as a hydroxy acid 394 by Jones and Halsall [233]. This acid, as would be expected on the basis of either structure, was oxidized further by lead tetraacetate to the lower acid 395 and carbon dioxide. Although spectral evidence was advanced in support of the keto-acid structure 393, the hydroxy-acid formulation 394 is preferable since such a structure would be expected to be stable to alkaline hydrogen peroxide, unlike the α-keto acid 393 which should suffer further oxidation with this reagent to the acid 395. A mechanism for the production of the keto acid 393 would be:

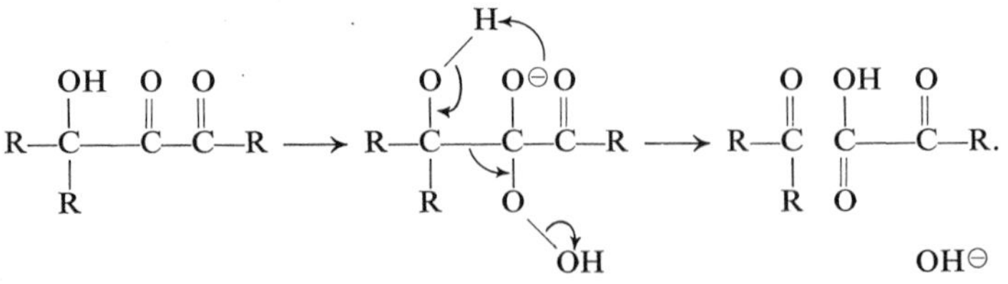

The hydrolysis of the acid 395 followed by oxidation with chromic acid afforded a triketone with loss of carbon dioxide. This decarboxylation must involve a β-keto acid, and the activating carbonyl group must be located at C-17 and must be derived from the acetoxy group at that position in the diketone 392, and that acetoxy group was introduced at the point of attachment of the side chain in lanosterol [231, 232].

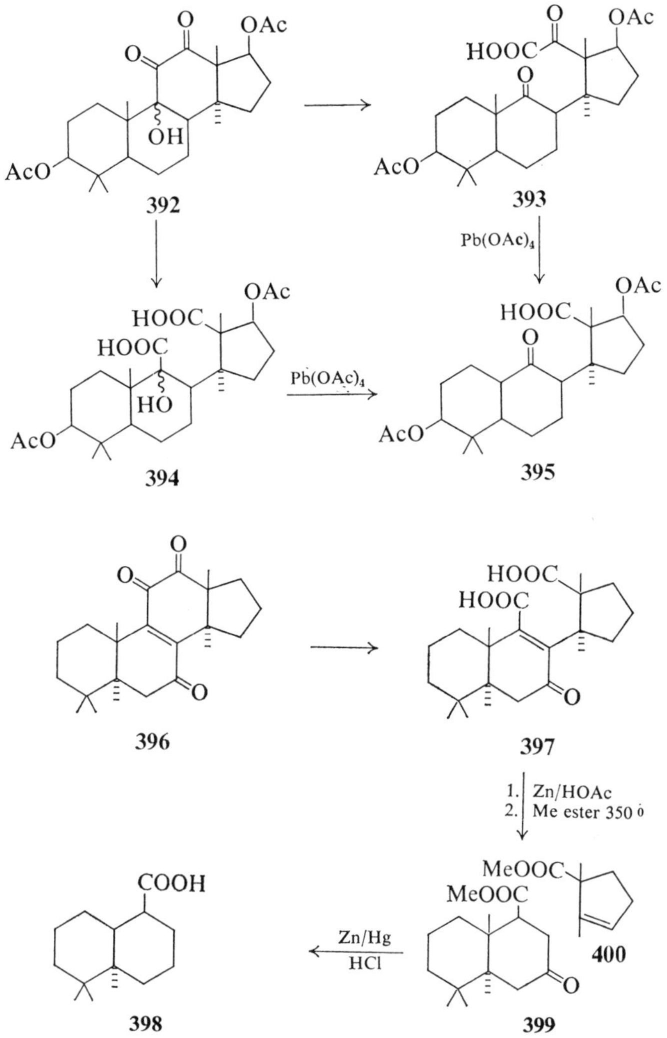

An oxidative fission of an α-diketone system in Ring C by alkaline hydrogen peroxide was also involved in the elucidation of the nature of Rings A and B of lanosterol. The triketone (**396**) on oxidation in this way gave the dibasic acid **397**, the methyl ester of which on reduction and pyrolysis afforded the esters **399** and **400**. Clemmensen reduction of the ester **399** gave the acid **398** of known structure, previously obtained from manoöl [234].

Both hydrogen peroxide and peracids generally in acidic media have been used for the conversion of ketones into esters by the Baeyer-Villiger oxidation

[235]. The reaction may be represented by the following mechanism: **401 → 403**. It has been utilized in the study of the structure of natural products. In the steroid and triterpene series side-chain keto-groups have been converted into esters, for example, the ketone **404**, derived from lanosterol,

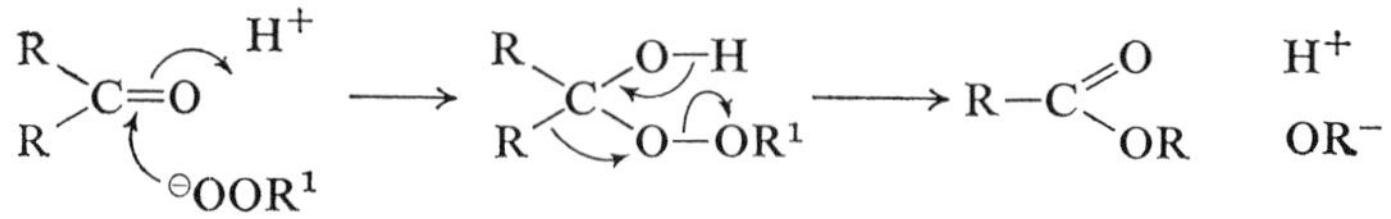

401   **402**   **403**

was oxidized to the ester **405** by perbenzoic acid, and this ester on further oxidation with selenium dioxide gave the α-diketone **392** already mentioned [231, 232]. Sarsapogenin (**406**) was related to pregnan-3,16,20-triol (**408**) by oxidation and hydrolysis as shown [236]. Application of the reaction to cyclic ketones yields lactones from which hydroxy acids can be obtained.

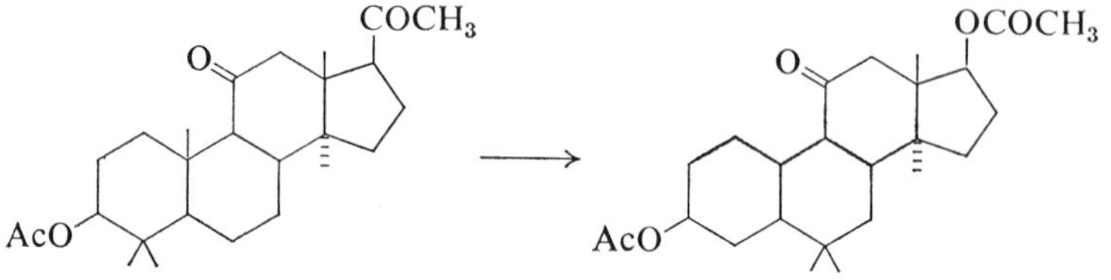

404   **405**

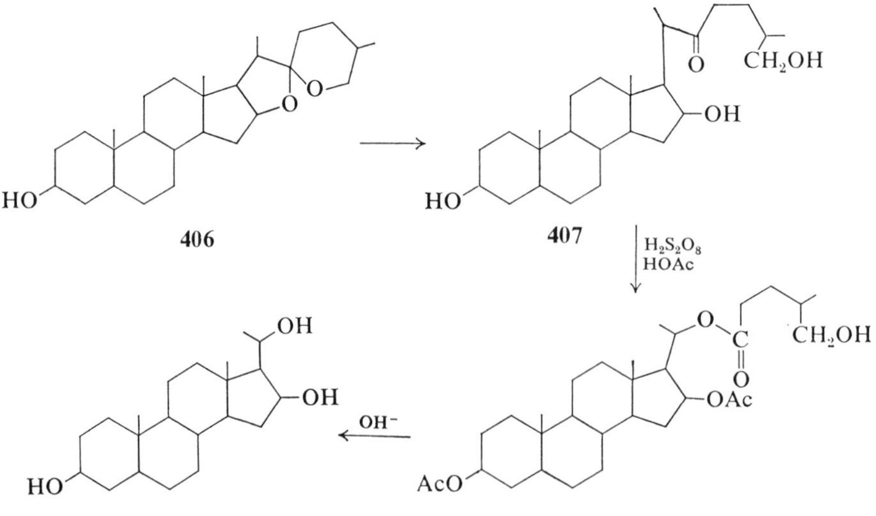

406   **407**

$\begin{array}{c} H_2S_2O_8 \\ HOAc \end{array}$

408   **409**

In the steroid series ketones with carbonyl groups at C-3, C-7, and C-17 are readily oxidized, whereas C-12 ketones are oxidized with difficulty. The reaction was utilized in studies of the biogenesis of labeled steroids, for example the ketone **410** was converted via the lactone into the hydroxy acid **411** which could be decarboxylated, and in material incorporating ^{14}C from $CH_3^{14}COOH$, radioactive carbon was lost during the decarboxylation [237].

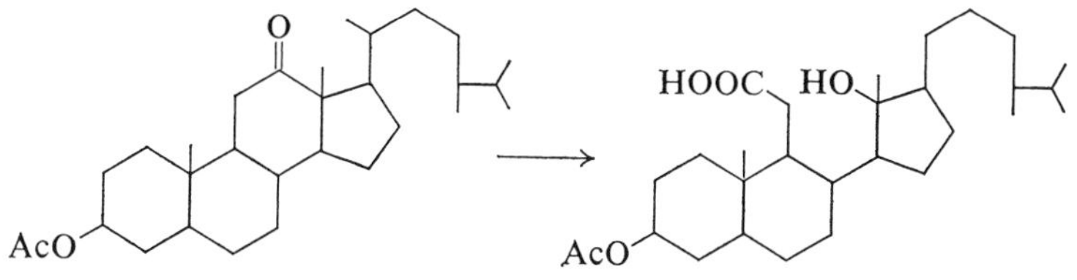

The direction of the fission was not rigorously proved [238], but does not affect the evidence of incorporation of ^{14}C at C-12 in the ketone. In a similar manner it has been shown that carbon atom C-7 in Ring B of cholesterol originates from the methyl group of acetic acid [239, 240].

Apoaromadendrone (**412**) has been oxidized to the hydroxy acid **413** which was converted by two successive Barbier-Wieland degradations (see p. 239) into the acid **414**, thus establishing the relationship of the three- and five-membered rings. The oxidation of the hydroxy acids **413** and **414** to derivatives of cyclopentanone showed that the parent carbonyl group is attached to the cyclopentane **412** and not the cyclopropane ring [241].

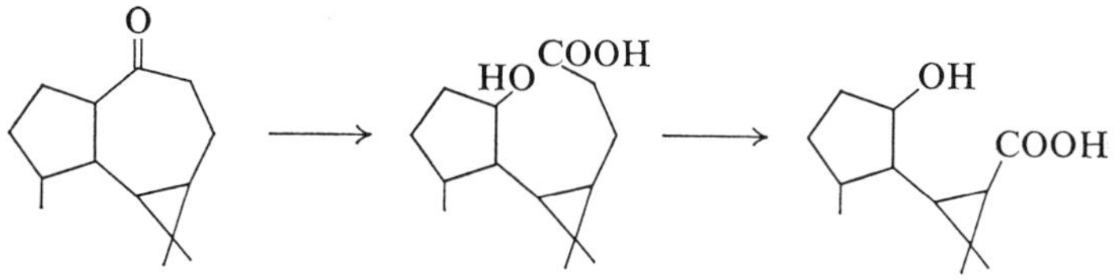

An unusual loss of a methyl group in the Baeyer-Villiger oxidation of 4,4-dimethylcholestan-3-one (**415**) to the lactone **416** has been observed [242].

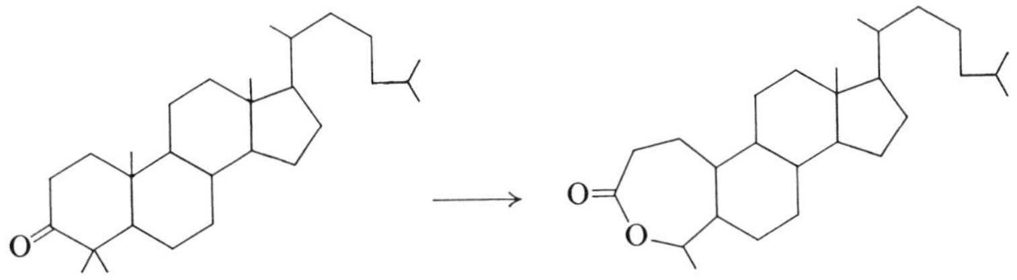

Diaryl ketones **417** are generally oxidized with ease by the Baeyer-Villiger procedures, and the reaction is to be commended in structural studies as being less drastic than alkali fusion or hydrolytic fission with acids (see Section 3). In general the migrating group is the more electron-releasing of the two. With aryl alkyl ketones no general rule can be given for predicting the direction of oxidation, and fissions on both sides of the carbonyl group

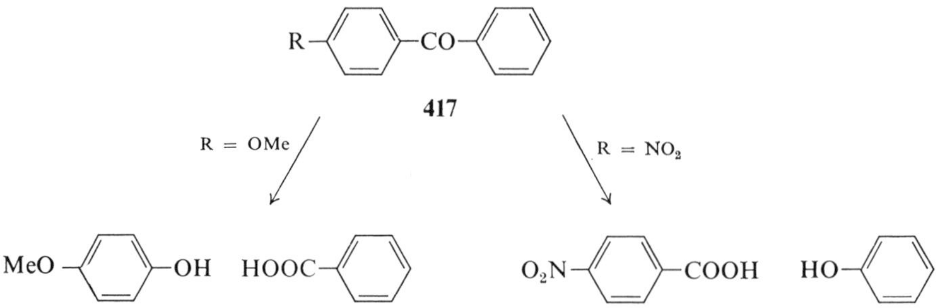

417

R = OMe

R = NO₂

are common [243–245]. The reaction has been widely used in the oxidation of *o*- and *p*-amino or hydroxy aromatic ketones and aldehydes under the general name of the Dakin reaction. It has especially been applied in structural studies of naturally occurring phenols. For example, clavatol (**418**), which shows the properties of a monohydric phenol, was proved to contain a carbonyl group and a second hydroxyl group by oxidation to 3-hydroxy-2,6-dimethylbenzoquinone (**419**) [246].

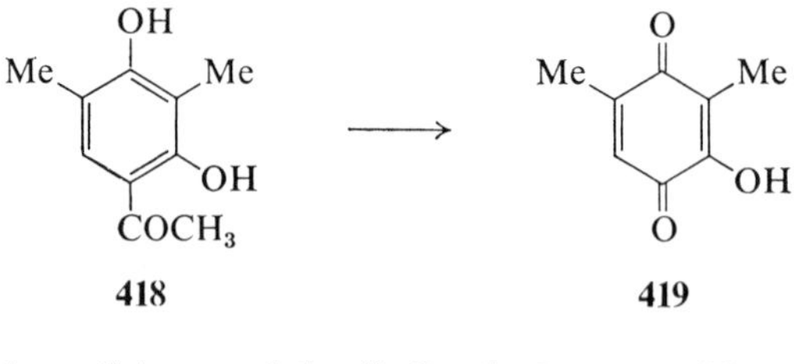

418   **419**

Experimental conditions and details for the Baeyer-Villiger-Dakin reaction are given in a review by Hassall [235].

The Haloform Reaction

Methyl ketones and compounds that can generate methyl ketones under the alkaline oxidation conditions of the reaction are oxidized to carboxylic acids and a haloform, $CHCl_3$, $CHBr_3$, or CHI_3, under a variety of experimental conditions that can be essentially represented as the use of alkaline hypohalites. The reaction proceeds with the formation of a trihaloketone by carbanion generation and reaction with halogen, followed by alkaline cleavage of the ketone:

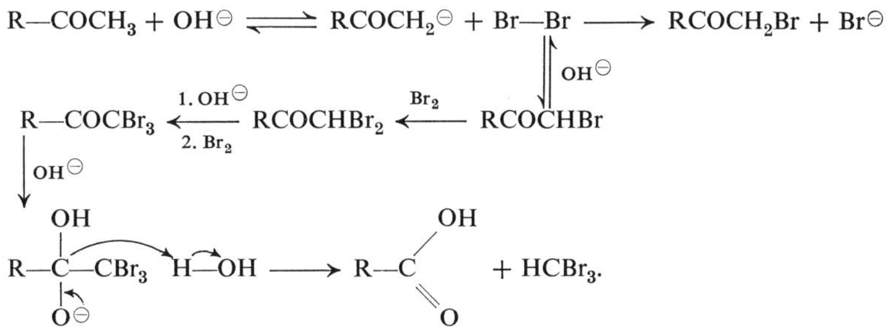

$$R{-}COCH_3 + OH^\ominus \rightleftharpoons RCOCH_2^\ominus + Br{-}Br \longrightarrow RCOCH_2Br + Br^\ominus$$

$$R{-}COCBr_3 \xleftarrow[\text{2. Br}_2]{\text{1. OH}^\ominus} RCOCHBr_2 \xleftarrow{\text{Br}_2} RCOCHBr$$

$$R{-}\underset{\underset{O^\ominus}{|}}{\overset{\overset{OH}{|}}{C}}{-}CBr_3 \quad H{-}OH \longrightarrow R{-}\overset{OH}{C}\diagdown_{O} \quad + HCBr_3.$$

When the reaction is carried out in aqueous methanol the initial product is a methyl ester:

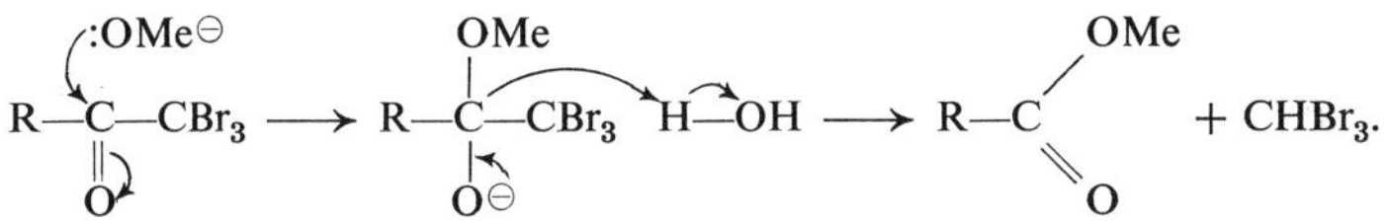

$$R{-}\underset{O}{\overset{:OMe^\ominus}{C}}{-}CBr_3 \longrightarrow R{-}\underset{\underset{O^\ominus}{|}}{\overset{\overset{OMe}{|}}{C}}{-}CBr_3 \quad H{-}OH \longrightarrow R{-}\overset{OMe}{C}\diagdown_{O} \quad + CHBr_3.$$

In structural studies the reaction has been most used in the detection and oxidation of methyl ketones obtained as a result of the oxidation of olefins bearing a methyl group on the double bond, particularly in the terpene series, for example, as in the conversion of dihydrocarveol (**420**) into the acid **421** which was dehydrogenated to 3-hydroxy-4-methylbenzoic acid

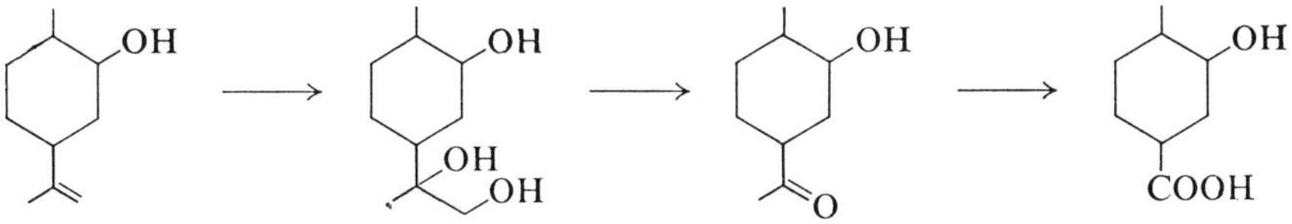

420   **421**

[247–249], and the oxidation to β,β-dimethyladipic acid (**24**) of isogeronic acid (**21**) obtained from α-ionone (**19**). The structure of carotol (**422**) was demonstrated by the sequence of reactions shown in formulas **422** → **430**. The double bond was hydroxylated and the vicinal glycol **423** protected during dehydration and reduction. Oxidation of the glycol **425** with periodate gave the aldehydic ketone **426** which was oxidized by permanganate and a haloform reaction to the dibasic acid **427**. This on pyrolysis gave a ketone (**428**) which was oxidized to a dibasic acid that gave a cyclopentanone derivative (**430**) on pyrolysis [250–252].

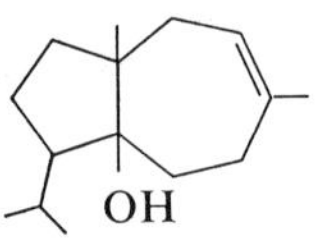

422 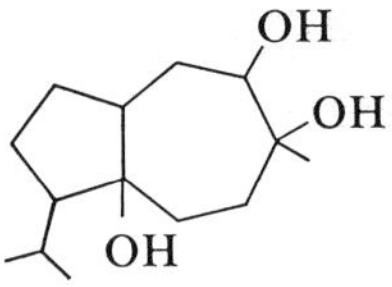 → **423** 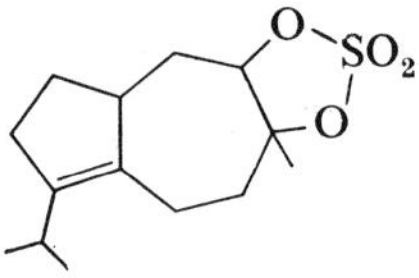 → **424**

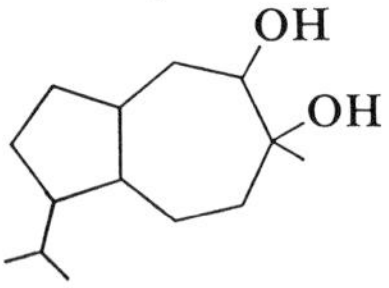

424 →

425 ← **426** ← **427**

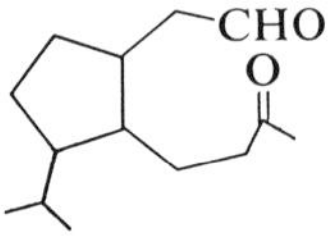

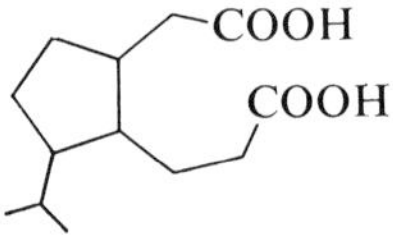

427 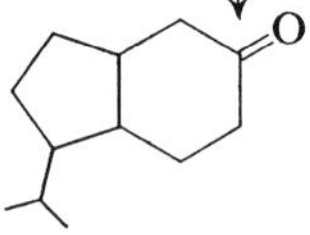 → **428** 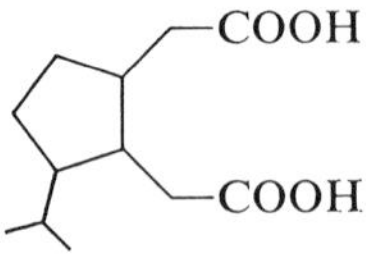 → **429** 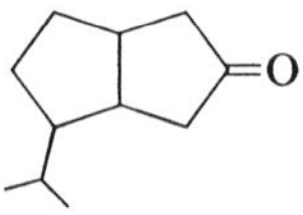 → **430**

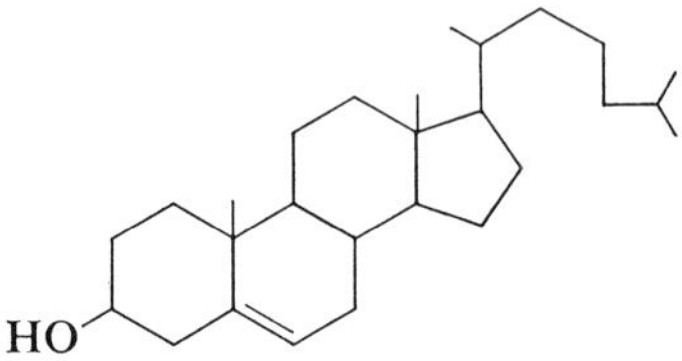

431

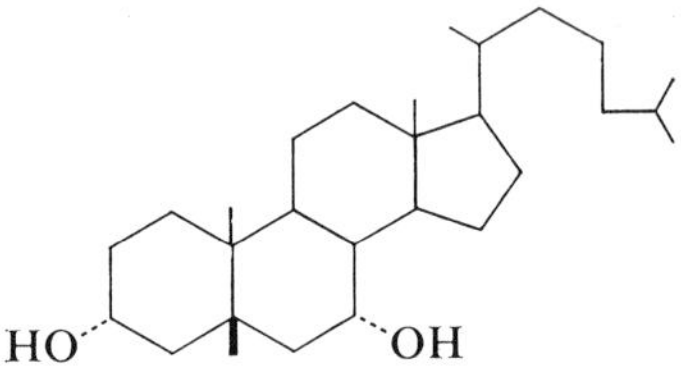

432

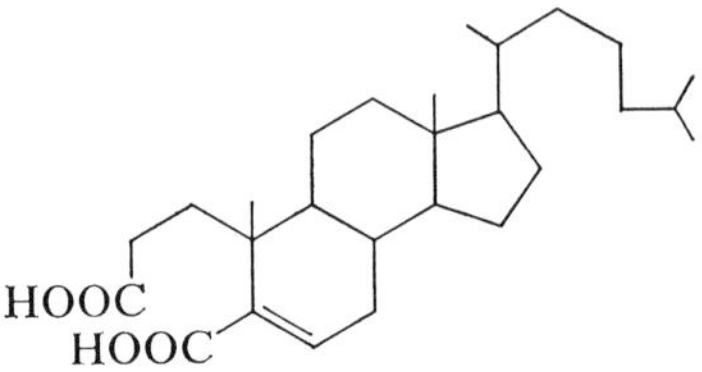

433

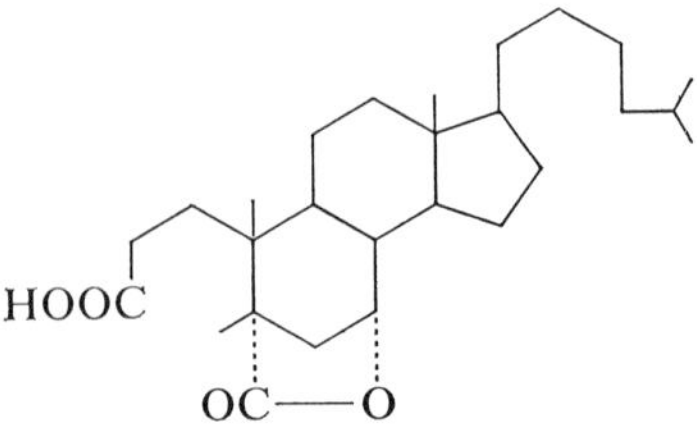

434

The haloform reaction is generally regarded as specific for the oxidation of R–COCH$_3$ to R–COOH, but other reactions can occur, for example, Ring A alcohols of the steroid series can be oxidized to dibasic acids, cholesterol (431) [253], and chenodesoxycholic, and 432 [254] being oxidized to Diels acid 433 and the lactone 434, respectively. This process may be regarded as

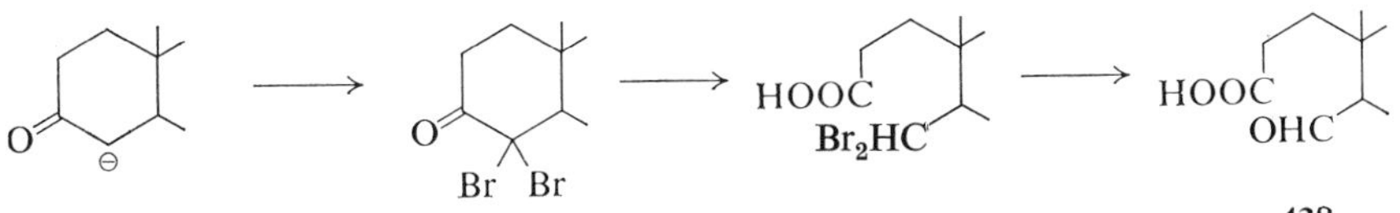

proceeding by essentially the same mechanism as that set out above, namely oxidation to a ketone, bromination through the carbanion 435 to a *gem*-dibromoketone (436), cleavage by alkali (437), hydrolysis to an aldehyde (438), and further oxidation.

The haloform reaction may be carried out in aqueous solution or in an alcohol that is not oxidized by the reagent, for example, methanol. An alternative process that has been used in some cases for the conversion of R–COCH$_3$ to R–COOH involves the bromination of the ketone and the treatment of the bromo compound with pyridine. The pyridinium salt thus formed can then be hydrolyzed by dilute alcoholic sodium hydroxide [255, 256]. The process can be represented as in 439–441. The reaction has been simplified by the use of the methyl ketone, iodine, and pyridine or

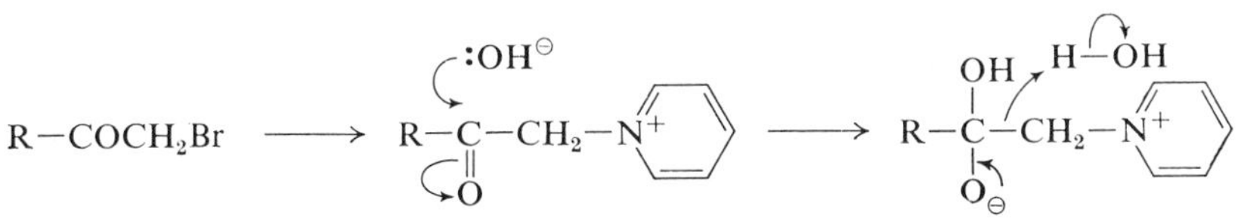

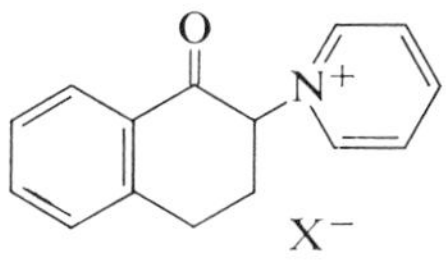

related bases, which give good yields of quaternary salts directly [257, 258]. The salts (442, X = Br or I) on prolonged heating with sodium hydroxide give the acids (443, X = Br or I) [259, 260], a reaction that resembles the ring fission of derivatives of cholesterol 431 and 432.

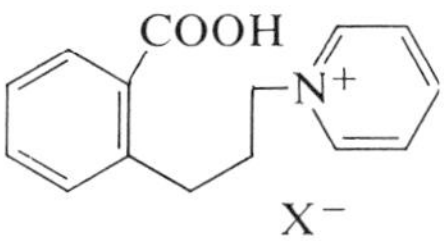

Oxides of Nitrogen

Nitric acid was used quite widely as an oxidizing agent in early work in organic chemistry, but it is an unspecific oxidizing agent and unless used under carefully controlled conditions it frequently leads to widespread degradation with the formation of acids of low molecular weight, and aromatic nuclei often suffer nitration with the result that its use has largely been superseded by the use of other oxidizing agents of more predictable behavior. Its oxidizing power depends on the conditions under which it is used. In concentrated solution it is reduced to nitrogen dioxide and in dilute solution to nitric oxide:

$$NO_3^- + 2H^+ + \varepsilon^- \rightarrow NO_2 + H_2O$$
$$(2HNO_3 \rightarrow 2NO_2 + H_2O + O)$$
$$NO_3^- + 4H^+ + 3\varepsilon^- \rightarrow NO + 2H_2O$$
$$(2HNO_3 \rightarrow 2NO + H_2O + 2O).$$

A wide range of oxidations has been effected in this way, and several examples are given here. Strychnine (**444**) with dilute nitric acid gives picric

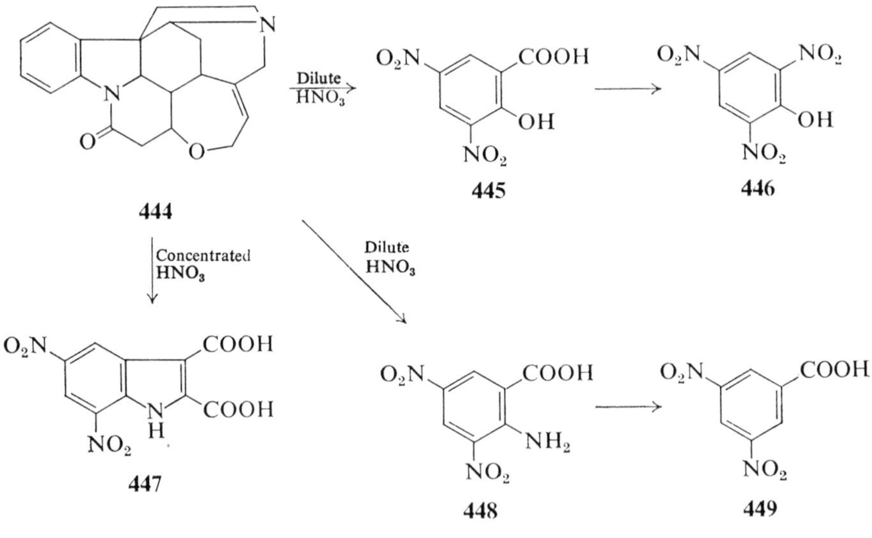

acid (**446**) [261] and 3,5-dinitrobenzoic acid (**449**) [262], doubtless via the acids **445** and **448** [263], and with concentrated nitric acid gives dinitrostrychol carboxylic acid (**447**) [264].

Trimethylbrazilone (**450**) is oxidized by nitric acid to the ketone hydrate lactone **451**, the structure of which was proved by its conversion into 4-methoxysalicylic acid (**452**) and the osazone **453** on treatment with phenylhydrazine [265].

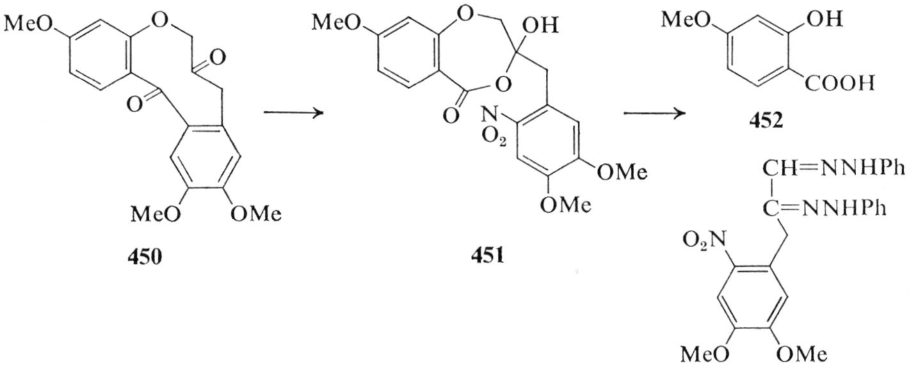

450　　　　**451**　　　　**452**

453

The acid **455** is obtained from both helminthosporin (**454**) [266] and kermesic acid (**456**) [267] on oxidation with nitric acid.

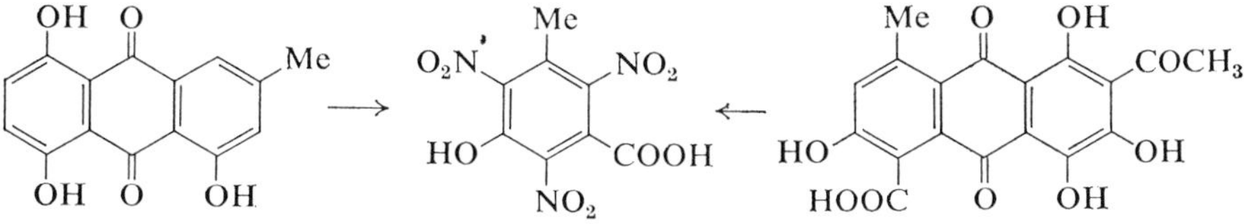

454　　　　**455**　　　　**456**

Dilute nitric acid oxidizes magnamycin (**457**) very rapidly and gives *cis*-ethylene oxide dicarboxylic acid (**458**), thus confirming the presence in the molecule of an epoxide ring inferred from other less direct evidence [268].

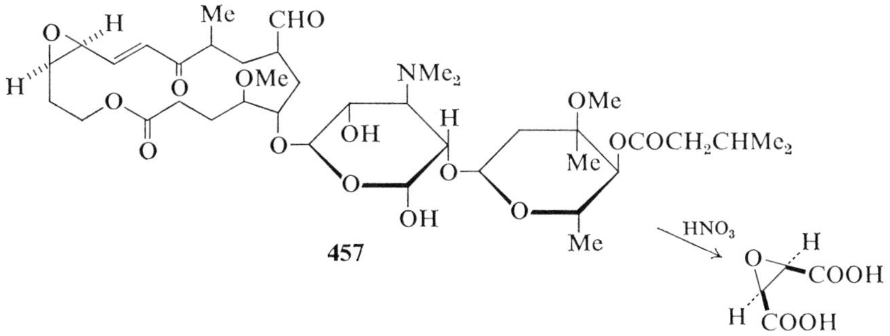

457

458

The survival of the ethylene oxide system in this reaction under acid hydrolyzing conditions is remarkable.

Cyclic ketones and the related secondary alcohols may be oxidized to dicarboxylic acids, for example, cyclohexanone and cyclohexanol both give adipic acid, and the ketones **459** [269, 270], **460** [271, 272], and **461** [273] are all oxidized to the corresponding dibasic acids. In such cases where other functional groups, double bonds, or aromatic nuclei are not present, yields

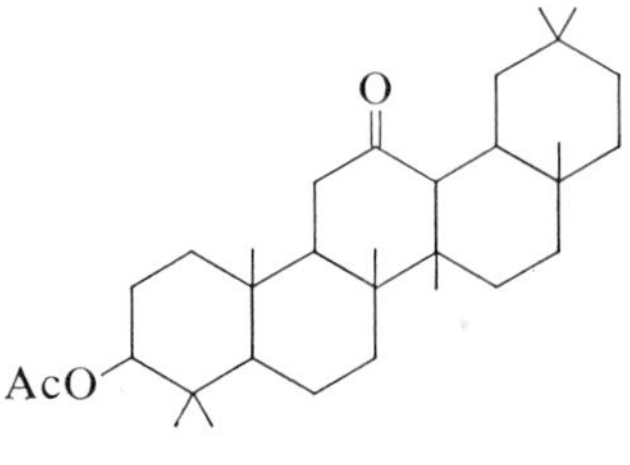

459

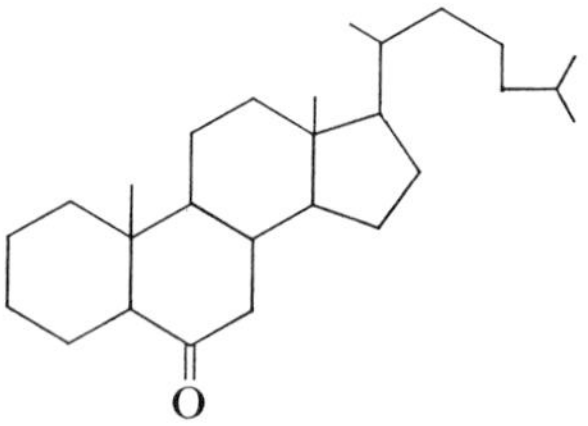

460

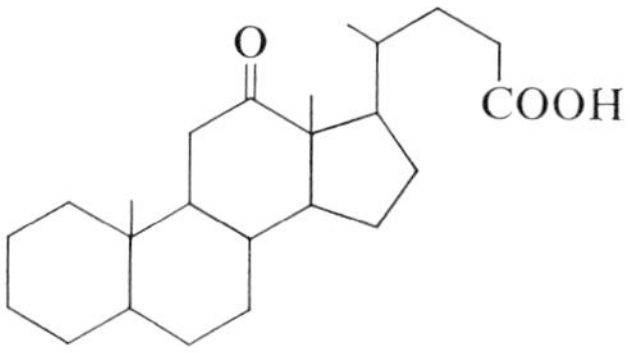

461

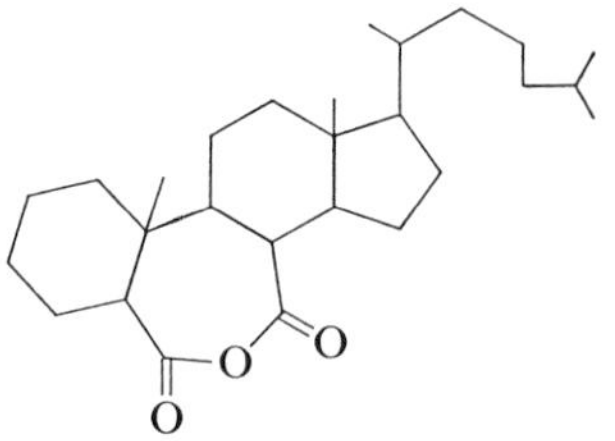

462

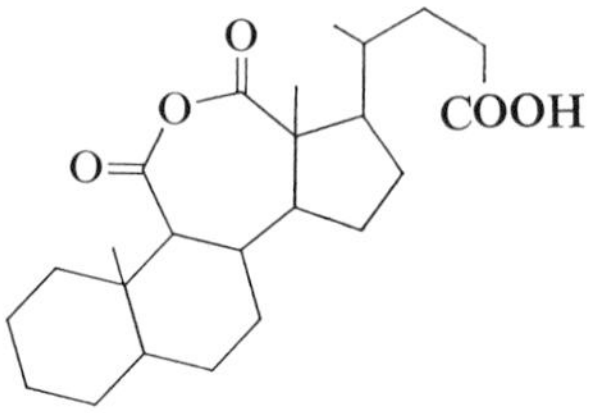

463

are good and overoxidation is not a problem. It may be noted that the acids obtained from the ketones **460** and **461**, although derived from six-membered cyclic ketones, give anhydrides **462** and **463** rather than ketones on heating, in apparent contravention of the Blanc rule (see page 148). It is now known that the Blanc rule can only be *reliably* applied to dicarboxylic acids where the carboxyl groups are not themselves attached to two different rings [274],

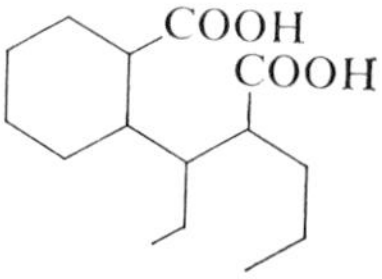

464

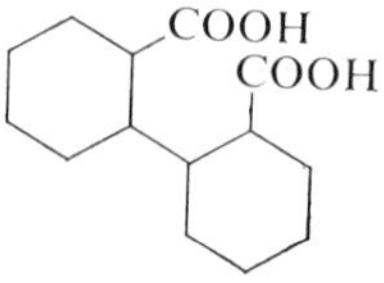

465

that is, the rule may be applied to acids of the type **464** but not to those of the type **465**. This is only a generalization however, and acids of the general

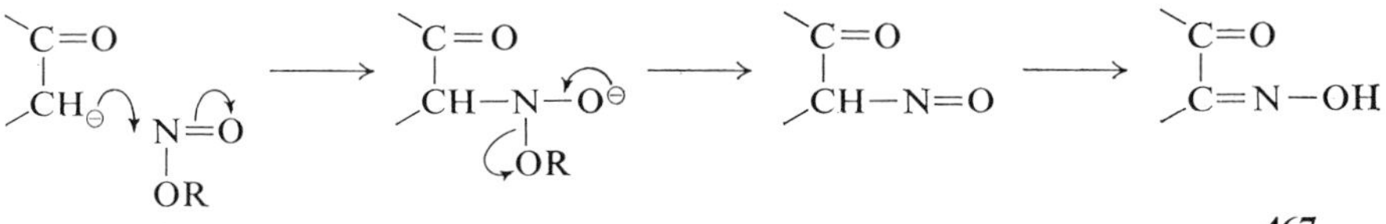

466   **467**

type **465**, for example, that derived from the ketone **459**, are known that give cyclic ketones on pyrolysis in agreement with the original Blanc rule.

Nitrous acid has not been utilized directly as a reagent for the fission of carbon–carbon bonds, but by virtue of the ability of the acid and its esters to attack reactive $-CH_2-$ and $-\overset{|}{C}H-$ groups, it has been used to convert ketones into derivatives, generally α-diketones, suitable for further degradation.

Esters of nitrous acid are generally used in basic media where carbanions formed from the reactive organic compound can react with the ester (**466** → **467**). The α-oximino ketone so formed can be hydrolyzed to an α-diketone. These reactions have been used in structural studies both to prove the presence of a reactive $-CO-CH_2-$ system [275] and to facilitate oxidation,

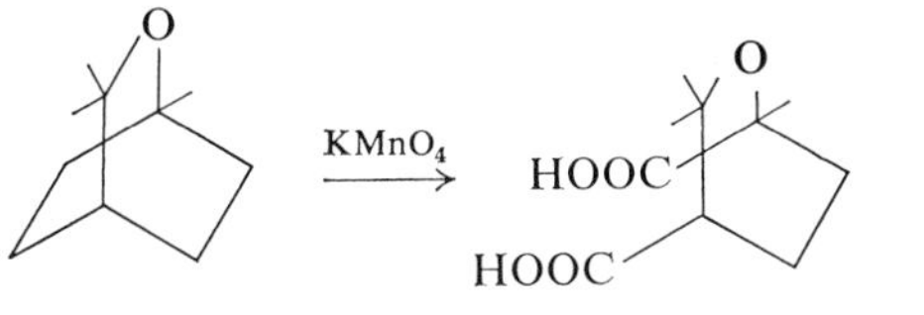

468   **469**

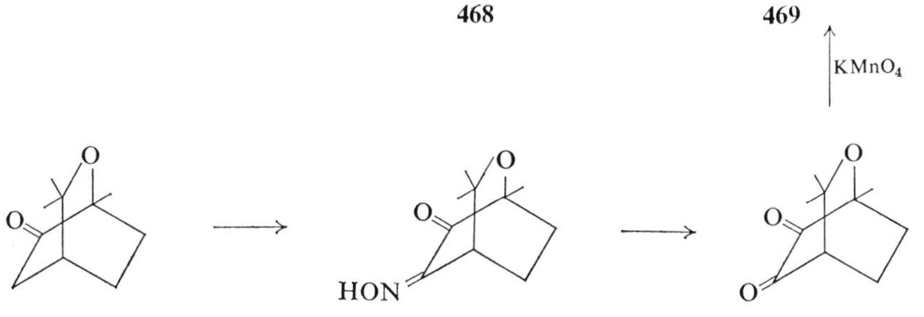

470   **471**   **472**

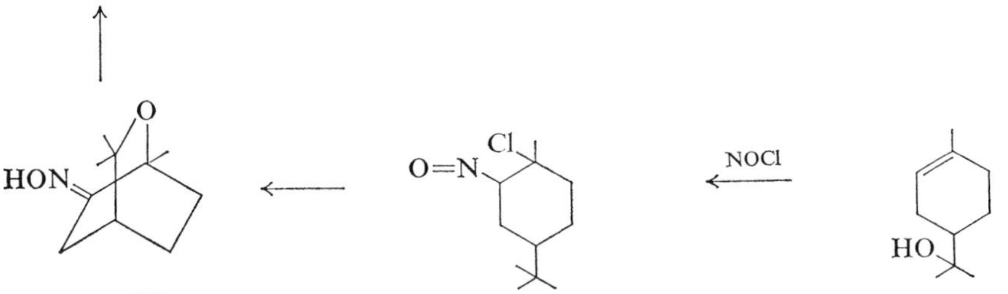

473   **474**   **475**

for example, in the degradation of ketocineole (**470**) through diketocineole (**472**) to the dicarboxylic acid **469** [276] which is obtainable directly from cineole (**468**), thus establishing the relationship between cineole and α-terpineol, from which the ketone **470** can be prepared. The conversion of α-terpineol (**475**) into ketocinole (**470**) involves the addition of nitrosyl chloride to the double bond to give the nitroso compound **473**, a general reaction of considerable value in the chemistry of the terpenoids. From the general mechanism (**466**) → (**467**) it will be seen that –CH–CO– groups will also react with esters of nitrous acid to give tertiary nitroso compounds (**476**) that cannot isomerize to oximes, but in such cases attack of the system by base **478** can effect carbon–carbon bond fission with the production of an oximino ester (**479**). In this way menthone (**480**) was

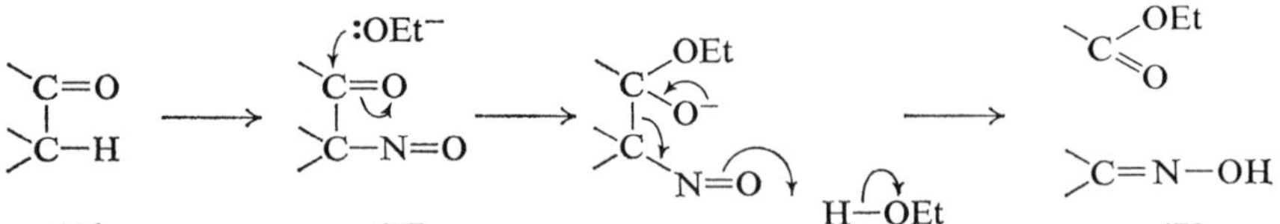

converted into the ester **481** [277] and apoaromadendrome (**412**) was converted into the oxime **482** [278], the reactions in each demonstrating the presence of a tertiary carbon atom adjacent to the carbonyl group.

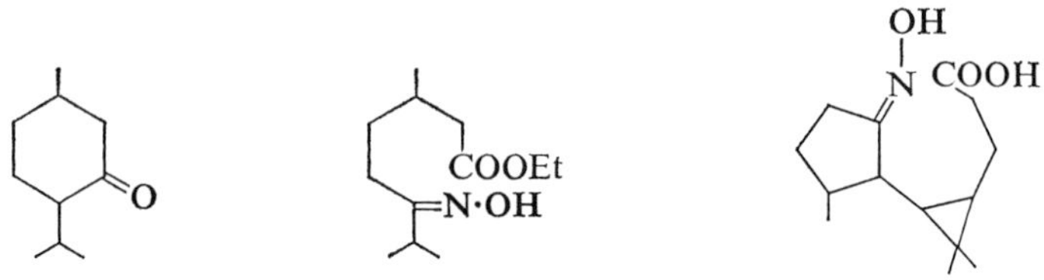

Other Oxidation Processes

Some other reagents have been used occasionally for oxidative fission, and these may be summarized as follows.

Nickel Peroxide

Cleavage of vicinal glycols and α-hydroxy acids to aldehydes or ketones has been satisfactorily effected with nickel peroxide in aprotic solvents. In aqueous alkaline solution the peroxide cleaves α-keto alcohols and α-keto acids to carboxylic acids [279]. In aqueous alkalies the peroxide oxidizes primary alcohols to carboxylic acids and secondary alcohols to ketones [280], and it may also be used in the presence of ammonia to oxidize aldehydes to acid amides [281]. Carbon–carbon bond fission with this reagent has also been observed in the oxidation of *o*-phenylenediamines (483) to *cis,cis*-1,4-dicyanobuta-1,3-dienes (484) [282].

483   484

Oxygen

Molecular oxygen has recently been shown to be capable of effecting the oxidation of cyclic ketones to dibasic acids (489) in good yield in the presence of base, provided conditions are carefully controlled. Best results are obtained with potassium or sodium hydroxide in hexamethylphosphoramide [283]. The reaction proceeds through a carbanion (485) and hydroperoxide ion (486) to an aldehyde (488) which suffers further oxidation.

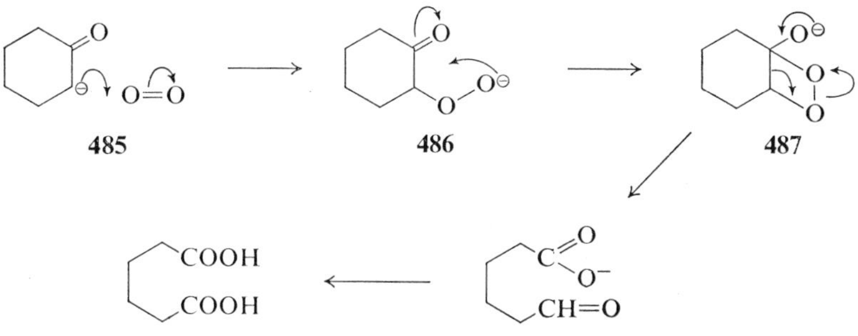

485   486   487

489   488

Oxygen in methanol in the presence of acids converts partially hydrogenated polynuclear aromatic hydrocarbons into ketals with carbon–carbon bond cleavage, for example, 1,2,3,4,4a,9a-hexahydrofluorene (490) is oxidized to the ketal 491 [284].

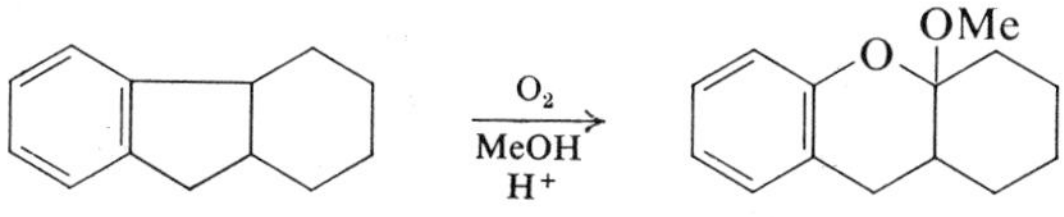

490   491

α,β and β,γ-Unsaturated ketones capable of forming conjugated dienols are oxidized by oxygen in the presence of cupric amine complexes with initial formation of a γ-hydroperoxide (493), the subsequent fate of which depends on the substituents at the γ-position. Primary, secondary, and tertiary peroxides giving aldehydes, ketones, and alcohols, respectively [285, 286].

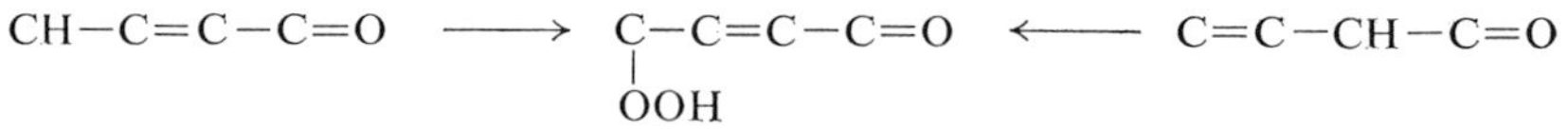

$$CH-C=C-C=O \longrightarrow \underset{\underset{OOH}{|}}{C}-C=C-C=O \longleftarrow C=C-CH-C=O$$

492   **493**   **494**

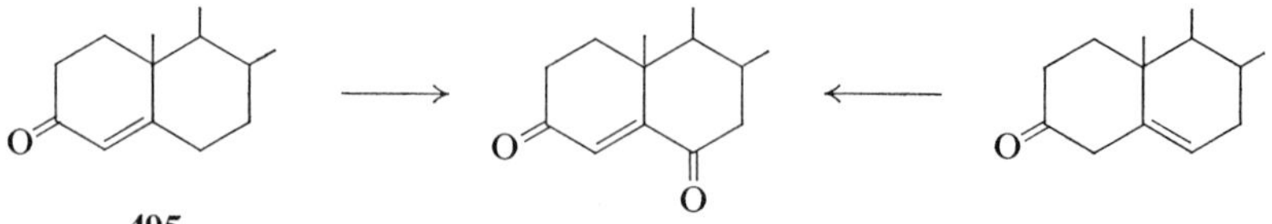

495   **496**   **497**

For example derivatives of cholest-4-ene-3-one (**495**) and cholest-5-ene-3-one (**497**) give derivatives of cholest-4-ene-3,6-dione (**496**). In this way ketones may be modified prior to other oxidations.

Oxidative decarboxylation of aromatic acids has been observed with the copper salts of the acids or with the acids in the presence of other copper salts, with or without oxygen. Good yields of phenols are obtained in aqueous media or of esters under anhydrous conditions [287, 288].

Dimethylsulfoxide

Dimethylsulfoxide, in the presence of dicyclohexylcarbodiimide [289–292] or diethylcarbodiimide [293] and a proton source, has recently been used as an oxidizing agent. It is principally used to oxidize secondary alcohols to ketones, which may then be subjected to other degradative processes. The use of this system has been reviewed [294].

4-Phenyl-1,2,4-triazolin-3,5-dione

This reagent has been shown to oxidize secondary alcohols to ketones in excellent yield at room temperature in neutral solution [295], and may be of great value in structural studies.

Partial Oxidative Cleavage of Olefins

Certain oxidation processes can be used to convert carbon–carbon double bonds into vicinal glycol systems or their derivatives which may subsequently be subjected to cleavage under the relatively mild conditions of oxidation with periodic acid or lead tetraacetate. The use of potassium permanganate for this purpose has already been mentioned on p. 140. Several of these processes have been discussed in a review on hydroxylation methods [296].

Oxidation with Peracids

The treatment of olefins with organic peracids (e.g., perbenzoic acid, monoperphthalic acid, peracetic acid, per-p-nitrobenzoic acid, or performic acid) affords either an oxirane or a *trans*-α-glycol according to the peracid used and the conditions of the reaction (**498** → **500**). The process was first investigated by Prileschajew [297] and has been widely used in synthesis. A review is given by Swern [298].

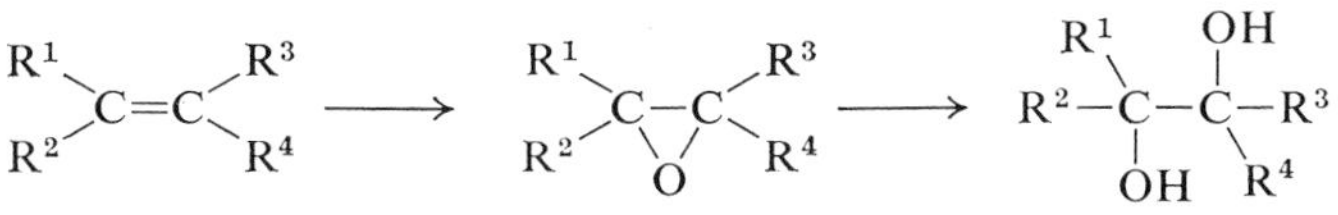

498   **499**   **500**

The precise mode of action of peracids in epoxidizing double bonds is not known. It may involve—and may certainly be satisfactorily represented as involving—direct attack of the double bond by OH^+ as shown in **501–504**

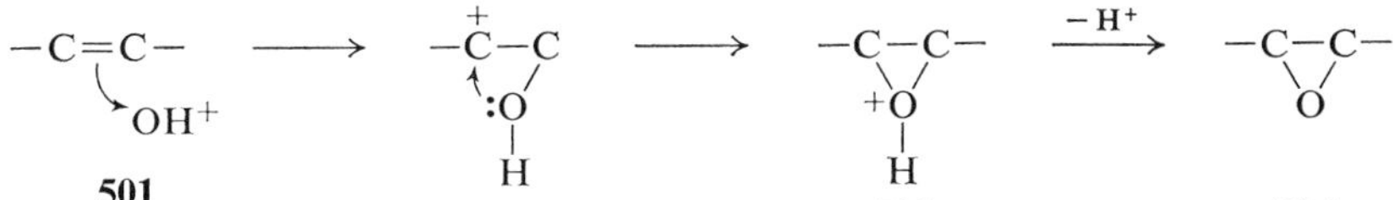

501   **502**   **503**   **504**

[299, 300]. Alternatively the peracid molecule may be directly involved in some process such as **505** → **506**. It is easy to understand why the relatively

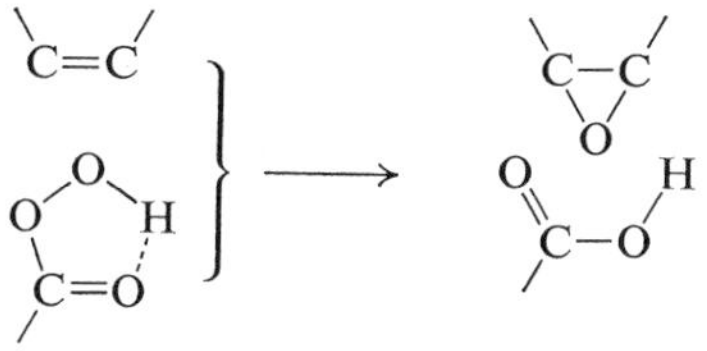

505   **506**

electron-deficient double bond of an α,β-unsaturated ketone is less susceptible than an isolated double bond to attack by the electron-deficient OH^+.

The hydrolysis of the epoxide to the *trans*-α-glycol may be simply represented as in **507** → **509**.

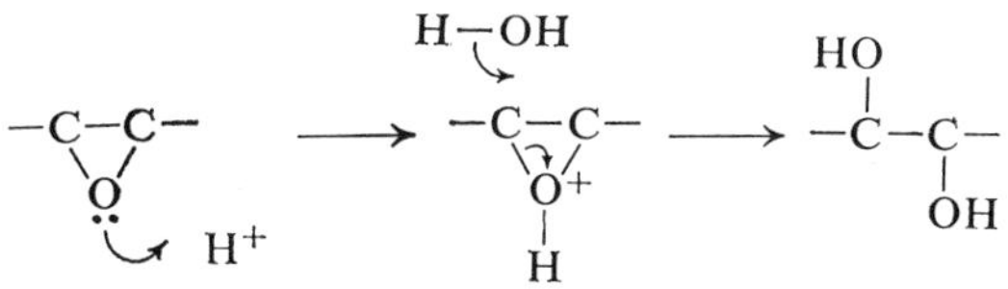

507   **508**   **509**

The reaction between olefin and peracid takes place under mild conditions in inert solvents (chloroform, ether, benzene, dioxane, acetone, etc.). The reaction time is generally relatively short and the yields high.

For epoxide formation perbenzoic acid and monoperphthalic acid are most frequently used. Of these perbenzoic acid is perhaps the most convenient to make, and may be prepared from benzoyl peroxide and sodium methoxide [301], from benzoyl chloride and sodium peroxide [302], or by the autoxidation of benzaldehyde *in situ* by air or oxygen [303–305]. (It may be noted that autoxidation of aliphatic aldehydes under similar conditions gives only very poor yields of the corresponding peracids [304, 305].)

More recently details have been published of a method of preparing organic peracids in 85–97% yield of 93–99% purity by the action of 90–95% hydrogen peroxide on a solution or suspension of the organic acid in methanesulfonic acid [306]. Trifluoroperacetic acid [307] has been widely used but frequently has to be in carefully buffered solution if epoxides are required. Permaleic acid, from maleic anhydride and hydrogen peroxide, gives epoxides in excellent yield in unbuffered solution [308].

Monoperphthalic acid is somewhat more stable than perbenzoic acid, and its use may be preferred if, and when, an extended reaction time is necessary for epoxidation [97].

Peracetic acid generally oxidizes olefins to α-glycols or to their monoacetates. Good yields of epoxides may, if required, be obtained by working at 20–25° for as short a time as possible with the exclusion of strong acids which catalyze the opening of epoxide rings to glycols. Peracetic acid may be prepared by the action of hydrogen peroxide on acetic anhydride [309], or *in situ* from acetic acid and hydrogen peroxide [310]. Excellent yields of α-glycols may be obtained from olefins by the use of acetic acid and stoichiometric quantities of 25–30% hydrogen peroxide at 40° in the presence of sulfuric acid as catalyst.

Performic acid is particularly useful for glycol formation; it reacts rapidly and completely with olefins and even with the relatively less reactive double bonds of α,β-unsaturated acids, which give α,β-dihydroxy acids. α,β-Unsaturated alcohols give approximately 50% of the triol together with cleavage products [311]. Performic acid may be prepared by the interaction of formic acid and 25–90% hydrogen peroxide [312] or may be generated *in situ* in the same way. This peracid has also been used for the preparation of *cis*-glycols [313].

The ease of *trans*-glycol formation with these various reagents is in the following order: hydrogen peroxide–formic acid > hydrogen peroxide–acetic acid–sulfuric acid > preformed performic or peracetic acid–sulfuric acid > perbenzoic acid followed by acid hydrolysis of the isolated epoxide.

The rate of the reaction of the olefin with the peracid may be determined by the measurement of the amount of unconsumed peroxide left after specific intervals of time [314].

Peracids will epoxidize or hydroxylate isolated double bonds readily at room temperature with a reaction time of 8–48 hr; electron-releasing groups attached to the double bond cause a considerable acceleration of the reaction whereas electron-attracting groups greatly reduced the speed of reaction. Isolated double bonds may be selectively oxidized in the presence of de-activated double bonds; for example, **510** may be oxidized to **511** by per-benzoic acid [315] and **512** to **513** by monoperphthalic acid [316].

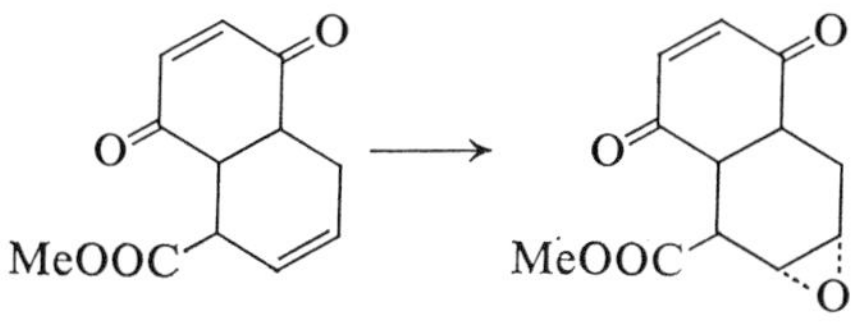

Stereospecific epoxidation of α,β-unsaturated esters has been observed with *m*-chloroperbenzoic acid [317], and this reagent has also been found to cleave certain enol ethers, for example the enol ether **514** is oxidized to the keto ester **515** [318].

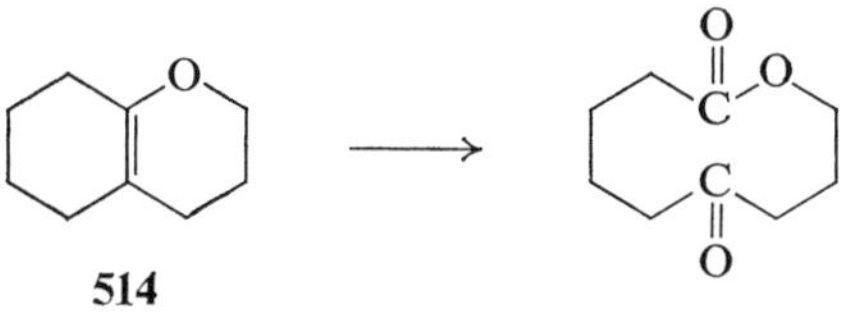

When glycol formation is attempted by peracid oxidation of olefins, or by the acid-catalyzed hydrolysis of epoxides, pinacol–pinacolone rearrangement or other transformations may occur in the acid medium. For example the peracid oxidation of camphene (**516**) affords camphenilanaldehyde (**519**) as well as the epoxide **517**, and this doubtless arises from the epoxide as shown [318]. The epoxide **520** when treated with acid rearranges through the non-classical carbonium ion **521** to the diol **522** ≡ **523** [319]. Vicinal glycol production in this case proved impossible. A similar rearrangement of nor-bornene epoxide has been noted [320, 321].

It has been found that hydrogen peroxide in methanol in the presence of benzonitrile or acetonitrile at pH 7–8 is an effective oxidizing agent [322, 323].

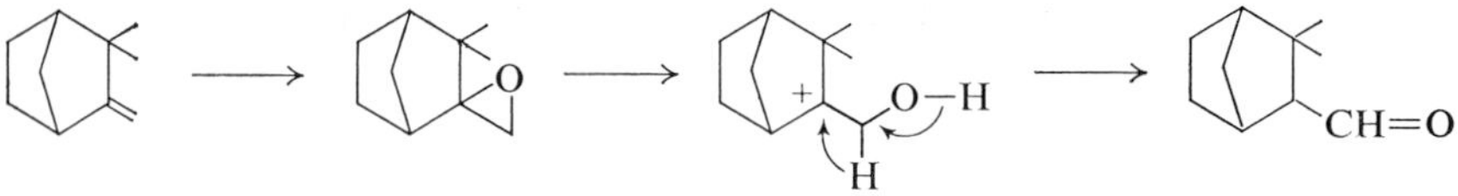

516 **517** **518** **519**

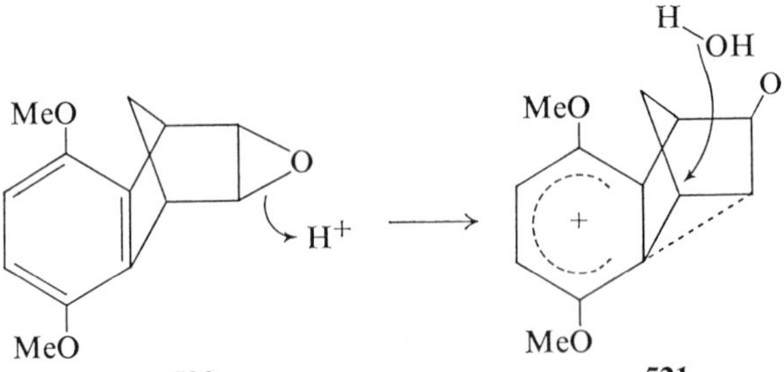

520 **521**

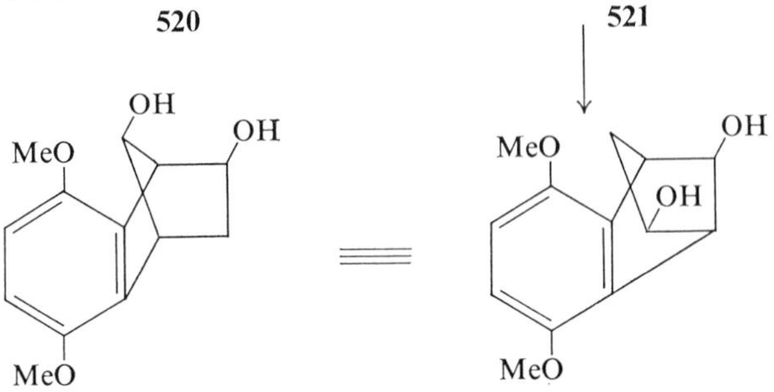

523 **522**

The active species are believed to be the peroxyimidic acids (**524**) which, however, are not stable entities. In the absence of an oxidizable substrate they

$$\mathrm{R{-}C{\equiv}N} + \mathrm{HO{-}OH} \longrightarrow \mathrm{R{-}C}\begin{matrix}\mathrm{{=}NH} \\ \mathrm{OOH}\end{matrix}$$

524

oxidize hydrogen peroxide to oxygen [323]. Cyclohexene, styrene, acrolein dimethyl acetal, and methylenecyclohexane are oxidized to the corresponding epoxides in 60–70% yield [322], and 2-allylcyclohexanone (**526**) is oxidized to the epoxide **525** by benzonitrile/hydrogen peroxide at pH 8 but, by a Baeyer-Villiger reaction, to the lactone **527** by peracetic acid at the same pH [323].

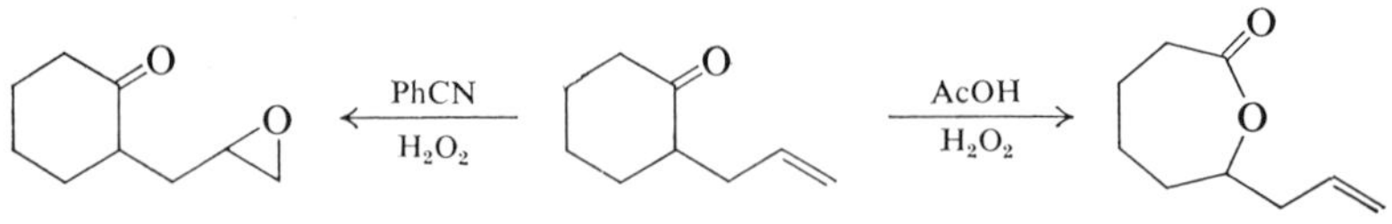

525 **526** **527**

Oxidation with Osmium Tetroxide

The reaction of olefins with osmium tetroxide in an inert solvent (ether or dioxane) results in the direct *cis* addition of the tetroxide to the double bond, as shown in **528–529**, to give the osmate ester of the *cis-α-glycol*, from which the α-glycol itself may be recovered by hydrolysis, preferably in a reducing medium, by reduction with lithium aluminum hydride or by decomposition with hydrogen sulfide. The geometry of the addition reaction is clearly such as to favor *cis* addition, which can take place with little distortion of the Os–O valencies.

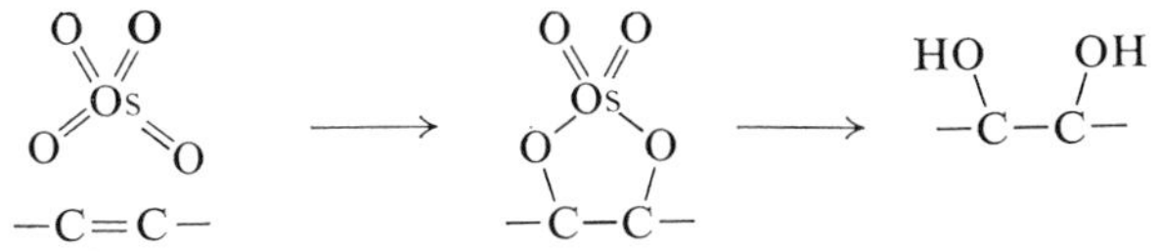

528   **529**

The yields are generally very high, and the process relatively convenient. Owing to the extremely noxious nature of osmium tetroxide, however, stringent precautions must be taken to avoid inhalation of the vapor and contact of the vapor with the eyes for deposition of osmium can rapidly cause blindness. The lower oxidation state of osmium produced during the reaction may, however, be reoxidized *in situ* to OsO_4 by hydrogen peroxide, sodium chlorate, or potassium permanganate and in this way only small catalytic quantities of the tetroxide need be used [324, 325], and the hazards and costs accordingly greatly reduced.

The process has proved of considerable value in the steroid series. For example, the olefins **530** and **532** have been oxidized to the *cis*-glycols **531** and

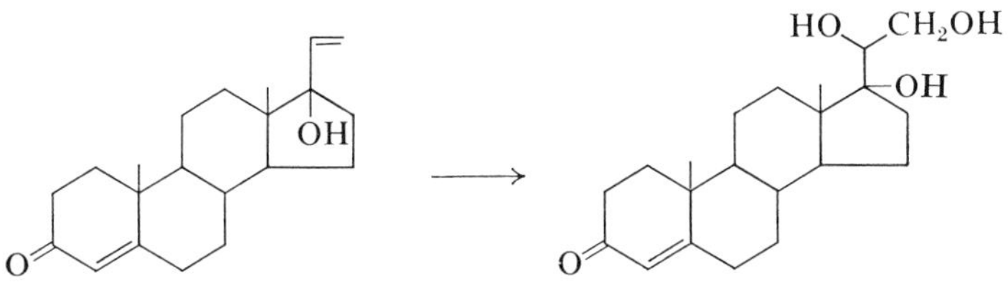

530   **531**

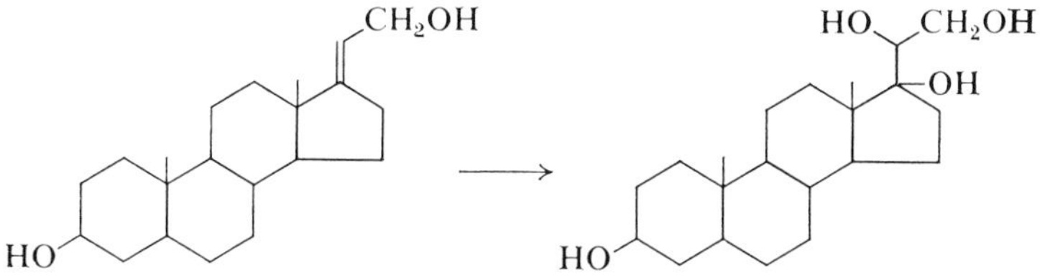

532   **533**

533 [326–331], and the unsaturated compound **534** was oxidized to the glycol **535** during Woodward's total synthesis of the steroids [332]. This

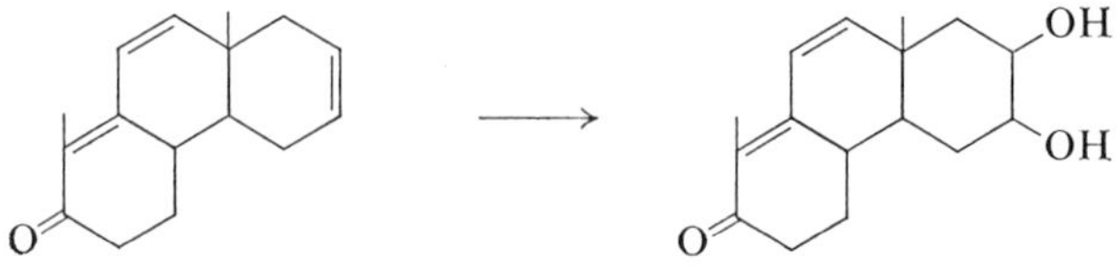

534   535

oxidation procedure is particularly valuable giving, as it does, the *cis*-glycol, which is more rapidly cleaved by periodic acid oxidation than is the corresponding *trans*-α-glycol obtained by hydroxylation with peracids. Using this process certain nitrogen-free exhaustive methylation products from codeine have been converted through the glycol into the corresponding aldehyde (**536 → 538**) [333]. Derivatives of codeine having a 7,8-double bond also react with osmium tetroxide to give the corresponding 7,8-glycol, but the

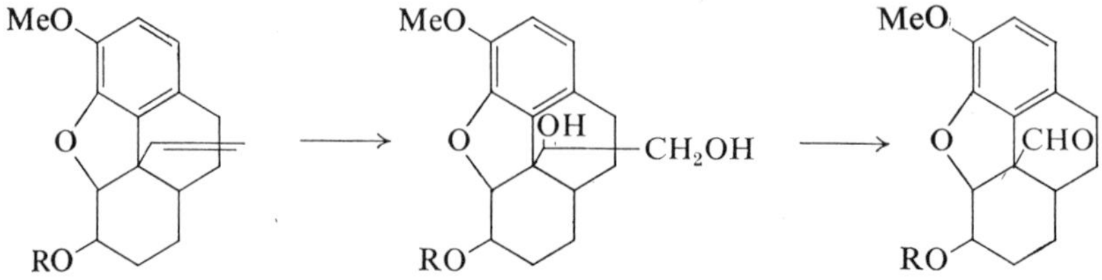

536   537   538

yields have been found to be sensitive to the substituent at position 6, and with codeinone dimethyl ketal oxidation of the N–Me group to N–CHO occurs [334].

Osmium tetroxide will attack the double bonds of α,β-unsaturated ketones, which may be unaffected by peracids. For example, the unsaturated ketone **539**, an intermediate in the synthesis of reserpine, readily afforded the glycol **540**. Similarly, compound **510** gave only the epoxide **511** on oxidation with perbenzoic acid, but the epoxide could be further oxidized to the glycol **541** by means of hydrogen peroxide and osmium tetroxide [315].

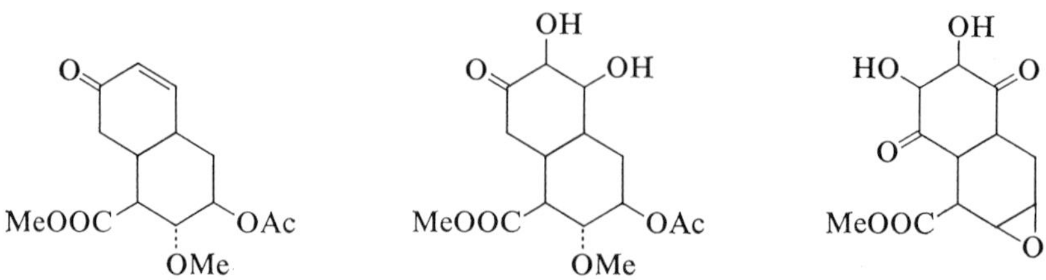

539   540   541

The use of hydrogen peroxide with oxmium tetroxide may, however, lead to compounds other than the corresponding glycol. The olefin **542**, for example, gives a 48% yield of the hydroxy ketone **543** with these two reagents [335]. β-Carotene (**544**) with osmium tetroxide and hydrogen peroxide undergoes cleavage of the molecule, giving retinene **545** and β-ionylideneacetaldehyde **546** [336].

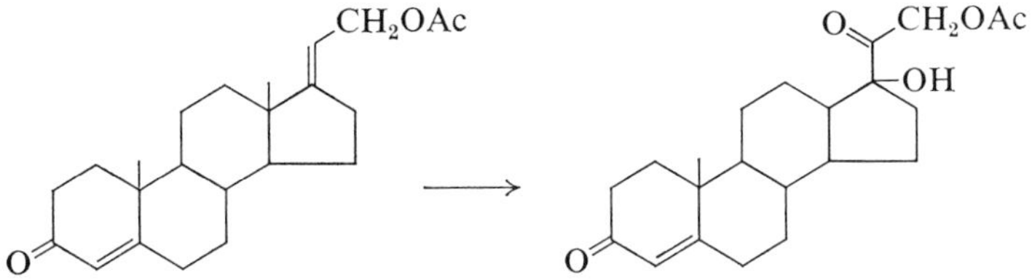

542   543

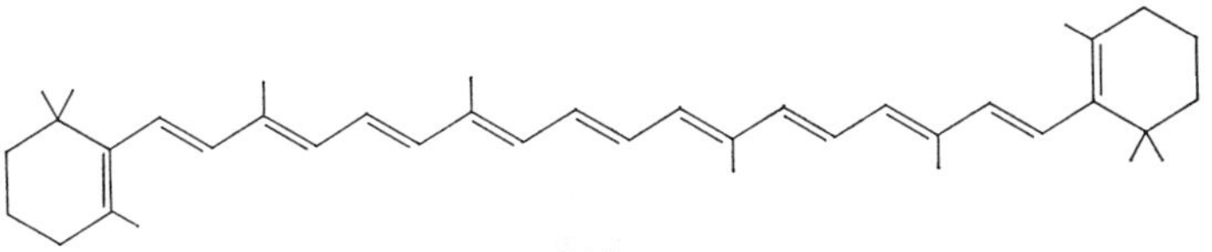

544

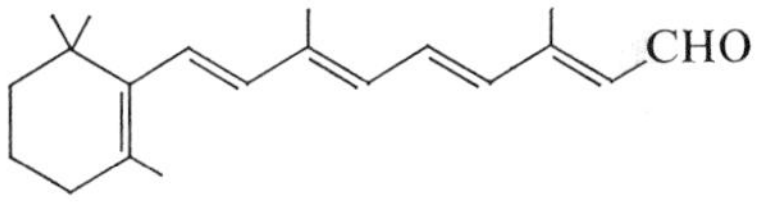

545 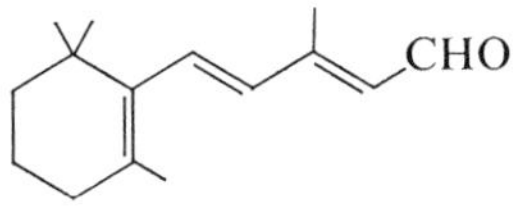   546

Osmium tetroxide in pyridine–benzene will oxidize 1-acetyl and 1-benzoyl derivatives of 2,3-dimethylindole, 2,3-diphenylindole, 2,3,5-trimethylindole, and tetrahydrocarbazole (**547**, R = Ac and PhCO) to osmate esters hydrolyzable to the corresponding 2,3-dihydroxydihydroindoles (e.g., **548**) [134]. No crystalline materials have been isolated from the oxidation of the corresponding indoles with free –NH– groups.

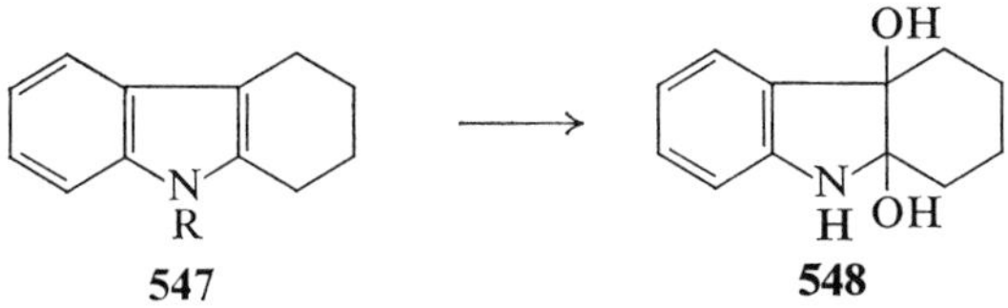

547   548

Hydroxylation and fission of a double bond in one operation have been achieved in the oxidation of cyclohexene (**549**) to adipaldehyde (**550**) by the addition of sodium metaperiodate to a mixture of cyclohexene, osmium tetroxide, ether, and water, the yield of aldehyde being 77% [207].

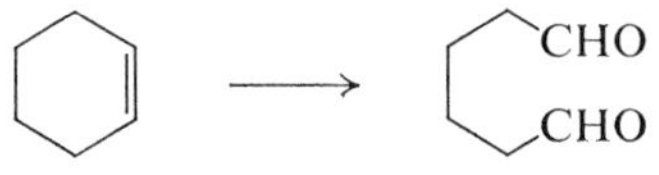

549   **550**

Hydroxylation with Silver Acetate and Iodine

An alternative method of obtaining *cis*-α-glycols from olefins is the treatment of the unsaturated compound with silver acetate and iodine in acetic acid containing water [337]. This is a modification of the Prevost reaction [338], which is the conversion of olefins to *trans*-α-glycol benzoates by iodine and benzoic acid in anhydrous solvents and was developed from the observation that whereas the interaction of *trans*-1-acetoxy-2-bromocyclohexane, *trans*-1,2-dibromocyclohexane, *threo*- and *erythro*-2-acetoxy-3-bromobutanes, and (+)(−)- and *meso*-1,2-dibromobutanes with dry silver acetate in dry acetic acid resulted in the corresponding diacetoxy compounds with almost complete retention of configuration, in wet acetic acid there was formed 65–75% of the corresponding glycol monoacetate with nearly 100% inversion. The monoacetate produced in this way is directly formed and does not arise from hydrolysis of a diacetate, and the percentage inversion is slightly greater than the percentage of water added to the acetic acid [339].

In these processes where there is retention of configuration during the reaction in dry solvent, there must be a participating neighboring group effect of some such type as that shown in **551** → **553**, whereas inversion with the direct formation of a glycol monoacetate in wet solvent must result from attack of the intermediate **552** by water as shown in **554**, giving the ortho-acetate **556** from which the monoacetate **557** is derived [339].

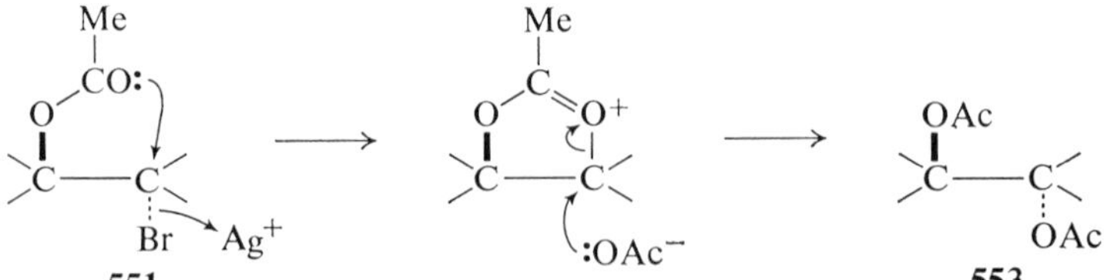

551   **552**   **553**

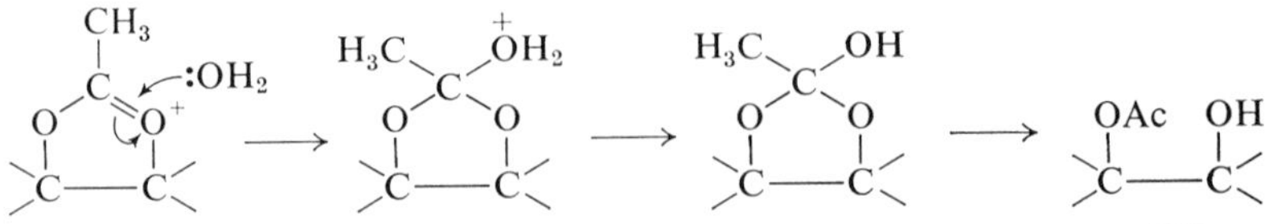

554   **555**   **556**   **557**

From these observations there has been developed the use of silver acetate and iodine in wet acetic acid for the formation of a *cis*-α-glycol monoacetate that may be hydrolyzed to the corresponding glycol in good yield; for example, the olefin **558** has been oxidized in this way to the glycol **559** in very good yield [340]. The general utility of this process in the study of natural products has not been investigated, but the potentialities are obvious. The

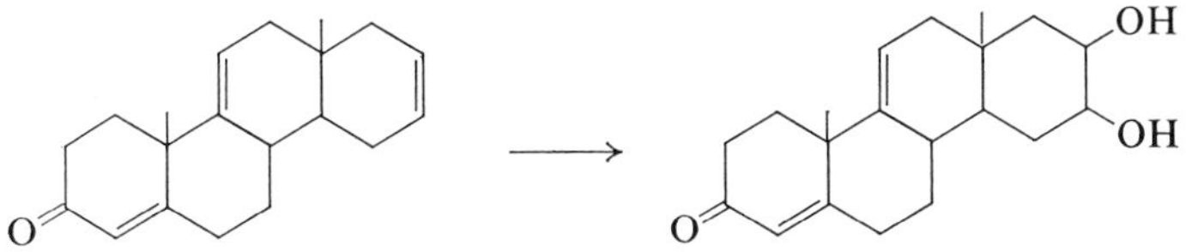

reaction has generally been carried out at 45–95° for 1–20 hr [296, 337, 341], but better yields are obtained at room temperature [342].

Miscellaneous Processes

Oxidation of double bonds to vicinal glycols may be accomplished by means of potassium permanganate, and complete fission may be achieved by potassium periodate and small amounts of permanganate.

Olefins may be converted into the corresponding bromohydrins by treatment with hypobromous acid (*N*-bromosuccinimide in dilute sulfuric acid is a useful source of this reagent), as in the oxidation of **560** to the bromohydrin **561** [315], and the bromohydrin may then be oxidized to the

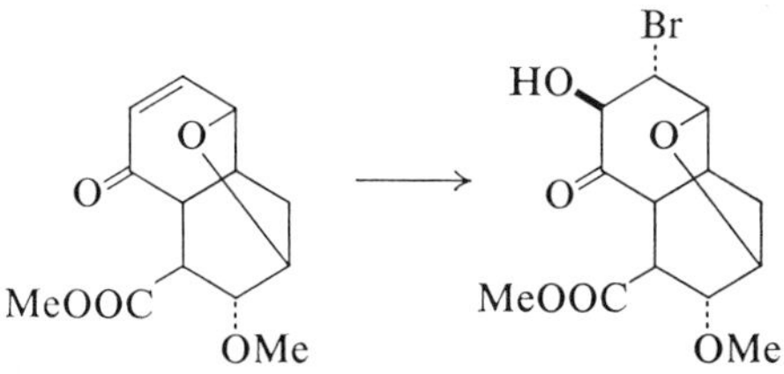

α-bromoketone, which can be reduced to the saturated ketone, or dehydrobrominated to the α,β-unsaturated ketone, prior to further oxidation. In cases where, for steric reasons, reaction of the olefin with hypobromous acid produces only one of the two possible diaxial bromohydrins, the other may be obtained by epoxidation with peracids and hydrolysis of the epoxide. For example, the olefin **565** gives the bromohydrin **569** by direct attack by Br⁺ and the epoxide **562** by oxidation with peracids. The epoxide **562** gives the isomeric bromohydrin **563** when hydrolyzed by hydrobromic acid, and the bromohydrin **569** is cyclized to the epoxide **568** by alkalis. Both bromohydrins **563** and **569** have diaxial substituents and both may be oxidized to bromo-ketones **568** and **570** in which the bromine adopts the equatorial position, and the ketone **566** can be reduced to the saturated ketone **564** and dehydro-brominated, by boiling 2,4,6-collidine, to the unsaturated ketone (**567**).

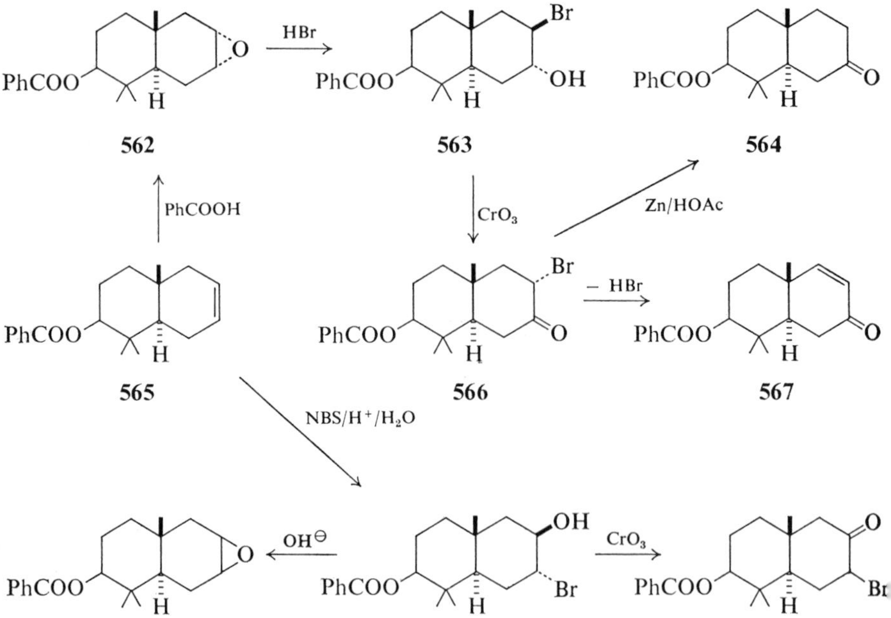

Hydration of Olefins

The conversion of olefins into easily oxidized substances may also be achieved by hydration of the double bond. This process affords alcohols which may be oxidized directly with fission of a carbon–carbon bond or may first be converted into ketones.

Hydration may be accomplished by sulfuric acid or halogen hydracid via the hydrogen sulfate or halide and hydrolysis. In these cases hydration follows the Markowinkoff rule and the hydroxyl group appears on the most substituted carbon atom. This process was used by Gates in his synthesis of morphine for the conversion of β-dihydrodesoxycodeine methyl ether (**571**) through the alcohols (**572**, R = Me and H) into β-dihydrothebainone (**573**) [343].

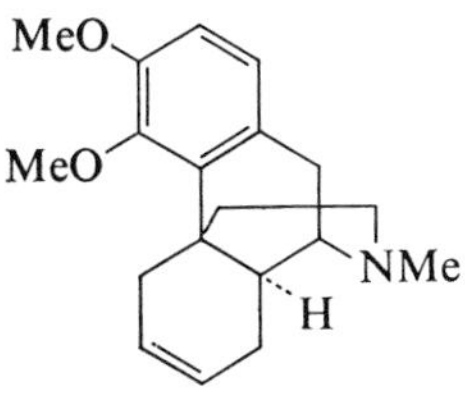

571 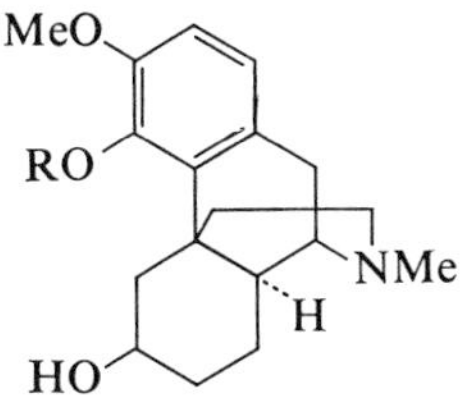   **572** 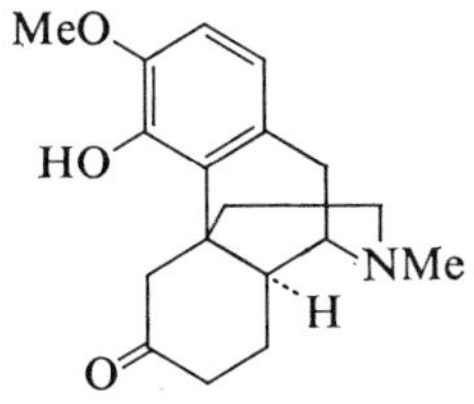   **573**

Hydration via hydroboration provides a method of stereospecific *cis* anti-Markowinkoff hydration of double bonds [344–347]. The reaction involves the treatment of the olefin with diborane (which may be generated *in situ* or externally by the action of boron trifluoride etherate on an alkali metal borohydride) when a boron–hydrogen bond adds in a *cis* fashion to the carbon–carbon double bond. Of the three hydrogen atoms of diborane BH_3, all are utilized by mono and disubstituted olefins, two by trisubstituted and

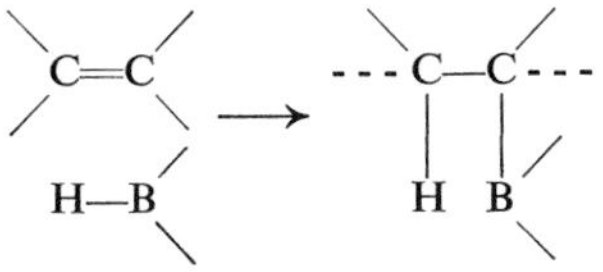

$$\diagdown C=C \diagup \longrightarrow ---\diagdown C-C \diagup --- $$
$$H-B \qquad\qquad H \quad B$$

one by tetrasubstituted olefins, giving products with three, two, or one carbon–boron bonds. In all cases the boron atom becomes linked predominantly to the *least* substituted carbon atom.

The resulting tri-, di-, or monoalkylboranes can be hydrolyzed by organic acids, with replacement of boron by hydrogen, the net effect being a reduction of the double bond:

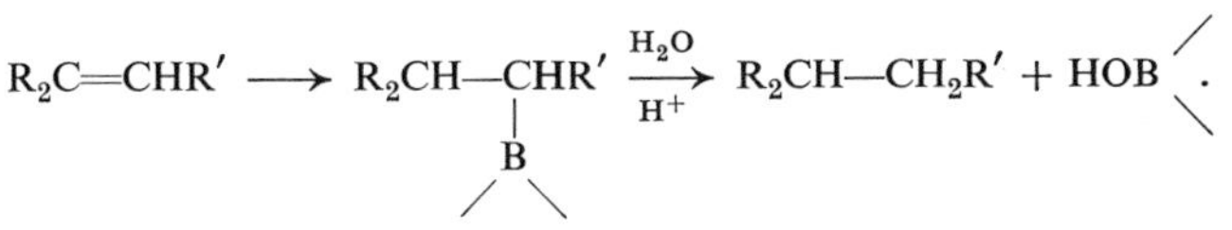

$$R_2C{=}CHR' \longrightarrow R_2CH{-}\underset{\underset{B}{|}}{C}HR' \xrightarrow[H^+]{H_2O} R_2CH{-}CH_2R' + HOB\diagup$$

Oxidation of the alkylboranes with hydrogen peroxide and alkali, however, results in replacement of the boron by a hydroxyl group *with retention of configuration*, the net result being hydration of the double bond:

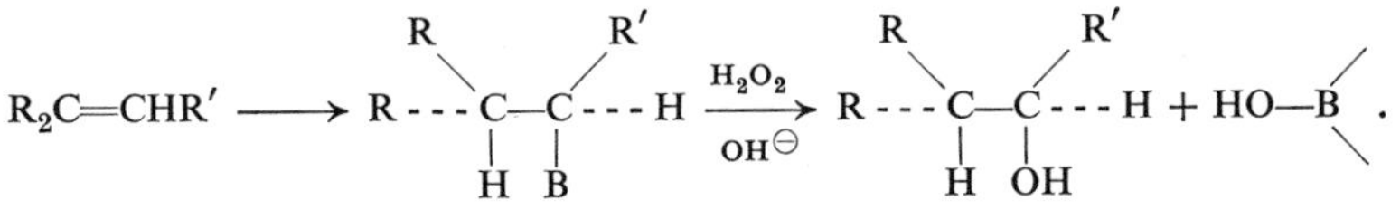

$$R_2C{=}CHR' \longrightarrow R---\underset{\underset{H}{|}}{\overset{\overset{R}{|}}{C}}-\underset{\underset{B}{|}}{\overset{\overset{R'}{|}}{C}}---H \xrightarrow[OH^\ominus]{H_2O_2} R---\underset{\underset{H}{|}}{\overset{\overset{R}{|}}{C}}-\underset{\underset{OH}{|}}{\overset{\overset{R'}{|}}{C}}---H + HO{-}B\diagup$$

The alcohols produced in this way can be separately oxidized to ketones, but the two oxidation processes can be combined if the alkylborane is oxidized with chromic acid, the product then being the ketone:

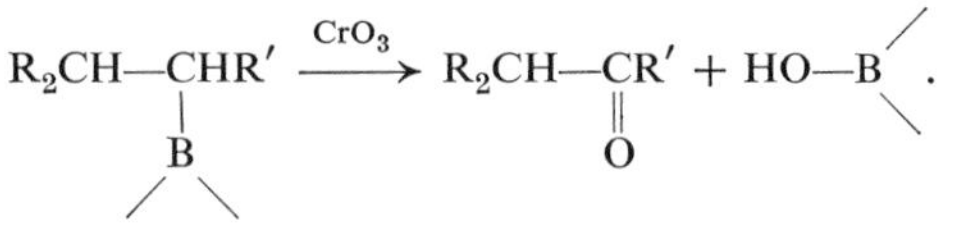

$$R_2CH{-}CHR' \xrightarrow{\text{CrO}_3} R_2CH{-}CR' + HO{-}B\diagdown\ .$$

The process of hydroboration in general and its use for the hydration of double bonds in particular has been reviewed [346, 347].

Hydroboration generally involves addition of diborane to the less hindered side of a double bond, for example, norbornene (**574**) gives *exo*-norborneol (**576**) almost exclusively [348] and isodrin (**577**) gives the alcohol (**578**) [349].

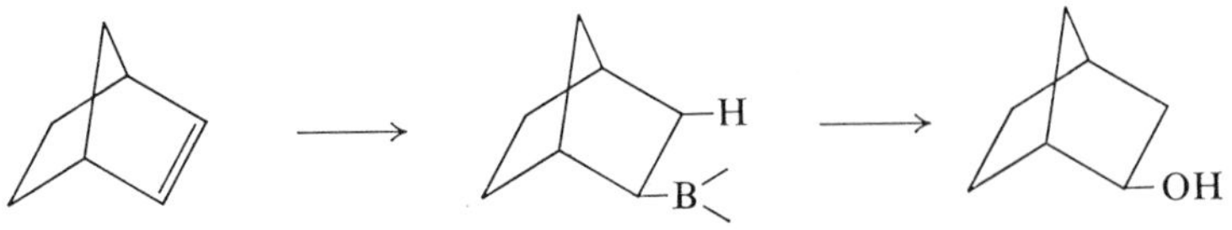

| 574 | 575 | 576 |

| 577 | 578 |

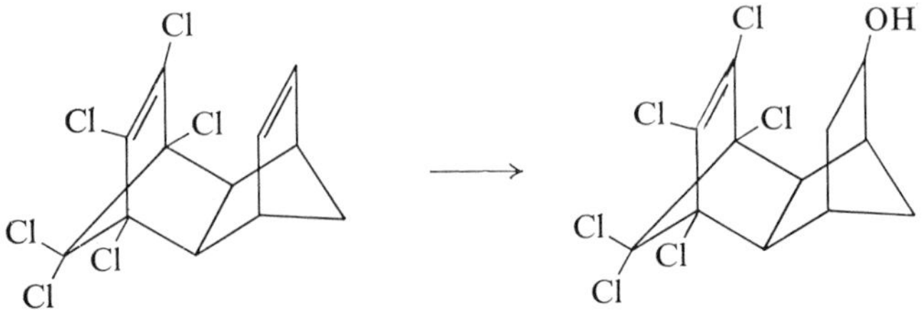

In some cases, however, addition occurs on the more hindered face of the molecule, for example, 8,14-unsaturated compounds in the morphine series give principally the 14α-substituted compounds of the B/C *trans* or isomorphinan series, Δ⁸-desoxycodeine (**579**, R = H) giving *trans*-8α-hydroxy-dihydrodesoxycodeine (**580**, R = H) [350], isoneopine (**579**, R = OH) the diol (**580**, R = OH), and thebaine (**582**) a mixture of the diastereoisomers **583** and **584** [351]. Catalytic reduction of all 8,14-unsaturated compounds in the morphine series, and the reaction of thebaine with hydrogen peroxide

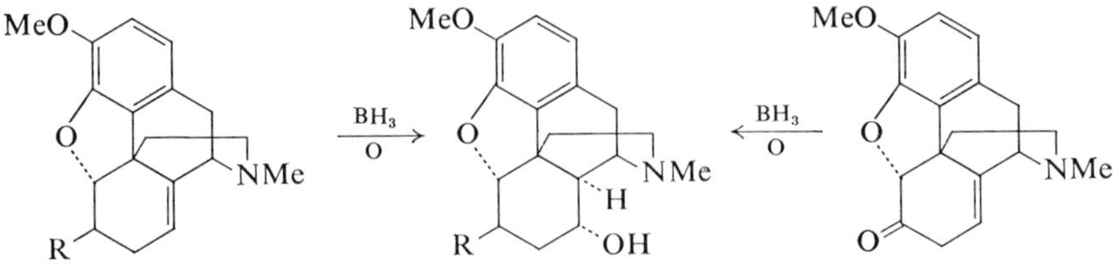

| 579 | 580 | 581 |

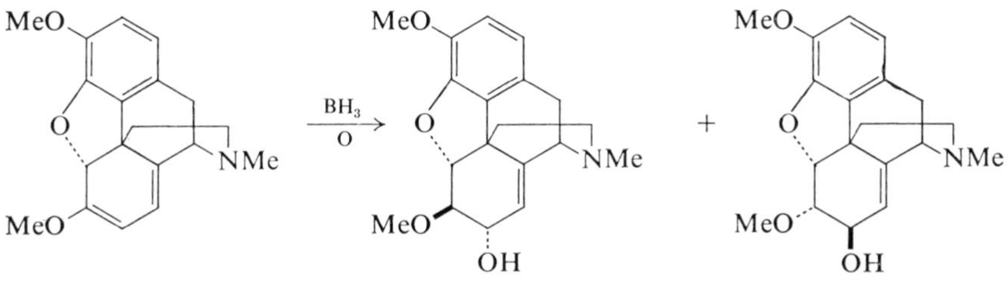

582   583   584

and with bromine, give exclusively 14β-substituted compounds of the normal B/C *cis*-morphine series. These reactions have provided a route to B/C *trans*-morphine itself [352].

Diborane is a reducing agent and will reduce carbonyl groups, for example, neopinone (**581**) affords the diol (**580**, R = OH) with diborane followed by oxidation [351].

Under controlled conditions partial hydration of dienes has been observed, for example, bicycloheptadiene (norbornadiene **585**) gives *exo*-dehydro-norborneol (**588**) [353, 354], caryophyllene (**586**) reacts initially at the highly reactive *trans*-cyclic double bond to give the alcohol (**589**) [347], and myrcene (**591**) with restricted amounts of diisoamyldiborane gives first the alcohol (**592**), then the diol (**593**), before giving the triol (**594**) [355].

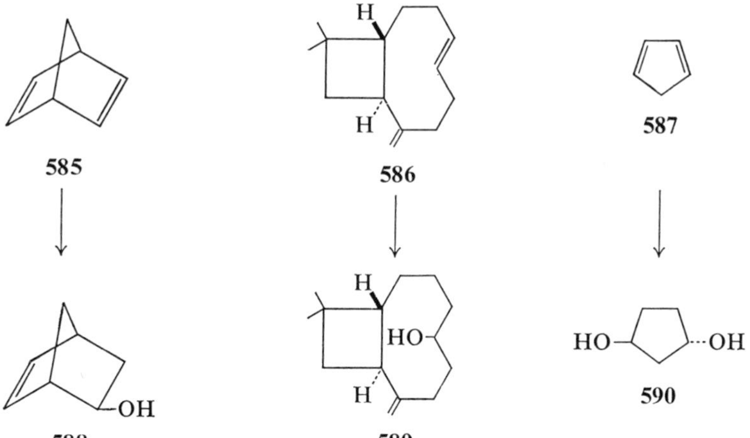

585   586   587

588   589   590

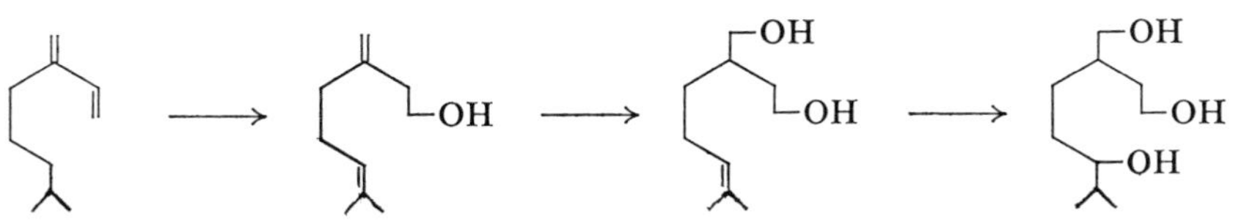

591   592   593   594

Complete reaction of polyenes can be accomplished of course as with myrcene, and cyclopentadiene (587) gives *trans*-1,3-cyclopentandiol (590) [356].

3 HYDROLYTIC PROCESSES

Carbon–carbon bonds of certain types may be broken by hydrolytic processes with either acids or bases. Reactions that may be effected are generally the fission of β-dicarbonyl systems, the dealdolization of β-hydroxy or α,β-unsaturated carbonyl compounds, the fission of aromatic ketones, and certain types of phenyl ether. Such processes are involved in alkali fusions and many examples are given in Chapter XIV, but may also occur under much milder conditions. β-Diketones, for example, are cleaved readily by aqueous alkalis to carboxylic acids and ketones. The process may be represented as in formulas 595 → 596 and 597 from which it will be seen that

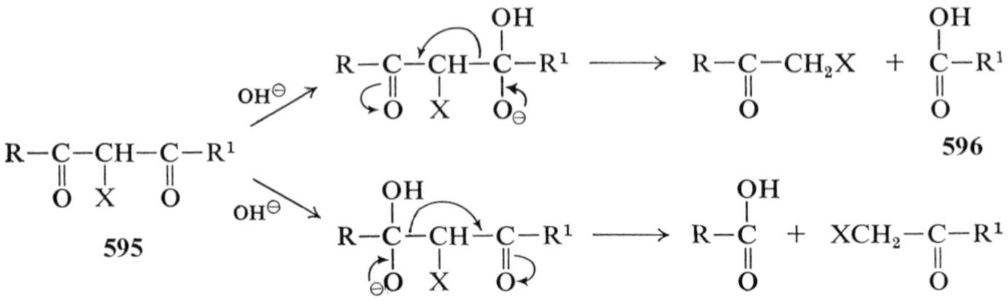

unsymmetrical diketones can be cleaved to two different acid/ketone pairs [357, 358]. In general the cleavage appears to take place preferentially to give the stronger of the two acids in greater yield, and exceptions to this are probably explicable on steric grounds.

Such degradation played a major part in the elucidation of the structure of Terramycin [359]. The antibiotic on alkali fusion gives succinic acid and *o*- and *m*-hydroxybenzoic acids, but on mild treatment with alkali it gives terracinoic acid (605), whereas treatment with alkali in the presence of zinc dust, that is, under reducing conditions, yields terranaphthol (600). Terracinoic acid and terranaphthol have certain structural features in common, and since they are produced under similar basic conditions it is reasonable to suppose that they represent stable end-products of degradation of the same part of the Terramycin molecule. If all of the bonds in the five-membered ring of terracinoic acid were present in Terramycin, the antibiotic would have to contain the system (607) and, in order that the carbon marked with an asterisk could become a carbonyl group as in 608 to permit base-catalyzed dehydration followed by aromatization, it would have to be linked

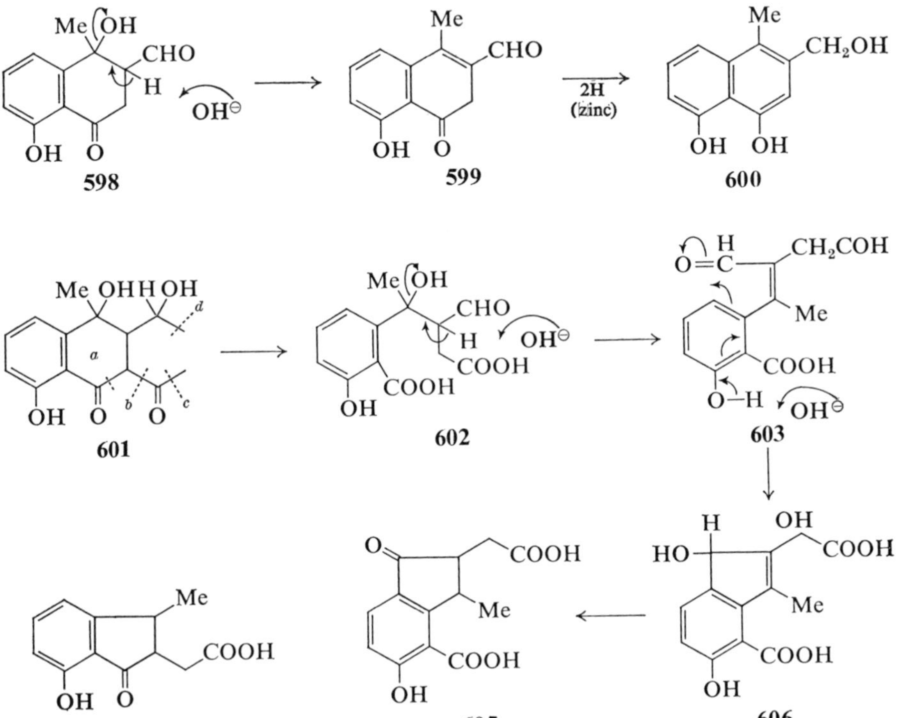

Me OH
CHO
H
OH⊖
OH O
598

Me
CHO
OH O
599

$\xrightarrow{\underset{(zinc)}{2H}}$

Me
CH₂OH
OH OH
600

Me OHH OH
a
a
OH O O
601

Me OH
CHO
H OH⊖
COOH
COOH
OH
602

H CH₂COH
O=C
Me
COOH
O–H OH⊖
603

H OH
HO COOH
Me
COOH
OH
606

O COOH
Me
COOH
OH
605

Me
COOH
OH O
604

to oxygen, and hence could bear no hydrogen. In the products of degradation of Terramycin in acids, that carbon atom is found to bear hydrogen, and hence part-structure **607** can be discounted.

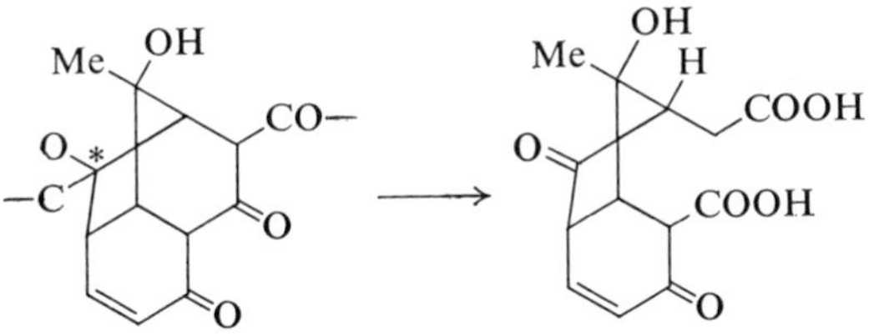

Me OH
CO–
O *
–C
O
O
607

Me OH H
COOH
O
COOH
O
608

The simplest assumption is that the five-membered ring of terracinoic acid is absent in Terramycin and that the latter contains the part structure **601** in which the base-catalyzed fission of the bond *d* generates an aldehyde system **598** and **602**. Simultaneous fission of bond *b* would give the intermediate **598**, the base-catalyzed dehydration of which, followed by aromatization and reduction, would produce terranaphthol (**600**). Cleavage of the β-diketone system of **601** in the alternative sense *a*, accompanied by bond cleavages *d* and *c*, would generate the acid **602** which by dehydration and cyclization *para* to the phenolic hydroxyl and isomerization would give terracinoic acid (**602** → **605**). On the basis of this reasoning it was predicated that cyclization of the acid **603** *ortho* to the phenolic hydroxyl should also occur, and that the product of decarboxylation of the resulting β-keto acid, that is, isodecarboxy-terracinoic acid (**604**), should also be produced during the alkaline degradation of Terramycin, and a search for this among the products resulted in its isolation.

Hydrolytic cleavage of a β-dicarbonyl system occurs in the acid-catalyzed degradations of Terramycin, and careful interpretation of these degradations allowed the expansion of the part-structure **601** for the antibiotic to the complete structure (**609**). In acid the first product is anhydroterramycin (**610**), which is very smoothly converted into two diastereoisomeric apo-terramycins (**612**). These are lactones and the facile conversion of ketone into lactone indicates that the ketonic carbonyl group is part of a β-diketone system as shown. The apoterramycins yield 2,5-dihydroxybenzoquinone on alkali fusion and the molecular fission involved in this may be represented as

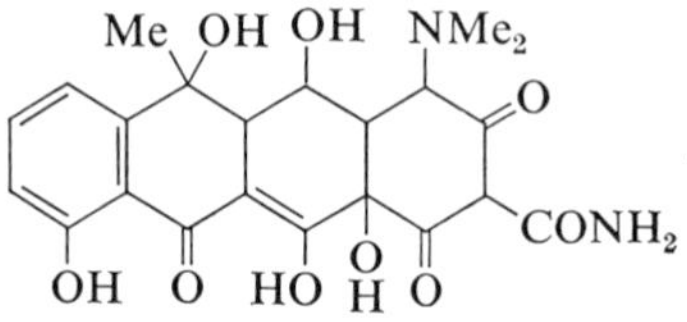

609

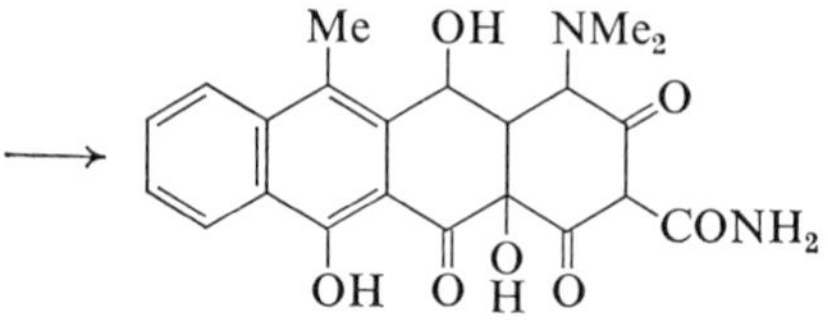

610

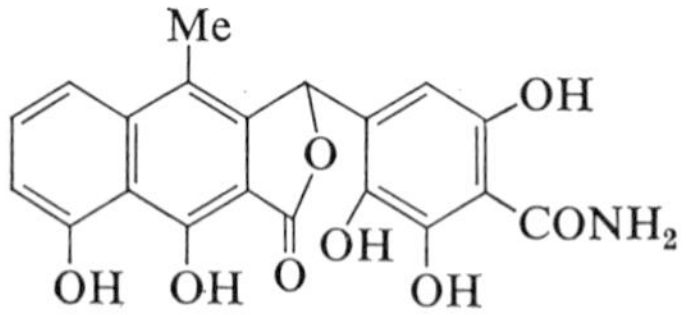

611

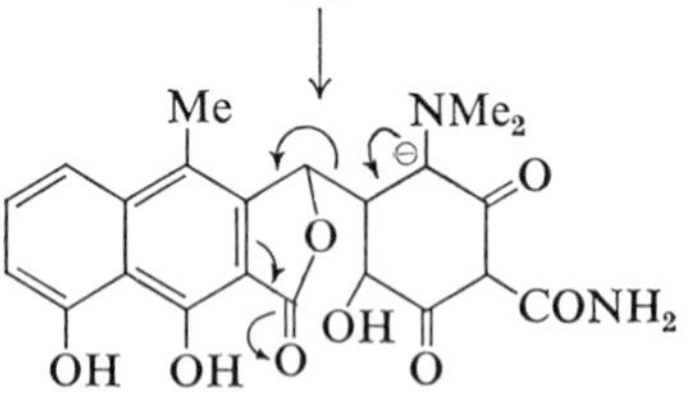

612

in formula **612**. Further acid treatment of the apoterramycins gives terrinolide (**611**), and the elimination of dimethylamine may be represented as a β-elimination in the diketone (**614**), which is related to **612** through the common enol.

On the basis of the complete structure for Terramycin (**609**), the generation of an aldehyde group in the part structure **601** by the bond cleavage *d* may be represented as initial cleavage of the two β-diketone systems to give first the α-hydroxyketone **613**, which is convertible through the enol **615** into the

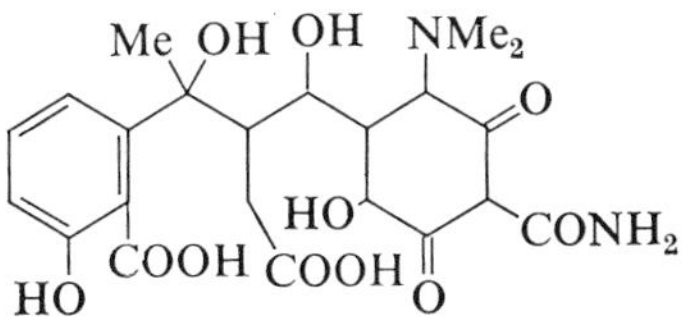

613

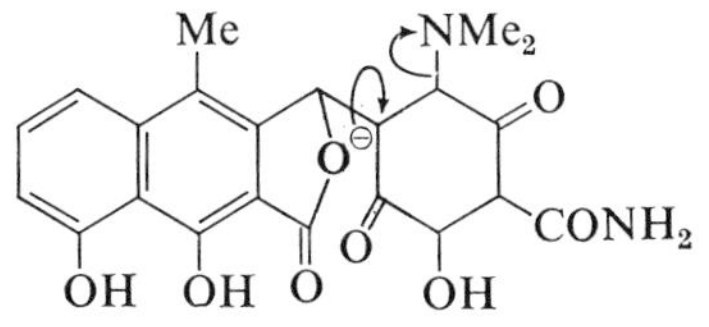

614

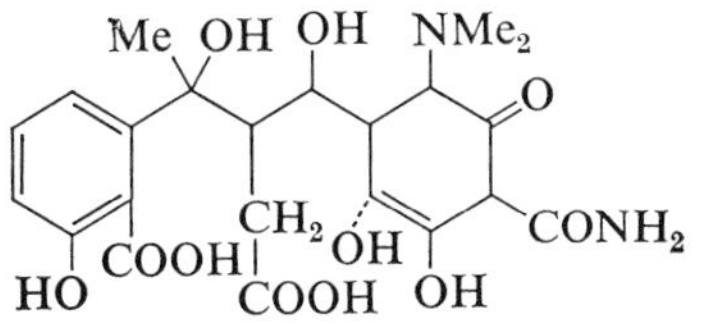

615

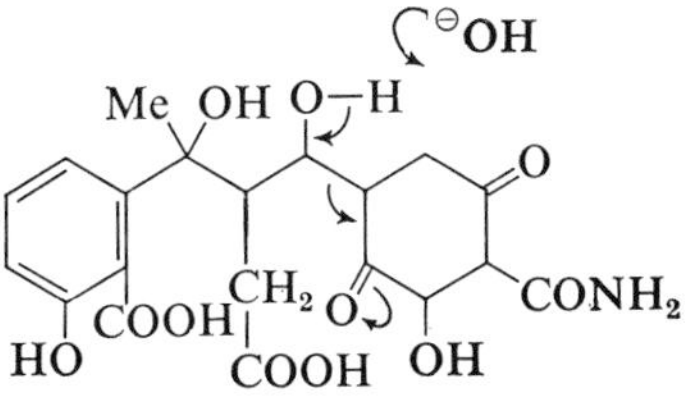

616

isomeric α-hydroxyketone **616** in which a dealdolization can be accomplished under the influence of base as shown.

Dealdolization reactions are involved in a number of other interesting degradations and transformations, and the following examples may be given.

(a) The conversion of flavothebaone trimethyl ether methine (**617**) into the ψ-methine (**621**). This involves initial hydration of the α,β-unsaturated

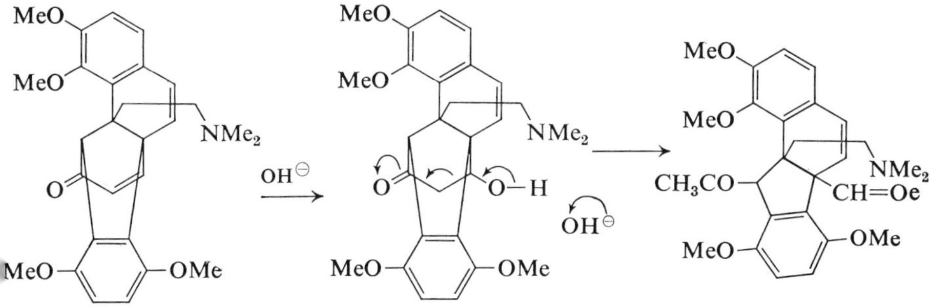

617

618

619

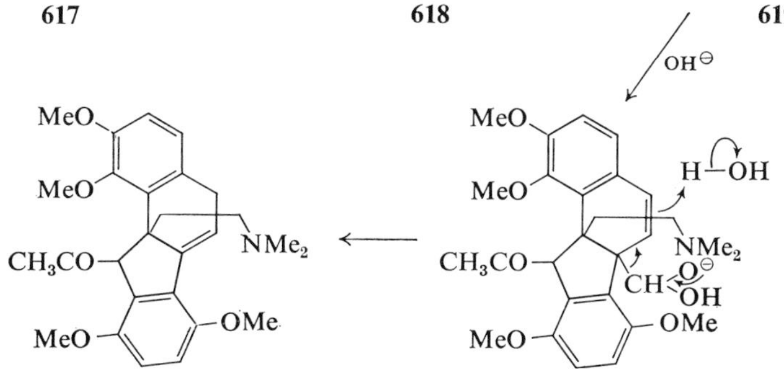

621

620

ketone and a final interesting elimination of an angular formyl group [360], which may be compared with the easy decarboxylation with associated double bond shift in morolic acid and related compounds (p. 229).

(b) The isomerization of tenulin (622) to isotenulin (623) under the influence of very mild basic conditions, for example, boiling with hard water [361].

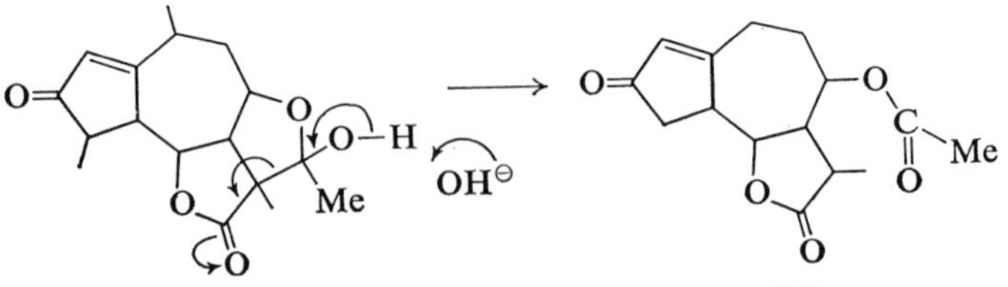

622   623

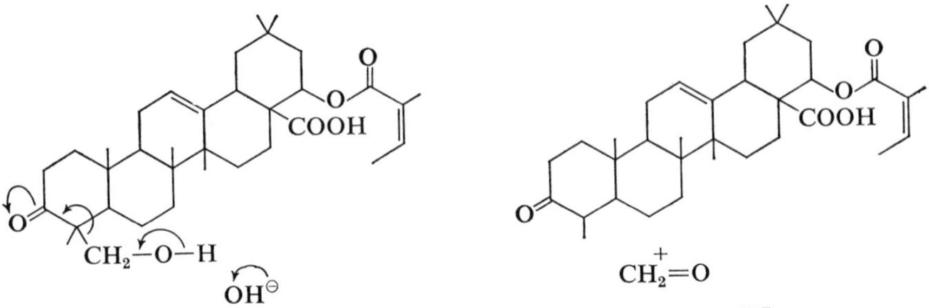

624   625

(c) The loss of formaldehyde from icterogenin (624)–(625) on mild treatment with alkali [362].

(d) The Clemmensen reduction of eleutherin (626) to the furan derivative (627) as well as the related 2-ethyl compound [363].

It may be noted in this context that the β-keto aldehyde (629) derived from isodihydroiresin (628) is very stable towards alkalis but is readily hydrolyzed to the ketone 630 by acids [364].

The elucidation of the structures of geodin (636, R = Me) and erdin (636, R = H) further illustrates the use of hydrolytic processes that gave vital structural information [365–368]. These mold metabolites may be catalytically reduced to benzophenones (635, R = Me and R = H) which on reductive fission with hydriodic acid and red phosphorus afford methyl iodide, carbon dioxide, orcinol (631), and α-resorcylic acid (632). Hydrolytic fission of the ketones with 80% sulfuric acid, which hydrolyzes other benzophenones, gives α-resorcylic acid monomethyl ether (639) and 3,5-dichloro-2,6-dihydroxy-4-methyl benzoic acid (638). The carboxyl group of the acid (638) must represent the carbonyl group of dihydrogeodin and erdin, which must

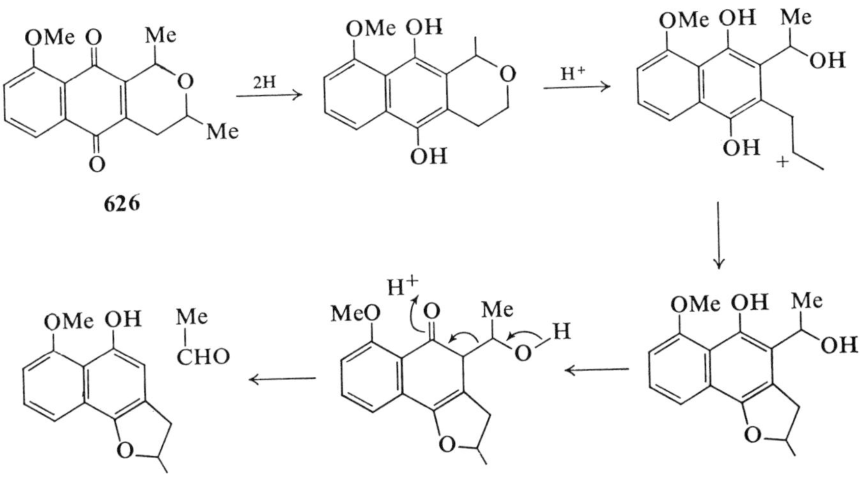

626

627

therefore have the structures (**635**), (**633**), or (**634**). A synthesis of dihydro-geodin trimethyl ether eliminated **634**, leaving the two possibilities **635** and **633** for dihydrogeodin and erdin, and a distinction between these was obtained by the action of mild alkali on dihydroerdin. This compound **635** (R = H) is an acylresorcinol and suffers hydrolysis in the diketone form **640** to dichloroorcinol (**641**) and 5-hydroxy-3-methoxyphthalic acid (**642**). The alternative structure (**634**, R = H) for dihydroerdin would have given 3-hydroxy-5-methoxyphthalic acid on hydrolysis [368]. This base-catalyzed hydrolysis in addition afforded the xanthone **645**, which presumably arose from dihydroerdin as shown in formulas **643** → **645**.

Both geodin and erdin are also hydrolyzed at the carbon–carbon linkage by 80% sulfuric acid, the products being geodin and erdin hydrates (**636** → **637**) [367].

Geodin (**636**, R = Me) is optically active, but is easily racemized by acids, whereas the acetyl ester and methyl ether are optically stable. The hydroxyl group is clearly necessary for racemization, which probably involves the reversible carbon–carbon bond fission of an extended diketone system represented in **646** ⇌ **647** [368].

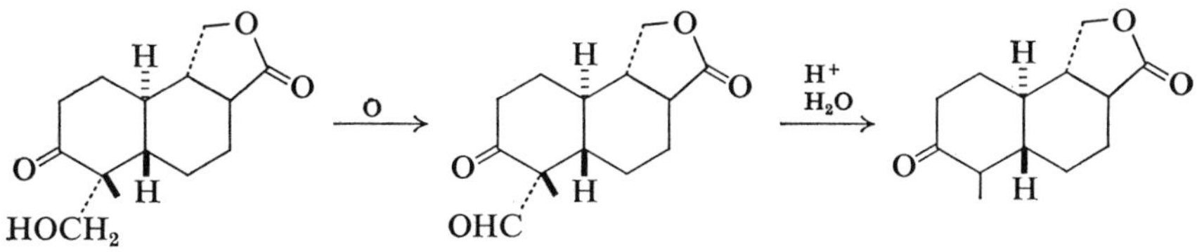

628   **629**   **630**

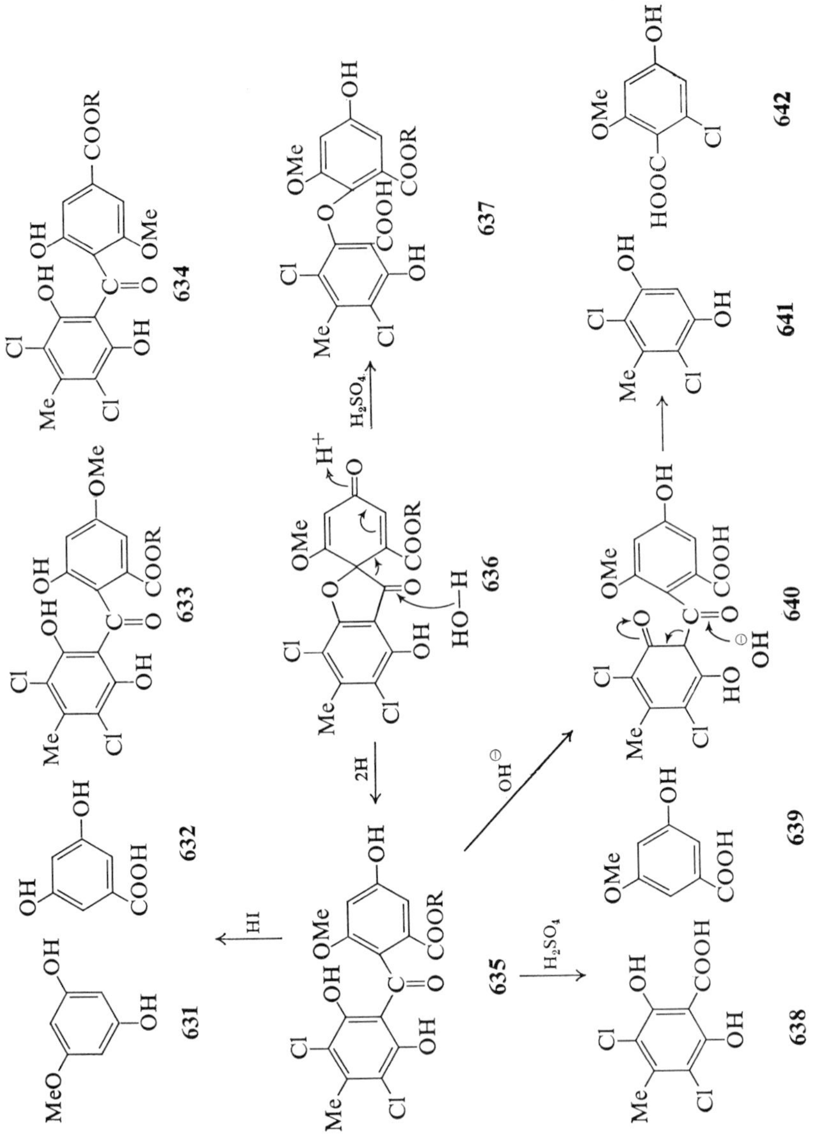

COOR

OH OH

Cl

Me OMe

Cl

634

OMe

OH

COOH

Cl

OH

Me

Cl

637

OH

OMe

HOOC

Cl

642

OMe

OH OH

Cl

Me C=O

Cl COOR

633

OMe

OH

C

Cl

Me

Cl

OH

COOH

641

OH

COOH

OH

632

OMe

OH O COOR

Cl

Me

Cl

OH

636

+H⁺

HO—H

2H

OMe

OH

Cl

Me

Cl

C O

COOH

OH

640

OH⁻

OMe

OH

MeO

OH

631

HI

OMe

OH

Cl

Me

Cl

C=O

O COOR

OH

635

OH⁻

H₂SO₄

OMe

OH

Cl

Me

Cl

COOH

OH

639

OMe

COOH

OH

Cl

Me

Cl

OH

OH

638

224

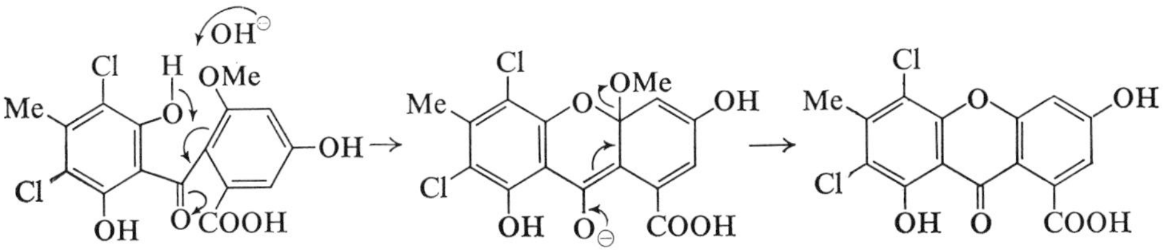

643 **644** **645**

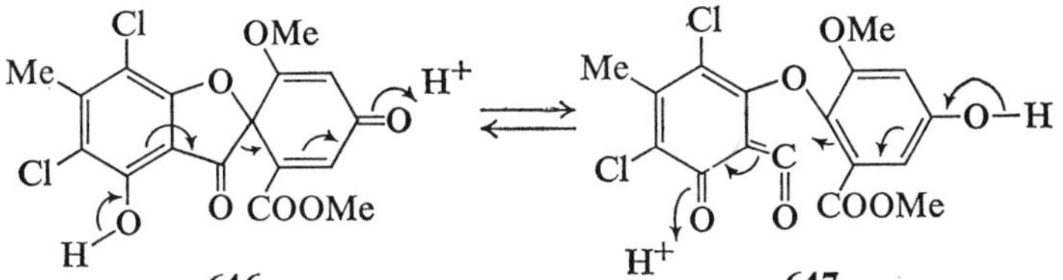

646 **647**

The hydrolysis of a benzophenone by sulfuric acid is involved in the conversion of trimethylbrazilone (**648**) into ψ-trimethylbrazilone (**650**) [265], and the reaction may be represented as proceeding through the spirointermediate **649**.

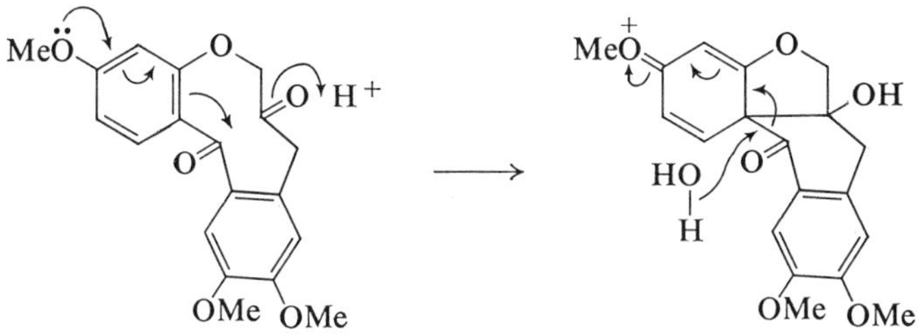

648 **649**

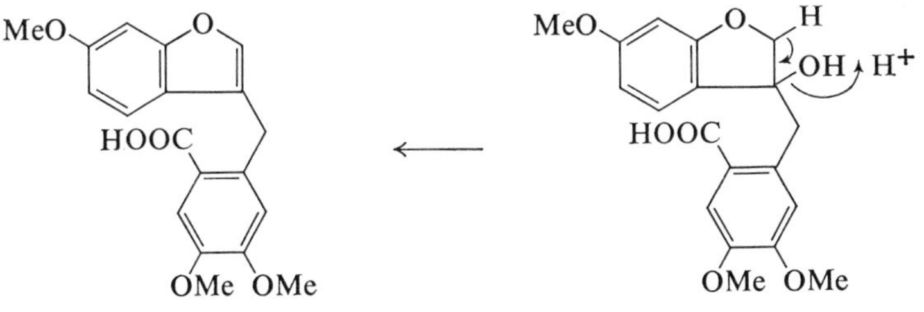

650 **651**

4 DECARBOXYLATION, DECARBONYLATION

The decarboxylation of simple acids of the type RCOOH and the decarbonylation of aldehydes RCHO proceed by the elimination of the group R as an anion $R^{\ominus}$ with its pair of bonding electrons. Ionization of acids by proton removal with bases (hydroxyl ions, organic bases such as quinoline and collidine), gives the carboxylate anion in which elimination occurs, and the reaction proceeds with greatest ease when a relatively stable ion $R^{\ominus}$ is generated, for example, with β-resorcylic acid [369, 370]:

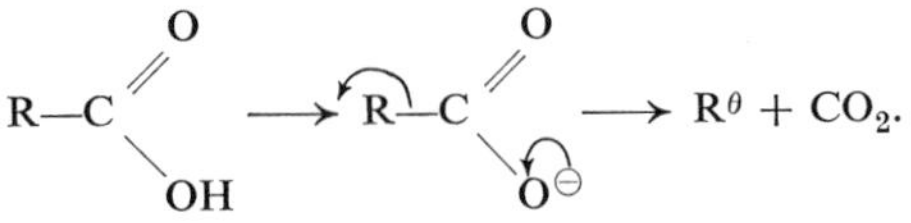

$$R-C \overset{O}{\underset{OH}{\big\langle}} \longrightarrow R-C \overset{O}{\underset{O^{\ominus}}{\big\langle}} \longrightarrow R^{\theta} + CO_2.$$

With aldehydes the process may be written in the form

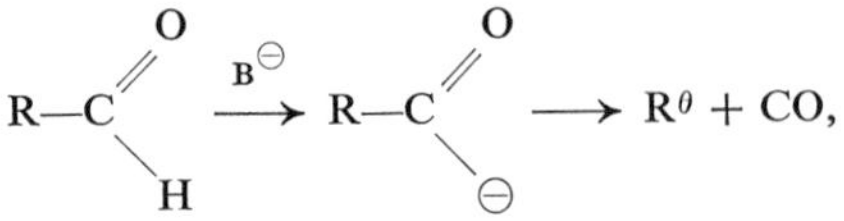

$$R-C \overset{O}{\underset{H}{\big\langle}} \xrightarrow{B^{\ominus}} R-C \overset{O}{\underset{\ominus}{\big\langle}} \longrightarrow R^{\theta} + CO,$$

and when a relatively stable ion is generated this process, like decarboxylation, may occur with surprising ease, for example ajmaline (652), which is a masked aldehyde, loses carbon monoxide and gives decarbonochanoajmaline (654) simply on boiling with Raney nickel in xylene [371].

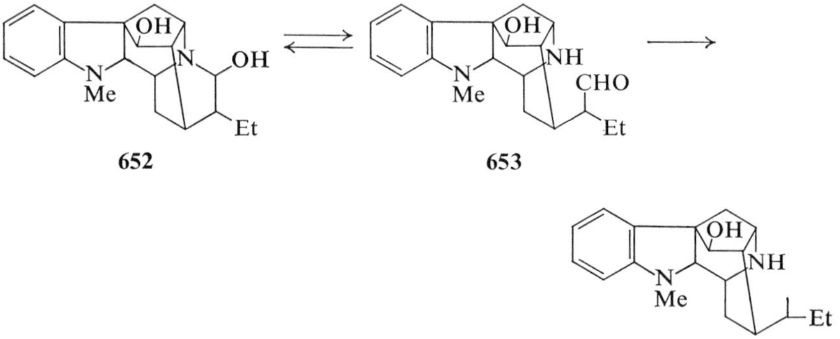

652   653

654

Decarboxylations of this type are readily effected by the Hunsdiecker reaction which involves treating the silver salt of the acid with bromine, the products being silver bromide, carbon dioxide, and the alkyl bromide [372–375]. The reaction must be carried out in the absence of water, generally

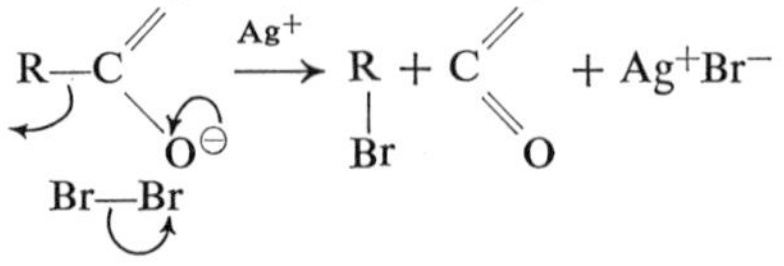

$$R-C \overset{O}{\underset{O^{\ominus}}{\big\langle}} \underset{Br-Br}{\overset{Ag^{+}}{\longrightarrow}} R + C \overset{O}{\underset{Br}{\big\langle}} + Ag^{+}Br^{-}$$

in carbon tetrachloride, though ethyl bromide has been recommended [376].

An improved procedure for achieving the same result has been introduced in which the acid is treated with red mercuric oxide and bromine in carbon tetrachloride:

$$2RCOOH + HgO + 2Br_2 \rightarrow 2RBr + HgBr_2 + H_2O + CO_2.$$

Yields are generally superior to those obtained in the Hunsdiecker reaction and the process has the advantage that it is not necessary to exclude moisture. The reaction is effective even in complex systems, for example, 5,6,7,8,9,9-hexachloro-1,2,3,4,4a,5,8,8a-octahydro-*exo*,*endo*-1,4,5,8-dimethano-2-naphthoic acid (**655**) gives 71 % *exo*-bromide (**656**) and 29 % *endo*-bromide [377].

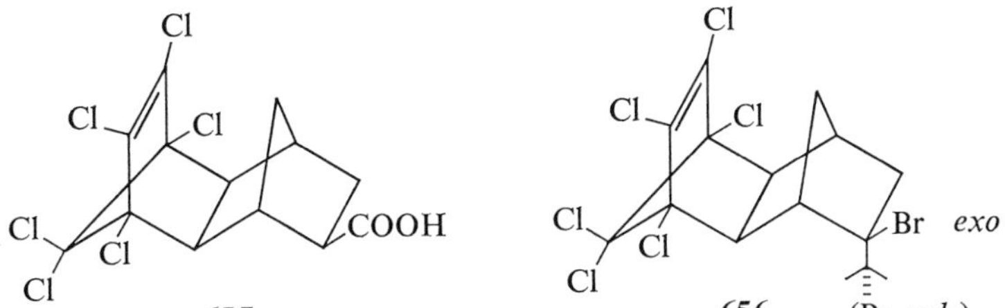

Many aromatic acids and their vinylogues are very readily decarboxylated, with the formation of bromo compounds, merely by treating an aqueous solution of the sodium salt with bromine. For example, piperonylic acid (**657**) affords methylenedioxybromobenzene, and dimethoxycinnamic acid (**658**) gives the ω-bromostyrene. The latter process is not stereospecific and a mixture of *cis* and *trans* isomers is obtained. The dibromide of the acid (**658**), having an excellent leaving group, readily undergoes stereospecific elimination of carbon dioxide and bromide ion (**659**), simply on warming with aqueous sodium carbonate, to give *trans*-ω-bromo-3,4-dimethoxy-stryene [378, 379].

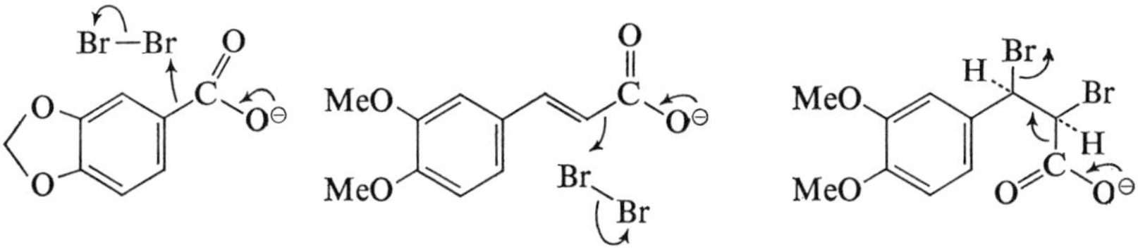

Decarboxylation of β-keto acids is generally a very smooth and simple process. The reaction is generally regarded as proceeding through a six-membered hydrogen bonded cyclic transition state (**660**) in which elimination

affords initially the enol form (662) of the resulting ketone (662) [380]. As
will be expected from such a mechanism, if the double bond of the enol 661

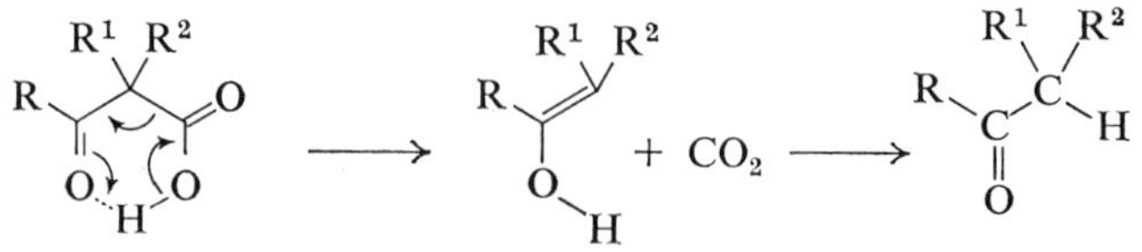

660   661   662

cannot be accommodated in a reasonably strain-free system, the acid will
resist decarboxylation as in the case of bridgehead carboxylic acids such as
ketopinic acid (663) and camphenonic acid (664), which are stable even above
300°.

663    664

A precisely similar process is involved in the conversion of voacangic acid
(665) into ibogaine (667) [383], and it was on the basis of such a mechanism
that the carbomethoxy group in voacangine (665, OH = OMe) was assigned

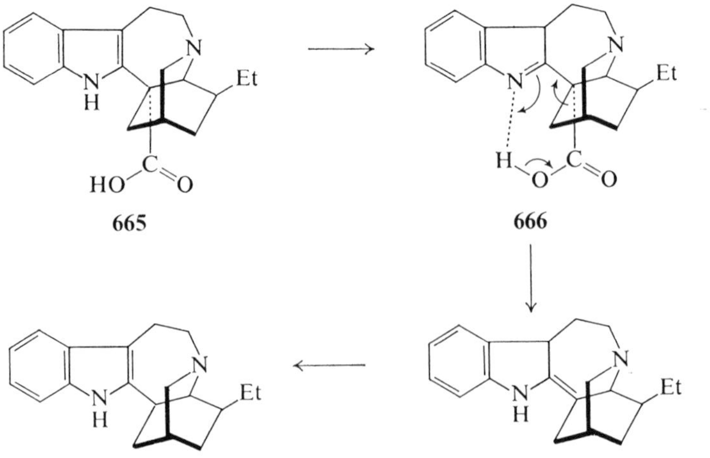

665   666

667   668

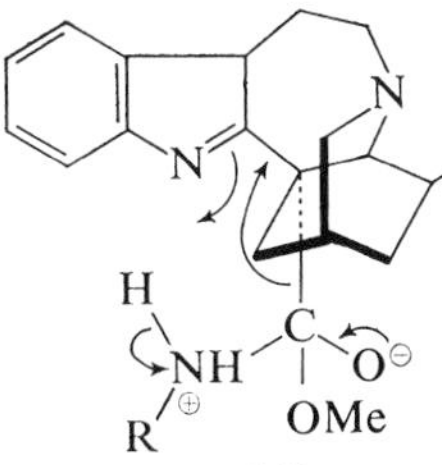

669

the position indicated [384]. In this respect voacangic acid resembles indole 2-acetic acid [385]. A similar degradation occurs when voacangine is heated with bases such as hydrazine or ethanolamine [386], and this doubtless proceeds by the mechanism shown in **669**.

A similar cyclic process is involved in the facile decarboxylation of β,γ-unsaturated acids with migration of the double bond (**670**), as in the decarboxylation of morolic acid (**672**) to the olefin **673** [387], which was a reaction of structural significance in the elucidation of the structure of the acid, and of dehydrooleanolic acid (**674**) [387, 388] and of the acid **675**,

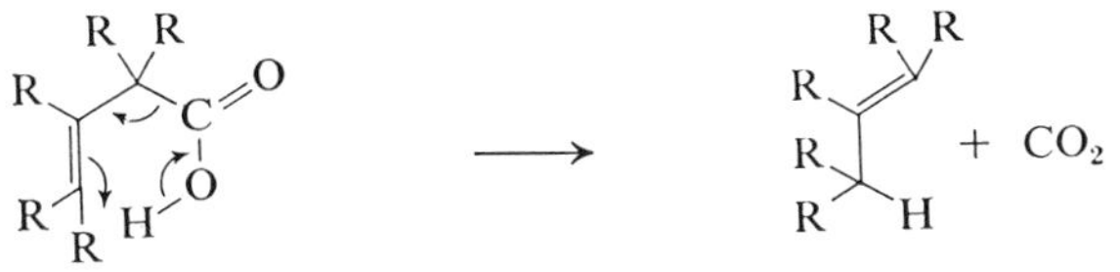

670 $\longrightarrow$ **671** $+ CO_2$

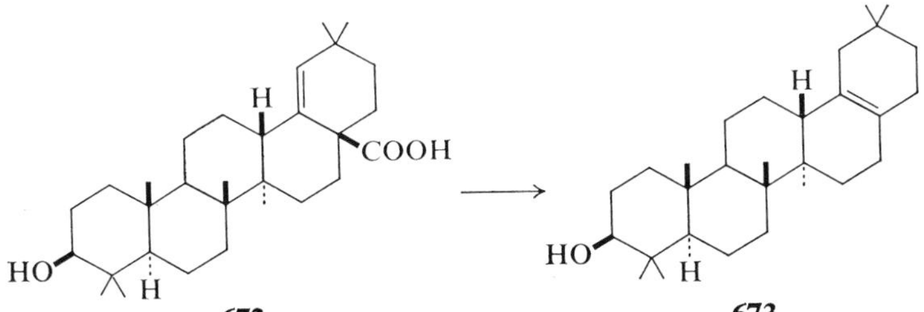

672 $\longrightarrow$ **673**

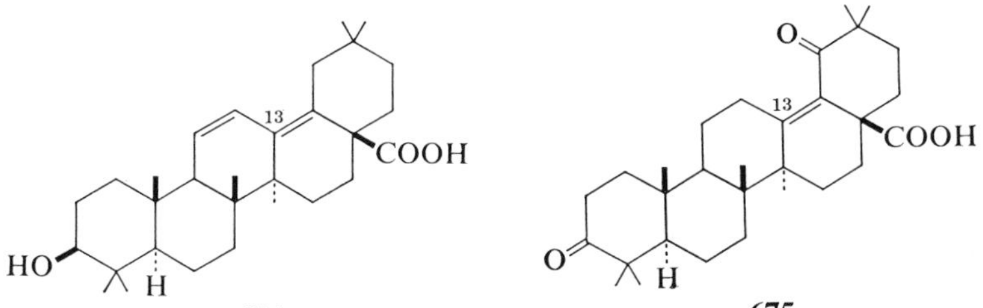

674 **675**

derived from siaresinolic acid. In the last two cases it is a consequence of the cyclic mechanism 670 → 671 for decarboxylation that the hydrogen atom deposited at C-13 must be on the same side of the molecule as the carboxyl group in the parent acid and, since the product of decarboxylation of the acid 675 was related to morolic acid, the stereochemistry at C-13 in this acid 672 was made clear [387]. A synthesis of morolic acid from siaresinolic confirmed this point [389].

An interesting cleavage of vincamine (676) to carbon monoxide and eburnamonine (677) by lithium aluminum hydride has been observed, and this has been rationalized by the mechanism shown (676) by analogy with the

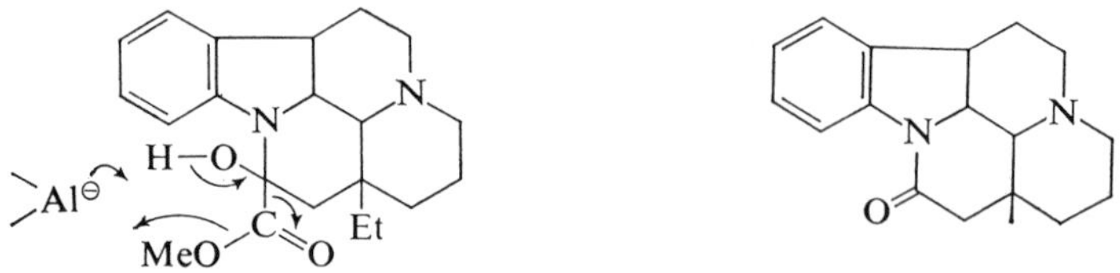

676   677

decomposition of formic esters to carbon monoxide and alkoxide ion [390].

5   MISCELLANEOUS REACTIONS

Beckmann Transformation

This process, which involves either a 1,2-shift from carbon to nitrogen with formation of an amide:

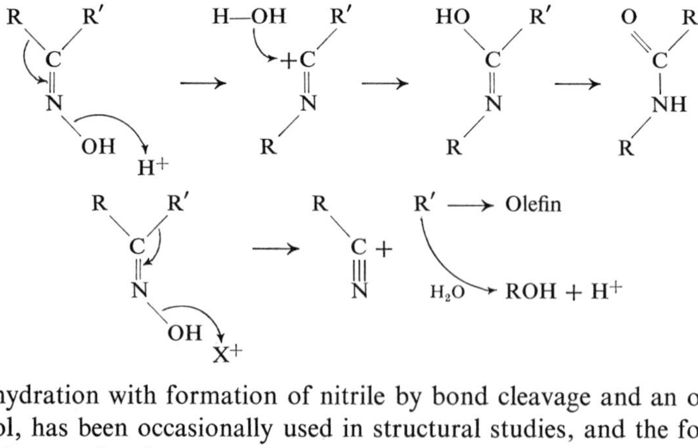

or dehydration with formation of nitrile by bond cleavage and an olefin or alcohol, has been occasionally used in structural studies, and the following examples may be given.

(a) The rearrangement of the oxime 678 to the amide 679 [391], also obtained by the Schmidt reaction on the diketone from which the oxime was derived [392], in structural studies with lanostadienol [393].

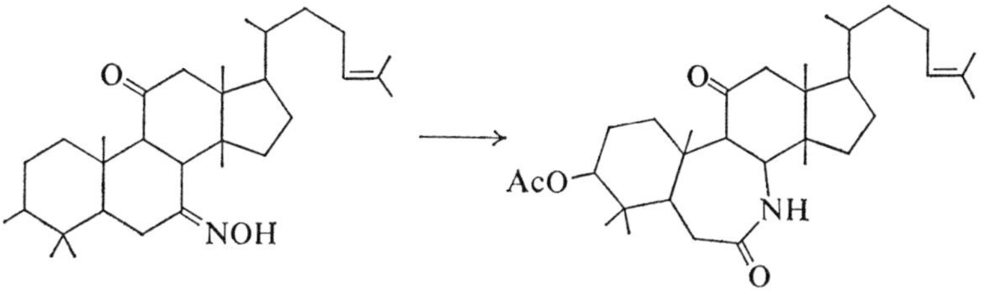

678   679

(b) The degradation of camphor oxime (680) to α-camphenonitrile (681) [3].

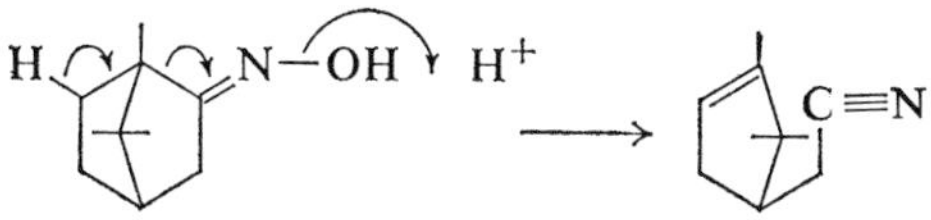

680   681

(c) The degradation of flavothebaone trimethyl ether ψ-methine oxime (682) to acetonitrile and the alcohol 683 and the combined loss of acetonitrile, group migration, and aromatization of the related angular vinyl compound (685 → 686). Dehydration group migration and aromatization of the alcohol 683 can be achieved by the more vigorous action of acids, and the difference

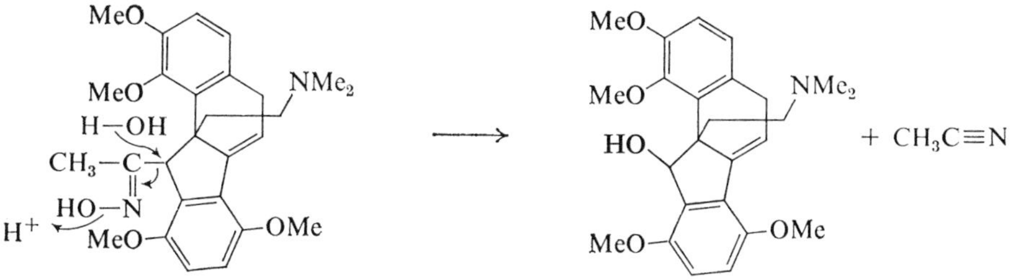

682   683

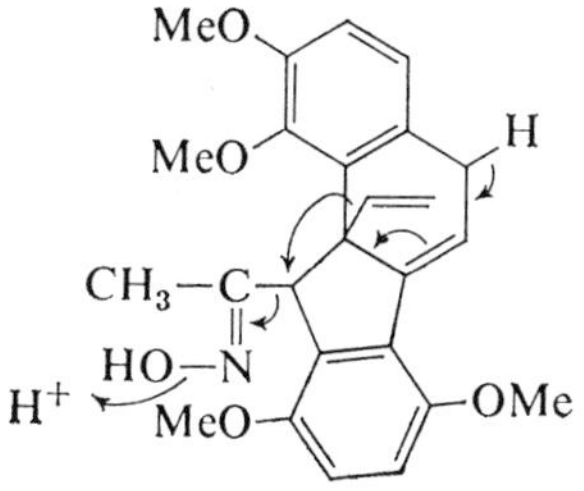

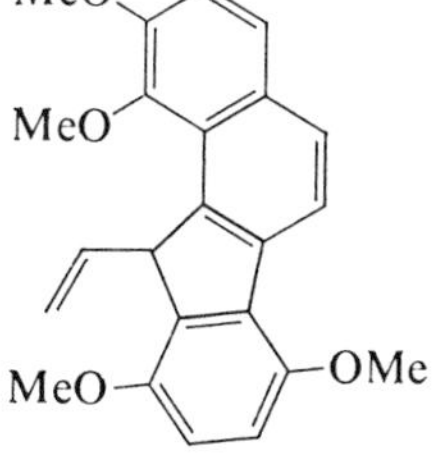

684   685

in behavior of the two oximes may be attributed to differences in the migratory aptitudes of alkyl and vinyl groups [394]. These rearrangements are represented by concerted mechanisms, but may well proceed through carbonium ions.

(*d*) Rearrangement of isonitrosostrychnine (**686**) to the lactam (**688**) and the Wieland-Gumlich aldehyde (**689**), carbon dioxide, and hydrogen cyanide via the cyanhydrin **687** [395–397].

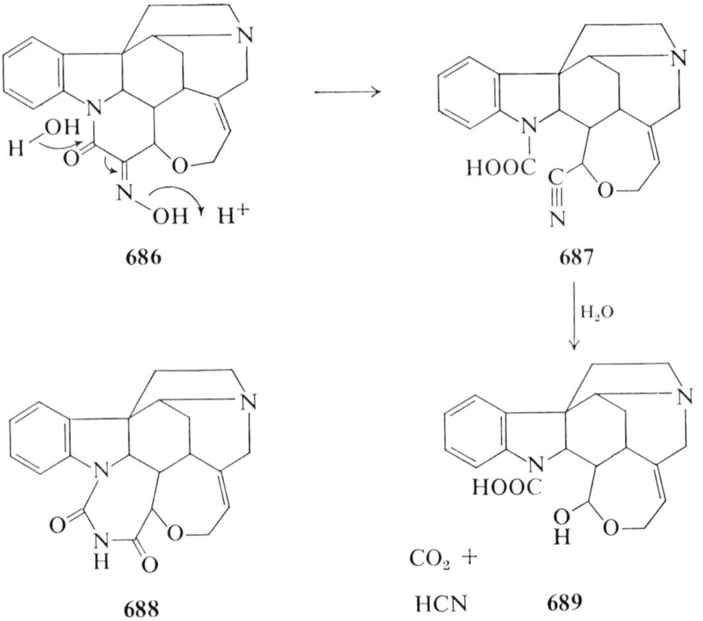

686

687

H_2O

688

$CO_2 +$

HCN   **689**

(*e*) The rearrangement of bisisonitrosothebenone (**690**) to the dinitrile **691**. Hydrolysis of this gave an amido acid **692** which could be cyclized to an imide, whereas the C-14 epimer of the isonitroso compound gave an amido acid that could not be so cyclized [398]. These reactions were of importance in the determination of the stereochemistry of morphine.

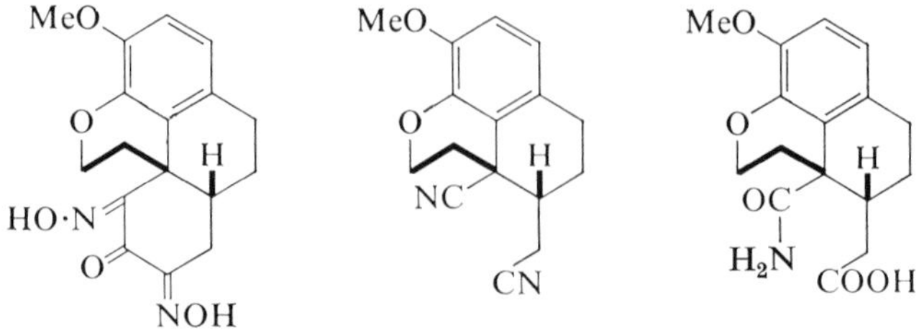

690

691

692

(*f*) There arrangement of dihydrocodeinone oxime (**693**) to the aldehydo-nitrile **694**. This reaction was of importance in proving a C-13 attachment of the nitrogen-containing side-chain in the morphine group, as a C-5 attachment would have given a keto-nitrile (**695**) [399]. The reaction is one that is often observed in Beckmann rearrangements of oximes bearing an oxygen substituent on the α-carbon, for example, benzoin oxime (**696**).

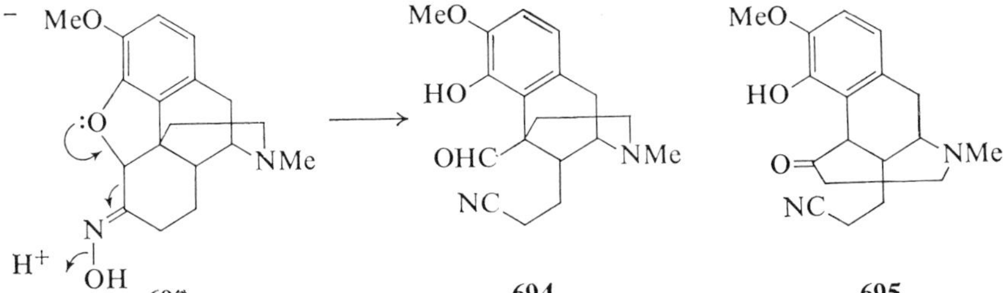

693   694   695

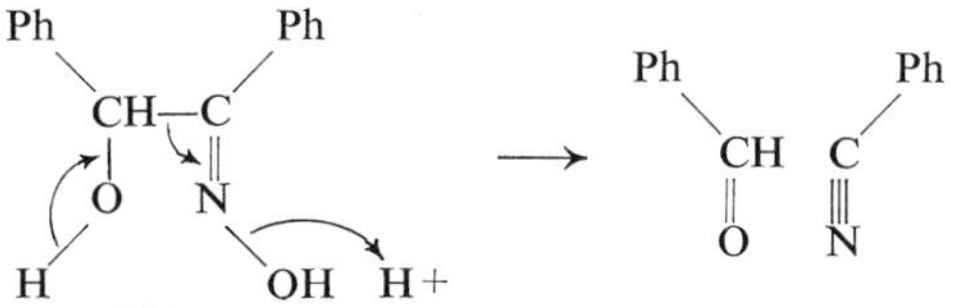

696

(*g*) Ibogaine, (**697**, R = OMe, R′ = H), tabernanthine (**697**, R = H, R′ = OMe), and ibogamine (**697**, R = R′ = H) were related by conversion into the pseudoindoxyls (**698**), the oxime tosylates of which (**701**) were rearranged and hydrolyzed to the same ketone (**700**) and the appropriate anthranilonitriles (**699**) [384].

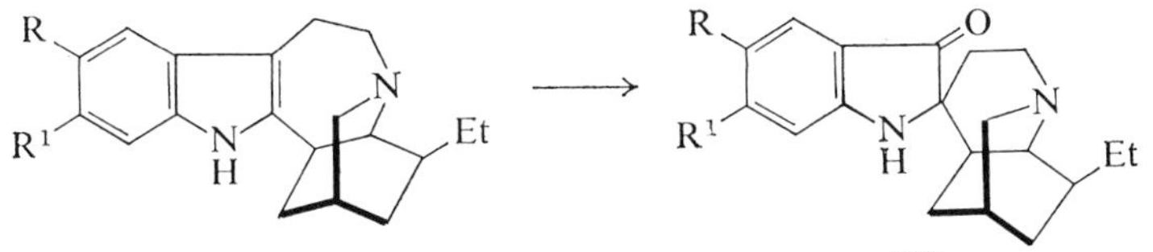

697   698

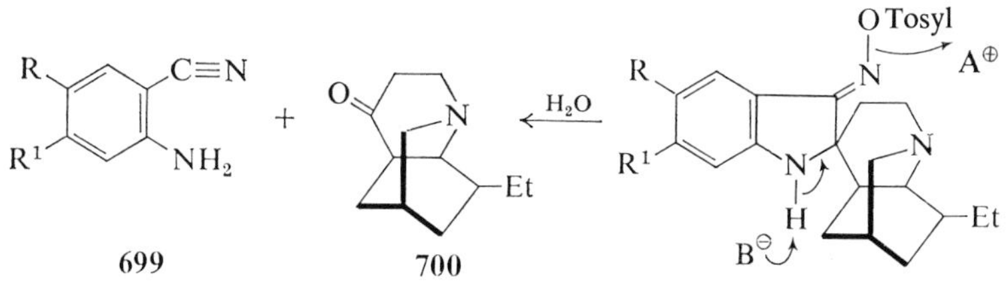

699   700   701

Fission of Hydroxydiarylmethanes

Tetrahydroallorottlerin (**702**) has been cleaved with benzene-diazonium chloride to give the azo compounds **704** and **706** and the benzylic alcohols **707** and **708** [400]. The mechanism of one mode of fission is shown in formula **702**, and this leads as shown to the products **704** and **708**. The attack of the other nucleus would lead to the alternative fission products **706** and **707**.

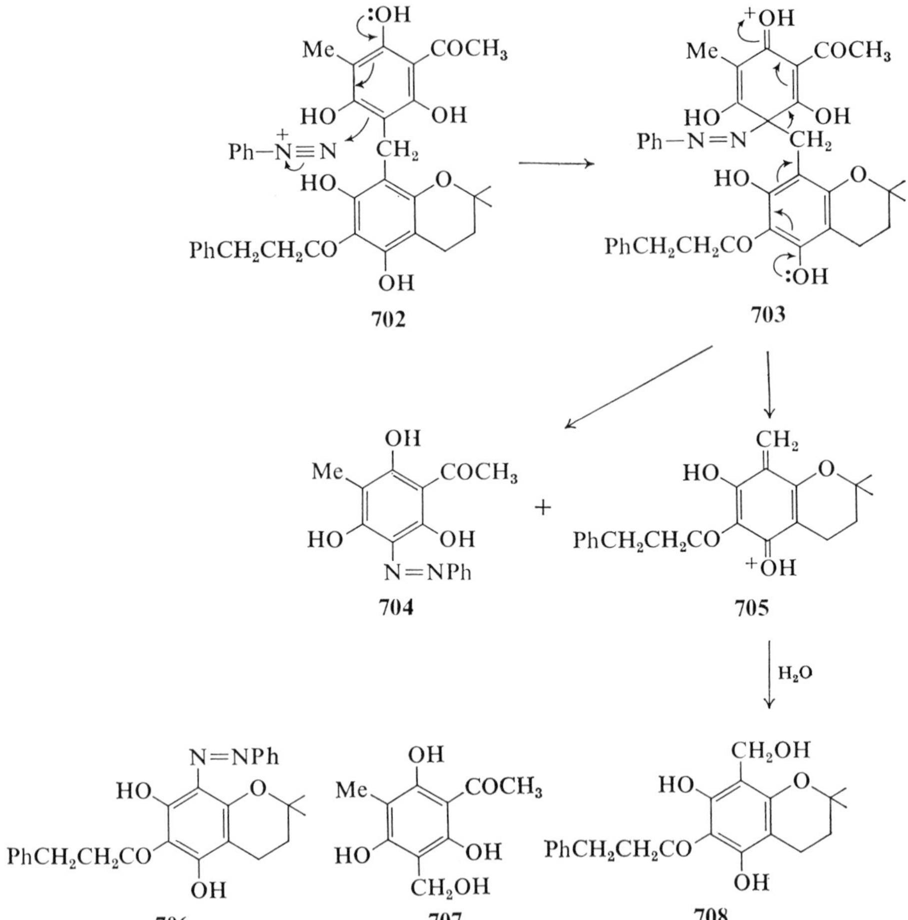

702   **703**

704   **705**

706   **707**   **708**

Similar fissions of other polyhydroxydiphenylmethanes have been observed [401].

Fission of 1,3-Glycols and Their Ethers

1,3-Glycols and their ethers can undergo a type of "retro-Prins" reaction under dehydrating conditions. For example, certain complex alcohols in the morphine series are very readily dehydrated to give ketones (**709**) [402], and the same reactions have been observed in simpler analogous bridged-ring systems [403]. Quinamine (**710**) and dihydroquinamine eliminate formaldehyde on heating above their melting points, presumably as shown [404], and

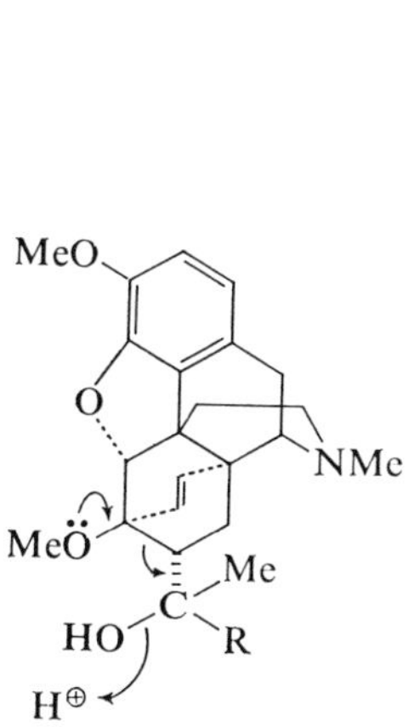

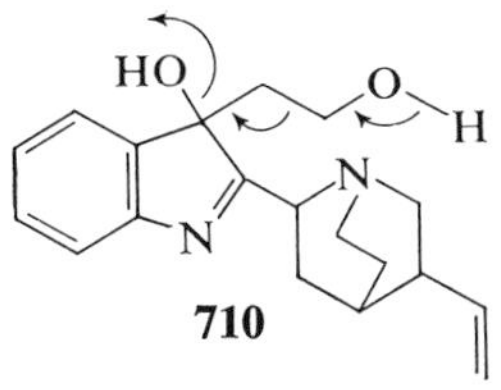

710

709

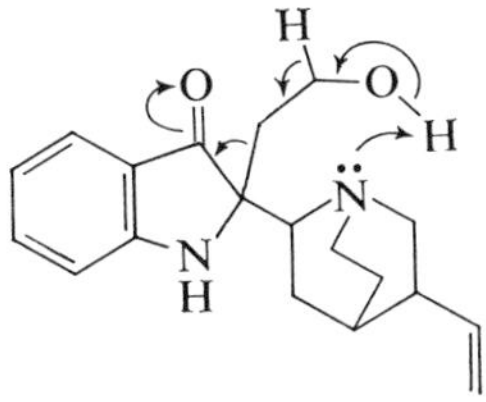

711

the reaction was originally incorrectly interpreted structurally as evidence for the location of a 2-hydroxymethyl group at the indole α-position [405]. The related pyrolysis of isoquinamine with the elimination of acetaldehyde [406] may be formulated as in **711** [404].

Photochemical Cleavages

Photochemical reactions have not been widely used in structural studies, and useful results are mainly confined to the cleavage of ketones and unsaturated ketones in solution, when the choice of solvent is frequently the critical factor since it often influences the reaction path. The reaction initially generally involves the fission of the CO–C bond to generate the more substituted radical **712** → **713**, which can revert to a stable species in a variety of ways, involving hydrogen transfer to form an unsaturated aldehyde (**714**) or a

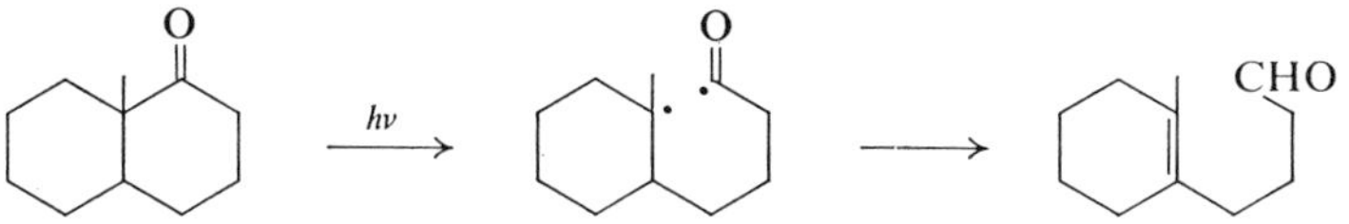

712   **713**   **714**

ketene (**716**), and the ketene, if formed in the presence of water, alcohols, or amines, gives acids (**715**, R = H), esters (**715**, R = alkyl), or amide (**717**).

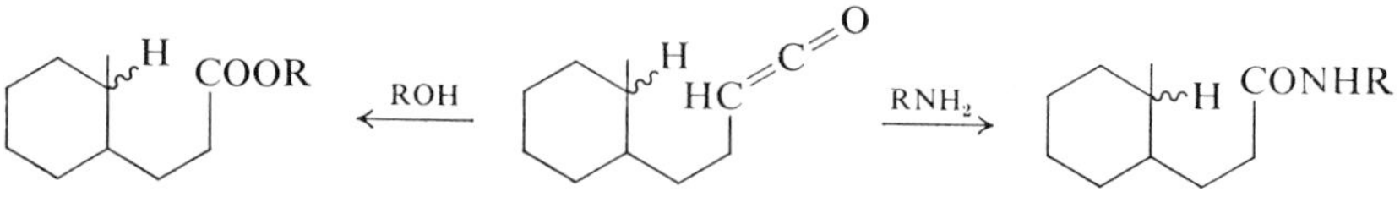

715   716   717

The choice between the stabilization pathways leading to aldehyde or ketene depends on the availability of hydrogen at the two relevant centers α to the electron-deficient carbon atoms, on the steric requirements of the two transition states, and on the availability of nucleophiles that can react with the ketene when generated [407].

The processes may be illustrated by the following examples:

(*a*) Menthone (**718**) affords the aldehydes **719** and **720** and the acid **721** when irradiated in aqueous ethanol [408].

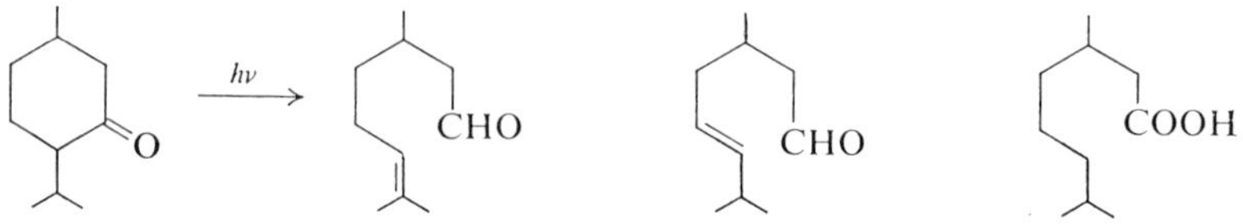

718   719   720   721

(*b*) The conversion of cholesta-3β-ol-6-one (**722**) into the acid **723** [409].

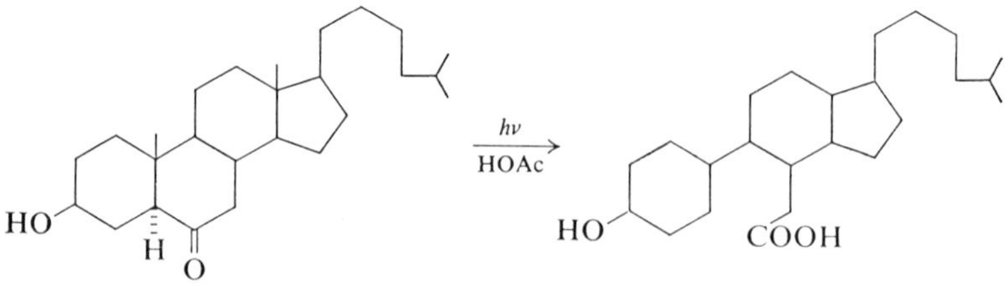

722   723

(*c*) The conversion of the lactone **724**, prepared from hydroxydammaren-one-II, of known structure, into the acid **725**, which on reduction with lithium aluminum hydride gave the triol **727**, already obtained by a sequence of reactions from dammarenolic acid (726), which was of importance in the determination of the structure of the last-named acid [410].

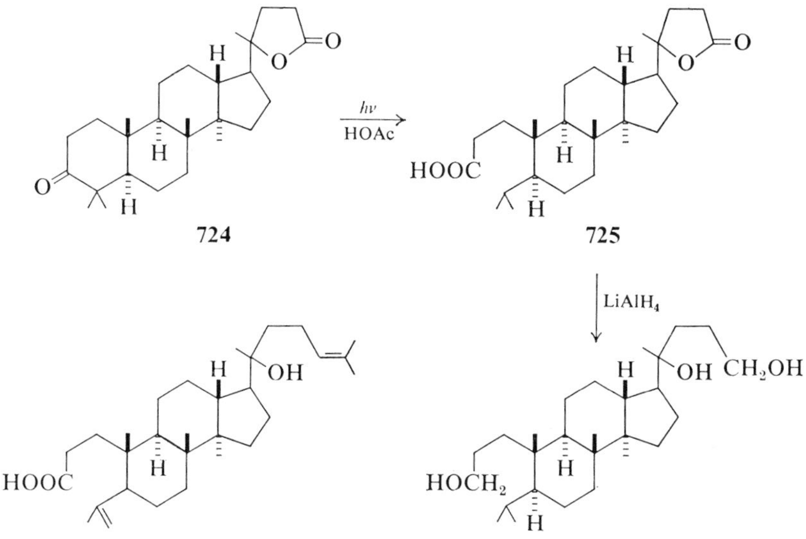

(d) The dienone 1-oxosanta-2,4-dien-8α;12-olide (728) is very readily converted into the dienoic acid 729, together with the other geometrical isomer [411]. Unsaturated ketones are more easily attacked than saturated ketones, but a variety of rearrangements and dimerizations is also possible in such cases.

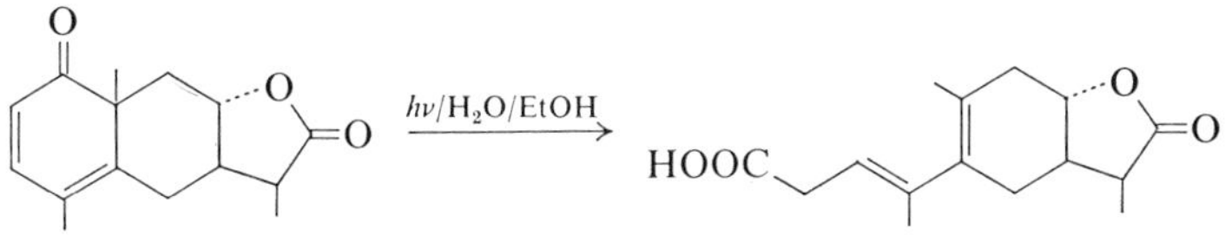

Cyclic conjugated olefins may also suffer cleavage of a carbon–carbon bond on photolysis, and examples of this are the conversion of ergosterol (730) into precalciferol (731) [412] and of methyl dehydroursolate acetate (733) to the triene (734) [413]. The initial products in both cases are readily isomerized to the trienes 732 and 735, respectively.

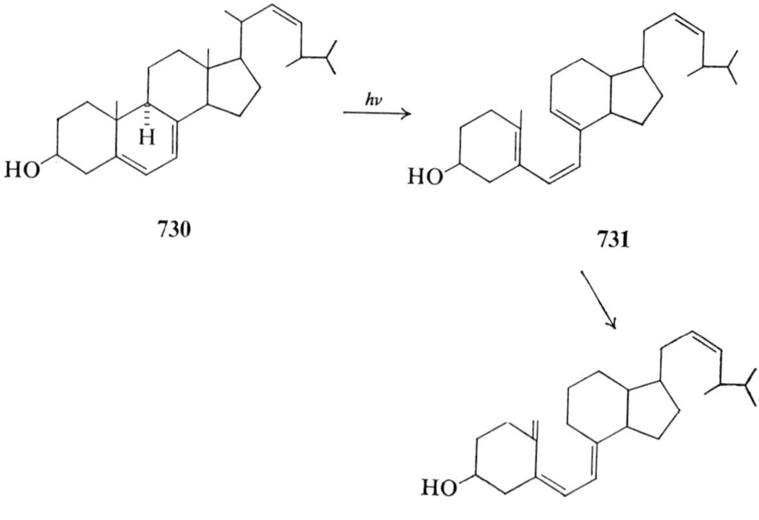

730

731

732

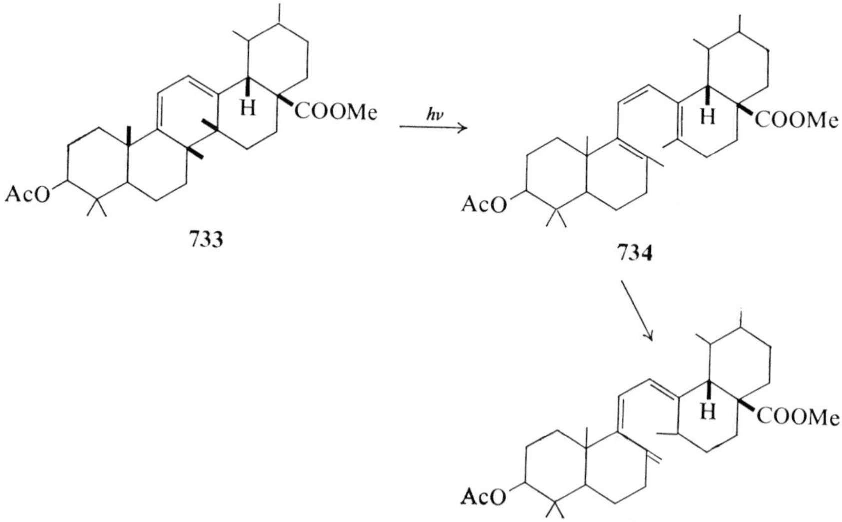

733

734

735

6 SPECIAL REACTION SEQUENCES APPLICABLE TO CERTAIN SYSTEMS

Certain reaction sequences have been devised for the stepwise degradation of long-chain acids, ketones, amines, and so on, which may be summarized as follows:

(a) The Barbier-Wieland degradation [414, 415] for the degradation of acids involves the stepwise removal of one carbon atom at a time, except where a point of chain branching is reached, when more than one carbon atom is removed, the number depending on the length of the branch in the chain, though in such cases the reaction sequence may be terminated in a ketone instead of an acid. The reaction sequence is summarized here. The

Step One

$$\underset{\substack{|\\ \text{RCHCH}_2\text{COOH}}}{\overset{\text{CH}_3}{|}} \xrightarrow[\text{PhMgBr}]{\text{Ester}} \underset{\substack{|\quad\ |\\ \text{RCHCH}_2\text{CPh}_2}}{\overset{\text{CH}_3\ \ \text{OH}}{|\quad\ |}} \xrightarrow{-\text{H}_2\text{O}} \underset{\substack{|\\ \text{RCHCH}{=}\text{CPh}_2}}{\overset{\text{CH}_3}{|}} \xrightarrow{\text{O}} \underset{\substack{|\\ \text{RCHCOOH}}}{\overset{\text{CH}_3}{|}}$$

$$\downarrow \begin{array}{l}\text{Ester}\\\text{PhMgBr}\end{array}$$

Step Two

$$\underset{\text{RCOOH}}{} \longleftarrow \overset{\text{O}}{} \text{R·CO} \longleftarrow \underset{\substack{|\\ \text{R—C}{=}\text{CPh}_2}}{\overset{\text{CH}_3}{|}} \overset{\text{O}}{} \xleftarrow{-\text{H}_2\text{O}} \underset{\substack{|\ \ |\\ \text{CH CPh}_2}}{\overset{\text{CH}_3\text{OH}}{|\ \ |}}$$

process is clearly applicable to the degradation of long-chain ketones, which will give tertiary alcohols with phenylmagnesium bromide, and these may be dehydrated to olefins as in the sequence shown.

Chromic acid is the reagent of choice for the oxidation step [416–418], but the double bond fission has also been accomplished with ozone [419–421], sodium periodate, and ruthenium dioxide or osmium tetroxide [422]. The use of phenylmagnesium bromide as Grignard reagent ensures that the dehydration can proceed in one direction only. Adequate experimental details are available [415–423].

(b) The Hunsdiecker reaction (see Section 4) can be utilized for a similar stepwise degradation of acids through bromides and alcohols. The sequence

$$\underset{\substack{|\\ \text{R·CHCH}_2\text{COOH}}}{\overset{\text{CH}_3}{|}} \xrightarrow[\text{Br}_2]{\text{Ag salt}} \underset{\substack{|\\ \text{RCHCH}_2\text{Br}}}{\overset{\text{CH}_3}{|}} \xrightarrow{\text{H}_2\text{O}} \underset{\substack{|\\ \text{R—CH·CH}_2\text{OH}}}{\overset{\text{CH}_3}{|}} \xrightarrow{\text{O}} \underset{\substack{|\\ \text{R—CHCOOH}}}{\overset{\text{CH}_3}{|}}$$

$$\downarrow \text{Acid}$$

$$\underset{\substack{|\\ \text{R·CO}}}{\overset{\text{CH}_3}{|}} \overset{\text{O}}{} \longleftarrow \text{R·CH·OH} \xleftarrow{\text{H}_2\text{O}} \underset{\substack{|\\ \text{R·CHBr}}}{\overset{\text{CH}_3}{|}}$$

Ketone

may be represented as shown. Ketones are formed on oxidation where chain-branching occurs, but further oxidation can convert these into acids for a continuation of the process.

(c) The Hofman rearrangement and Hofmann degradation may be utilized for removing two carbon atoms per step in a straight-chain acid, though more are removed where chain branching is encountered [424, 425]. The reaction sequence involves the following processes: Hofmann re-arrangement of the amide of an acid, quaternization of the primary amine, elimination in the quaternary salt, and oxidation of the resulting olefin, and

$$\underset{\substack{\text{rearrange-}\\\text{ment}}}{\overset{\text{Amide}}{\longrightarrow}}$$

$$RCH_2\overset{\overset{\displaystyle CH_3}{|}}{C}HCH_2CH_2COOH \xrightarrow[\substack{\text{rearrange-}\\\text{ment}}]{\text{Amide}} RCH_2\overset{\overset{\displaystyle CH_3}{|}}{C}HCH_2CH_2NH_2 \longrightarrow RCH_2\overset{\overset{\displaystyle CH_3}{|}}{C}HCH_2CH_2\overset{+}{N}Me_3$$

$$RCH_2\overset{\overset{\displaystyle CH_3}{|}}{C}HNMe_2 \xleftarrow{\text{Repeat}} RCH_2\overset{\overset{\displaystyle CH_3}{|}}{C}HCOOH \xleftarrow{O} RCH_2\overset{\overset{\displaystyle CH_3}{|}}{C}HCH{=}CH_2$$

$$RCH{=}\overset{\overset{\displaystyle CH_3}{|}}{C}H \xrightarrow{O} RCOOH$$

may be summarized as shown. Where chain branching occurs, more than one olefin may be formed and the products of oxidation may be more complex. The reaction has been modified to achieve one-carbon removal by the reduction of the acid amide to an amine containing the same number of carbon atoms as the parent acid [426]:

$$R{-}CH_2COOH \xrightarrow[\text{LiAlH}_4]{\text{Amide}} RCH_2CH_2NH_2 \xrightarrow[\text{above}]{\text{As}} RCH{=}CH_2 \xrightarrow{O} RCOOH.$$

Pyrolysis of the tertiary amine oxide may be used for the generation of the olefin [426]. The reaction sequence is obviously applicable to the degradation of long-chain amines.

(d) Acids may be brominated at the α-carbon atom and then either hydrolyzed to α-hydroxy acids and oxidized or pyrolyzed to aldehydes with the loss of one carbon atom [427–429], or dehydrobrominated to α,β-unsaturated acids which on fusion with alkali lose two carbon atoms [430]. Where chain branching occurs in the first of these processes the product is a ketone, which can be further oxidized to an acid with the loss of more carbon. The second process has been adapted to the degradation of amino and keto acids. Keto acids can be converted, via oximes or hydrazones, into

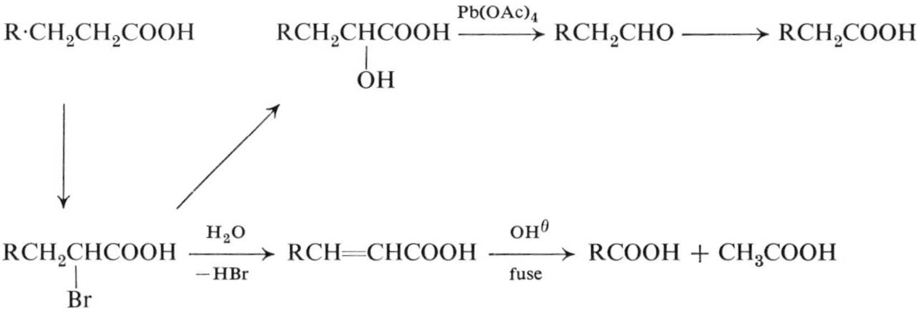

$$R \cdot CH_2CH_2COOH \qquad RCH_2CHCOOH \xrightarrow{Pb(OAc)_4} RCH_2CHO \longrightarrow RCH_2COOH$$
$$\underset{OH}{|}$$

$$RCH_2CHCOOH \xrightarrow[-HBr]{H_2O} RCH=CHCOOH \xrightarrow[fuse]{OH^\theta} RCOOH + CH_3COOH$$
$$\underset{Br}{|}$$

amino acids, which on quaternization and Hofmann degradation give olefinic acids, and these on alkali fusion suffer double bond migration into the α,β-position before degradation as shown [431].

(e) The removal of three carbon atoms (more if chain branching at the α or β position is encountered) from an acid may be achieved by a modification of the Barbier-Wieland process in which the olefin is brominated at the allylic position by N-bromosuccinimide, and the resulting bromide is dehydrobrominated to a diene which is then oxidized [433–438]. The sequence of reactions is shown.

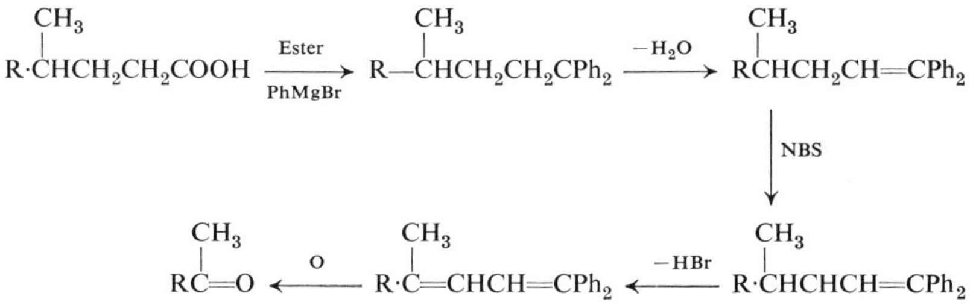

$$\underset{CH_3}{\overset{|}{R \cdot CHCH_2CH_2COOH}} \xrightarrow[PhMgBr]{Ester} \underset{CH_3}{\overset{|}{R-CHCH_2CH_2CPh_2}} \xrightarrow{-H_2O} \underset{CH_3}{\overset{|}{RCHCH_2CH=CPh_2}}$$

$$\downarrow NBS$$

$$\underset{CH_3}{\overset{|}{RC=O}} \xleftarrow{\quad O \quad} \underset{CH_3}{\overset{|}{R \cdot C=CHCH=CPh_2}} \xleftarrow{-HBr} \underset{CH_3}{\overset{|}{R \cdot CHCHCH=CPh_2}}$$

References

1. E. Bayer, *Chem. Ber.*, **13**, 2254 (1880).
2. A. W. Gilbody, W. H. Perkin, and J. Yates, *J. Chem. Soc.*, **79**, 1396 (1901).
3. W. H. Perkin, *J. Chem. Soc.*, **81**, 221, 1008 (1902).
4. W. H. Perkin and R. Robinson, *J. Chem. Soc.*, **93**, 489 (1908).
5. W. H. Perkin, *J. Chem. Soc.*, **57**, 992 (1890).
6. W. H. Perkin and R. Robinson, *J. Chem. Soc.*, **97**, 305 (1910).
7. H. Leuchs, *Chem. Ber.*, **41**, 1711 (1908).
8. H. Leuchs and G. Schwaebel, *Chem. Ber.*, **3693** (1913).
9. E. Späth and H. Bretschneider, *Chem. Ber.*, **63**, 2997 (1911).

10. L. F. Fieser and M. Fieser, *Steroids*, Reinhold, New York, 1959, and references there given.
11. W. A. Waters, *Quart. Rev.* (London), **12**, 277 (1958).
12. J. W. Ladbury and C. F. Cullis, *Chem. Rev.*, **58**, 403 (1958).
13. R. Stewart, *Oxidation in Organic Chemistry* (K. B. Wiberg, ed.), Academic Press, New York and London, 1965, pp. 1–68.
14. J. Kenyon and M. C. R. Symons, *J. Chem. Soc.*, **1953**, 2129, 3580.
15. K. B. Wiberg and A. S. Fox, *J. Amer. Chem. Soc.*, **85**, 3487 (1963).
16. R. H. Eastman and R. A. Quinn, *J. Amer. Chem. Soc.*, **82**, 4249 (1960).
17. K. B. Wiberg and K. A. Saegebarth, *J. Amer. Chem. Soc.*, **79**, 2822 (1957).
18. F. Tiemann and P. Kruger, *Chem. Ber.*, **26**, 2693 (1893).
19. G. Wagner, *Chem. Ber.*, **23**, 2311 (1890).
20. G. Wagner, *Chem. Ber.*, **29**, 124 (1897).
21. G. Goldschmiedt, *Monatsch. Chem.*, **9**, 327 (1888).
22. G. Goldschmiedt and E. Ostersetzer, *Monatsh. Chem.*, **9**, 762 (1888).
23. K. W. Bentley and R. Robinson, *J. Chem. Soc.*, **1952**, 947.
24. R. Kuhn and H. Roth, *Chem. Ber.*, **66**, 1274 (1933).
25. W. J. Hickinbottom, D. Peters, and D. G. M. Wood, *J. Chem. Soc.*, **1955**, 1360.
26. K. B. Wiberg, *Oxidation in Organic Chemistry* (K. B. Wiberg, ed.), Academic Press, New York and London, 1965, pp. 69–184.
27. W. J. Hickinbottom, D. R. Hogg, D. Peters, and D. G. M. Wood, *J. Chem. Soc.*, **1954**, 4400.
28. W. J. Hickinbottom and D. G. M. Wood, *J. Chem. Soc.*, **1953**, 1906.
29. W. Treibs and H. Schmidt, *Chem. Ber.*, **61**, 459 (1928).
30. A. A. Berkin, *J. Appl. Chem. USSR*, **18**, 217 (1945).
31. H. H. Zeiss and F. R. Zwanzig, *J. Amer. Chem. Soc.*, **79**, 1733 (1957).
32. S. J. Cristol and K. R. Eilar, *J. Amer. Chem. Soc.*, **72**, 4353 (1950).
33. R. A. Stairs, D. G. M. Diaper, and A. L. Gatska, *J. Amer. Chem. Soc.*, **41**, 1059 (1963).
34. C. E. Groves, *Justus Liebig's Ann. Chem.*, **167**, 357 (1873).
35. F. Beilstein and A. Kurbatow, *Justus Liebig's Ann. Chem.*, **202**, 213 (1880).
36. J. Hertzenberg and S. Ruhemann, *Chem. Ber.*, **60**, 893 (1927).
37. J. Fritzsche, *J. Prakt. Chem.*, **106**, 287 (1869).
38. R. P. Linstead and W. v. E. Doering, *J. Amer. Chem. Soc.*, **64**, 1998 (1942).
39. R. Pschorr, *Chem. Ber.*, **39**, 3128 (1906).
40. M. Freund and E. Göbel, *Chem. Ber.*, **28**, 941 (1895).
41. J. Hampton, A. Leo, and F. Westheimer, *J. Amer. Chem. Soc.*, **78**, 306 (1956).
42. J. J. Crawley and F. H. Westheimer, *J. Amer. Chem. Soc.*, **85**, 1771 (1963).
43. K. Bowden, I. M. Heilbron, E. R. H. Jones, and B. C. L. Weedon, *J. Chem. Soc.*, **1946**, 39.
44. H. H. Sisler, J. D. Bush, and O. E. Accountius, *J. Amer. Chem. Soc.*, **70**, 3827 (1948).
45. G. I. Poos, G. E. Arth, R. E. Beyler, and L. M. Sarett, *J. Amer. Chem. Soc.*, **75**, 422 (1953).
46. R. V. Oppenauer and H. Oberrauch, *An. Asoc. Quim. Arg.*, **37**, 246 (1949).

47. H. H. Inhoffen, H. Pommer, K. Winkelmann, and H. J. Aldag, *Helv. Chim. Acta*, **84**, 87 (1951).
48. T. Suga, *Nippon Kagaki Zasshi*, **80**, 918 (1959); *Chem. Abstr.*, **55**, 3464 (1961).
49. T. Matsuura and T. Suga, *Korzo*, **62**, 13 (1961).
50. T. Suga, K. Kihara, and T. Matsuura, *Bull. Chem. Soc. Japan*, **38**, 893, 1141 (1965).
51. T. Suga and T. Matsuura, *Bull. Chem. Soc. Japan*, **38**, 1503 (1965).
52. F. C. Whitmore and G. W. Pedlow, *J. Amer. Chem. Soc.*, **63**, 758 (1941).
53. H. E. Stavely and G. N. Bollenback, *J. Amer. Chem. Soc.*, **65**, 1285 (1943).
54. W. F. Sager and A. Bradley, *J. Amer. Chem. Soc.*, **78**, 1187 (1956).
55. G. Snatzke, *Z. Anal. Chem.*, **209**, 412 (1965).
56. C. D. Harries, *Untersuchungen uber das Ozon und seine Einwirkung auf organische Verbindung*, Springer, Berlin, 1916.
57. H. Staudinger, *Chem. Ber.*, **58**, 1088 (1925).
58. R. Criegee, *Justus Liebig's Ann. Chem.*, **560**, 131 (1948).
59. R. Criegee and G. Wenner, *Justus Liebig's Ann. Chem.*, **564**, 9 (1949).
60. R. Criegee and G. Lohaus, *Justus Liebig's Ann. Chem.*, **583**, 6 (1953).
61. R. Criegee and G. Lohaus, *Justus Liebig's Ann. Chem.*, **583**, 12 (1953).
62. R. Criegee, G. Blust, and H. Zinke, *Chem. Ber.*, **87**, 766 (1954).
63. R. Criegee, A. Kerckow, and H. Zinke, *Chem. Ber.*, **88**, 1878 (1955).
64. R. Criegee, *Rec. Chem. Progr.*, **18**, 111 (1957).
65. L. Long, *Chem. Rev.*, **27**, 437 (1940).
66. P. S. Bailey, *Chem. Rev.*, **58**, 925 (1958).
67. P. S. Bailey, *Chem. Ber.*, **88**, 795 (1955).
68. P. S. Bailey, *J. Amer. Chem. Soc.*, **78**, 3811 (1956); *J. Org. Chem.*, **22**, 1548 (1957).
69. E. Briner and E. Dallwigk, *C.R. Acad. Sci., Paris*, **243**, 360 (1956).
70. E. Briner and E. Dallwigk, *C.R. Acad. Sci., Paris*, **244**, 1695 (1957).
71. E. Dallwigk and E. Briner, *Helv. Chim. Acta*, **39**, 1826 (1956).
72. G. Slomp and J. C. Johnson, *J. Amer. Chem. Soc.*, **80**, 915 (1958).
73. A. Serini and W. Logemann, *Chem. Ber.*, **71**, 1362 (1938).
74. T. S. Work, *J. Chem. Soc.*, **1944**, 334.
75. P. S. Bailey and A. G. Lane, *J. Amer. Chem. Soc.*, **89**, 4473 (1967).
76. R. W. Murray, R. D. Youssefyeh, and P. R. Story, *J. Amer. Chem. Soc.*, **88**, 3143 (1966).
77. P. R. Story, R. W. Murray, and R. D. Youssefyeh, *J. Amer. Chem. Soc.*, **88**, 3144 (1966).
78. R. Criegee and G. Schröder, *Chem. Ber.*, **93**, 689 (1960).
79. F. L. Greenwood and S. Cohen, *J. Org. Chem.*, **28**, 1159 (1963).
80. F. L. Greenwood, *J. Org. Chem.*, **30**, 3108 (1965).
81. L. J. Durham and F. L. Greenwood, *Chem. Commun.*, **1967**, 843.
82. R. Criegee and G. Lohaus, *Chem. Ber.*, **86**, 1 (1953).
83. R. Criegee, *Products of Ozonisation of Some Olefins*, in *Advances in Chemistry Series, No. 21*, American Chemical Society, Washington, D.C., 1958, pp. 133.

84. A. C. Andrews, B. L. Mickel, and K. C. Klassen, *J. Chem. Educ.*, **32**, 154 (1955).
85. L. I. Smith, F. L. Greenwood, and O. Hudrlik, *Org. Syn.*, Coll. Vol. 3, 673 (1955).
86. H. Reich, M. Sutter, and T. Reichstein, *Helv. Chim. Acta*, **23**, 170 (1940).
87. W. A. Bonner, *J. Chem. Educ.*, **30**, 452 (1953).
88. R. C. Elderfield and A. P. Gray, *J. Org. Chem.*, **16**, 520 (1951).
89. A. L. Henne and W. L. Perilstein, *J. Amer. Chem. Soc.*, **65**, 2183 (1943).
90. J. E. Ransford, *J. Chem. Educ.*, **28**, 477 (1951).
91. H. Boer, *Rec. Trav. Chim. Pays-Bas*, **67**, 217 (1948); **70**, 1020 (1951).
92. F. L. Greenwood and M. G. Wolkowich, *J. Org. Chem.*, **17**, 1511 (1952).
93. M. Tits and A. Bruylants, *Bull. Soc. Chim. Belges*, **57**, 50 (1948).
94. C. R. Noller and R. Adams, *J. Amer. Chem. Soc.*, **48**, 1074 (1926).
95. E. Briner and S. de Nemitz, *Helv. Chim. Acta*, **21**, 748 (1938).
96. B. Witkop and S. Goodwin, *J. Amer. Chem. Soc.*, **75**, 3371 (1953).
97. R. Criegee and M. Lederer, *Justus Liebig's Ann. Chem.*, **583**, 29 (1953).
98. M. Stoll and A. Rouve, *Helv. Chim. Acta*, **37**, 950 (1944).
99. G. M. Badger and N. K. Wilson, *J. Chem. Phys.*, **18**, 998 (1950).
100. B. Witkop and J. B. Patrick, *J. Amer. Chem. Soc.*, **74**, 3855 (1952).
101. W. F. Symes and C. R. Dawson, *J. Amer. Chem. Soc.*, **75**, 4952 (1953).
102. M. Hinder and M. Stoll, *Helv. Chim. Acta*, **33**, 1308 (1950).
103. J. B. Patrick and B. Witkop, *J. Org. Chem.*, **19**, 1824 (1954).
104. F. L. Greenwood, *J. Org. Chem.*, **20**, 803 (1955).
105. H. Pleininger and K. Suhr, *Chem. Ber.*, **89**, 270 (1956).
106. J. L. Warnell and R. L. Shriner, *J. Amer. Chem. Soc.*, **80**, 915 (1958).
107. R. B. Woodward and E. G. Kovach, *J. Amer. Chem. Soc.*, **72**, 100 (1950).
108. W. G. Young, A. C. McKinnis, I. D. Webb, and J. D. Roberts, *J. Amer. Chem. Soc.*, **68**, 293 (1946).
109. D. H. R. Barton and E. J. Seoane, *J. Chem. Soc.*, **1956**, 4150.
110. E. Späth and M. Pailer, *Chem. Ber.*, **73**, 238 (1940).
111. E. Späth, M. Pailer, and G. Gergely, *Chem. Ber.*, **73**, 795 (1940).
112. G. S. Fisher and J. S. Stinson, *Ind. Eng. Chem.*, **47**, 1569 (1955).
113. F. Holloway, H. J. Anderson and W. Rodin, *Ind. Eng. Chem.*, **47**, 2111 (1955).
114. K. Ziegler, W. Hechelhammer and H. Wilms, *Justus Liebig's Ann. Chem.*, **567**, 99 (1950).
115. C. S. Marvel, W. M. Schilling, D. J. Shields, C. Bluenstein, O. R. Irwin, P. G. Sheth, and J. Honig, *J. Org. Chem.*, **16**, 838 (1951).
116. A. Riechee, R. Meister, and H. Sauthoff, *Justus Liebig's Ann. Chem.*, **553**, 187 (1942).
117. G. A. Howard and A. R. Tatchell, *J. Chem. Soc.*, **1954**, 2400.
118. N. A. Milas and J. T. Nolan, *Advances in Chemistry Series*, No. 21, American Chemical Society, Washington, D.C., 1958, p. 136.
119. J. P. Wibaut, F. L. J. Sixma, L. W. F. Kampschmidt, and H. Boer, *Rec. Trav. Chim. Pays-Bas*, 1355 (1950).
120. F. L. J. Sixma, H. Boer, and J. P. Wibaut, *Rec. Trav. Chim. Pays-Bas*, **70**, 1005 (1951).

121. E. Speyer and A. Popp, *Chem. Ber.*, **59**, 390 (1926).

122. H. Rapoport and G. B. Payne, *J. Org. Chem.*, **15**, 1093 (1950).

123. E. Speyer, *Chem. Ber.*, **62**, 209 (1929).

124. K. W. Bentley, D. G. Hardy, and P. A. Mayor, *J. Chem. Soc. C*, **1969**, 2385.

125. R. Pschorr and H. Einbeck, *Chem. Ber.*, **40**, 3652 (1907).

126. H. Wieland and L. F. Small, *Justus Liebig's Ann. Chem.*, **467**, 17 (1928).

127. G. M. Badger, J. E. Campbell, J. W. Cook, R. A. Raphael, and A. I. Scott, *J. Chem. Soc.*, **1950**, 2326.

128. P. S. Bailey and J. B. Ashton, *J. Org. Chem.*, **22**, 98 (1957).

129. A. F. Lindenstruth and C. A. van der Wert, *J. Amer. Chem. Soc.*, **71**, 3020 (1949).

130. H. Boer, F. L. J. Sixma, and J. P. Wibaut, *Rec. Trav. Chim. Pays-Bas*, **70**, 509 (1951).

131. R. L. Shriner and R. B. Moffett, *J. Amer. Chem. Soc.*, **62**, 2711 (1940).

132. B. Witkop, *Justus Liebig's Ann. Chem.*, **556**, 103 (1944).

133. Y. Berguer, D. Molho, and C. Mentzer, *C.R. Acad. Sci., Paris*, **230**, 760 (1950).

134. D. W. Ockenden and K. Schofield, *J. Chem. Soc.*, **1953**, 612.

135. R. Criegee, *Newer Methods of Preparative Organic Chemistry*, Interscience, New York, 1948, p. 1.

136. W. A. Waters, in *Organic Chemistry* Vol. 4 (H. Gilman, ed.), Wiley, New York, 1953, p. 1185.

137. R. Criegee, *Angew. Chem.*, **67**, 752 (1955); **70**, 173 (1958).

138. R. Criegee, in *Oxidation in Organic Chemistry* (K. B. Wiberg, ed.), Academic Press, New York, 1965, p. 277.

139. C. A. Burton, in *Oxidation in Organic Chemistry* (K. B. Wiberg, ed.), Academic Press, New York, 1965, p. 367.

140. R. Criegee, L. Kraft, and B. Rank, *Justus Liebig's Ann. Chem.*, **507**, 159 (1933).

141. R. Criegee, E. Höger, G. Huber, and P. Kruck, F. Marktscheffel, and H. Schnellenberger, *Justus Liebig's Ann. Chem.*, **599**, 81 (1956).

142. R. Criegee, *Angew. Chem.*, **50**, 153 (1937).

143. R. Criegee, B. Marchand, and H. Wannowins, *Justus Liebig's Ann. Chem.*, **550**, 99 (1942).

144. C. A. Burton and M. D. Carr, *J. Chem. Soc.*, **1963**, 770.

145. S. J. Angyal and R. J. Young, *J. Amer. Chem. Soc.*, **81**, 5251, 5467 (1959).

146. R. J. Dimler, H. A. Davis, and G. E. Hilbert, *J. Amer. Chem. Soc.*, **68**, 1377 (1946).

147. B. H. Alexander, R. J. Dimler, and C. L. Mehltretter, *J. Amer. Chem. Soc.*, **73**, 4658 (1951).

148. H. H. Wasserman, in *Steric Effects in Organic Chemistry* (M. S. Newman, ed.), Wiley, New York, 1956, p. 378.

149. H. R. Goldschmid and A. S. Perlin, *Can. J. Chem.*, **38**, 2280 (1960).

150. E. Baer, R. Grosheintz, and H. O. L. Fischer, *J. Amer. Chem. Soc.*, **61**, 2607 (1939).

151. R. Criegee and E. Büchner, *Chem. Ber.*, **73**, 563 (1940).

152. J. P. Cordiner and K. H. Pausacker, *J. Chem. Soc.*, **1953**, 102; L. K. Dyall and K. H. Pausacker, *J. Chem. Soc.*, **1958**, 3950.
153. C. Djerassi and J. A. Zderic, *J. Amer. Chem. Soc.*, **78**, 2907, 6390 (1956).
154. E. Baer, *J. Amer. Chem. Soc.*, **62**, 1597, 1600 (1940); **64**, 1416 (1942).
155. R. B. Bradbury and S. Masamune, *J. Amer. Chem. Soc.*, **81**, 5201 (1959).
156. V. M. Mićović and M. L. Mihailović, *Rec. Trav. Chim. Pays-Bas*, **71**, 970 (1952).
157. R. D. Partch, *Tetrahedron Lett.*, **1964**, 3071.
158. Baker Castor Oil Co., British Patent 759416 (1956).
159. E. A. Braude and O. H. Wheeler, *J. Chem. Soc.*, **1955**, 320.
160. A. Amorosa, L. Caglioti, G. Cainelli, H. Immer, J. Keller, H. Wehrli, M. L. Mihailović, K. Schaffner, D. Arigoni, and O. Jeger, *Helv. Chim. Acta*, **45**, 2674 (1962).
161. D. Hauser, K. Heusler, J. Kalvoda, K. Schaffner, and O. Jeger, *Helv. Chim. Acta*, **47**, 1961 (1964).
162. M. L. Mihailović, M. Stefanović, L. Lorenc, and M. Gašić, *Tetrahedron Lett.*, **1964**, 1867.
163. V. M. Micović, R. J. Masmizić, D. Jeremic, and M. L. Mihailović, *Tetrahedron*, **20**, 2279 (1964).
164. M. Stefanović, L. Lorenc, and M. L. Mihailović, *Tetrahedron*, **20**, 2289 (1964).
165. W. A. Mosher, C. L. Kehr, and L. W. Wright, *J. Org. Chem.*, **26**, 1044 (1961).
166. R. B. Woodward, T. Fukunaga, and R. C. Kelly, *J. Amer. Chem. Soc.*, **86**, 3162 (1964).
167. J. Jacques, C. Weidmann, and A. Horeau, *Bull. Soc. Chim. Fr.*, **1959**, 424.
168. C. A. Grob, M. Ohta, and A. Weiss, *Angew. Chem.*, **70**, 343 (1958).
169. E. E. van Tamelen and S. P. Pappas, *J. Amer. Chem. Soc.*, **85**, 3297 (1963).
170. N. B. Chapman, S. Sotheeswaren, and K. J. Toyne, *Chem. Commun.*, **1965**, 214.
171. S. Borćić and J. D. Roberts, *J. Amer. Chem. Soc.*, **87**, 1056 (1965).
172. K. Alder and S. Schneider, *Justus Liebig's Ann. Chem.*, **524**, 189 (1936).
173. R. Partch, *J. Org. Chem.*, **28**, 276 (1963).
174. L. Malaprade, *C.R. Acad. Sci.*, *Paris*, **186**, 382 (1928); L. Malaprade, *Bull. Soc. Chim. Fr.*, [4]**43**, 683 (1928); [5]**1**, 833 (1933).
175. E. L. Jackson, *Org. Reactions*, **2**, 341 (1944); B. Sklarz, *Quart. Rev.*, **21**, 3 (1967).
176. A. K. Qureshi and B. Sklarz, *J. Chem. Soc.*, *C*, **1966**, 412; *Chem. Abstr.*, **29**, 6820 (1935).
177. R. Criegee, *Sitzber. Ges. Beförder. Ges. Naturw. Marburg*, **69**, 25 (1934).
178. C. J. Buist, C. A. Bunton, and J. H. Miles, *J. Chem. Soc.*, **1959**, 743. F. R. Duke, *J. Amer. Chem. Soc.*, **69**, 3054 (1947).
179. P. Fleury and M. Joly, *J. Pharm. Chim.*, [8]**26**, 341, 397 (1937).
180. J. G. P. Schwarz, *Chem. Ind.* (London), **1955**, 1388.
181. G. R. Barker, *J. Chem. Soc.*, **1960**, 624.
182. E. L. Jackson and C. S. Hudson, *J. Amer. Chem. Soc.*, **58**, 378 (1936); **59**, 994 (1937).
183. P. Fleury and J. Lange, *J. Pharm. Chim.*, [8]**17**, 313 (1933).

184. P. Fleury and G. Bon-Bernatets, *J. Pharm. Chim.*, [8]23, 85 (1936).
185. O. Schmidt and P. Günthert, *Chem. Ber.*, 71, 493 (1938).
186. R. E. Reeves, *J. Amer. Chem. Soc.*, 61, 664 (1939).
187. G. W. Perdd and K. Pachler, *Proc. Chem. Soc.*, 1964, 62.
188. R. D. Diamand and D. Rogers, *Proc. Chem. Soc.*, 1964, 63.
189. W. S. Rapson, *J. Chem. Soc.*, 1940, 1271.
190. E. A. Kaczka, C. H. Shunk, J. W. Richter, F. J. Wott, M. M. Gasser, and K. Folkers, *J. Amer. Chem. Soc.*, 78, 4125 (1956).
191. C. F. Spencer, C. H. Stammer, J. O. Rodin, E. Walton, F. W. Holly, and K. Folkers, *J. Amer. Chem. Soc.*, 78, 2655 (1956).
192. J. W. Hinman, E. L. Caron, and H. Hoeksema, *J. Amer. Chem. Soc.*, 79, 3789 (1957).
193. E. Walton, J. O. Rodin, F. W. Holly, J. W. Richter, C. H. Shunk, and K. Folkers, *J. Amer. Chem. Soc.*, 82, 1489 (1960).
194. N. G. Brink, F. A. Kuehl, E. H. Flynn, and K. Folkers, *J. Amer. Chem. Soc.*, 68, 2405 (1946); 70, 2085 (1948).
195. F. A. Kuehl, E. H. Flynn, N. G. Brink, and K. Folkers, *J. Amer. Chem. Soc.*, 68, 2679 (1946).
196. P. F. Wiley, *J. Amer. Chem. Soc.*, 84, 1514 (1962).
197. E. H. Flynn, M. V. Sigal, P. F. Wiley, and K. Gerzon, *J. Amer. Chem. Soc.*, 76, 3121 (1954).
198. R. K. Clark, *Antibiot. Chemotherapy*, 3, 663 (1953).
199. T. Reichstein, *Helv. Chim. Acta*, 20, 978 (1937).
200. T. Reichstein, C. Meystre, and J. v. Euw, *Helv. Chim. Acta*, 22, 1107 (1939).
201. R. U. Lemieux and E. v. Rudloff, *Can. J. Chem.*, 33, 1701, 1710 (1955).
202. E. V. Rudloff, *Can. J. Chem.*, 33, 1714 (1955); 34, 1413 (1956).
203. E. v. Rudloff, *J. Am. Oil Chem. Soc.*, 33, 126 (1956).
204. A. P. Tulloch and G. A. Ledingham, *Can. J. Microbiol.*, 6, 625 (1960).
205. F. D. Gunstone and P. J. Sykes, *J. Chem. Soc.*, 1962, 3058.
206. D. A. Martin and A. Weise, *Angew. Chem. Int. Ed.*, 6, 168 (1967).
207. R. Pappo, O. S. Allen, R. U. Lemieux, and W. S. Johnson, *J. Org. Chem.*, 21, 478 (1956).
208. L. Marion and K. Sargent, *J. Amer. Chem. Soc.*, 78, 5127 (1956).
209. J. C. P. Schwarz and M. MacDougall, *J. Chem. Soc.*, 1956, 3065.
210. M. Cantley, L. Hough, and A. O. Pettit, *J. Chem. Soc.*, 1963, 2527.
211. E. Adler, J. Dahlen, and G. Westin, *Acta Chem. Scand.*, 14, 1580 (1960).
212. E. Adler, I. Falkehag, and B. Smith, *Acta Chem. Scand.*, 16, 529 (1962).
213. W. Rigby, *J. Chem. Soc.*, 1950, 1907.
214. H. Beucker and R. Criegee, *Justus Liebig's Ann. Chem.*, 541, 218 (1939).
215. K. H. Pausacker, *J. Chem. Soc.*, 1953, 107.
216. L. K. Dyall and K. H. Pausacker, *Aust. J. Chem.*, 11, 485 (1958).
217. W. Rigby, *J. Chem. Soc.*, 1951, 793.
218. T. P. C. Mulholland, *J. Chem. Soc.*, 1958, 2693.
219. W. H. Richardson, in *Oxidation in Organic Chemistry* (K. Wiberg, ed.) Academic Press, New York and London, 1965, p. 243.

220. W. A. Waters and J. S. Littler, in *Oxidation in Organic Chemistry* (K. Wiberg, ed.), Academic Press, New York and London, 1965, p. 185.

221. B. Krishna and K. C. Tewari, *J. Chem. Soc.*, **1961**, 3097.

222. E. Weitz and A. Scheffer, *Chem. Ber.*, **54**, 2327 (1921).

223. C. A. Bunton, *Nature*, **163**, 444 (1949).

224. C. A. Bunton and G. J. Minkoff, *J. Chem. Soc.*, **1949**, 665.

225. R. E. Corbett, A. W. Johnson, and A. R. Todd, *J. Chem. Soc.*, **1950**, 6; **1951**, 1139.

226. G. Barger and O. Derrer, *Biochem. J.*, **28**, 11 (1934).

227. T. Nozoe, S. Ito, K. Matsui, and T. Ozeki, *Proc. Japan. Acad.*, **30**, 599, 603 (1954).

227. T. Nozoe, *Sci. Rep. Tohoku Univ.*, [1]**34**, 199 (1950).

229 T. Nozoe, M. Sato, S. Ito, K. Matsui, and T. Ozeki, *Sci. Rep. Tohoku Univ.*, [1]**39**, 190 (1955).

230. T. Nozoe, *Experientia Suppl.*, **7**, 306 (1957).

231. W. Voser, Hs. H. Günthard, H. Heusser, O. Jeger, and L. Ruzicka, *Helv. Chim. Acta*, **35**, 2065 (1952).

232. W. Voser, M. V. Mijovic, O. Jeger, and L. Ruzicka, *Helv. Chim. Acta*, **35**, 2414 (1952).

233. E. R. H. Jones and T. G. Halsall, *Fortschr. Chem. Org. Naturstoffe*, **12**, 61 (1955).

234. E. Kyburg, B. Riniker, H. R. Schenk, H. Heusser, and O. Jeger, *Helv. Chim. Acta*, **36**, 1891 (1953).

235. C. H. Hassall, *Org. Reactions*, **9**, 73 (1957).

236. R. E. Marker, E. Rohrmann, H. M. Crooks, E. L. Whittle, E. M. Jones, and D. L. Turner, *J. Amer. Chem. Soc.*, **62**, 525 (1940).

237. W. G. Dauben, T. W. Hutton, and G. A. Boswell, *J. Amer. Chem. Soc.*, **81**, 403 (1959).

238. E. S. Rothman, M. E. Wall, and C. L. Eddy, *J. Amer. Chem. Soc.*, **76**, 526 (1954).

239. K. Block, *Helv. Chim. Acta*, **36**, 1611 (1953).

240. W. G. Dauben and K. H. Takemura, *J. Amer. Chem. Soc.*, **75**, 6302 (1953).

241. L. Doleiš and F. Šorm, *Tetrahedron Lett.*, **10**, 1 (1959); **17**, 1 (1959).

242. J. S. E. Holker, W. R. Jones, and P. J. Ramm, *Chem. Commun.*, **1965**, 435.

243. W. v. E. Doering and L. Speers, *J. Amer. Chem. Soc.*, **72**, 5515 (1950).

244. S. L. Friess and M. Farnham, *J. Amer. Chem. Soc.*, **72**, 5518 (1950).

245. S. L. Friess and A. H. Soloway, *J. Amer. Chem. Soc.*, **73**, 3968 (1951).

246. C. H. Hassall and A. R. Todd, *J. Chem. Soc.*, **1947**, 611.

247. O. Wallach, *Chem. Ber.*, **24**, 3984 (1891).

248. O. Wallach, *Justus Liebig's Ann. Chem.*, **275**, 111, 155 (1893).

249. F. Tiemann and F. Semmler, *Chem. Ber.*, **28**, 2142 (1895).

250. V. Sýkora, L. Novotný, and F. Šorm, *Tetrahedron Lett.*, **14**, 24 (1959).

251. G. Chiurdogen and M. Descampo, *Tetrahedron*, **8**, 271 (1960).

252. V. Sýkora, L. Novotrý, M. Holub, V. Herout, and F. Šorm, *Collect. Czech. Chem. Commun.*, **36**, 788 (1961).

253. O. Diels and E. Aberhalden, *Chem. Ber.*, **36**, 3177 (1903); **37**, 3092 (1904).
254. A. Windaus and A. van Schoor, *Z. Physiol. Chem.*, **148**, 225 (1925); **157,** 175 (1926).
255. F. Kröhnke, *Chem. Ber.*, **66**, 604 (1933).
256. L. C. King, *J. Amer. Chem. Soc.*, **66**, 894, 1612 (1944).
257. L. C. King, M. McWhirter, and D. M. Barton, *J. Amer. Chem. Soc.*, **67,** 2089 (1945).
258. L. C. King and M. McWhirter, *J. Amer. Chem. Soc.*, **68,** 46 (1946).
259. F. Krollpfeiffer and A. Müller, *Chem. Ber.*, **68,** 1169 (1935).
260. L. C. King, M. McWhirter, and R. L. Rowland, *J. Amer. Chem. Soc.*, **70,** 239 (1948).
261. W. A. Shenstone, *J. Chem. Soc.*, **47**, 139 (1885).
262. K. N. Menon, W. H. Perkin, and R. Robinson, *J. Chem. Soc.*, **1930**, 842.
263. R. Robinson, *Proc. Roy. Soc.*, **130A**, 431 (1931).
264. J. Täfel, *Justus Liebig's Ann. Chem.*, **268**, 229 (1892); **301,** 285 (1898).
265. W. H. Perkin and R. Robinson, *J. Chem. Soc.*, **95,** 381 (1909).
266. J. H. V. Charles, H. Raistrick, R. Robinson, and A. R. Todd, *Biochem. J.*, **27,** 499 (1933).
267. O. Dimroth, *Chem. Ber.*, **43**, 1387 (1910).
268. R. B. Woodward, *Angew. Chem.*, **69,** 50 (1957).
269. A. Meyer, O. Jeger, V. Prelog, and L. Ruzicka, *Helv. Chim. Acta*, **34,** 747 (1951).
270. H. Menard and O. Jeger, *Helv. Chim. Acta*, **36**, 355 (1953).
271. A. Windaus and O. Dalmer, *Chem. Ber.*, **52,** 612 (1919).
272. A. Windaus, *Chem. Ber.*, **53,** 488 (1920).
273. H. Wieland and P. Weyland, *Z. Physiol. Chem.*, **110,** 123 (1920).
274. H. Wieland and E. Dane, *Z. Physiol. Chem.*, **210,** 268 (1932).
275. K. W. Bentley, D. G. Hardy, C. F. Howell, W. Fulmor, J. E. Lancaster, J. J. Brown, G. O. Morton, and R. A. Hardy, *J. Amer. Chem. Soc.*, **89,** 3303 (1967).
276. O. Wallach, *Justus Liebig's Ann. Chem.*, **277,** 120 (1893).
277. R. W. Clarke, L. A. Lapworth, and E. Wechsler, *J. Chem. Soc.*, **93,** 90 (1908).
278. A. J. Birch and F. N. Lahey, *Aust. J. Chem.*, **6,** 379 (1953).
279. K. Nakagawa, K. Igano, and J. Sugita, *Chem. Pharm. Bull.* (Tokyo) **12,** 603 (1964).
280. N. J. Leonard and C. R. Johnson, *J. Org. Chem.*, **27,** 282 (1962).
281. K. Nakagawa, H. Onoue, and K. Minami, *Chem. Commun.*, **1966,** 17.
282. K. Nakagawa and H. Onoue, *Tetrahedron Lett.*, **1965,** 1433.
283. T. J. Wallace, H. Pobiner, and A. Schriesheim, *J. Org. Chem.*, **30,** 3768 (1965).
284. W. Triebs and R. Schöllner, *Chem. Ber.*, **94,** 42, 2983 (1961).
285. H. C. Volger and W. Brackmann, *Rec. Trav. Chim. Pays-Bas*, **84,** 579, 1233 (1965).
286. H. C. Volger, W. Brackmann, and J. W. F. M. Lemmers, *Rec. Trav. Chim. Pays-Bas*, **84,** 1203 (1965).
287. W. G. Toland, *J. Amer. Chem. Soc.*, **83,** 2507 (1961).
288. W. W. Kaeding, *J. Org. Chem.*, **26,** 3144 (1961).
289. K. E. Pfitzner and J. G. Moffatt, *J. Amer. Chem. Soc.*, **85,** 3027 (1963).

290. J. D. Albright and L. Goldman, *J. Amer. Chem. Soc.*, **87**, 4214 (1965); **89**, 2416 (1967); *J. Org. Chem.*, **30**, 1107 (1965).

291. J. B. Jones and D. C. Wigfeld, *Can. J. Chem.*, **44**, 2517 (1966).

292. A. J. Fatiadi, *Chem. Commun.*, **1967**, 441.

293. C. W. L. Bevan, M. P. Patel, A. H. Rees, and A. G. Loudon, *Tetrahedron*, **23**, 3809 (1967).

294. F. W. Sweet, *Chem. Rev.*, **67**, 247 (1967).

295. R. C. Cookson, I. D. R. Stevens, and C. T. Watts, *Chem. Commun.*, **1966**, 744; R. C. Cookson, S. S. H. Gilani, and I. D. R. Stevens, *J. Chem. Soc.*, C, **1967**, 1905.

296. F. D. Gunstone, *Advan. Org. Chem.*, **1**, 103 (1960).

297. N. Prileschajew, *Chem. Ber.*, **42**, 4811 (1902); *J. Russ. Phys. Chem. Soc.*, **42**, 1387 (1910); **43**, 609 (1911); **44**, 613 (1912).

298. D. Swern, *Org. Reactions*, **7**, 378 (1953).

299. I. Roit and W. A. Waters, *J. Chem. Soc.*, **1949**, 3060.

300. F. L. Weisenborn and D. Taub, *J. Amer. Chem. Soc.*, **74**, 1329 (1952).

301. G. Braun, *Org. Syntheses*, Coll. Vol. **1**, 431 (1948).

302. B. T. Brooks and W. B. Brooks, *J. Amer. Chem. Soc.*, **55**, 4310.

303. E. Raymond, *J. Chim. Phys.*, **28**, 480 (1931).

304. V. V. Pigulevskii, *J. Gen. Chem. USSR*, **4**, 616 (1934).

305. D. Swern and T. W. Findlay, *J. Amer. Chem. Soc.*, **72**, 4315 (1950).

306. L. S. Gilbert, E. Siegel, and D. Swern, *J. Org. Chem.*, **27**, 1336 (1962).

307. W. D. Emmons and G. B. Lucas, *J. Amer. Chem. Soc.*, **77**, 2287 (1955).

308. R. W. White and W. D. Emmons, *Tetrahedron*, **17**, 31 (1962).

309. F. P. Greenspan, *J. Amer. Chem. Soc.*, **68**, 907 (1946).

310. F. P. Greenspan, *Ind. Eng. Chem.*, **39**, 847 (1947).

311. J. A. Ross, A. B. Gebhart, and J. F. Gerecht, *J. Amer. Chem. Soc.*, **71**, 282 (1949).

312. J. T. Scanlan and D. Swern, *J. Amer. Chem. Soc.*, **62**, 2711 (1940).

313. H. Riviere, *Bull. Soc. Chim. Fr.*, **1964**, 97.

314. D. Swern, *J. Amer. Chem. Soc.*, **69**, 1692 (1947).

315. R. B. Woodward, F. E. Bader, H. Bickel, A. J. Frey, and R. W. Kierstead, *Tetrahedron*, **2**, 1 (1958).

316. D. H. R. Barton and M. Mousseron-Canet, *J. Chem. Soc.*, **1960**, 271.

317. I. J. Borowitz, G. Gonis, R. Kelsey, R. Rapp, and G. J. Williams, *J. Org. Chem.*, **31**, 3032 (1966).

318. A. Endō, M. Saitō, and Y. Fusizaki, *J. Chem. Soc. Japan*, **85**, 593 (1964).

319. J. Meinwald and G. A. Wiley, *J. Amer. Chem. Soc.*, **80**, 3667 (1958).

320. H. M. Walborsky and D. F. Loncrini, *J. Amer. Chem. Soc.*, **76**, 5396 (1954).

321. H. Kwart and W. G. Vosburgh, *J. Amer. Chem. Soc.*, **76**, 5400 (1954).

322. G. B. Payne, P. H. Denning, and P. H. Williams, *J. Org. Chem.*, **26**, 659 (1961).

323. G. B. Payne, *Tetrahedron*, **18**, 763 (1962).

324. N. A. Milas and E. M. Terry, *J. Amer. Chem. Soc.*, **47**, 1415 (1925).

325. N. A. Milas and S. Sussman, *J. Amer. Chem. Soc.*, **58**, 1302 (1936).

326. A. Butenandt and J Schmid-Thomé, *Chem Ber.*, **71**, 1490 (1938).

327. A. Butenandt, J. Schmid-Thomé, and H. Paul, *Chem. Ber.*, **72**, 1112 (1939).

328. P. N. Chakravarty and R. H. Levin, *J. Amer. Chem. Soc.*, **64**, 2317 (1942).

329. A. Serini and W. Logemann, *Chem. Ber.*, **71**, 1362 (1938).

330. A. Serini, W. Logemann, and W. Hildebrand, *Chem. Ber.*, **72**, 391 (1939).

331. H. Reich, M. Sutter, and T. Reichstein, *Helv. Chim. Acta*, **23**, 170 (1940).

332. R. B. Woodward, F. Sondheimer, D. Taub, K. Heusler, and W. M. McLamore, *J. Amer. Chem. Soc.*, **74**, 4223 (1952).

333. H. Rapoport, A. D. Batcho, and J. E. Gordon, *J. Amer. Chem. Soc.*, **80**, 5767 (1958).

334. H. Rapoport, M. S. Chadha, and C. H. Lovell, *J. Amer. Chem. Soc.*, **79**, 4694 (1957).

335. K. Miescher and J. Schmidlin, *Helv. Chim. Acta*, **33**, 1840 (1950).

336. E. C. Grob and R. Butler, *Helv. Chim. Acta*, **38**, 1840 (1955).

337. R. B. Woodward and F. V. Brutcher, *J. Amer. Chem. Soc.*, **80**, 209 (1958).

338. C. Prevost, *C.R. Acad. Sci.*, *Paris*, **196**, 1129 (1933); **197**, 1661 (1933).

339. S. Winstein and R. E. Buckles, *J. Amer. Chem. Soc.*, **64**, 2780 (1942).

340. L. B. Barkley, M. W. Farrer, W. S. Knowles, H. Rattelson, and Q. E. Thompson, *J. Amer. Chem. Soc.*, **76**, 5014 (1954).

341. W. S. Knowles and Q. E. Thompson, *J. Amer. Chem. Soc.*, **79**, 3212 (1957).

342. P. E. Ellington, D. G. Hey, and G. D. Meakins, *J. Chem. Soc.*, C, **1966**, 1327.

343. M. Gates and G. Tschudi, *J. Amer. Chem. Soc.*, **78**, 1380 (1956).

344. H. C. Brown and B. C. SubbaRao, *J. Amer. Chem. Soc.*, **78**, 5694 (1956); **81**, 6423, 6428 (1959); *J. Org. Chem.*, **22**, 1136 (1957)

345. H. C. Brown and G. Zweifel, *J. Amer. Chem. Soc.*, **81**, 1512 (1959); **83**, 3834 (1961).

346. H. C. Brown, *Hydroboration*, Benjamin, New York, 1962.

347. G. Zweifel and H. C. Brown, *Org. Reactions*, **13**, 1 (1963).

348. C. W. Bird, R. C. Cookson, and E. Crundwell, *J. Chem. Soc.*, **1961**, 4809.

349. P. Bruck, D. Thompson, and S. Winstein, *Chem. Ind.* (London), **1960**, 405.

350. H. Kugita and M. Takeda, *Chem. Pharm. Bull.* (Tokyo), **13**, 1422 (1965).

351. M. Takeda, H. Inoue, and H. Kugita, *Tetrahedron*, **25**, 1839 (1969).

352. H. Kugita, M. Takeda, and H. Inoue, *Tetrahedron*, **25**, 1851 (1969).

353. H. C. Brown and G. Zweifel, *J. Amer. Chem. Soc.*, **81**, 5832 (1959).

354. G. Zweifel, K. Nagase, and H. C. Brown, *J. Amer. Chem. Soc.*, **84**. 183 (1962).

355. H. C. Brown, K. P. Singh, and B. J. Garner, *J. Organometal. Chem.*, **1**, 2 (1963).

356. K. A. Saegebarth, *J. Amer. Chem. Soc.*, **82**, 2081 (1960); *J. Org. Chem.*, **25**, 2212 (1960).

357. W. Bradley and R. Robinson, *J. Chem. Soc.*, **1926**, 2356.

358. C. L. Bickel, *J. Amer. Chem. Soc.*, **67**, 2204 (1945); **68**, 865 (1946).

359. F. A. Hochstein, C. R. Stephens, L. H. Conover, P. P. Regina, R. Pasternack, P. N. Gordon, F. J. Pilgrim, K. J. Brunings, and R. B. Woodward, *J. Amer. Chem. Soc.*, **75**, 5455 (1953).

360. K. W. Bentley, J. Dominguez, and J. P. Ringe, *J. Org. Chem.*, **22**, 409 (1957).

361. D. H. R. Barton and P. de Mayo, *J. Chem. Soc.*, **1956**, 142.

362. D. H. R. Barton and P. de Mayo, *J. Chem. Soc.*, **1954**, 887.
363. H. Schmid, A. Ebnöther, and Th. M. Meijer, *Helv. Chim. Acta*, **33**, 1751 (1950).
364. C. Djerassi and S. Burstein, *Tetrahedron*, **7**, 37 (1959).
365. H. Raistrick and G. Smith, *Biochem. J.*, **30**, 1315 (1936).
366. P. W. Clutterbuck, W. Koerber, and H. Raistrick, *Biochem. J.*, **31**, 1089 (1937).
367. C. T. Calam, P. W. Clutterbuck, A. E. Oxford, and H. Raistrick, *Biochem. J.*, **33**, 579 (1939); **41**, 458 (1947).
368. D. H. R. Barton and A. I. Scott, *J. Chem. Soc.*, **1958**, 1767.
369. K. J. Pedersen, *J. Phys. Chem.*, **38**, 559 (1934).
370. B. R. Brown, *Quart. Rev.* (London), **5**, 131 (1951).
371. Sir Robert Robinson and J. D. Hobson, *Chem. Ind.* (London), **1955**, 285.
372. H. Hunsdiecker and C. Hunsdiecker, *Chem. Ber.*, **75**, 291 (1942).
373. J. Kleinberg, *Chem. Rev.*, **40**, 381 (1947).
374. R. G. Johnston and R. K. Ingham, *Chem. Rev.*, **56**, 219 (1956).
375. C. V. Wilson, *Org. Reactions*, **9**, 332 (1957).
376. M. Rottenberg, *Helv. Chim. Acta*, **35**, 1286 (1952).
377. S. J. Cristol and W. C. Firth, *J. Org. Chem.*, **26**, 280 (1961).
378. M. Reimer, E. Tobin, and M. Schaffner, *J. Amer. Chem. Soc.*, **57**, 211 (1933).
379. K. W. Bentley, S. F. Dyke, and A. R. Marshall, *J. Chem. Soc.*, **1963**, 3914.
380. F. H. Westheimer and W. A. Jones, *J. Amer. Chem. Soc.*, **63**, 3283 (1941).
381. O. Aschan, *Justus Liebig's Ann. Chem.*, **410**, 243 (1915).
382. G. Komppa, *Chem. Ber.*, **44**, 1537 (1911).
383. M.-M. Janot and R. Goutarel, *C.R. Acad. Sci., Paris*, **241**, 986 (1955).
384. M. F. Bartlett, D. F. Dickel, and W. I. Taylor, *J. Amer. Chem. Soc.*, **80**, 126 (1958).
385. W. Schindler, *Helv. Chim. Acta*, **41**, 1441 (1958).
386. U. Renner, D. Prins, and W. G. Stoll, *Helv. Chim. Acta*, **42**, 1572 (1959).
387. D. H. R. Barton and C. J. W. Brooks, *J. Chem. Soc.*, **1951**, 257.
388. P. Bilham, G. A. R. Kon, and W. C. J. Ross, *J. Chem. Soc.*, **1942**, 532.
389. D. H. R. Barton, C. J. W. Brooks, and N. J. Holmes, *J. Chem. Soc.*, **1951**, 278.
390. P. Plat, D. D. Manh, J. Le Men, M.-M. Janot, H. Budzikiewicz, J. M. Wilson, L. J. Durham, and C. Djerassi, *Bull. Soc. Chim. Fr.*, **1962**, 1082.
391. W. Voser, M. Montavon, Hs. H. Günthard, O. Jeger, and L. Ruzicka, *Helv. Chim. Acta*, **33**, 1893 (1950).
392. C. S. Barnes, D. H. R. Barton, J. S. Fawcett, and B. R. Thomas, *J. Chem. Soc.*, **1952**, 2339.
393. W. Thiele, *Chem. Ber.*, **20**, 922 (1893).
394. K. W. Bentley and J. P. Ringe, *J. Org. Chem.*, **22**, 424 (1957).
395. H. Wieland and W. Gumlich, *Justu's Liebigs Ann. Chem.*, **494**, 191 (1932).
396. H. Wieland and K. Kaziro, *Justus Liebig's Ann. Chem.*, **506**, 60 (1933).
397. F. A. L. Anet and Sir Robert Robinson, *J. Chem. Soc.*, **1955**, 2260.
398. H. Rapoport and J. B. Lavigne, *J. Amer. Chem. Soc.*, **75**, 5329 (1953).
399. C. Schöpf, *Justus Liebig's Ann. Chem.*, **452**, 411 (1927).

400. A. McGookin, A. Robertson, and E. Tittensor, *J. Chem. Soc.*, **1939**, 1587.
401. R. Boehm, *Justus Liebig's Ann. Chem.*, **318**, 230 (1901); **329**, 269 (1903).
402. K. W. Bentley, D. G. Hardy, and B. Meek, *J. Amer. Chem. Soc.*, **89**, 3293 (1967).
403. A. J. Birch and J. S. Hill, *J. Chem. Soc.*, **1966**, 419.
404. W. I. Taylor, in *The Alkaloids*, Vol. 8 (R. H. F. Mauske, ed.), Academic Press, New York, 1965, p. 243.
405. K. S. Kirby, *J. Chem. Soc.*, **1945**, 528.
406. K. S. Kirby, *J. Chem. Soc.*, **1949**, 735.
407. R. O. Kan, *Organic Photochemistry*, McGraw-Hill, New York, 1966.
408. G. Ciamician and P. Silber, *Chem. Ber.*, **40**, 2415 (1907).
409. G. Quinkert, B. Wegemund, F. Hemburg, and G. Cimbollek, *Chem. Ber.*, **97**, 958 (1964).
410. D. Arigoni, D. H. R. Barton, R. Bernasconi, C. Djerassi, J. S. Mills, and R. Wolf, *J. Chem. Soc.*, **1960**, 1900.
411. W. G. Dauben, D. A. Lightner, and W. K. Hayes, *J. Org. Chem.*, **27**, 1897 (1962).
412. E. Havinga, R. J. de Koch, and M. P. Rappoldt, *Tetrahedron*, **11**, 276 (1960).
413. R. L. Autry, D. H. R. Barton, A. K. Ganguli, and W. H. Reusch, *J. Chem. Soc.*, **1961**, 3313.
414. P. Barbier and R. Locquin, *C.R. Acad. Sci.*, Paris, **156**, 1443 (1913).
415. H. Wieland, O. Schlichting, and R. Jacobi, *Z. Physiol. Chem.*, **161**, 80 (1926).
416. W. M. Hoelm and H. L. Mason, *J. Amer. Chem. Soc.*, **60**, 1493 (1938); **62**, 569 (1940).
417. T. Reichstein and E. v. Arx, *Helv. Chim. Acta*, **23**, 747 (1940).
418. E. R. Stadtman, T. C. Stadtman, and H. A. Barker, *J. Biol. Chem.*, **178**, 677 (1949).
419. J. v. Euw, A. Lardon, and T. Reichstein, *Helv. Chim. Acta*, **27**, 821 (1944).
420. C. A. Vodoz and H. Schinz, *Helv. Chim. Acta*, **33**, 1040 (1950).
421. R. B. Turner, V. R. Mattox, W. F. McGuckin, and E. C. Kendall, *J. Amer. Chem. Soc.*, **74**, 5814 (1952).
422. Y. Yanuka and S. Sarel, *Bull. Res. Council Israel, Sect. A*, **6**, 286 (1957).
423. B. Riegel, R. B. Moffett, and A. V. McIntosh, *Organic Syntheses*, Coll. Vol. 3, 234, 237 (1955).
424. J. v. Braun, *Justus Liebig's Ann. Chem.*, **490**, 100 (1931).
425. H. B. Macphillamy and C. R. Scholz, *J. Org. Chem.*, **14**, 643 (1949).
426. H. Baumgarten, F. A. Bower, and T. T. Okamoto, *J. Amer. Chem. Soc.*, **79**, 3145 (1957).
427. E. E. Blaise, *C.R. Acad. Sci.*, Paris, **138**, 697 (1904).
428. C. Darzens and A. Levy, *C.R. Acad. Sci.*, Paris, **196**, 348 (1933).
429. H. Mendel and J. Coops, *Rec. Trav. Chim. Pays-Bas*, **58**, 1133 (1939).
430. G. D. Hunter and G. Popják, *Biochem. J.*, **50**, 163 (1951).
431. J. W. Cornforth, G. D. Hunter, and G. Popják, *Biochem. J.*, **54**, 590 (1953).
432. J. W. Cornforth, and G. Popják, *Biochem. J.*, **58**, 403 (1954).
433. C. Meystre, H. Frey, A. Wettstein, and K. Miescher, *Helv. Chim. Acta*, **27**, 1815 (1944).

434. C. Meystre and K. Miescher, *Helv. Chim. Acta*, **28,** 1497 (1945).
435. C. Meystre, H. Frey, R. Neher, A. Wettstein, and K. Miescher, *Helv. Chim. Acta*, **29,** 627 (1946).
436. C. Meystre and A. Wettstein, *Helv. Chim. Acta*, **30,** 1037, 1256 (1947).
437. A. Wettstein and C. Meystre, *Helv. Chim. Acta*, **30,** 1262 (1947).
438. K. Miescher and J. Schmidlin, *Helv. Chim. Acta*, **30,** 1405 (1947).

Chapter **XIII**

THE FISSION OF CARBON–NITROGEN AND CARBON–OXYGEN BONDS

K. W. Bentley

I CARBON–NITROGEN BOND CLEAVAGE

Exhaustive Methylation

The reaction whereby an olefin and a tertiary base are obtained by the degradation of a quaternary ammonium hydroxide, generally by pyrolysis, known as the Hofmann degradation [1, 2], is one of the most widely used processes in degradative studies of alkaloids and is also of considerable value in synthetic organic chemistry. In degradative studies, however, the results obtained in this way must not infrequently be interpreted with care.

The normal Hofmann degradation involves the attack of a hydrogen atom on the carbon β to the nitrogen atom by a base, usually a hydroxyl ion, and subsequent elimination of the base and water in a one-step process [3–6]:

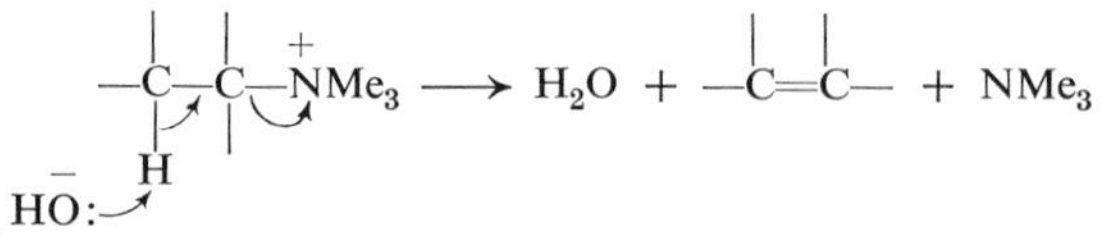

$$-\overset{|}{\underset{|}{C}}-\overset{|}{\underset{\substack{| \\ H \\ HO:}}{C}}-\overset{+}{N}Me_3 \longrightarrow H_2O + -\overset{|}{C}=\overset{|}{C}- + NMe_3$$

In the absence of a β-hydrogen, abstraction of a proton further removed from the nitrogen atom with a similar net result is occasionally observed, but the most common reaction in such cases is displacement by attack of one of the carbon atoms attached to nitrogen (see p. 277).

If the parent base is an open-chain amine (1), degradation of the salt (2) will afford a nonbasic olefin (3) and a simple tertiary amine. Heterocyclic amine salts (5), however, give olefinic bases (6) which can be subjected to

$$CH_3CH_2CH_2CH_2NMe_2 \rightarrow CH_3CH_2CH_2CH_2\overset{+}{N}Me_3 \rightarrow CH_3CH_2CH=CH_2 + NMe_3 + H_2O$$
$$OH^-$$

$$\quad\quad 1 \quad\quad\quad\quad\quad\quad 2 \quad\quad\quad\quad\quad\quad 3 \quad\quad\quad\quad 4$$

N-methylation (7) and further Hofmann degradation to give a nonbasic diene (8).

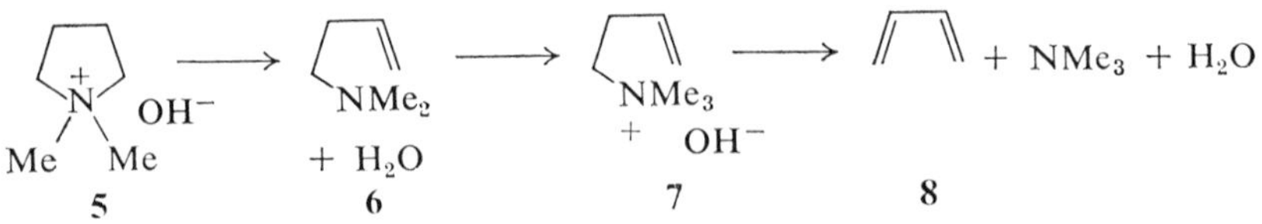

If the nitrogen atom is bound in two rings (9), three stages of Hofmann degradation are required before it is eliminated from the molecule (9 → 12). This sequence of *N*-methylations and Hofmann degradations leading to complete elimination of the basic nitrogen atom from the molecule is generally referred to as exhaustive methylation.

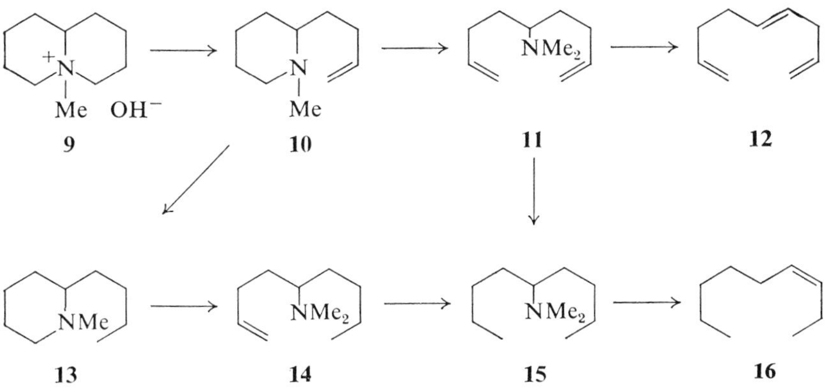

In some cases the first products of degradation of cyclic amines undergo rearrangement, fragmentation, or decomposition if further degradation is attempted (see p. 284), but in such cases reduction of the olefin first obtained yields products suitable for further reactions. Reduction may take place at any stage in a multiple step exhaustive methylation (9 → 16) to give a series of products of varying degrees of unsaturation.

The technique of exhaustive methylation has been popular in the study of alkaloids, both because of the facility with which the processes can be accomplished and because of the valuable information that may be obtained regarding the environment of the nitrogen atom from the number of degradations required completely to eliminate this atom from the molecule, and from a study of the products that are obtained with and without reduction at each stage of the process. The olefinic nature of the products makes them particularly suitable for spectral studies and for further chemical degradation, for example by oxidation.

The use of the Hofmann degradation in the elucidation of structures may be illustrated by the following examples.

(a) Apo-β-erythroidine (17) methohydroxide affords the methine base 18 on degradation, and this can be shown to contain the system –CH=CH₂ by the presence of characteristic absorption bands in its infrared spectrum and by the production of formaldehyde when it is ozonized. Reduction of the methine to the base 19 followed by a further Hofmann degradation affords the aromatic amine 20, which is resistant to further degradation and which can be shown to contain the system –CH=CH₂ by spectral studies and by ozonolysis. Apo-β-erythroidine therefore contains a nitrogen atom common to two rings and flanked by two –CH₂–CH₂– groups [7].

(b) Deoxynupharidine (21), which can be dehydrogenated with palladium to a base $C_{15}H_{19}NO$ oxidizable to pyridine-2,5-dicarboxylic acid and hence must contain a 2,5-substituted piperidine ring, on Hofmann degradation gives a methine base (22) which yields furan-3-aldehyde (23) on ozonolysis [8]. Reduction of the alkaloid involves saturation of the furan nucleus and scission of the quinzolizidine ring system to give a secondary base, which can be methylated to a quaternary salt (27), also obtainable from the methine base 22 [9, 10]. This quaternary salt, when subjected to two successive Hofmann degradations, gave a mixture of nitrogen-free products (25 and 29),

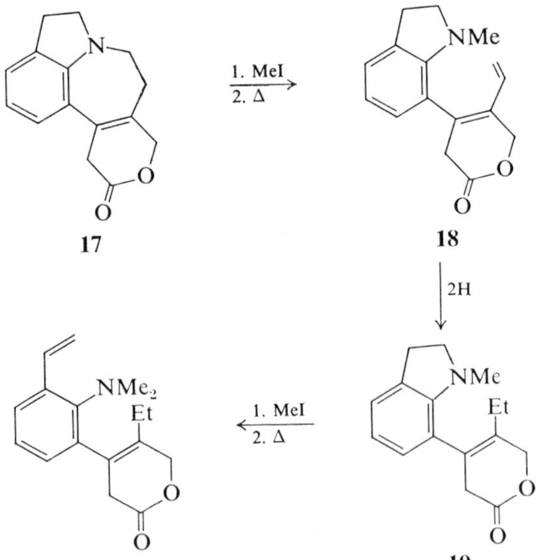

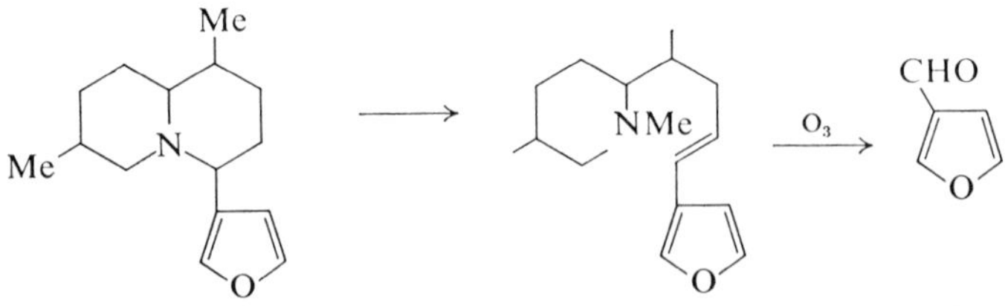

derived from the bases 24, 28, and 30, and ozonolysis of this mixture of dienes yielded the ketone 26 and the acid 32. Catalytic reduction of the mixture of unsaturated bases 24, 28, and 30 produced during the first of these Hofmann degradations, followed by further degradation produced a mixture of olefins 31 and 33, and this mixture on ozonolysis afforded the acid 32, isovaleric acid (35), the ketone 34, and formic acid [9–11]. These

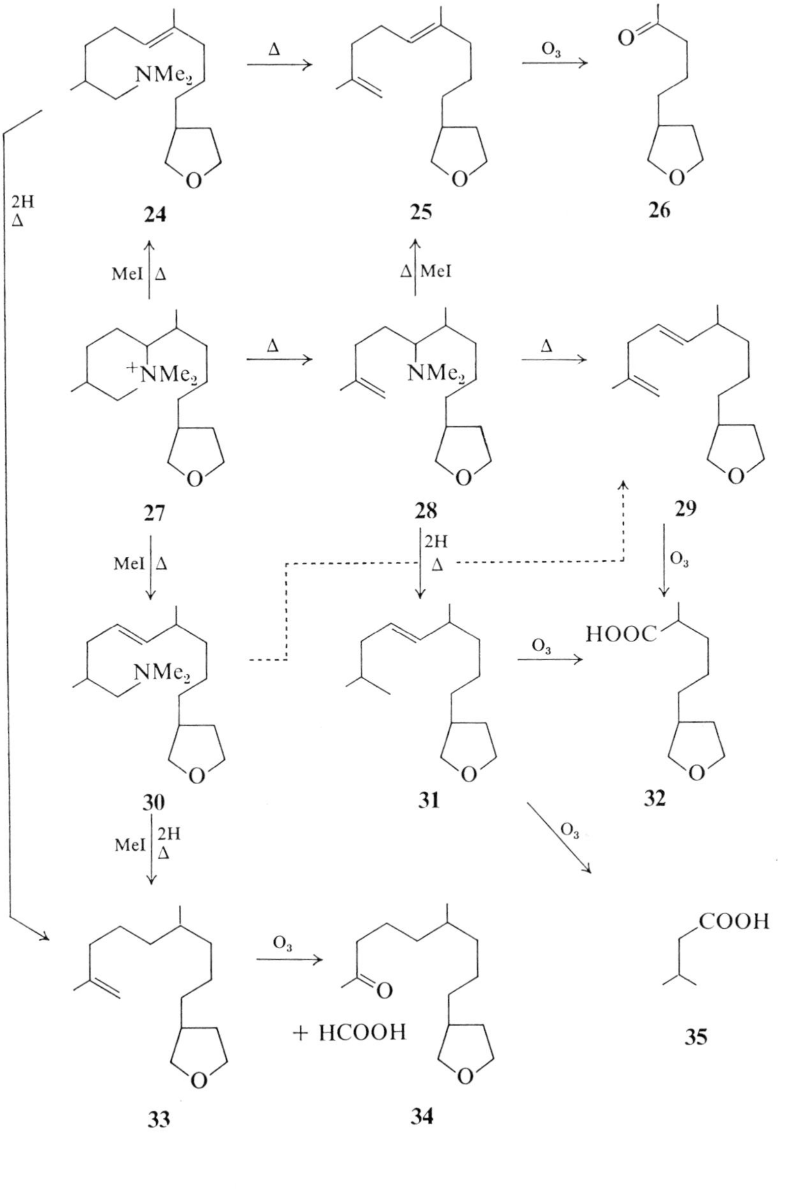

24 **25** **26**

2H
Δ

MeI Δ   Δ MeI

27 **28** **29**

MeI Δ   2H
Δ

O₃

30 **31** **32**

MeI 2H
Δ

O₃

HOOC

.COOH

33 **34** **35**

O₃

+ HCOOH

259

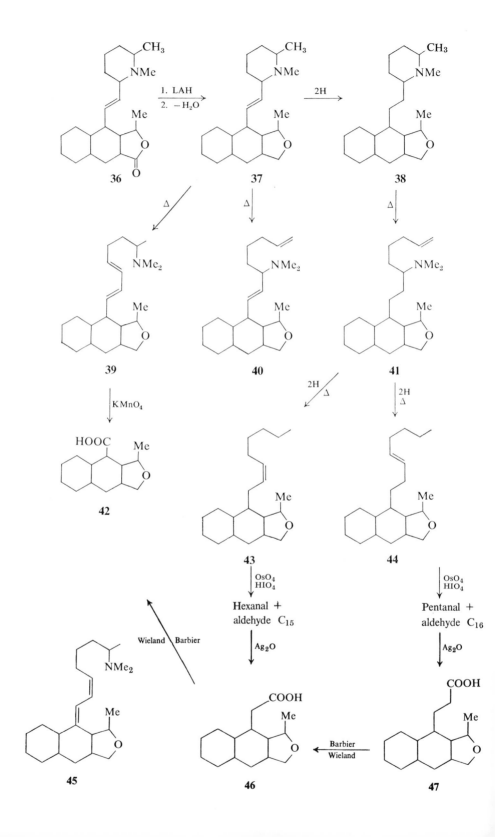

findings are only explicable on the basis of the formula **21** for deoxynuphari-dine.

(*c*) The alkaloid himbacine (**36**), which resists Hofmann degradation, gave anhydrohimbacine diol (**37**) on reduction with lithium aluminum hydride and dehydration, and further reduction of this yielded the dihydro-anhydro compound (**38**). Hofmann degradation of the last-named compound yielded a methine base (**41**) which gave formaldehyde on ozonolysis, and catalytic reduction of this followed by a second Hofmann degradation gave a mixture of olefinic nitrogen-free products **43** and **44**. Oxidation of this mixture gave pentanol, hexanol, and a mixture of nonvolatile aldehydes, and this mixture was oxidized further to the acids **47** and **46**, and of these the former was converted into the latter and further into the acid **42** by Barbier-Wieland degradation. The only rational explanation of these findings is that given in the accompanying formulas.

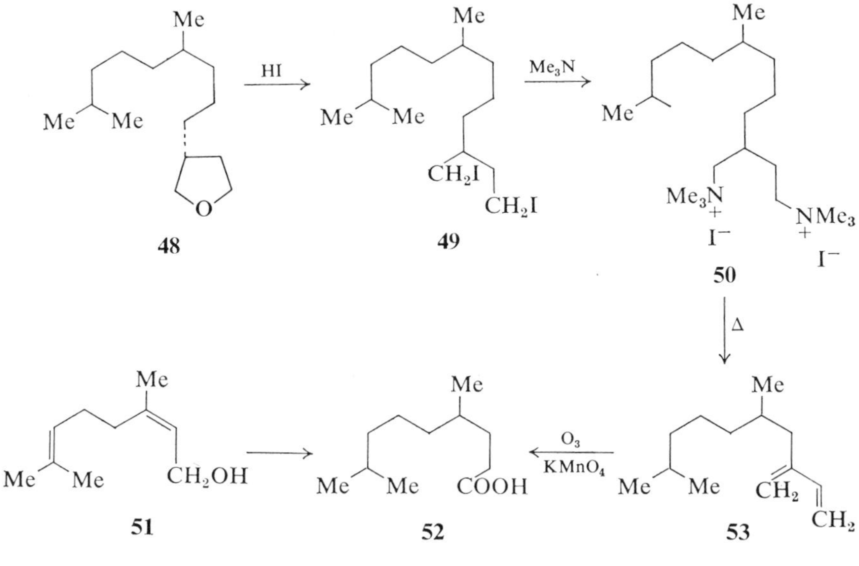

+ COOH—COOH

The anhydrodiol **37** on Hofmann degradation yielded a mixture of methine bases **40** and **39**, of which the latter was shown by its ultraviolet spectrum to be a conjugated diene. Since the conjugated diene gave the acid **42** on oxidation with permanganate, the alternative diene structure **45** was untenable and the structure of anhydrohimbacine diol (**37**) and hence of himbacine itself became clear [12].

(*d*) In some cases a tertiary amino group may be deliberately introduced into a molecule in order to achieve degradation to olefinic substances which can be subjected to oxidation during the elucidation of structures. For example, anhydronupharane diol (**48**) has been degraded via the diiodide **49** and bisquaternary salt **50** to the diene **53**, a two-step oxidation of which afforded oxalic acid and the acid **52**, obtainable from geraniol (**51**) [13].

Conditions of the Reaction

The Hofmann degradation is normally carried out by dry distillation of the quaternary methohydroxide at temperatures around 200°C [14]. Isolation of the quaternary hydroxide is frequently not necessary, pyrolysis of a mixture of the chloride or iodide and potassium hydroxide being a satisfactory process [14–16]. In a number of cases, particularly when the nitrogen atom is in the β-position relative to an aromatic nucleus so that elimination gives a styrene or stilbene derivative, the degradation may be simply accomplished by boiling one of the quaternary salts, most conveniently the methiodide or methosulfate, in aqueous or alcoholic sodium or potassium hydroxide. The use of alkoxides rather than hydroxides for the degradation of the quaternary salts in solution appears to have little effect on the yield of olefin [14].

The methohydroxide, when required, may be prepared by the classical but tedious method of heating the methiodide with a suspension of silver oxide in water or, much more conveniently, by the passage of an aqueous solution of the methiodide through a column of a suitable ion exchange resin [17]. Volatile materials produced during the degradation, for example trimethylamine or other simple base, may be collected in a suitable cooled trap, and nonvolatile materials remaining after pyrolysis may be extracted from the residue with ether or chloroform. Nitrogen-free products may be separated from tertiary bases (which are formed by the loss of methanol from the quaternary hydroxide, see below, p. 277) by extracting the latter with dilute acids. When the Hofmann degradation is carried out in aqueous solution the product, base or nitrogen-free olefin, frequently separates in a crystalline or semisolid state and may be readily collected.

Mechanism of Elimination

Most of the Hofmann eliminations of tertiary base from quaternary ammonium compounds are bimolecular elimination processes ($E2$), but other processes are involved in a minority of cases. For successful elimination the presence of β-hydrogen, quaternary nitrogen and moderately strong base are all generally required and the presence or absence of α-hydrogen is of no consequence. Either a concerted $E2$ or stepwise $E1cb$ process can be involved. In the latter a carbanion intermediate is regarded as being formed, but the mechanisms can be regarded as extremes of the same process, differing only in the lifetime of the carbanion, which in the $E2$ mechanism is zero [18, 19].

The $E2$ mechanism requires the β-hydrogen and quaternary nitrogen atom to be coplanar and in the *trans* relationship conformationally. This has been demonstrated in the degradation of *erythro*- and *threo*-trimethyl-1,2-diphenylpropylammonium iodide in ethanol in the presence of ethoxide ion,

$$\begin{array}{c}
B^{\ominus} \\
\searrow H \\
-\overset{|}{\underset{|}{C}}\!-\overset{|}{\underset{|}{C}}- \\
\underset{+}{\overset{|}{N}R_3}
\end{array} \quad \xrightarrow{\ E2\ } \quad \overset{}{\underset{}{>}}C\!=\!C\overset{}{\underset{}{<}} \ + \ NR_3 \ + \ BH$$

$$\begin{array}{c}
B^{\ominus} \\
\searrow H \\
-\overset{|}{\underset{|}{C}}\!-\overset{|}{\underset{|}{C}}- \\
\underset{+}{NR_3}
\end{array} \quad \xrightarrow{\ E1cb\ } \quad
\begin{array}{c}
\overset{|}{} \\
-\overset{\ominus}{\underset{|}{C}}\!-\overset{|}{\underset{|}{C}}- \\
\underset{+}{NR_3}
\end{array} \quad \longrightarrow \quad
\begin{array}{c}
\overset{|}{} \\
-C\!=\!\overset{|}{\underset{|}{C}}- \\
NR_3
\end{array}$$

when the *erythro* form (54) gave *cis*-1,2-diphenylpropene (56) and the *threo* form (57) gave the *trans*-olefin (59). As will be seen from the Newman projection formulas, steric hindrance in the transition state 58 for the *threo* isomer is less than in the corresponding conformation (55) for the *erythro* isomer, and in accord with expectation a more rapid degradation of the former than the latter was observed [20].

There are no such steric requirements for the $E1cb$ mechanism, and degradation of the two quaternary salts 54 and 57 with the more basic t-butoxide ion in t-butanol affords the *trans* olefin (59) only, the reaction rate being the same with both isomers [20]. These findings are what would be expected for a two-step reaction mechanism involving a carbanion (60–62) having an appreciable lifetime in which the more stable configuration (61) can be assumed before completion of the elimination.

The degradation of quaternary salts of menthylamine and neomenthyl-amine presents interesting evidence of the importance of *trans* elimination. With neomenthylamine salts the conformation 63 would permit *trans* elimination of nitrogen and hydrogen at C-2 or C-4; the predominance of menth-3-ene (66) (92%) in the product indicates that, in this case, the removal of a proton from R_2CH- is preferred to removal from $R-CH_2-$. With menthylamine salts the conformation 65 does not permit *trans* elimi-nation, and degradation involves the conformation 67 in which the only β-proton *trans* to nitrogen is at C-2 and 86% of the product is menth-2-ene (64) [21, 22].

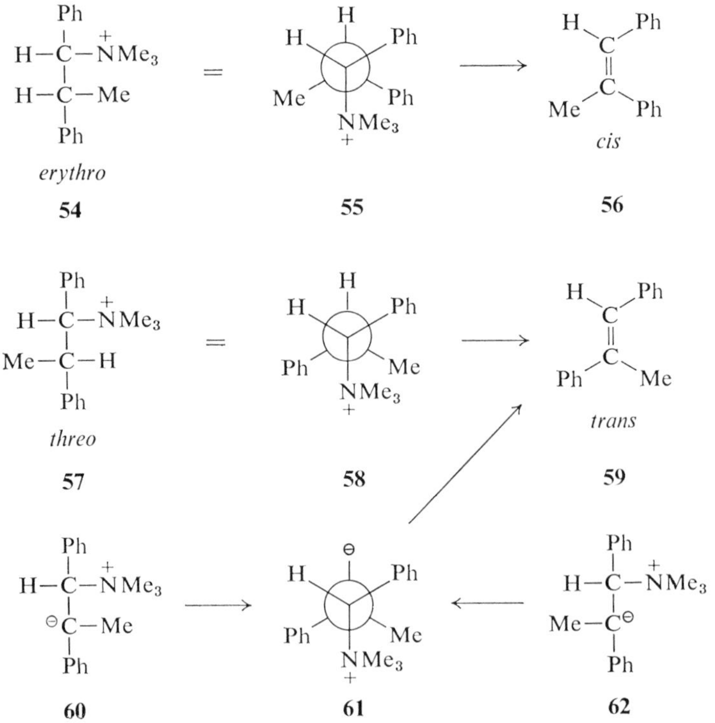

Ph
| +
H—C—NMe₃
|
H—C—Me   =
|
Ph

erythro

54

55

$\longrightarrow$

cis

56

Ph
| +
H—C—NMe₃
|
Me—C—H   =
|
Ph

threo

57

58

$\longrightarrow$

trans

59

Ph
| +
H—C—NMe₃
|
⊖C—Me
|
Ph

60

$\longrightarrow$

61

$\longleftarrow$

Ph
| +
H—C—NMe₃
|
Me—C⊖
|
Ph

62

The preferential formation of menth-3-ene by the *E*2 degradation of the quaternary salts of neomenthylamine was until recently regarded as unusual. It was believed that in general the direction of elimination was determined only by the acidity of the proton removed [4, 5], and the nature of the olefin produced during Hofmann degradation was believed to be predictable by the so-called Hofmann rule that the least substituted olefin is formed by

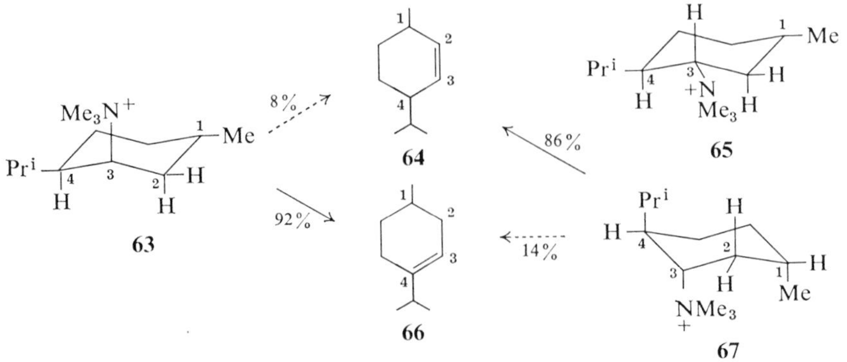

63   8% **64**   **65**

92%   **66**   14%   86%   **67**

removal of a proton from the least substituted β-carbon atom. The degradation of the salt **63** to the olefin **66**, when degradation to the less substituted olefin is not sterically inhibited, clearly provides an exception to this rule, which is an oversimplification. In general ethylene is preferentially eliminated from salts containing an N-ethyl group, but the salt **69** degrades preferentially to isobutene (**68**), only 7.2% of ethylene (**70**) being formed. Certainly elimination involves an ethyl group more easily than the other primary alkyl groups, but secondary alkyl groups are somewhat more readily and tertiary alkyl groups generally much more readily involved than ethyl [23].

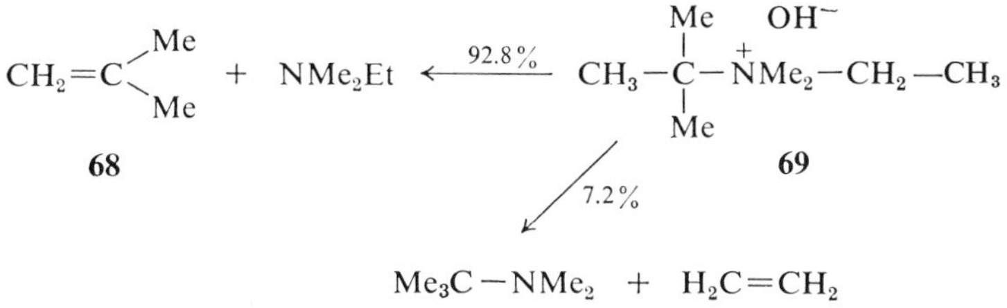

$$CH_2\!=\!C\diagdown\begin{matrix}Me\\Me\end{matrix} \;+\; NMe_2Et \;\xleftarrow{\;92.8\%\;}\; CH_3\!-\!\underset{\underset{Me}{|}}{\overset{\overset{Me}{|}}{C}}\!-\!\overset{+}{N}Me_2\!-\!CH_2\!-\!CH_3 \quad OH^-$$

68   **69**

$$\diagdown\;7.2\%$$

$$Me_3C\!-\!NMe_2 \;+\; H_2C\!=\!CH_2$$

70

In $E2$ eliminations in general the steric requirements in the transition state of both base and leaving group are important factors, and a degree of control of Hofmann over Saytzeff elimination may be obtained [24]. Other examples of Saytzeff rather than Hofmann rule elimination are provided by the degradations of quaternary salts of 6β-aminocholestane (**71**) which give cholest-5-ene (**72**) exclusively, in spite of the fact that in the salt (**73**) β-hydrogen atoms 180° *trans* to the quaternary nitrogen are available for

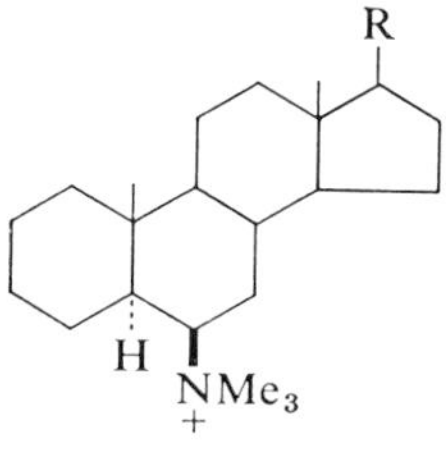

71 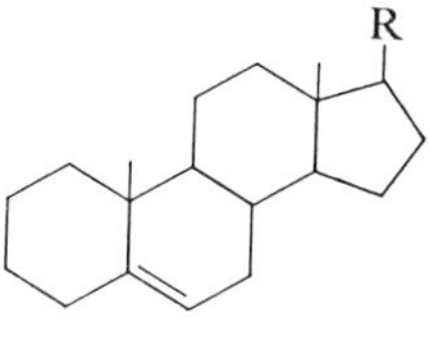   **72** 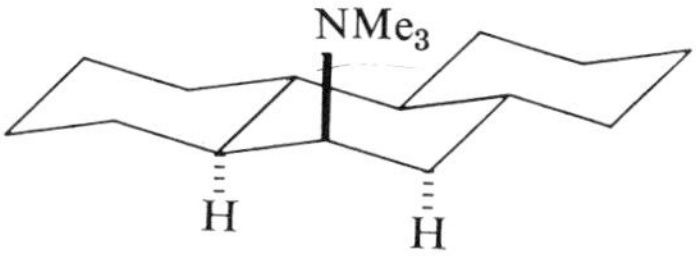   **73**

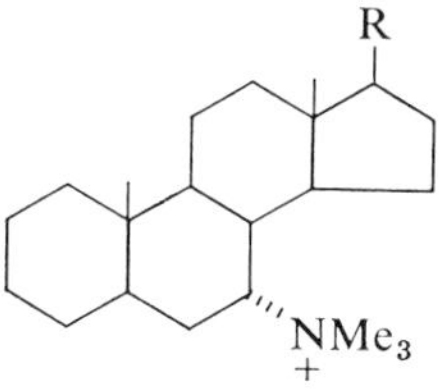

74 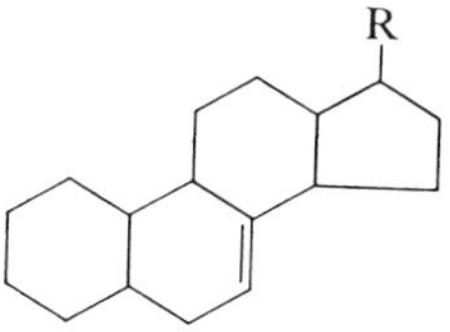   **75** 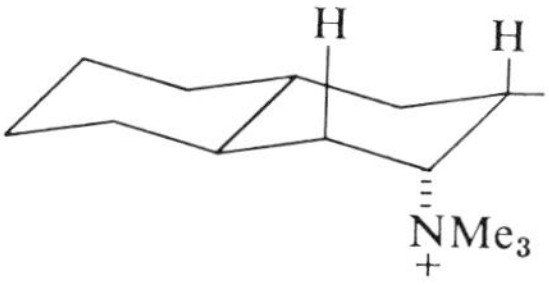   **76**

elimination on both the more substituted C-5 and the less substituted C-7 [25]. The quaternary salts of 7α-amino cholestane (**74**), in which the nitrogen and β-hydrogen atoms are not 180° *trans* (**76**), on boiling in neutral ethanol give almost pure cholest-7-ene (**75**), undoubtedly by an *E*1 mechanism, but in alkali the product is cholest-6-ene [26]. The *E*1 mechanism involves a two-step process in which the C–N bond is broken first to give an amine and a carbonium ion (**77 → 79**). No base is required for the reaction.

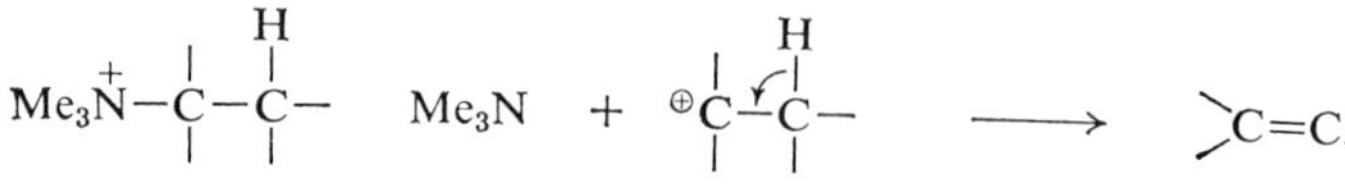

$$
\underset{\textbf{77}}{Me_3\overset{+}{N}-\overset{\overset{\displaystyle H}{|}}{\underset{|}{C}}-\overset{|}{\underset{|}{C}}-} \qquad Me_3N \;+\; \underset{\textbf{78}}{\overset{\oplus}{C}}\overset{\overset{\displaystyle H}{|}}{\underset{|}{C}}- \qquad\longrightarrow\qquad \underset{\textbf{79}}{\diagdown C=C \diagup}
$$

Hofmann degradations that cannot proceed by a *trans* elimination mechanism occur by the two-step, nonsterospecific *E*1*cb* process and give the most stable olefin. Many of these reactions involve the production of stabilized benzylic or allylic β-carbanions (**80 → 82**) [27]. Such reactions also occur, however, in salts in which the β-hydrogen atoms are activated only by the

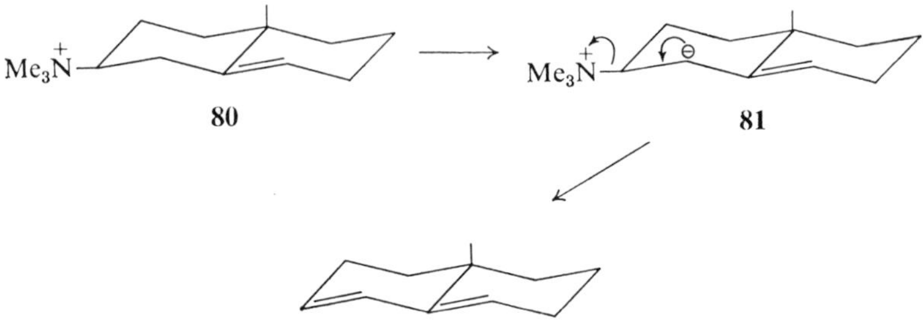

$$
\underset{\textbf{80}}{Me_3\overset{+}{N}} \qquad\longrightarrow\qquad \underset{\textbf{81}}{Me_3\overset{+}{N}} \qquad\qquad \underset{\textbf{82}}{}
$$

positively charged nitrogen and, although the *E*1*cb* process via a β-carbanion may still be operative, an alternative mechanism involving an ylid may be involved (**83 → 84 → 86** or **83 → 85 → 86**). The cyclic α,β-elimination mechanism (**85**) is similar to that involved in the pyrolysis of amine oxides (see p. 292).

It has been shown that *E*2 β-elimination and α,β-elimination by the ylid mechanism (**85**) are competing reactions in the degradation of β-tritium labeled ethyltrimethylammonium hydroxide (**87**), by the detection of tritium labeled trimethylamine in the products of the degradation [28]. Mass spectrometric detection of deuterium labeled trimethylamine in the products of

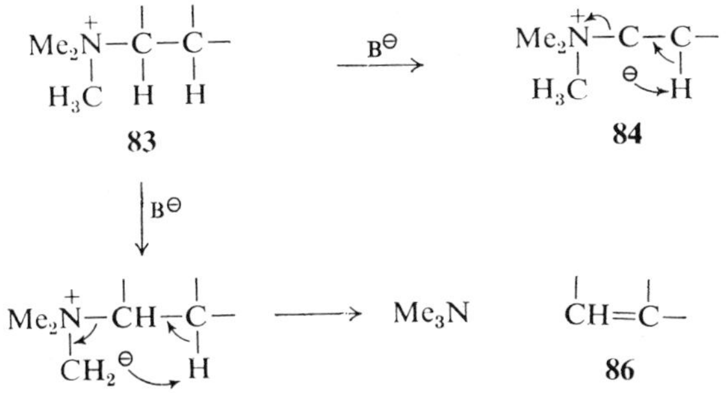

$$Me_2\overset{+}{N}-\overset{|}{\underset{H_3C}{C}}-\overset{|}{\underset{H}{C}}- \qquad \xrightarrow{\;B^{\ominus}\;} \qquad Me_2\overset{+}{N}\overset{\frown}{-}\overset{|}{C}\overset{}{\underset{\underset{H}{\ominus}}{-}}\overset{|}{C}-$$

83   84

$$\Big\downarrow B^{\ominus}$$

$$Me_2\overset{+}{N}\overset{\frown}{\underset{\underset{CH_2}{\overset{|}{\underset{\ominus}{\,}}}}{-}CH}\overset{\frown}{\underset{H}{-}}\overset{|}{C}- \quad\longrightarrow\quad Me_3N \qquad \overset{|}{CH}=\overset{|}{C}-$$

85   86

degradation of the salt (**88**) showed that this decomposition proceeds principally by the ylid mechanism (**85**) [29]. In contrast with these findings, however, the degradations of the salts **89**, **90**, and **91** apparently do not

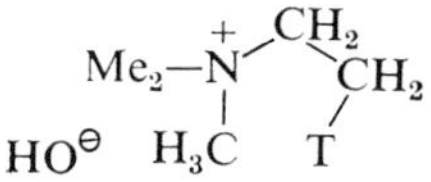

87 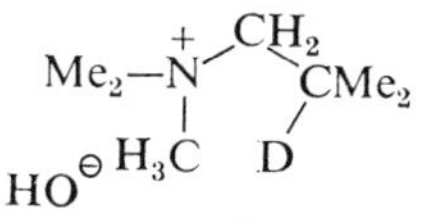   88

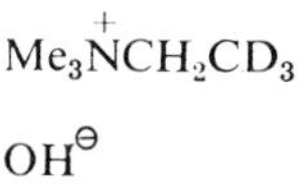

$$Me_3\overset{+}{N}CH_2CD_3$$
$$OH^{\ominus}$$
89

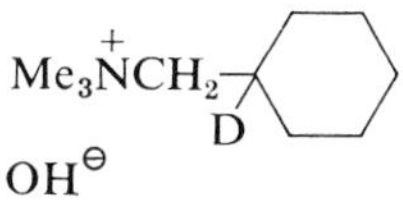

$$Me_3\overset{+}{N}CH_2{-}\overset{\text{cyclohexyl}}{\underset{D}{}}$$
$$OH^{\ominus}$$
90

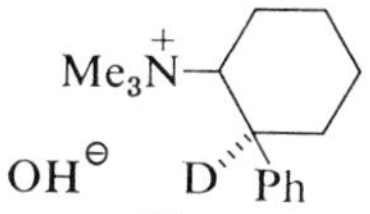

$$Me_3\overset{+}{N}{-}$$
$$OH^{\ominus}\quad D^{\,\backslash}Ph$$
91

proceed significantly by the ylid mechanism [30, 31], which is probably of most importance in cases where *trans E2* elimination is either impossible or highly unfavored.

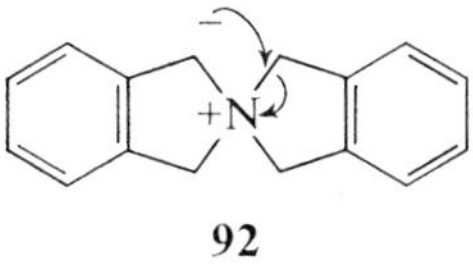

92 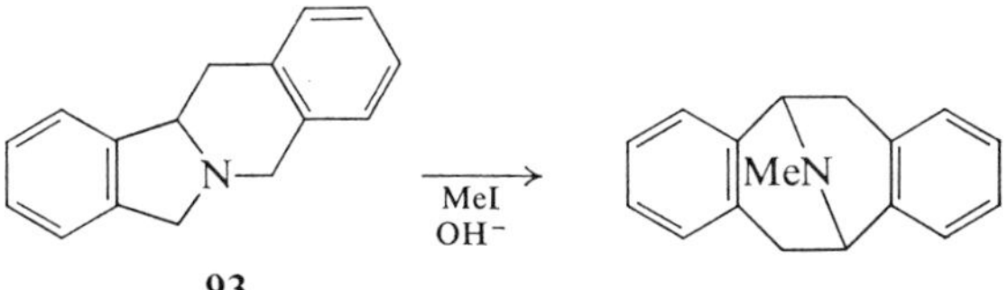   93   94

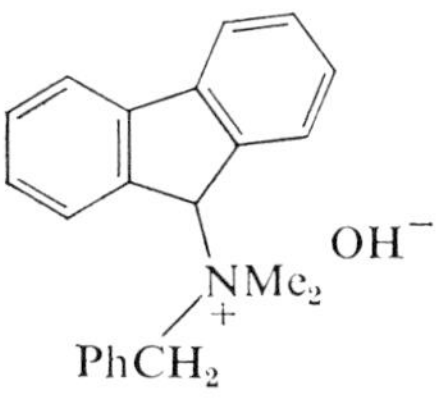

$$\xrightarrow[\text{OH}^-]{\text{MeI}}$$

95 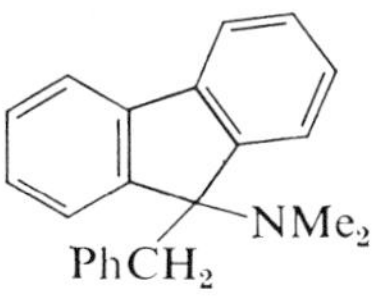   96 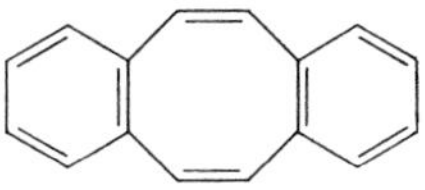   97

Ylid involvement is obvious in the exhaustive methylation of the quaternary salt **92** to dibenzocyclooctatetraene (**97**) via the bases **93** and **94** [32, 33], and the conversion of the quaternary salt **95** into the tertiary base **96** [34].

Direction of Elimination

The stereochemical factors governing *E*2 elimination in the Hofmann degradation have been outlined above. In general in aliphatic systems elimination of β-hydrogen is easiest when the hydrogen is part of a $-CH_3$ group and least easy when it is part of a $-CHR_2$ group, and in many cases this is the factor governing the direction of elimination in such systems [14]. The possibility of effecting degradation by other than the *E*2 mechanism may, however, affect the direction of elimination.

In alicyclic systems the most important factor in the elimination reaction is the availability in one conformation or the other where two possible conformations exist, of *trans* diaxial $\overset{+}{N}R_3$ and β-hydrogen (see p. 265). Salts of both menthyl and neomenthylamine can be arranged in conformations where such dispositions exist (**63** and **67**) as a result of the interconvertibility of the two chair forms of cyclohexane, but the conformational rigidity of the *trans*-A/B ring junction in cholestane derivatives permits such a disposition of $\overset{+}{N}Me_3$ and β-hydrogen only in derivatives of the 3α-amines (**98**), which are easily degraded, whereas the 3β-compounds, in which only *trans* diequatorial groups can be involved (**99**), give only poor yields of elimination products [27].

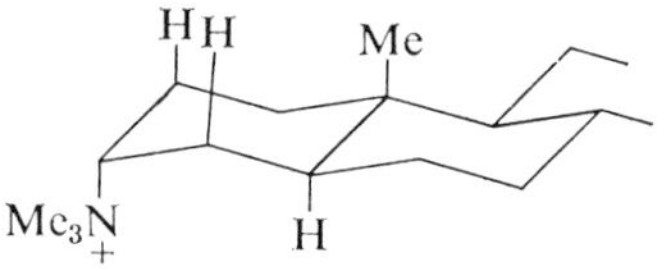

                          98 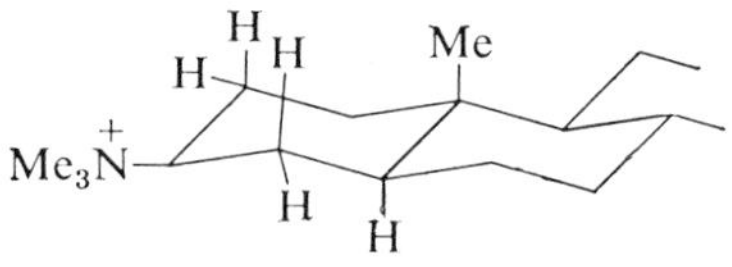   99

Most of the useful applications of the Hofmann degradation in the elucidation of structures, however, has been in the alkaloid field, with heterocyclic bases, most frequently in five- or six-membered rings.

Pyrrolidinium and piperidinium salts readily undergo Hofmann degradation with apparently equal ease since the spiro salt **101** gives the products of alternative fission of the two systems, **100** and **102** [35]. When

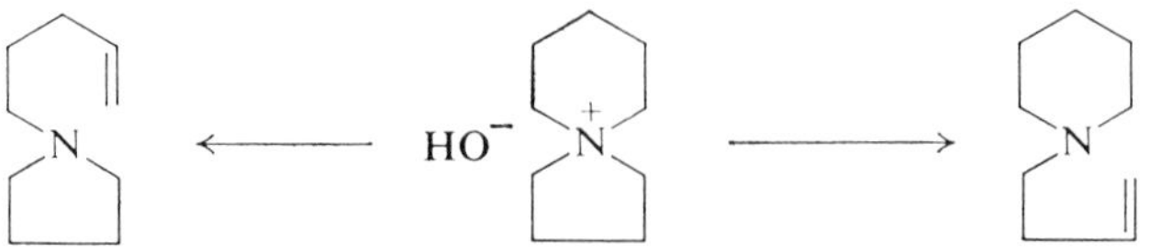

        100   101   102

the pyrrolidinium ring is part of a more complex system, the nature of that system can determine the mode of fission of the ring. For example, salts of *cis*-octahydroindole in which the nitrogen can be accommodated in an axial (**104**) or equatorial (**103**) position by a conformational flip, are degraded to the base **105** [36], whereas salts of *trans*-octahydroindole have the nitrogen locked in an equatorial position (**106**) and are degraded to the base **107** [37].

103

104

105

106

107

108

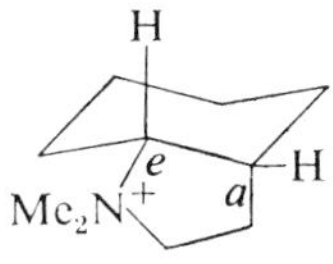

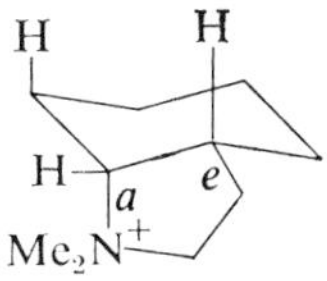

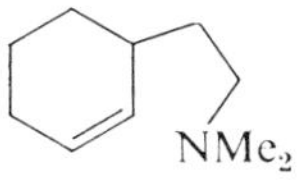

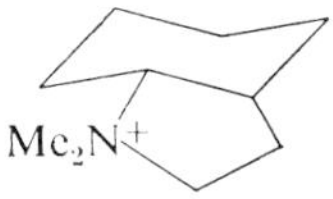

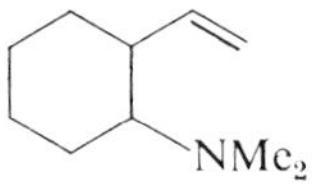

Where the pyrrolidine ring bears an α-methyl group, both the acidity of the CH_3 protons and steric factors in the transition state favor elimination involving the methyl group and salts of 2-methyl-*cis*-octahydroindole degrade to the base **108** rather than to an analog of the base **105** [14].

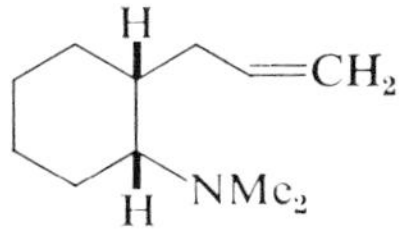

Salts of α-methylpiperidine behave similarly (**109**), giving terminal vinyl compounds (**110**). Exhaustive methylation of such compounds gives the

109

110

111

112

113

114

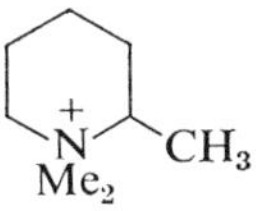

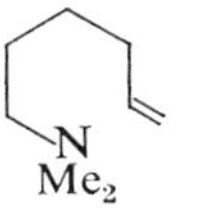

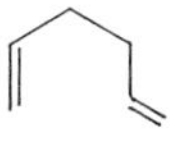

unconjugated 1,5-diene **111** as end product [38], whereas the exhaustive methylation of piperidines lacking such an α-substituent (**112**) gives a conjugated 1,3-diene (**114**) as a result of double bond migration under the strongly alkaline conditions of the reaction [2].

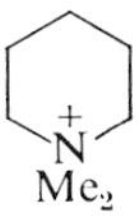

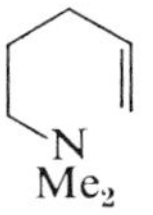

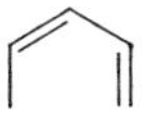

Bridged piperidinium salts are equally easily degraded, for example the salt **115** affords the olefin **115a** [39], tropidine methohydroxide (**116**) can be degraded to the base **116a** [40], *N*-methyl granatanine salts (**117**) to the amine **117a** [41, 42], and the 3,9-diazabicyclo[4.2.1]nonan-4-one salt (**118**) yields

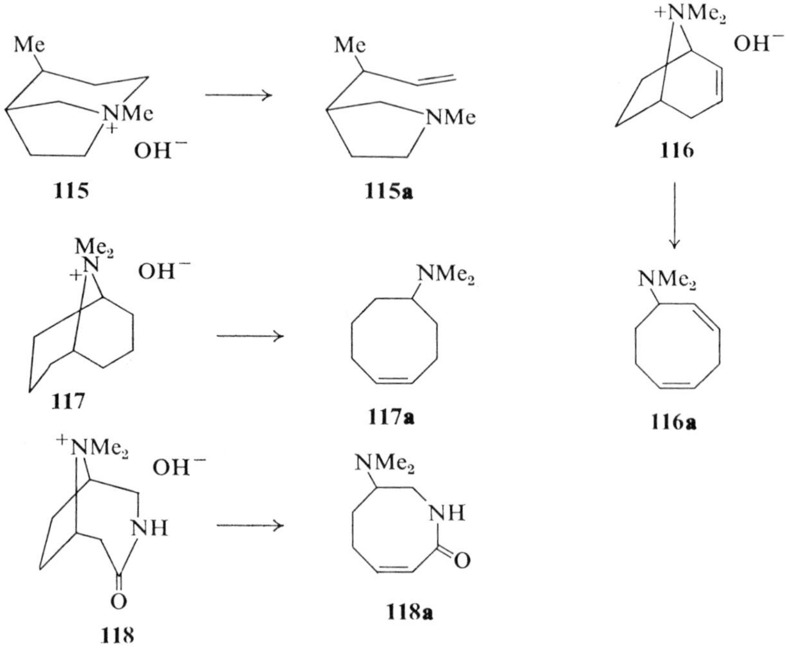

the α,β-unsaturated amide **118a** [43]. The ketone **119**, obtained from dioscorine, suffers Hofmann degradation very readily under the influence of sodium bicarbonate to give the amine **119a** and, by β-elimination of dimethylamine from this, the neutral product **119b** [44].

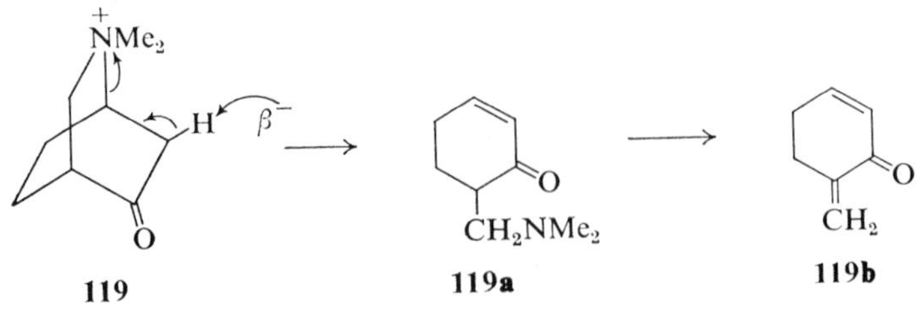

It has been repeatedly stated that derivatives of tetrahydroquinoline (**120**) do not suffer Hofmann degradation, even in the presence of a 2-methyl group, but that the quaternary hydroxide loses methanol to give the parent tertiary base [14]. However, successful degradation to the olefin **120a** has been reported [45]. Decahydroquinoline salts (**122**) degrade preferentially to bases of the type **122a** [14]. By contrast, tetrahydroisoquinolines (**121**) are degraded with very great ease to styrenes (**121a**).

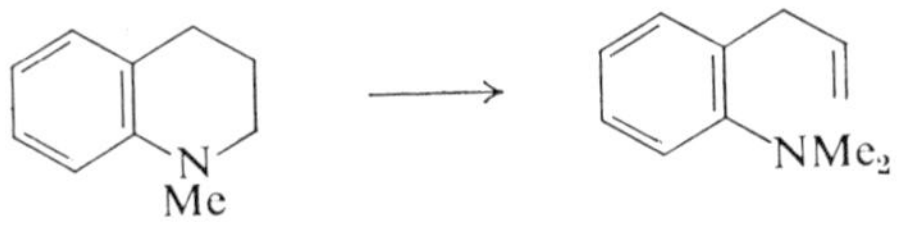

The aromatic nucleus has a great activating effect on the ease with which the β-hydrogen atoms may be removed under attack by bases, and when alternative modes of fission would permit formation either of a styrene or an unconjugated olefin the styrene is always formed. For example, the bases of the morphine group invariably give styrenes in the first step [44], dihydro-codeine (**123**) giving the methine base **124** with ease in the first stage [47, 48], and the nitrogen-free product **125** only under much more vigorous conditions in the second stage [49].

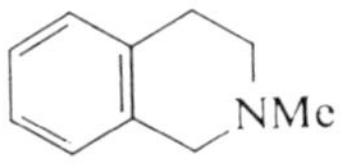

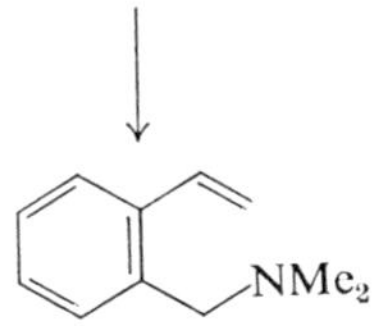

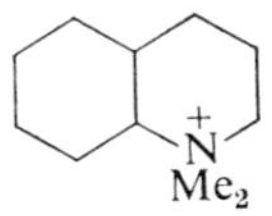

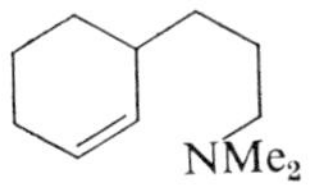

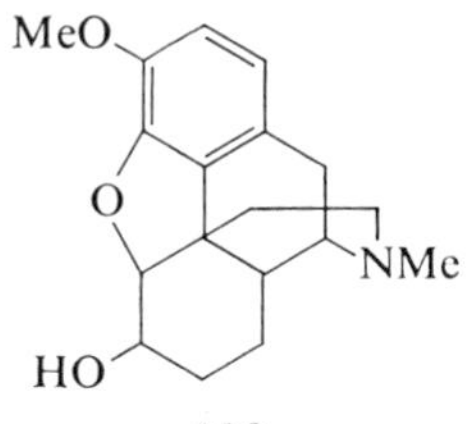

Bases of the aporphine group in which β-hydrogen atoms on both sides of the nitrogen atom are activated by aromatic nuclei generally give mixtures of products of both alternative modes of fission, for example, isothebaine methyl ether (**126**) gives the optically inactive methine **127** which is a phen-anthrene derivative, and the optically active isomethine **128**, both of which give the trimethoxyvinylphenanthrene **129** on further degradation [50, 51].

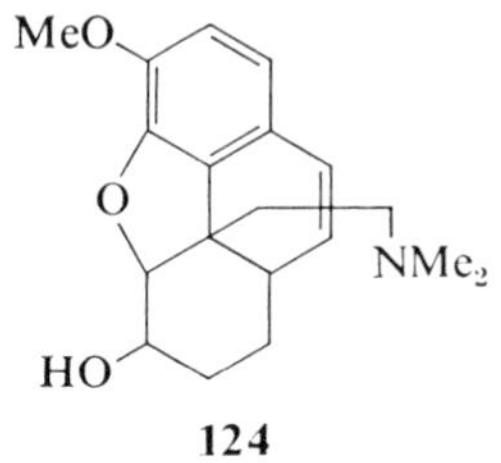

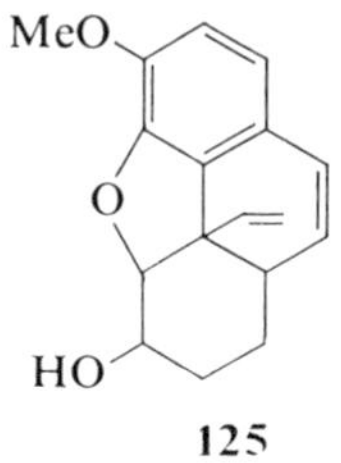

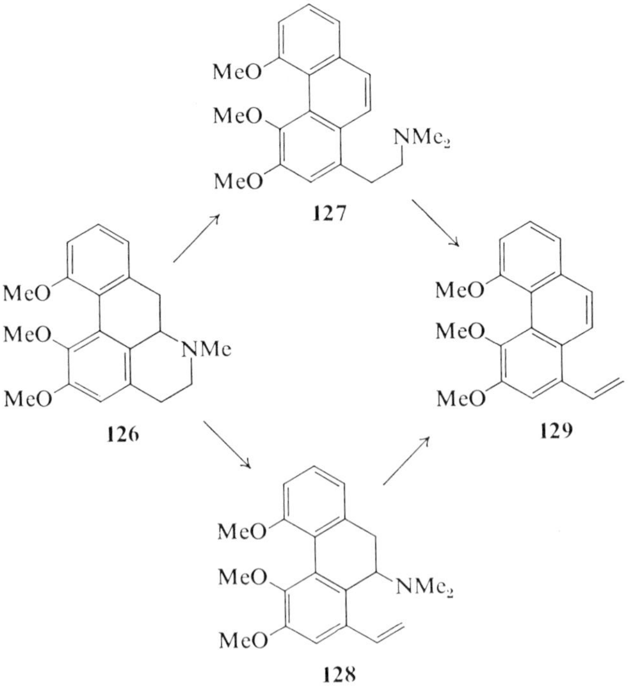

Phenyldihydrothebaine (**131**, R = Ph) and methyldihydrothebaine (**131**, R = Me) likewise yield mixtures of methine base **130** and isomethine **132** [52–54].

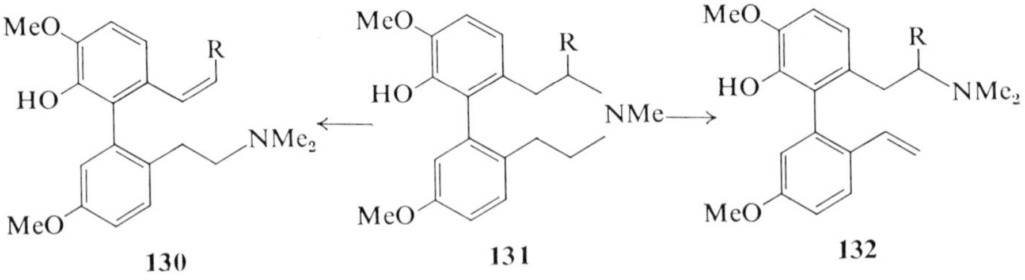

Benzylisoquinoline alkaloids, such as laudanosine (**133**), degrade preferentially to the stilbenes (**134**), and finally, with equal ease, to the nitrogen-free olefins (**135**) [55, 56]. Hydroxylaudanosine, however, does not behave in this way, molecular cleavage preceding degradation (see p. 288).

The alkaloid (−)-canadine provides an interesting case. The optically active quaternary salt **136** on degradation affords a mixture of the two methine bases **137** (optically inactive) and **138** (optically active), together

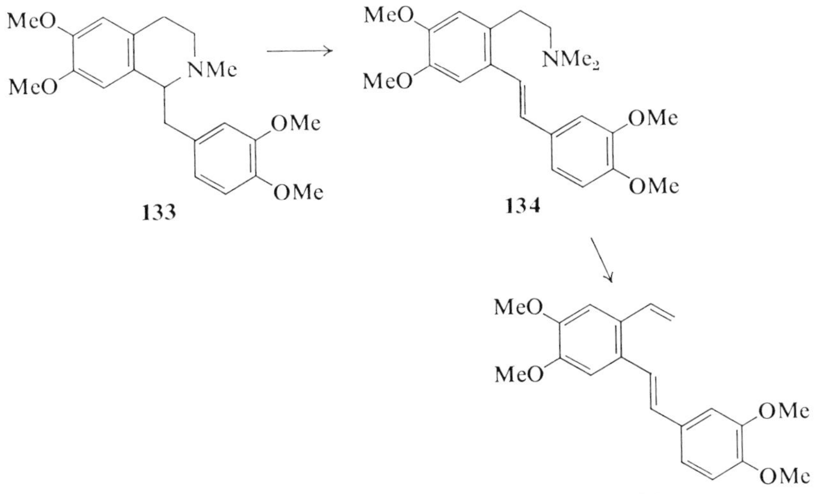

133   **134**

135

with the racemate of the original quaternary salt **136** and the racemate of the
methine base **138**. This could be accounted for by postulating reversal of
the Hofmann degradation of **136** to **137**, which would inevitably give the
racemate of the salt **136**, irreversible degradation of which in the alternative
manner would give the racemic methine **138**. An alternative explanation
would be the racemization of the salt **136** via the ylid **139**, but although such

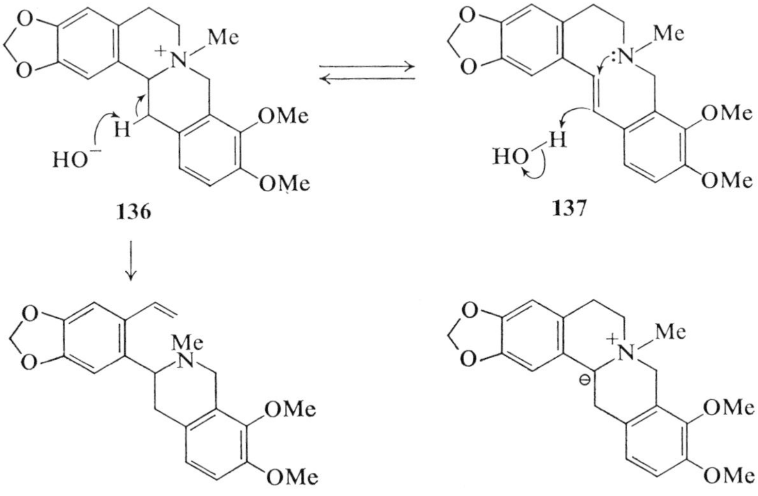

136   **137**

138   **139**

a process may contribute to the racemization, reversal of the Hofmann is undoubtedly a factor in the reaction since the pure methine base **137** yields the racemic quaternary hydroxide **136** when it is boiled with aqueous ethanol [57]. The reversal of the reaction is unusual, and analogous intermolecular reactions can be accomplished only with great difficulty [58]; it is probably facilitated by the favorable steric arrangement of the methine base **137** in which the double bond and tertiary amino groups are held in close proximity by the ten-membered ring. The recombination of Hofmann degradation products has also been observed in the strychnine [59] and flavothebaone [60] series.

β-Amino ketones always split in the Hofmann degradation with great ease to give the *α*,*β*-unsaturated ketone; for example, the salt **140** very readily affords **141** in spite of the presence of *β*-hydrogens at three positions other than that marked by an asterisk [61]. The ease of degradation of the salt **119** with bicarbonate ion (see above) further illustrates this point.

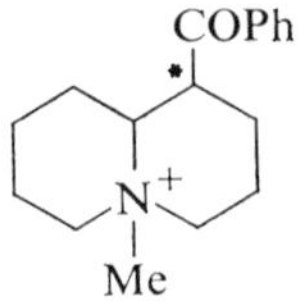

140

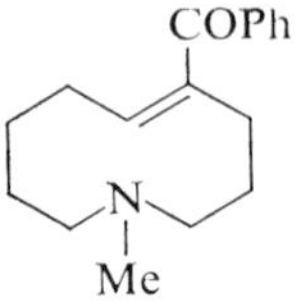

141

When two quaternary nitrogens are present in the same molecule, both basic centers are usually affected during Hofmann degradation; for example, *N*-methylemetine dimethiodide (**142**) gives the bismethine **143** [62]. Advantage may be taken of the differing reactivity of two tertiary nitrogens, however, to obtain products degraded at one center only, for example, the tetrahydro compound of **143** can be converted via a monomethiodide into the base

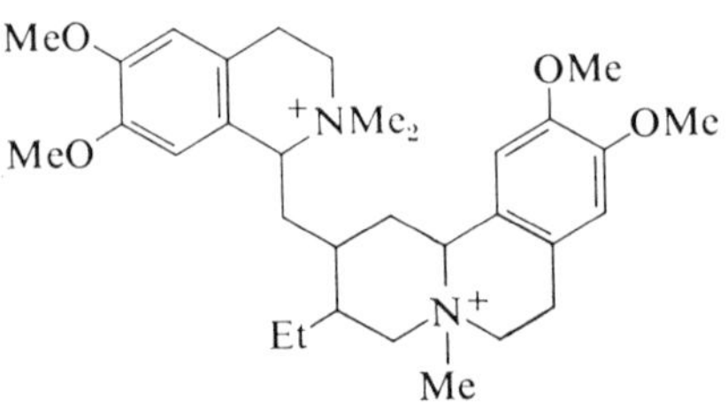

142

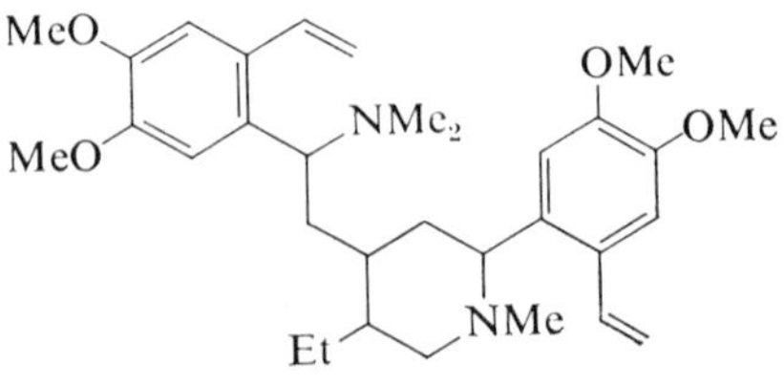

143

144 [62]. Secondary nitrogens may be protected during the *N*-methylation and degradation by acylation; for example, by working on the *N*-acetyl compound only the tertiary nitrogen of tetrahydrodeoxycytisine (**145**) may be eliminated [63].

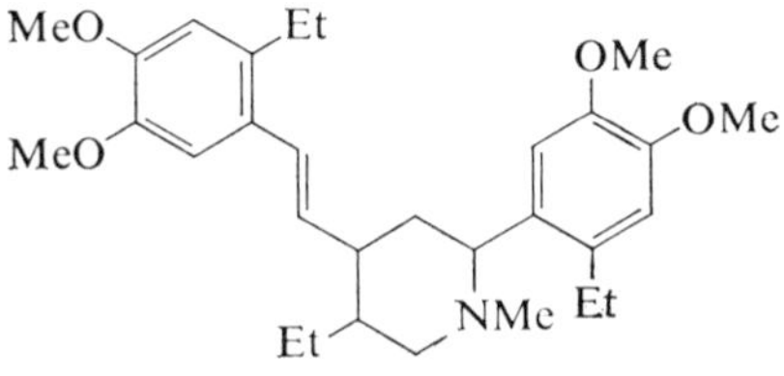

144

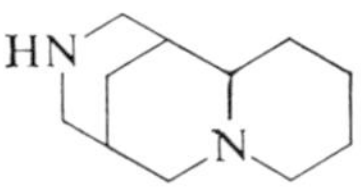

145

The degradation of pyrazoline methiodide (**146**) occurs with ease and takes the expected course, giving unsaturated substituted hydrazones [64]. The degradation of N,N,N'-trimethyl-1,2,3,4-tetrahydroquinoxalinium iodide (**148**) gives the N,N,N'-trimethyl-o-phenylenediamine (**150**), presumably via the enamine **149**, which would be the normal Hofmann degradation product [65].

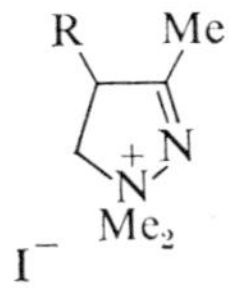

146

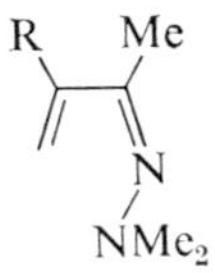

147

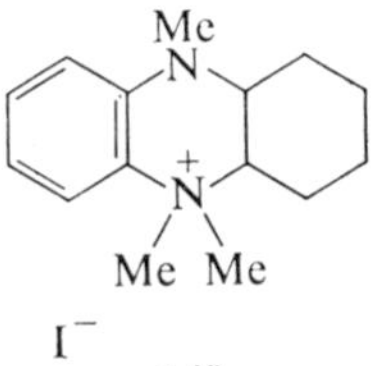

148

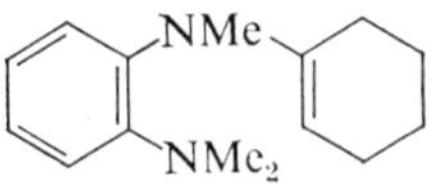

149

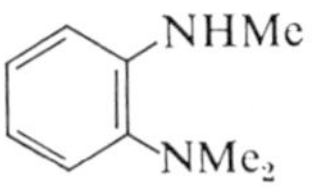

150

The degradation of salts of quinuclidine (**151**) is inhibited by obvious steric factors, but the presence of a suitably placed alkyl group outside the ring system facilitates ring fission, and 2-methyl-1-azabicyclo[2.2.2]octane methohydroxide (**152**) readily gives the piperidine derivative **153** on pyrolysis [66].

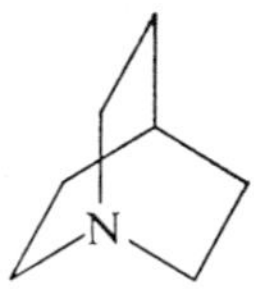

151

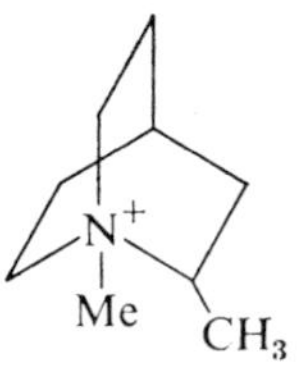

152

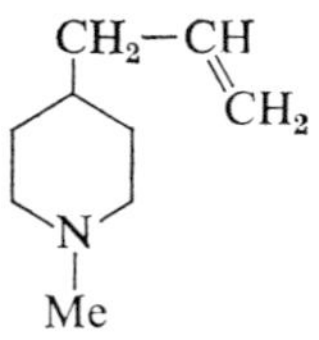

153

Migration of Double Bonds during Degradation

Where the exhaustive methylation of cyclic compounds might be expected to give 1,4-dienes, the alkaline conditions of the reaction frequently result in the migration of one of the double bonds to give a 1,3-diene, and double bonds already present in the molecule may also be moved into conjugation with those produced during the reaction or already present. For example, the exhaustive methylation of dimethylpiperidinium hydroxide (**112**) yields first the unsaturated base **113** and finally the conjugated piperylene **114**, not the 1,4-diene. Degradation of the methiodide of dihydrothebaine-ϕ (**154**) gives the conjugated triene **155** [67], and the nitrogen-free substance **157** is the product of the degradation of de-$N(a)$-emetinehexahydrobismethine (**156**) [68].

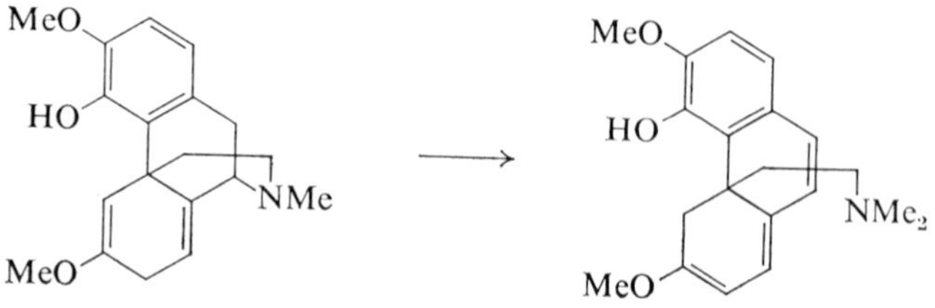

154   155

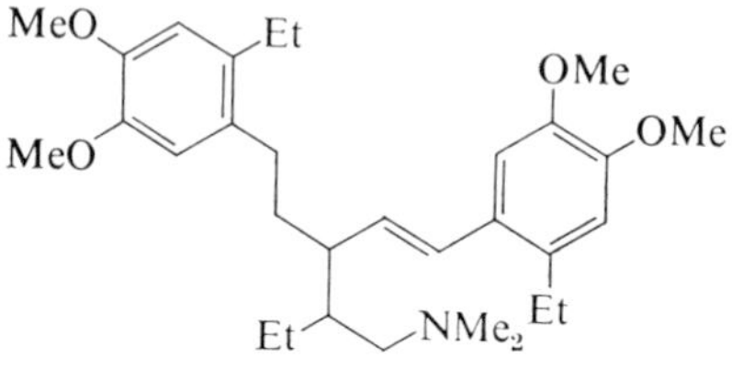

156 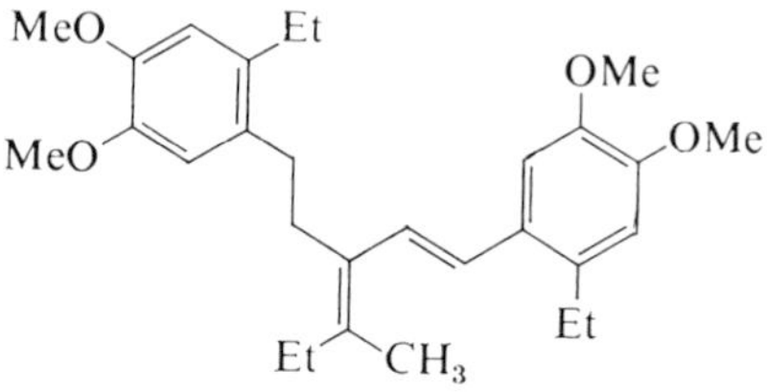   157

Aromatization occasionally becomes possible as a result of fission of the carbon–nitrogen bond by Hofmann degradation, as in the conversion of uleine methiodide (**158**) into the carbazole **159** [69].

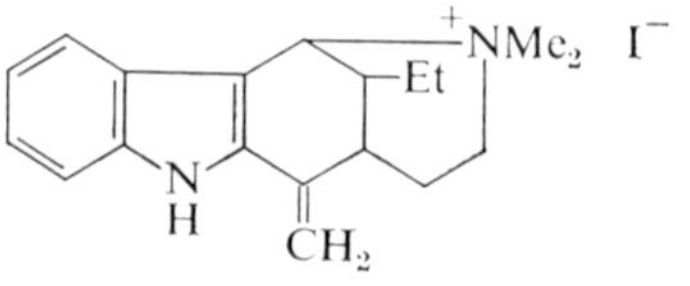

158 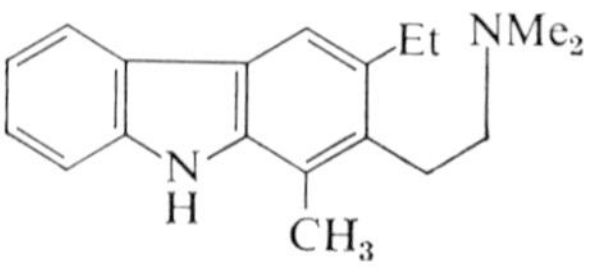   159

In a few cases olefin formation takes place during Hofmann degradation by removal of a proton on the carbon δ or ζ rather than β to the nitrogen.

In such cases the effect is transmitted through double bonds and elimination occurs in the usual manner, though of course such reactions are not governed by steric factors operative in β-elimination. An example of such a reaction is the degradation of des-N-methyltylophorine methine methiodide (160) to the nitrogen-free product 161 as a result of attack by base at the doubly allylic position relative to the nitrogen [70].

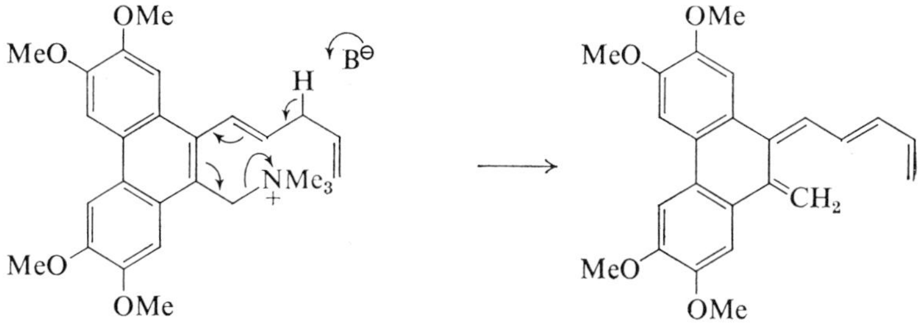

160   161

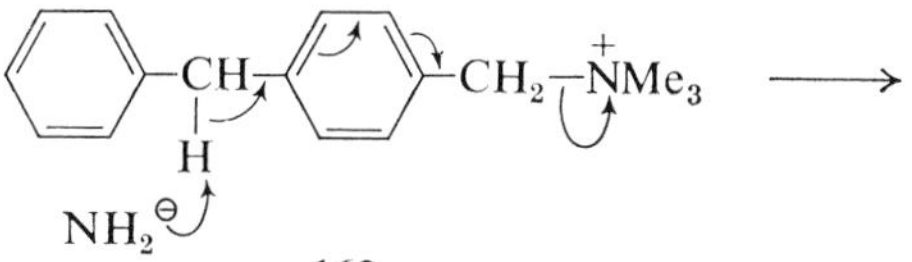

162 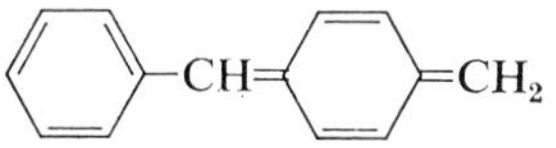   163

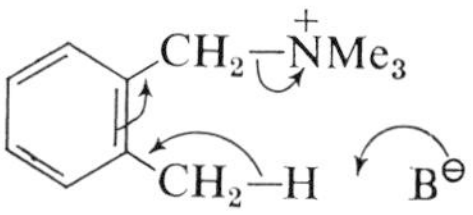

164 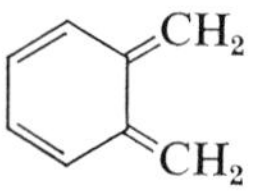   165

Similar processes are presumably involved in the elimination of trimethylamine from quaternary salts of p-benzylbenzylamine (162) [71] and o- and p-methylbenzylamine (164) [72]. The neutral end-products of these eliminations are polymers, doubtless formed from the highly reactive primary products of elimination 163 and 165.

Displacement Reactions

The action of bases on quaternary salts can result in the production of tertiary bases in a manner different from that involved in the Hofmann degradation, and one particular variant of this alternative process almost always accompanies Hofmann degradation and is not infrequently the principal reaction involved. The reaction involves the direct attack by base of one of the carbon atoms directly attached to the nitrogen (167) rather than of a proton attached to the β-carbon (166).

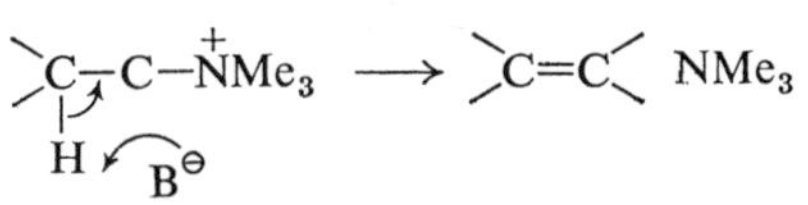

166

Hofmann degradation

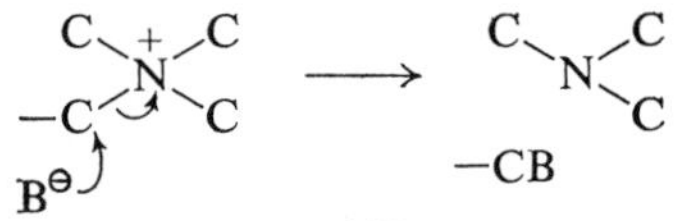

167

displacement

168

$$-CH_2CH_2N\begin{matrix}CH_3\\CH_3\end{matrix} + CH_3OH$$

169

The most common variant of this reaction is the displacement of methanol from a salt bearing a methyl group attached to nitrogen on dry distillation of the methohydroxide **168**, the products being the parent tertiary base **169**, from which the salt was derived, and methanol. This process, where possible, almost invariably accompanies normal Hofmann degradation and is sometimes the only reaction involved. Dihydrothebainone methine methyl ether **(170)** and its dihydro compound behave in this way [73], and N-methyltetrahydroquinoline methohydroxide **(171)** affords only the parent base **172** on attempted degradation under most conditions [14], though conditions giving up to 48% of the elimination product **120a** have been discovered [45].

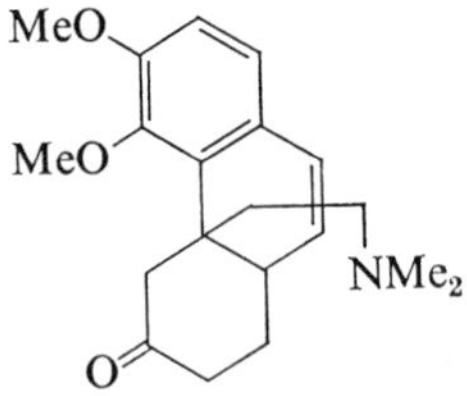

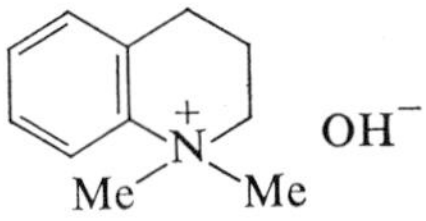

171

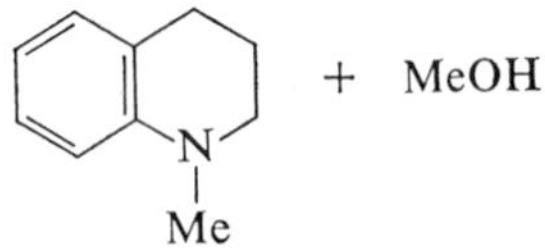

170

172

+ MeOH

The same reaction can, of course, involve attack of a carbon directly attached to nitrogen which is not part of a methyl group, even when methyl groups are present, and in open chain compounds this can lead **(173)** to alcohols or ethers **(174)** and the production of trimethylamine as if successful Hofmann degradation had occurred.

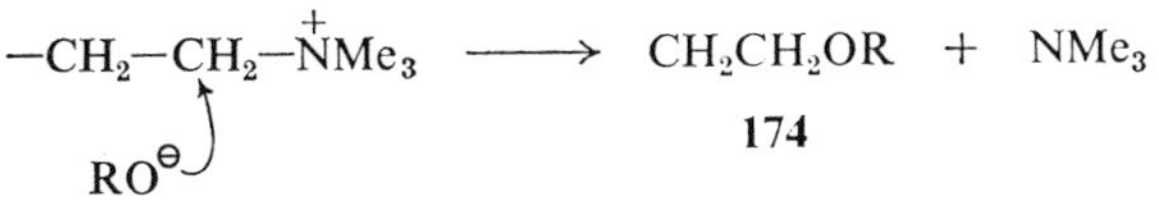

$$-CH_2-CH_2-\overset{+}{N}Me_3 \longrightarrow CH_2CH_2OR + NMe_3$$

$$RO^{\ominus}$$

173   174

Alcoholic, phenolic, or enolized ketonic groups within the molecule can become ionized under the influence of base and then participate in the displacement reaction (**175** → **176**).

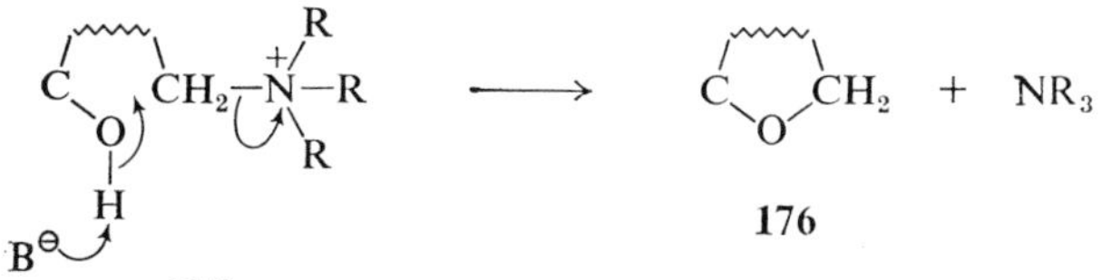

The following are examples of processes of this type:

(*a*) Apparicine methiodide (**177**) on heating with sodium methoxide suffers nucleophilic attack of OMe$^{\ominus}$ to give the tertiary base **178** [74]. Reduction of the quaternary salt with lithium aluminum hydride involves fission of the same C–N bond.

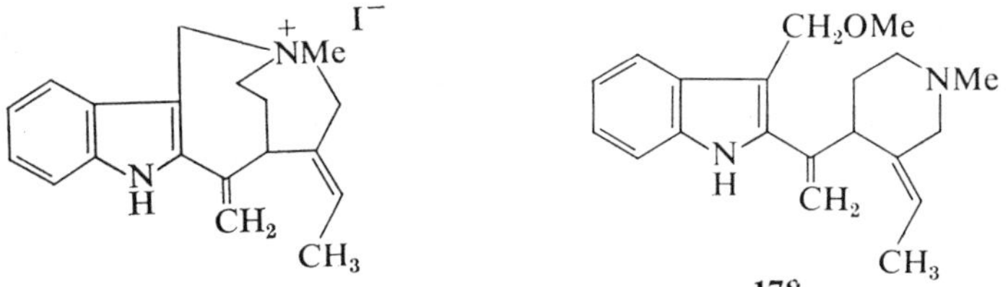

(*b*) γ-Tetrahydrocodeimethine (**179**, R = H) presents an interesting case. On dry distillation the methohydroxide of this base affords the olefins **180** (R = H, Me) and the cyclic ether **181**, together with undegraded base **179** (R = H) and the methyl ether **179** (R = Me) of this. The origin of these substances may be explained by postulating the production of the internal salt **182** under the vigorous alkaline conditions of the degradation; internal attack by the negative oxygen at *a* would afford the cyclic ether **181**, while attack at *b* would give the base methyl ether **179** (R = Me). The quaternary salt of **179** (R = Me) would result from the intermolecular attack by the negative oxygen at the point *b* of another molecule, and normal Hofmann elimination of trimethylamine from the product would

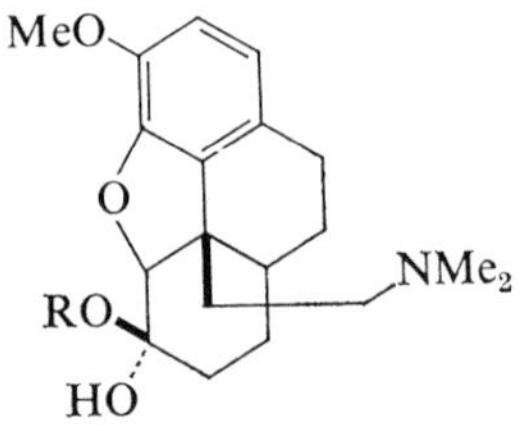

179

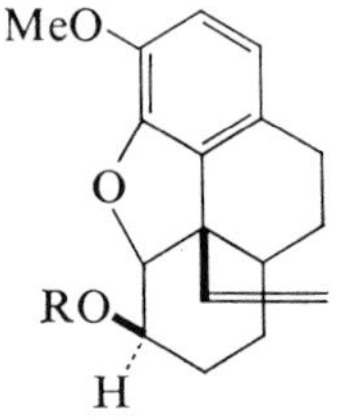

180

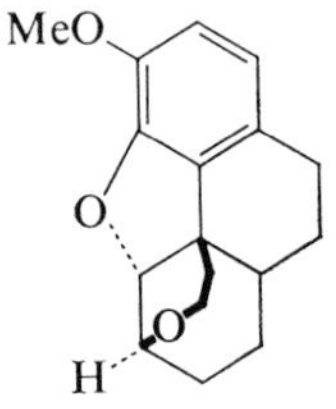

181

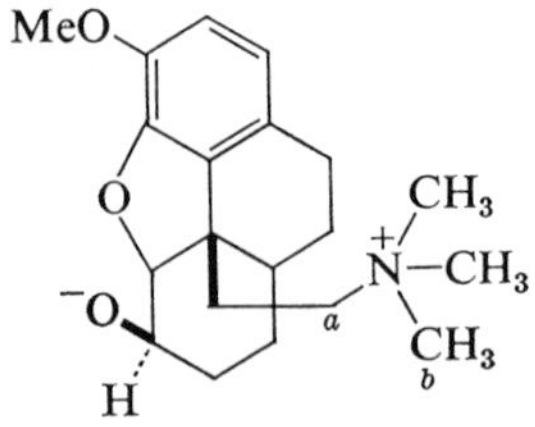

182

then give rise to the olefin **180** (R = Me). It is significant that the degradation of the methohydroxide of **179** (R = H) gives about 30% of undegraded bases, whereas the methohydroxide of **179** (R = Me) gives only about 2–3% of undegraded base [49, 75].

(c) 14-Hydroxydihydrocodeinone methine (**183**) on degradation affords the cyclic ether **184** by a similar process [76]; no careful search of the products of this reaction for compounds, other than the ether **184**, methylated at the hydroxyl group has been made.

(d) In like manner the dihydromethine **185** derived from chelidonine gives the ether **186** on degradation (the production of such an ether has been cited

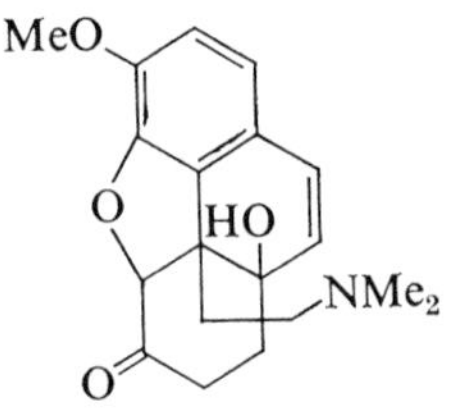

183

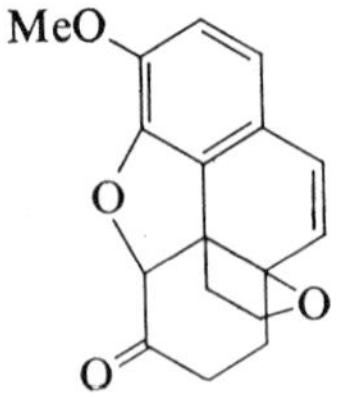

184

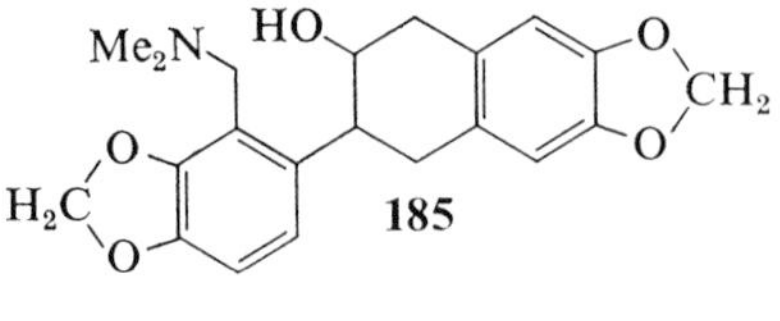

185

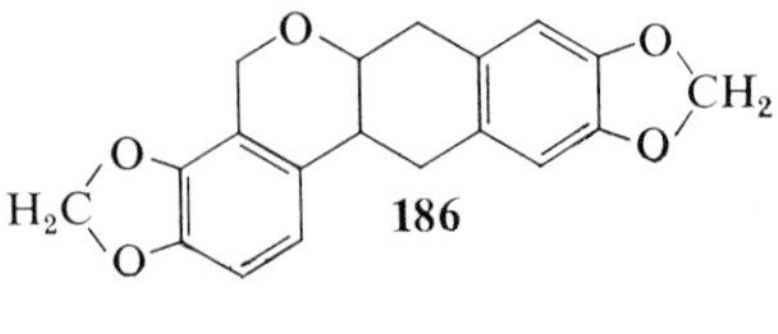

186

as evidence for the position of the hydroxyl group in the parent alkaloid [77])
and the base **187** from α-erythroidine yields the ether **188** [78].

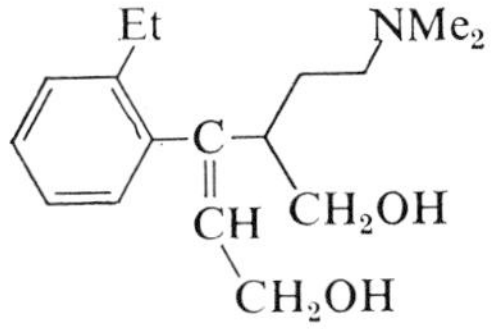

187 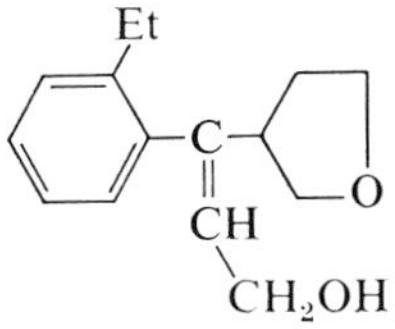   **188**

(*e*) Degradation of N_a-methylnorgelsemine carbinol methohydroxide (**189**)
gives only the unstable trimethylene oxide derivative **190**, which spontaneously
reverts to the salt **189** in the presence of moisture [79].

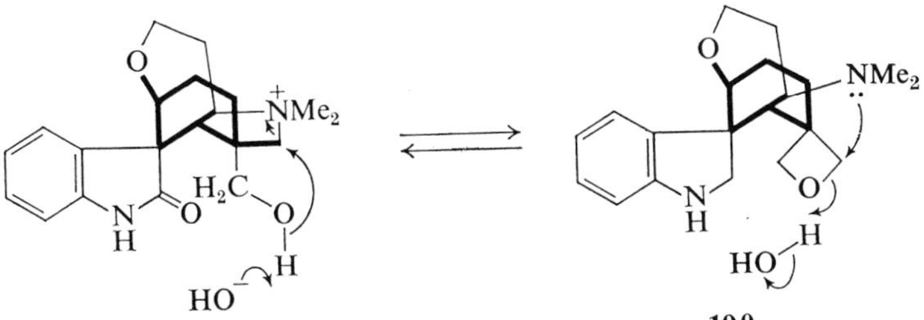

189   **190**

(*f*) When a phenolic hydroxyl group in a molecule can participate in such
a process the degradation may be accomplished very simply. For example,
derivatives of alkaloids of the morphine group bearing a hydroxyl group at
position 4 give cyclic ethers with very great ease in the final stage of exhaustive
methylation, dihydrothebainone dihydromethine methiodide (**191**) giving
thebenone (**192**) simply by boiling with aqueous alkali. This process does not

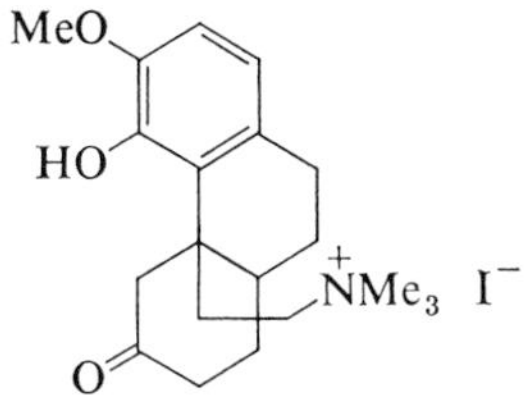

191 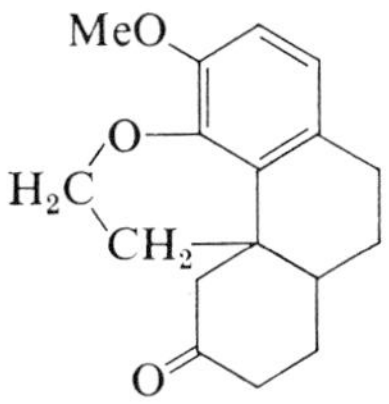   **192**

appear to be accompanied by the displacement of methanol to any extent,
and the yields are very high [46, 67, 80]. The tendency for this process to
occur is so great that a methoxyl group at position 4 is reported to suffer
demethylation [81], a statement that conflicts with an earlier one that
dihydrothebainone methyl ether methine (**170**) and its dihydro compound

suffer only displacement of methanol when subjected to Hofmann degradation [73].

(*g*) Participation of carbonyl groups in such displacement reactions is exemplified by the degradation of quaternary salts of 3β,20α-bisdimethylamino-5α-pregnan-18-al (part structure **193**) to the hemiacetal **194** and of the homologous methyl ketone **195** to the enol ether **196** [82], and by the degradation of cryptopine methosulfate (**197**) to the enol ether **198** as well as to the normal Hofmann degradation product (**199**) [83].

193

194

195

196

197

198

199

200

201

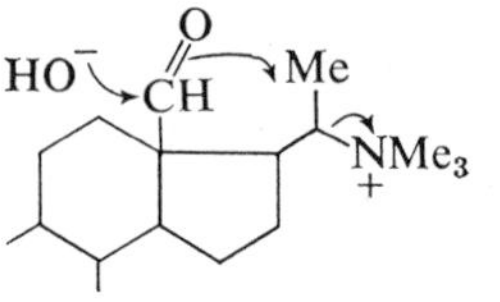

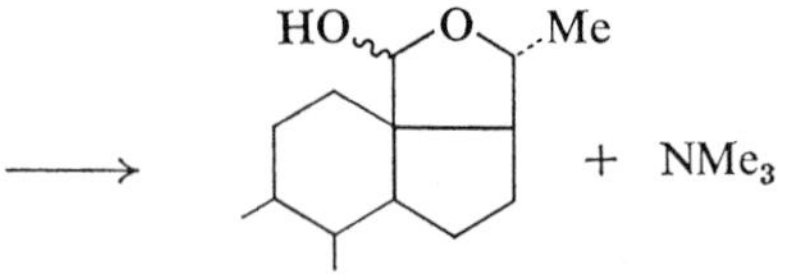

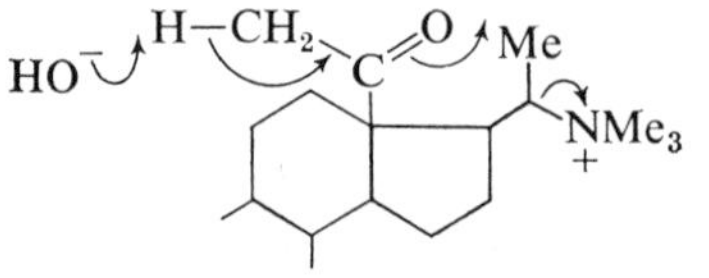

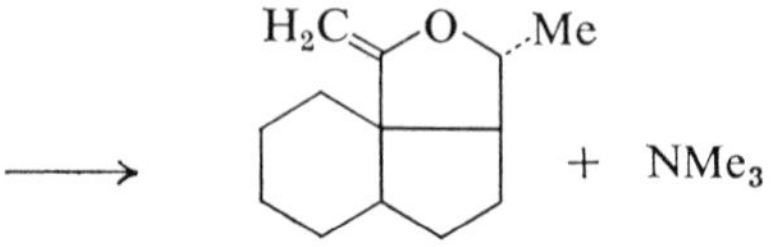

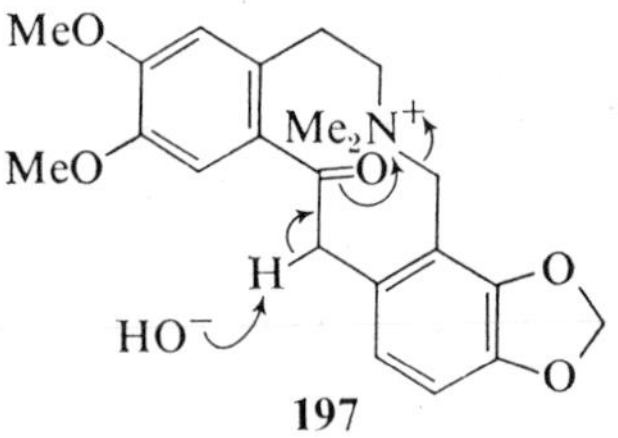

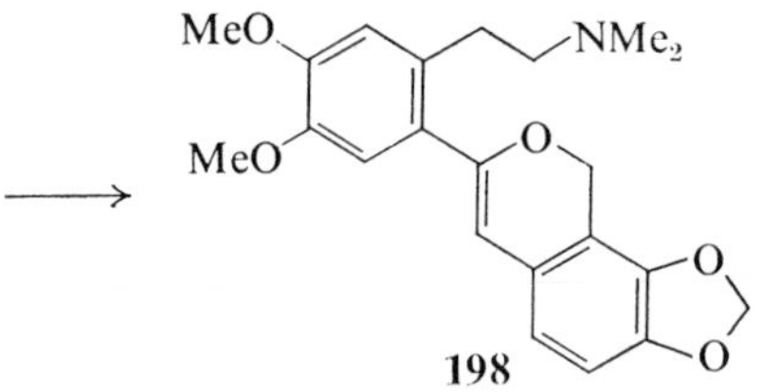

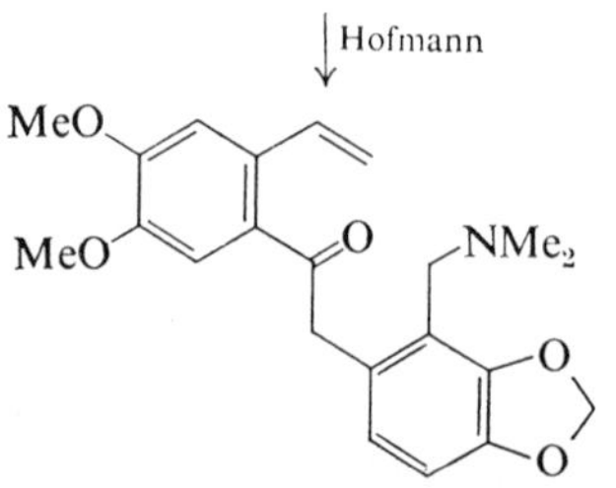

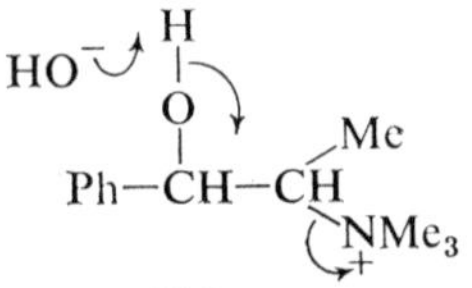

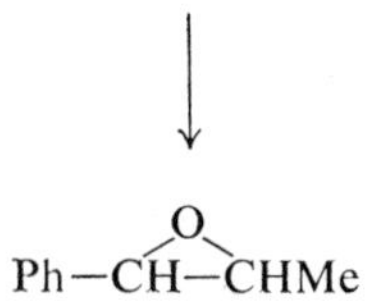

(*h*) Epoxide formation is observed with displacement of trimethylamine when ephedrine quaternary hydroxide (**200**) is heated [84].

(*i*) In many cases, displacement of tertiary base, instead of being the result of intramolecular attack of the α-carbon atom by a negative oxygen, is the result of direct attack of the α-carbon by a hydroxyl ion, the products being a tertiary base and an alcohol. The displacement of methanol from the methohydroxide, with which this process competes, is simply a particular case of attack in this way of one of the four carbon atoms directly attached to the nitrogen. This α-carbon attack is particularly observed when there is no β-hydrogen; for example, the exhaustive methylation of anhydrodihydro-cryptopine-A (**202**) gives the olefin **203** in the first step and the alcohol

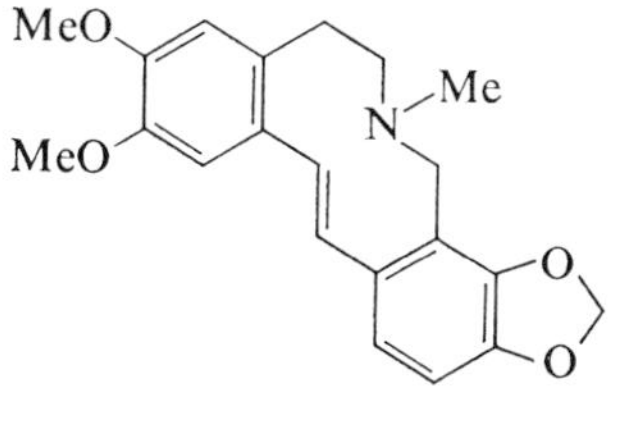

202

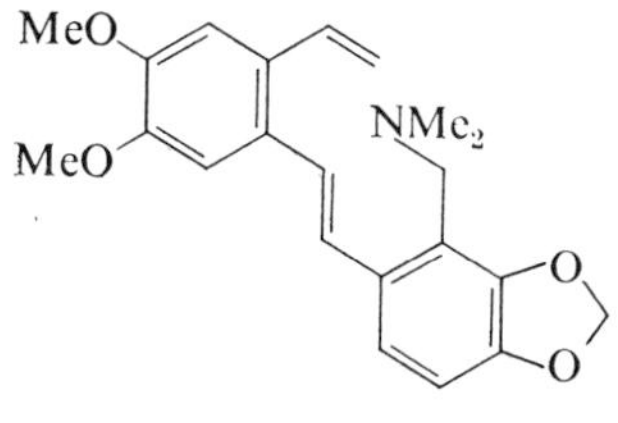

203

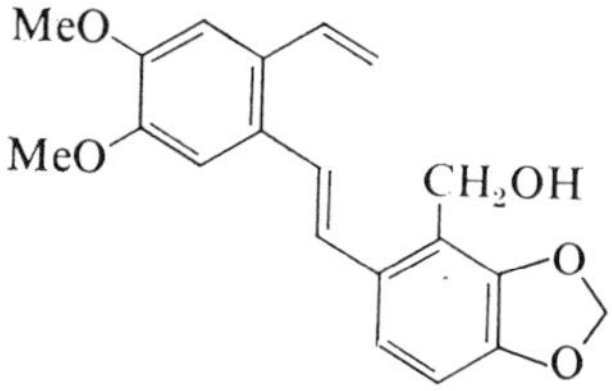

204

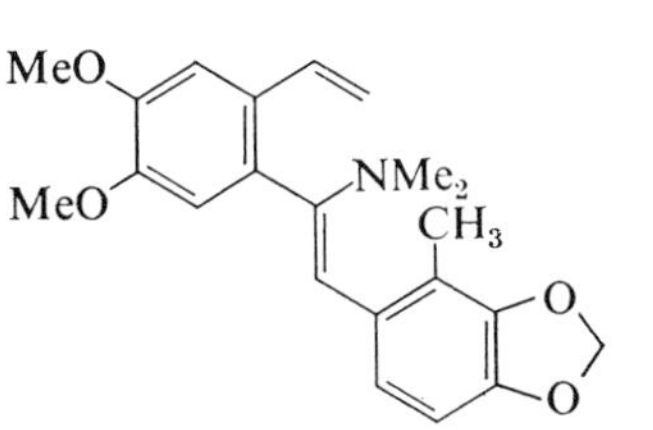

205

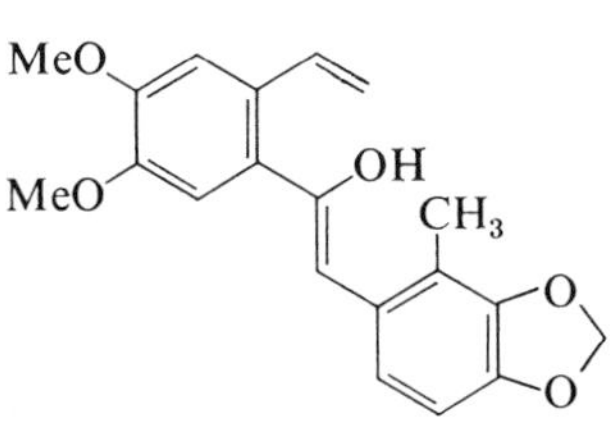

206

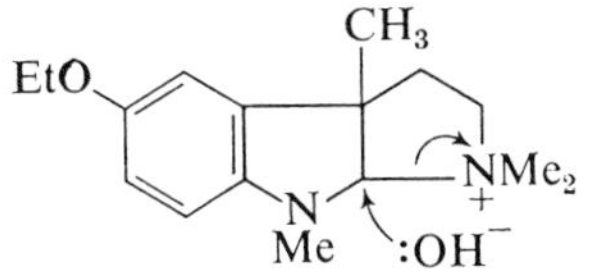

207

207

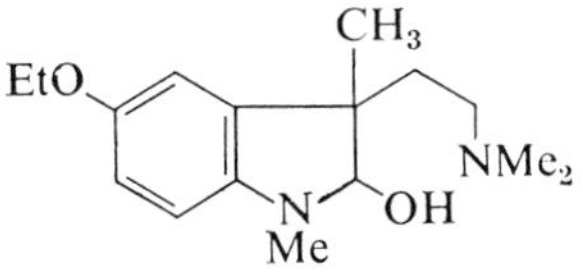

208

α-isocryptopidol (**204**) in the final step [85, 86]. The degradation of dihydro-anhydromethylcryptopine (**205**) yields hydroxycryptopidine (**206**) [83, 85], but this may be regarded as the hydrolysis of a vinylamine. Eserethole methiodide (**207**), on heating with alkali, is also degraded by attack of the α-carbon atom to eseretholemethine (**208**) [87].

Other Pathways for Elimination

It may well be that the products of a single stage Hofmann degradation bear no superficial resemblance to the base from which they were derived, for under the strongly alkaline and vigorous conditions of the reaction other changes may precede or succeed the simple elimination stage, or pathways different from those already mentioned may be available for elimination of tertiary base, involving other reactive centers in the molecule and resulting in molecular rearrangement or the loss of fragments of the molecule.

The group of morphine alkaloids and their derivatives, containing as it does a plethora of reactive centers, provides a number of interesting examples of rearrangements resulting from novel pathways of elimination in the Hofmann degradation. In this series all the rearrangements are observed during the second stage of exhaustive methylation.

α-Codeimethine methohydroxide (**209**) on distillation gives the fully aromatic phenanthrene derivative methylmorphenol (**211**) [88-90] by dehydration and elimination of ethylene, trimethylamine, and water as shown in **209 → 211**. Flavothebaone trimethyl ether methine methohydroxide (**212**) under the strongly alkaline conditions is first transformed into the ψ-methine

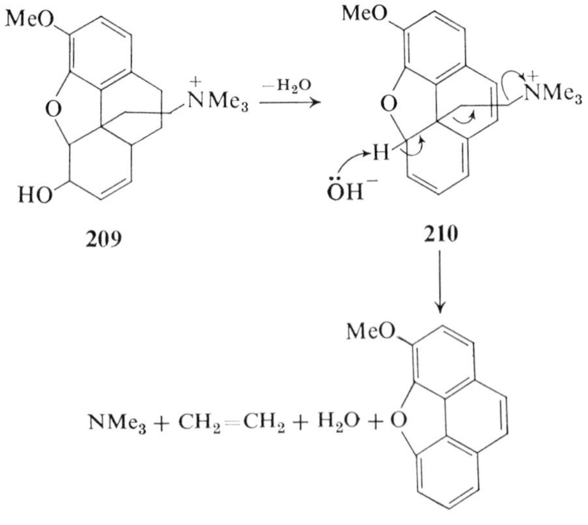

$$NMe_3 + CH_2\!=\!CH_2 + H_2O + O$$

211

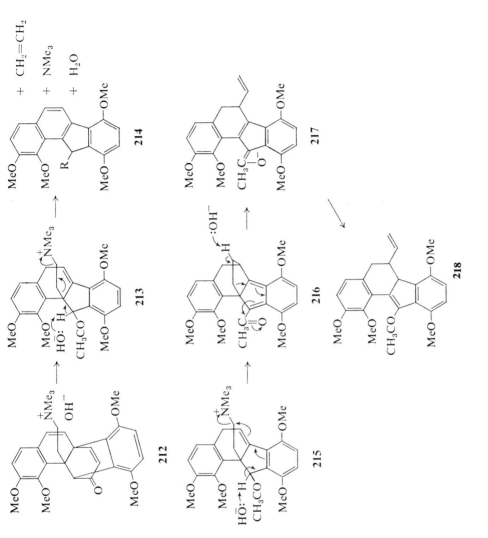

derivative 213, which then suffers elimination of ethylene, trimethylamine, and water to give the benzofluorene 214 (R = $COCH_3$), which under the alkaline conditions is converted partly into 214 (R = H) and, by base-catalyzed methylation with the methanol displaced from the methohydroxide 213, partly into 214 (R = CH_3) [91]. When the conditions of this degradation are modified slightly, the alternative elimination of trimethylamine outlined in 215 → 216 occurs, with the participation of the aromatic nucleus, and the initial product 216 is further transformed through 217 into the olefin 218 [91].

The degradation of tazettine (219) provides a further interesting example of processes preceding and succeeding normal Hofmann elimination. Degradation of the methiodide affords the dimethylglycine ester 221 of 6-phenylpiperonyl alcohol and the alcohol itself. These can also arise by normal Hofmann degradation to the methine base 220, followed by base-catalyzed elimination of the methoxide ion to give the ester 221, hydrolysis of which would then give the free alcohol. The elimination of methoxide ion could, of course, precede the Hofmann reaction. The exhaustive methylation of O-methyltazettine (222), which lacks the hemiketal hydroxyl group, yields the ester 227 and the alcohol 225, and these may be regarded as arising by normal Hofmann degradation to a methine base 223 which, after quaternization, suffers removal of the most acidic proton in a base-catalyzed elimination of trimethylamine 224 to give the enol ether 226, which can be hydrolyzed by acids during isolation to give the ester 227 and/or the alcohol 225 [92, 93]. Although other explanations have been suggested [94], it is clear that all of the Hofmann degradations of tazettine and its derivatives can be explained by assuming normal Hofmann elimination to methine bases that react with acidic or basic media to give further elimination, hydrolysis, or rearrangement products [93].

Whereas degradation of laudanosine gives a stilbene as stated above, and the degradation of narcotine methiodide (228) affords the keto acid narceine (231) via the enol lactone 229 [95], degradation of the methiodides of the two diastereoisomeric hydroxylaudanosines (232) affords veratric aldehyde

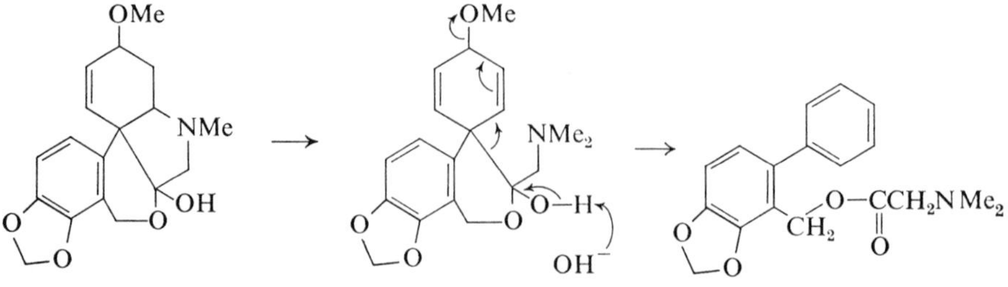

219   220   221

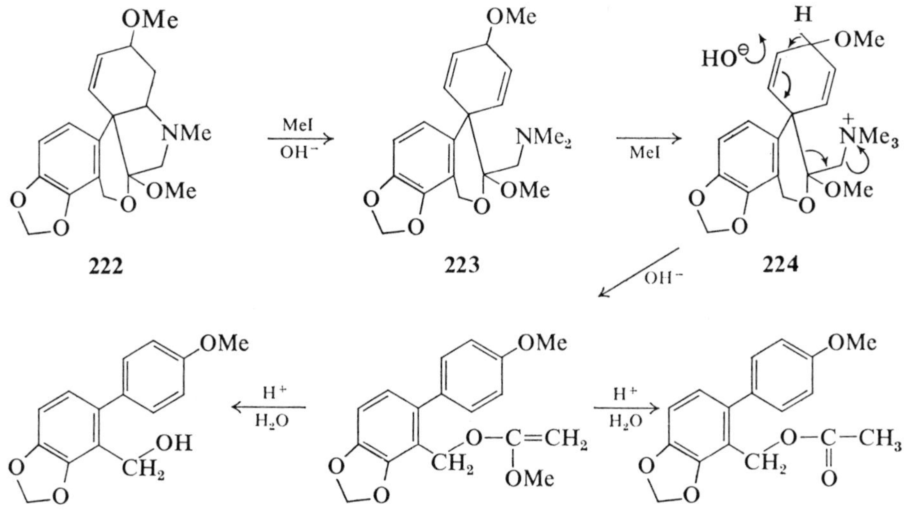

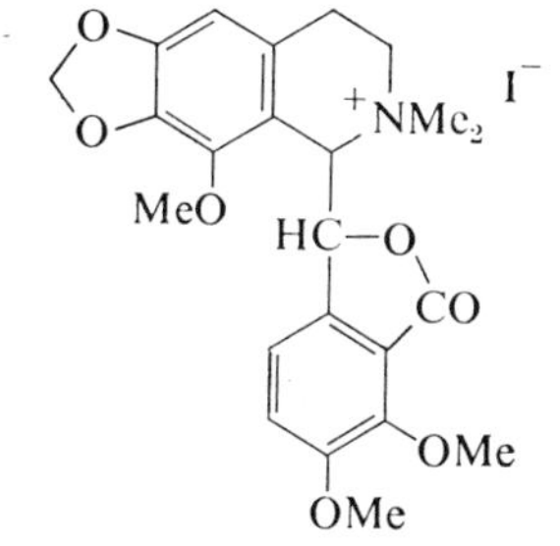

222     $\xrightarrow[\text{OH}^-]{\text{MeI}}$ 223 $\xrightarrow{\text{MeI}}$ 224

$\text{HO}^{\ominus}$

225     $\xleftarrow[\text{H}_2\text{O}]{\text{H}^+}$ 226 $\xrightarrow[\text{H}_2\text{O}]{\text{H}^+}$ 227

$\xrightarrow{\text{OH}^-}$

228

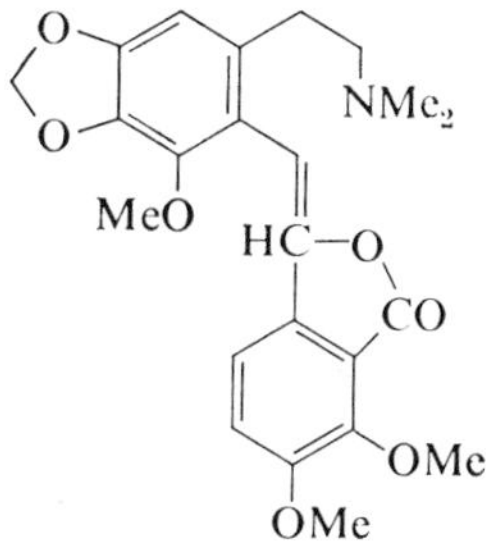

229

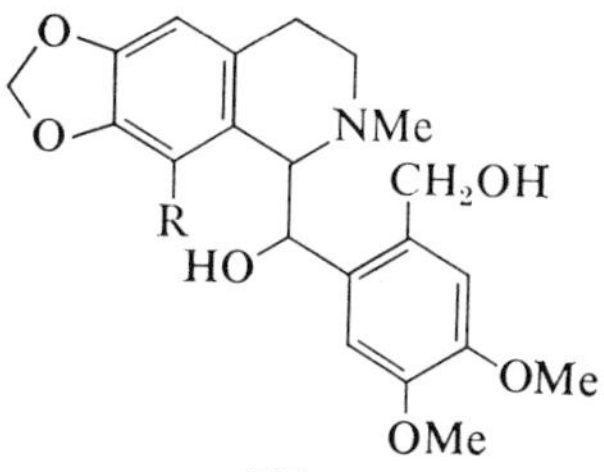

230

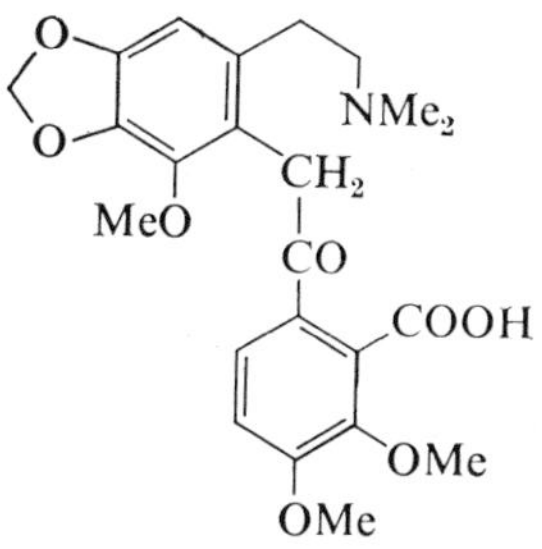

231

(234) and the styrenoid base 235 [96]. The difference in behavior of narcotine and the hydroxylaudanosine methiodides probably arises from the fact that the first proton to be removed from 232 would be proton a, leaving a negative oxygen that would effectively inhibit removal of proton b on the neighboring carbon atom. Removal of proton b is a necessary first step in a degradation analogous to that of narcotine methiodide, in which removal of proton b is

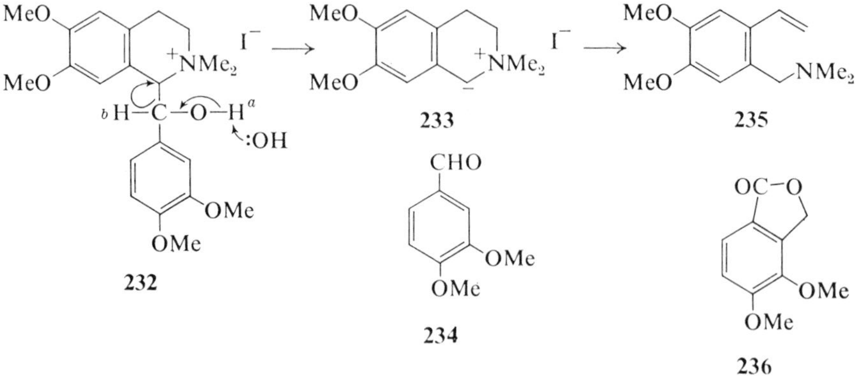

232   233   234   235   236

not inhibited, since the oxygen atom can only acquire a negative charge in that case by opening of the lactone ring, a process that occurs less rapidly than the Hofmann elimination. The whole process is governed by the ease of formation of the transitory ylid 233, which is increased in the quaternary salt, relative to the base, by the positive charge on the nitrogen atom; β-hydroxylaudanosine itself is stable in refluxing methanolic potassium hydroxide [97]. Entirely similar fissions of the methiodides of narcotine diol 230 (R = OMe) [97–99] and hydrastine diol 230 (R = H) [97] also occur on Hofmann degradation though in these cases the initially-formed aldehyde suffers Cannizzaro reaction or aerial oxidation to m-meconine (236) [97, 98]. Ketolaudanosine methiodide is degraded to the base 235 and veratric acid [97].

An interesting example of the complexity of mechanism that may be involved in an apparently simple elimination of trimethylamine from a quaternary salt under extremely mild conditions is provided by the degradation of the methohydroxide 237, which occurs spontaneously at room temperature [100]. The product is an optically active phenylnaphthalene (243) which is racemized by heat. Two essentially similar mechanisms involving attack of protons α or β to nitrogen (237 → 242 → 240 and 245 → 244 → 240) can be advanced for the rearrangement. In each case the rearrangement is an essential feature of the elimination of base, and elimination is an essential feature of rearrangement. The parent base is stable to alkali under a wide variety of conditions.

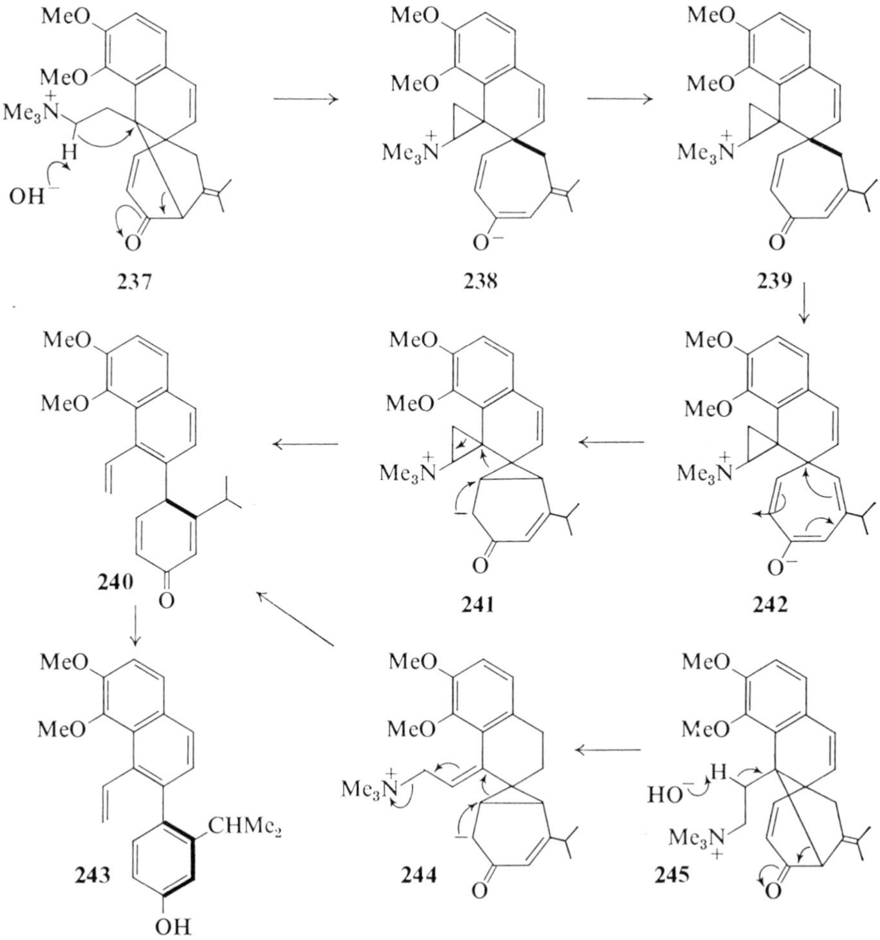

237   238   239

240   241   242

243   244   245

Fragmentation Reactions

Quaternary salts of pyrrolidine undergo fragmentation when treated with strong bases such as phenyllithium. In simple cases the heterocyclic ring is cleaved with the production of an olefin and an enamine, which may be hydrolyzed to an aldehyde and a secondary base. For example, the salt **246** gives ethylene and vinyldimethylamine (**247**), which gives acetaldehyde (**248**) on hydrolysis. With increasing substitution other reactions compete more

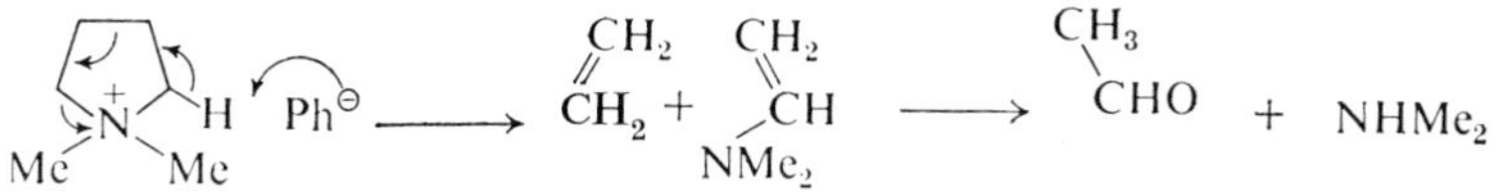

246   247   248

successfully with this fragmentation process, these being Hofmann elimination, S_N2 substitution and, in suitably substituted compounds, Sommelet-
Hauser rearrangement. Three of these processes occurring when 1,1-dimethyl-

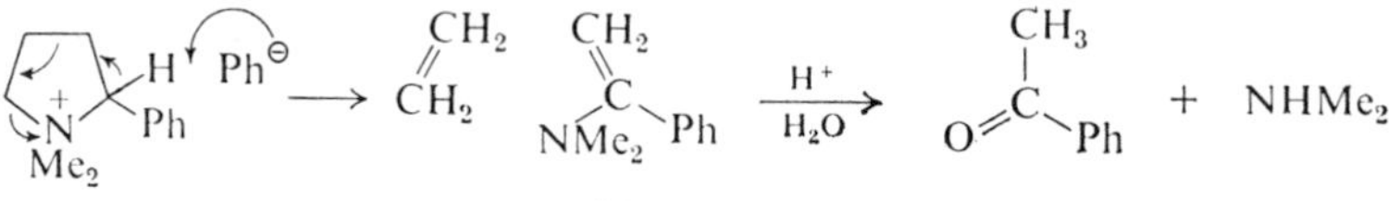

249   **250**
                              Fragmentation

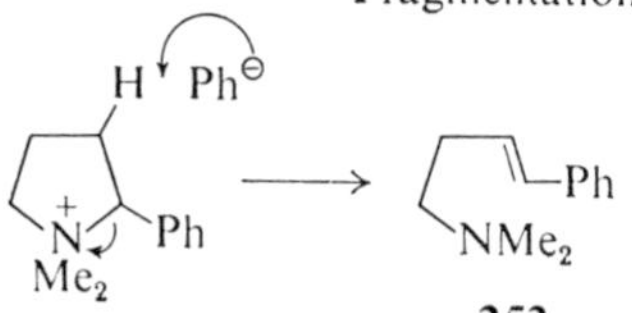

251
                   Hofmann degradation

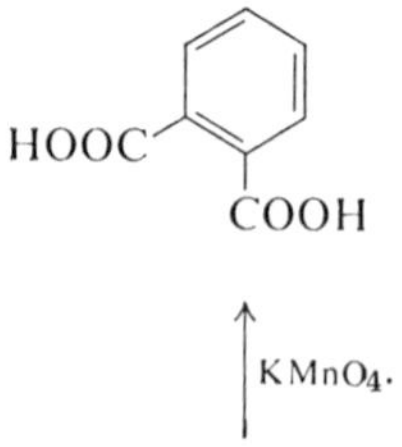

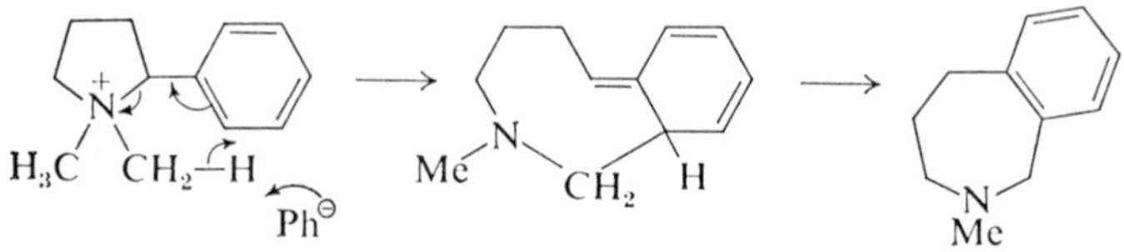

253
                Sommelet-Hauser rearrangement   **254**

2-phenylpyrrolidinium iodide is treated with phenyllithium are illustrated in
formula **249–254**.

The presence of a second bond in the 3,4-position in quaternary salts of
Δ^3-pyrrolene ensures that breakage of the carbon chain at that point does not
occur when such salts are treated with bases, the products being unsaturated
enamines as shown in formulas **255 → 257** [101–103].

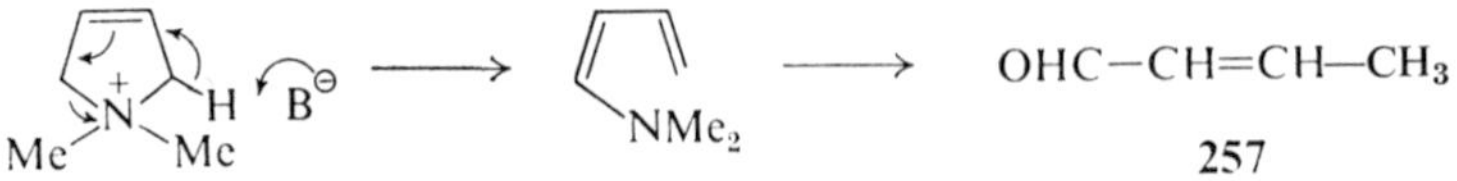

255   **256**

Sommelet-Hauser Rearrangement

Quaternary salts of benzylamines frequently undergo rearrangement in
the presence of strong bases (NH_2^-, Ph^-) in a process that involves ylid
formation and nuclear substitution (**258 → 260**) [71, 104–107]. In most

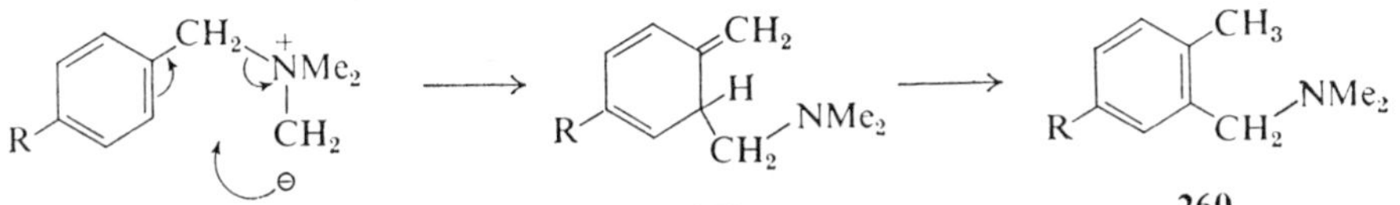

258   259   260

cases, where possible, the reaction competes with Hofmann elimination though exceptions to this generalization are not uncommon. For example, the salt **262** (R = H) gives only the rearrangement product **263** whereas its methyl ether **262** (R = Me) gives the elimination product **261** [108].

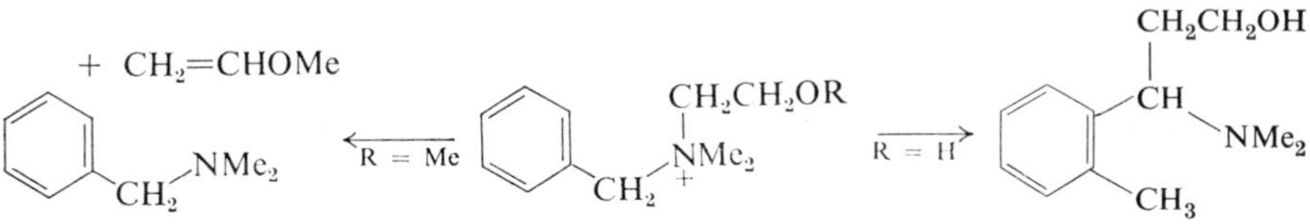

261   262   263

The closely related derivatives of piperidine (**264**) and morpholine (**265**) differ in that the former undergoes rearrangement and the latter β-elimination [109]. The salt **266** suffers β-elimination to give amine and the olefin **268**, S_N2 substitution to give amine and the alcohol **267**, and rearrangement

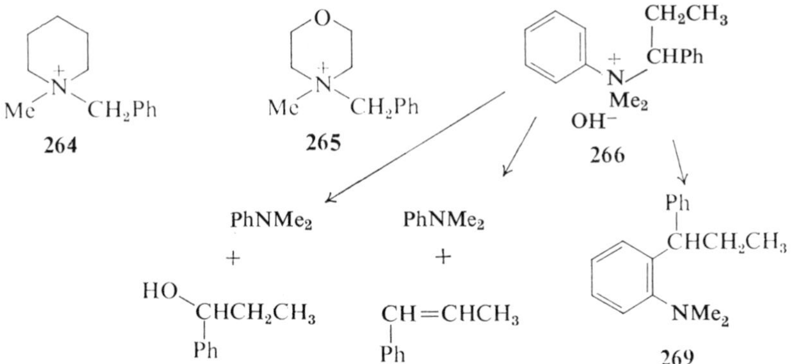

264   265   266

PhNMe₂   PhNMe₂
+   +

267   268   269

to the amine **269** through the most easily formed ylid [110]. The effect of substitution in the benzene ring on the relative ease of elimination and rearrangement has been studied. In general, methoxyl groups in the 2- and 4-position favor rearrangement almost to the exclusion of elimination, 3-methoxyl has no effect, and 2- or 3-fluorine substitution favors elimination [111].

Degradation of Amine Oxides

Tertiary amine oxides on pyrolysis [14], or at room temperature in dimethyl sulfoxide in some cases [112], readily give olefins and substituted hydroxylamines as a result of a cyclic elimination process involving the β-hydrogen

270 **271**

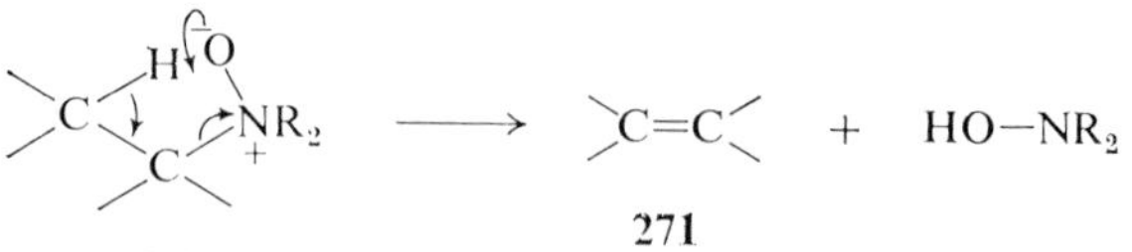

(**270**–**271**). The mechanism, as would be expected, involves *cis* elimination, as is shown by the production of *cis*- (**273**) and *trans*- (**275**) 2-phenylbut-2-ene by the degradation of the oxides of *threo*- (**272**) and *erythro*- (**274**) 2-dimethylamino-3-phenyl butane [112, 113]. The eclipsed transition state **272** from the

272 **273**

274 **275**

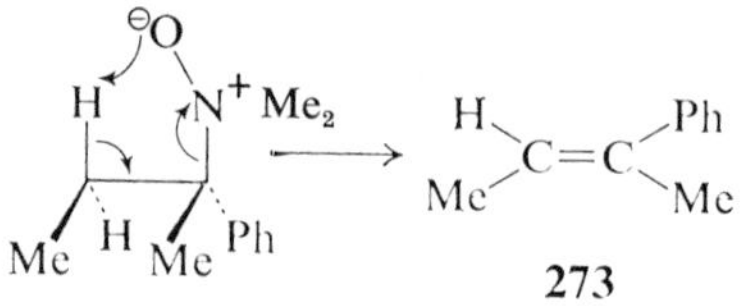

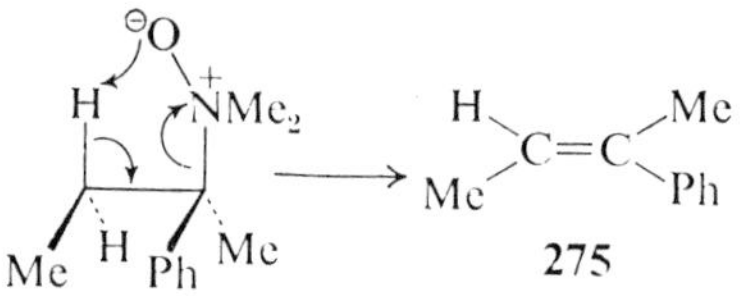

threo form involves less steric hindrance than that from the *erythro* isomer **274**, and degradation of the former takes place more readily than that of the latter [113].

In open-chain systems the direction of elimination is governed by the number of β-hydrogen atoms; for example, the oxide **276** gives more than twice as much propylene (**278**) as ethylene (**277**) on pyrolysis [114].

276

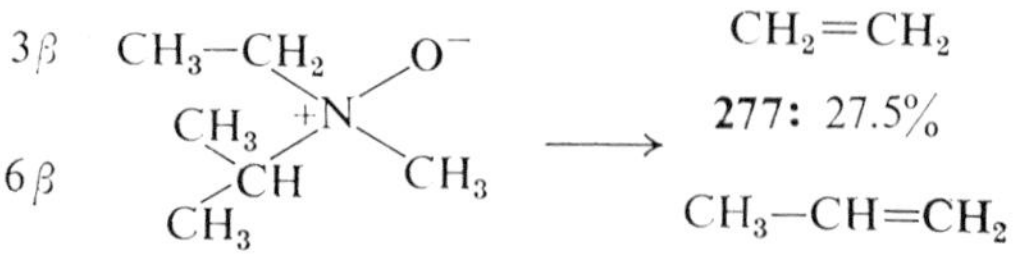

In cyclic systems the direction of elimination is determined by steric factors and the availability of *cis*-β-hydrogen. For example, *trans*-1-dimethyl-amino-2-phenylcyclohexane oxide (**279**), in which *cis* elimination can occur on both sides of the nitrogen atom, affords mainly 1-phenylcyclohexene (**280**), whereas the *cis*-amine oxide **282** can only suffer *cis* elimination to give 3-phenylcyclohexene (**281**), and in this case the 2% of isomeric olefin **280** probably arose from traces of the *trans*-amine oxide **279** in the starting materials [115].

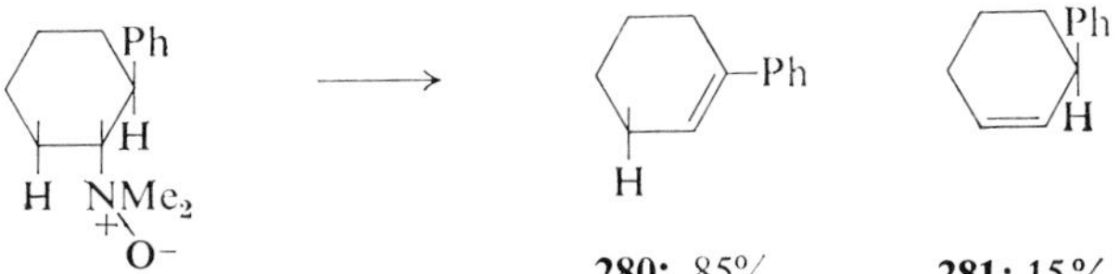

279   280: 85%   281: 15%

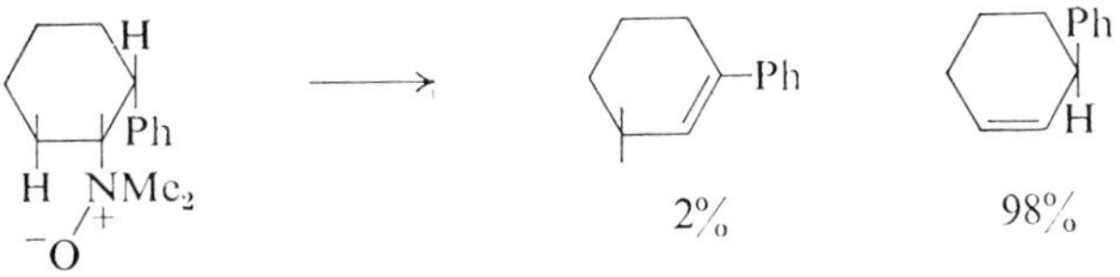

282   2%   98%

In the steroid field elimination occurs in the oxides of 6β- and 7α-dimethyl-amino cholestanes (**283**) to give exclusively cholest-6-ene, since with the axial 6β- and 7α-amines the alternative elimination to cholest-5-ene and cholest-7-ene could not proceed through a *syn* periplanar transition state [26]. The 6α- and 7β-amine oxides can, however, suffer elimination by a *cis* cyclic mechanism on either side **283a** of the nitrogen.

The conditions of the reaction are relatively mild and side reactions are very rare. The migration of double bonds into conjugation with the unsaturated systems is not generally observed.

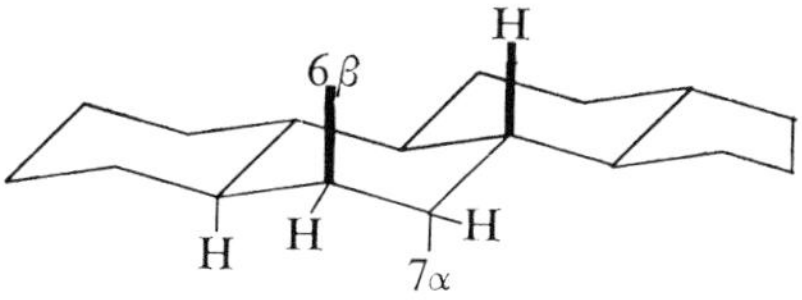

283 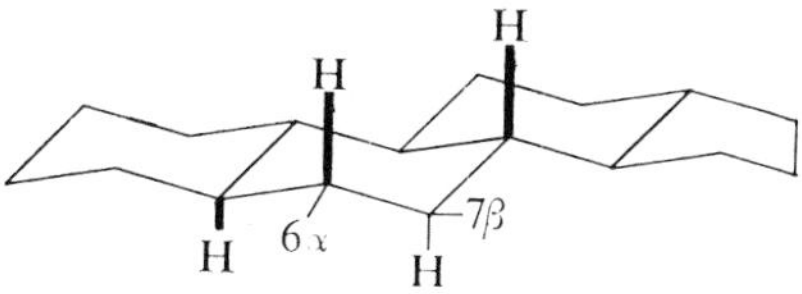   283a

Six-membered heterocyclic amine oxides do not suffer elimination unless there is a suitable β-hydrogen in a side chain. For example, *N*-methyltetra-hydroisoquinoline *N*-oxide (**284**) gives tetrahydroisoquinoline and formaldehyde on pyrolysis [14], whereas the *N*-oxide of laudanosine (**133**) yields the olefinic hydroxylamine **285** [97]. Hydroxylaudanosine *N*-oxide (**232**, NMe$_2$ = NMe → O) suffers fission to veratric aldehyde (**234**) by a process similar to that involved in Hofmann degradation [97].

Owing to the limitations imposed on this reaction by the mechanism, a study of the application of amine oxide pyrolysis to the cases in which Hofmann degradation involves rearrangement or fragmentation was made. It was hoped that the cyclic elimination involved would, on geometric grounds, be likely to leave these reactive centers unattacked. The utility of this process appears to be limited to open-chain tertiary amines, since such internal attack of the β-hydrogen in cyclic bases is geometrically prohibited,

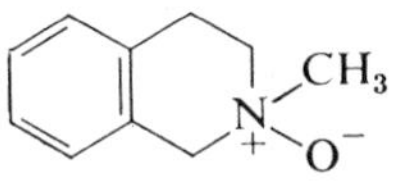

284

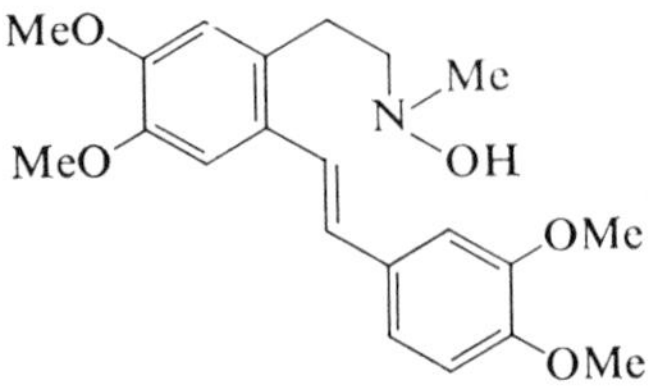

285

and it has so far been applied in the alkaloid series mainly to derivatives of the morphine group. The alkaloids of this group do, however, provide a particularly suitable field for the study of this process in view of the facility with which they undergo molecular rearrangement under the conditions of the normal Hofmann degradation.

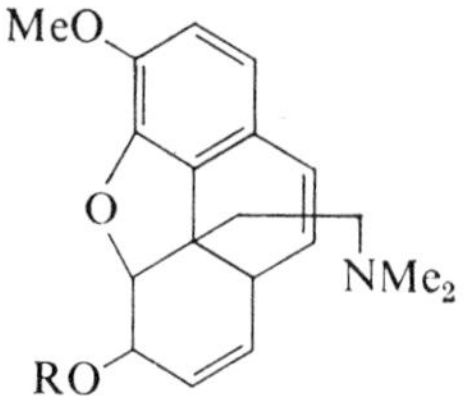

286

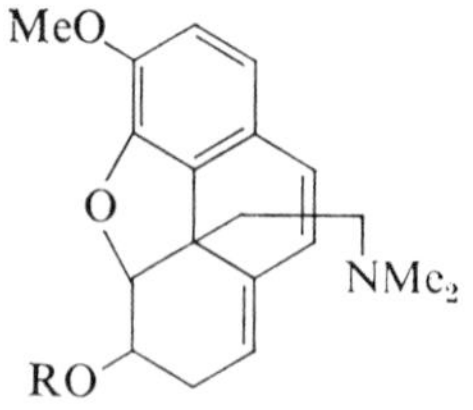

287

Degradation of the *N*-oxide of α-codeimethine (**286**, R = H), β-codei-methine (**287**, R = H), and their methyl ethers (**286**, R = Me; **287**, R = Me) gives, respectively, the vinyl compounds **288** and **289** (R = H) and

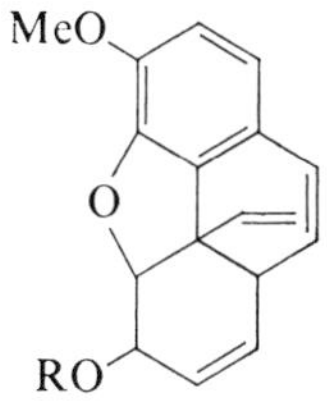

288

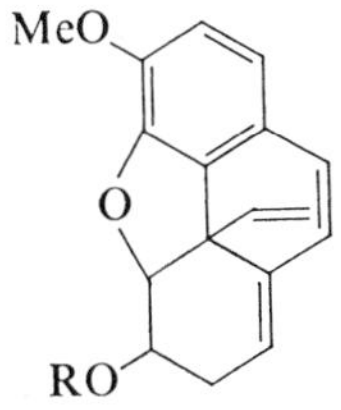

289

288 and **289** (R = Me), and the degradation of flavothebaone trimethyl ether methine (**290**) and the ψ-methine **292** via the *N*-oxide likewise yielded the vinyl compounds **291** and **293** [116]. None of these nitrogen-free substances is accessible by the standard Hofmann degradation procedure. The degradation of reduced bases derived from the above also proceeded satisfactorily, the yield in all cases being good.

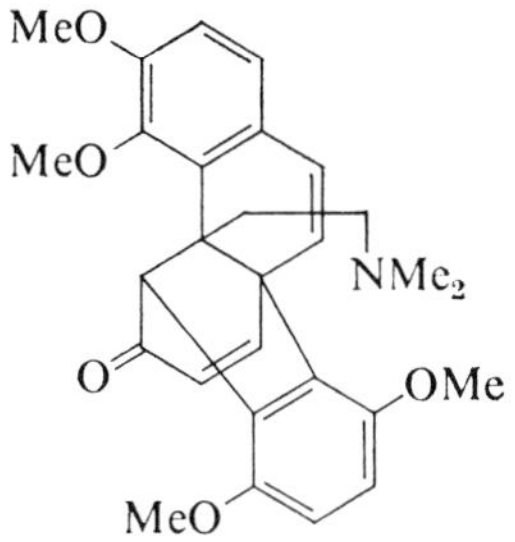

290

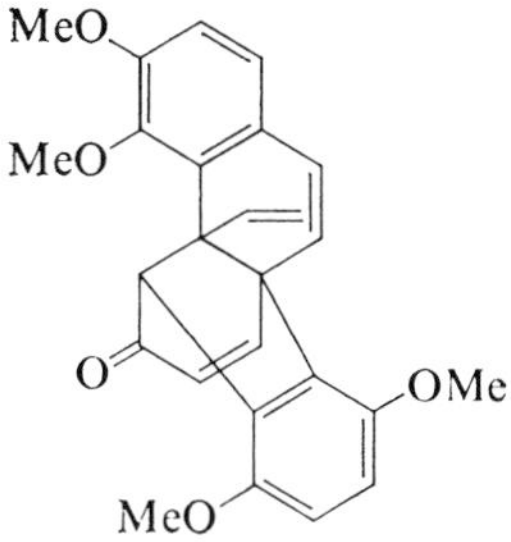

291

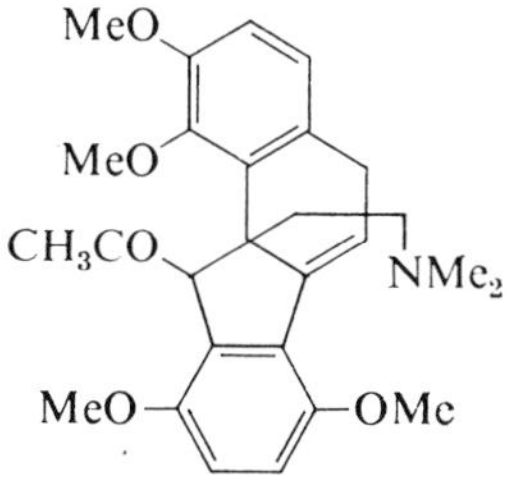

292

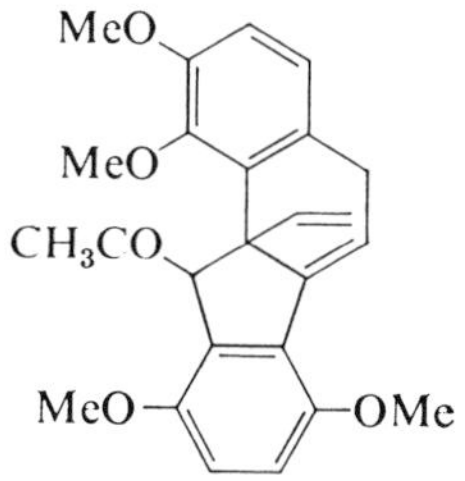

293

14-Hydroxydihydrocodeinone methine (**294**) degraded via the *N*-oxide affords the vinyl compound **295** instead of the cyclic ether **184** which is the product of the classical process; dihydrothebainone dihydromethine, however, furnished thebenone (**192**), as in the Hofmann method, this being the only case so far encountered in which the process yielded a recognizable product other than the expected olefin. The method is not generally suitable for the degradation of phenols, which are extensively oxidized during the pyrolysis of the amine oxide [116].

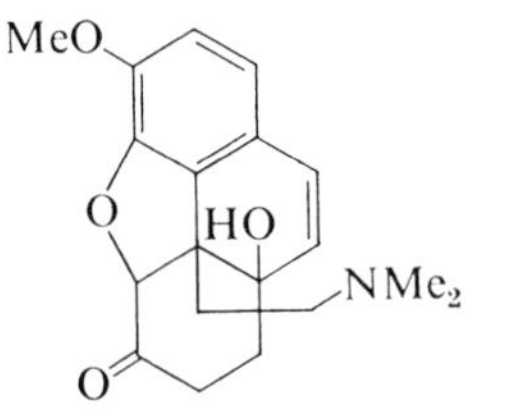

294 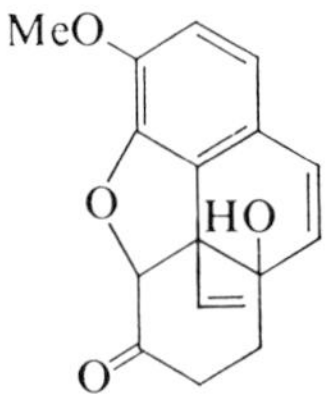   295

Emde Degradation

In a number of cases the normal Hofmann degradation fails to bring about ring fission of cyclic amines, and in such instances reduction of the quaternary salt can often achieve this end. This reaction, generally known as the Emde degradation [117, 118], may be achieved by sodium amalgam, catalytically, or by sodium and liquid ammonia. For example, Hofmann degradation of

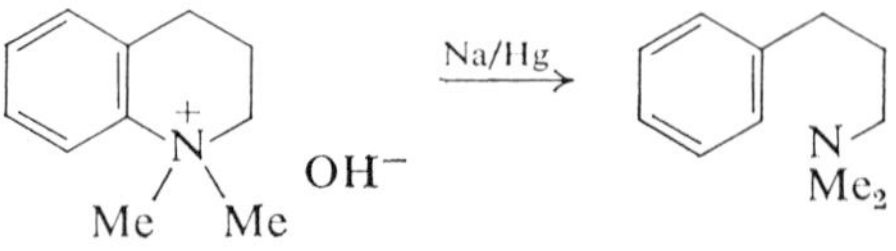

296   297

N-methyltetrahydroquinoline methohydroxide (296) generally results in regeneration of the parent base with displacement of methanol, but reduction of the salt with sodium amalgam affords the ring-opened amine 297. Reduction of N-methyltetrahydroisoquinoline methiodide (298) with sodium amalgam yields the styrene 299 produced by the Hofmann degradation, but catalytic reduction of the same salt gives the β-tolylethylamine (300), which

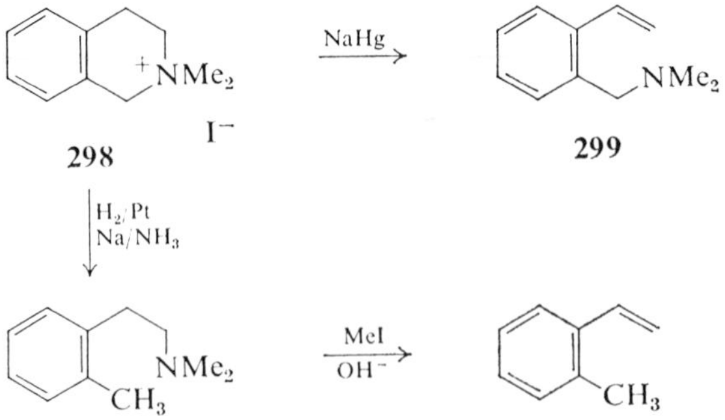

298   299

300   301

can subsequently be converted into the nitrogen-free product 301 by Hofmann degradation [117, 118].

The reduction of quaternary salts of tetrahydroisoquinolines by sodium and liquid ammonia takes the same course as the catalytic reduction, yielding β-tolylethylamines. But if other reactive systems are present, additional changes may occur; for example, hydrocotarnine methiodide (**302**) affords the base **303**, the methylenedioxy group being affected at the same time [119].

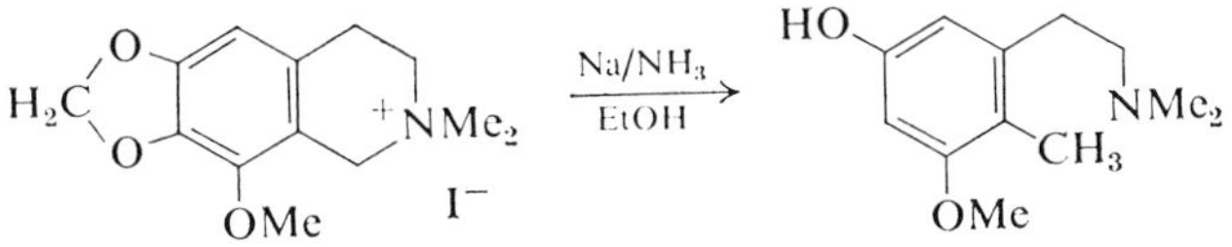

302

303

Reduction of the thebaine methiodide (**304**) with sodium and liquid ammonia yields the base **305**, the methiodide of which on further reduction with the same reagent gives the nitrogen-free substance **306** [120]. Successive

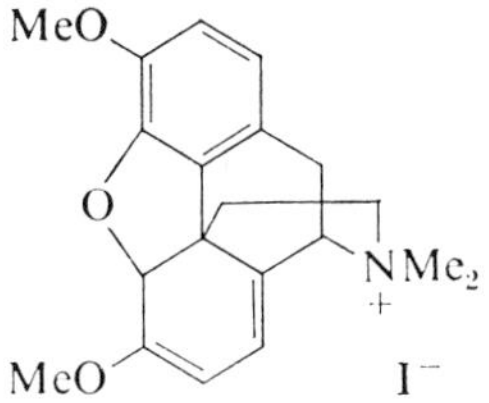

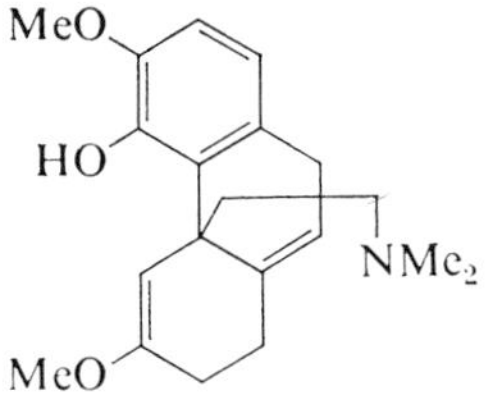

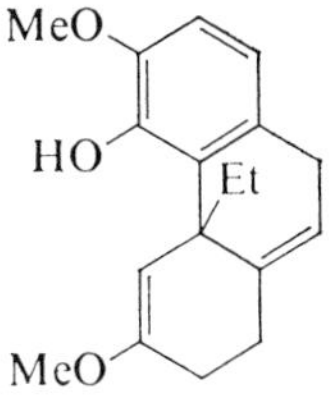

304

305

306

degradations of anhydrolycorine methiodide (**307**) by the Emde process yielded the bases **308** and **309** and the nitrogen-free substance **310** [121].

For Emde reactions of β-carbolines see the following subsection.

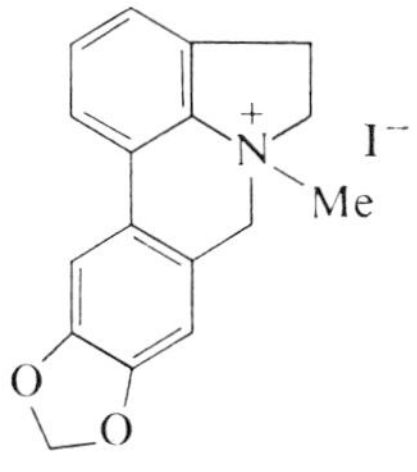

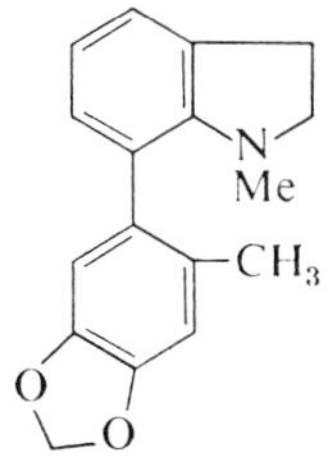

307

308

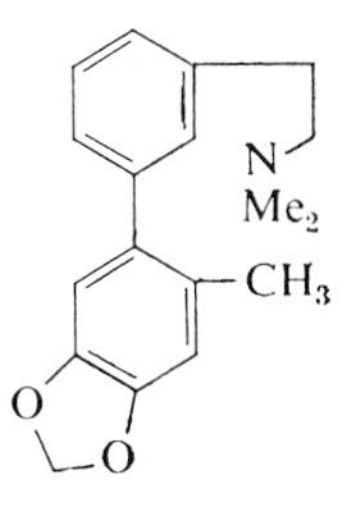

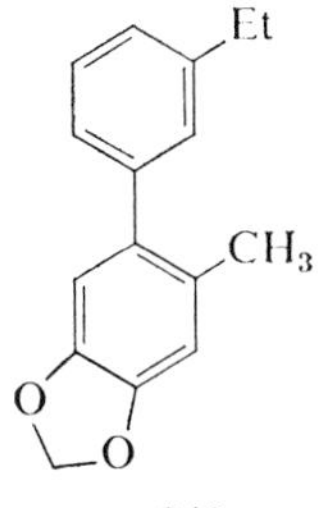

309

310

Carbon–Nitrogen Fission in β-Carbolines

A recent publication [122] reports the successful reduction of bases in the tetrahydro-β-carboline series by zinc and acetic acid. Fission of the bases between the aliphatic nitrogen atom and the carbon atom joining this to the indole α-position was observed. The reaction path could be either of those shown in formulas **311–317**. Bases reduced in this way include reserpine, 1,2-butano-1,2,3,4-tetrahydro-β-carboline (**311**), and ajmalicine (**318**). The reduction of pleiocarpamine (**319**) with zinc and hydrochloric acid gives the corresponding 2,3-dihydroindole, dihydropleiocarpamine, and isodihydro-pleiocarpamine (**320**), which is doubtless formed from the *seco*-product (**321**) [123].

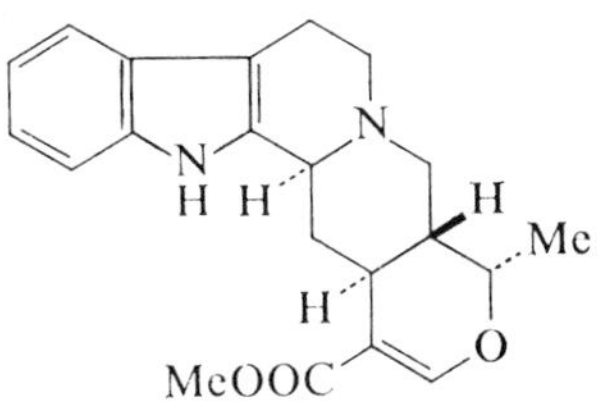

318

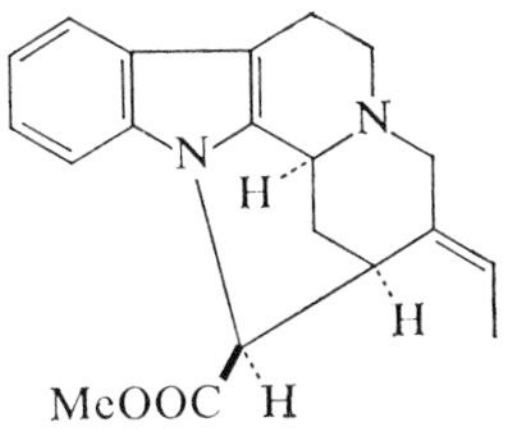

319

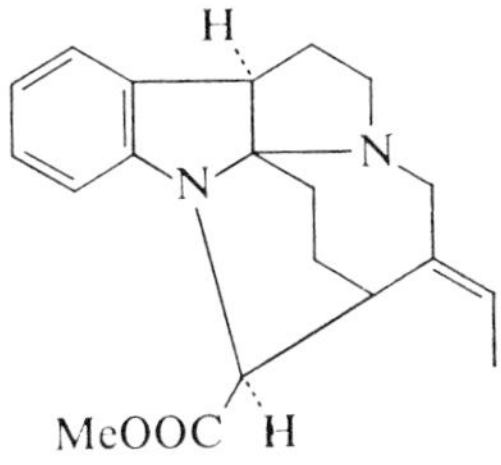

320

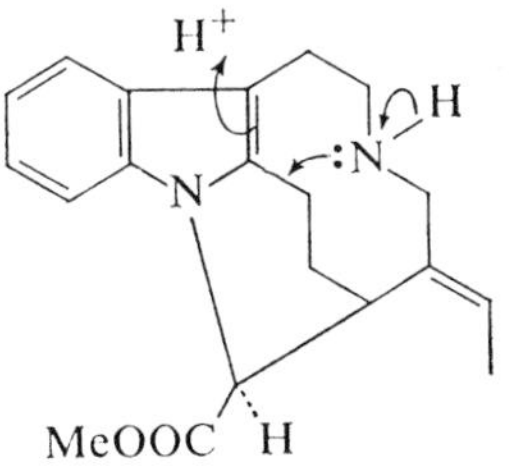

321

Other methods of fission of the same carbon–nitrogen bond in β-carbolines have been of importance in the synthesis and degradation of indole alkaloids, and these may be summarized as follows:

(*a*) Emde reduction of quaternary salts (**322–323**) [124, 125].

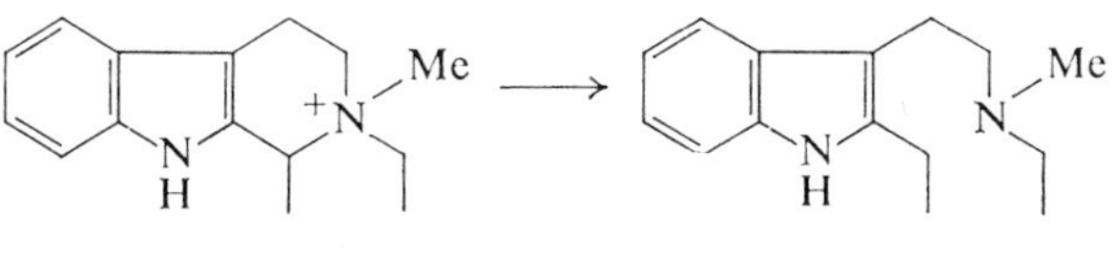

322

323

(*b*) The treatment of quaternary salts with potassium cyanide (**324–325**) [126–128].

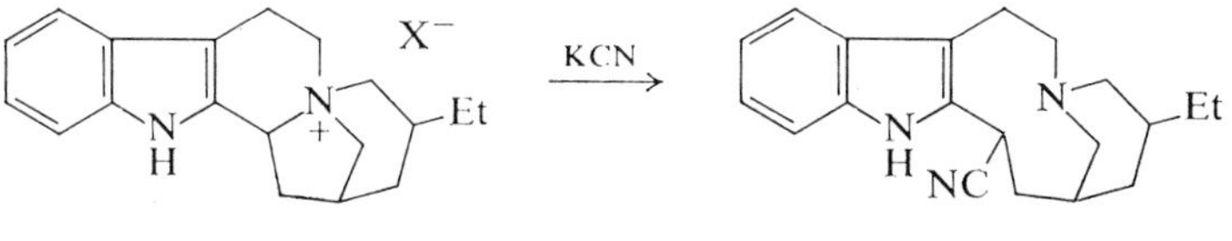

324

325

(c) The treatment of bases with acetic anhydride and sodium acetate (326–327) [126, 129, 130].

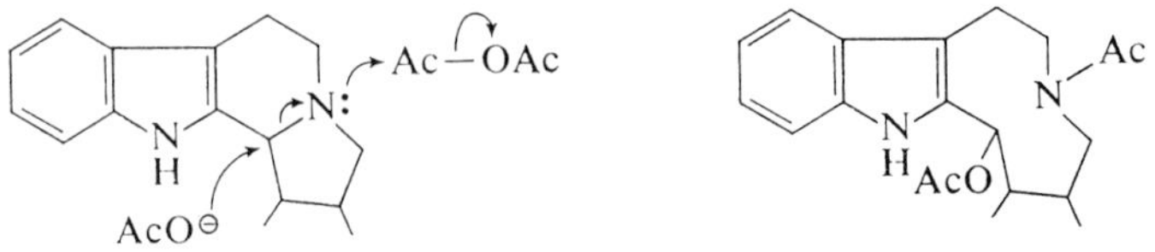

326   327

(d) The treatment of quaternary salts of indolenines of the general type **328**, where X = OAc or Cl, with acetic anhydride and sodium acetate, followed by hydrolysis (**328–333**) [131, 132]. The last step in this sequence is a normal reaction of carbinolamine salts.

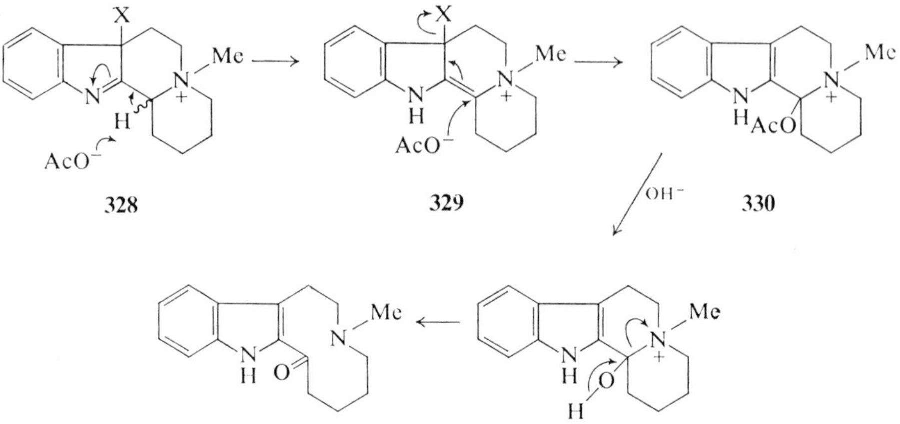

328   329   330

332   331

The Von Braun Reaction

Tertiary amines in general react with cyanogen bromide to give an alkyl bromide and a disubstituted cyanamide as shown in **333 → 334** [133]. If the groups attached to the nitrogen atom are all different, then it is clearly possible for three disubstituted cyanamides to be formed in this way by the

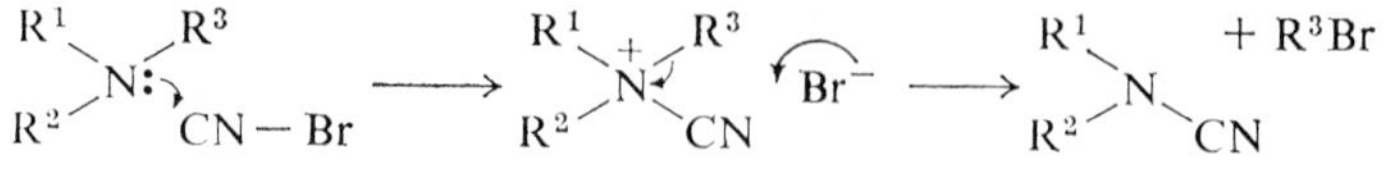

333   334

elimination of three different alkyl bromides, although formation of one product generally predominates. When the nitrogen atom is part of a ring

(**336**), the bromide and the substituted cyanamide may be two distinct entities (**335**), or the groups may be part of the same molecule (**337**). This reaction, which has been reviewed by Hageman [134], has proved to be of

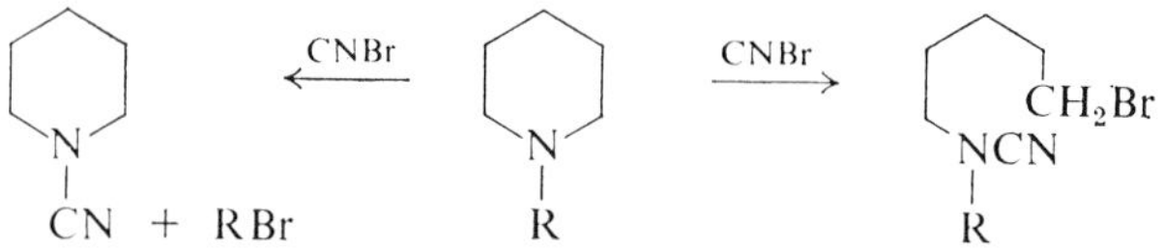

considerable value in degradative studies of the alkaloids, wherein it can give information about the environment of the nitrogen atom.

Conditions of the Reaction

The interaction of cyanogen bromide with amines is best carried out in an inert solvent, as this results in adequate mixing of the reactants and more accurate temperature control than is obtainable if no solvent is used. The physical properties of cyanogen bromide (m.p. 52°, b.p. 62°) make control of the reaction in the absence of a solvent difficult. Nonpolar solvents such as chloroform, benzene, and the hydrocarbons are generally recommended because they are immiscible with water and also cause precipitation of such by-products as salts of amines, which are then easily removed by filtration. The exclusion of water from the reaction mixture is also desirable in order to avoid interference resulting from the production of hydrobromic acid.

It is generally preferable to add the amine to the cyanogen bromide, which is thus always in excess, thereby reducing the risk of formation of quaternary salts by interaction of the amine with generated alkyl bromide, since cyanogen bromide reacts with amines more rapidly than do alkyl bromides.

The products of the reaction can usually be separated by filtration of the solution to remove any precipitated salts, followed by extraction of un-reacted amine, if any, from the solution by dilute acids, and finally separation of the residual mixture of cyanamide and alkyl bromide, after removal of solvent, by fractional crystallization or fractional distillation.

Recently the use of cyanogen bromide in water or methanol has been studied and found to be advantageous in certain cases (see below).

Products of the Reaction

The cleavage of an unsymmetrically substituted amine of low molecular weight occurs predominantly in the direction involving the displacement of the smallest group as the alkyl bromide; in alkaloid work this frequently means the displacement of methyl bromide with the replacement of N–Me by N–CN; for example, cocaine (**338**) on treatment with cyanogen bromide affords the cyanonor compound **339** in about 60% yield [135], and diacetylmorphine (**340**) gives cyanonordiacetylmorphine (**341**) in 75% yield [136].

338 **339**

340 **341**

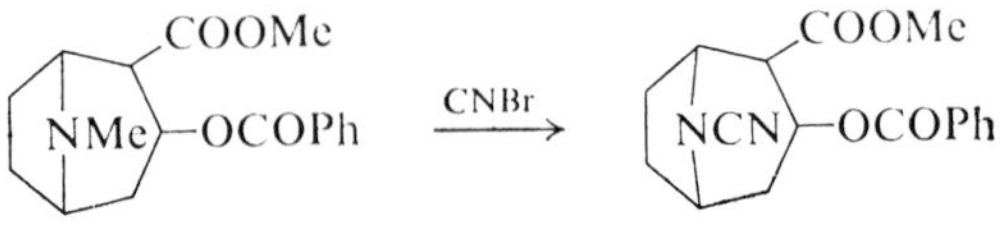

Other structural features, however, are more important than mere size of group attached to the nitrogen atom in determining the direction of fission of the molecule, for example, chain branching and the presence or absence of β,γ-unsaturation. In general, those groups whose halides are known to be reactive, such as allyl and benzyl, are displaced most readily in the reaction with cyanogen bromide [137]. For example, anhydroecgonine methyl ester (**342**), unlike cocaine (**338**), gives only a small amount of the cyanonor compound **343**, together with a considerable amount of the ring-cleaved product **344** [135]; hydrohydrastinine (**345**) gives only the substituted

342

343

344

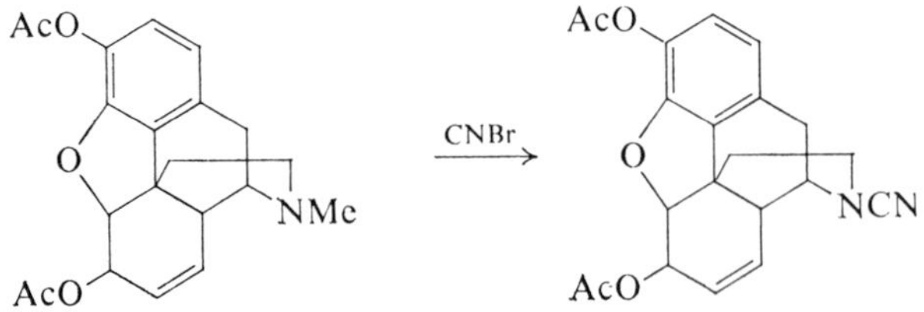

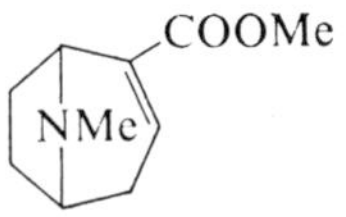

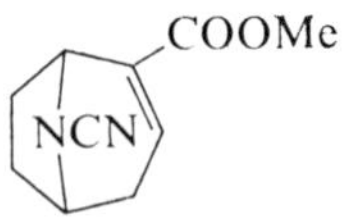

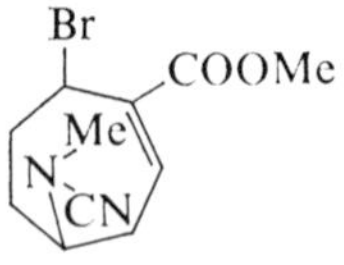

benzyl bromide **346** in this process [138], and thebaine (**347**), which also contains a double bond β,γ to the nitrogen, undergoes rearrangement and

345

346

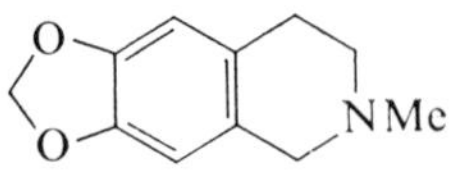

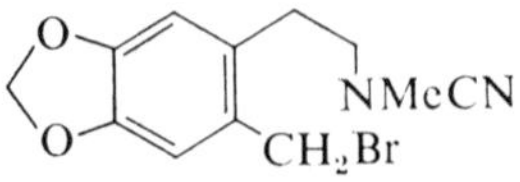

affords cyanonorthebaine (**348**) [139], together with a small amount of an ill-defined bromide-containing substance that may be the simple product of ring fission [136].

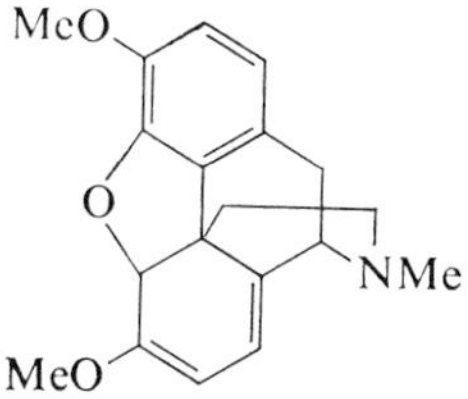

347

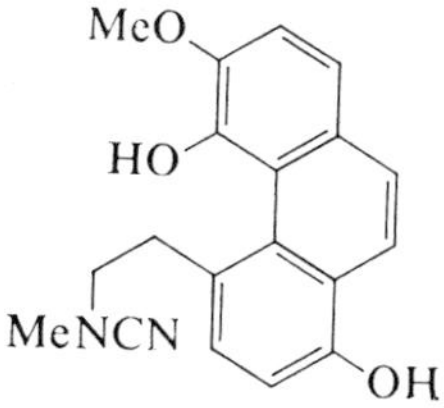

348

In the aporphine series it has been found that diacetylapomorphine (**349**) undergoes ring fission in the normal manner, but the product spontaneously loses hydrogen bromide to give the phenanthrene derivative **350** [140].

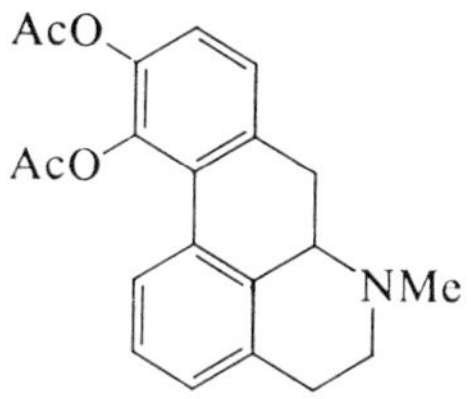

349

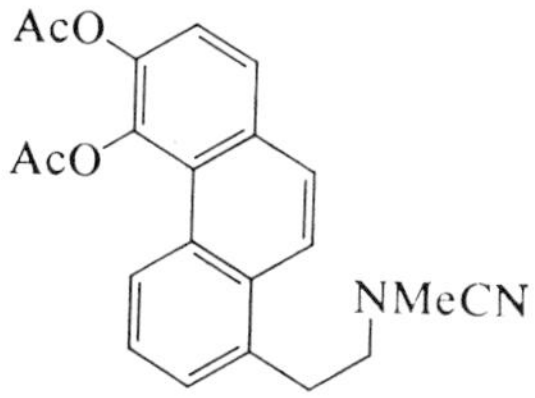

350

As with all generalizations, there are exceptions to the rule that where possible benzylic C–N bonds are broken by cyanogen bromide. For example, diacetyldihydrolycorine (**351**) is converted by this reagent into the *N*-cyano compound **352** [141].

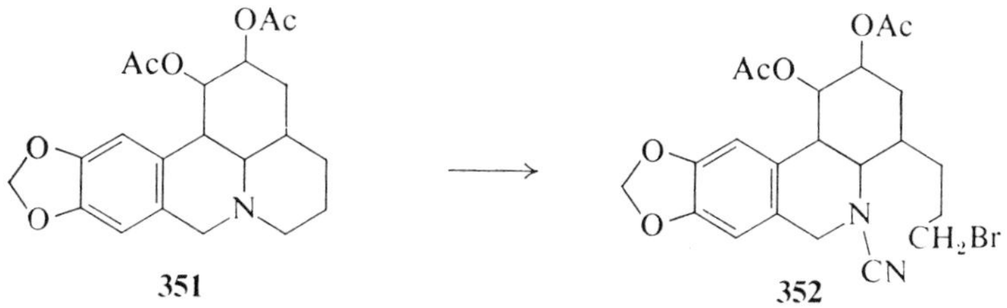

351

352

Canadine (tetrahydroberberine, **353**) with cyanogen bromide gives two products, the unsaturated cyanamide **356** and the bromo compound **355** in the ratio of 1:2. Of these, the olefin **356** doubtless arises as a result of the dehydrobromination of the primary product **354** and fission of the N–C-6 rather than the N–C-8 bond, to give the bromo compound **355**, can be explained on the basis of steric hindrance by the C-9 methoxyl group to the approach of bromide ion along a line colinear with the N–C8 bond in the transition state, there being no such hindrance of attack at C-6 [142]. The conversion of NCH$_3$ into NCN rather than fission of the benzylic C–N bond

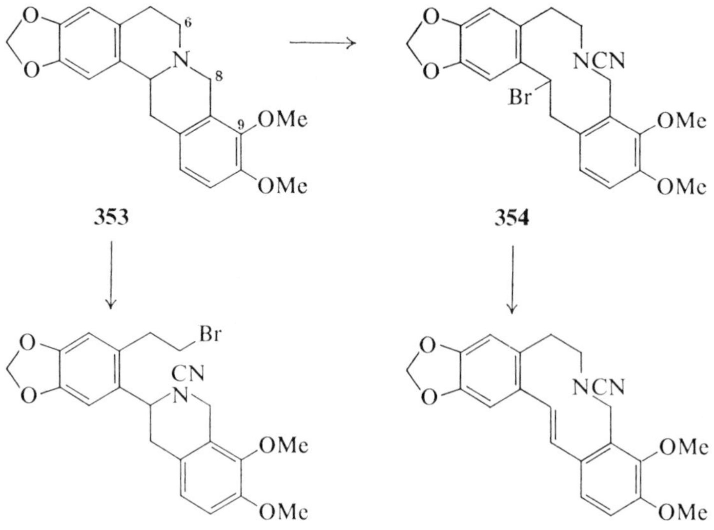

in cryptopine (**197**, $^+$NMe$_2$ = NMe) can be explained in a similar manner [143].

As examples of the use of the von Braun reaction in the elucidation of the structures of alkaloids the following may be mentioned:

(*a*) The lactam **357**, derived from tuberostemonine, on reduction with lithium aluminum hydride and treatment with cyanogen bromide gave the *N*-cyano compound **358** which was reduced to the base **359**. The methohydroxide of this base on oxidation with alkaline permanganate gave one of the optical isomers of 1-ethyl-2-butyl succinic acid (**361**), thus allowing a distinction between the structure **357** and the alternative **360** for the original lactam and hence a choice between two possible structures for tuberostemonine [144].

(*b*) Lycopodine (**362**) on treatment with cyanogen bromide gives the bromo compound **363** which may be converted into the diamine **365**, and this on cyclization and dehydrogenation yields the alkaloid lycodine **364**, thus establishing the structure of the last-named base [145].

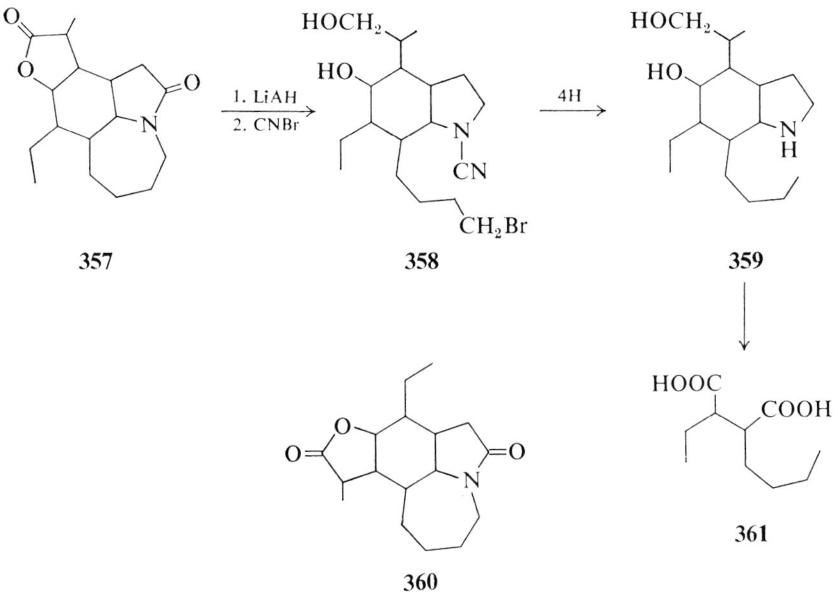

357   358   359

360   361

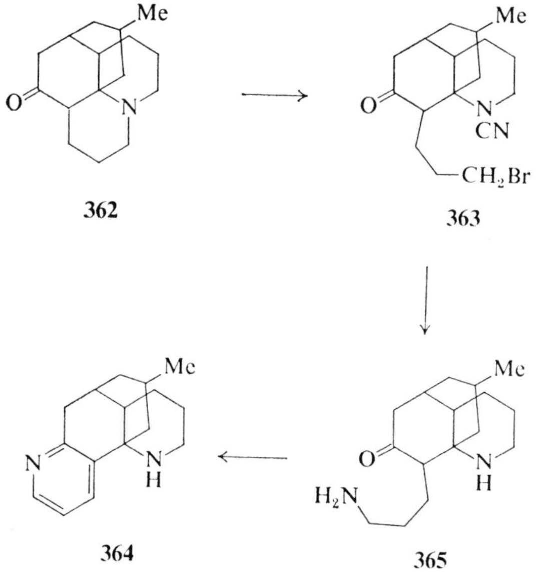

362   363

364   365

Recently an investigation has been made into the use of hydroxylic solvents in the reaction between cyanogen bromide and alkaloids. Bases in the tetrahydro-β-carboline and tetrahydroisoquinoline series give high yields of hydroxy or ethoxy N-cyano compounds when the reactions are carried out in aqueous tetrahydrofuran or ethanol–chloroform solutions. For example, yohimbine (366) gives 81% of the hydroxycyanamide 367 (R = H) in tetrahydrofuran–water and 94% of the ethoxy compound 367 (R = Et) in ethanol–chloroform, and canadine (353), in contrast to the reaction in dry benzene (see above), gives only the ethoxy compound 368 in ethanol chloroform. The reaction in hydroxylic solvents avoids the formation of reactive bromides and, by facilitating the introduction of hydroxyl groups in particular, affords compounds suitable for further transformation [146].

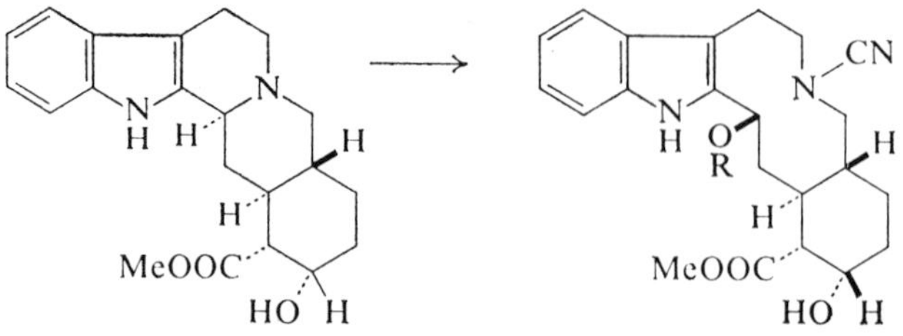

366   367

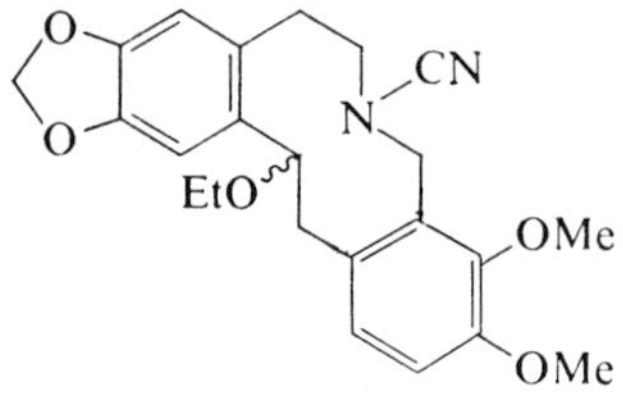

368

N-Demethylation

Several methods are available for the demethylation of N–CH$_3$ to NH.

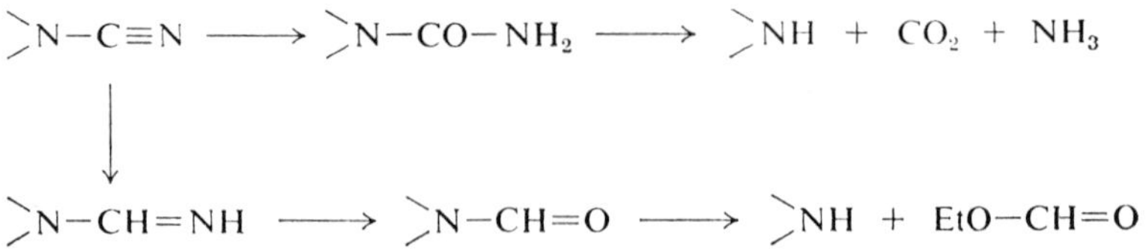

(a) The von Braun reaction (p. 300) is applicable in general in cases that do not involve ring fission or preferential displacement of a group other than methyl. The substituted cyanamide resulting from displacement of CH$_3$ may be readily hydrolyzed to the secondary amine by acids [147] or alkalis [148] or, if the molecule contains other hydrolyzable groups, by reduction, hydrolysis with water, and transacylation [149].

The reaction with cyanogen bromide, however, sometimes fails to give the expected *N*-cyano compound. For example, bases in the methadone series lose $Me_2N–CN$ rather than MeBr, methadol (369) giving the tetrahydrofuran 370 [150] and methadone (372) giving the related product 371 [151].

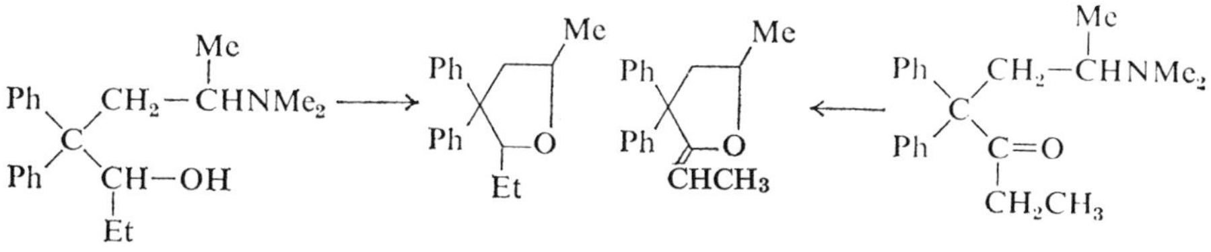

| 369 | 370 | 371 | 372 |

| 373 | 374 |

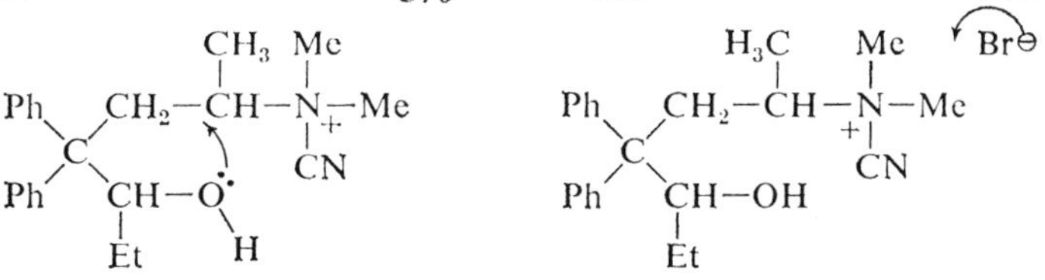

Presumably in this case steric proximity favors the internal displacement (373) in the quaternary ion rather than displacement as in 374 by a bromide ion as in the normal von Braun reaction (333 → 334).

(*b*) Tertiary amines bearing an *N*-methyl group (375) react readily with esters of azodicarboxylic acid to give addition products 376 [148, 149, 152–154] which are stable enough in simple cases to be isolated and purified [154].

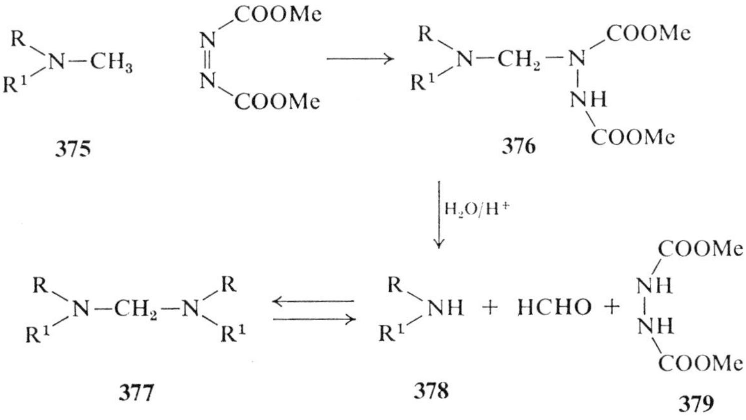

| 375 | 376 |

| 377 | 378 | 379 |

These addition products are unstable to acids, and even in some cases to water or alcohols [148], which readily hydrolyze them to secondary amines (**378**), formaldehyde, and esters of hydrazodicarboxylic acid (**379**). In some cases if the formaldehyde is not removed from the reaction mixture in some way (e.g., evaporation or reaction with a reactive methylene compound such as dimedone) it may react with the secondary base to give N,N'-methylenebis compounds **377**, which like the addition products **376** may be hydrolyzed to secondary bases [148].

(*c*) Oxidation of tertiary alkaloidal bases with hydrogen peroxide affords the corresponding amine oxides in good yield, and further oxidation of these oxides with potassium chromate affords yet another method of converting N–Me into N–H. The CH_3 group is oxidized to $-CH_2OH$, and the resulting carbinolamine readily loses formaldehyde to give the secondary amine. For example, codeine (**380**, R = Me) has been transformed in this way into norcodeine (**380**, R = H), identical with the compound produced by demethylation by the processes (*a*) and (*b*) above [155].

By analogy with this preparation of norcodeine, Bailey and Robinson [156] developed a convenient and reliable preparation of ψ-strychnine (**382**) by the oxidation of strychnine N-oxide (**381**) with potassium chromate. Strychnine

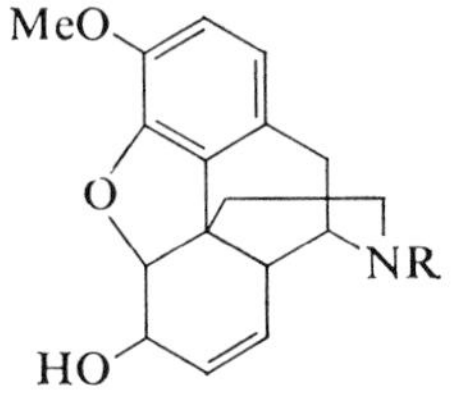

MeO ... NR   380 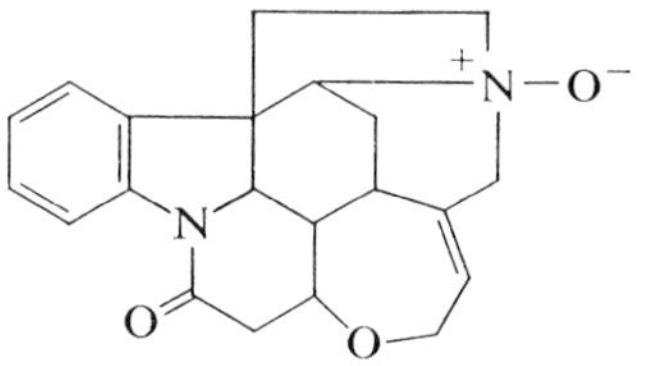   381 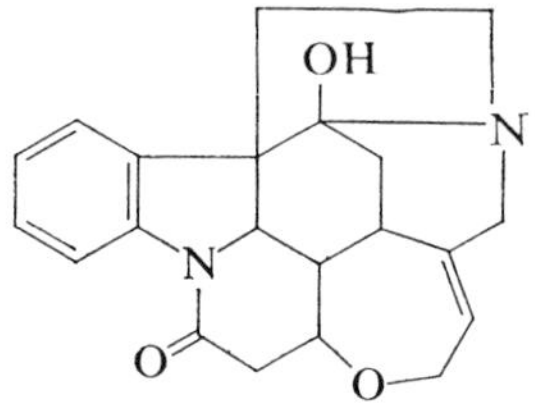   382

N-oxide contains no $-CH_3$ group, so oxidation to a carbinolamine containing the group $-CH_2OH$, which could lose formaldehyde, cannot occur. The oxide could, however, theoretically afford three different carbinolamines, of which ψ-strychnine (**382**) is one; the other two compounds do not appear to be formed in the reaction. The formation of ψ-strychnine from strychnine by the oxidation of the alkaloid with oxygen in the presence of coordinated cupric salts [157] presumably proceeds via strychnine N-oxide, a view supported by the isolation of this compound in small amount from the reaction mixture.

The same process has been used for the conversion of alkaloids of the tetrahydroberberine group into analogs of protopine, for example, the conversion of the N-oxide **383** into cryptopine (**385**) [143].

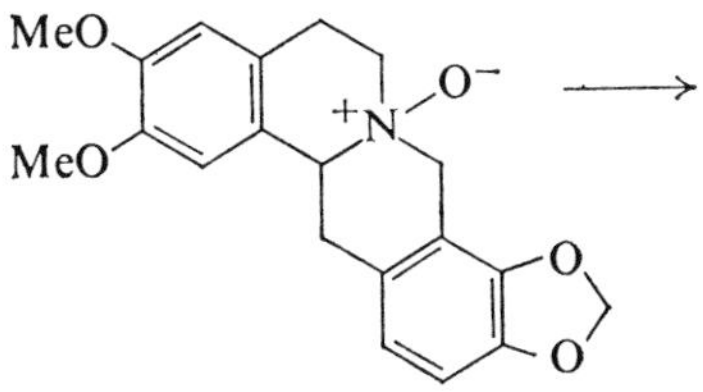

383

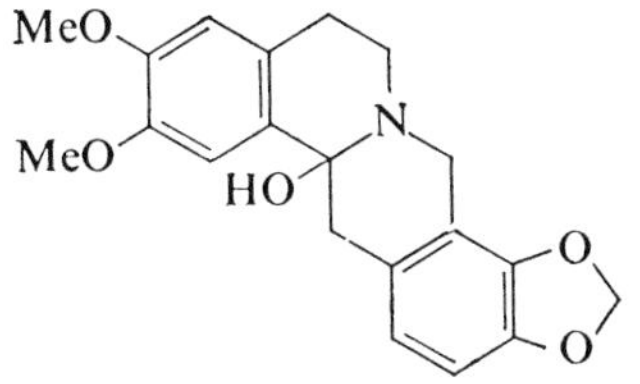

384

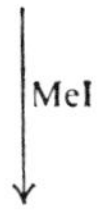

MeI

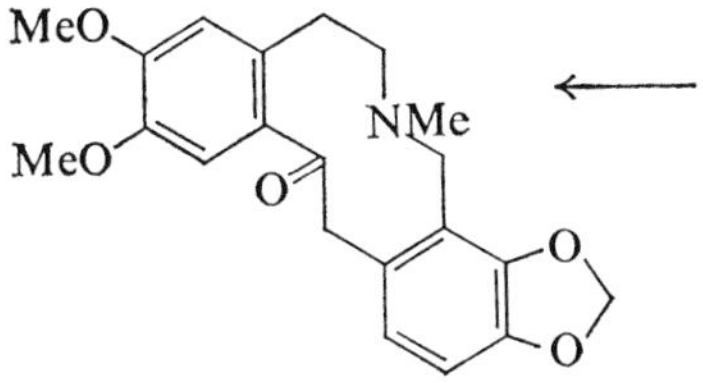

385

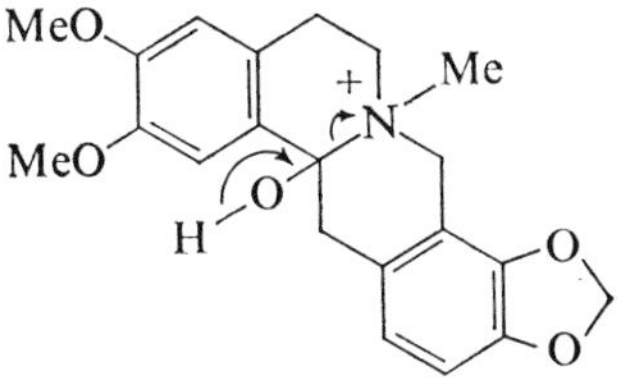

386

2 CARBON–OXYGEN BOND CLEAVAGE

Examples of carbon–oxygen cleavage by hydrolysis are given in Chapters XV and XVI of this volume.

Other fissions of carbon oxygen bonds may be subdivided into (*1*) the cleavage of ethers and (*2*) the reduction of –CH(OH)– and –CO– to –CH$_2$–.

Cleavage of Ethers

The carbon–oxygen bond of ethers may be cleaved under either acidic or basic conditions. Phenol methyl ethers are frequently demethylated under alkaline conditions at high temperatures, for example, during the Huang-Minlon modification of the Wolff-Kishner reduction [67]. This process has

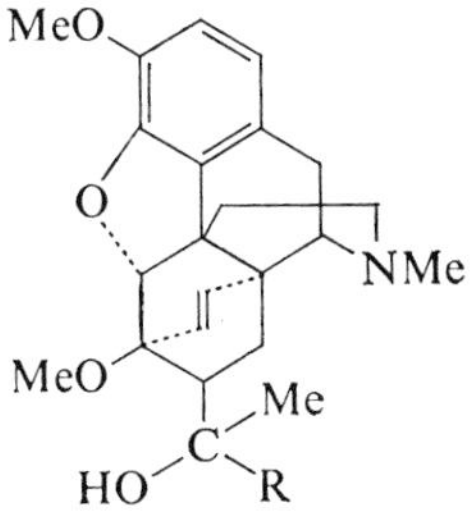

387

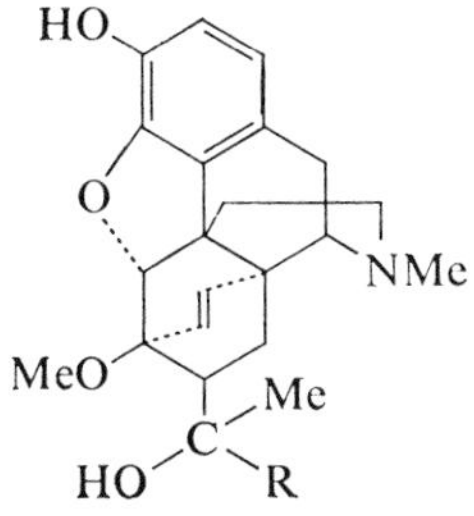

388

been utilized in the total synthesis of morphine [158], and in the demethylation of complex bases in the 6,14-*endo*-ethenotetrahydrothebaine series (**387**), which are very readily rearranged under acidic conditions [148]. Alkaline demethylation of bases of this type affects only the phenyl methyl ether group, the products being phenols (**388**) which are soluble in alkalis, and the progress of the reaction can easily be followed, completion being indicated by the production of homogeneous solutions on dilution.

A number of acidic reagents have been applied to the cleavage of ethers— for example, hydriodic acid, aluminum chloride, pyridine hydrochloride, boron tribromide, and Grignard reagents—although no widespread use has been made of any of these reagents in the study of natural products.

Hydriodic Acid

Hydriodic acid is usually used in solution, with or without the addition of red phosphorus to combine with any liberated iodine; it is one of the classical reagents for demethylation and deethylation of ethers. The general reaction may be represented as R–OMe + HI → R–OH + MeI and is the basis of the Zeisel method of estimation of methoxyl and ethoxyl groups in analysis.

The process is a vigorous one and cannot, of course, be applied to the demethylation of substances that are decomposed or rearranged by acids, such as morphine, and bases such as those of the series **387**. It has found considerable use in organic synthesis, for example, in the syntheses of flavones, flavonols, and anthocyanidins, in the course of which phenolic hydroxyl groups are protected by methylation, the methyl groups being removed in a final step by heating with hydriodic acid, as in the syntheses of quercetin (**389**) [160], gentisein (**390**) [161], and delphinidin (**391**) [162].

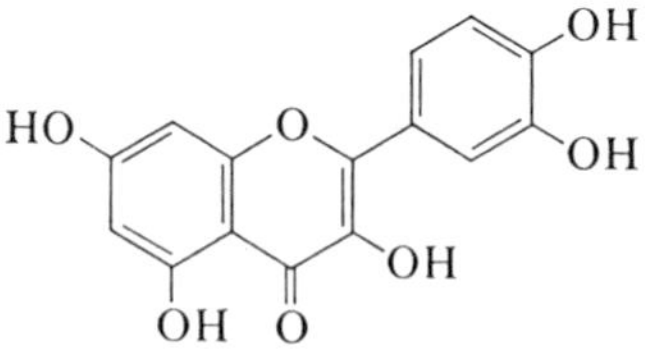

389

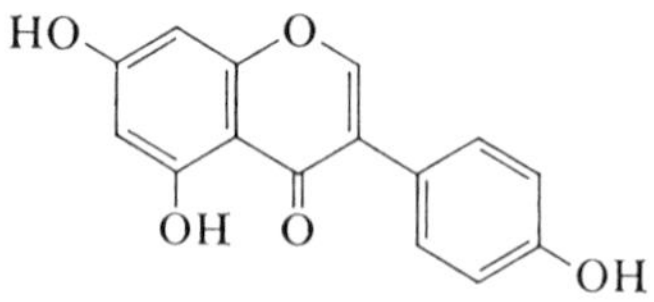

390

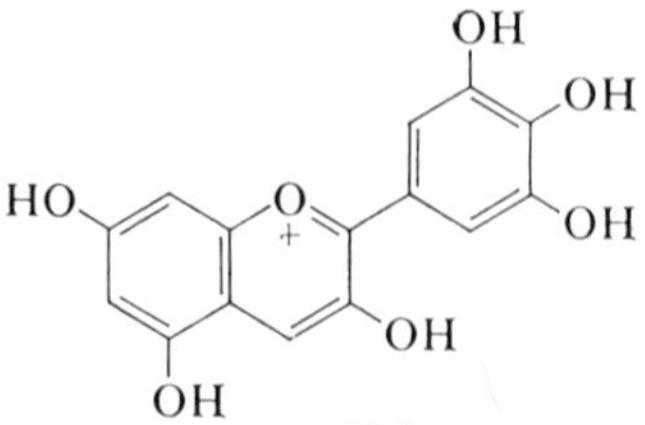

391

Other simple changes may take place, of course, during the reaction. These include the establishment of new ether linkages, as is illustrated by the demethylation of mangostin (**392**) with hydriodic acid. This affords demethylmangostin (**393**), of which only a monomethyl ether can be prepared, indicating that cyclization of the olefinic side chains with the adjacent phenolic hydroxyl groups must have taken place. The isomeric isodemethylmangostin (**394**) is the product when the demethylation is effected by hydriodic acid and iodine with subsequent bisulfite treatment [162].

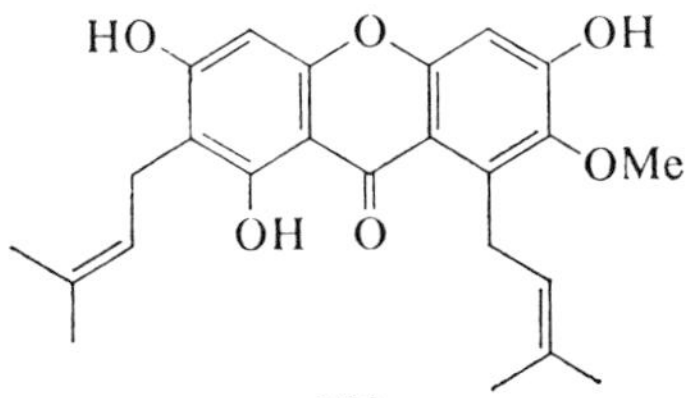

HO. ...O... ...OH
...OMe
OH O

392

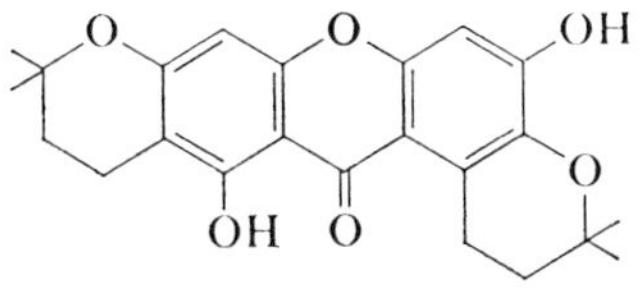

O... ...O... ...OH
O
OH O

393

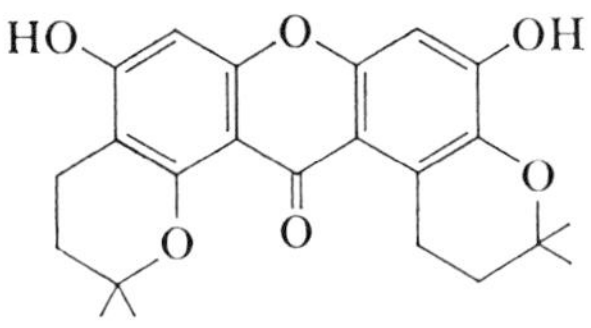

HO. ...O... ...OH
O
O O

394

Hydriodic acid is equally effective as a reagent for the cleavage of methylenedioxy groups, as instanced by the conversion of 3,4-methylenedioxyphenylalanine (**395**) into 3,4-dihydroxyphenylalanine (**396**) [163].

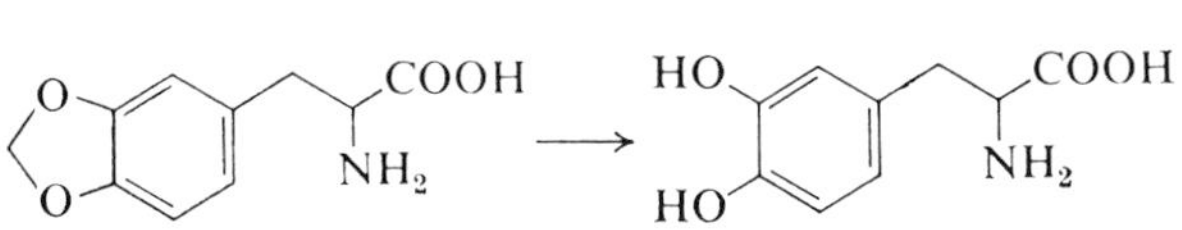

O ...COOH
O ...NH₂

395

HO ...COOH
HO ...NH₂

396

Hydrobromic Acid

Hydrobromic acid is an alternative to hydriodic acid for the process of demethylation and has been widely used for this purpose in organic synthesis, principally for the demethylation of phenol methyl ethers. In the natural product field it has been used successfully for the demethylation of substituted dihydrocodeinones to the corresponding derivatives of dihydromorphinone [164].

8-Methoxyquinoline may be demethylated in very good yield in this way [165], and details of other typical demethylation reactions are given by Long and Burger [166] and by Burton and Hoggarth [167].

Aluminum Chloride

Aluminum chloride is another acidic reagent frequently used for demethylating phenol methyl ethers in general organic chemistry [168, 169], but only infrequently used in the natural product field. Polymethoxyflavones having a methoxyl group and a carbonyl group in the *ortho* relationship may be demethylated, at this methoxyl group only, by the action of aluminum chloride under relatively mild conditions [170]. Thus 7-hydroxy-5,8-dimethoxyflavone (**397**) yields 5,7-dihydroxy-8-methoxyflavone (**398**) with 0.75

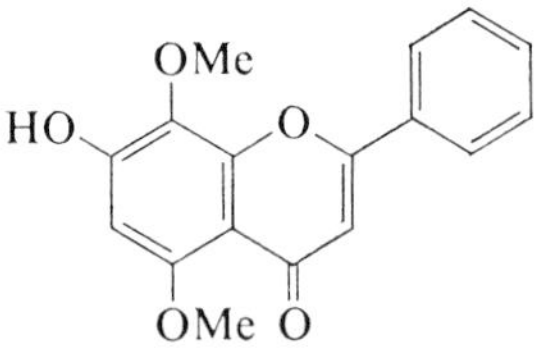

397

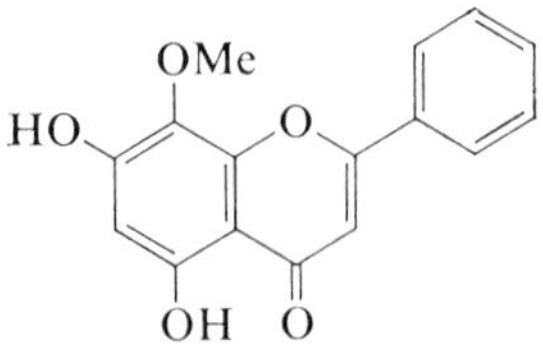

398

mole of aluminum chloride, but with 1.5 mole demethylation is complete and the trihydroxyflavone is the product [171]. Chrysin dimethyl ether (**399**) may in like manner be demethylated in nitrobenzene solution to tectochrysin (**400**) [172], and rubiadin methyl ether (**401**, R = Me) yields rubiadin (**401**, R = H) on treatment with aluminum chloride at 200° [173].

Aluminum tribromide has also been used for demethylations [174, 175].

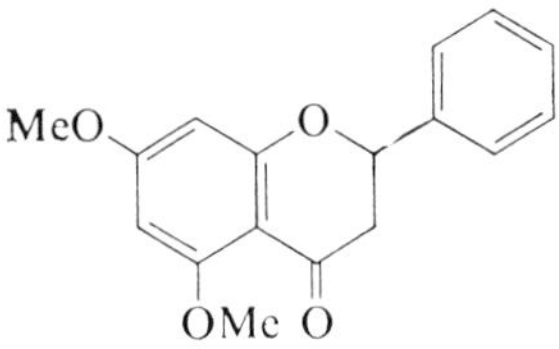

399

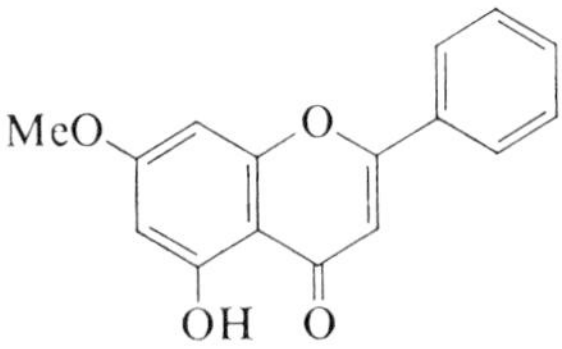

400

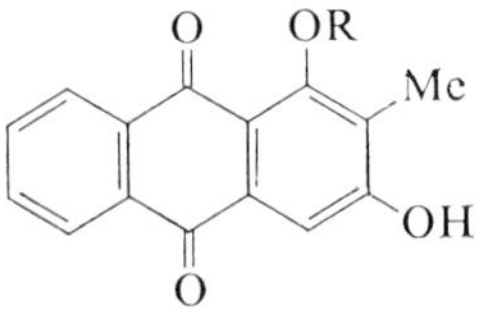

401

Pyridine Hydrochloride

Pyridine hydrochloride has been used for the demethylation of phenol ethers with considerable success, even with compounds that are sensitive to other demethylation procedures. The process was first introduced by Prey [174, 176] as an improvement of an earlier method using aniline hydrochloride or *p*-toluididine hydrochloride [177] and gives good yields. The reaction is carried out at temperatures within the range 170–240°; for example, 1,2,10,11-tetrahydro-4,13-dimethoxychrysene may be demethylated by heating with the reagent at 230–240° for 3 hr [178], whereas isobisdehydrodoisynolic acid (402) may be obtained in good yield from the corresponding methyl ether after 4 hr at only 170° [179]. Estrone methyl ether (403) has also been successfully demethylated in this way [180], and the process has even been successfully applied to the demethylation of codeine (404, R = Me), to morphine (404, R = H) [181, 182] (the latter is extremely unstable to acids and is not obtained by the demethylation of codeine with other reagents), and also to the demethylation of desoxycodeines D and E [183].

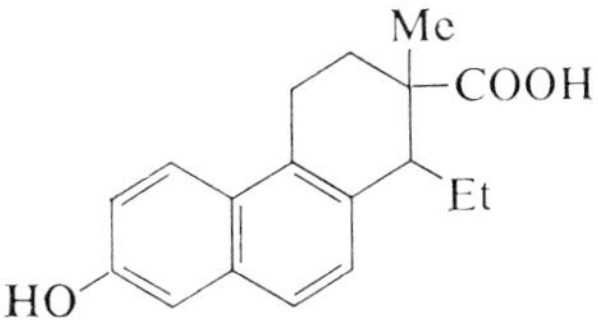

402 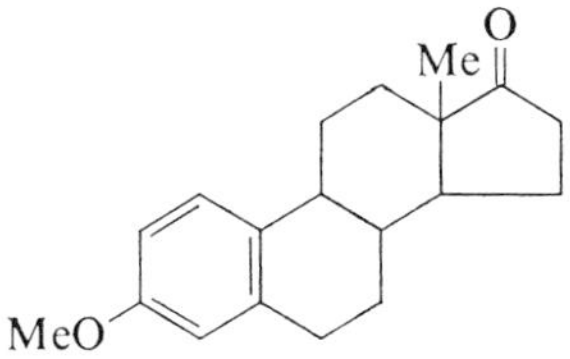   403 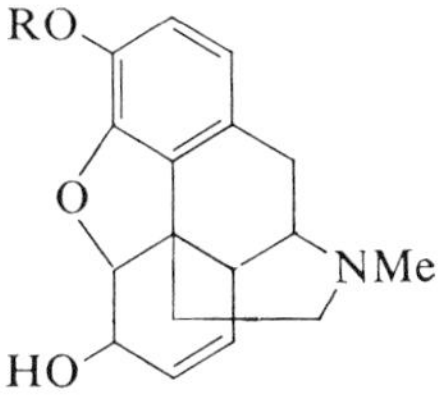   404

Grignard Reagents

When 1,2,10,11-tetrahydro-4,13-dimethoxychrysene is heated with methylmagnesium iodide at 180°, it gives the corresponding dihydroxy compound in very good yield, and the temperature required is considerably less than that needed with pyridine hydrochloride [178]. This reaction is an example of ether cleavage by Grignard reagents at high temperatures, first noticed by Grignard [184, 185] and subsequently studied by Späth [186]. It is a fairly general reaction that may be written as

$$R-O-R' + R''MgX \rightarrow R-OMgX + R'-R'',$$

although unsaturated hydrocarbons have been found among the products of the reaction [186].

The reaction does not in general proceed at a measurable rate at temperatures below about 150°. For example, anisole may be used as a solvent for the preparation of Grignard reagents from relatively unreactive halides, as a much higher temperature can be obtained with it than with diethyl ether, but at the boiling point of this solvent the ether group is attacked and phenol is formed [187]. Demethylation of a phenol in appreciable amounts has, however, been detected at the temperature of boiling benzene, for in the reaction of dihydrothebaine (405, R = Me) with phenylmagnesium bromide an appreciable quantity of dihydromorphinone Δ^6-enol methyl ether (405, R = H) is obtained [164], and demethylation (of the C-3 methoxyl group) has even been observed in boiling ether in the reaction between the ketone **406** and Grignard reagents [148].

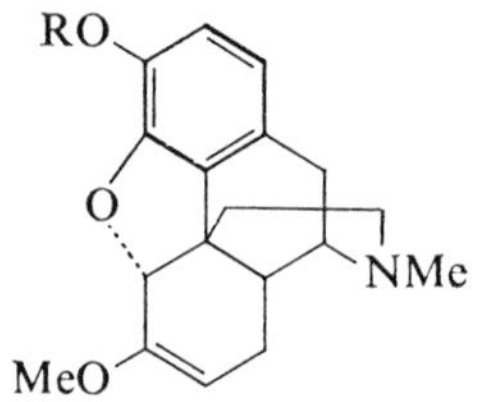

405

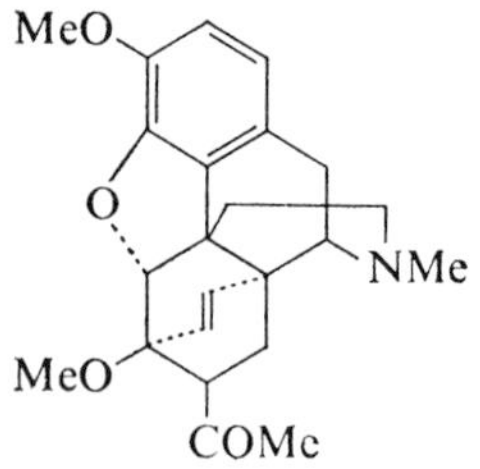

406

Some Reduction Processes

Sodium and Liquid Ammonia

Allylic ether linkages are reductively split by sodium and ammonia, generally in the presence of an additional proton source such as ethanol. However, in a number of cases this is not required; for example, thebaine is reduced by sodium and ammonia alone to give an excellent yield of dihydro-thebaine-ϕ (**408** → **407**) [67], whereas strychnine (**409**) gives the base **410** only in the presence of ethanol [188], the allylamine system suffering fission in addition to the allylic ether system. Methylenedioxy groups are particularly prone to undergo fission with this reagent; for example, hydrocotarnine

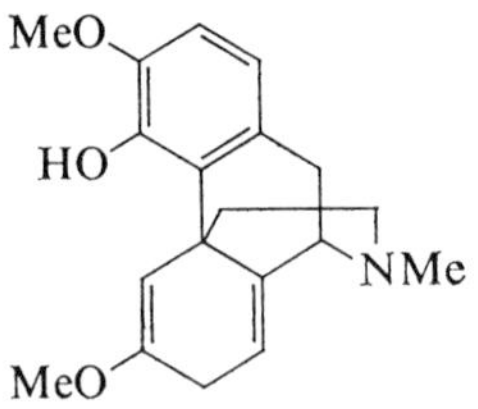

407

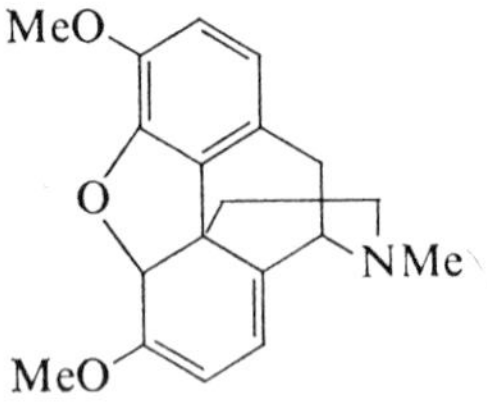

408

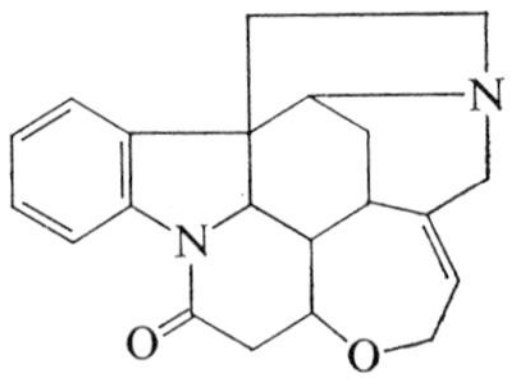

409

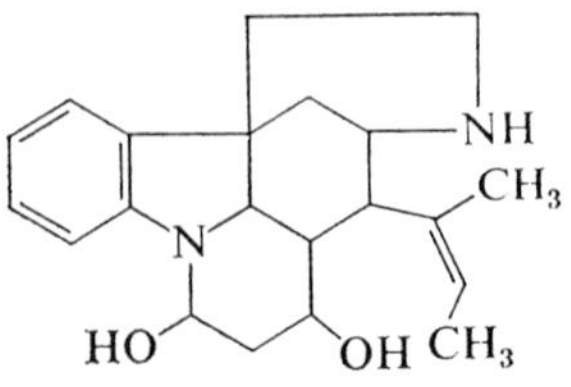

410

(411) with sodium and ammonia plus ammonium chloride gives the phenolic base 412 [119]. With sodium and alcohol, on the other hand, hydrocotarnine gives hydrohydrastinine (413) [84, 119], and a similar reductive elimination of a methoxyl group has been noted during the sodium–liquid ammonia–ethanol reduction of flavothebaone trimethyl ether ψ-methine (414), which is converted into the base 415 [91].

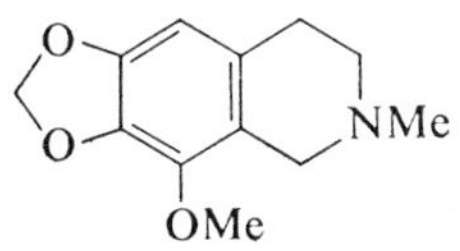

411 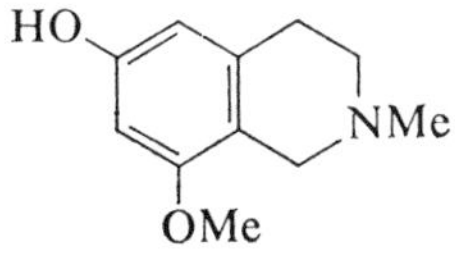   412 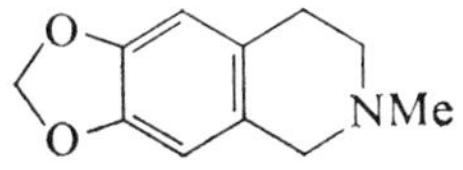   413

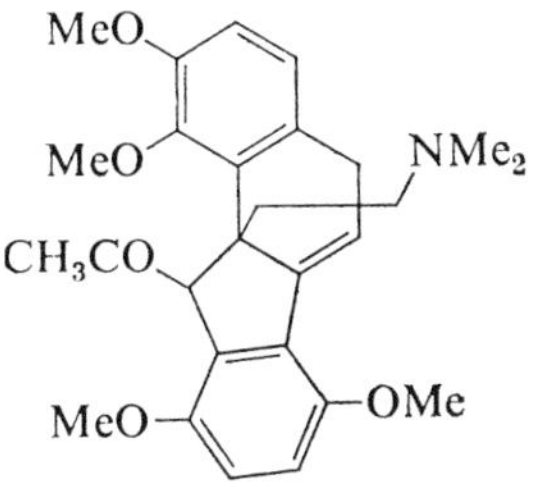

414 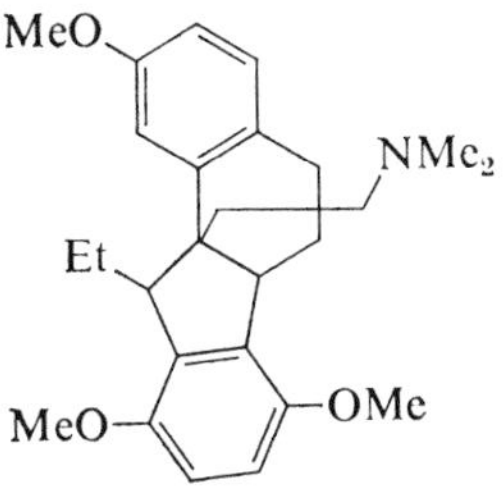   415

Halogens and Borohydrides

Alcohols and ethers can be converted into halides on treatment with halogens and diborane, higher boranes, or alkali metal borohydrides [189, 190].

$$6PhOMe + 3I_2 + B_2H_6 \rightarrow 6MeI + 2B(OPh)_3 + 3H_2$$
$$3ROH + X_2 + NaBH_4 \rightarrow 3RX + NaX + B(OH)_3 + 2H_2$$
$$3ROR' + X_2 + NaBH_4 \rightarrow 3RX + NaX + B(OR')_3 + 2H_2.$$

(Hydrogen chloride replaces hydrogen as a product when chlorine is used.) The reaction is applicable to the fission of cyclic ethers, including epoxides; for example, ethylene oxide (416) gives the borate ester 417, which can be hydrolyzed to 2-iodoethanol (418) and tetrahydrofuran can similarly be

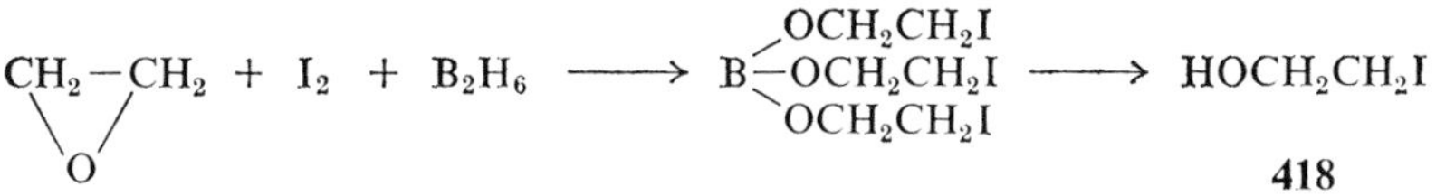

416   417   418

converted into 4-iodobutanol. The reaction is effected under mild conditions and yields are very good, though with the more reactive chlorine other reactions may occur as well.

Reduction of Toluenesulfonyl Esters

The C–OH link may be reduced to C–H by the reduction of the corresponding toluenesulfonyl ester with lithium aluminum hydride in a number of cases. This process was developed as a logical extension [191] of the observation of Gilman [192, 193] that Grignard reagents will react with toluenesulfonyl esters with the production of hydrocarbons and toluenesulfonic acid salts according to

$$R—O—SO_2C_6H_4—CH_3 + R'MgX \rightarrow R—R' + CH_3—C_6H_4—SO_2OMgX.$$

The reduction of toluenesulfonyl esters with lithium aluminum hydride follows an analogous course in a number of cases [182, 194], as shown by

$$R—O—SO_2C_6H_4—CH_3 + H^- \rightarrow R—H + CH_2—C_6H_4—SO_2O^-,$$

although in other cases the reaction takes a different course and gives the parent alcohol [191]:

$$R—O—SO_2C_6H_4—CH_3 + H^- \rightarrow R—OH + CH_3—C_6H_4—SO_2^-.$$

For example, (−)-menthyltoluenesulfonate (419, R = O-tosyl) in this way affords p-menthane (419, R = H) [191], 6-tolylsulfonyldiacetone-D-galac-

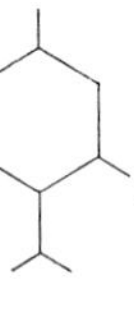

419 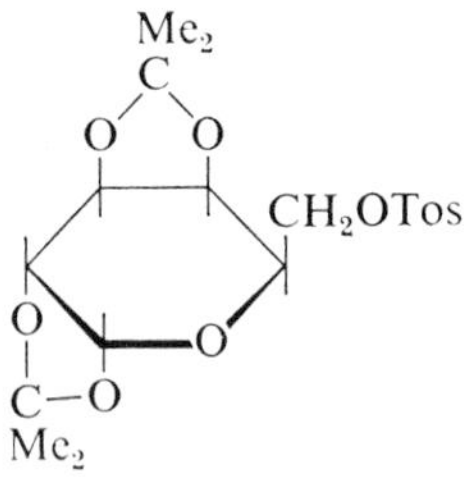   420 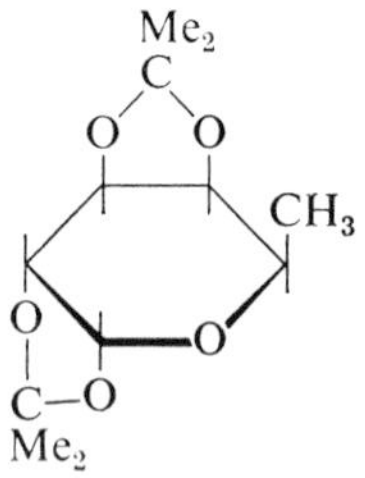   421

tose ⟨1,5⟩ (420) is converted into diacetone-D-fucose (421) [191], and yohimbyl alcohol tritoluenesulfonate (422) is reduced to 16-methylyohimbane (423) [195].

Toluenesulfonyl esters of primary alcohols are reduced very rapidly (see also Chapter XI), while esters of secondary alcohols are reduced more slowly and generally require several hours under reflux with lithium aluminum hydride in tetrahydrofuran or ether [196]. For example, 2,4,6-tritolylsulfonyl-8-methyl-D-idoside 3-methyl ether (424) in boiling tetrahydrofuran

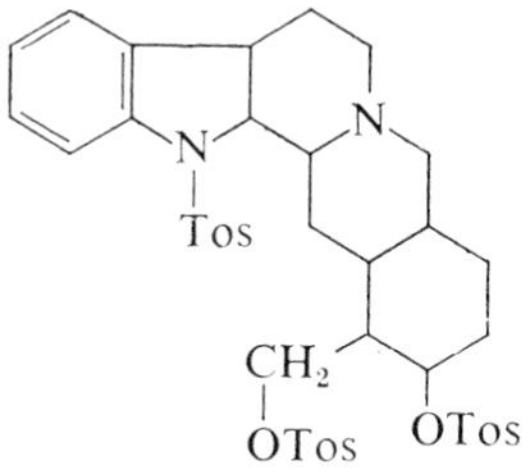

422 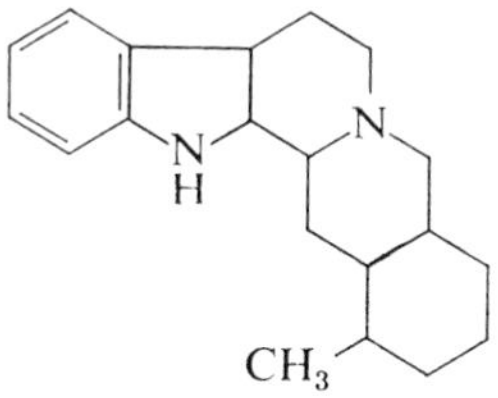   423

is transformed by lithium aluminum hydride into **425** in 30 min. 4,6-Ditolyl-sulfonyl-2,3-anhydro-α-methyl-D-alloside (**428**) suffers reductive opening of the epoxide ring on standing for 6 hr at 18° with lithium aluminum hydride in tetrahydrofuran, the product being 4,6-ditolylsulfonyl-α-methyl-D-alloside (**426**); in addition to epoxide ring opening the primary toluene-sulfonyloxy group is reduced after 1 hr under reflux in tetrahydrofuran, the product being 4-tolylsulfonyl-α-methyl-D-digitoxoside (**427**) whereas after 6 hr under reflux the secondary ester group is also attacked, with displacement of a toluenesulfinate ion, the final product being α-methyl-D-digitoxoside (**429**) [197].

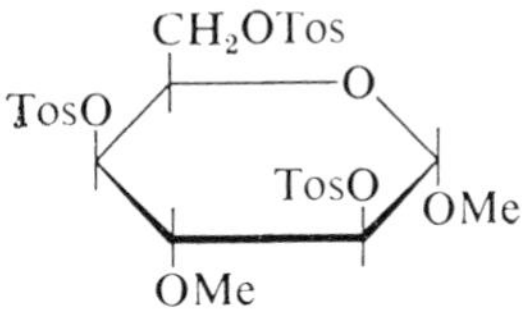

424 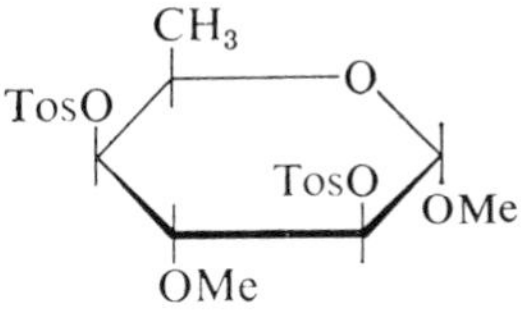   425 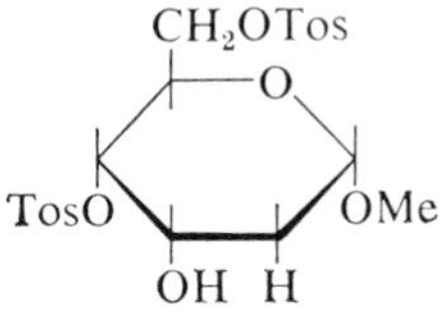   426

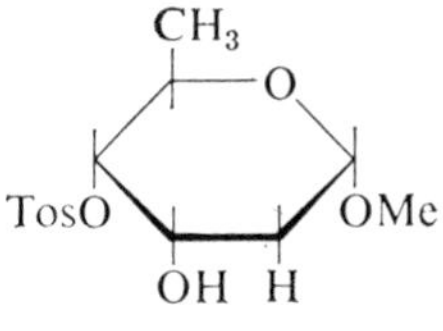

427 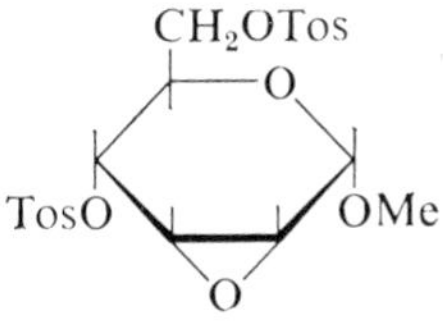   428 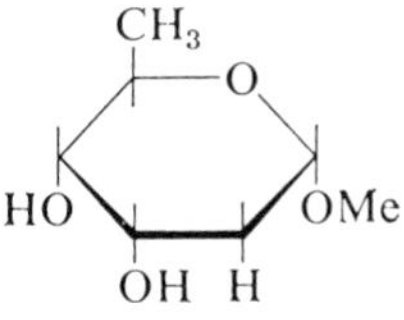   429

In some cases different stereoisomers may be obtained by the reduction of toluene sulfonates and by elimination reactions followed by reduction of the resulting olefin, as in part structures **430–434** [198].

An elimination reaction with the production of an olefin may occur in some cases as an alternative to reduction of the ester, and this process has been studied by Cram [199], who has shown that L-(+)-*threo*-3-phenylbutan-2-oltoluenesulfonate (**435**) gives in this way *trans*-2-phenylbut-2-ene (**436**),

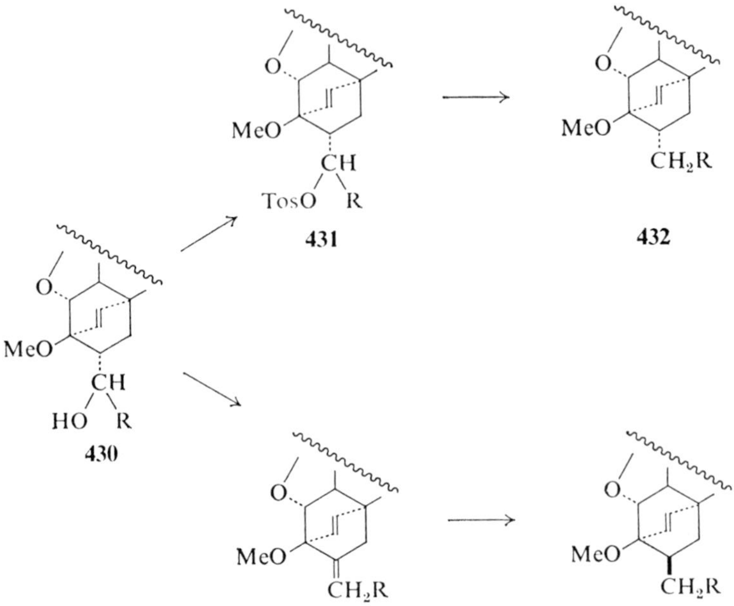

431

432

430

433

434

whereas L-(+)-*erythro*-3-phenylbutan-2-oltotoluenesulfonate (**437**) gives the *cis*-olefin **438**, the saturated hydrocarbon (+)-2-phenylbutane being obtained in addition to the olefin in both cases. For reduction there is a simple nucleophilic displacement of the toluene–sulfonate ion by an aluminohydride ion, although migration of the phenyl group occurs to a limited extent, leading to the production of racemic 2-phenylbutane. The elimination process is predominantly a *trans* one, as seen from the above results, and is bimolecular.

This process of reduction of toluenesulfonates may be used for the production of labeled compounds if lithium aluminum hydride is replaced by lithium aluminum hydride-*d* [200].

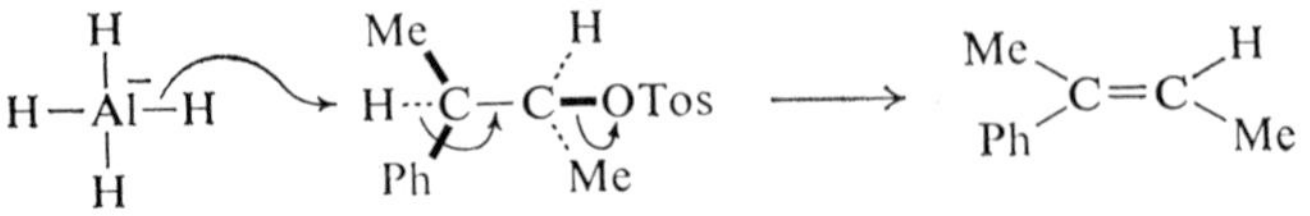

435

436

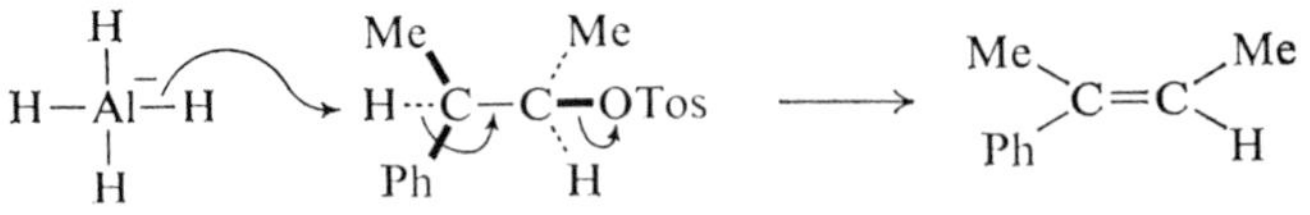

437

438

References

1. W. W. Hofmann, *Justus Liebig's Ann. Chem.*, **78**, 253 (1851); **79**, 11 (1851).
2. W. W. Hofmann, *Chem. Ber.*, **14**, 494, 659 (1881).
3. W. Hanhart and C. K. Ingold, *J. Chem. Soc.*, **1927**, 997.
4. C. K. Ingold and C. C. H. Voss, *J. Chem. Soc.*, **1928**, 3125.
5. M. L. Dhar, E. D. Hughes, C. K. Ingold, A. M. M. Mandour, C. A. Maw, and I. Woolf, *J. Chem. Soc.*, **1948**, 2093.
6. W. v. E. Doering and H. Meislich, *J. Amer. Chem. Soc.*, **74**, 2099 (1952).
7. M. F. Grundon and V. Boekelheide, *J. Amer. Chem. Soc.*, **75**, 2537 (1953).
8. M. Kotake, S. Kusumoto, and T. Ohara, *Justus Liebig's Ann. Chem.*, **606**, 148 (1957).
9. Y. Arata, *J. Pharm. Soc. Japan*, **66**, 138 (1946).
10. Y. Arata and T. Ohashi, *J. Pharm. Soc. Japan*, **77**, 229 (1957).
11. Y. Arata, *J. Pharm. Soc. Japan*, **76**, 1447 (1956).
12. J. T. Pinkey, E. Ritchie, W. C. Taylor, and S. D. Binns, *Aust. J. Chem.*, **14**, 106 (1961).
13. Y. Arata, M. Koseki, and K. Sakai, *J. Pharm. Soc. Japan*, **77**, 232 (1957).
14. A. C. Cope and E. R. Trumbull, *Org. Reactions*, **11**, 317 (1960).
15. R. B. Woodward and W. v. E. Doering, *J. Amer. Chem. Soc.*, **67**, 860 (1945).
16. R. M. F. Manske, *J. Amer. Chem. Soc.*, **72**, 55 (1950).
17. J. Weinstock and V. Boekelheide, *J. Amer. Chem. Soc.*, **75**, 2546 (1953).
18. E. D. Hughes, C. K. Ingold, and C. S. Patel, *J. Chem. Soc.*, **1933**, 526.
19. W. H. Saunders and R. A. Williams, *J. Amer. Chem. Soc.*, **79**, 3712 (1957).
20. D. J. Cramm, F. D. Greene, and C. H. Depuy, *J. Amer. Chem. Soc.*, **78**, 790 (1956).
21. A. C. Cope and E. M. Acton, *J. Amer. Chem. Soc.*, **80**, 355 (1958).
22. N. L. McNiven and J. Read, *J. Chem. Soc.*, **1952**, 153.
23. L. D. Freedman, *J. Chem. Educ.*, **43**, 662 (1966).
24. H. C. Brown and I. Moritani, *J. Amer. Chem. Soc.*, **78**, 2203 (1956).
25. B. B. Gent and J. McKenna, *J. Chem. Soc.*, **1959**, 137.
26. R. Ledger, A. J. Smith, and J. McKenna, *Tetrahedron*, **20**, 2413 (1964).
27. R. D. Haworth, J. McKenna, and R. G. Powell, *J. Chem. Soc.*, **1953**, 1110.
28. F. Weygand, H. Daniel, and H. Simon, *Chem. Ber.*, **91**, 1691 (1958).
29. A. C. Cope and A. S. Mehta, *J. Amer. Chem. Soc.*, **85**, 1949 (1963).
30. A. C. Cope, N. A. LeBel, P. T. Moore, and W. Moore, *J. Amer. Chem. Soc.*, **83**, 3861 (1961).
31. G. Ayrey, E. Buncel, and A. N. Bourns, *Proc. Chem. Soc.*, **1961**, 458.
32. G. Wittig, G. Koenig, and K. Clauss, *Justus Liebig's Ann. Chem.*, **593**, 127 (1955).
33. G. Wittig and R. Polster, *Justus Liebig's Ann. Chem.*, **599**, 13 (1956).
34. G. Wittig, H. Tenhaeff, W. Schoch, and G. Koenig, *Justus Liebig's Ann. Chem.* **572**, 1 (1951).

35. K. Jewers and J. McKenna, *J. Chem. Soc.*, **1958**, 2209.
36. F. E. King, D. M. Bovey, K. G. Mason, and R. L. S. Whitehead, *J. Chem. Soc.*, **1953**, 250.
37. H. Booth, F. E. King, and J. Parrick, *J. Chem. Soc.*, **1958**, 2302.
38. G. Merling, *Justus Liebig's Ann. Chem.*, **264**, 310 (1891).
39. A. D. Yanina and M. V. Rubtsov, *Zh. Obshch. Khim.*, **32**, 1789 (1962).
40. G. Merling, *Chem. Ber.*, **24**, 3108 (1891).
41. R. Willstätter and H. Veraguth, *Chem. Ber.*, **40**, 957 (1907).
42. A. C. Cope, C. F. Howell, and A. Knowles, *J. Amer. Chem. Soc.*, **84**, 3190 (1962).
43. L. A. Paquette and L. D. Wise, *J. Org. Chem.*, **30**, 228 (1965).
44. W. A. M. Davies, I. G. Morris, and A. R. Pinder, *Chem. Ind.* (London), **1958**, 1000.
45. D. A. Archer and H. Booth, *Chem. Ind.* (London), **1962**, 894.
46. K. W. Bentley, *The Chemistry of the Morphine Alkaloids*, Clarendon Press, Oxford, 1954.
47. M. Freund, W. W. Melber, and E. Schlesinger, *J. Prakt. Chem.*, **101**, 1 (1920).
48. E. Speyer and W. Krausz, *Justus Liebig's Ann. Chem.*, **432**, 233 (1923).
49. H. Rapoport, *J. Org. Chem.*, **13**, 714 (1948).
50. W. Klee, *Arch. Pharm.* (Weinheim), **252**, 211 (1914).
51. E. Schlittler and J. Müller, *Helv. Chim. Acta*, **31**, 1119 (1948).
52. K. W. Bentley and R. Robinson, *J. Chem. Soc.*, **1952**, 947.
53. L. Small, L. J. Sargent, and J. A. Bralley, *J. Org. Chem.*, **12**, 847 (1947).
54. L. Small and E. M. Fry, *J. Org. Chem.*, **3**, 509 (1939).
55. J. von Braun, *Chem. Ber.*, **50**, 45 (1917).
56. V. Bruckner, J. Kovacs, and J. Nagy, *Chem. Ber.*, **77**, 710 (1944).
57. F. L. Pyman, *J. Chem. Soc.*, **103**, 817 (1913).
58. R. Wegler and G. Pieper, *Chem. Ber.*, **83**, 1 (1950).
59. O. Achmatowicz and R. Robinson, *J. Chem. Soc.*, **1934**, 581.
60. K. W. Bentley and J. Dominguez, *J. Org. Chem.*, **21**, 1348 (1956).
61. C. Schöpf, E. Schmidt, and W. Braun, *Chem. Ber.*, **64**, 683 (1931).
62. A. R. Battersby and H. T. Openshaw, *J. Chem. Soc.*, **1949**, 62.
63. E. Späth and F. Galinovsky, *Chem. Ber.*, **66**, 1338 (1933).
64. B. V. Ioffe and K. N. Zelenin, *Dokl. Akad. Nauk USSR*, **154**, 864 (1964).
65. A. S. Elina and I. S. Musatova, *Khim. Geterotsikl. Soedin.*, **1965**, 291.
66. J. Palecek, *Collect. Czech. Chem. Commun.*, **31**, 1340 (1966).
67. K. W. Bentley, R. Robinson, and A. E. Wain, *J. Chem. Soc.*, **1952**, 958.
68. A. R. Battersby and H. T. Openshaw, *J. Chem. Soc.*, **1949**, 3207.
69. J. Schmutz, F. Hunzicker, and R. Hirt, *Helv. Chim. Acta*, **40**, 1189 (1957).
70. T. R. Govindachari, M. V. Lakshmikantham, K. Nagarajan, and B. R. Pai, *Tetrahedron*, **4**, 311 (1958).
71. C. R. Hauser, W. Q. Beard, and F. N. Jones, *J. Org. Chem.*, **26**, 4789 (1961).
72. R. Oda and K. Nakaoka, *Kogyo Kagaku Zasshi*, **64**, 1992 (1961).
73. R. S. Cahn, *J. Chem. Soc.*, **1926**, 2562.
74. J. A. Joule, H. Monteiro, L. J. Durham, B. Gilbert, and C. Djerassi, *J. Chem. Soc.*, 4773 (1965).

75. H. Rapoport and G. B. Payne, *J. Chem. Soc.*, **74**, 2630 (1952).

76. C. Schöpf and F. Borkowsky, *Justus Liebig's Ann. Chem.*, **452**, 249 (1927).

77. E. Späth and F. Kuffner, *Chem. Ber.*, **64**, 370 (1931).

78. J. C. Godfrey, D. S. Tarbell, and V. Boekelheide, *J. Amer. Chem. Soc.*, **77**, 3342 (1955).

79. A. M. Roe and M. Gates, *Tetrahedron*, **11**, 148 (1960).

80. L. Small and G. L. Browning, *J. Org. Chem.*, **3**, 618 (1939).

81. L. J. Sargent and L. Small, *J. Org. Chem.*, **16**, 1031 (1951).

82. V. Černý and F. Šorm, *The Alkaloids*, Vol. 9, Academic Press, New York, 1967, p. 373.

83. W. H. Perkin, *J. Chem. Soc.*, **109**, 815 (1916).

84. F. L. Pyman and F. G. P. Remfrey, *J. Chem. Soc.*, **101**, 1595 (1912).

85. W. H. Perkin, *J. Chem. Soc.*, **115**, 713 (1919).

86. P. Rabe, *Chem. Ber.*, **44**, 824 (1911).

87. G. Barger and E. Stedman, *J. Chem. Soc.*, **123**, 758 (1923).

88. E. v. Gerichten, *Chem. Ber.*, **29**, 65 (1896).

89. L. Knorr, *Chem. Ber.*, **32**, 181 (1899).

90. E. Mosettig and E. Meitzner, *J. Amer. Chem. Soc.*, **56**, 2738 (1934).

91. K. W. Bentley, J. Dominguez, and J. P. Ringe, *J. Org. Chem.*, **22**, 409 (1957).

92. T. Ikeda, W. I. Taylor, Y. Tsuda, and S. Uyeo, *Chem. Ind.* (London), **1956**, 411.

93. W. I. Taylor and S. Uyeo, *J. Chem. Soc.*, **1956**, 4750.

94. K. Wiesner and Z. Valenta, *Chem. Ind.* (London), **1956**, R 36.

95. M. Freund and G. B. Frankforter, *Justus Liebig's Ann. Chem.*, **277**, 20 (1893).

96. F. E. King and P. l'Ecuyer, *J. Chem. Soc.*, **1937**, 427.

97. K. W. Bentley and A. W. Murray, *J. Chem. Soc.*, **1963**, 2491.

98. W. Wiegrebe, *Arch. Pharm.* (Weinheim), **296**, 801 (1963).

99. W. Awe and W. Wiegrebe, *Arch. Pharm.* (Weinheim), **296**, 370 (1963).

100. K. W. Bentley, H. P. Crocker, R. Walser, W. Fulmor, and G. O. Morton, *J. Chem. Soc.*, **1969**, 2225.

101. F. Weygand and H. Daniel, *Chem. Ber.*, **94**, 1688 (1961).

102. H. Daniel, *Justus Liebig's Ann. Chem.*, **673**, 92 (1964).

103. H. Daniel and F. Weygand, *Justus Liebig's Ann. Chem.*, **671**, 111 (1964).

104. S. W. Kantor and C. R. Hauser, *J. Amer. Chem. Soc.*, **73**, 4122 (1951).

105. C. R. Hauser and A. J. Weinheimer, *J. Amer. Chem. Soc.*, **76**, 1264 (1954).

106. W. Q. Beard and C. R. Hauser, *J. Org. Chem.*, **25**, 334 (1960).

107. W. Q. Beard, D. N. Van Eenam, and C. R. Hauser, *J. Org. Chem.*, **26**, 2301 (1961).

108. G. C. Jones and C. R. Hauser, *J. Org. Chem.*, **27**, 806 (1962).

109. L. P. A. Fery and L. van Hove, *Bull. Soc. Chim. Belges.*, **68**, 65 (1959); **69**, 79 (1960).

110. D. A. Archer and H. Booth, *Chem. Ind.* (London), **1962**, 1570.

111. F. N. Jones and C. R. Hauser, *J. Org. Chem.*, **27**, 4020 (1962).

112. D. J. Cram, M. R. V. Sahyun, and G. R. Knox, *J. Amer. Chem. Soc.*, **84**, 1734 (1962).

113. D. J. Cram and J. E. McCarty, *J. Amer. Chem. Soc.*, **76**, 5740 (1954).

114. A. C. Cope, N. A. le Bel, H.-H. Lee, and W. R. Moore, *J. Amer. Chem. Soc.*, **79**, 4720 (1937).
115. A. C. Cope and C. L. Bumgardner, *J. Amer. Chem. Soc.*, **79**, 960 (1957).
116. K. W. Bentley, J. C. Ball, and J. P. Ringe, *J. Chem. Soc.*, **1956**, 1963; K. W. Bentley, J. Dominguez, and J. P. Ringe, *J. Org. Chem.*, **22**, 422 (1957).
117. H. Emde, *Justus Liebig's Ann. Chem.*, **391**, 88 (1912).
118. H. Emde, *Helv. Chim. Acta*, **15**, 130 (1932).
119. D. B. Clayson, *J. Chem. Soc.*, **1949**, 2016.
120. K. W. Bentley and A. E. Wain, *J. Chem. Soc.*, **1952**, 972.
121. L. G. Humber, H. Kondo, K. Kotera, S. Takagi, K. Takeda, W. I. Taylor, B. R. Thomas, Y. Tsuda, K. Tsukamoto, H. Uyeo, H. Yajima, and N. Yanaihara, *J. Chem. Soc.*, **1954**, 4622.
122. A. J. Gaskell and J. A. Joule, *Tetrahedron*, **24**, 5115 (1968).
123. A. A. Gorman and H. Schmid, *Monatsh. Chem.*, **98**, 1554 (1967).
124. D. Herbst, R. Rees, G. A. Hughes, and H. Smith, *J. Med. Chem.*, **9**, 864 (1966).
125. J. P. Kutney, N. Abdurahman, P. LeQuesne, E. Piers, and I. Vlattas, *J. Amer. Chem. Soc.*, **88**, 3656 (1966).
126. G. H. Foster, J. Harley-Mason, and W. R. Waterfield, *Chem. Commun.*, **1967**, 21.
127. J. Harley-Mason and Atta-ur-Rahman, *Chem. Commun.*, **1967**, 208.
128. J. P. Kutney, Ka Kong Chan, A. Failli, J. M. Fromson, C. Gletsos, and V. R. Nelson, *J. Amer. Chem. Soc.*, **90**, 389 (1968).
129. J. Harley-Mason and Atta-ur-Rahman, *Chem. Commun.*, **1967**, 1048.
130. K. Freter and K. Zeile, *Chem. Commun.*, 416 (1967).
131. L. J. Dolby and S. Sakai, *Tetrahedron*, **23**, 1 (1967).
132. L. J. Dolby and G. W. Gribble, *J. Org. Chem.*, **32**, 1391 (1967).
133. J. von Braun, *Chem. Ber.*, **33**, 1438 (1900).
134. H. A. Hageman, *Org. Reactions*, **7**, 198 (1953).
135. J. von Braun and E. Müller, *Chem. Ber.*, **51**, 235 (1918).
136. J. von Braun, *Chem. Ber.*, **47**, 2312 (1914).
137. J. von Braun and H. Engle, *Justus Liebig's Ann. Chem.*, **436**, 299 (1924).
138. J. von Braun, *Chem. Ber.*, **49**, 2624 (1916).
139. E. Speyer and H. Rosenfeld, *Chem. Ber.*, **58**, 1125 (1925).
140. J. von Braun and E. Aust, *Chem. Ber.*, **50**, 43 (1917).
141. S. Takagi, W. I. Taylor, S. Uyeo, and H. Yajima, *J. Chem. Soc.*, **1955**, 4003.
142. I. Sallay and R. H. Ayers, *Tetrahedron*, **19**, 1397 (1963).
143. K. W. Bentley and A. W. Murray, *J. Chem. Soc.*, **1963**, 2497.
144. T. Kaneko, *Ann. Rept. Itsuu Lab.*, **14**, 49 (1965).
145. D. Dvornck and O. E. Edwards, *Chem. Ind.* (London), **1958**, 623.
146. J. D. Albright and L. Goldman, *J. Amer. Chem. Soc.*, **91**, 4317 (1969).
147. J. von Braun, *Chem. Ber.*, **49**, 977 (1916).
148. K. W. Bentley and D. G. Hardy, *J. Amer. Chem. Soc.*, **89**, 3281 (1967).
149. K. W. Bentley, J. D. Bower, J. W. Lewis, M. J. Readhead, A. C. B. Smith, and G. R. Young, *J. Chem. Soc.*, **1969**, 2237.
150. A. F. Casy and M. M. A. Hassan, *Tetrahedron*, **23**, 4075 (1967).
151. A. F. Casy and M. M. A. Hassan, *J. Chem. Soc.*, **1966**, 683.

152. O. Diels and E. Fischer, *Chem. Ber.*, **47**, 2043 (1914).
153. O. Diels and M. Paquin, *Chem. Ber.*, **46**, 2000 (1913).
154. G. W. Kenner and R. J. Stedman, *J. Chem. Soc.*, **1952**, 2089.
155. O. Diels and E. Fischer, *Chem. Ber.*, **49**, 1721 (1916).
156. A. S. Bailey and R. Robinson, *J. Chem. Soc.*, **1948**, 703.
157. H. Leuchs, *Chem. Ber.*, **70**, 1543 (1937).
158. M. Gates and G. Tschudi, *J. Amer. Chem. Soc.*, **78**, 1380 (1956).
159. St. v. Kostanecki, V. Lampe, and J. Tambor, *Chem. Ber.*, **37**, 1402 (1904).
160. W. Baker and R. Robinson, *J. Chem. Soc.*, **1928**, 3115.
161. D. D. Pratt and R. Robinson, *J. Chem. Soc.*, **127**, 166 (1925).
162. P. Yates and G. H. Stout, *J. Amer. Chem. Soc.*, **80**, 1691 (1958).
163. R. H. Barry, A. M. Mattocks, and W. H. Hartung, *J. Amer. Chem. Soc.*, **70**, 693 (1948).
164. L. Small, S. G. Turnbull, and H. M. Fitch, *J. Org. Chem.*, **3**, 204 (1938).
165. F. E. King and J. A. Sherred, *J. Chem. Soc.*, **1942**, 415.
166. L. Long and A. Burger, *J. Org. Chem.*, **6**, 852 (1941).
167. H. Burton and E. Hoggarth, *J. Chem. Soc.*, **1945**, 14.
168. C. Hartmann and L. Gatterman, *Chem. Ber.*, **25**, 3531 (1892).
169. H. Shapiro and K. A. Smith, *J. Chem. Soc.*, **1946**, 143.
170. K. Venkataraman, *Curr. Sci.*, **2**, 50 (1934).
171. D. Chakravarti and B. C. Banerjee, *J. Indian Chem. Soc.*, **14**, 37 (1937).
172. K. D. Bulati and K. Venkataraman, *J. Chem. Soc.*, **1936**, 267.
173. P. C. Mitter and H. Biswas, *J. Indian Chem. Soc.*, **7**, 839 (1930).
174. V. Prey, *Chem. Ber.*, **75**, 537 (1942).
175. R. Adams and M. Mathieu, *J. Amer. Chem. Soc.*, **70**, 2120 (1948).
176. V. Prey, *Chem. Ber.*, **75**, 350 (1942).
177. A. Klemenc, *Chem. Ber.*, **49**, 1371 (1916).
178. E. Hardegger, D. Redlich, and A. Gal, *Helv. Chim. Acta*, **28**, 628 (1945).
179. J. Heer, J. R. Billeter, and K. Miescher, *Helv. Chim. Acta*, **28**, 991 (1945).
180. W. S. Johnson, D. K. Banerjee, W. P. Schneider, C. D. Gutsche, W. E. Schleberg, and L. J. Chinn, *J. Amer. Chem. Soc.*, **74**, 2832 (1952).
181. M. Gates and G. Tschudi, *J. Amer. Chem. Soc.*, **78**, 1380 (1956).
182. H. Rapoport, C. H. Lovell, and B. M. Tolbert, *J. Amer. Chem. Soc.*, **73**, 5900 (1951).
183. H. Rapoport and R. M. Bonner, *J. Amer. Chem. Soc.*, **73**, 2872, 5845 (1951).
184. V. Grignard, *C.R. Acad Sci.*, Paris, **138**, 1048 (1904).
185. V. Grignard, *C.R. Acad Sci.*, *Paris*, **151**, 322 (1910).
186. E. Späth, *Montash. Chem.*, **35**, 319 (1914).
187. H. Simons and P. Remmert, *Chem. Ber.*, **47**, 269 (1914).
188. G. R. Clemo and T. J. King, *J. Chem. Soc.*, **1948**, 1661.
189. G. F. Freeguard and L. H. Long, *Chem. Ind.* (London), **1964**, 1582.
190. L. H. Long and G. F. Freeguard, *Chem. Ind.* (London), **1965**, 223.
191. H. Schmid and P. Karrer, *Helv. Chim. Acta*, **32**, 1371 (1949).
192. H. Gilman and J. J. Beaber, *J. Amer. Chem. Soc.*, **45**, 839 (1923).
193. H. Gilman and J. J. Beaber, *J. Amer. Chem. Soc.*, **47**, 578 (1925).
194. P. Karrer and G. Widmark, *Helv. Chim. Acta*, **34**, 34 (1951).

195. P. Karrer and R. Aeman, *Helv. Chim. Acta*, **35**, 1371 (1949).

196. J. Strating and H. J. Backer, *Rec. Trav. Chim. Pays-Bas*, **69**, 638 (1950).

197. H. R. Bollinger and P. Ulrich, *Helv. Chim. Acta*, **35**, 93 (1952).

198. K. W. Bentley, D. G. Hardy, J. W. Lewis, W. I. Rushworth, and M. J. Readhead, *J. Chem. Soc.*, **1969**, 826.

199. D. J. Cram, *J. Amer. Chem. Soc.*, **74**, 2149 (1952).

200. H. L. Yale, E. J. Pribyl, W. Brake, J. Bernstein, and W. A. Lott, *J. Amer. Chem. Soc.*, **72**, 3716 (1950).

Chapter XIV

ALKALI FUSION AND SOME RELATED PROCESSES*

B. C. L. Weedon

1   INTRODUCTION

Alkali fusion is one of the classical methods of degrading organic compounds and has provided valuable information in the study of many natural products. The reactions which occur under such drastic conditions are sometimes complex, and it is scarcely surprising that the results observed have at times been misinterpreted. Moreover, it is only recently that any

* This review is reprinted from the first edition; a few new references have been added.

systematic studies have been carried out on the reactions of organic compounds under strongly alkaline conditions, and therefore many of the mechanisms which are proposed in this review must be regarded as speculative. It is, however, hoped that they will serve a useful purpose by helping to correlate much of the available information and by suggesting possible interpretations of new results.

In addition to its wide use for degradation, alkali fusion has a number of synthetic uses. As is well known, it provided a general method for the preparation of phenols and was used to effect a number of important syntheses in the dyestuffs field, including the first synthesis of a natural dye (alizarin). Although this review is concerned primarily with degradation, brief reference is also made to the synthetic applications since these help to give a fuller picture of the reactions which are likely to be encountered in alkali fusions. For obvious reasons, no attempt has been made to cover all examples of a particular reaction.

In surveying the extensive literature on alkali fusions, two main difficulties have been encountered. First, many authors have failed to specify fully the experimental conditions used, in particular whether the reaction was performed in air or in an inert atmosphere (e.g., nitrogen or hydrogen). It is now known that with some compounds—for example, alcohols, ketones [1–3] phenols [4], and unsaturated acids [5]—the composition and yield of the product is altered markedly if the reaction is carried out in air. The second difficulty arises from the fact that many studies on alkali fusion have obviously been made with compounds of questionable purity and without adequate identification of the products. Much of the early work was, of course, carried out before convenient methods of analysis and identification had been developed. Claims which in the light of more recent information are considered to be of doubtful significance, or to be based on erroneous reports, are excluded from the following survey.

The term "alkali fusion" has been used very loosely in the literature, but can generally be taken to refer to a reaction with an alkali metal hydroxide in which the reagent is liquid at the reaction temperature but solid at room temperature. Potassium hydroxide is often preferred to sodium hydroxide because of its superior solvent properties and lower melting point; the comparatively low melting eutectic of sodium and potassium hydroxides is sometimes used when it is desired to moderate the reaction. The temperature at which the reagent liquefies depends greatly on the amount of water present. Commercial potassium hydroxide pellets contain ca. 15% water, and more is sometimes added to the reaction mixture. Even molten alkalies retain water tenaciously, and the water almost invariably present during alkali fusions is believed to play an important role in many of the reactions which take place.

Glass apparatus is obviously unsuitable for carrying out alkali fusions, but nickel, stainless steel, iron, copper, or silver apparatus has proved satisfactory. Except possibly with some dehydrogenations, there is little evidence to suggest that the material of the vessel influences the reaction to any significant extent.

A wide variety of compounds has been submitted to alkali fusion; it is obviously an advantage, though not a prerequisite, for the starting material to have a carboxyl, phenolic hydroxyl, or other group conferring solubility in alkali. Unless directly concerned with the reaction under consideration, these solubilizing groups, if any, are ignored in classifying compounds in the following review.

2   CARBOXYLIC ACIDS

Ethylenic Fatty Acids (The Varrentrapp Reaction)

In 1840 Varrentrapp heated oleic acid (*cis*-**1**) with molten potassium hydroxide and obtained palmitic acid (**2**), acetic acid, and hydrogen [6].

$$CH_3(CH_2)_7CH\!=\!CH(CH_2)_7CO_2H \rightarrow CH_3(CH_2)_{14}CO_2H + CH_3CO_2H + H_2$$
$$\mathbf{1} \qquad\qquad\qquad\qquad\qquad \mathbf{2}$$

For many years oleic acid was formulated, largely on this evidence, as octadec-2-enoic acid, and the nature of the transformation involved in the Varrentrapp reaction was not appreciated. Other similar fatty acids which have been converted by alkali fusion at 300–360° into saturated acids containing two fewer carbon atoms include elaidic (*trans*-**1**) [6], petroselinic (*cis*-**3**) [5], erucic (*cis*-**4**) [7]; brassidic (*trans*-**4**) [8], and undecylenic acid (**5**) [5, 9–11]. The reaction is clearly general for both *cis*- and *trans*-ethylenic fatty acids and is independent of the position of the double bond in the

$$CH_3(CH_2)_{10}CH\!=\!CH(CH_2)_4CO_2H$$
$$\mathbf{3}$$
$$CH_3(CH_2)_7CH\!=\!CH(CH_2)_{11}CO_2H$$
$$\mathbf{4}$$
$$CH_2\!=\!CH(CH_2)_8CO_2H$$
$$\mathbf{5}$$

starting material. Provided that the reactions are carried out in an inert atmosphere, yields of ca. 80% are readily obtained. An interesting variant of the Varrentrapp reaction is the formation of pimelic acid (**7**) from cyclohex-3-enecarboxylic acid (**6**), the loss of two carbon atoms being avoided with this cyclic starting material [12, 13].

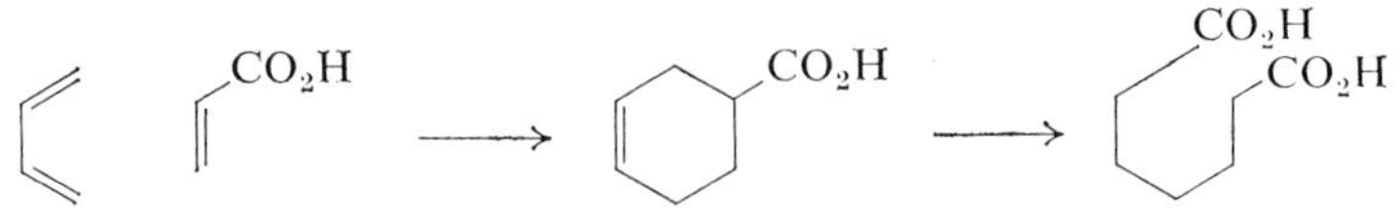

$$\text{(structure)} \quad \text{(structure with } CO_2H) \longrightarrow \text{(cyclohexene with } CO_2H) \longrightarrow \text{(cyclohexane with } CO_2H, CO_2H)$$

$$\qquad\qquad\qquad\qquad\qquad\qquad\qquad 6 \qquad\qquad\qquad\qquad 7$$

The course of the Varrentrapp reaction has been the subject of much speculation. It has now been shown that initially the double bond in, for example, oleic acid migrates in both directions along the fatty acid chain and that the newly formed double bonds are mixtures of *cis* and *trans* isomers [5, 5a, 11, 14]. With the reservation that proton removal and proton return may be either consecutive or concerted processes, this reversible migration of the double bond may be represented by a prototropic mechanism [15] (Scheme 14.1). The mobility of the double bond seems to increase as it

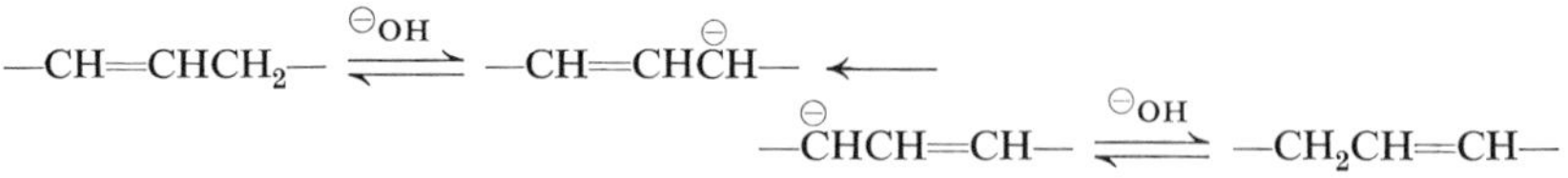

$$-CH{=}CHCH_2{-} \; \underset{}{\overset{\ominus OH}{\rightleftharpoons}} \; -CH{=}CH\overset{\ominus}{CH}{-} \longleftarrow$$

$$-\overset{\ominus}{C}HCH{=}CH{-} \; \underset{}{\overset{\ominus OH}{\rightleftharpoons}} \; -CH_2CH{=}CH{-}$$

Scheme 14.1

approaches the carboxylate group, and it is conceivable that reaction of the Δ^4 and Δ^5 isomers is facilitated by intramolecular processes [5] (e.g., Scheme 14.2).

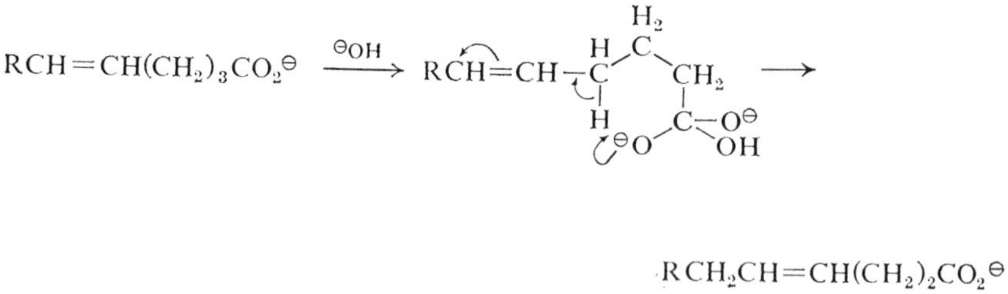

$$RCH{=}CH(CH_2)_3CO_2{}^{\ominus} \; \overset{\ominus OH}{\longrightarrow} \; RCH{=}CH{-}\overset{H}{\underset{H}{C}}\overset{H_2}{\underset{}{C}}\;\;CH_2 \longrightarrow$$

$$R\,CH_2CH{=}CH(CH_2)_2CO_2{}^{\ominus}$$

Scheme 14.2

By a sequence of reversible prototropic rearrangements, the double bond in the starting material may be regarded as migrating into the α,β-position where it then undergoes fission [5, 11, 13, 16, 17]. Strong support for this view comes from the observation that neither 1-methylcyclohex-3-ene-1-carboxylic acid (8) [12] nor 3,3-dimethylnon-8-en-1-oic acid (9) [18], in

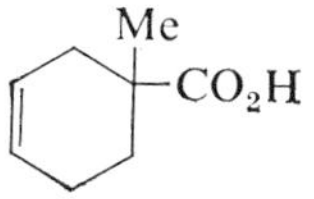

Me

CO_2H

8

$CH_2=CH(CH_2)_4CMe_2CH_2CO_2H$

9

which the α- and β-positions, respectively, are blocked, undergo the normal Varrentrapp reaction. However, α,β-unsaturated acids—for example, **10**, $n = 2$, 4, 5, 8, and 14, and (**11**)—are readily split under similar conditions [5, 12, 19–22].

$$CH_3(CH_2)_nCH=CHCO_2H \rightarrow CH_3(CH_2)_nCO_2H + CH_3CO_2H$$
10

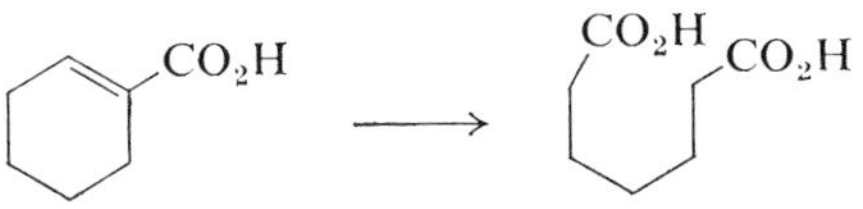

CO_2H $\longrightarrow$ CO_2H CO_2H

11

Concerning the fission of the α,β-ethylenic acid (**12**) (Scheme 14.3), it seems probable that this involves a nucleophilic attack by hydroxyl ion to give a β-hydroxy acid (**13**) [5, 11, 23]. The formation of β-hydroxy acids during the equilibration of α,β- and β,γ-ethylenic acids in aqueous alkali has been reported [24–26]. Such hydrations are reversible; when heated with alkali, crotonic acid gives β-hydroxybutyric acid [5], and this yields crotonic acid together with small amounts of vinylacetic acid [24]. The conversion of 2-hydroxycyclohexane-1-carboxylic acid into cyclohexene-1-carboxylic acid in concentrated alkalies has also been described [12, 13]. Although β-hydroxy acids have not been isolated from alkali fusions of ethylenic acids, this is probably due to their ease of fission. Alkali fusion of β-hydroxystearic and β-hydroxybutyric acids gives palmitic and (1.9 mole) acetic acid, respectively [5].

There are two plausible mechanisms for the breakdown of the β-hydroxy acid (**13**) into the final products [5, 11]. Either dehydrogenation occurs (Route A, Scheme 14.3) followed by normal "acid fission" of the resulting β-keto acid (**14**); or fission by a reaction of the "retro-aldol" or "retro Claisen" type gives acetate and an aldehyde, **15**, which is then converted into the corresponding acid (Route B). The dehydrogenation in alkalies of other hydroxy acids to keto acids has been demonstrated [3], and it is known that aldehydes under similar conditions are converted into the corresponding

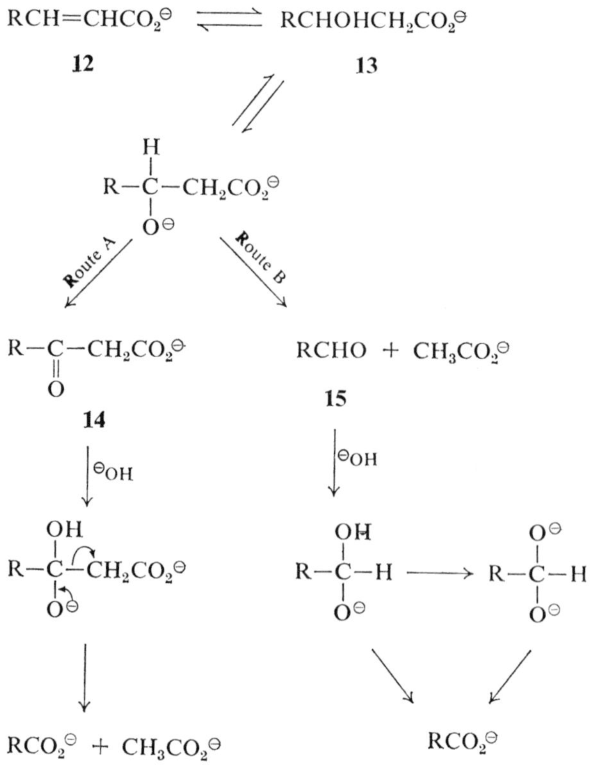

$$RCH=CHCO_2^{\ominus} \rightleftharpoons RCHOHCH_2CO_2^{\ominus}$$

12   **13**

$$R-\overset{\overset{\textstyle H}{|}}{\underset{\underset{\textstyle O^{\ominus}}{|}}{C}}-CH_2CO_2^{\ominus}$$

Route A   Route B

$$R-\overset{\overset{\textstyle}{\|}}{\underset{\underset{\textstyle O}{}}{C}}-CH_2CO_2^{\ominus} \qquad RCHO + CH_3CO_2^{\ominus}$$

14   **15**

$\downarrow {}^{\ominus}OH$ $\downarrow {}^{\ominus}OH$

$$R-\overset{\overset{\textstyle OH}{|}}{\underset{\underset{\textstyle O^{\ominus}}{|}}{C}}-CH_2CO_2^{\ominus} \qquad R-\overset{\overset{\textstyle OH}{|}}{\underset{\underset{\textstyle O^{\ominus}}{|}}{C}}-H \longrightarrow R-\overset{\overset{\textstyle O^{\ominus}}{|}}{\underset{\underset{\textstyle O^{\ominus}}{|}}{C}}-H$$

$$RCO_2^{\ominus} + CH_3CO_2^{\ominus} \qquad\qquad\qquad RCO_2^{\ominus}$$

Scheme 14.3

acids (p. 342); both processes are believed to involve hydride ion transfers with formation of molecular hydrogen. Although the evidence is as yet inconclusive, Route B is favored for the fission stage of the Varrentrapp reaction [5].

Fission of ethylenic acids is accompanied by a number of minor side reactions. These include further reaction of the main product with alkali (see p. 333); the formation of the saturated analog of the starting material by transfer of a hydride ion from a donor such as **16** to the α,β-unsaturated carboxylate (**17**) (Scheme 14.4); and the formation of "polymers," probably

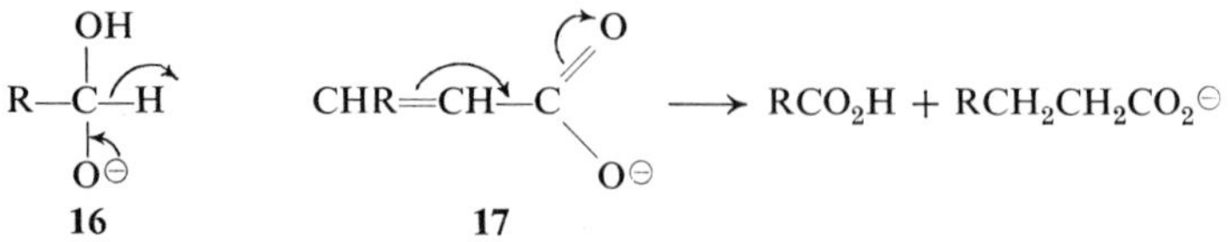

$$R-\overset{\overset{\textstyle OH}{|}}{\underset{\underset{\textstyle O^{\ominus}}{|}}{C}}-H \qquad CHR=CH-C\overset{\overset{\textstyle O}{\|}}{\underset{\underset{\textstyle O^{\ominus}}{}}{}} \longrightarrow RCO_2H + RCH_2CH_2CO_2^{\ominus}$$

16   **17**

Scheme 14.4

by addition of carbanions to the α,β-unsaturated carboxylate (17) in reactions of the Michael type [5].

Diethylenic and Acetylenic Acids

Few examples of alkali fusions on these acids have been reported. It is nevertheless clear that the fission products to be expected by analogy with the simple ethylenic acids are obtained at 250–300°, although polymerization and other side reactions become significant with these more highly unsaturated starting materials.

Alkali fusion of 6,6-diallylhexanoic acid (18) gives di-n-propylacetic (19) and acetic acids [18].

$$(CH_2{=}CHCH_2)_2CH(CH_2)_4CO_2H \rightarrow (CH_3CH_2CH_2)_2CHCO_2H + CH_3CO_2H$$
$$\mathbf{18} \qquad\qquad \mathbf{19}$$

Linoleic (20) [27] and *trans,trans*-octadeca-10,12-dienoic (21, $n = 4$) [28] acids both give (40–60%) myristic acid (22), acetic acid, and hydrogen. Linoleic acid is known [14, 29–31] to isomerize readily on treatment with alkalies to a mixture of 9,11- and 10,12-dienoic acids (21, $n = 4$ and 5),

$$CH_3(CH_2)_4CH{=}CHCH_2CH{=}CH(CH_2)_7CO_2H \rightarrow$$
$$\mathbf{20}$$

$$CH_3(CH_2)_nCH{=}CHCH{=}CH(CH_2)_{12-n}CO_2H \rightarrow$$
$$\mathbf{21}$$

$$CH_3(CH_2)_{12}CO_2H + CH_3CO_2H \;[+CH_3(CH_2)_{14}CO_2H]$$
$$\mathbf{22} \qquad\qquad\qquad \mathbf{2}$$

and there is no doubt that alkali fusion involves a reversible migration of the diene system in both directions along the fatty acid chain and subsequent fission of a 2,4-dienoate intermediate [5a, 27]. It is of interest that in addition to myristic acid (22), both linoleic acid and its conjugated isomer (21, $n = 4$) also give small amounts (ca. 5%) of palmitic acid (2) [27, 28]. The formation of this byproduct may be attributed to a hydride ion transfer from an intermediate of the type 16, involved in the main fission reaction, to the 2,4-dienoate (23) (Scheme 14.5). Nucleophilic attack by hydride ion

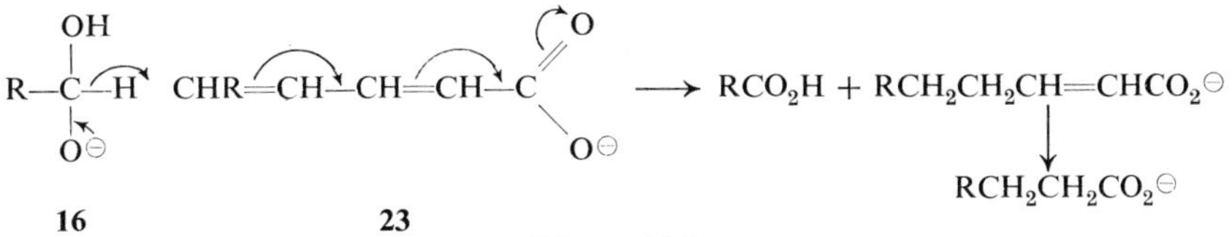

$$R{-}\overset{\overset{\displaystyle OH}{|}}{\underset{\underset{\displaystyle O^\ominus}{|}}{C}}{-}H \quad CHR{=}CH{-}CH{=}CH{-}C\overset{\displaystyle O}{\underset{\displaystyle O^\ominus}{\diagup}} \longrightarrow RCO_2H + RCH_2CH_2CH{=}CHCO_2{}^\ominus$$
$$\mathbf{16} \qquad\qquad \mathbf{23} \qquad\qquad\qquad\qquad\qquad RCH_2CH_2CO_2{}^\ominus$$

Scheme 14.5

apparently assumes greater importance with the dienoates than with the enoates from monoethylenic acids [27].

Linolenic acid (24) gives cyclic isomers in high yield on prolonged treatment with glycolic potassium hydroxide [32]. It has, however, been claimed that linolenic acid is converted into lauric acid (25) during alkali fusion of mixtures of fatty acids [33], but no experimental details have been published.

$$CH_3CH_2CH{=}CHCH_2CH{=}CHCH_2CH{=}CH(CH_2)_7CO_2H \xrightarrow{?} CH_3(CH_2)_{10}CO_2H$$
$$\mathbf{24} \qquad\qquad\qquad\qquad\qquad\qquad\qquad\qquad\qquad \mathbf{25}$$

Alkali fusion of stearolic (26), undecynoic (27), and dodec-6-ynedioic (28) acids shows that acetylenic acids, like the isomeric diene acids, give mainly

$$CH_3(CH_2)_7C{\equiv}C(CH_2)_7CO_2H \rightarrow CH_3(CH_2)_{12}CO_2H\ [+CH_3(CH_2)_{14}CO_2H]$$
$$\mathbf{26}$$

$$HC{\equiv}C(CH_2)_8CO_2H \rightarrow CH_3(CH_2)_5CO_2H\ [+CH_3(CH_2)_7CO_2H]$$
$$\mathbf{27}$$

$$HO_2C(CH_2)_4C{\equiv}C(CH_2)_4CO_2H \rightarrow HO_2C(CH_2)_6CO_2H$$
$$\mathbf{28}$$

saturated products with loss of four carbon atoms [27, 34, 35]. With stearolic acid it has been demonstrated that reaction occurs by an initial migration of the triple bond in both directions along the fatty acid chain, presumably by a sequence of reversible acetylene → allene → acetylene rearrangements [5a, 27] (Scheme 14.6). An octadec-4-ynoic acid formed in this way would then

$$-C{\equiv}CCH_2- \rightleftharpoons -CH{=}C{=}CH- \rightleftharpoons -CH_2C{\equiv}C-$$

Scheme 14.6

be expected to isomerize to the 2,4-dienoic acid, since such conversions are well authenticated with other Δ^4-acetylenic acids under milder alkaline conditions [36], and finally undergo fission of the type discussed for linoleic acid. As is to be expected if a common intermediate is involved, the side reaction leading to the saturated acid with only two carbon atoms less than the starting material is also observed [27].

The transformations discussed above for the fission of acetylenic acids will not apply to the α,β- and β,γ-acetylenic compounds. These are known to decompose in strong alkali, probably by an initial hydration to a β-keto acid, and alkali fusion of dec-2-ynoic acid gives octanoic acid as the main acidic fission product [27].

Other Carboxylic Acids

Under sufficiently drastic conditions, the saturated carboxylic acids will also react with molten alkali. It is claimed that, at 370–430°, 4-cyclohexyl-butyric acid (29) gives cyclohexylacetic acid (30) (ca. 50%), acetic acid, and hydrogen, and that the main products from the γ-alkylhexanoic acids (31, R = CMe_3, Me, or Et) are the corresponding α-alkylbutyric acids (32)

$$\text{29} \qquad\qquad \text{30}$$

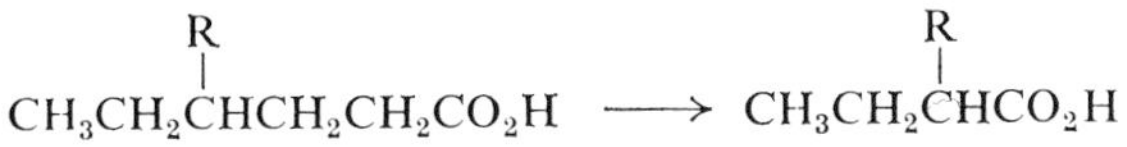

$$\text{31} \qquad\qquad \text{32}$$

[37, 38]. Similar transformations have been observed as minor side reactions during the alkali fusion of unsaturated acids under much milder conditions [5]. It has been tentatively suggested that these degradations proceed (Scheme 14.7) by an initial slow dehydrogenation, followed by a rapid fission of the resulting α,β-ethylenic acid [5a, 5].

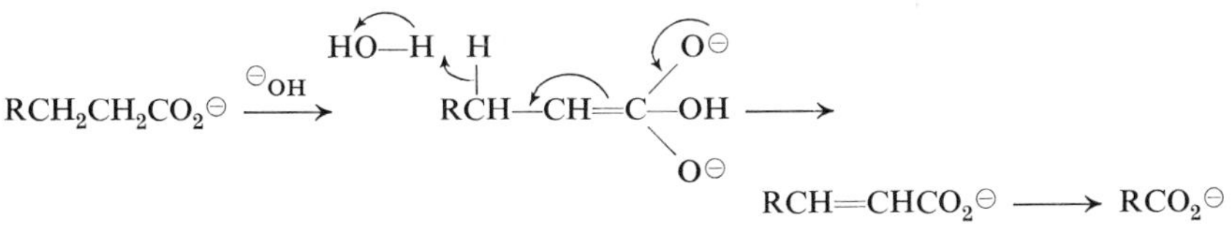

Scheme 14.7

Reactions of a different type are reported to occur during the alkali fusion of highly alkylated acids. α,α-Dimethyloctanoic acid (33) is stated to give oct-1-ene, 2-methyloct-1-ene (34), and 2-methyloct-2-ene (35); β,β-dimethyl-nonanoic acid (36), a mixture of the same methyloctenes together with acetic acid: γ,γ-dimethyldecanoic acid (37), mainly 2-methyloct-1-ene (34) [18]; and camphoric acid (38), a complex mixture of aliphatic mono- and di-carboxylic acids [39]. A careful re-examination of these and related reactions seems desirable.

$$CH_3(CH_2)_5CH=CH_2$$

$$CH_3(CH_2)_5CMe_2CO_2H$$

33

$$CH_3(CH_2)_5CMe=CH_2$$

34

+

$$CH_3(CH_2)_4CH=CMe_2$$

35

$$CH_3(CH_2)_5CMe_2CH_2CO_2H$$

36

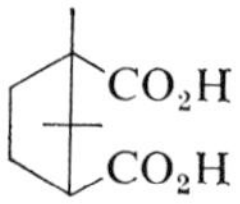

38

$$CH_3(CH_2)_5CMe_2(CH_2)_2CO_2H$$

37

Under drastic conditions the decarboxylation of acids is also to be expected [40–44]. Monocarboxylic acids have been obtained from adipic and β-alkyladipic acids, but these reactions may well involve cyclic intermediates [3].

The cyclization of phenylglycine (**39**) to indoxyl (**40**) in molten alkali is of importance in the manufacture of indigo. This route has long since superseded an earlier process using phenylglycine-o-carboxylic acid (**41**) [45].

Before concluding this section, it is of interest to recall that sodium formate, which may be prepared by the action of carbon monoxide on

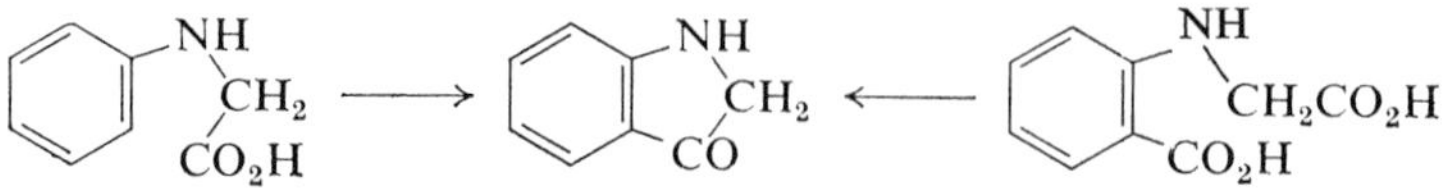

39

40

41

sodium hydroxide at ca. 200°, is converted into sodium oxalate and hydrogen when heated at 400°, or at 300° in the presence of a small excess of alkali [40].

$$NaOH + CO \rightarrow HCO_2Na \rightarrow (CO_2Na)_2 + H_2$$

These processes have found industrial applications [46].

3 KETONES

A variety of products has been reported from alkali fusions of ketones, but recent model studies with fatty acid derivatives show that the reactions involved fall mainly into three categories: (1) hydrolysis of a carbon–carbon bond adjacent to the carbonyl group, (2) nucleophilic attack of the carbonyl group by a carbanion derived from a second molecule of the same, or another, ketone, and (3) dehydrogenation to give an α,β-unsaturated ketone, followed by a fission of the retro-aldol type. Of these reactions hydrolysis is the commonest and was first utilized by Wallach in his classical studies on terpenes at the turn of the century. Dehydrogenation is normally only a minor side reaction.

The hydrolysis of a ketone (42) on alkali fusion may be attributed (Scheme 14.8) to an initial attack of the carbonyl group by hydroxyl ion to give

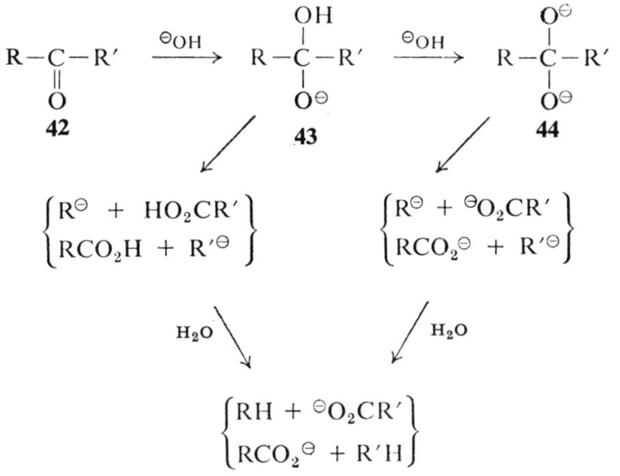

$$
\begin{array}{ccccc}
& & \mathrm{OH} & & \mathrm{O}^{\ominus}\\
& & | & & |\\
\mathrm{R-C-R'} & \xrightarrow{\ominus\mathrm{OH}} & \mathrm{R-C-R'} & \xrightarrow{\ominus\mathrm{OH}} & \mathrm{R-C-R'}\\
\| & & | & & |\\
\mathrm{O} & & \mathrm{O}^{\ominus} & & \mathrm{O}^{\ominus}\\
\mathbf{42} & & \mathbf{43} & & \mathbf{44}
\end{array}
$$

$$
\left\{ \begin{array}{l} \mathrm{R}^{\ominus}\ +\ \mathrm{HO_2CR'}\\ \mathrm{RCO_2H}\ +\ \mathrm{R'}^{\ominus} \end{array} \right\} \qquad \left\{ \begin{array}{l} \mathrm{R}^{\ominus}\ +\ {}^{\ominus}\mathrm{O_2CR'}\\ \mathrm{RCO_2}^{\ominus}\ +\ \mathrm{R'}^{\ominus} \end{array} \right\}
$$

$$
\mathrm{H_2O} \searrow \qquad \swarrow \mathrm{H_2O}
$$

$$
\left\{ \begin{array}{l} \mathrm{RH}\ +\ {}^{\ominus}\mathrm{O_2CR'}\\ \mathrm{RCO_2}^{\ominus}\ +\ \mathrm{R'H} \end{array} \right\}
$$

Scheme 14.8

either **43** or, by analogy with the Cannizzaro reaction [47], the doubly charged species **44**, followed by fission of one of the adjacent carbon–carbon bonds [3, 48] (cf. Haller-Bauer reaction, i.e., cleavage by sodamide [49]). It is unlikely that the carbanions indicated in this scheme have a free existence [50]; they are probably transferred directly to an acceptor such as water (e.g., Scheme 14.9) [3]. The simplest reaction of this type is that of acetone to give acetic acid and methane [42].

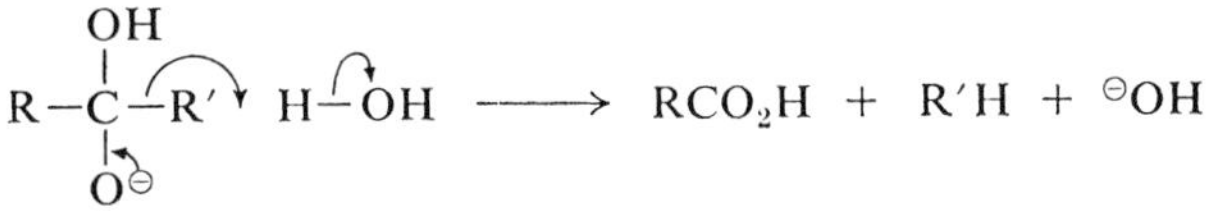

$$
\begin{array}{c}
\mathrm{OH}\\
|\\
\mathrm{R-C-R'} \quad \mathrm{H-OH} \longrightarrow \mathrm{RCO_2H}\ +\ \mathrm{R'H}\ +\ {}^{\ominus}\mathrm{OH}\\
|\\
\mathrm{O}^{\ominus}
\end{array}
$$

Scheme 14.9

With unsymmetrical ketones, hydrolysis can occur in two ways, depending on which carbon–carbon bond is broken, and a mixture of products is usually obtained. The relative importance of these two hydrolyses depends on the nature of the groups R and R'. The influence of a number of substituents on the direction of fission of diaryl ketones has been studied [51]. Thus it is reported that p-methylbenzophenone gives benzoic and p-toluic acids in the proportions 1:1.4, whereas o-methylbenzophenone gives benzoic and o-toluic acids in the proportions 1:0.2. Alkali fusion of fluorene-2-carboxylic acid (45) is stated to give diphenyl 2,4'-dicarboxylic acid (46) [52]. There

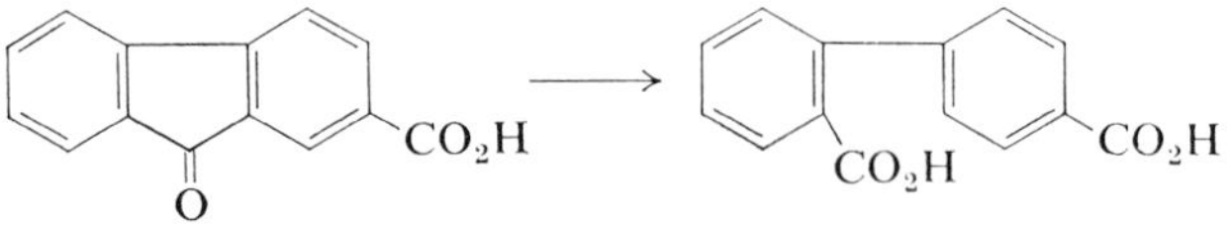

45   46

is some evidence that, with unsymmetrical alicyclic ketones, cleavage of the bond joining the carbonyl group to the least substituted α-carbon atom predominates. Thus alkali fusion of camphor (47) gives campholic acid (48) together with smaller amounts of isocampholic acid (49) [53–55]; and menthone oxime (50) is believed to give mainly 2-isopropyl-5-methylhexanoic

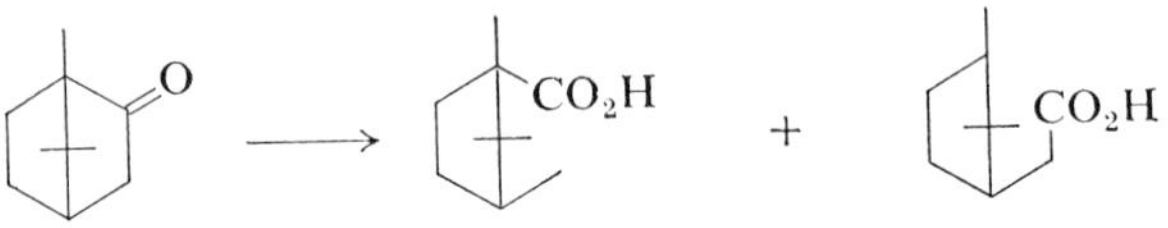

47   48   +   49

acid (51) [56]. By analogy, it is likely that isofencholic acid, formed from isofenchone (52), is 53 rather than 54 [57]. Fenchone (55) itself gives fencholic acid (56) [55]. It is probable that steric factors, as well as the relative stability of ionic intermediates, help to determine the preferred direction of fission.

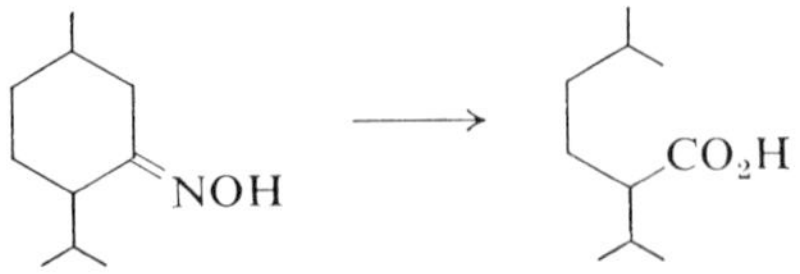

50   51

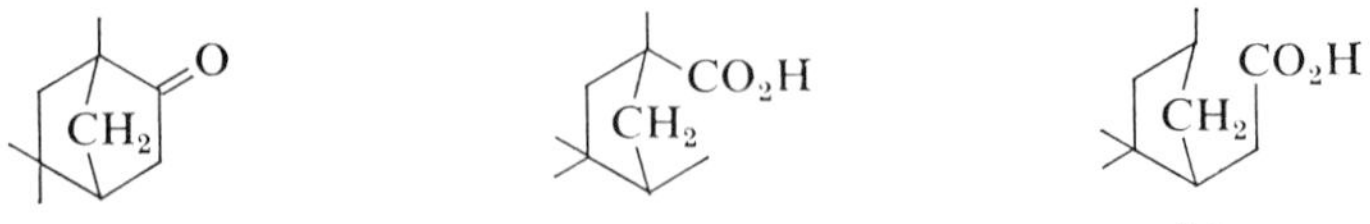

52   53   54

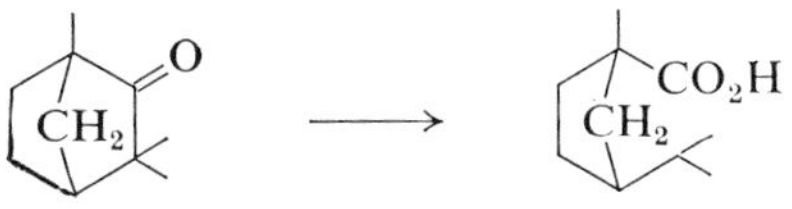

55   56

In a careful study of the fission of some estrogenic hormones, it has been clearly established that doisynolic acid, formed (ca. 50%) by alkali fusion of estrone (57), has the structure 58, in accordance with the above generalization

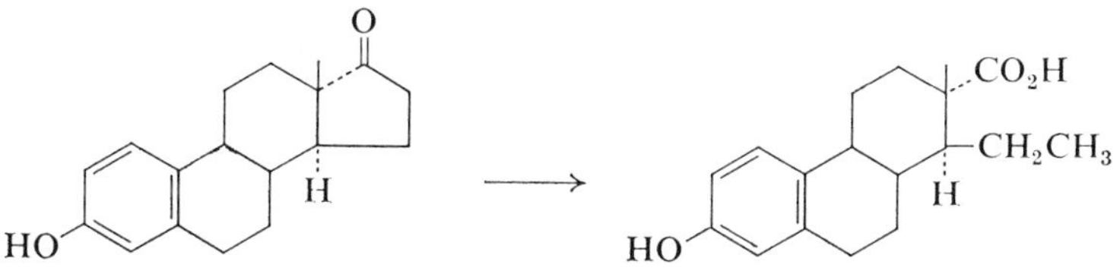

57   58

[58–60]. With equilenin (59), (+)-*trans*-bisdehydrodoisynolic acid (60) is obtained together with a biologically active (−)-*cis*-isomer (61) by inversion at the "benzylic" position [59, 60]. The production of a potent estrogenic acid by alkali fusion of equilin has been claimed but the constitution of the product has not been fully established [60, 61].

In the absence of substituent or steric effects, fission normally occurs to the same extent in both directions. Alkali fusion of 12-oxooctadecanoic

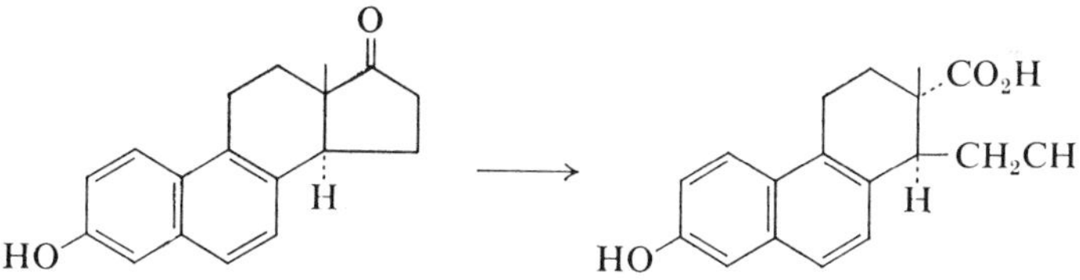

59   60

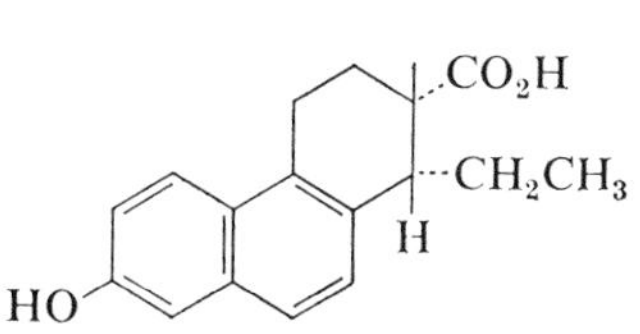

61

acid (62) gives heptanoic and undecanoic acids, by fission of the 11,12 bond, and dodecanedioic acid, by fission of the 12,13 bond, in approximately equimolar amounts [3]. However, the yield of undecanoic acid (0.25 mole)

$$CH_3(CH_2)_5CO(CH_2)_{10}CO_2H \rightarrow \begin{cases} CH_3(CH_2)_5CO_2H \\ CH_3(CH_2)_9CO_2H \\ HO_2C(CH_2)_{10}CO_2H \end{cases}$$

62

is somewhat lower than that of heptanoic and dodecanedioic acids (0.3–0.4 mole), and no large amount of hexane is isolated. This indicates that the hydrolysis of a ketone in molten alkali (Schemes 14.8 and 14.9) is accompanied by a reaction in which a carbanion is transferred from an intermediate of the type **43** or **44** to an acceptor other than water, presumably the ketone itself (Scheme 14.10) [3, 48]. Although the resulting tertiary alcohols do not

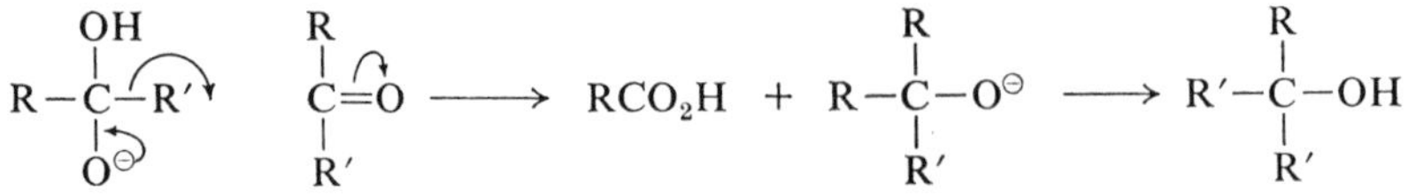

Scheme 14.10

$$CH_3(CH_2)_{10}CO(CH_2)_{10}CH_3 \longrightarrow CH_3(CH_2)_{10}CO_2H +$$
63   **25**

$$\begin{array}{c} CH_3(CH_2)_{10} \\ \diagdown \\ CH_3(CH_2)_{10} \end{array} C=CH(CH_2)_9CH_3$$

64

survive, evidence for reactions of this type can be inferred from the detection of dehydration and other transformation products of the alcohols. The fission of laurone (**63**) to lauric acid (**25**) is accompanied by the formation of the olefin **64** [48]. Production of 6-cyclohexenylhexanoic acid (**65**) and

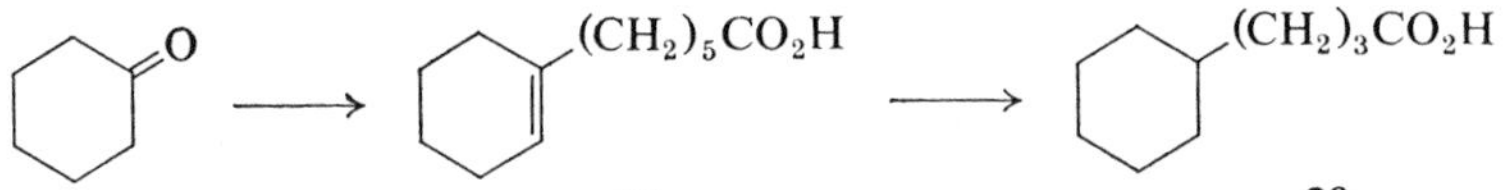

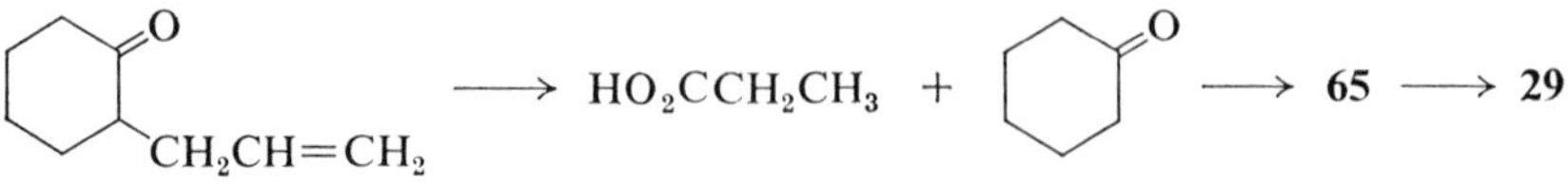

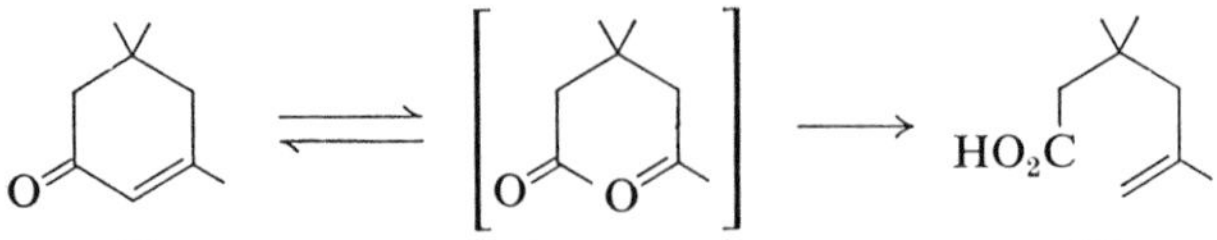

its isomers or, by a further reaction of the Varrentrapp type, of 4-cyclo-hexylbutyric acid (**29**) can largely replace hydrolysis to hexanoic acid during alkali fusion of cyclohexanone [62, 63]. The same products are also obtained from 2-allylcyclohexanone (66); clearly, cyclohexanone and propionic acid are first formed by a Varrentrapp fission [62]. The conversion of isophorone (67) to an acid regarded as 68 can be attributed to a retro-aldol reaction followed by processes of the type now under consideration [64]. (The initial reaction can be regarded as reversible. Treatment of isophorone with concentrated alkali yields a dimer [65] which probably has a constitution analogous to that (70) proposed [66] for the corresponding product from piperitone (69).)

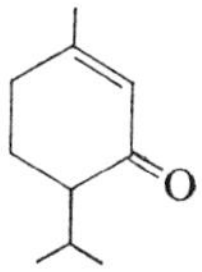

69

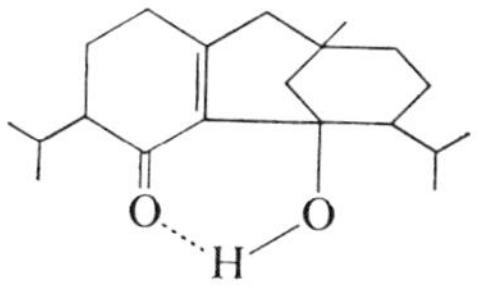

70

Model studies on the alkali fusion of ketones [3, 3a] indicate a minor side reaction (Scheme 14.11) in which an initial dehydrogenation, similar to, but occurring more rapidly than, that postulated for acids (Scheme 14.7), gives rise to methyl ketones. The latter then react further, either by hydrolysis (Scheme 14.8) or by processes similar to that shown in Scheme 14.10. The reported [67] formation from stearone (71) of palmitic acid (2) in addition to stearic acid (72) may be due, at least partially, to an initial dehydrogenation of the type shown in Scheme 14.11. A similar process may also

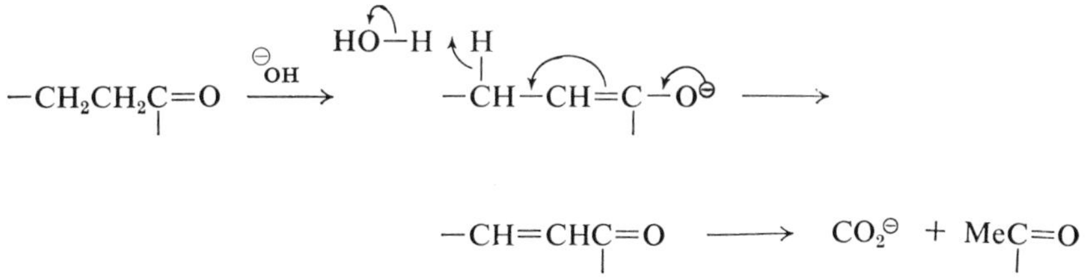

Scheme 14.11

account for the formation of thymol during the fission of menthone oxime [56].

$$CH_3(CH_2)_{16}CO(CH_2)_{16}CH_3 \rightarrow CH_3(CH_2)_{16}CO_2H + CH_3(CH_2)_{14}CO_2H$$

71   72   2

Results analogous to those from 12-oxooctadecanoic acid (72) are also obtained with 10-oxooctadecanoic acid, 10-oxoheptadecanoic acid, and nonadecan-10-onedioic acid [3]. Departures from this general pattern of ketone fission are, however, encountered with appropriately substituted compounds. Thus fission of a 5-keto acid seems to occur mainly by reactions of the retro-Michael type (Scheme 14.12) [3].

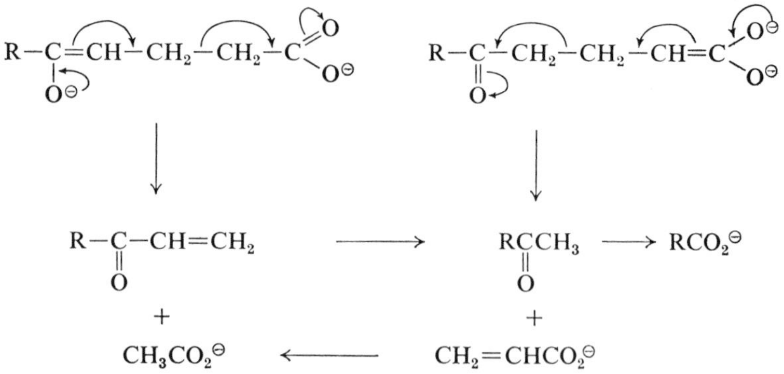

Scheme 14.12

A more striking example is the alkali fusion of 6-oxooctadecanoic acid (**73**). This gives only small amounts of tridecanoic and adipic acids (Scheme 14.8). The main products are the cyclopentenyl acid (**74**) and, by further reaction with molten alkali, β-dodecyladipic acid (**76**) [3, 3a]. The initial production of the cyclopentenyl acid may be attributed (Scheme 14.13) to an intramolecular reaction of the aldol or Claisen type, followed by dehydration of the resulting β-hydroxy acid (cf. conversion of 2-hydroxycyclohexanecarboxylic acid into cyclohexenecarboxylic acid [12]). The isolation of an α,β-unsaturated acid from an alkali fusion is at first sight surprising, but is no doubt associated

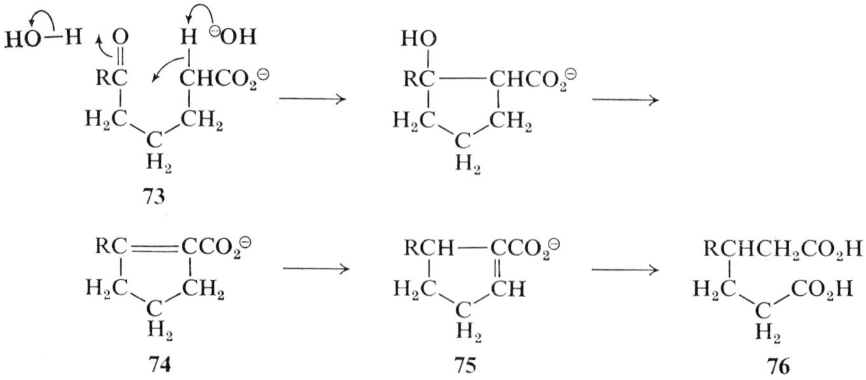

Scheme 14.13

with the fact that direct fission of **74** merely regenerates 6-oxooctadecanoic acid (**73**). Isomerization of **74** to **75** by a prototropic migration of the double bond, followed by a fission of the Varrentrapp type, accounts for the formation of the dicarboxylic acid (**76**) (cf. isomerization and fission of octadec-2-enoic acid [124] and the conversion of cyclohexenecarboxylic acid into pimelic acid [12, 13]).

A further example of the interaction of functional groups is provided by the alkali fusion of methymycin (**77**) when 2,4,6-trimethylcyclohex-2-en-1-one (**79**) is obtained as a volatile product [68]. The formation of this ketone is ascribed (Scheme 14.14) to fission of the macrocyclic ring (hydrolysis

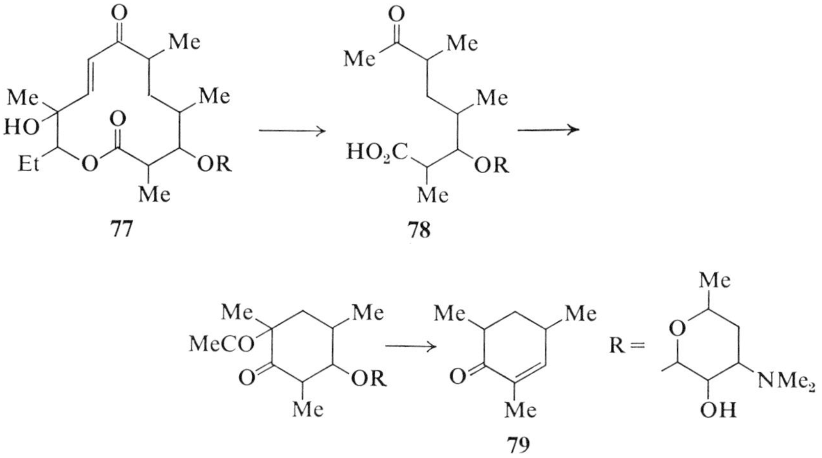

Scheme 14.14

and retro-aldol) to give the 7-keto acid (**78**), followed by intramolecular acylation, β-diketone cleavage, and β-elimination. The recognition of the transformation involved provided an important clue to the structure of the antibiotic.

4 ALCOHOLS AND RELATED COMPOUNDS

Primary Alcohols (The Dumas-Stass Reaction)

In the same year that Varrentrapp discovered the fission of ethylenic acids, Dumas and Stass made an observation which also has an important bearing on the interpretation of reactions of organic compounds under strongly alkaline conditions [69]. They reported that primary alcohols are dehydrogenated to the corresponding acids when heated with potassium hydroxide. This reaction is now regarded [3, 70] as involving two successive hydride ion transfers with the intermediate formation of the corresponding aldehyde (Scheme 14.15).

Both solid and concentrated (ca. 40%) aqueous alkali metal hydroxides can be used to effect these dehydrogenations [22, 69, 71–73], and the addition of cadmium oxide or cadmium to the reaction mixture is claimed to be

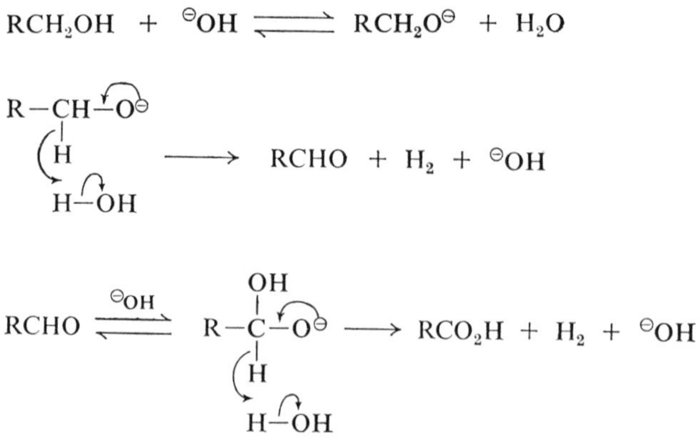

$$RCH_2OH + {}^{\ominus}OH \rightleftharpoons RCH_2O^{\ominus} + H_2O$$

$$R-CH{-}O^{\ominus} \atop \left(\begin{matrix} H \\ | \\ H{-}OH \end{matrix}\right) \longrightarrow RCHO + H_2 + {}^{\ominus}OH$$

$$RCHO \underset{{}^{\ominus}OH}{\rightleftharpoons} R-\overset{\displaystyle OH}{\underset{\displaystyle \left(\begin{matrix} H \\ | \\ H{-}OH \end{matrix}\right)}{\overset{|}{C}}}-O^{\ominus} \longrightarrow RCO_2H + H_2 + {}^{\ominus}OH$$

Scheme 14.15

beneficial [74, 75]. With many primary alcohols excellent yields of the corresponding acids have been reported; at one time the measurement of the volume of hydrogen evolved was even suggested as a method of determining the molecular weight of primary alcohols [76]. It is, however, stated that aliphatic alcohols with fewer than six carbon atoms are partly dehydrated to olefins during the reaction [77] and that the hydrogen liberated in the conversion of benzyl alcohol to benzoic acid (ca. 65%) effects partial reduction (ca. 13%) of the starting material to toluene [78]. As required by Scheme 14.15, both aliphatic [42, 79] and aromatic aldehydes [80, 81] give acids in good yield on alkali fusion, although complex condensation products may be formed on treatment of aliphatic aldehydes with concentrated aqueous alkalies [82, 83].

A few applications of the Dumas-Stass reaction may be cited. The constitution of homoisopilopic acid (80), a degradation product of the jaborandi alkaloid isopilocarpine, was deduced from its transformation into ethyltricarballylic acid (81) on alkali fusion [84]. β-Anilinoethanol (82), which is readily obtained from aniline and ethylene chlorohydrin, may be substituted for phenylglycine (39) in the manufacture of indoxyl (40) [45]. Aliphatic α-amino acids can be prepared from the corresponding amino alcohols by dehydrogenation with alkali [74].

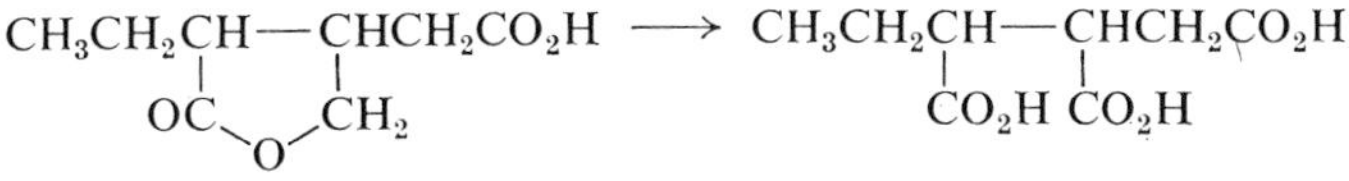

$$CH_3CH_2CH\!-\!CHCH_2CO_2H \longrightarrow CH_3CH_2CH\!-\!CHCH_2CO_2H$$

80   **81**

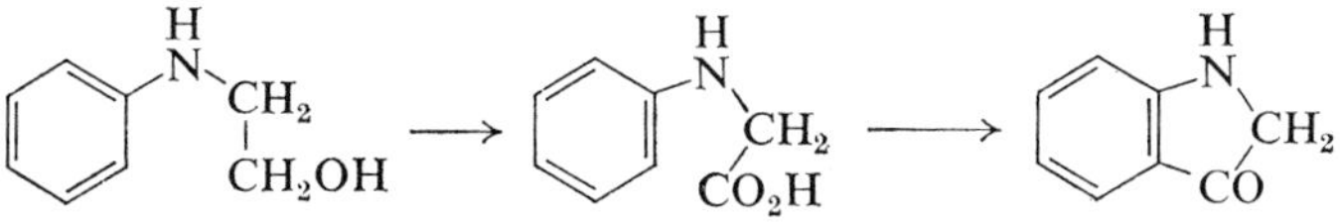

82   **39**   **40**

Many examples exist of the reduction of aliphatic and aromatic ketones to secondary alcohols by treatment with a primary alcohol and concentrated alkali [85, 86]. The reduction of diacetyldeuteroporphyrin to hematoporphyrin in Fischer's synthesis of hemin provides a particularly elegant example [87]. Hydrogen transfers of this type show that ketones can compete efficiently with water or an alcohol as acceptors of a hydride ion from the intermediates involved in the Dumas-Stass dehydrogenation of primary alcohols to acids (e.g., Scheme 14.16). Under alkaline conditions formaldehyde, and presumably other aldehydes, can also serve as hydride ion donors for the reduction of ketones [88–91].

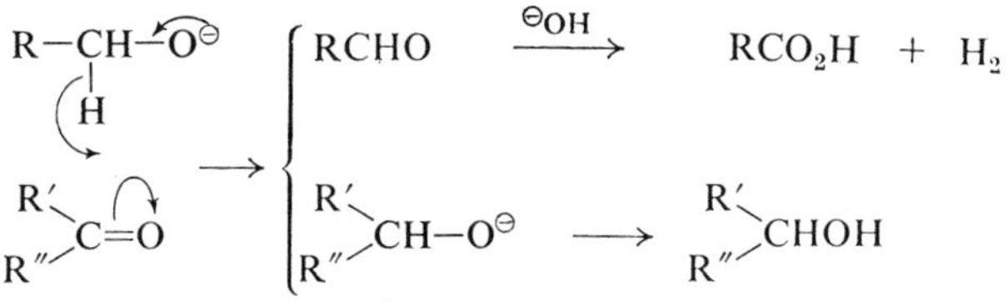

Scheme 14.16

Secondary alcohols formed during alkaline fission of natural products are often the result of a reduction, of the type generalized in Scheme 14.16, of a ketone produced initially. The formation of 6-methylhept-5-en-2-ol from geraniol [92–94] and mangostin [95], and of octan-2-ol from ricinoleic acid [79, 96], can be readily explained in this way (see pp. 349, 361, and 348).

The Guerbet Reaction (Primary and Secondary Alcohols)

When heated with alkalies, both primary (**83**) and secondary alcohols possessing an unsubstituted methylene group at C-2 can condense to give alcohols (e.g., **84**) with two or more times as many carbon atoms as the starting material. The reaction was discovered independently by Markownikow

[97] and Guerbet [77], but the latter worker was mainly responsible for its early development.

The condensation is greatly favored, and the Dumas-Stass reaction suppressed, by essentially anhydrous conditions. These are best achieved by removing the water formed during the condensation, either by azeotropic distillation or by adding a dehydrating agent to the reaction mixture. The addition of dehydrogenation catalysts is also claimed to be beneficial [70, 98–103]. A common procedure for carrying out the reaction consists in heating the alcohol with the corresponding alkali metal alkoxide rather than potassium hydroxide [73, 77, 104].

The Guerbet reaction may be regarded as involving an aldol condensation, followed by hydride ion transfers from either the alcoholate, or a derived

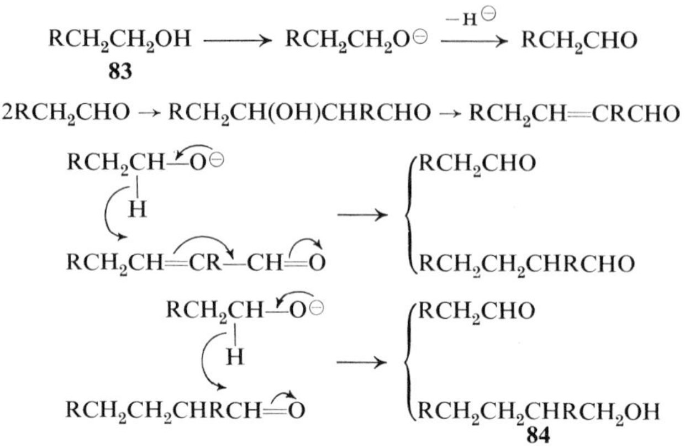

$$RCH_2CH_2OH \longrightarrow RCH_2CH_2O^{\ominus} \xrightarrow{-H^{\ominus}} RCH_2CHO$$
$$\textbf{83}$$

$$2RCH_2CHO \rightarrow RCH_2CH(OH)CHRCHO \rightarrow RCH_2CH{=}CRCHO$$

$$RCH_2CH{-}O^{\ominus} \qquad\qquad \left\{ \begin{array}{l} RCH_2CHO \\ \\ RCH_2CH_2CHRCHO \end{array} \right.$$
$$\overset{|}{C}{-}H$$
$$RCH_2CH{=}CR{-}CH{=}O$$

$$RCH_2CH{-}O^{\ominus} \qquad\qquad \left\{ \begin{array}{l} RCH_2CHO \\ \\ RCH_2CH_2CHRCH_2OH \\ \hspace{3.5cm}\textbf{84} \end{array} \right.$$
$$\overset{|}{C}{-}H$$
$$RCH_2CH_2CHRCH{=}O$$

Scheme 14.17

aldehyde, to the α,β-unsaturated aldehyde and its saturated analogue (e.g., Scheme 14.17) [3, 70, 103–105–109]. As required by this interpretation, benzyl alcohol, methyl alcohol, and 2-alkyl aliphatic alcohols do not exhibit the Guerbet reaction [77, 103, 108]. However, as expected, a mixture of benzyl and ethyl alcohols gives 3-phenylpropanol (**85**) [110] and a mixture of isobutanol and cyclohexanol gives 2-isobutylcyclohexanol (**86**) [103]. Moreover a mixture of n-butanol and cyclohexanol is reported to give 2-butylcyclohexanol (**87**), but no 2-cyclohexylbutanol, in good agreement with the course of the reaction known to occur between the two postulated intermediates, butyraldehyde and cyclohexanone [109].

With primary alcohols the Guerbet reaction is frequently accompanied, or followed, by a Dumas-Stass dehydrogenation [72, 73, 77]. Thus, depending on the experimental conditions, octadecanol (**88**) can give mainly 2-hexadecyleicosanol (**89**) [106] or 2-hexadecyleicosanoic acid (**90**) [111].

$$PhCH_2OH + CH_3CH_2OH \longrightarrow PhCH_2CH_2CH_2OH$$

85

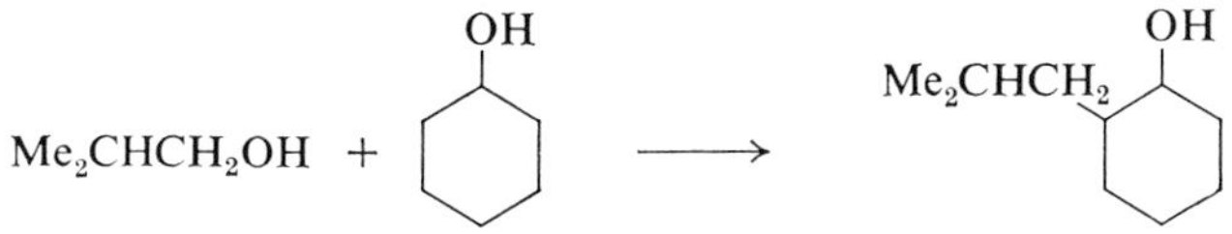

86

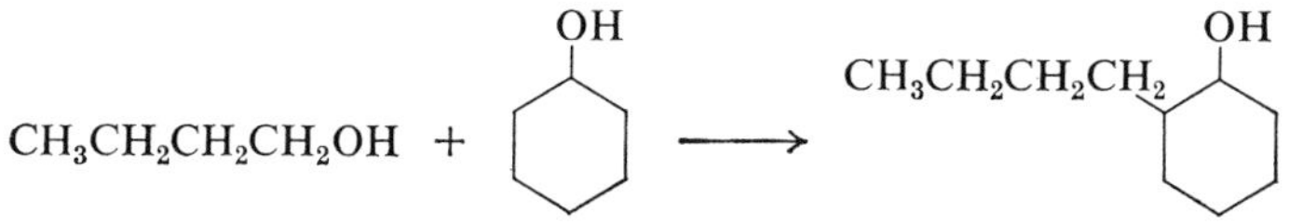

87

Related to the Guerbet reaction is the formation of fatty acids with two extra carbon atoms on heating a sodium alkoxide with either sodium acetate [112] or *N*-ethylacetanilide [113].

$$CH_3(CH_2)_{16}CH_2OH$$

88

$$(CH_3(CH_2)_{15})(CH_3(CH_2)_{17})CHCH_2OH \qquad (CH_3(CH_2)_{15})(CH_3(CH_2)_{17})CHCO_2H$$

89 **90**

Secondary Alcohols

As mentioned above, some secondary alcohols can undergo the Guerbet reaction if heated with alkali under essentially anhydrous conditions. However, in most alkali fusions the conditions are not favorable for this condensation, and fissions to give products very similar in composition to those obtained from the corresponding ketones are far more frequently encountered [77]. Thus borneol (**91**), like camphor (**47**) gives a mixture of campholic (**48**) and isocampholic (**49**) acids [53, 54]. Estradiol (**92**), like estrone (**57**), gives doisynolic acid (**58**) [61]. Cyclohexanol can be substituted for cyclohexanone in the reactions outlined on page 338 [63], and the fission of 4-alkylcyclohexanols (**93**) gives 4-alkylhexanoic acids (**31**) [38]. Both 10- and 12-hydroxyoctadecanoic acids (**94** and **95**) give mixtures of products almost

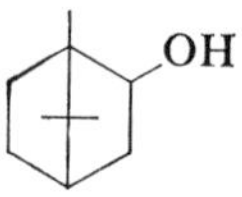

91

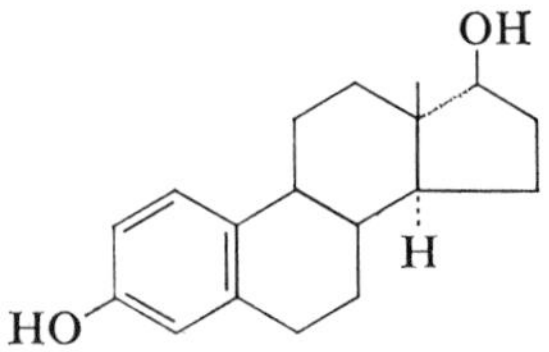

92

93

31

$$CH_3(CH_2)_7CHOH(CH_2)_8CO_2H$$
94

$$CH_3(CH_2)_5CHOH(CH_2)_{10}CO_2H$$
95

identical in composition with those from the corresponding keto acids [3]. γ-Lactones undergo a similar fission to give the expected fatty acid and propionic acid [114, 115]. On alkali fusion 6-hydroxyoctadecanoic acid, like the 6-keto acid (73), gives both the cyclopentenyl acid (74) and β-dodecyl-adipic acid (76) [3]. α-hydroxy fatty acids undergo fission of the 1,2-bond to give the nonacids [3a].

Although fission by a mechanism of the type shown in Scheme 14.18 cannot be excluded, and is even favored (p. 331) for β-hydroxy acids [5] and some phenolic derivatives (p. 364), the close similarity between the

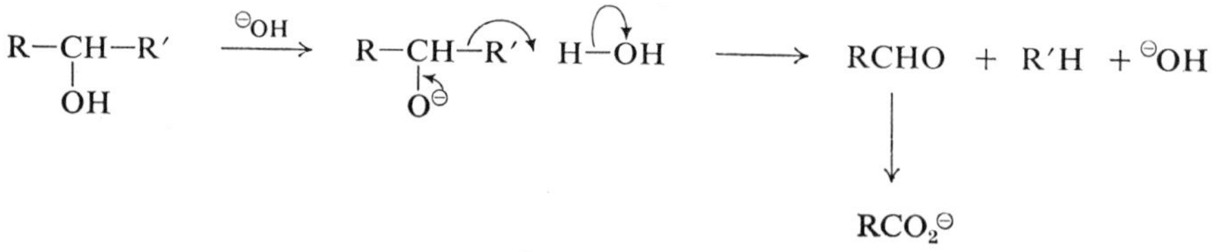

Scheme 14.18

products from ketones and the corresponding secondary alcohols strongly suggests that reaction of the latter compounds proceeds by an initial dehydrogenation to the ketone (Scheme 14.19). The feasibility of such a transformation was recently demonstrated by the conversion in good yield of 10-hydroxyoctadecanoic acid (94) and 12-hydroxyoctadecanoic (dihydroricinoleic) acid (95) into the corresponding keto acids by controlled alkali fusion at 200–250° or by treatment with 30% aqueous potassium hydroxide at 300–360° [3].

The reported formation of camphor (97) during the fission of borneol (91) with molten alkali is also of interest in this connection [53, 54].

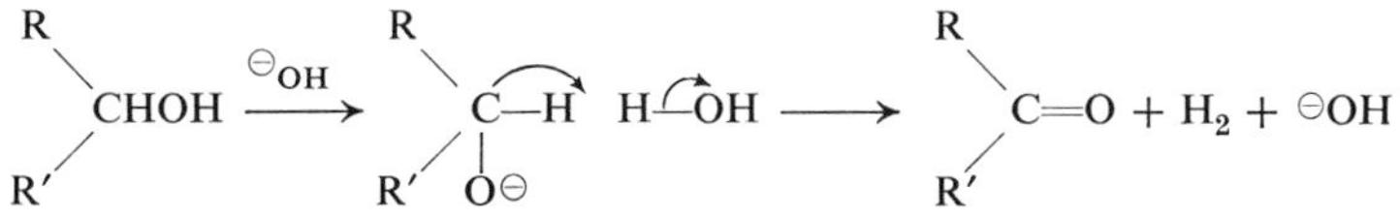

Scheme 14.19

Early information on the nature of the cinchona alkaloids was furnished by the observation that quinoline (97) is produced on alkali fusion of cinchonine (96) [13, 116, 117] (quinine similarly gives 6-methoxyquinoline [117, 118]. This degradation can also be interpreted as involving the initial formation of a ketone, either by dehydrogenation or by rearrangement. Dehydration to cinchene, followed by a retro-aldol fission, may occur

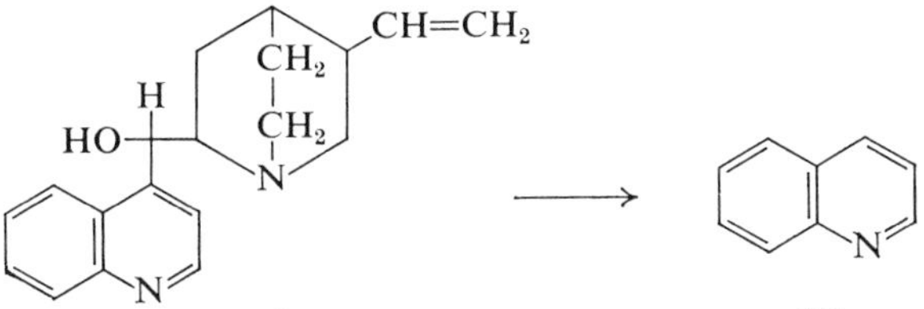

simultaneously, since lepidine (98) is also formed. Other bases detected include 3-ethylpyridine and 3-ethyl-4-methylpyridine, which presumably arise from the quinuclidine moiety by reactions of the type discussed on page 369.

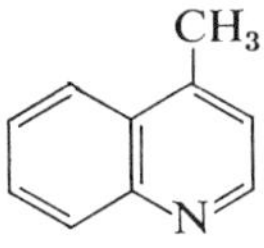

Tertiary Alcohols

The behavior of tertiary alcohols under strongly alkaline conditions has received very little study, but it seems probable that the principal reaction is dehydration. Olefins produced during alkali fusions are frequently attributed to the dehydration of a tertiary alcohol formed initially [3, 48, 62, 63, 119].

As mentioned previously (p. 338), there is good reason to believe that tertiary alcohols are formed *in situ* during the alkali fusion of ketones owing to attack of the carbonyl group by a carbanion. The converse of this type of reaction (Scheme 14.20) may occur under suitable conditions [93, 120, 121].

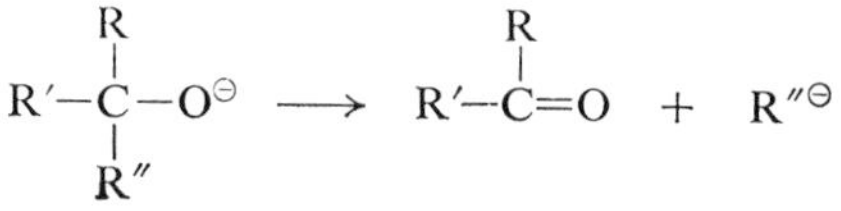

$$R'-\underset{\underset{R''}{|}}{\overset{\overset{R}{|}}{C}}-O^{\ominus} \longrightarrow R'-\underset{}{\overset{\overset{R}{|}}{C}}=O + R''^{\ominus}$$

Scheme 14.20

Unsaturated Alcohols

Of the alkaline fissions of natural products, that of ricinoleic acid (**100**) has achieved considerable commercial importance. It is reported that treatment of the acid or castor oil with concentrated alkalies at ca. 200° gives mainly octan-2-one (**103**) and 10-hydroxydecanoic acid (**104**), but that octan-2-ol (**105**), sebacic acid (**106**), and hydrogen are the principal products from reactions at temperatures above 240° [79, 96]. To explain the formation of the low temperature products, it has been suggested [3] that ricinoleic acid, like dihydroricinoleic acid (**95**), first undergoes dehydrogenation (Scheme 14.21). Isomerization of the resulting β,γ-unsaturated ketone (**99**) into the

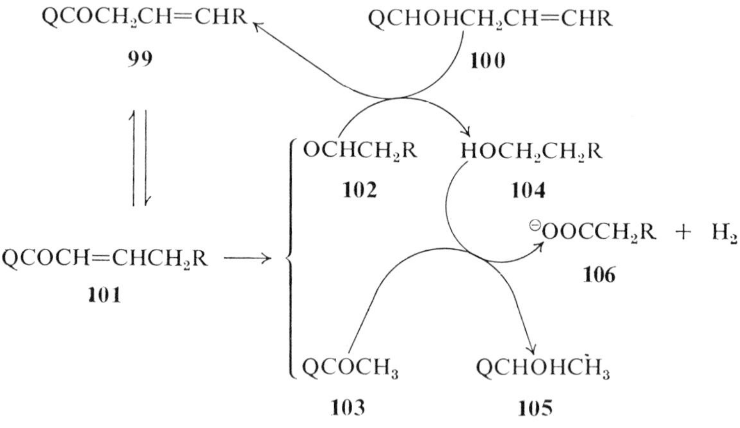

QCOCH$_2$CH=CHR   QCHOHCH$_2$CH=CHR

99   100

OCHCH$_2$R   HOCH$_2$CH$_2$R

102   104

$^{\ominus}$OOCCH$_2$R + H$_2$

106

QCOCH=CHCH$_2$R $\longrightarrow$

101

QCOCH$_3$ QCHOHCH$_3$

103   105

$$Q = CH_3(CH_2)_5-; \quad R = (CH_2)_7CO_2^{\ominus}$$

Scheme 14.21

α,β isomer (**101**), followed by a retro-aldol fission, would then give octan-2-one (**103**) and the aldehydic acid **102**. The isolation of the hydroxy acid **104** rather than **102** is attributed to the latter's acting as a hydride ion acceptor in the initial dehydrogenation (cf. Cannizzaro and Guerbet reactions in which hydride ions are also believed to be transferred to aldehydes). It has been shown that the products from the high temperature fission of ricinoleic acid are formed by further interaction of the low temperature products [79]. This is clearly an instance of the reduction of a ketone by a primary alcohol in a reaction of the type generalized by Scheme 14.16. Glycerol liberated in reactions using castor oil as the starting material may also be implicated in the reduction of the aldehydic acid **102** and the ketone **103**.

One feature of the above explanation of the fission of ricinoleic acid (**100**) deserves further comment. It will be noted that the low temperature products are regarded as being formed by interaction of a secondary alcohol (ricinoleic acid) and an aldehyde (**102**) to give a ketone (**99**) and a primary alcohol (**104**), and the high temperature products as resulting from a subsequent reduction of a ketone (**103**) by the primary alcohol (**104**). This apparent contradiction is resolved by the assumption [3] that a hydrogen transfer of the type shown in Scheme 14.22 is reversible (cf. Meerwein-Ponndorff and Oppenauer

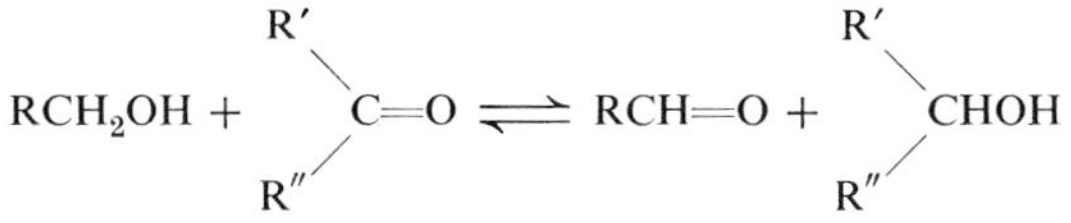

$$RCH_2OH + \underset{R''}{\overset{R'}{\diagdown}}C{=}O \rightleftharpoons RCH{=}O + \underset{R''}{\overset{R'}{\diagdown}}CHOH$$

Scheme 14.22

reactions). In most reactions involving this scheme, equilibrium is disturbed by the (nonreversible) conversion of aldehyde into carboxylate, and the formation of secondary alcohol is therefore favored (Scheme 14.16). In the initial stage of the ricinoleic acid reaction, however, the ketone formed is unsaturated and hence destroyed by fission; the reaction therefore proceeds in the reverse direction with reduction of the aldehyde. The ability of secondary alcohols to reduce aldehydes under appropriate alkaline conditions is also implicit in the observation that (−)-active amyl alcohol

$$(CH_3CH_2CH(Me)CH_2OH)$$

is racemized in the presence of both alkalies and ketones, but not of alkalies alone. This result is attributed to dehydrogenation, and re-formation, of the primary alcohol by reactions of the type given in Scheme 14.22, and to racemization of the intermediate aldehyde by enolization [122].

The mechanisms discussed above for ricinoleic acid can obviously be extended to embrace the alkaline fission not only of other β,γ-unsaturated alcohols—for example of isopulegol (**107**) to 3-methylcyclohexanol (**108**) [123] and of hept-3-ene-1,7-diol (**109**) to glutaric and acetic acids [22]—but also of allylic alcohols—for example, of geraniol (**110**) to methylheptenone (**111**) and methylheptenol (**112**) [22, 92–94].

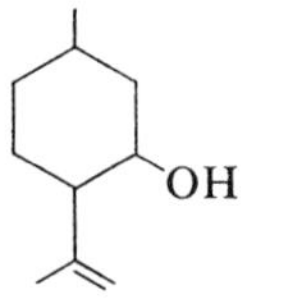

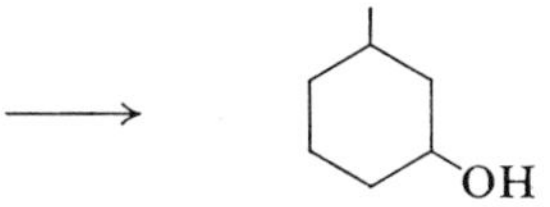

107 → **108**

$$HOCH_2CH_2CH{=}CH(CH_2)_2CH_2OH \rightarrow CH_3CO_2H + HO_2C(CH_2)_3CO_2H$$
109

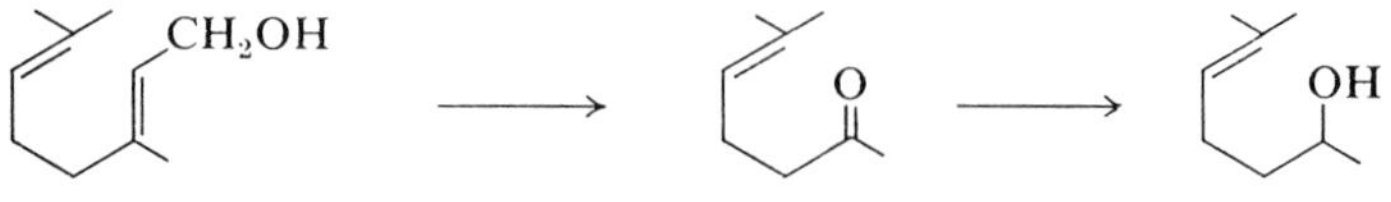

110 → **111** → **112**

Glycols and Some Related Oxygen Functions

Studies on the behavior of α-glycol groupings during alkali fusion have been carried out principally with the dihydroxy derivatives (**113**) of oleic

$$CH_3(CH_2)_7CHOHCHOH(CH_2)_7CO_2H$$
113

acid [2, 124–126]. Both *erythro-* and *threo-*9,10-dihydroxyoctadecanoic acids give the same products [2, 125], and the conclusions concerning the transformations involved are summarized in Scheme 14.23. It is probable that an

RCHOHCHOHR′
113

↓

RCOCHOHR′ ⟶ $R\overset{\displaystyle OH}{\underset{\displaystyle O^{\ominus}}{C}}CHOHR'$ ⟶ $RCO_2H + HOCH_2R'$
116

↓ ↓

RCOCOR′ ⟶ $R\overset{\displaystyle OH}{\underset{\displaystyle O^{\ominus}}{C}}COR'$ ⟶ $RCO_2H + OCHR'$
117

↓ ↓

$HO_2C\overset{\displaystyle R}{\underset{\displaystyle OH}{C}}R'$ ⟶ RCOR′ ⟶ $RCO_2H + HR'$

114 $\overset{\displaystyle HO_2CR}{}$ **115**

R and *R*′ represent the two different end groups, viz.,

—$(CH_2)_7CH_3$ or   —$(CH_2)_7CO_2H$

Scheme 14.23

initial dehydrogenation occurs to give successively the α-hydroxy ketones (**116**) and the α-diketone (**117**) and that the latter then undergoes a benzilic acid type of rearrangement (Scheme 14.24) to give α-hydroxy-α-octylsebacic

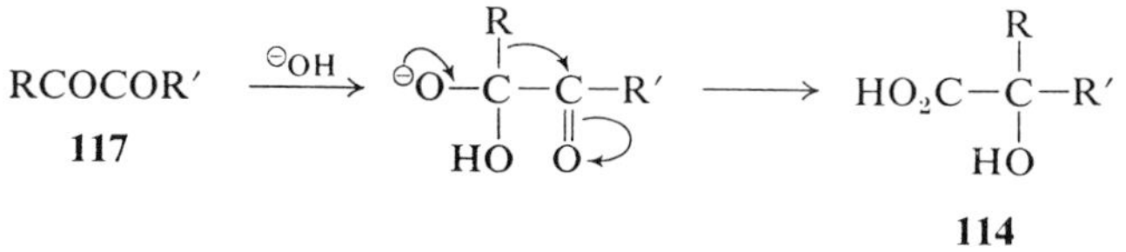

Scheme 14.24

acid (**114**). This hydroxy diacid can be obtained in high yield from controlled alkali fusion of the dihydroxy acids and the hydroxy-keto acids at 230–240° and of the 9,10-dioxooctadecanoic acid at 160° [2, 127, 128] (an analogous transformation has been observed with *erythro*-13,14-dihydroxybehenic acid derived from erucic acid [125]). Under more drastic conditions (ca. 270°) the hydroxy diacid (**114**) is converted into 9-oxoheptadecanoic acid (**115**), which can be isolated in yields up to 30%; two mechanisms (Scheme 14.25)

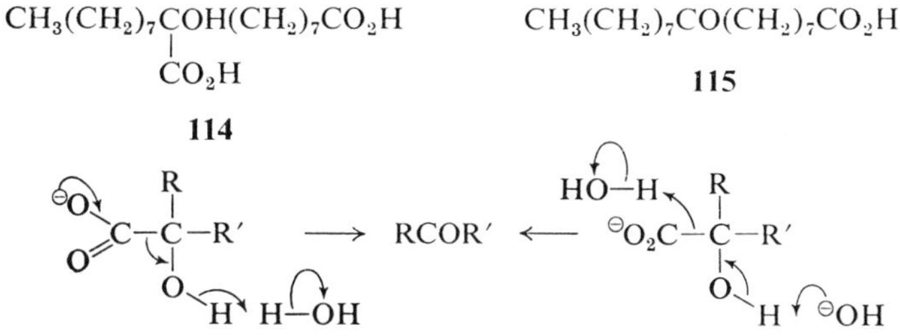

Scheme 14.25

have been tentatively suggested for the transformation [2]. It is of interest that no products resulting from the dehydration of the hydroxy diacid (**114**) are observed, although alkali fission of 1-hydroxycyclohexanecarboxylic acid gives some pimelic acid (**7**) [12] and there is little doubt that cyclohex-1-enecarboxylic acid is an intermediate.

At reaction temperatures of ca. 300° the keto acid (**115**) undergoes the expected fission, the principal products being azelaic acid, by hydrolysis of the 9,10 bond, and both octanoic and nonanoic acids by hydrolysis of the 8,9 bond [2]. The direct hydrolysis of the hydroxyketo (**116**) and diketo (**117**) acids, giving initially alcohols or aldehydes, occurs to only a small extent during alkali fusion, but may become more important under milder conditions when fission without the formation of octanoic acid is sometimes observed [129, 130, 130a].

Alkali fusion of 10,11-dihydroxyundecanoic acid (118) gives sebacic acid (90%) and small amounts of suberic and nonanoic acids [2, 126]. With

$$HOCH_2CHOH(CH_2)_8CO_2H \rightarrow HO_2C(CH_2)_8CO_2H$$
118

the natural trihydroxy acid, aleuritic acid (119), fission and simultaneous

$$HOCH_2(CH_2)_5\underset{\underset{OH}{|}}{CH}\underset{\underset{OH}{|}}{CH}(CH_2)_7CO_2H \longrightarrow \begin{cases} CH_3(CH_2)_4CO_2H \\ CH_3(CH_2)_6CO_2H \\ HO_2C(CH_2)_5CO_2H \\ HO_2C(CH_2)_7CO_2H \end{cases}$$
119

dehydrogenation of the primary alcohol function occur, together with other reactions, and give hexanoic, octanoic, pimelic, and azelaic acids in approximately equal amounts (0.25 mole) [2, 3a]. Fission of the dihydroxy derivative (120) of ricinoleic acid, which contains both 1,2- and 1,3-diol systems, gives

$$CH_3(CH_2)_5\underset{\underset{OH}{|}}{CH}CH_2\underset{\underset{OH}{|}}{CH}\underset{\underset{OH}{|}}{CH}(CH_2)_7CO_2H \dashrightarrow \begin{cases} CH_3(CH_2)_5CO_2H \\ HO_2C(CH_2)_7CO_2H \end{cases}$$
120

azelaic (0.45 mole) and heptanoic (0.25 mole) acids, in addition to a number of minor by-products [2, 126]; as with other 1,3-glycols [131], fission probably involves β-hydroxy ketone or β-diketone intermediates. Sativic (121) and linusic (122) acids, derived from linoleic and linolenic acids, are stated to

$$CH_3(CH_2)_4\underset{\underset{OH}{|}}{CH}\underset{\underset{OH}{|}}{CH}CH_2\underset{\underset{OH}{|}}{CH}\underset{\underset{OH}{|}}{CH}(CH_2)_7CO_2H \longrightarrow \begin{cases} CH_3(CH_2)_4CO_2H \\ HO_2C(CH_2)_7CO_2H \end{cases}$$
121

$$CH_3CH_2\underset{\underset{OH}{|}}{CH}-\underset{\underset{OH}{|}}{CH}CH_2\underset{\underset{OH}{|}}{CH}-\underset{\underset{OH}{|}}{CH}CH_2\underset{\underset{OH}{|}}{CH}-\underset{\underset{OH}{|}}{CH}(CH_2)_7CO_2H \rightarrow$$
122

$$CH_3CH_2CO_2H + HO_2C(CH_2)_7CO_2H$$

give azelaic acid, acetic acid, and either hexanoic or propionic acid, respectively [132] (other acids are to be expected as by-products). Alkaline fission of dihydroxydihydrocinnamic acid (123) at 250° gives both benzoic and oxalic acids; at 165–175° the products include toluene and hydrobenzoin [133].

$$PhCHOHCHOHCO_2H$$
123

With the estrogenic hormone estriol (**124**), the possibility of a benzilic acid rearrangement is precluded by steric considerations, and alkali fusion gives (60%) marrianolic acid (**125**) without loss of one carbon atom; this transformation played an important role in studies on the structure of the hormone [60, 134].

Similar reactions to those considered in this section must also be involved in the manufacture of oxalic acid by alkali fusion of sawdust (cellulose) at 240–250° [135].

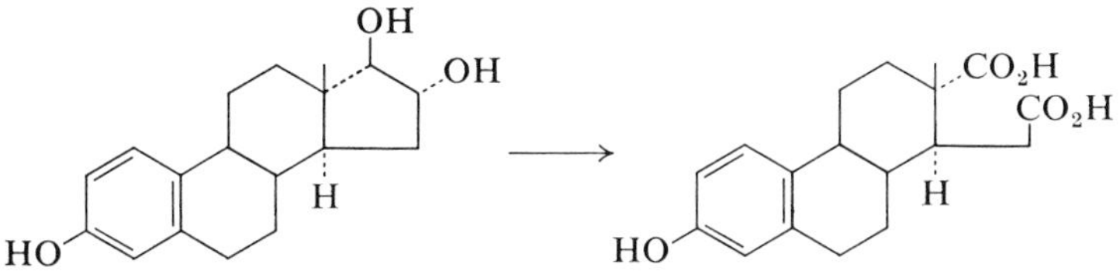

124   **125**

Ethers

The main products from the alkali fusion of epoxy fatty acids resemble those from the corresponding dihydroxy derivatives [135a]. Under similar conditions simple alkyl ethers undergo an initial β-elimination [135a].

Several examples have been reported of the fission of tetrahydrofuran rings during alkali fusion [22]. Thus the main product (55%) from 3-α-tetra-hydrofurylpropionic acid (**126**) is pimelic acid (**7**). Little information is available concerning the course of the reaction, but a few speculative suggestions can be made (Scheme 14.26). It is likely that unsaturated hydroxy

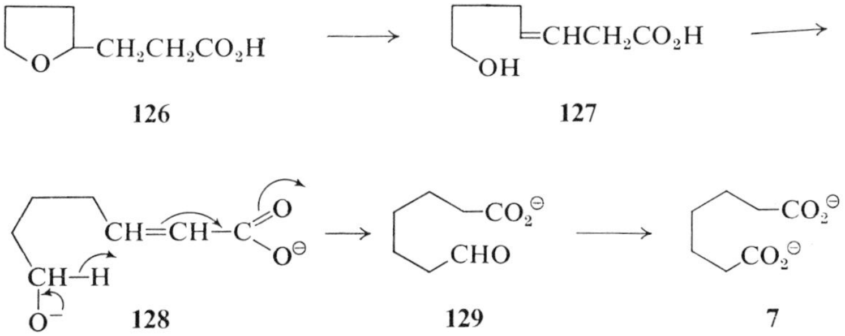

126   **127**

128   **129**   **7**

Scheme 14.26

acids (e.g., **127**) are first produced and that these rearrange to the α,β isomer (**128**). The formation of pimelic acid can then be attributed to a Dumas-Stass dehydrogenation accompanied by reduction of the α,β double bond by inter- and intramolecular processes of the type indicated in Scheme 14.4. The intramolecular process would be favored here by steric factors, and it is of interest that the homolog (130) of the starting material gives a comparatively low yield (27%) of suberic acid (131). The by-products observed in the

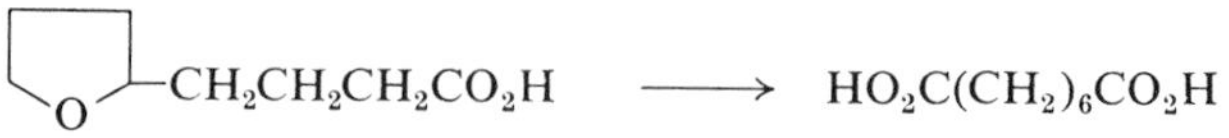

$$\text{[structure]}\text{---CH}_2\text{CH}_2\text{CH}_2\text{CO}_2\text{H} \longrightarrow \text{HO}_2\text{C(CH}_2)_6\text{CO}_2\text{H}$$

130   131

reaction of 126 are readily explicable in terms of the above proposals. Glutaric and acetic acids can be attributed to a Varrentrapp fission of the unsaturated intermediates, and cyclohex-1-ene-carboxylic acid to cyclization of the aldehydic acid (129) (cf. Scheme 14.13).

The conversion (20–35%) of the 3-α-tetrahydrofurylpropanols (132, R = H, Me, and Et) into pimelic acids (133) probably involves a preliminary

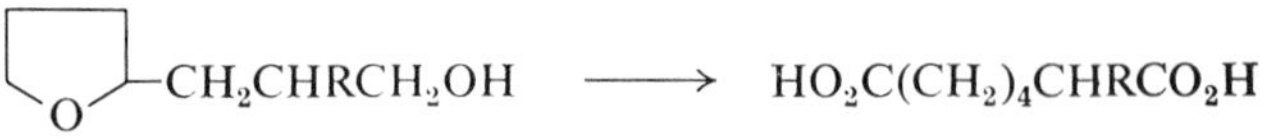

$$\text{[structure]}\text{---CH}_2\text{CHRCH}_2\text{OH} \longrightarrow \text{HO}_2\text{C(CH}_2)_4\text{CHRCO}_2\text{H}$$

132   133

Dumas-Stass dehydrogenation. This cannot be a key stage in the conversion of tetrahydrofuryl alcohol (134) into glutaric acid (135), since the yield (60%) is much greater than that (17%) obtained starting with the corresponding acid. Glutaric acid (45%) is also formed on alkali fusion of 2-hydroxytetrahydropyran (136) [22].

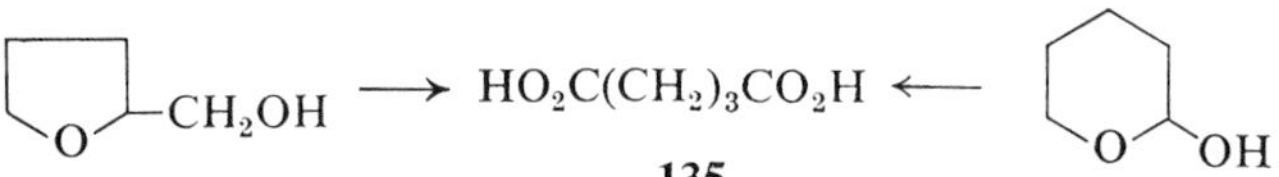

$$\text{[structure]}\text{---CH}_2\text{OH} \longrightarrow \text{HO}_2\text{C(CH}_2)_3\text{CO}_2\text{H} \longleftarrow \text{[structure]}\text{OH}$$

134   135   136

5   ARYLSULFONIC ACIDS AND ARYL HALIDES

The alkali fusion of arylsulfonic acids, which was introduced independently by Kekulé [136], Wurtz [137], and Dusart [138] in 1867, is still of importance in the preparation of phenols. This nucleophilic aromatic substitution is often accompanied by oxidation of the initial product by processes which may be fundamentally homolytic in character [139]. Phenol can be converted by alkali fusion into catechol and resorcinol (137), and resorcinol into phloroglucinol (138) [140]. Alkali fusion of mesitylenesulfonic acid (139) gives a carboxylic acid (141), and, by decarboxylation of the latter, a xylenol (142), in addition to the main product, mesitol (140) [141]. It has been shown that many of these various side reactions depend on the absorption of atmospheric oxygen, and can be largely excluded by carrying out the fusion in an inert atmosphere [4].

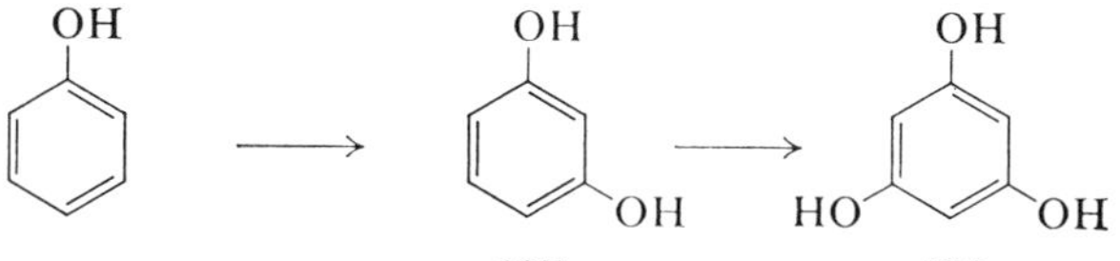

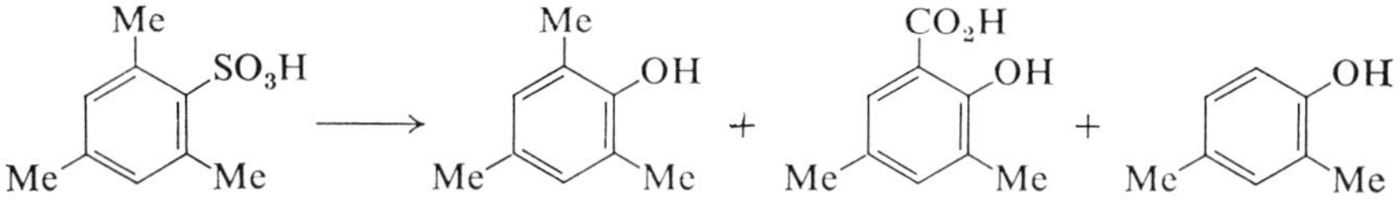

Nucleophilic substitution of aryl halides is also used in the preparation of phenols. However, reactions in which either a halogen atom or a sulfonic acid group is replaced by hydroxyl in the alkali fusion of substituted phenoxides or sulfonates often, but not always, give rearranged products [139]. Thus resorcinol is produced not only from halophenols, halosulfonic acids, and the disulfonic acid of the meta series, but also from their ortho and para isomers [142–144]. Before the possibility of rearrangement was recognized, the use of these reactions in attempts to correlate orientation in different benzene derivatives led to much confusion.

Many of these abnormal substitutions give mixtures of products and are reminiscent of reactions proceeding by benzyne intermediates. However, having regard to what occurs to phenol itself in fused alkali, it has been suggested that the above abnormal substitutions may result from the successive loss and gain of hydroxyl substituents in the alkali melt, presumably by mechanisms that are at least partly homolytic [139]. The gain and loss of hydrogen that would necessarily accompany this loss and gain of hydroxyl can, in fact, be observed separately. Thus there is extensive exchange between aryl hydrogen and hydrogen of the alkali during hydrolysis of a deuterated benzanilide with fused alkali [145]. Again deuterium is introduced into aromatic nuclear positions when decarboxylating sodium benzoate with sodium deuteroxide, and calcium benzoate and calcium trimesate (benzene-1,3,5-tricarboxylate) with calcium hydroxide-d [44]. Moreover, not only the original salts but also the benzene formed undergo hydrogen exchange with the alkali. Clearly, the aromatic ring is attacked by fused alkali in comprehensive ways which are as yet little understood[5a].

A rearrangement of a different type may have been involved in the original synthesis of alizarin (**144**) [146]. Though not realized at the time, the dibromoanthraquinone submitted to alkali fusion was the 2,3 (**143**) and not the

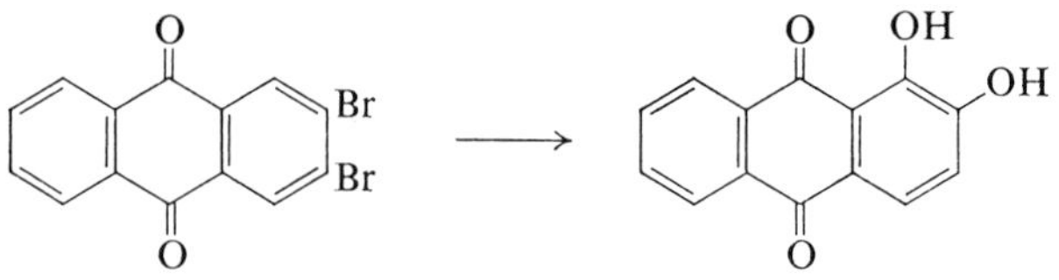

143   144

1,2 compound; it is conceivable that this spectacular first synthesis of a natural dyestuff involved a fission of the quinone ring and ring closure in an alternate position [147].

6 PHENOLS, PYRONES, AND THEIR DERIVATIVES

The degradation of these compounds constitutes one of the principal uses of alkali fusion. Though the reactions which take place are very similar to those already discussed, three of them deserve special mention: (1) conversion of phenol ethers to the parent phenols; (2) elimination of carboxyl or acyl groups, or of side chains capable of giving rise to these groups in the alkali melt; and (3) elimination of alkenyl, or potential alkenyl, groups which are situated in the ortho or para position with respect to phenolic hydroxyl groups.

Many examples are known of the replacement of a phenol ether group by hydroxyl during alkali fusion. p-Methoxybenzoic acid (**145**, R = Me) gives p-hydroxybenzoic acid (**145**, R = H) [148]. Diphenylene oxide (**146**, R = H) and its methyl derivative (**146**, R = Me) give 2,2′-dihydroxydiphenyls

$$RO-\langle\ \rangle-CO_2H$$

145

146   147

148   149

(147) [52, 149]. Methylmorphenol (148) gives 3,4,5-trihydroxyphenanthrene (149), a reaction which revealed the presence of a phenanthrene nucleus in morphine and the location of two of the oxygen functions [150]. Both galipine (150, R = R' = Me) and cusparin (150, RR' = CH$_2$) give proto-catechuic acid (151) [151]. No rearrangements have been reported to accompany these substitution reactions of the phenolic ethers, in contrast to

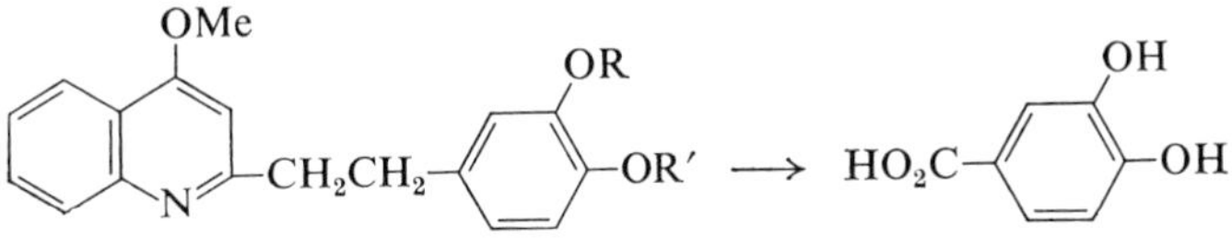

150   151

those of the arylsulfonic acids and halides. The analogous conversion [152] of xanthone (152, R = H) to 1,1'-dihydroxybenzophenone (153) may proceed by 1,4 addition to the formal α,β-unsaturated ketone system, followed by ring fission (cf. Scheme 14.27). Under more vigorous conditions phenol and salicylic acid are obtained.

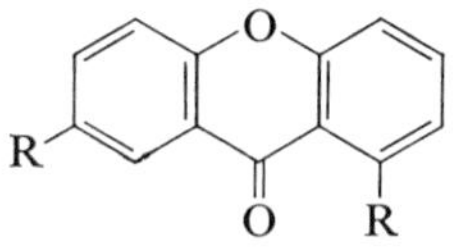

152 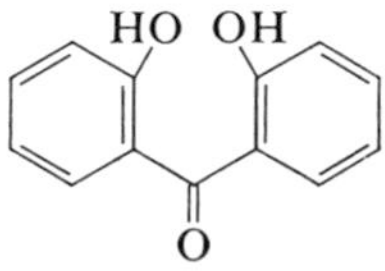   153

Degradations which involve elimination of a carboxyl or acyl group are also common. These appear to be facilitated by ortho and para hydroxyl sub-stituents, and it is conceivable that reaction proceeds via the keto form of the phenol (cf. decarboxylation of β-keto acids and hydrolysis of β-diketones). Thus protocatechuic acid (151), formed by alkali fusion of lignin, is converted by prolonged reaction with alkali into catechol [153]. Alkali fusion of the natural methyl ether of alternariol (154) gives the tetrahydroxymethyl-diphenyl (155) [154] and euxanthone (152, R = OH) gives a mixture of hydroquinone and resorcinol [155].

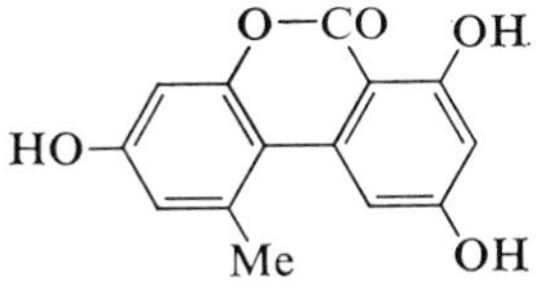

154 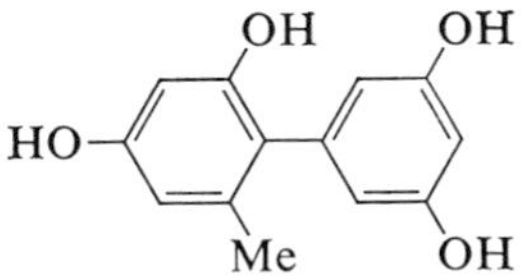   155

Fission of α- and γ-pyrone rings, present in many natural products, occurs readily on alkali fusion (cf. α,β-unsaturated acids and ketones). With coumarins and chromones the o-acylphenols, which may be regarded as the initial products, are usually converted *in situ* into the parent phenols, presumably either by direct hydrolysis (cf. β-diketones) or by formation and decarboxylation of the corresponding carboxylic acid (cf. β-keto acids). The phenolic acid is sometimes isolated; for example, salicylic acid (157) can

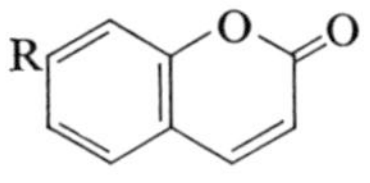

156 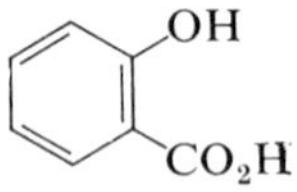   157

be obtained from coumarin (156, R = H) [156], and β-resorcylic acid (159) from oreoselone (158) [157]. However, elimination of the ortho carboxyl or acyl group is common. Thus umbelliferone (156, R = OH) gives both β-resorcyclic acid and resorcinol [158], and bergapten (160) gives phloroglucinol (161) [159].

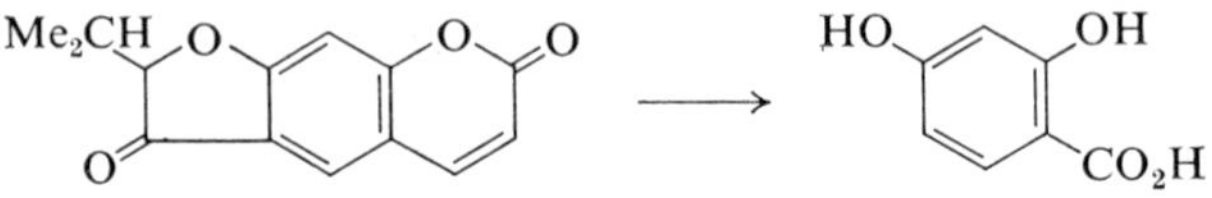

158   159

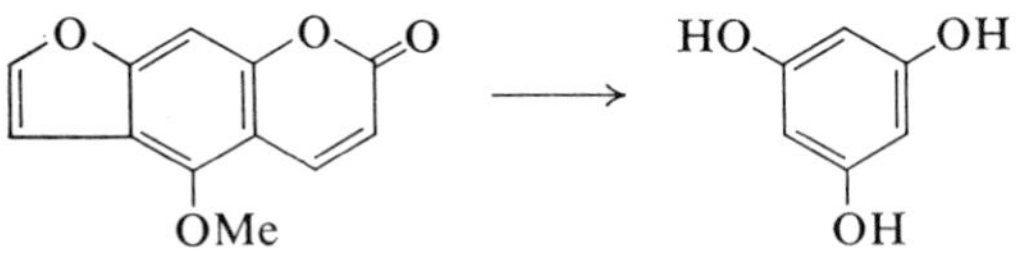

160   161

A similar fission is exhibited by flavones and flavonols. Treatment of chrysin (162) with boiling concentrated alkali gives phloroglucinol (161), benzoic acid, acetic acid, and traces of acetophenone (Scheme 14.27) [160], and alkali fusion of quercetin (163, R = H) and rhamnetin (163, R = Me) gives phloroglucinol (161) and protocatechuic acid (151) [153, 161]. Alkali fusion of the related isoflavone, biochanin A (164), gives phloroglucinol (161) and p-methoxyphenylacetic acid (Scheme 14.28) [162]. It will be obvious from these examples that alkali fusion cannot distinguish between coumarin and chromone derivatives.

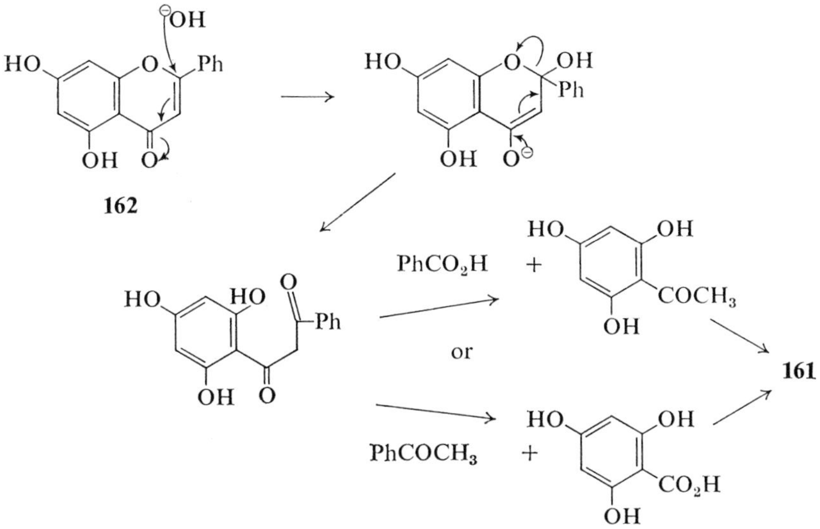

Scheme 14.27

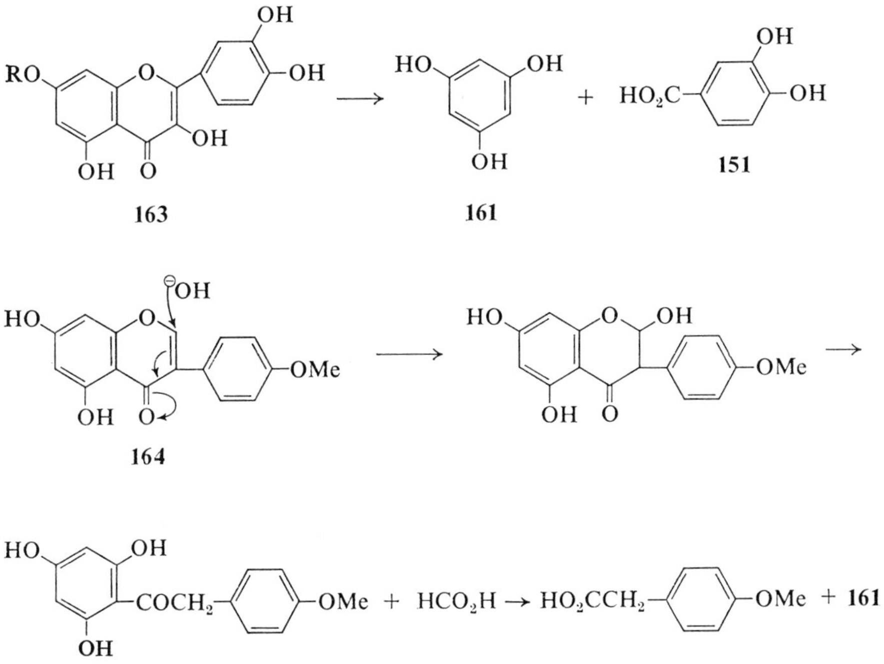

Scheme 14.28

Willstätter's classical work on the structures of the anthocyanidins constitutes another well-known application of alkali fusion [163]. The chromylium ring system is converted into phloroglucinol, and the various C-2-aryl substituents are isolated as the corresponding benzoic acids (168). Thus pelargonidin (165, R = R' = H) gives p-hydroxybenzoic acid, cyanidin (165, R = OH, R' = H) gives protocatechuic acid, and delphinidin (165, R = R' = OH) gives gallic acid. The course of these degradations is unknown, but it seems reasonable to suppose (Scheme 14.29) that there is an initial attack by hydroxyl ion at C-4 of the color base (166) of the anthocyanidin to give a 4-hydroxyflavenol (167). Several plausible mechanisms can then be envisaged for the subsequent conversion of 167 to the final products; for example, dehydrogenation to the flavonol and fission (cf. Scheme 14.27).

As expected, methoxyl groups present in some anthocyanidins are converted into hydroxyl during alkali fusion. They can, however, be preserved by carrying out the degradations under much milder alkaline conditions [164]; such procedures are also to be preferred for the degradation of some flavones, flavonols, and isoflavones [165].

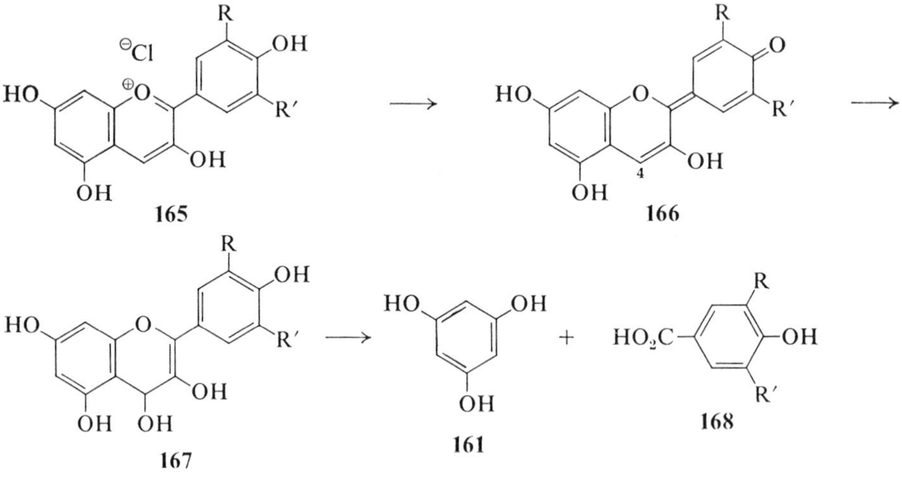

Scheme 14.29

The elimination of acyl or carboxyl groups produced during alkali fusion provides an explanation of the formation of salicylic acid and its meta isomer from Terramycin (169); this observation enabled the tentative conclusion to be drawn that the antibiotic contains a phenol ring linked to carbon at both C-2 and C-3 [166]. The formation of 5-chloro-2-hydroxybenzoic acid from aureomycin (170) is similar [167].

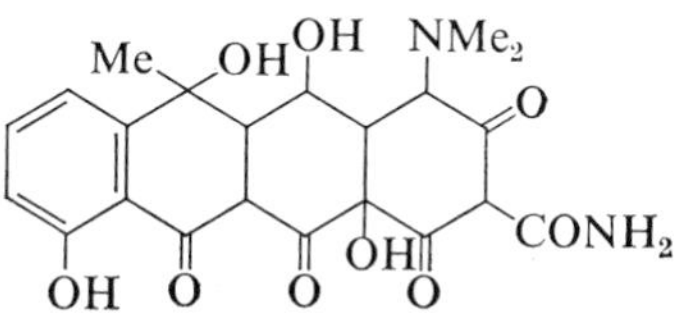

169 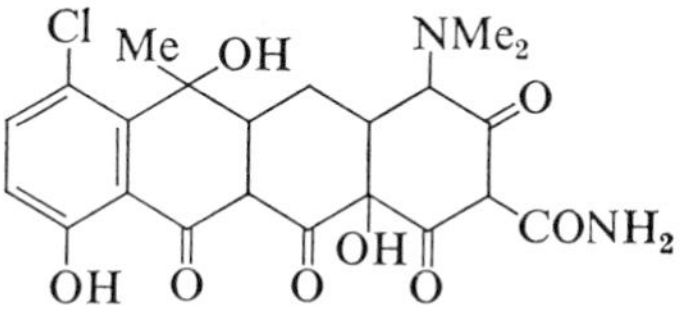   **170**

Concerning the third type of reaction singled out above for special mention, the alkali fusion of mangostin (**171**) provides an instructive example [95, 168, 169]. The initial products include acetic acid, isovaleric acid, and a pigment, **172**, which on further alkali fusion gives isovaleric acid and phloroglucinol. These transformations can be interpreted in the following way (Scheme 14.30). Migration of the double bond, hydration, and retro-aldol cleavage

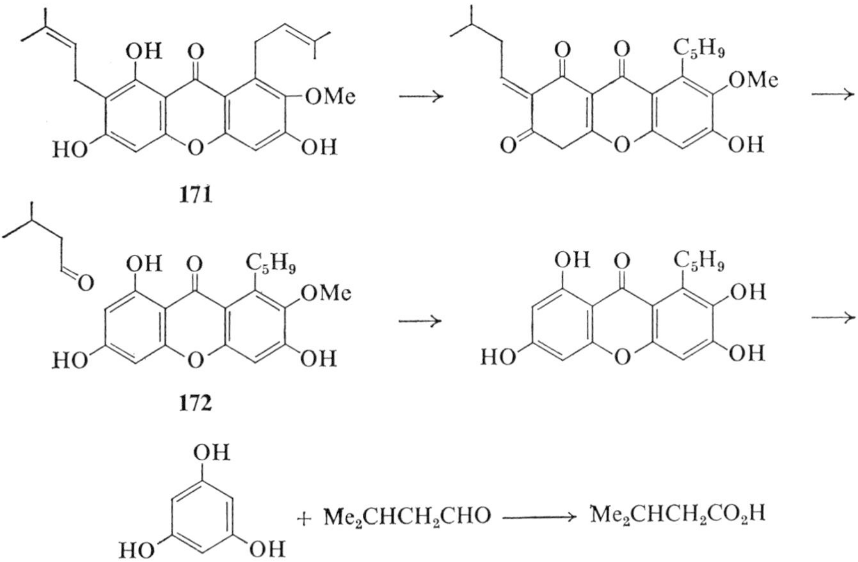

Scheme 14.30

(cf. Varrentrapp reaction) give the pigment **172** and isovaleraldehyde, which is subsequently dehydrogenated to isovaleric acid (cf. Scheme 14.15). The second alkenyl side chain has no free hydroxyl group in either the ortho or para position which would permit the retro-aldol elimination and therefore survives in the pigment **172**. The formation of isovaleric acid on further alkali fusion of **172** may be attributed to substitution of the methoxyl group by a free hydroxyl group in the ortho position to the side chain, which is then eliminated.

An alternative route (Scheme 14.31) for the degradation of mangostin also seems feasible. Rupture of the potential phloroglucinol ring by hydration and retro-aldol cleavage and fission of the resulting β-diketone would give **173** and methylheptenone (**111**). The latter, in molten alkali, would be degraded to isovaleric acid and acetic acid (cf. Varrentrapp reaction and fission of ketones). However, it is of interest to note that treatment of mangostin with

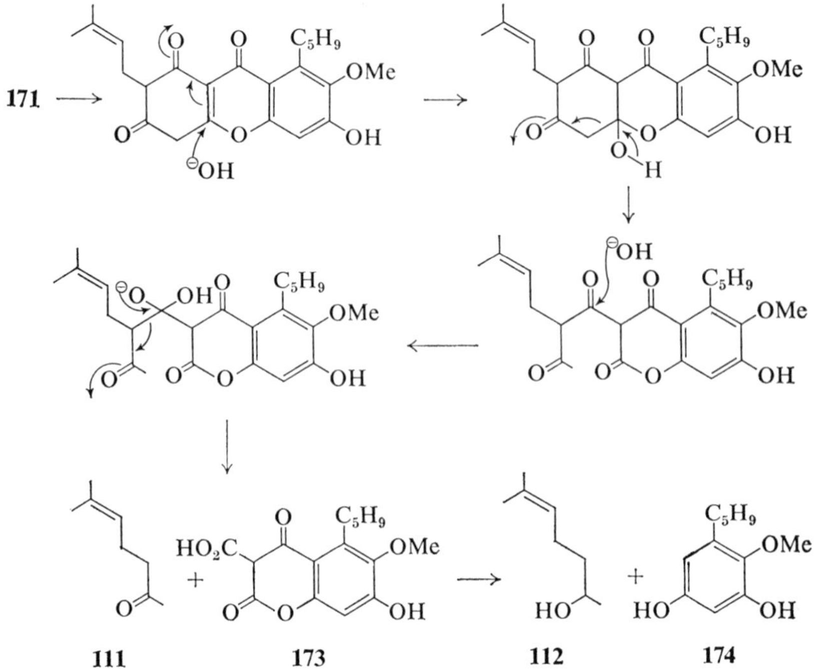

Scheme 14.31

ethanolic potassium hydroxide at 170–180° gives an isopentenylmethoxy-resorcinol (**174**), which can be regarded as a fission product of **173**, and both isoamyl alcohol and methylheptenol (**162**) [170]. These alcohols doubt-less arise by reduction of isovaleraldehyde and methylheptenone by the ethanolic alkali (cf. Scheme 14.16).

Similar reactions to those discussed above may account for the formation of isovaleric acid from other isopentenyl-phenol derivatives—e.g., the tri-methyl ether of pomiferin (**175**) [171]—and the elimination of alkenyl substituents, as in the production of resorcinol from ostruthin (**176**) [172]. The removal of the furanoid ring in the formation of phloroglucinol (**161**) from furocoumarins, such as bergapten (**160**) [159], and oxypeucedanin (**177**) [173], and from visnaginone (**179**) [174], the product of mild hydrolysis

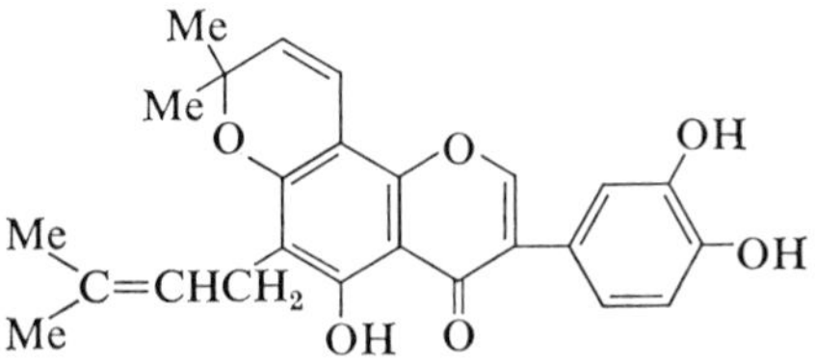

175

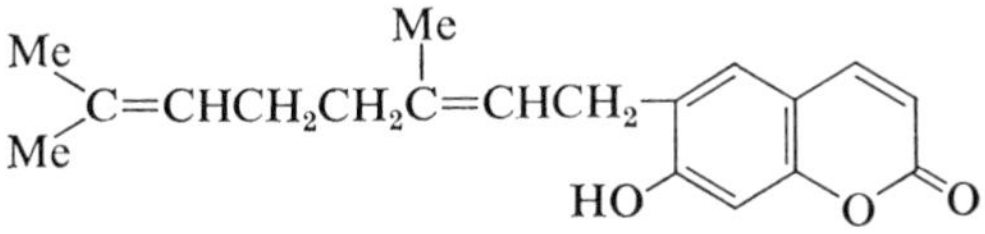

176

of visnagin (**178**), may follow the same basic procedure (Scheme 14.32); presumably fission of the carbon–carbon link precedes that of the carbon–oxygen link.

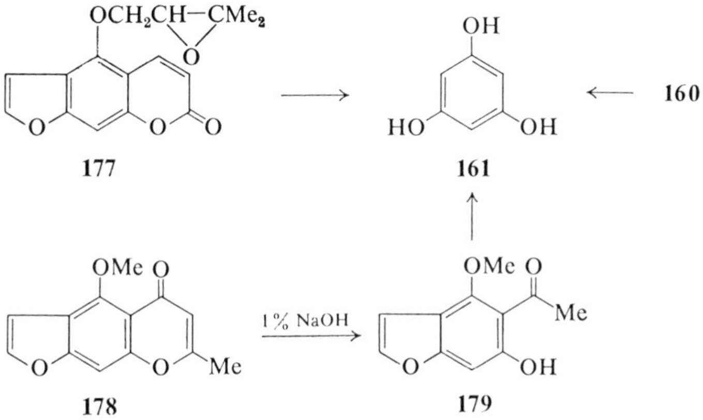

Scheme 14.32

The formation (Scheme 14.33) of acetic acid, isovaleric acid, and 3,4,5-trihydroxyphenylacetic acid (**183**) on alkali fusion of the dihydro derivative (**181**) of fuscin (**182**) [175] can also be regarded as a simple extension of the same basic process in which the alkenyl group is formed *in situ* by cleavage of the phenolic ether and dehydration of the resulting alcohol. Since mild alkaline hydrolysis of dihydrofuscin gives acetaldehyde and fuscinic acid (**183**) (cf. retro-aldol reaction), it is probable that a similar reaction occurs initially during alkali fusion.

The migration of a double bond, postulated in a number of the above reactions, can be observed in the conversion of tubaic acid (**184**) into isotubaic

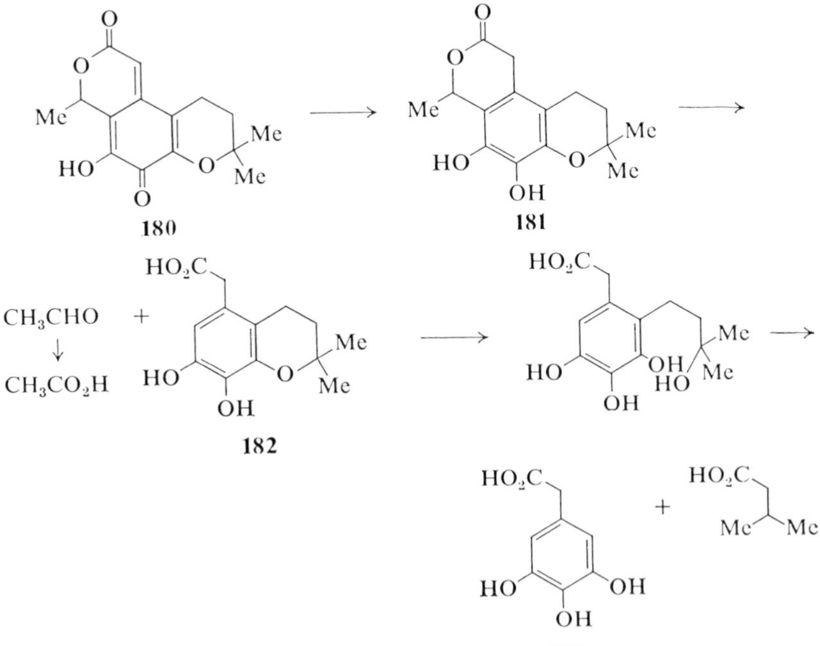

180   181

CH₃CHO +

↓

CH₃CO₂H

182   183

Scheme 14.33

acid (185) in molten alkali [176] and of safrole (186, RR' = CH₂) and eugenol (186, R = Me, R' = H) into their conjugated isomers on treatment with concentrated alkalies [177, 178]. Alkali fusion of rotenone (187) and a number of its derivatives at 200–210° also gives isotubaic acid, but at higher temperatures (260–300°) this acid is converted into resorcinol [176] (cf. elimination of the furanoid ring from, for example, bergapten).

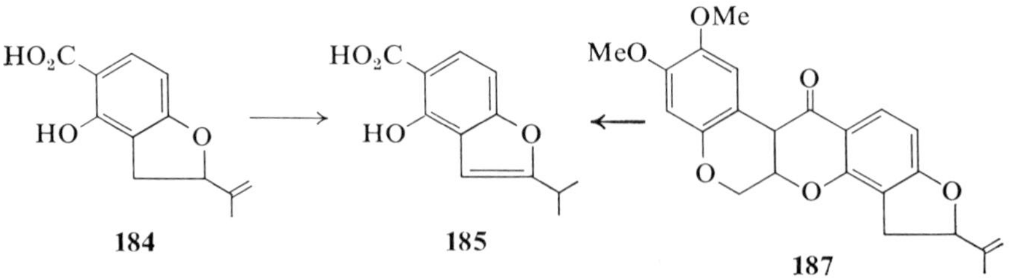

184   185   187

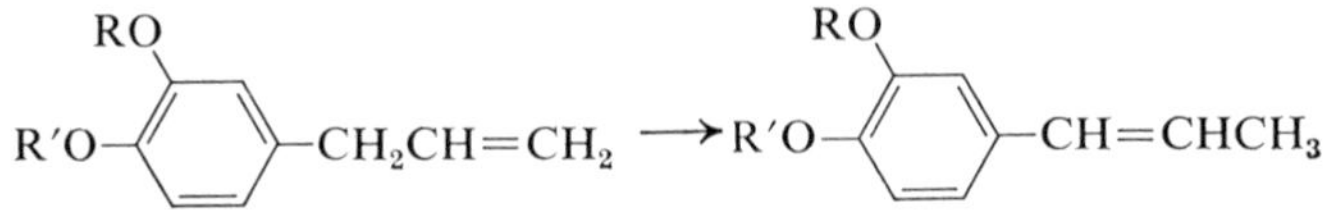

186

7 ANTHRAQUINONES

Under strongly alkaline conditions, anthraquinones are prone to nucleophilic substitution. When this involves replacement of a hydrogen atom, the reaction can often be facilitated by adding an oxidizing agent to the reaction mixture. The manufacture of alizarin (144) by alkali fusion of anthraquinone-2-sulfonic acid (188) in the presence of a nitrate or chlorate [179] provides a good illustration (Scheme 14.34). Alizarin is also formed on alkali fusion of

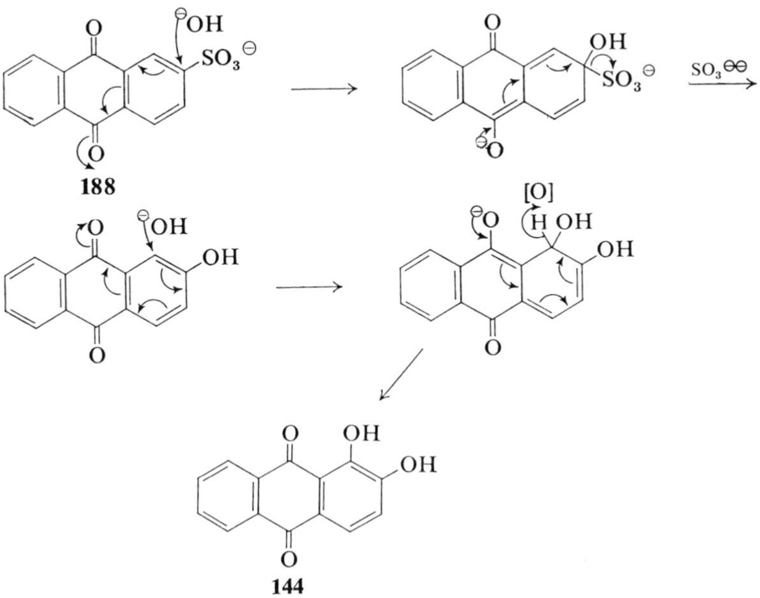

Scheme 14.34

β-hydroxy- and β-anilinoanthraquinone [180], and by the reaction of anthraquinone itself with concentrated sodium hydroxide and an oxidant [181]; alkali fusion of anthraquinone yields benzoic acid [182] (cf. fission of aryl ketones, p. 336).

The observation by Bohn in 1901 that alkali fusion of β-aminoanthraquinone (189) gives indanthrone (Indanthrene Blue) (193) inaugurated an important new era in the chemistry of vat dyes [183, 184]. Yields of 50% are obtainable by fusion with potassium hydroxide, containing potassium nitrate as an oxidizing agent, at 150–200°; alizarin is formed as a by-product. Fluoranthrone (Indanthrone Yellow G) (194) is obtained in addition to indanthrone [183, 184], especially if the temperature of the alkali melt is allowed to rise too high (>270°). The formation of these polycyclic quinones has been attributed [165, 185] to nucleophilic attack of a molecule of β-aminoanthraquinone by an ion, **190**, formed from a second molecule (Scheme 14.35). Depending on whether this leads to a C or N derivative (cf. C- and O-alkylation of enols) the intermediate would be **191** or **192**, and subsequent cyclization would give indanthrone (**193**) or fluoranthrone (**194**), respectively. The related pyranthrone (**196**) can be prepared by heating the dianthraquinonyl (**195**) with solid potassium hydroxide [186].

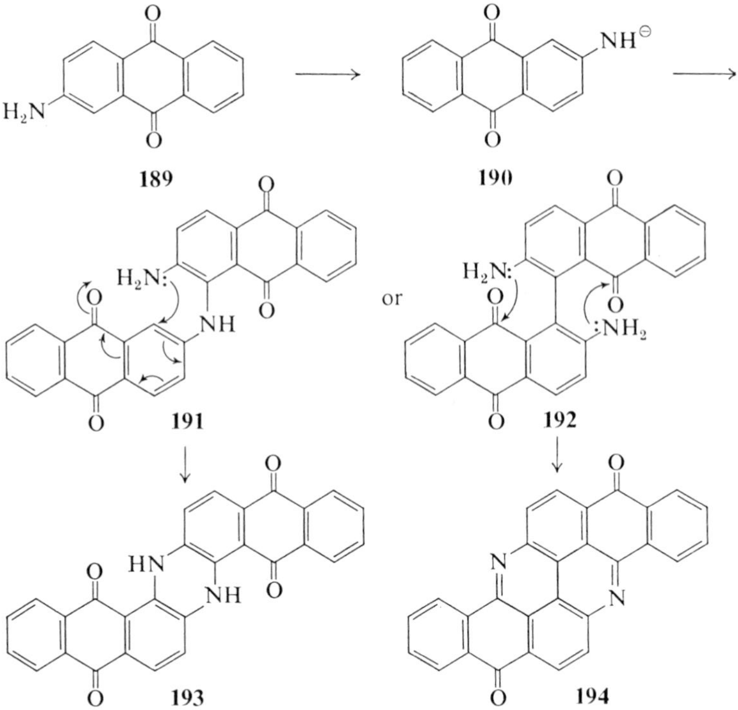

Scheme 14.35

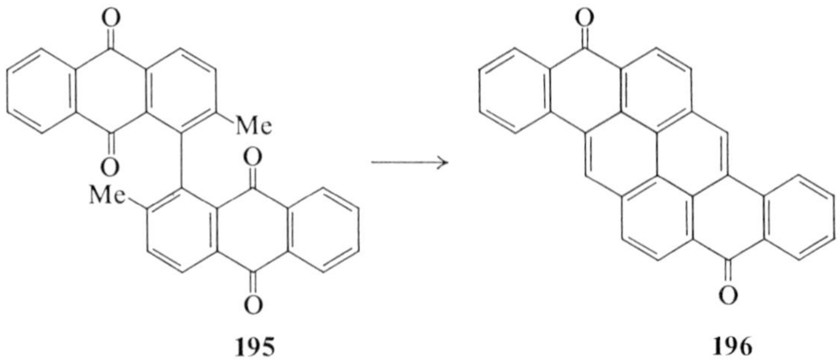

Alkali fusion of carminic acid (197) yields coccinin (198) [187]. The degradation of the sugarlike side chain to methyl doubtless involves hydride ion transfers which are responsible for the reductive removal of a phenolic hydroxyl group and the conversion of the anthraquinone to the anthrone system; both transformations can readily be effected by conventional chemical means (Scheme 14.36), and similar processes are observed on

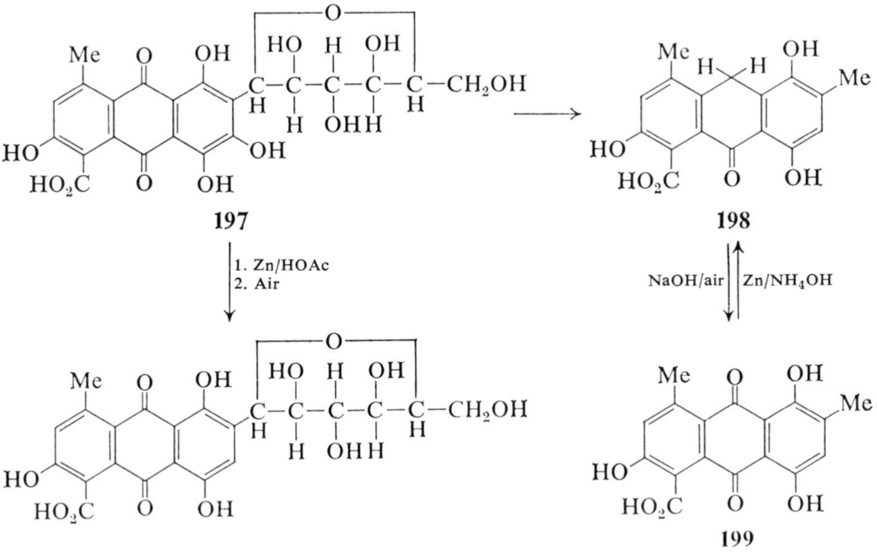

Scheme 14.36

heating other anthraquinone derivatives with a mixture of alkali and glycerol. The recognition that coccinin is related to its oxidation product coccinone (199) in the same way as anthrone to anthraquinone provided valuable evidence for the anthraquinonoid structure of carminic acid.

8 BENZANTHRONES

Following Bohn's remarkable syntheses, Bally [188] submitted benzanthrone (200) to alkali fusion and obtained an excellent vat dye, violanthrone (202). By analogy with the hydroxylation which takes place when benzanthrone is fused with alkalies in the presence of an oxidant, it has been suggested [181] that the self-condensation results from an attack of benzanthrone by an anion of benzanthrone to give the dibenzanthronyl (201), which subsequently cyclizes (Scheme 14.37). It is interesting to note that the

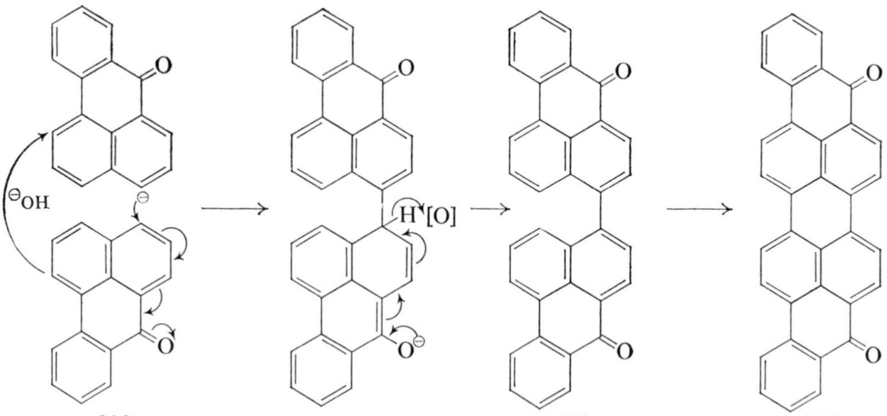

200 201 202

Scheme 14.37

postulated intermediate (**201**) can be obtained by treatment of benzanthrone with alcoholic alkali at 100°, and that it gives violanthrone on alkali fusion [189].

A number of substituted benzanthrones undergo similar self-condensation on alkali fusion [165, 181]. The configuration of the important vat dye, Caledon Jade Green (Indanthrone Brilliant Green B) (**203**) was first established by the synthesis of the dye by alkali fusion of 12-methoxybenzanthrone [190].

The symmetrical isomer (**205**) of violanthrone can be prepared by intermolecular dehydrochlorination of 13-chlorobenzanthrone (**204**) with alkali [191].

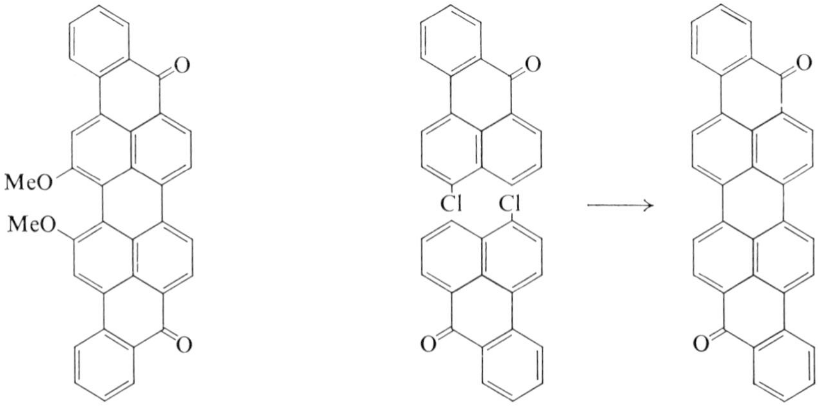

203 204 205

9 MISCELLANEOUS COMPOUNDS

The results which have been reported from the alkali fusions of many natural products indicate the occurrence of a variety of dehydrogenations, and of carbon–carbon and carbon–nitrogen fissions, which are difficult to rationalize from the scant information at present available. Whether these reactions require the presence of alkali, or air, or are purely thermal in character, is not clear in many instances. Thus the carbon–carbon fission which occurs in the conversion of galipine (**150**, R = R' = Me) and cusparine (**150**, RR' = CH$_2$) into protocatechuic acid (**151**) [151] may conceivably proceed by elimination of a hydride ion from a benzylic carbanion, followed by a retro-aldol fission of the resulting benzylidene quinaldine (e.g., Scheme 14.38; cf. Scheme 14.7), or, alternatively, by aerial oxidation of the benzylic carbanion.

$$QCH_2CH_2R \rightarrow QCH_2\overset{\ominus}{C}HR \rightarrow QCH{=}CHR \rightarrow QCH_2CH(OH)R \rightarrow$$

$$QCH_3 + OCHR \rightarrow HO_2CR$$

Q = 4-methoxy-2-quinolinyl;   R = 3,4-dimethoxyphenyl

Scheme 14.38

Sometimes the products isolated afford little useful information concerning the starting material; for example, the formation of *o*-cresol and isophthalic acid from limonin (**206**) [192], or of a terephthalic acid derivative from

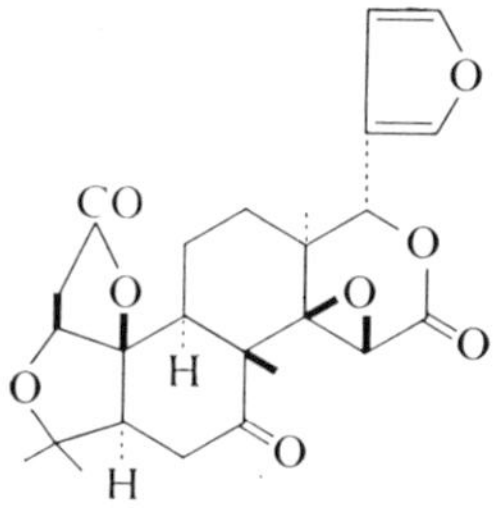

206

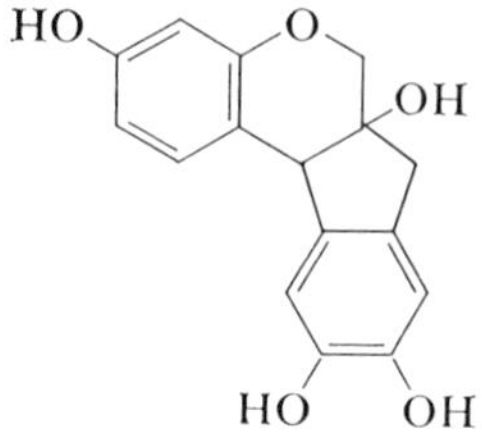

207

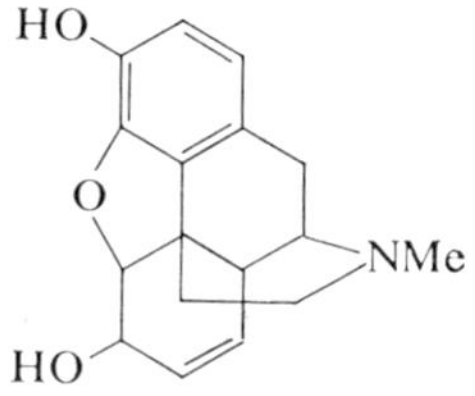

208

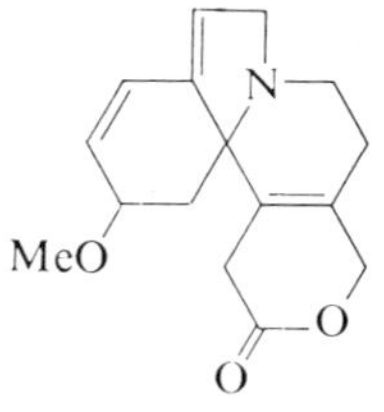

209

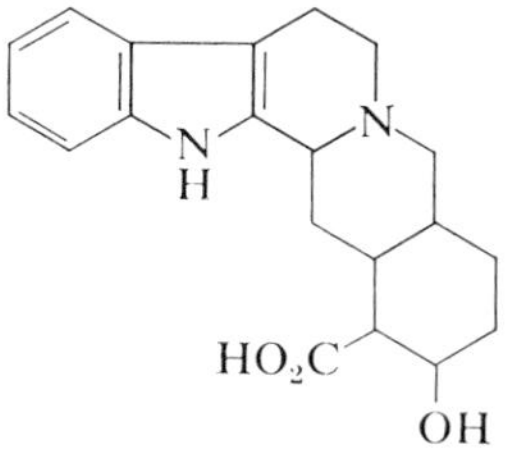

210

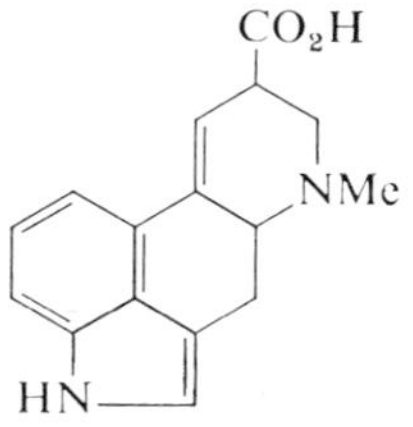

211

colchicine [193], is of little value. Frequently, however, the products provide a clue to structural features present in the starting material. The formation of protocatechuic acid (151) from brazilin (207) [194] and morphine (208) [195] indicates the presence in both compounds of a benzene ring with two oxygen substituents in adjacent positions. The isolation of indole from β-erythroidine (209) and other erythrina alkaloids [196], of 3-ethylindole and indole-2-carboxylic acid from yohimbic acid (210) [197], and of 3,4-dimethylindole from the dihydro derivative of lysergic acid (211) [198] indicates the presence of an indole or hydroindole ring system. The formation of 1-methyl-5-aminonaphthalene and methylamine in the last-mentioned degradation is also significant. Important information on the structure of papaverine (212) was obtained [199] by the isolation of 6,7-dimethoxyiso-quinoline (213) and 3,4-dimethoxytoluene (214) after alkali fusion. Although these two products together account for all 20 carbon atoms of the alkaloid, it seems likely that they are formed in different reactions.

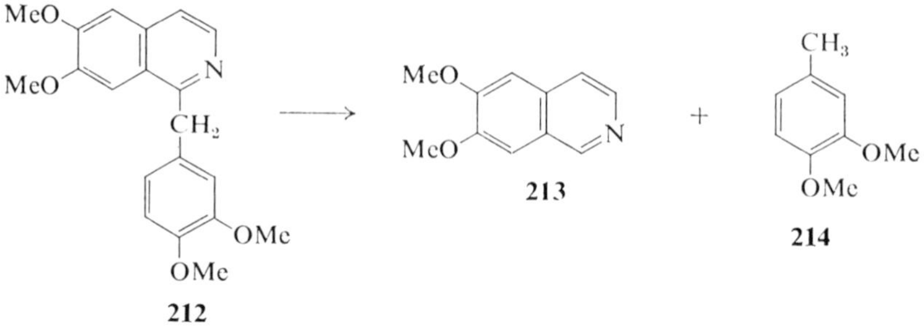

The examples given in this section also serve to re-emphasize the caution which must be exercised in drawing conclusions from the results of alkali fusions.

References

1. W. v. E. Doering and R. M. Haines, *J. Amer. Chem. Soc.*, **76**, 482 (1954).
2. R. A. Dytham and B. C. L. Weedon, *Tetrahedron*, **9**, 246 (1960).
3. R. A. Dytham and B. C. L. Weedon, *Tetrahedron*, **8**, 246 (1960).
3a. M. F. Ansell, D. J. Redshaw, and B. C. L. Weedon, *J. Chem. Soc.* C, **1971**, 1846.
4. M. C. Boswell and J. V. Dickson, *J. Amer. Chem. Soc.*, **40**, 1786 (1918).
5. R. G. Ackman, R. P. Linstead, B. J. Wakefield, and B. C. L. Weedon, *Tetrahedron*, **8**, 221 (1960).
5a. M. F. Ansell, A. N. Radziwill, and B. C. L. Weedon, *J. Chem. Soc.* C. **1971**, 1851.

6. F. Varrentrapp, *Justus Liebigs Ann. Chem.*, **35**, 196 (1840).

7. A. Fitz, *Chem. Ber.*, **4**, 442 (1871).

8. G. Goldschmiedt, *Jahresber.*, **1877**, 522.

9. F. Becker, *Chem. Ber.*, **11**, 1412 (1878); cf. I. Jegorow, *J. Russ. Phys.-Chem. Soc.*, **46**, 975 (1914).

10. P. Chuit, F. Boelsing, J. Hausser, and G. Malet, *Helv. Chim. Acta*, **9**, 1074 (1926), **10**, 113 (1927).

11. A. Lüttringhaus and W. Reif, *Justus Liebigs Ann. Chem.*, **618**, 221 (1958).

12. H. J. Pistor and H. Plieninger, *Justus Liebigs Ann. Chem.*, **562**, 239 (1949).

13. F. X. Werber, J. E. Jansen, and T. L. Gresham, *J. Amer. Chem. Soc.*, **74**, 532 (1952).

14. J. J. A. Blekkingh, H. J. J. Janssen, and J. G. Keppler, *Rec. Trav. Chim. Pays-Bas*, **76**, 35, 42 (1957).

15. E. H. Farmer, *Trans. Faraday Soc.*, **38**, 356 (1942).

16. F. G. Edmed, *J. Chem. Soc.*, **73**, 627 (1898).

17. M. Saytzeff, C. Saytzeff, and A. Saytzeff, *J. Prakt. Chem.*, **37**, 269 (1888).

18. R. Lukeš and J. Hofman, *Chem. Listy*, **52**, 1747 (1958).

19. G. Hunter and G. Popják, *Biochem. J.*, **50**, 163 (1951).

20. C. H. Kao and S.-Y. Ma, *J. Chem. Soc.*, **1931**, 2046.

21. G. Ponzio, *Gazz. Chim. Ital.*, **34**, 77 (1904).

22. F. Runge, R. Hueter, and H.-D. Wulf, *Chem. Ber.*, **87**, 1430 (1954).

23. M. M. Shemyakin and I. A. Red'kin, *J. Gen. Chem. (USSR)*, **11**, 1142, 1169 (1941); *Chem. Abstr.*, **37**, 4053, 4054 (1943); M. M. Shemyakin and L. A. Shchukina, *Quart. Rev.* (London), **10**, 261 (1956).

24. F. Fichter and F. Sonneborn, *Chem. Ber.*, **35**, 938 (1902).

25. R. Fittig and J. G. Spenzer, *Justus Liebigs Ann. Chem.*, **283**, 66, 80 (1894).

26. D. J. G. Ives and R. H. Kerlogue, *J. Chem. Soc.*, **1940**, 1362.

27. R. G. Ackman, R. A. Dytham, B. J. Wakefield, and B. C. L. Weedon, *Tetrahedron*, **8**, 239 (1960).

28. R. A. Dytham and B. C. L. Weedon, unpublished results.

29. J. E. Jackson, R. F. Paschke, W. Tolberg, H. M. Boyd, and D. H. Wheeler, *J. Amer. Oil Chem. Soc.*, **29**, 229 (1952).

30. P. L. Nichols, S. F. Herb, and R. W. Riemenschneider, *J. Amer. Chem. Soc.*, **73**, 247 (1951).

31. A. W. Ralston, *Fatty Acids and Their Derivatives*, Wiley, New York, 1948.

32. C. R. Scholfield and J. C. Cowan, *J. Amer. Oil Chem. Soc.*, **36**, 631 (1959).

33. H. G. Kirschenbauer, U.S. Patent 2,682,549 (1945); *Chem. Abstr.*, **48**, 11819 (1954).

34. M. Bodenstein, *Chem. Ber.*, **27**, 3397 (1894).

35. S. Marasse, *Chem. Ber.*, **2**, 359 (1869).

36. E. R. H. Jones, G. H. Whitham, and M. C. Whiting, *J. Chem. Soc.*, **1954**, 3201.

37. E. L. Pelton and A. A. Holzschuh, U.S. Patent 2,425,343 (1947); *Chem. Abstr.*, **41**, 7414 (1947).

38. E. L. Pelton and A. A. Holzschuh, U.S. Patent 2,531,363 (1950); *Chem. Abstr.* **45**, 2969 (1951).

39. A. W. Crossley and W. H. Perkin, Jr., *J. Chem. Soc.*, **73**, 1 (1898).
40. M. C. Boswell and J. V. Dickson, *J. Amer. Chem. Soc.*, **40**, 1779 (1918).
41. H. S. Fry and E. Otto, *J. Amer. Chem. Soc.*, **50**, 1122 (1928).
42. H. S. Fry and E. L. Schulze, *J. Amer. Chem. Soc.*, **48**, 958 (1926).
43. H. S. Fry, E. L. Schulze, and H. Weitkamp, *J. Amer. Chem. Soc.*, **46**, 2268 (1924).
44. L. H. P. Weldon and C. L. Wilson, *J. Chem. Soc.*, **1946**, 244.
45. E. H. Rodd, *Thorpe's Dictionary of Applied Chemistry*, 4th ed., Vol. 6, Longmans, London, 1943, pp. 445, 446.
46. J. F. Thorpe and M. A. Whiteley, *Thorpe's Dictionary of Applied Chemistry*, 4th ed., Vol. 5, Longmans, London, 1941, p. 323.
47. L. P. Hammett, *Physical Organic Chemistry*, McGraw-Hill, New York, 1940.
48. K. Takeshita, *J. Chem. Soc. Japan, Ind. Chem. Sect.*, **55**, 223, 279 (1952); *Chem. Abstr.*, **48**, 7545 (1954); **47**, 12228 (1953).
49. K. E. Hamlin and A. W. Weston, *Org. Reactions*, **9**, 1 (1957).
50. H. M. Walborsky and F. J. Impastato, *Chem. Ind.* (London), **1958**, 1690.
51. W. E. Bachmann, *J. Amer. Chem. Soc.*, **57**, 737 (1935).
52. O. Kruber, *Chem. Ber.*, **65**, 1382 (1932).
53. M. Guerbet, *Bull. Soc. Chim. Fr.*, **5**, 420 (1909).
54. M. Guerbet, *C.R. Acad. Sci., Paris*, **148**, 720 (1909).
55. O. Wallach and M. Behnke, *Justus Liebigs Ann. Chem.*, **369**, 98 (1909).
56. O. Wallach, *Justus Liebigs Ann. Chem.*, **296**, 120 (1897); **389**, 197 (1912).
57. O. Wallach and A. Holmberger, *Justus Liebigs Ann. Chem.*, **369**, 97 (1909).
58. D. W. MacCorquodale, S. A. Thayer, and E. A. Doisy, *J. Biol. Chem.*, **99**, 327 (1933), **101**, 753 (1933).
59. K. Miescher, *Helv. Chim. Acta*, **27**, 1727 (1944); *Experientia*, **2**, 237 (1946).
60. K. Miescher, *Chem. Rev.*, **43**, 367 (1948).
61. W. Hohlweg and H. H. Inhoffen, German Patents 705,862 (1941) and 719,572 (1942).
62. T. L. Cairns, R. M. Joyce, and R. S. Schreiber, *J. Amer. Chem. Soc.*, **70**, 1689 (1948).
63. E. L. Pelton, C. J. Starnes, and S. A. Shrader, *J. Amer. Chem. Soc.*, **72**, 2039 (1950).
64. H. Finch, K. E. Furman, and S. A. Ballard, *J. Amer. Chem. Soc.*, **73**, 4299 (1951).
65. J. G. Fife, British Patent 560,598 (1944); *Chem. Abstr.*, **40**, 3130 (1946).
66. W. A. Ayer and W. I. Taylor, *J. Chem. Soc.*, **1955**, 2227; cf. G. Büchi, J. H. Hansen, D. Knutson, and E. Koller, *J. Amer. Chem. Soc.*, **80**, 5517 (1958).
67. K. Takeshita, *J. Chem. Soc. Japan, Ind. Chem. Sect.*, **56**, 28 (1953); *Chem. Abstr.*, **48**, 7545 (1954).
68. C. Djerassi and J. A. Zderic, *J. Amer. Chem. Soc.*, **78**, 6390 (1956).
69. J. Dumas and J. S. Stass, *Justus Liebigs Ann. Chem.*, 35, 129 (1840).
70. H. Machemer, *Angew. Chem.*, **64**, 213 (1952).
71. M. Guerbet, *Bull. Soc. Chim. Fr.*, **11**, 164 (1912).
72. E. E. Reid, H. Worthington, and A. W. Larchar, *J. Amer. Chem. Soc.*, **61**, 99 (1939).

73. C. Weizmann and S. F. Garrard, *J. Chem. Soc.*, **117**, 324 (1920).

74. H. C. Chitwood, U.S. Patent 2,384,817 (1945); *Chem. Abstr.*, **40**, 354 (1946); British Patent 601,817 (1948); *Chem. Abstr.*, **42**, 7325 (1948).

75. W. Stein, British Patent 698,154 (1953); *Chem. Abstr.*, **48**, 6147 (1954); U.S. Patent 2,696,501 (1954); *Chem. Abstr.*, **49**, 2760 (1955).

76. C. Hell, *Justus Liebigs Ann. Chem.*, **223**, 269 (1884).

77. M. Guerbet, *Bull. Soc. Chim. Fr.*, **5**, 420 (1909), **7**, 212 (1910), **11**, 164, 279 (1912); *Ann. Chim. Phys.*, **27**, 67 (1902); *C.R. Acad. Sci.*, *Paris*, **128**, 511, 1002 (1899), **132**, 207, 685 (1901), **133**, 300, 1220 (1902), **134**, 467 (1902), **135**, 172 (1902), **149**, 129 (1909), **150**, 183, 979 (1910), **154**, 222 (1912), **155**, 1156 (1912); cf. E. G. E. Hawkins and W. E. Nelson, *J. Chem. Soc.*, **1954**, 4704.

78. G. Lock, *Chem. Ber.*, **63**, 551 (1930).

79. G. H. Hargreaves and L. N. Owen, *J. Chem. Soc.*, **1947**, 753.

80. G. Lock, *Chem. Ber.*, **61**, 2234 (1928), **62**, 1177 (1929).

81. R. Piria, *Justus Liebigs Ann. Chem.*, **30**, 165 (1839).

82. A. T. Nielsen, *J. Amer. Chem. Soc.*, **79**, 2518, 2524 (1957).

83. R. Pummerer and J. Smidt, *Justus Liebigs Ann. Chem.*, **610**, 192 (1957).

84. H. A. D. Jowett, *J. Chem. Soc.*, **79**, 1331 (1901).

85. C. S. Barnes and A. Palmer, *Aust. J. Chem.*, **10**, 334 (1957).

86. G. H. Hargreaves and L. N. Owen, *J. Chem. Soc.*, **1947**, 750.

87. H. Fischer and K. Zeile, *Justus Liebigs Ann. Chem.*, **468**, 98 (1928).

88. M. Apel and B. Tollens, *Chem. Ber.*, **27**, 1087 (1894).

89. C. Mannich and W. Brose, *Chem. Ber.*, **56**, 833 (1923).

90. J. R. Roach, H. Wittcoff, and S. E. Miller, *J. Amer. Chem. Soc.*, **69**, 2651 (1947).

91. H. Witticoff, *Org. Syntheses*, **31**, 101 (1951).

92. P. Barbier, *C.R. Acad. Sci.*, *Paris*, **126**, 1423 (1898).

93. J. Doeuvre, *Bull. Soc. Chim. Fr.*, **45**, 351 (1929).

94. F. Tiemann, *Chem. Ber.*, **31**, 2989 (1898).

95. P. Yates and G. H. Stout, *J. Amer. Chem. Soc.*, **80**, 1691 (1958).

96. M. J. Bouis, *C.R. Acad. Sci.*, *Paris*, **33**, 141 (1851), **41**, 603 (1859).

97. W. Markownikow and P. Zubow, *Chem. Ber.*, **34**, 3246 (1901); cf. *J. Russ. Phys.-Chem. Soc.*, **21**, 128 (1889).

98. J. Bolle and L. Bourgeois, *C.R. Acad. Sci.*, *Paris*, **233**, 1466 (1951).

99. C. A. Carter, U.S. Patent 2,457,866 (1949); *Chem. Abstr.*, **43**, 3437 (1949).

100. R. E. Miller and G. E. Bennett, U.S. Patent 2,762,847 (1956); *Chem. Abstr.*, **51**, 5112 (1957).

101. E. F. Pratt and D. G. Kubler, *J. Amer. Chem. Soc.*, **76**, 53 (1954).

102. M. Sulzbacher, *J. Appl. Chem.*, **5**, 637 (1955); British Patent 655, 864 (1951); *Chem. Abstr.*, **46**, 7580 (1952).

103. C. Weizmann, E. Bergmann, and M. Sulzbacher, *J. Org. Chem.*, **15**, 54 (1950).

104. C. Weizmann, M. Sulzbacher, and E. Bergmann, *J. Chem. Soc.*, **1947**, 772.

105. J. Bolle, *C.R. Acad. Sci.*, *Paris*, **233**, 1629 (1951).

106. L. E. Gast, E. D. Bitner, J. C. Cowan, and H. M. Teeter, *J. Amer. Oil Chem. Soc.*, **35**, 703 (1958).

107. W. Hückel and H. Naab. *Chem. Ber.*, **64**, 2137 (1931).
108. A. Lüttringhaus, *Angew. Chem.*, **62**, 87 (1950), **63**, 244 (1951).
109. C. Weizmann, E. Bergmann, and L. Haskelberg, *Chem. Ind.* (London), **56**, 589 (1937).
110. M. Guerbet, *C.R. Acad. Sci., Paris*, **146**, 298 (1908).
111. B. W. Howk, U.S. Patent 2,293,649 (1942); *Chem. Abstr.*, **37**, 978 (1943).
112. W. Poetsch, *Justus Liebigs Ann. Chem.*, **218**, 66 (1883).
113. J. U. Nef, *Justus Liebigs Ann. Chem.*, **318**, 143 (1901).
114. A. J. Birch, O. C. Musgrave, R. W. Rickards, and H. Smith, *J. Chem. Soc.*, **1959**, 3146.
115. J. D. Bu'Lock and H. Gregory, *Biochem. J.*, **69**, 35P (1958).
116. C. Gerhardt, *Justus Liebigs Ann. Chem.*, **44**, 279 (1842).
117. R. B. Turner and R. B. Woodward, *The Alkaloids*, Vol. 3, Academic Press, New York, 1955, p. 2.
118. A. Butlerow and A. Wischnegradsky, *Chem. Ber.*, **12**, 2093 (1879).
119. M. Guerbet, *C.R. Acad. Sci.*, **154**, 713 (1912).
120. H. Pines and L. Schaap, *J. Amer. Chem. Soc.*, **79**, 2956 (1957).
121. H. D. Zook, J. March, and D. F. Smith, *J. Amer. Chem. Soc.*, **81**, 1617 (1959).
122. W. v. E. Doering and T. C. Aschner, *J. Amer. Chem. Soc.*, **71**, 839 (1949), **75**, 393 (1953).
123. G. H. Hargreaves and L. N. Owen, *J. Chem. Soc.*, **1947**, 756.
124. T. Asahara and M. Tomita, *J. Oil Chem. Soc. Japan*, **2**, 105 (1953); *Chem. Abstr.*, **48**, 11341 (1954).
125. H. R. Le Sueur and J. C. Withers, *J. Chem. Soc.*, **105**, 2801 (1913); cf. H. R. Le Sueur, *J. Chem. Soc.*, **79**, 1313 (1901).
126. R. L. Logan, U.S. Patent 2,625,558.
127. J. F. McGhie, unpublished results.
128. B. H. Nicolet and A. E. Jurist, *J. Amer. Chem. Soc.*, **44**, 1136 (1922).
129. T. P. Hilditch and H. Plimmer, *J. Chem. Soc.*, **1942**, 204.
130. R. S. Morrell and E. O. Phillips, *J. Soc. Chem. Ind.*, **57**, 245 (1938).
130a. M. F. Ansell, I. S. Shepherd, and B. C. L. Weedon, *J. Chem. Soc. C*, **1971**, 1857.
131. J. U. Nef, *Justus Liebigs Ann. Chem.*, **335**, 302 (1904); cf. K. C. Brannock and G. R. Lappin, *J. Amer. Chem. Soc.*, **77**, 6053 (1955).
132. A. Eckert, *Monatsh. Chem.*, **38**, 1 (1917).
133. H. R. Le Sueur and C. C. Wood, *J. Chem. Soc.*, **123**, 1697 (1921).
134. G. F. Marrian and G. Haslewood, *J. Soc. Chem. Ind.*, **51, II**, 279 T (1932).
135. J. Baddiley, *Thorpe's Dictionary of Applied Chemistry*, 4th ed., Vol. 9, Longmans, London, 1949, p. 146.
135a. M. F. Ansell, I. S. Shepherd, and B. C. L. Weedon, *J. Chem. Soc. C*, **1971**, 1840.
136. A. Kekulé, *C.R. Acad. Sci., Paris*, **64**, 752 (1867); N. N. Vorozhtsov, *Bull. Acad. Sci. URSS, Classe Sci. Chim.*, **1940**, 107; I. A. Makolkin, *Acta Physicochim. URSS*, **16**, 88 (1942).
137. A. Wurtz, *C.R. Acad. Sci., Paris*, **64**, 749 (1867).

138. J. Dusart, *C.R. Acad. Sci., Paris*, **64**, 795 (1867).
139. C. K. Ingold, *Structure and Mechanism in Organic Chemistry*, Bell, London, 1953; J. F. Bunnett and R. E. Zahler, *Chem. Rev.*, **49**, 273 (1951).
140. L. Barth and J. Schreder, *Chem. Ber.*, **12**, 417, 503 (1879); R. Lemberg, *Chem. Ber.*, **62**, 592 (1929); M. V. Troitskii, *Bull. Acad. Sci. URSS, Classe Sci. Chim.*, **1940**, 127; *Chem. Abstr.*, **35**, 2484 (1941).
141. O. Jacobsen, *Justus Liebigs Ann. Chem.*, **206**, 200 (1881).
142. A. Faust, *Chem. Ber.*, **6**, 1022 (1873).
143. H. E. Fierz-David and G. Stamm, *Helv. Chim. Acta*, **25**, 364 (1942).
144. R. Fittig and E. Mager, *Chem. Ber.*, **7**, 1175 (1874), **8**, 362 (1875).
145. H. Erlenmeyer, H. Lobeck, and A. Epprecht, *Helv. Chim. Acta*, **19**, 546 (1936).
146. C. Graebe and C. Liebermann, *Chem. Ber.*, **2**, 332 (1869); cf. M. Phillips, *J. Amer. Chem. Soc.*, **49**, 473 (1927).
147. L. F. Fieser and M. Fieser, *Organic Chemistry*, Heath, Boston, 1944.
148. L. Barth, *Z. Chem.*, 650 (1866); S. Komatsu and S. Tanaka, *J. Chem. Soc. Japan*, **51**, 138 (1930); *Chem. Abstr.*, **26**, 706 (1932); R. Amatatsu and S. Araki, *J. Chem. Soc. Japan*, **52**, 484 (1931); *Chem. Abstr.*, **26**, 5083 (1932).
149. G. Krämer and R. Weissgerber, *Chem. Ber.*, **34**, 1662 (1901).
150. E. v. Gerichten and O. Dittmer, *Chem. Ber.*, **39**, 1718 (1906).
151. H. Bechurts and G. Frerichs, *Apotheker Ztg.*, **18**, 697 (1903); *Chem. Zentr.*, **74**, II, 1010 (1903); *Arch. Pharm.*, **243**, 470 (1905); cf. J. Tröger and O. Müller, *Apotheker Ztg.*, **24**, 678 (1909); *Chem. Zentr.*, **80**, II, 1570 (1909); *Arch. Pharm.*, **248**, 1 (1910); W. G. Körner and C. Böhringer, *Gazzetta*, **13**, 363 (1883).
152. R. Richter, *J. Prakt. Chem.*, **28**, 273 (1883); cf. V. Mierz and W. Weith, *Chem. Ber.*, **14**, 192 (1881); G. Goldschmiedt, *Monatsh. Chem.*, **4**, 124, 129 (1883).
153. H. Hlasiewetz, *Justus Liebigs Ann. Chem.*, **112**, 96 (1859).
154. H. Raistrick, C. E. Stickings, and R. Thomas, *Biochem. J.*, **55**, 421 (1953).
155. C. Graebe, *Justus Liebigs Ann. Chem.*, **254**, 265 (1889).
156. Z. Delalande, *Justus Liebigs Ann. Chem.*, **45**, 333, 336 (1843); cf. H. Bleibtreu, *Justus Liebigs Ann. Chem.*, **59**, 189 (1846).
157. E. Späth and K. Klager, *Chem. Ber.*, **66**, 749 (1933).
158. F. Tiemann, and C. L. Reimer, *Chem. Ber.*, **12**, 993 (1879); F. Tiemann and A. Parrisius, *Chem. Ber.*, **13**, 2359 (1880).
159. C. Pomeranz, *Monatsh. Chem.*, **12**, 379 (1891).
160. J. Piccard, *Chem. Ber.*, **6**, 884 (1873); cf. S. v. Kostanecki, *Chem. Ber.*, **26**, 2901 (1893).
161. S. Smorawski, *Chem. Ber.*, **12**, 1595 (1879).
162. J. Bose and S. Siddiqui, *J. Sci. Ind. Res.*, **4**, 231 (1945); *Chem. Abstr.*, **40**, 2832 (1946).
163. R. Willstätter and H. Mallison, *Justus Liebigs Ann. Chem.*, **408**, 40 (1915); R. Willstätter and E. K. Bolton, *Justus Liebigs Ann. Chem.*, **408**, 59 (1915); R. Willstätter and W. Mieg, *Justus Liebigs Ann. Chem.*, **408**, 61 (1915).

164. P. Karrer and R. Widmer, *Helv. Chim. Acta*, **10**, 5 (1927).

165. K. Venkataraman, *Fortschr. Chem. Org. Naturstoffe*, **17**, 28 (1959); *The Chemistry of Synthetic Dyes*, Academic Press, New York, 1952.

166. R. Pasternack, A. Bavley, R. L. Wagner, F. A. Hochstein, P. P. Regna, and K. J. Brunings, *J. Amer. Chem. Soc.*, **74**, 1926 (1952); F. A. Hochstein, C. R. Stephens, L. H. Conover, P. P. Regna, R. Pasternack, P. N. Gordon, F. J. Pilgrim, K. J. Brunings, and R. B. Woodward, *J. Amer. Chem. Soc.*, **75**, 5455 (1953).

167. B. L. Hutchings, C. W. Waller, S. Gordon, R. W. Broschard, C. F. Wolf, A. A. Goldman, and J. H. Williams, *J. Amer. Chem. Soc.*, **74**, 3710 (1952).

168. O. Dragendorff, *Justus Liebigs Ann. Chem.*, **482**, 280 (1930), **487**, 62 (1931).

169. S. Yamashiro, *Bull. Chem. Soc. Japan*, **7**, 1 (1932).

170. M. Murakami, *Proc. Imp. Acad. (Tokyo)*, **7**, 254, 311 (1931); *Ann.*, **496**, 122 (1932); *J. Chem. Soc. Japan*, **53**, 150, 162 (1932).

171. M. L. Wolfrom and J. Mahan, *J. Amer. Chem. Soc.*, **64**, 308 (1942); M. L. Wolfrom, W. D. Harris, G. F. Johnson, J. E. Mahan, S. M. Moffett, and B. Wildi, *J. Amer. Chem. Soc.*, **68**, 406 (1946).

172. E. Gorup-Besanez, *Chem. Ber.*, **7**, 564 (1874); *Justus Liebigs Ann. Chem.*, **183**, 321 (1876); cf. E. Späth and K. Klager, *Chem. Ber.*, **67**, 859 (1934).

173. E. Späth and K. Klager, *Chem. Ber.*, **66**, 914 (1933).

174. E. Späth and W. Gruber, *Chem. Ber.*, **74**, 1492 (1941).

175. J. H. Birkinshaw, A. Bracken, S. E. Michael, and H. Raistrick, *Biochem. J.*, **48**, 67 (1951); cf. D. H. R. Barton and J. B. Hendrickson, *Chem. Ind.* (London), **1955**, 682.

176. S. Takei, *Biochem. Z.*, **157**, 1 (1925); *Bull. Inst. Phys. Chem. Research* (Tokyo), **2**, 485 (1923), **3**, 673 (1924); cf. F. B. La Forge, H. L. Haller, and L. E. Smith, *Chem. Rev.*, **12**, 181 (1933); H. L. Haller and F. B. La Forge, *J. Amer. Chem. Soc.*, **52**, 4505 (1930); T. Kariyone and M. Hadano, *J. Pharm. Soc. Japan*, **50**, 542 (1930); *Chem. Abstr.*, **24**, 4785 (1930).

177. J. F. Eykman, *Chem. Ber.*, **23**, 859 (1890); E. Grimaux and J. Ruotte, *Justus Liebigs Ann. Chem.*, **152**, 91 (1869); A Einhorn and C. Frey, *Chem. Ber.*, **27**, 2455 (1894).

178. A. R. Bader, *J. Amer. Chem. Soc.*, **78**, 1709 (1956).

179. D. Richter, *Thorpe's Dictionary of Applied Chemistry*, 4th ed., Vol. 1, Longmans, London, 1949, p. 203.

180. W. Bradley and E. Leete, *J. Chem. Soc.*, **1951**, 2129.

181. W. Bradley and F. K. Sutcliffe, *J. Chem. Soc.*, **1951**, 2118, **1952**, 1247; W. Bradley and G. V. Jadhav, *J. Chem. Soc.*, **1948**, 1622.

182. C. Graebe and C. Liebermann, *Justus Liebigs Ann. Chem.*, **160**, 129 (1871).

183. R. Bohn, German Patent, 129,845 (1902).

184. R. Scholl, *Chem. Ber.*, **36**, 3410 (1903).

185. W. Bradley and H. E. Nursten, *J. Chem. Soc.*, **1951**, 2170.

186. R. Scholl, *Chem. Ber.*, **43**, 246 (1910).

187. O. Dimroth, *Justus Liebigs Ann. Chem.*, **399**, 1 (1913); M. A. Ali and L. J. Haynes, *J. Chem. Soc.*, **1959**, 1033; R. Oda, K. Tamura, and K. Maeda, *J. Soc. Chem. Ind. Japan*, **41**, *Suppl.*, 193 (1938); *Chem. Abstr.*, **32**, 7447 (1938).

188. O. Bally, *Chem. Ber.*, **38,** 196 (1905), German Patent 185,221 (1907); cf. R. Scholl and C. Seer, *Justus Liebigs Ann. Chem.*, **394,** 111 (1912).
189. A. Lüttringhaus and H. Neresheimer, *Justus Liebigs Ann. Chem.*, **473,** 259 (1929).
190. R. F. Thomson, *Thorpe's Dictionary of Applied Chemistry*, 4th ed., Vol. 1, Longmans, London, p. 425.
191. O. Bally, German Patent 194,252 (1908); cf. A. Zinke, F. Linner, and O. Wolfbauer, *Chem. Ber.*, **58,** 323 (1925).
192. A. Melera, K. Schaffner, D. Arigoni, and O. Jeger, *Helv. Chim. Acta*, **40,** 1420 (1957); D. Arigoni, D. H. R. Barton, E. J. Corey, O. Jeger, L. Cagliotti, S. Dev, P. G. Ferini, E. R. Glazier, A. Melera, S. K. Pradhan, K. Schaffner, S. Stenhell, J. Templeton, and S. Tobinago, *Experientia*, **16,** 41 (1960).
193. A. Windaus, *Sitzber. Heidelberg. Akad. Wiss., Math.-Naturw. Kl., Abhandl.*, **1914,** 18.
194. J. Herzig, *Monatsh. Chem.*, **19,** 738 (1898); W. Feuerstein and S. v. Konstanecki, *Chem. Ber.*, **22,** 1024 (1899).
195. L. Barth and H. Weidel, *Monatsh. Chem.*, **4,** 700 (1883).
196. K. Folkers, F. Koniuszy, and J. Shavel, *J. Amer. Chem. Soc.*, **64,** 2146 (1942).
197. G. Barger and C. Scholz, *J. Chem. Soc.*, **1933,** 614.
198. W. A. Jacobs and L. C. Craig, *J. Biol. Chem.*, **111,** 455 (1935), **113,** 767 (1936), **128,** 715 (1939).
199. G. Goldschmiedt, *Monatsh. Chem.*, **8,** 510 (1887).

Chapter **XV**

DEGRADATION OF POLYSACCHARIDES

G. O. Aspinall

I INTRODUCTION

Polysaccharides may be defined as condensation polymers of high molecular weight in which glycose units are joined by glycosidic linkages. There is no rigorously defined dividing line between oligo- and polysaccharides, but the term polysaccharide is normally used for materials containing more than 10 sugar residues. Other natural macromolecules, which are not composed entirely of sugar units, contain blocks of monosaccharide residues as part of the molecular structure. This chapter will include some reference to glycoproteins, teichoic acids, and related polymeric carbohydrate phosphates, but nucleic acids will not be considered.

Investigations directed towards the determination of the fine structure of polysaccharides seek information on (*1*) the nature and proportions of glycose units, (*2*) the position and configuration of the glycosidic linkages, (*3*) the sequence of sugar units along the chain, (*4*) the type of branching, and (*5*) the location of ester, ketal, and other substituents.

Knowledge of the structure of polysaccharides [1] has been derived primarily from degradation studies on the original macromolecule leading to the formation of simpler fragments. Hydrolysis and similar solvolytic procedures form the commonest means for effecting degradation, and are of greatest value when incomplete depolymerization leads to the isolation of oligosaccharides or their derivatives. The hydrolytic degradation of structurally modified polysaccharides similarly furnishes the corresponding modified mono- or oligosaccharides. Pre-eminent among structural modification procedures is the methylation method in which hydrolysis of the fully etherified derivative gives methylated or partially methylated sugars in which hydroxyl groups indicate sites of linkage from one glycose residue to another or positions from which acid-labile substituents, for example, ester groups, were removed during depolymerization. Among oxidative degradation procedures that involving glycol cleavage with periodate is by far the most commonly used. The application of each degradation procedure gives certain information, but no single procedure provides sufficient data to uniquely define the structure of a polysaccharide. Usually structural information is sought by as many different and complementary procedures as possible, but in the case of complex polysaccharides it is often still not possible to put forward completely unambiguous primary structures.

Since polysaccharides differ widely in detailed structure, many variations of these general procedures are used in particular cases. In this chapter it will only be possible to mention some of the special cases, but the more important procedures and their variations, which are used in studies on neutral, acidic, and amino polysaccharides, and on carbohydrate polymers containing phosphate and sulfate ester functions, will be outlined.

2 DEGRADATION STUDIES OF THE NATIVE POLYSACCHARIDE

Hydrolysis

Discussion

Acidic or enzymic hydrolysis of the glycosidic linkages in a polysaccharide may lead to the production of the free glycose components, which may be qualitatively and quantitatively analyzed, thus establishing the composition of the glycan. Graded acidic or enzymic hydrolysis leads to the production of a degraded polysaccharide or oligosaccharides of constitutional significance since it is assumed (but see later) that these fragments arise from direct breakdown of the polysaccharide without modification of the linkages. Thus the isolation of maltose (4-O-α-D-glucopyranosyl-D-glucopyranose) in high yield from the hydrolysis of starch [2] furnished early unequivocal proof of the $(1 \rightarrow 4)$ α-D-linkages in starch. The subsequent isolation of a polymer-homologous series of maltodextrins from the partial hydrolysis of amylose [3] provided detailed evidence for regularly repeating linkages of the same type. The characterization of tri- and higher oligosaccharides containing more than one type of linkage, for example, panose (O-α-D-glucopyranosyl-$(1 \rightarrow 6)$-O-α-D-$(1 \rightarrow 4)$-D-glucopyranose) from amylopectin [4], provides unambiguous evidence for these linkages in a single polysaccharide. Similarly, the identification of oligosaccharides with more than one glycose component, for example, 6-O-α-D-galactopyranosyl-D-mannose from the galactomannan guaran [5], clearly establishes their origin in a heteropolysaccharide.

If all the linkages in a glycan are split at approximately the same rate, incomplete or partial hydrolysis will give rise to a selection of oligosaccharide fragments representative of all regions in the parent macromolecule. Since this situation is rarely encountered in practice, the failure to isolate a particular oligosaccharide on partial hydrolysis *may indicate* but, without independent supporting evidence, *does not prove* the absence of that particular linkage in the polysaccharide. Many glycans contain a number of types of linkage which are hydrolyzed at markedly different rates. With such complex polysaccharides the ready isolation on partial hydrolysis of some only of the theoretically possible oligosaccharides is a simplifying factor experimentally which necessarily excludes the isolation of oligosaccharides containing the other linkages. Evidence for the presence of these other linkages in the polysaccharide must be sought independently, for example, by using a depolymerization procedure of different relative selectivity or by structural modification of the polysaccharide so as to alter the relative rates of cleavage of glycosidic bonds.

The products which may be isolated from the controlled acid hydrolysis of complex polysaccharides will be determined by the location of the more acid-sensitive linkages. The acid-labile linkages are often found in the outer chains of branched polysaccharides, for example, many plant gums [6, 7], and mild hydrolysis gives rise to structurally simpler degraded polysaccharides, frequently the "core" or "backbone" of the macromolecule, which are more amenable to structural analysis. In the case of acid hydrolysis, however large the differences in rates of cleavage of glycosidic bonds may be, selectivity of hydrolysis is relative rather than absolute and some cleavage of the more stable linkages necessarily accompanies the complete hydrolysis of the labile linkages. Only in some, but by no means all, cases of enzymic hydrolysis is complete specificity encountered.

Provided that the enzyme preparation has been adequately purified and is free from contaminating enzymic activities, hydrolysis with enzymes, in contrast to acid hydrolysis, shows some measure of absolute specificity. Thus β-D-glucosides will not be hydrolyzed by α-D-glucosides. The enzyme hydrolysing α-D-glucopyranosyl bonds will normally cleave bonds of a certain type only, for example, $(1 \rightarrow 4)$ but not $(1 \rightarrow 6)$ bonds. Furthermore, an enzyme hydrolyzing α-D-$(1 \rightarrow 4)$ bonds in a glucan may be limited as to the bonds of this type which will be split, and even in a polymer of uniform linkage type a minimum substrate size may be required for effective catalysis. Again, if other linkages are present a minimum number of bonds of the given type in sequence may be necessary before hydrolysis will occur. Furthermore, the ability of an enzyme to hydrolyze large sequences of bonds of a specific type will depend on the mode of action of the enzyme, and further hydrolysis by the enzyme may be completely stopped by a single foreign linkage if action takes place from the end of the chain only. Some examples which relate to enzyme specificity are given later.

The yields of oligosaccharides which are isolated during the partial hydrolysis of glycans are often low because small fragments liberated in the process are unprotected and are progressively degraded to monosaccharides as reaction proceeds. Painter has obtained high yields of oligosaccharides during enzymic hydrolysis by carrying out the reaction in a dialysis membrane so that dialyzable fragments of low molecular weight are removed as rapidly as they are formed [8]. The procedure has been used in studies on Jack pine glucomannan [9]. Essentially the same method has been proposed for partial acid hydrolysis using water-soluble polystyrenesulfonic acid as the catalyst [10]. The technique, however, is only applicable to very acid-labile polysaccharides such as inulin since the cellophane membrane is attacked under the conditions required for the hydrolysis of more resistant materials. As an alternative an apparatus has been designed to enable hydrolysis to be carried out in a stepwise manner followed by dialysis at the end of each period of hydrolysis to remove products of low molecular weight until no nondialyzable polysaccharide is left [11].

Acid Hydrolysis

Glycosidic linkages in polysaccharides are more or less randomly hydrolyzed by acid, and the over-all rate of hydrolysis of a particular polysaccharide depends largely on its structure. α-D-Glycosidic linkages are usually more easily hydrolyzed than β-D-linkages, and bonds of the $(1 \rightarrow 6)$ type are generally more resistant to hydrolysis than others [12]. The rates of hydrolysis of different neutral glycopyranosidic linkages (except those of deoxyglycosides) probably differ by no more than a factor of 10. In the case of simple glycopyranosides these differences have been rationalized in terms of steric factors [13]. On the other hand, the rates of hydrolysis of furanosides may be between 10 and 10^3 times greater than those of the corresponding pyranosides (for a discussion of the mechanism of hydrolysis of furanosides, see Ref. 16). Polysaccharides containing glycose units in their furanose form, such as the fructoglycans [14] and many arabinose-containing polysaccharides [6], are hydrolyzed under very mild conditions.

Glycosiduronic acids are much more resistant than the corresponding neutral glycosides to hydrolysis at low pH values, and graded hydrolysis of glycuronic acid-containing polysaccharides frequently leads to the isolation of aldobiouronic acids of structural significance. The marked differences between entropies of activation in the hydrolysis of aldobiouronic acids and of the corresponding neutral disaccharides [15] suggest that different mechanisms may be involved. At higher pH values, however, these differences in rates are much smaller [16] and in some instances glycosiduronic acid linkages may be *more readily* hydrolyzed than neutral glycosidic linkages [17]. In the case of glycuronans the relatively high rates of hydrolysis above pH 2 have been shown to be due to intramolecular autocatalysis [18]. Hydrolyses at different pH values have been used to isolate overlapping sequences of sugar units in the so-called linkage region of heparin (see Appendix).

The introduction of electron-withdrawing toluene-*p*-sulfonyl substituents in primary positions has been shown to stabilize glycosidic linkages. Model experiments with dextran have been used to assess the potential of this type of structural modification of glycans in linkage analysis [19].

Another type of relative resistance to acid hydrolysis is encountered with 2-amine-2-deoxyglycopyranosides, when the NH_3^+ group formed in acid solution electrostatically shields the neighboring glycosidic substituents from attack by hydrions [20]. This effect is most pronounced in these amino sugar derivatives and amino groups in the "aglycone," for example, in 2-amino-ethyl β-D-glucopyranoside [21] and glycopeptides from collagen containing D-galactose units glycosidically linked to δ-hydroxylysine [22–24], retard hydrolysis to a much smaller extent. 2-Amino-2-deoxyglycose residues in polysaccharides and other carbohydrate polymers are most frequently found naturally as the N-acetyl derivatives whose glycosidic linkages are not unusually resistant to acid hydrolysis. Provided that conditions can be established under which glycoside hydrolysis is more rapid than hydrolysis of the N-acetyl substituents [25], the preferred sites of hydrolysis of 2-acetamido-2-deoxyglycose-containing heteropolysaccharides may be quite different from those in the polysaccharide from which the N-acetyl substituents have been removed, for example, by basic hydrolysis. Thus hydrolysis of carboxyl-reduced chondroitin with and without N-acetyl substituents have been used to isolate oligosaccharides with different sequences of sugar residues (see Appendix).

For the complete hydrolysis of polysaccharides as a prelude to quantitative analysis of the hydrolyzate, a balance must be struck between maximum depolymerization and minimum destruction of sugars. The most favorable conditions for the hydrolysis of the $(1 \rightarrow 4)$ α-D-linkages in starch are $1.5N$ sulfuric acid at $100°$ for 2 hr [26], which effect a minimum destruction (ca. 1 %) of the free glucose. Similar conditions may be used for the complete or graded hydrolysis of other hexoglycans and pentoglycans, although rather more destruction of pentoses than of hexoses takes place [27]. Mannoglycans in general require longer hydrolysis.

Some sugars, for example, 2-deoxyaldoses, sialic acids (O-acylneuraminic acids), and 3,6-anhydro-D- and -L-galactose, may be largely destroyed under the above-mentioned hydrolytic conditions, and an entire component may be lost. The total hydrolysis of fructoglycans may be effected under much milder conditions, but a substantial degree of decomposition of D-fructose would take place with more drastic treatment so that the sugar might be missed as a component of a heteropolysaccharide. Some of the rarer reducing sugars, for example, L-idose and L-gulose, which are formed from the reduction of the corresponding hexuronic acid constituents of dermatan sulfate [28] and alginic acid [29], respectively, may be considerably under-estimated since under the usual hydrolysis conditions they undergo sub-stantial conversion into the nonreducing 1,6-anhydrides. The failure until quite recently [30, 31] to detect L-iduronic acid as a significant constituent of heparin and of L-idose from the carboxyl-reduced polysaccharide provides a striking example. The possibility that artifacts may also be produced during the neutralization of hydrolyzates with alkali or basic ion exchange resins should be noted, and the neutralizations should be carried out at a low temperature and, if artifact formation is suspected, accompanied by concurrent control reactions.

Glycosidation of sugars and hydrolysis of glycosides are the forward and reverse steps of a reversible reaction. Since monosaccharides undergo acid-catalyzed condensation or "acid reversion," the validity of the method of linkage analysis by graded hydrolysis procedures depends upon the demonstration that the oligosaccharides isolated are not artifacts. The more important disaccharides formed from monosaccharides, presumably under conditions of equilibrium control, have been characterized [32–36]. It is therefore usually assumed that an oligosaccharide isolated from the graded hydrolysis of a polysaccharide has constitutional significance and is not a reversion artifact, if it is obtained in much higher yield than the same sugar isolated from the reversion of the component monosaccharide under comparable conditions. True equilibrium, however, is probably not established during partial hydrolysis, and the possibility of quite different disaccharides being formed by reversion under kinetic control or intramolecularly by transglycosylation has not been given the same consideration. It has been suggested by Peat and Whelan [37] that a better method of providing control data for reversion products occurring during linkage analysis is to study the formation of such substances from the disaccharide containing the main polymeric linkage of the polysaccharide rather than the monosaccharide, and further that reversion synthesis should be kept to a minimum by maintaining low temperature and low concentrations of polysaccharide and acid during hydrolysis. The possibility that small amounts of nigerose, 3-O-α-D-glucopyranosyl-D-glucose, obtained from partial hydrolyzates of waxy maize starch [38], beef liver glycogen [39], and Floridean starch [40], is an artifact has been examined by heating maltose with acid under comparable conditions. The experimental evidence for [41] and against [42] artifact formation is not sufficiently convincing to draw definite conclusions.

Kuhn, Freudenberg, and others [43–45] have used a statistical approach to predict that, in the random degradation of a uniform polymer, the yield of the n-membered fragment is given by:

$$n\alpha^2(1 - \alpha)^{n-1}, \tag{15.1}$$

where α is the degree of scission. The maximum yield of the fragment is then

$$n[2/(n + 1)]^2[n - 1)/(n + 1)]^{n-1} \tag{15.2}$$

and occurs when $\alpha = 2/(n + 1)$.

Table 15.1 shows the theoretical extent of hydrolysis required for the maximum yields of oligosaccharides from a linear polymer in which all the glycosidic bonds are equally susceptible to cleavage. Although the experimental results obtained from polysaccharide hydrolyses do not exactly fit the above calculations, the values recorded in Table 15.1 may often suggest suitable conditions for obtaining oligosaccharides in good yield. In practice

Table 15.1   Degradation of Long Chain Molecules

No. of Glycose Units	Extent of Hydrolysis for Maximum Yields	Yield at Maximum Concentration (%)	Total Amount Formed during Complete Hydrolysis (%)
2	2/3	29.6	66.7
3	1/2	18.7	50.0
4	2/5	13.8	40.0
5	1/3	11.0	33.3
6	2/7	9.1	28.6
7	1/4	7.8	25.0
10	2/11	6.4	18.2

Freudenberg [46] has observed that the velocity of hydrolysis is dependent not only on the particular type of glycosidic linkage but also on its place in the glycan chain.

The progress of a polysaccharide hydrolysis may be followed by (1) fall in viscosity, (2) increase in reducing power, (3) the change in optical rotation of the solutions, (4) the change in the pattern of the sugars liberated by paper chromatographic examination of the hydrolyzate (preferably after neutralization), or (5) changes in molecular weight distribution as shown by gel-permeation chromatography (46a).

The following acidic reagents and conditions have found general application in the hydrolysis of polysaccharides.

SULFURIC ACID

Sulfuric acid (0.01N) at 100° has been used for the rapid hydrolysis of fructoglycans [47] and pentofuranose and hexofuranose [48] glycosidic linkages in polysaccharides. In the partial hydrolysis of the furanosidic linkages in beet arabinan [49] and galactocarolose [50], the polysaccharide solution was adjusted to pH 2.2 with sulfuric acid and the reaction mixture was maintained at 75° for several hours.

Sulfuric acid (1–$2N$) at 100° is widely used for the hydrolysis of pento- and hexopyranose glycans. In the case of polysaccharides, for example, cellulose, which are not readily soluble in dilute acids, the time of hydrolysis is prolonged with consequent destruction of much of the liberated sugars before hydrolysis is complete. It is recommended that such insoluble polysaccharides be dissolved in cold 72% sulfuric acid, and upon solution the reaction mixture diluted with water to 1–$2N$ and hydrolysis completed by boiling the diluted acid solution.

At the conclusion of the hydrolysis the sulfuric acid may be removed with a weak base anion exchange resin [51]. Alternatively, the acid may be neutralized in the cold with barium carbonate, or better, neutralized partially (80–90%) by the careful addition of barium hydroxide followed by barium carbonate. The solution is then filtered and the filtrate is treated with a small amount of mixed cation and anion exchange resins to remove the last traces of ionic materials. The hydrolyzate may then be concentrated by evaporation under reduced pressure at a low temperature (30–35°).

If basic or acidic glycoses are present in a hydrolyzate, they are absorbed on the ion exchange resins and then eluted. In the case of acidic sugars, it is preferable not to absorb them on the resin with the mineral acid. The hydrolyzate is best neutralized with barium carbonate (or barium hydroxide followed by barium carbonate), filtered, treated with cation exchange resin to remove barium ions, and passed through a weak base anion exchange resin of the polystyrene, cellulose (e.g., diethylaminoethylcellulose or ECTEOLA-cellulose), or Sephadex types.

HYDROCHLORIC ACID

Hydrochloric acid has been used for the hydrolysis of polysaccharides under conditions similar to those described for sulfuric acid. However, destruction of the liberated glycoses is more apparent than with sulfuric acid, particularly in the case of the aldopentoses, which are readily degraded to furfuraldehyde, and the hexuronic acids, which are destructively decarboxylated.

Glycosaminoglycans usually require treatment with 2–$5N$ hydrochloric acid at 100–120° for several hours to effect hydrolysis, which is accompanied by considerable destruction of neutral and acid glycoses. The hydrochloric acid may be removed from these hydrolyzates by evaporation under reduced pressure at low temperature, by evaporation *in vacuo* over potassium hydroxide, or by treatment with weak base anion exchange resins. Alkaline conditions must be avoided in the neutralization to prevent decomposition of aminoglycoses. Hydrochloric acid may be removed with weak base anion exchange resins, or by neutralization with silver carbonate, the last traces of silver being removed from the filtered solution with hydrogen sulfide.

NITRIC ACID

Aqueous 3% nitric acid is a good hydrolytic reagent for glycans, such as plant cell wall polysaccharides, which contain high proportions of pentose and hexuronic acid residues. The hydrolysis of xyloglycans is normally carried out by treating the polysaccharide (ca. 0.5%) containing a little urea with 3% nitric acid, free from oxides of nitrogen, at 100° for about 4 hr [52]. Longer periods up to 12 hr may be required for a reasonably complete hydrolysis of glycopyranosyluronic acid linkages, but under these conditions some degradation, up to 10% in the case of galacturonic acid, takes place. The nitric acid may be removed with a weak base anion exchanger.

ORGANIC ACIDS

Dilute acetic and oxalic acids are effective mild hydrolysis agents which have been used for the cleavage of the more readily hydrolyzable glycose residues in polysaccharides, for example, in fructoglycans. The acid may be removed by basic ion exchange resins. Alternatively, acetic acid may be removed by evaporation under reduced pressure, and oxalic acid may be neutralized with calcium carbonate.

Trifluoroacetic acid is a convenient reagent for the hydrolysis of all but the most resistant glycosidic, for example, glycopyranosiduronic acid, linkages and has the advantage that it may be removed readily by distillation under reduced pressure [53].

Formic acid (70–100%) at 100° has been used in the hydrolysis of many otherwise insoluble polysaccharides [54]. The acid successfully hydrolyzes glycuronosyl linkages, and the liberated glycuronic acids are fairly stable in the hydrolytic medium. Formic acid has been used in the hydrolysis of hexuronic acid-containing polysaccharides [55, 56] and in the hydrolysis of mannoglycans [57, 58] with particular success.

The polysaccharide is usually heated with 70–100% formic acid on a boiling-water bath (4–24 hr), and then the formic acid is removed by evaporation under reduced pressure. Formyl esters of the glycoses present in the residue may be hydrolyzed by repeated evaporation of their solutions in water [59] or by hydrolysis of the residue with $1N$ sulfuric acid at 90–100° for 1 hr, followed by neutralization (BaCO$_3$), filtration, and concentration under reduced pressure.

AUTOHYDROLYSIS

Polysaccharides containing glycuronic acid residues may be submitted to graded hydrolysis by heating aqueous 1% solutions of the ash-free glycans. The ash-free polysaccharides can be obtained by precipitation from aqueous solution with acidified ethanol or by passing its solution down a column of Amberlite IR-120(H$^+$) cation exchange resin [6]. Autohydrolysis provides a convenient means of effecting the controlled hydrolysis of L-arabinofuranose and 6-deoxyhexopyranose residues from the outer chains of many plant polysaccharides with relatively little degradation of the more resistant D-galactopyranose and other sugar residues in the interior chains [60]. The resistant nucleus, often referred to as the degraded polysaccharide, may be recovered by precipitation from solution with ethanol or acetone, leaving the low molecular weight glycoses, often including both monosaccharides and oligosaccharides, in solution.

The apiogalacturonan from *Lemna minor*, which carries apibiose units attached to a galacturonan chain, undergoes autohydrolysis with remarkable ease, even at pH 4.5 [60a]. Since the apiosidic linkage in the liberated disaccharide is stable under conditions comparable to those used in its formation, intramolecular catalysis involving participation of the free carboxyl groups of D-galacturonic acid residues in the cleavage of glycosidic bonds between the apibiose side-chains and the galacturonan has been postulated.

CATION EXCHANGE RESINS

Excess sulfonated polystyrene resin [e.g., Dowex 50(H⁺), 200–400 mesh] suspended in $0.05N$ hydrochloric acid at 100° has been found to be a suitable catalyst for the hydrolysis of glycosaminoglycuronoglycans [61]. This technique protects the liberated glycuronic acids from excessive destruction during hydrolysis. The resin used in the hydrolysis may also be used for the separation of neutral sugars and glycuronic acids from glycosamines by differential elution. Water-soluble polystyrenesulfonic acid may be used as a non-dialyzable catalyst, but its use is limited to conditions under which the dialysis membrane is not attacked [10].

Enzymic Hydrolysis

Enzymic hydrolysis is an increasingly important degradative method for the structural analysis of polysaccharides [62, 63]. The most obvious application is in furnishing oligosaccharides and degraded polysaccharides as products of partial hydrolysis, particularly when these simpler fragments are inaccessible by other methods (see Appendix for examples of the isolation of oligosaccharides with different sequences of sugar units by various acid-catalyzed depolymerizations and by enzymic hydrolysis). To the extent that the specificity of an enzyme has been established, information of structural significances is derived from a knowledge of both the bonds which are broken and those which resist hydrolysis. The specificity of an enzyme may vary from those instances in which all bonds of a given type, for example, (1→4)-linked α-D-glucopyranosyl bonds, are split to those in which only certain bonds in a particular structural environment are broken. Thus an enzyme which hydrolyses a linear glycan of uniform linkage type may be much more selective towards apparently similar linkages in a structurally complex glycan. Phosphorylase, although a transferase rather than a glycan-hydrolase, provides an example. Phosphorylase cleaves all the (1→4)-linked α-D-glucopyranosyl bonds in amylose with the formation of α-D-glucopyranosyl phosphate but is able to approach to only 4 D-glucose residues beyond the branch points in both side and main chains of amylo-pectin and glycogen [64]. Furthermore the ability of an enzyme to split bonds in a linear polysaccharide of uniform type does not simply define the type of bond which it is able to hydrolyze. For example, the "laminaranase" from *Rhizopus arrhizus* QM 1032 does not hydrolyze the (1→3) bonds per se but rather 3-*O*-substituted β-D-glucopyranosyl linkages [65]. Thus in oat β-D-glucan it is the (1→4) rather than the (1→3) linkages which are hydrolyzed (see Fig. 15.1). The isolation of various oligosaccharides from the selective enzymic hydrolysis of oat β-D-glucan with the laminaranase and a cellulase is illustrated in the Appendix.

The extent of degradation of a polysaccharide by a glycanhydrolase also depends on whether the enzyme is an *endo*-enzyme acting randomly along the chain hydrolyzing bonds of a given type or an *exo*-enzyme acting in a stepwise manner from the (usually but not always) non-reducing ends of the chain. In the latter case no bonds beyond a branch point or other structural variation will be hydrolyzed. β-Amylase acts on starch-type polysaccharides with the liberation of maltose units from the outer chains but is unable to attack (1→4) linkages in the interior chains of amylopectin or glycogen. The two types of enzymic degradation may be distinguished experimentally since in the stepwise degradation, reducing groups are readily liberated and the viscosity of the substrate decreases slowly and regularly. On the other hand random degradation causes a rapid fall in the viscosity of the substrate and reducing sugars of low molecular weight are only slowly liberated.

Ideally the enzymic degradation of a polysaccharide should involve the use of highly purified enzymes, devoid of other enzymic activities, and of known

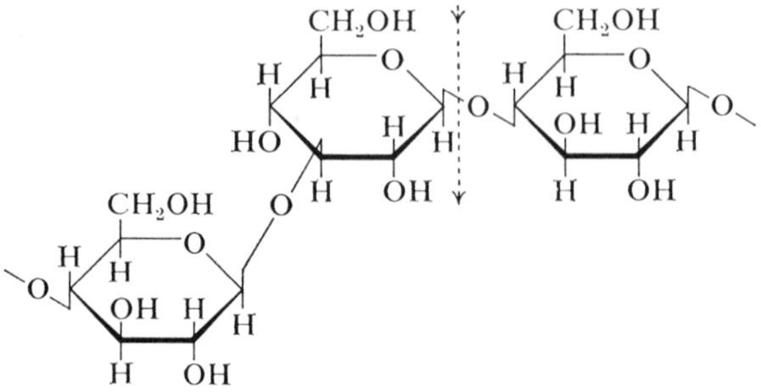

Fig. 15.1 Bond broken (↓) by the action of "laminaranase" on oat β-D-glucan.

specificity and action pattern. The specificity of an enzyme, however, is only defined by its mode of action on substrate of precisely known structures, and a full description of its action pattern requires *both* an analysis of the mono- and oligosaccharides which are liberated *and* an assessment of the changes in the molecular weight distribution of the polysaccharide as it undergoes breakdown. Among glycanhydrolases, this information has been most completely obtained for the enzymes which degrade the starch components, amylose and amylopectin, and glycogen [66].

From the many starch-degrading enzymes five types may be mentioned: (*1*) α-amylases, (*2*) glucoamylase, (*3*) β-amylase, (*4*) phosphorylase, and (*5*) pullulanase. The first two of these enzymes degrade starch and its components completely to soluble mono- and/or oligosaccharides and are useful for the removal of starch from other polysaccharide preparations. α-Amylases from different sources, for example, higher plants, mammalian secretions, and microorganisms, differ slightly in their action patterns, but all cause random hydrolysis of internal $(1\rightarrow4)$ bonds with the formation of maltose, malto-triose, and the so-called α-limit dextrins which retain the $(1\rightarrow6)$ linkages from amylopectin or glycogen. Maltotriose is not an end product but is further hydrolyzed to maltose and glucose. The nature of the α-limit dextrins vary with the source of the α-amylase. The α-limit dextrins produced by (*a*) salivary and pancreatic, and (*b*) *Bacillus subtilis* α-amylases are shown in Fig. 15.2.

Glucoamylases are of much commercial interest since they degrade starch almost completely to glucose. The enzymes, for example, from *Aspergillus niger*, act by the stepwise removal of D-glucose residues and appear to be capable of hydrolyzing both $(1\rightarrow4)$ and $(1\rightarrow6)$, and indeed also $(1\rightarrow3)$ linkages, and are thus able to hydrolyze branch points. The presence of a single enzyme, however, has not yet been rigorously established. The remaining three types of enzymes may be considered in greater detail since they are more selective in their action and are of particular value in probing the fine structure of the starch components and glycogen.

β-Amylases have been found only in plant sources, for example, sweet potato and common cereals, but appear to be essentially similar in their action. Stepwise action with the liberation of units of maltose (Fig. 15.3) is arrested by the presence of branch points or other structural variations, and in the case of amylopectin and glycogen appears to stop 2–3 glucose residues from the branch point. The so-called β-amylolysis limit, that is, the degree of conversion into maltose, expressed as a percentage, provides a measure of the average length of the outer chain, and the value is characteristic for various glycogens and amylopectins. Since only outer chains are attacked, the residual β-limit dextrin constitutes 42–44% of the parent amylopectin and 55–60% of the parent glycogen. Although amylose is an essentially linear

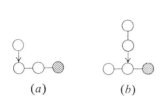

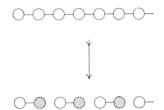

Fig. 15.2 Limit dextrins from the action of α-amylases on starch: ○, α-D-glucose residue; ●, reducing D-glucose residue; –, α-D-(1→4) bond; ↓, α-D-(1→6) bond.

Fig. 15.3 Stepwise liberation of maltose by the action of β-amylase on chains of (1→4)-linked α-D-glucopyranose residues: ○ = non-reducing; ● = reducing unit.

polysaccharide, not all amyloses are completely degraded by β-amylase. Since treatment of amylose β-limit dextrin with debranching enzymes (see below), such as yeast isoamylase [67] or pullulanase [68], is followed by complete breakdown by β-amylase, it has been concluded that some natural amyloses may contain occasional branch points of the (1→6) type.

Phosphorylases catalyze both the degradation to and the synthesis from α-D-glucopyranosyl phosphate of (1→4)-linked α-D-glucopyranose residues (Eq. 15.3). Little is known of the relative importance *in vivo* of the enzyme's two roles in starch metabolism, but there is evidence that its prime function in glycogen metabolism is in degradation [69]. Phosphorylase has no action on (1→6) linkages. The enzyme acts in a stepwise manner from the non-reducing ends of chains and approaches to 4 glucose residues short of the branch points in both the side and the main chains [64] leaving a phosphorylase-limit dextrin with intact interior chains.

$$x\text{-}\alpha\text{-D-Glucopyranosyl phosphate} + [\text{D-glucose}]_n \rightleftharpoons [\text{D-glucose}]_{n+x} + x\text{-phosphate}$$

$$(15.3)$$

Debranching enzymes capable of hydrolyzing (1→6) linkages between α-D-glucopyranose residues at the branch points in amylopectin and glycogen are found in plants (R-enzyme [70]) and in mammalian muscle (amylo-1,6-glucosidase [71]). Pullulanase, however, is potentially the most valuable debranching enzyme for structural studies. This extracellular enzyme from *Aerobacter aerogenes* effects essentially quantitative hydrolysis of pullulan to maltotriose [72]. Pullulanase does not hydrolyze all the branch linkages in amylopectin and glycogen, but is able to penetrate the highly branched structure to a greater degree than other debranching enzymes [73]. Complete debranching of these polysaccharides occurs under the concurrent action of pullulanase and β-amylase, and at high concentrations of the latter enzyme degradation to maltose and glucose only takes place. Assuming the statistical occurrence of equal numbers of even- and odd-numbered unit chains, glucose arises from chains containing an odd number of glucose residues. Estimation of glucose and maltose forms the basis of an enzymic assay of the average unit chain length of amylopectin and glycogen [74, 75]. The method is more reproducible than periodate oxidation and may be used on a microscale.

The successive action of an *exo*-enzyme, for example, β-amylase or phosphorylase, acting on the outer chains of glycogen or amylopectin, and of a debranching enzyme provides a valuable general approach to the determination of the degree of multiple branching. C. F. Cori and his collaborators [76], using phosphorylase and amylo-1,6-glucosidase, obtained the first convincing evidence for the multiply branched tree-like structure of glycogen in which branch points are arranged in tiers, with approximately half the branch points in the outermost tier and successively fewer branch points in the inner tiers towards the center of the molecule. A further approach to the quantitative determination of the degree of multiple branching in amylopectin was developed by Peat, Whelan, and Thomas [70] who treated amylopectin β-limit dextrin with the plant debranching enzyme, R-enzyme. The combined yield of maltose and maltotriose, which sugars could arise only from the stubs of A chains (see Fig. 15.4) gives an estimate of the degree of multiple branching. The observed yield (12.8%) of maltose and maltotriose is quite close to that calculated (10.4%) for a tree-like structure containing equal proportions of A and B chains. Rather similar results have been obtained from the debranching of amylopectin and glycogen β-limit dextrins with pullulanase [77].

Some examples of the use of enzymic hydrolysis to obtain structural information not readily obtainable otherwise for polysaccharides other than glucans have been reported recently. Certain $(1\rightarrow2)$ linkages in mannans from yeasts are hydrolyzed by an inducible enzyme preparation from a soil bacterium to give fragments containing predominantly the unattacked $(1\rightarrow6)$-linked α-D-mannopyranose residues [77a].

The distribution of $(1\rightarrow5)$ and $(1\rightarrow3)$ linkages in sugar beet arabinan has been established by the action of an α-L-arabinofuranosidase. This *exo*-enzyme, which is not linkage specific, acts by preferentially hydrolyzing terminal units. Since a linear $(1\rightarrow5)$-L-arabinan is formed, it follows that the $(1\rightarrow3)$-linked units are present in short side-chains [77b].

The technique of *sequential enzyme induction* takes advantage of the ability of microorganisms to elaborate enzymes when presented with the appropriate substrates. Figure 15.5 illustrates the successive removal of sugar residues from the nonreducing ends of carbohydrate chains of the complex α_1-acid glycoprotein (orosomucoid) by a series of enzymes induced when the natural polymer is introduced into a culture of a strain of *Klebsiella aerogenes* [78].

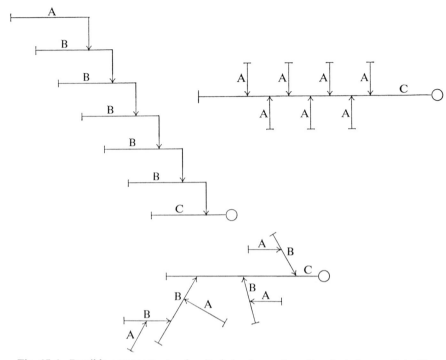

Fig. 15.4 Possible arrangements of unit chains in amylopectin. A chains are linked by (1→6) bonds to adjacent chains, B chains carry A chains and are themselves linked to other chains, and the single C chain contains the single reducing group: |- = nonreducing end group; → = (1→6) linkage at branch point; –○ = reducing end group.

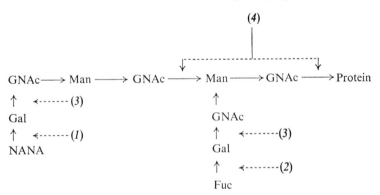

Fig. 15.5 Sequential induction of enzymes in the presence of α_1-acid glycoprotein: (*1*) neuraminidase action with liberation of *N*-acetyl neuraminic acid (NANA); (*2*) action of α-L-fucosidase with exposure of galactopyranose end groups and appearance of cross-reaction with Type XIV anti-*Pneumococcus sera*; (*3*) β-D-galactosidase results in loss of Type XIV cross-reactivity; (*4*) glucosaminidase action results in the liberation of a trisaccharide and a hexasaccharide from the cleavage of nonterminal linkages.

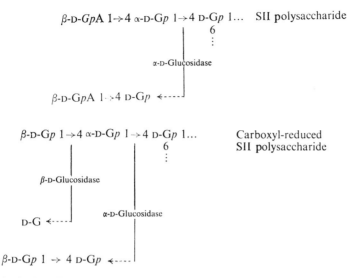

Fig. 15.6   Action of externally induced α- and β-D-glucosidases on Type II *Pneumococcus* specific polysaccharide (SII) and its carboxyl-reduced derivatives.

Enzymes may also be induced externally in the presence of simple glycosides of appropriate configuration, for example, methyl α-D-gluco-pyranoside for α-D-glucosidase, and then used for the hydrolysis of configurationally similar linkages in polysaccharides. Figure 15.6 illustrates the action of externally induced α- and β-D-glucosidases on a portion of the structure of the Type II *Pneumococcus* specific polysaccharide and on the corresponding portion of the structurally modified polysaccharide in which D-glucuronic acid have been converted into D-glucose residues [79].

The enzymic hydrolysis of glycosidic linkages, like acid hydrolysis, involves the scission of the glycosyl-oxygen [C-1–O] bond [80]. A few polysaccharide degrading enzymes, however, which are known as eliminases or lyases and are usually of bacterial origin, act by an entirely different mechanism. Hyaluronic acid (**1**), for example, is degraded by a bacterial enzyme with the formation of 2-acetamido-2-deoxy-3-*O*-(4-deoxy-α-L-*threo*-hexenopyrano-syluronic acid)-D-glucose (**2**) [81]. The reaction, which involves cleavage of the oxygen–aglycone [O–C-4] bond [82, 83], is probably mechanistically similar to the base-catalyzed β-eliminations of hexuronic acid derivatives.

Other enzymes, which effect transformations in polysaccharides and other carbohydrate-containing macromolecules and may be of value in the elucidation of structure, include: (*1*) esterases such as pectin esterase [84, 85]; (*2*) sulfatases, such as an enzyme from *Porphyra* species which catalyzes the transformation of L-galactose 6-sulfate to 3,6-anhydro-L-galactose residues in the glycan porphyran [86]; (*3*) peptidases which cleave the cross-bridges in bacterial cell wall peptidoglycans [87] to give the intact glycan; and (*4*) proteolytic enzymes, such as papain and pronase, which permit the isolation from glycoproteins of glycopeptides containing carbohydrate units attached to a limited number, sometimes only one, of amino acid units [88].

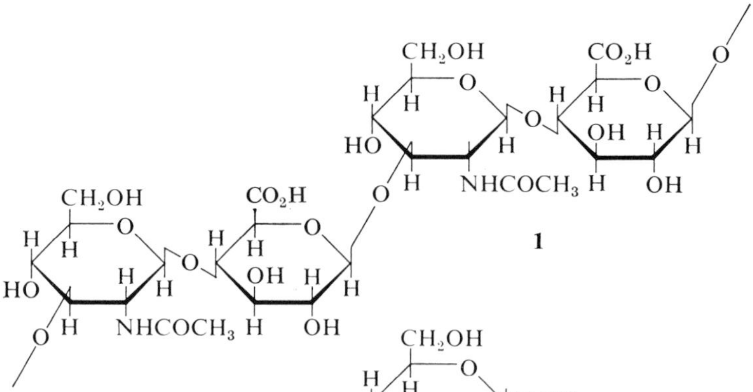

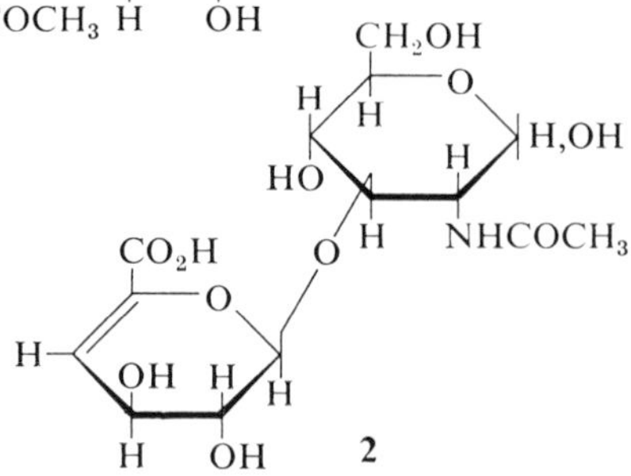

Examination of Products of Polysaccharide Hydrolysis

Complete hydrolysis of a polysaccharide is undertaken in order to determine the nature and proportions of its constituent glycose units. The qualitative analysis of the hydrolyzate may be performed by paper partition chromatography [89, 90], thin-layer chromatography [91], and gas–liquid chromatography [92]. Gas–liquid chromatography requires suitably volatile derivatives and examples include the use of alditol acetates [93], *O*-trimethylsilyl (TMS) derivatives of sugars themselves [94–96], or the derived aldonolactones [69, 97]. An extensive literature exists on the chromatography of carbohydrates and reference can only be made to some review articles and a few original papers. A positive identification of a sugar should not be made on the basis of chromatographic behavior alone, although the use of as many different chromatographic procedures as possible will limit the chances of a mistaken identification. Definite characterization of a glycose requires the isolation in a pure form followed by the determination of physical properties and the preparation of at least one suitable crystalline derivative. Quantities of material sufficient for characterization can usually be obtained after separation by filter sheet, preparative thin-layer [98] or column chromatography [99].

Methods for the quantitative analysis of mixtures of sugars in polysaccharide hydrolyzates are legion. Many of the chromatographic procedures mentioned above may be put on a quantitative basis, but it should be noted that calibration with known mixtures of sugars is often necessary. Gas–liquid chromatography gives direct results, but calibration of the detector response to different sugars is normally required. Separations by paper chromatography are followed by elution of the sugars from the paper and analysis by either reducing sugar [100] or colorimetric [101] methods. In some cases either individual sugars or groups of sugars of a given class may be analyzed in a hydrolyzate directly by specific colorimetric methods or enzymic procedures.

Ion exchange chromatography [102] is used for the separation of both acid [103] and basic [104] glycoses. Ion exchange resins are also used for the partition chromatography of glycoses and alditols [105]. This procedure forms the basis for the automated separation with a commercially available autoanalyzer of the common monosaccharides and the derived alditols. The aldoses may be analyzed by monitoring the eluant with the orcinol colorimetric reagent, and both aldoses and alditols may be analyzed by periodate oxidation and subsequent colorimetric determination of liberated formaldehyde, using pentane-2,4-dione.

Oligosaccharides of structural importance which are formed on graded hydrolysis of polysaccharides are examined by paper chromatography or thin-layer chromatography. Specimens required for identification or structural analysis may be obtained after separation by filter sheet chromatography or by one or more types of column chromatography. Complex mixtures of oligosaccharides frequently require the use of more than one type of separation procedure for the resolution of all components. For neutral oligosaccharides the most generally used column separations involve partition chromatography on cellulose [106] or gradient elution from charcoal–Celite with aqueous ethanol [107]. Mixtures of neutral oligosaccharides can be separated rapidly by chromatography on suitably cross-linked cation exchange resins [108] by development with water, or by molecular sieve chromatography (gel filtration) on Sephadex (a cross-linked dextran [109]). Complex mixtures of acidic oligosaccharides have been separated by displacement with increasing concentrations of acetic or formic acid from anion exchange resins in their acetate or formate form [103, 110]. Diethylaminoethylcellulose and diethylaminoethyl-Sephadex [111, 112] are convenient weak base ion exchangers for the separation of substantial quantities of acids. Mixtures of aldobiouronic acids have been separated by ion exchange chromatography using sodium tetraborate as eluant [113].

The structural analysis of oligosaccharides is determined by the use of classical procedures [114] involving hydrolysis, methylation, and periodate oxidation. The application of semimicro methods, for example, for the analysis of periodate oxidation [115] and lead tetraacetate oxidation [116] reactions, and gas–liquid chromatography for the separation of methylated and other sugar derivatives [92] has considerably reduced the quantities required for a reasonably complete structure determination.

Mass spectrometry promises to become a major tool in the structural analysis of disaccharides and higher oligosaccharides, and the mass spectra of a number of TMS derivatives of disaccharides have been reported [117]. The mass spectra of some di- to pentasaccharides as their 1-phenylflavazole peracetates have been measured, and oligosaccharides of uniform linkages of the (1→4) and (1→6) types have been distinguished [117a]. The determination of the structure of disaccharides as TMS derivatives of disaccharide alditols by combined gas–liquid chromatography–mass spectrometry has recently been described [118]. The structural symmetry involved in alditol formation may be avoided by reduction of the disaccharide with sodium borodeuteride. The mass spectra of the labeled disaccharide alditols enable a distinction to be made, for example, between (1→3) and (1→4) linkages. Analogous studies have been carried out on methylated disaccharide alditols [118a], from which derivatives the linkage position may also be assigned by identification of monosaccharide cleavage products. It must be borne in mind, however, that mass spectrometry is relatively insensitive to stereochemistry, and while this approach will provide a valuable method for linkage analysis, independent evidence must be sought on the configuration of glycosidic linkages, on the enantiomeric configuration of the sugar constituents, and, if stereoisomeric sugar sare present in oligosaccharides, on the sequence of the sugar units.

Acetolysis

Acetolysis of polysaccharides gives mixtures of fully acetylated monosaccharides and oligosaccharides. In certain cases small amounts of the acetates of monosaccharide aldehydrols are formed [119]. Acetolysis of cellulose [120] can give yields of over 40% of cellobiose octaacetate; the unusually high yield is probable due to the fact that the acetate is in part protected from degradation by crystallization during the reaction. In the case of essentially linear homoglycans, for example, ivory nut mannans A and B [121], acetolysis like partial acid hydrolysis affords a polymer-homologous series of oligosaccharides. In like manner linear heteroglycans furnish oligosaccharides containing the various constituent sugars. Thus Iles glucomannan on acetolysis [122] gave *inter alia* 4-*O*-β-D-glucopyranosyl-D-mannose and 4-*O*-β-mannopyranosyl-D-glucose.

Acetolysis provides a valuable alternative method to hydrolysis for the controlled fragmentation of complex polysaccharides since the relative rates of cleavage of glycosidic linkages are often quite different from those occurring with mineral acid in aqueous solution. Thus whereas (1→6) linkages between hexose residues are the most resistant to acid hydrolysis [123, 124], this type of linkage is most readily cleaved by acetolysis [125–128]. A notable example is provided by the isolation of nigerose in 20% by weight of parent dextran which contained 57, 35, and 8% respectively of (1→6), (1→3), and (1→4) linkages [125]. Another example of different "cracking" patterns has been reported in the case of desulfated λ-carrageenan where the (1→3) linkages are much more readily split by acid hydrolysis than the (1→4) linkages, whereas the reverse situation holds for acetolysis [129] (see Appendix).

The 6-deoxyhexopyranosyl linkages in complex glycans, especially those in terminal positions, tend to be preferentially hydrolyzed by dilute mineral acid, but the relative resistance of these linkages to acetolysis has enabled oligosaccharides with these linkages intact to be isolated from tragacanthic acid [130] and carboxyl-reduced gum arabic [131, 132] (see Appendix).

Although no examples have been reported of acid reversion during acetolysis, caution should be exercised in the assessment of the structural significance of oligosaccharides since a substantial degree of anomerization of glycosidic linkages may accompany depolymerization. Those oligosaccharides containing (1→6) linkages which were isolated from the acetolysis of carboxyl-reduced *Araucaria bidwillii* gum had the α-D-configuration whereas the corresponding oligosaccharides formed on partial acid hydrolysis had the β-D-configuration [133]. Artifacts resulting from anomerization appear to be minor products in the cleavage of the (1→4)-linked β-D-mannopyranosyl bonds in glucomannans [121, 134] and galactomannans [135].

As a method of linkage analysis acetolysis may have experimental advantages over partial hydrolysis with mineral acid for those polysaccharides which are insoluble in aqueous solution since either the polysaccharide itself may be or the acetylated polysaccharide almost certainly will be soluble in the acetolysis medium. Acetolysis is generally performed by dissolving the glycan or the acetylated glycan in either acetic anhydride or a mixture of acetic anhydride and acetic acid, containing 3–5% of sulfuric acid at 0°. The reaction mixture is usually kept at room temperature for 1–10 days, after which the reaction is stopped by pouring the solution into ice water and neutralizing to pH 3–4 with sodium hydrogen carbonate. The acetates may then be extracted with chloroform. The sugar acetates may be fractionated chromatographically on columns of Magnesol-Celite (hydrated magnesium acid silicate and diatomaceous silica) [99] or cellulose. Alternatively, the glycose acetate mixture may be easily deacetylated catalytically with sodium or barium methoxide in methanol to give the free glycoses [136] which can then be chromatographically separated and identified.

Mercaptolysis

Degradation of glycans by mercaptolysis has the advantage that otherwise acid-sensitive glycoses liberated in a hydrolysis are stabilized by protection of their potential carbonyl functions with thioacetal groups. Residues of 3,6-anhydrogalactose are very readily destroyed under normal acid hydrolysis conditions [137], but agarobiose, 3,6-anhydro-4-O-(β-D-galactopyranosyl)-L-galactose, and carrobiose, 3,6-anhydro-4-O-(β-D-galactopyranosyl)-D-galactose, have been isolated as their diethylthioacetals from the mercaptolysis of agarose [138], (see Appendix), and of κ- [139] and alkali-modified λ-carrageenan [140], respectively.

The degradation is normally carried out by dissolving the polysaccharide in concentrated hydrochloric acid at 0° followed by the slow addition of ethanethiol. After several hours at 0° the mixture is allowed to warm to room temperature, and the reaction is allowed to continue (6 hr to 7 days) until depolymerization has proceeded to the required extent. The cold solution is neutralized by the addition of a suspension of lead carbonate in methanol. The filtered reaction mixture is concentrated under reduced pressure, and the products may be isolated by some suitable form of column chromatography or sometimes by direct crystallization of the concentrate. The ethylthioacetal groups can be removed by treatment of the derivatives in warm aqueous solution with mercuric chloride and cadmium carbonate to yield the free glycoses.

Methanolysis

Methanolysis of polysaccharides, like mercaptolysis, renders acid-sensitive liberated glycoses more stable with the formation of methyl glycosides or dimethylacetals, and has found similar application in the controlled depolymerization of algal polysaccharides containing 3,6-anhydrogalactose residues. Agarose and related polysaccharides [141] furnish on methanolysis agarobiose, 3,6-anhydro-4-O-(β-D-galactopyranosyl)-L-galactose, as its di-methyl acetal. Milder conditions of methanolysis of agarose yield aragobiose dimethyl acetal linked in ketal form to pyruvic acid, and the structure **3**, 3,6-anhydro-4-O-(4,6-O-1'-carboxyethylidene-β-D-galactopyranosyl)-L-galac-tose dimethyl acetal, has been assigned to the compound.

The reactions of sulfated polysaccharides with aqueous mineral acid show little specificity in that hydrolysis of sulfate esters and of glycosidic linkages occur at comparable rates. In contrast complete desulfation (by transesterification) of certain polysaccharides has been achieved by treatment with

$$H_3C \quad H_3C \quad CH_2 \quad O \quad OH \quad H \quad O \quad H \quad HOCH(OMe)_2 \quad CH_2 \quad OH \quad H \quad OH$$

3

cold methanolic hydrogen chloride, a procedure unlikely to cause cleavage of any but the most acid-sensitive glycosidic linkages [142]. Comparison of the hydrolysis products from methylated desulfated polysaccharides with those from the methylated derivatives of the parent sulfated polysaccharides provide evidence for the location of sulfate ester substituents. This approach has been used in studies on λ-carrageenan [143] and heparin [144, 145].

Oxidative Glycol Cleavage of Polysaccharides

The cleavage of compounds containing 1,2-diol groups by periodate was introduced by Malaprade [146] and was generally applied to the analysis of carbohydrates by Hudson. The oxidation method has since been widely used for the cleavage of 1,2-diol and 1,2,3-triol groups in polysaccharides to provide information about the nature of their end groups and the types of glycosidic linkages present [115, 147]. The action (or lack of action) of periodate on some typical glycose residues is shown in the idealized Eqs. (15.4)–(15.8).

The essentially quantitative nature of periodate oxidations, particularly those of pyranosides unsubstituted other than at C-6 with the consumption of 2 moles of reagent and the liberation of 1 mole of formic acid and of other sugar residues containing 1,2-diol groups, readily lends itself to the analysis of certain aspects of structure. Furthermore the presence of sugar units, which are substituted in a manner that leaves no diol groups available for oxidation by the reagent, may be recognized by the detection of that sugar on hydrolysis of the periodate-oxidized polysaccharide or, preferably, by reduction of the polyaldehyde to the corresponding polyalcohol with sodium borohydride followed by hydrolysis.

Formaldehyde is less frequently encountered as a cleavage product from the periodate oxidation of polysaccharides, but is produced on oxidation of suitably substituted hexofuranose and heptopyranose residues. Insofar as acyclic diols are oxidized more rapidly than cyclic diols, glycans containing such units may be selectively oxidized using limited quantities of oxidant, and subsequent reduction with sodium borohydride gives a modified glycan with

1→2 link

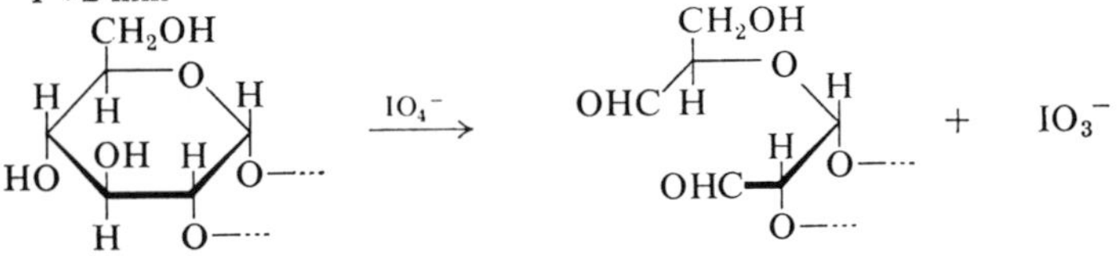

$$\xrightarrow{\text{IO}_4^-} \qquad + \quad \text{IO}_3^- \qquad (15.4)$$

1→3 link

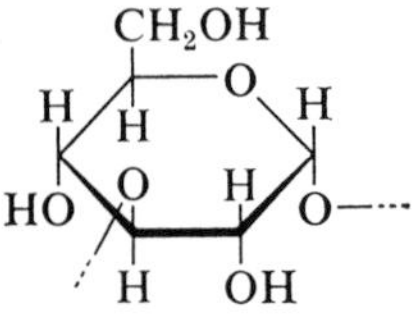

Not attacked by periodate (15.5)

1→4 link

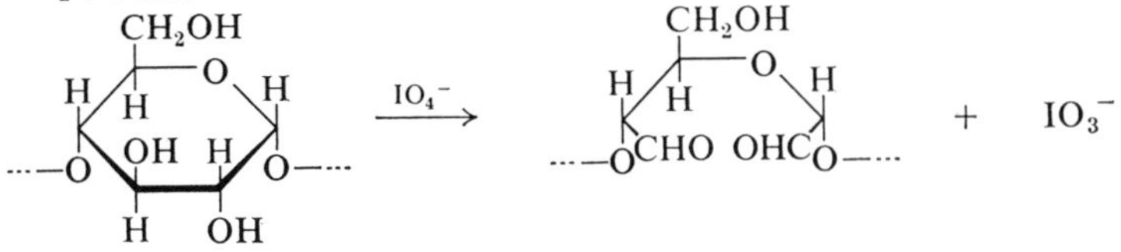

$$\xrightarrow{\text{IO}_4^-} \qquad + \quad \text{IO}_3^- \qquad (15.6)$$

1→5 link

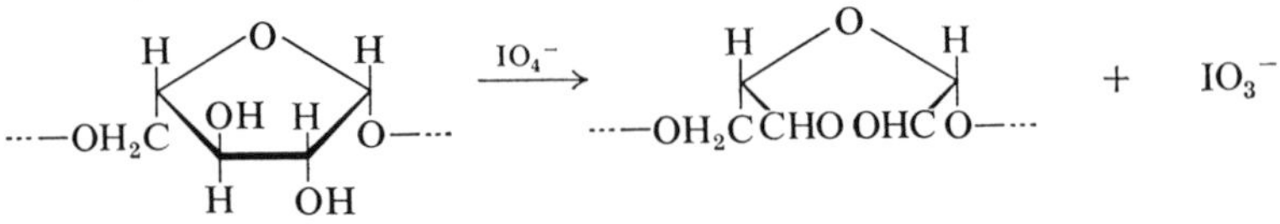

$$\xrightarrow{\text{IO}_4^-} \qquad + \quad \text{IO}_3^- \qquad (15.7)$$

1→6 link

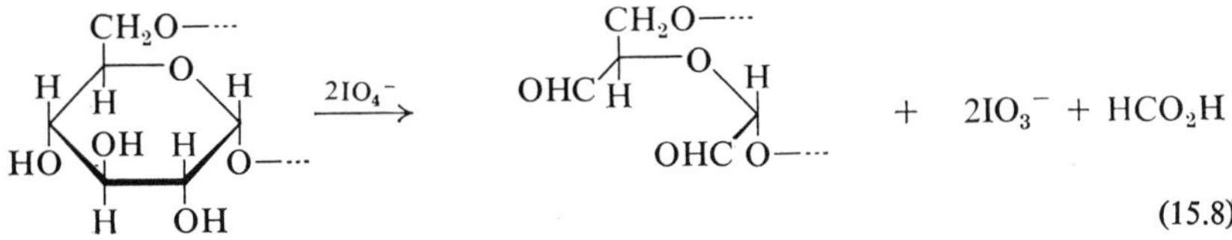

$$\xrightarrow{2\text{IO}_4^-} \qquad + \quad 2\text{IO}_3^- + \text{HCO}_2\text{H}$$

(15.8)

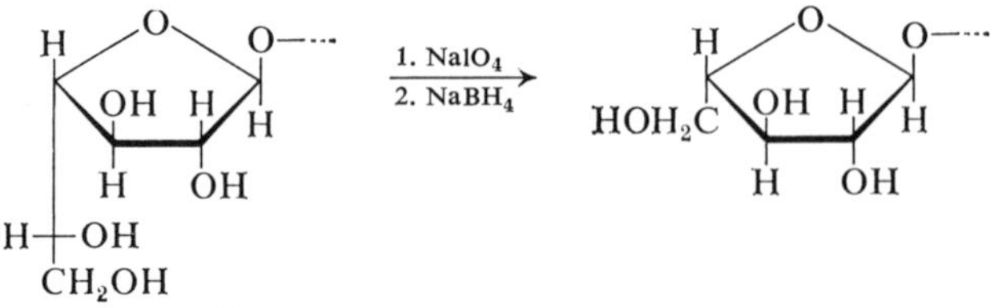

$$\xrightarrow[\text{2. NaBH}_4]{\text{1. NaIO}_4} \qquad (15.9)$$

$$\xrightarrow[\text{2. NaBH}_4]{\text{1. Na IO}_4} \qquad (15.10)$$

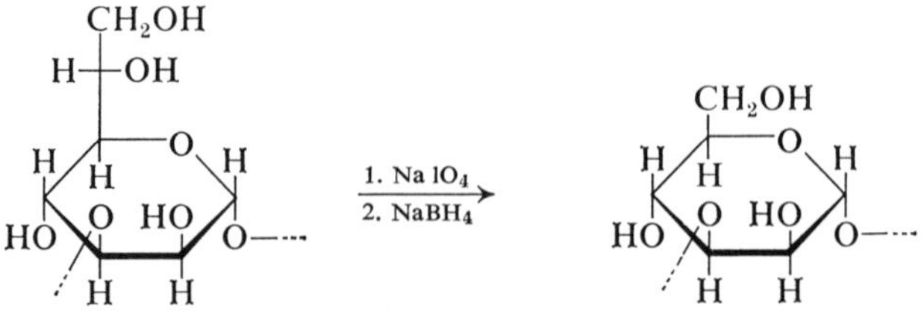

pentofuranose and hexopyranose units in place of the original sugar residues [11, 148]. Equations (15.9) and (15.10) show such transformations.

Periodate oxidation has been widely used for the quantitative estimation of certain structural features with glycans of known general structure and for giving a general indication of the nature of the linkages of polysaccharides of largely unknown structure. Even when the nonideality of behavior associated with the "overoxidation" or "underoxidation" of certain units is not taken into account, it must be borne in mind that the results from the above-mentioned analytical approach can be rarely interpreted unambiguously without at least qualitative information on structure derived by entirely independent methods. This is especially the case with heteropolysaccharides where epimeric asymmetric centers are destroyed during oxidation, for example, in 4-O-substituted D-glucose and D-mannose residues.

Periodate-oxidized polysaccharides do not undergo hydrolysis easily and the separation of the products often presents experimental difficulties. However, reduction of periodate-oxidized polysaccharides by catalytic hydrogenation or treatment with sodium borohydride followed by hydrolysis furnishes not only sugars from unoxidized units but also fragments, for example, ethylene glycol, glycerol, erythritol, and threitol, from the stubs of oxidized sugar units. Ethylene glycol has only been recognized from nonreducing pento-pyranose end groups, and glycerol may be derived from hexopyranose residues unsubstituted at C-3 and C-4 or from pentofuranose residues unsubstituted other than at C-5. Of the fragments of low molecular weight only erythritol and threitol from 4-O-substituted or 4,6-di-O-substituted units provide some information on the stereochemistry of the residues which have undergone cleavage. Estimations of the proportions of glycerol (from nonreducing end groups only) and erythritol formed from reduction followed by hydrolysis of periodate-oxidized amylopectin may be used to determine the average unit chain length [149, 150]. The alditols thus formed may be determined after paper chromatographic separation by periodate oxidation to formaldehyde which is determined by the chromotropic acid method [151]. Alternatively, alditols may be determined by gas chromatography of their acetates [152] or trimethylsilyl ethers [153].

The value of analytical uses of periodate oxidation may be limited by either "overoxidation" or less frequently by incomplete oxidation of units containing diol groups. Overoxidation is most frequently encountered when sugar residues give rise to malonic acid half aldehyde derivatives, for example, from hexuronic acid end groups, or to malondialdehyde derivatives, for example, from hexofuranosides. The degradative sequence in the former type of overoxidation is shown in Eq. (15.11). Unless the temperature and pH of the reaction mixtures are controlled, complete reaction involving over-oxidation of active hydrogens atoms in the intermediate oxidized residues leads to the formation of formic acid and oxalic acid [154]. Degradation during the oxidation of glycuronoglycans may be limited to normal cleavage if carried out with periodic acid or with periodate buffered at pH 2.2–4 [6].

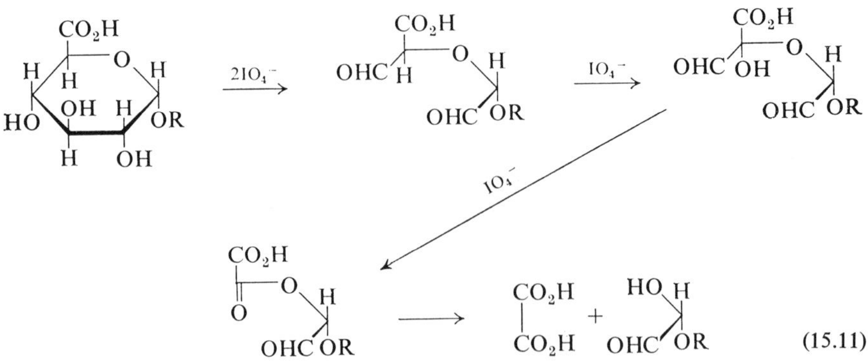

$$(15.11)$$

Another type of overoxidation may result from the periodate oxidation of reducing groups. The importance of this type of reaction will be inversely proportional to the molecular size of the polysaccharide. The reaction may be illustrated for (1→4)-linked glucoglycans as outlined in Eq. (15.13). Initial oxidation of the reducing group gives a formyl ester, which on hydrolysis undergoes further glycol cleavage to give a malondialdehyde derivative with an active hydrogen from which overoxidation is initiated. This further oxidation leads to cleavage of the glycosidic linkage from which point further erosion along the glycan chain may occur. Periodate oxidation of (1→2)- and (1→3)-linked glycans likewise results in overoxidation, but (1→6)-linked reducing groups afford substituted acetaldehyde derivatives which are not oxidized further (see Eq. 15.12).

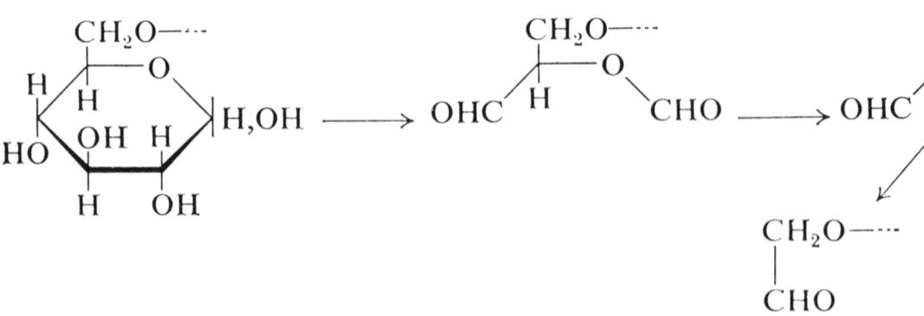

$$(15.12)$$

Overoxidation which leads to the erosion of the glycan chain from the reducing group may be minimized if the reaction medium contains only a slight excess of periodate and is buffered at pH 3.6 so that formyl esters are stabilized [155]. In practice overoxidation of this and other types cannot be entirely suppressed. Usually a fairly rapid initial consumption of reagent and release of formic acid is observed followed by a much slower reaction indicative of overoxidation. Extrapolation of the overoxidation part of a graphic plot to zero time gives a reasonably valid estimate of reagent consumed and formic acid liberated during the primary oxidation.

Deliberate overoxidation of polysaccharides, carried out at pH 8, has been used to determine the position of $(1\rightarrow6)$ linkages in hexoglycan chains otherwise containing only $(1\rightarrow2)$, $(1\rightarrow3)$, or $(1\rightarrow4)$ linkages. Measurement of the formaldehyde released from C-6 of glycose residues between the reducing group and the $(1\rightarrow6)$ linkage during periodate oxidation (see Eq. 15.13) indicates the relative position of the latter [156].

$$\begin{matrix} \cdots-\text{OCH}_2 \\ | \\ \text{CHO} \end{matrix} + n\text{HCHO} + n\text{CO}_2 + (4n - 4)\text{HCO}_2\text{H} \qquad (15.13)$$

The incomplete oxidation of sugar residues with 1,2-diol groups has been encountered and has been attributed to hemiacetal formation or similar interactions between the aldehyde groups of cleaved units and the hydroxyl group of a proximate and potentially oxidizable sugar unit. Alginic acid provides a striking example of an entirely $(1\rightarrow4)$-linked glycan [29, 157] which, under conditions which avoid overoxidation, consumes only ca. 0.50 mole of periodate per sugar residue. When the partially oxidized polysaccharide is reduced with sodium borohydride, further glycose units (but not all) are exposed to oxidation. Hemiacetal formation between oxidized and neighboring nonoxidized units provides a probable explanation for the resistance of the latter to oxidation, but the role of carboxyl groups in promoting the interaction is not yet known [158]. In the light of observations of this type, care must be exercised in attributing structural significance, for example, the presence of $(1\rightarrow3)$ linkages, to sugar residues which escape oxidation by periodate, and independent positive evidence should be sought for structural features thus implied.

Limited oxidation of methyl glycopyranosides and of dextran have also been observed in dimethyl sulfoxide [159]. In the case of methyl β-L-arabino-pyranoside the formation of the product of incomplete oxidation has been shown to be due to rapid cyclization to an intramolecular hemiacetal. Some stereoselectivity in oxidation in dimethyl sulfoxide has been attributed to the faster oxidation of vicinal glycol groups in an axial-equatorial orientation than of those in an equatorial-equatorial orientation.

The periodate oxidation of polysaccharides is normally performed in aqueous solution (ca. 0.1–0.5%) in the dark at a low constant temperature (4–20°) using a minimal excess of oxidant, which may be periodic acid, sodium metaperiodate, or potassium metaperiodate. The reaction may be adjusted to a known volume by the addition of distilled water or buffer solution. Aliquot portions are withdrawn at suitable intervals from both the reaction mixture and a reagent blank from which the polysaccharide is omitted, and analyzed for periodate reduced, and formic acid and formaldehyde liberated. The analytical methods may be carried out on a submicroscale [160].

The determination of periodate may be performed by a number of titrimetric methods [161]. The method most commonly employed involves reduction of periodate to iodate with excess standard arsenite in bicarbonate buffer followed by back titration with standard iodine solution [162]. Alternative methods which may be used on a microscale involve spectrophotometry at the ultraviolet maximum of periodate at 222.5 nm [163, 164] and polarographic techniques [165].

Acidity measurement is generally directed toward the determination of the formic acid released by periodate oxidation of the polysaccharide. Samples of the unbuffered reaction mixture and blank are neutralized by the addition of ethylene glycol and are then titrated with $0.01N$ sodium or barium hydroxide, preferably potentiometrically, to pH 6.25 [166]. Small-scale periodate oxidations of polysaccharide in sodium bicarbonate solution at pH 5.7 have been carried out in a Warburg respirometer in which the formic acid releases carbon dioxide, which is measured manometrically [167]. A spectrophotometric method in which the chromophore is formed on heating the 2-thiobarbituric acid has the advantages of high selectivity and avoidance of alkaline conditions [168].

Formaldehyde liberation during the periodate oxidation of polysaccharides is usually slight, and colorimetric methods are normally used for its determination. After removal of periodate, the formaldehyde may be determined (1) by the chromotropic acid method [151], (2) by the phenylhydrazine–ferric chloride method [169], or (3) by the measurement of the yellow color produced when formaldehyde reacts with pentane-2,5-dione in the presence of ammonium acetate [170]. Method (3), which forms the basis for an automated spectrofluorimetric determination [171], is not affected by the presence of large amounts of residual polysaccharide material.

In addition to procedures involving direct analysis of reagent consumed and of small molecules liberated during the oxidation, further transformations of periodate-oxidized polysaccharides provide valuable information on structure, and in the case of highly ramified polysaccharides with interior units resistant to oxidation, degradations based on sequential oxidations provide a means for effecting controlled stepwise depolymerization. The first procedure of this type was that developed by Barry and co-workers [172, 173] involving the reaction of periodate-oxidized polysaccharides with carbonyl-condensing reagents such as phenylhydrazine, isonicotinhydrazide, and thiosemicarbazide [174]. Further treatment of the product from reaction with phenylhydrazine in the presence of acetic acid results in cleavage of adjacent glycosidic linkages. Thus treatment of periodate-oxidized Floridean starch with phenylhydrazine gives glyoxal-bisphenylhydrazone (4) and D-erythrosephenylosazone (5). Reducing sugar units which are exposed by glycoside cleavage furnish the corresponding phenylosazones. The Barry degradation, which gave useful structural information when snail galactan [175] and arabic acid [176] were submitted to successive treatments with sodium periodate and phenylhydrazine, has now been largely superseded by the experimentally cleaner Smith degradation. A modification of the Barry degradation involving N,N-dimethylhydrazine in place of phenyl-hydrazine has been used to degrade the tetraglycosylglycerol (6), derived in three stages from the *Pneumococcus* type 10A substance [148], to 2-O-α-D-galactopyranosylglycerol (7).

Reduction of periodate-oxidized polysaccharides with sodium borohydride gives polyalcohols which when treated with dilute mineral acid at room temperature undergo essentially selective hydrolysis of acyclic linkages with the liberation of the unoxidized sugar residues as glycosides of polyhydric alcohols. This sequence of reactions, known as the Smith degradation [147], has found extensive application in probing the fine structure of polysaccharides since minimum cleavage of glycosidic linkages occurs during the mild acid hydrolysis. Glycans which contain single or groups of two or three adjacent sugar residues furnish chromatographically separable glycosides of glycerol, erythritol, or threitol Figure 15.7 illustrates the formation of 2-O-β-D-glucopyranosyl-D-erythritol from oat β-D-glucan [147]. Side reactions, however, may occur and in the example quoted a glycoealdehyde acetal (8) of 2-O-β-D-glucopyranosyl-D-erythritol is also formed [147]. Whereas β-D-glycans give rise preferentially to 6-membered O-2'-hydroxyethylidene acetals, the predominant by-products from α-D-glycans are 5-membered O-2'-hydroxyethylidene acetals [177].

In the case of many highly branched polysaccharides which contain substantial blocks of periodate-resistant sugar residues the Smith degradation leads to the formation of degraded polysaccharides of less complex structure. Successive applications of the sequence of reactions may be used to expose the interior chains of highly ramified polysaccharides. Thus Japanese larch

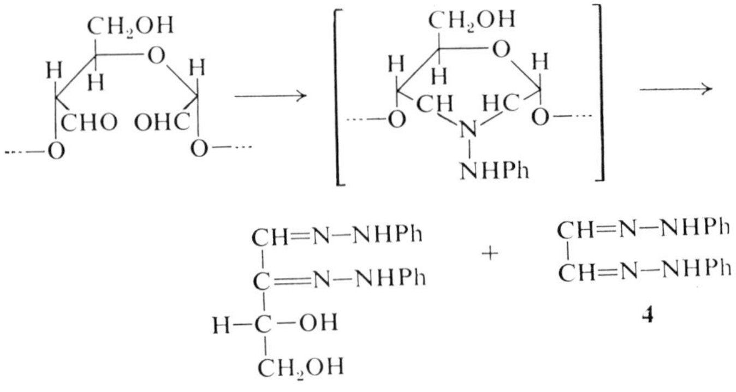

5

CH=N—NHPh
C=N—NHPh
H—C—OH
CH₂OH

+

CH=N—NHPh
CH=N—NHPh

4

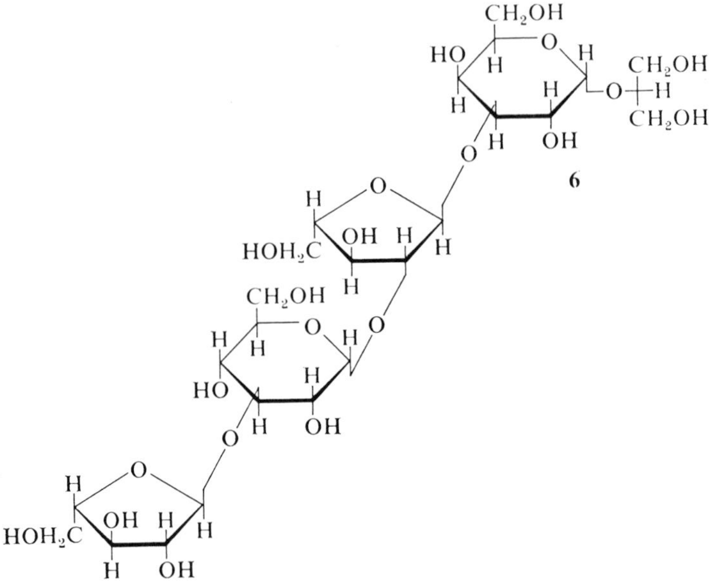

6

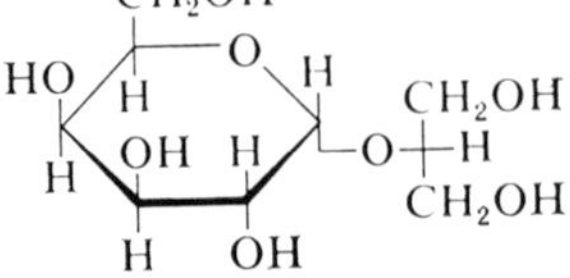

7

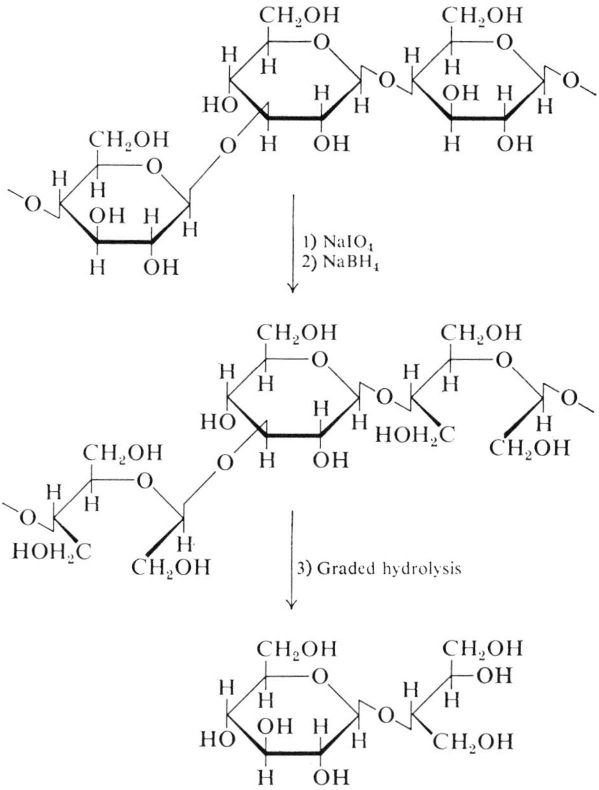

Fig. 15.7 Formation of 2-*O*-β-D-glucopyranosyl-D-erythritol by the Smith degradation of oat β-D-glucan.

arabinogalactan (**9**) after two such degradations furnishes a (1→3)-linked β-D-galactan with not more than 5% of branch points [178].

The oxidation of glycans with lead tetraacetate has been studied only to a small extent [116], mainly because of the insolubility of glycans in most nonaqueous solvents. Neutral glycans, however, are rapidly oxidized by lead tetraacetate in dimethyl sulfoxide [179]. Oxidized glycans, which are often

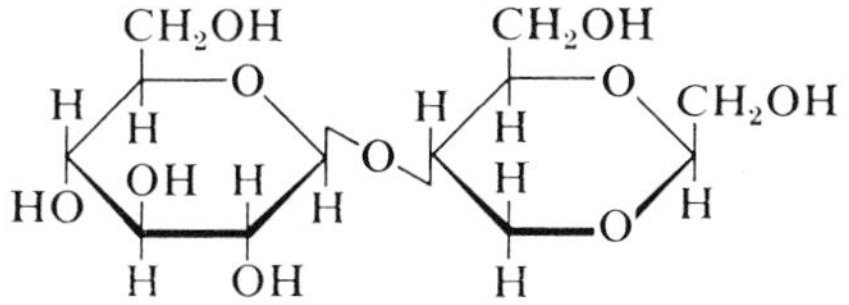

8

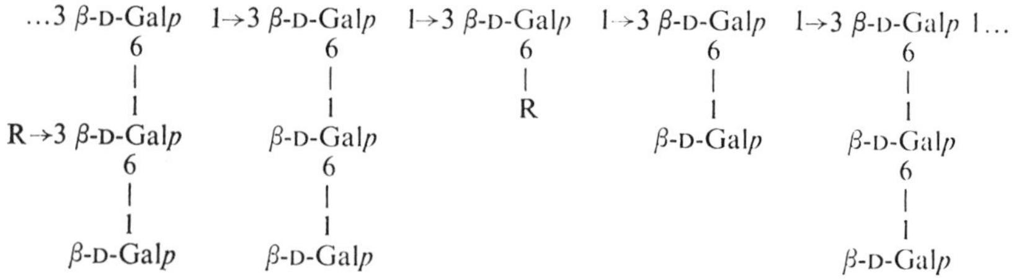

...3 β-D-Gal*p*   1→3 β-D-Gal*p*   1→3 β-D-Gal*p*   1→3 β-D-Gal*p*   1→3 β-D-Gal*p* 1...
  6   6   6   6   6
  | | | | |
  | | R | |
R→3 β-D-Gal*p*   β-D-Gal*p*   β-D-Gal*p*   β-D-Gal*p*
  6   6   6
  | | |
  | | |
β-D-Gal*p*   β-D-Gal*p*   β-D-Gal*p*

9: R = L-Ara*f* 1... or β-L-Ara*p* 1→3 L-Ara*f* 1...

difficult to isolate from aqueous solutions, are readily recoverable from dimethyl sulfoxide by precipitation with ethanol.

Alkaline Degradation of Polysaccharides

Glycosidic linkages are normally stable to alkali, but compounds containing β-glycosyloxy carbonyl groups undergo elimination when treated with alkali. The full potential of this type of reaction as a method for the determination of the fine structure of polysaccharides remains to be exploited. However, a number of structural units in glycans and glycoproteins which are susceptible to alkali have been recognized. These are (*1*) reducing glycose units especially those carrying 3- or 4-*O*-substituents, (*2*) esterified hexuronic acid residues carrying substituents at C-4, and (*3*) *O*-glycosides of serine and threonine residues in polypeptide chains. In addition, the removal of attached ester groups may result in modification of the glycan structure.

Investigations on model compounds have shown that 3- and 4-*O*-substituted glycoses undergo β-alkoxycarbonyl eliminations with the formation of metasaccharinic and isosaccharinic acids, respectively, as shown in the general mechanism of Fig. 15.8, which is based on the mechanism originally

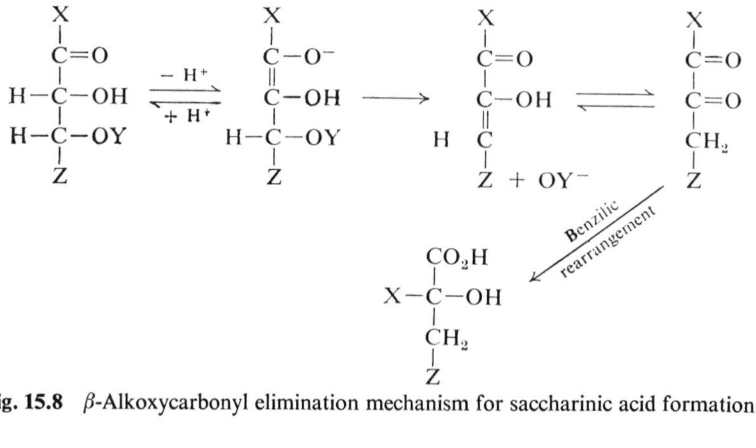

Fig. 15.8 β-Alkoxycarbonyl elimination mechanism for saccharinic acid formation.

proposed by H. S. Isbell [180]. 3-O-Substituted glycoses undergo rapid degradation, for example, with oxygen-free saturated lime-water at room temperature. 4-O-Substituted glycoses are degraded less rapidly since the dominant reaction involving elimination of the 4-glycosyloxy substituent occurs only after the Lobry de Bruyn-Van Ekenstein rearrangement of aldose to ketose has taken place. In general, the elimination of this substituent after rearrangement takes place more rapidly than the direct elimination of a 3-hydroxyl group from a reducing aldose. The formation of the particular saccharinic acid via the benzilic acid rearrangement of an intermediate 1,2-dicarbonyl compound is diagnostic for the substitution pattern of the reducing group. On the basis of studies on oligosaccharide model compounds it is probable that (1→2)-linked glycans will prove to be stable to dilute alkali at room temperature [181] but not necessarily at elevated temperatures [182]. The effect of alkali on (1→5)-linked glycans has not been studied but saccharinic acid formation is unlikely to occur. (1→6)-Linked glycans may be expected to be stable to alkali at room temperature but at high temperatures erosion of the glycan chain may take place if fragmentation of the reducing units by dealdolization to give three-carbon fragments [183, 184] occurs.

The reactions of oligo- and polysaccharides with alkali [185] which result in the cleavage of glycosidic linkages expose new reducing groups from which further degradation may take place. Stepwise erosion of the polysaccharide from the reducing end of the chain is referred to as the "peeling" reaction. The "peeling" reaction proceeds rapidly in the case of (1→3)-linked glycans. (1→4)-Linked glycans are degraded, although less rapidly, by the "peeling" mechanism, but a competing reaction involving elimination of hydroxide ion from C-3 of reducing glycose units has been recognized by the formation of an alkali-stable degraded polysaccharide terminated by metasaccharinic acid units (Fig. 15.9) when hydrocellulose [186] and amylose [187] are treated with alkali until saccharinic acid formation has ceased.

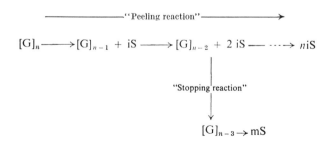

Fig. 15.9 Alkaline degradation of (1→4)-linked glucans. G = D-glucose residue, iS = isosaccharinic acid, and mS = metasaccharinic acid.

The term "stopping" reaction has been used for the chain termination step in the alkaline degradation.

Relatively little information is yet available on the effect of branching in a glycan chain on the course of alkaline degradation. O-Glycosyl substituents which are remote from the reactive sites in the reducing glycose units may remain attached to the degradation product. Thus, guaran, which consists of a main (1→4)-linked β-D-mannan chain with, on the average, every second unit substituted at C-6 with an α-D-galactopyranosyl unit, is degraded by alkali to give equal proportions of D-isosaccharinic acid and 5-O-(α-D-galactopyranosyl)-D-isosaccharinic acid [188]. The characterization of O-glycosylsaccharinic acids from branched glycans has not yet been explored in detail in relation to the determination of fine structure. However, some preliminary experiments [189] on the alkaline degradation of larch arabinogalactan, a polysaccharide containing interior chains of (1→3)-linked sugar residues carrying side chains at C-6, have indicated that the characterization of 6-O-glycosylmetasaccharinic acids may be facilitated if these are further degraded by hypochlorite solution at pH 5 to give the corresponding 5-O-glycosyl(2-deoxypentoses) (Fig. 15.10).

A number of speculations have been made [185] on the likely nature of the products to arise from the alkaline degradation of other branched glycans,

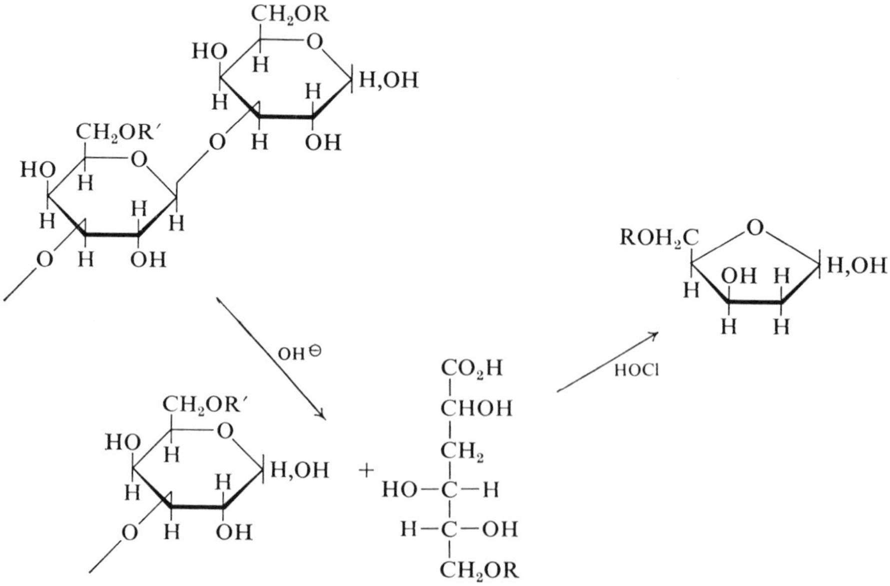

Fig. 15.10   Alkaline degradation of (1→3)-linked galactans.

but in some cases where experimental results are available the speculations have not been confirmed. For example, where there are branches at C-3 on (1→4)-linked glycans the elimination of the 3-O-substituent does not necessarily result in rearrangement of the reducing residue to give an alkali-stable metasaccharinic acid unit. Studies of the decrease in molecular weight of the highly branched rye flour arabinoxylan led to the suggestion [190] supported by parallel studies on a model compound [191] that both 3- and 4-O-substituents may be eliminated with exposure to a new reducing group in the main glycan chain. The degradation product from the original reducing glycose unit was not characterized, but the probable formation of glycosulo-3-enes may be inferred since the degradation of blood group specific substances containing 3,4-di-O-substituted glycose units with sodium hydroxide containing sodium borohydride gives rise to 3-hexenetetrols (Fig. 15.11) [192].

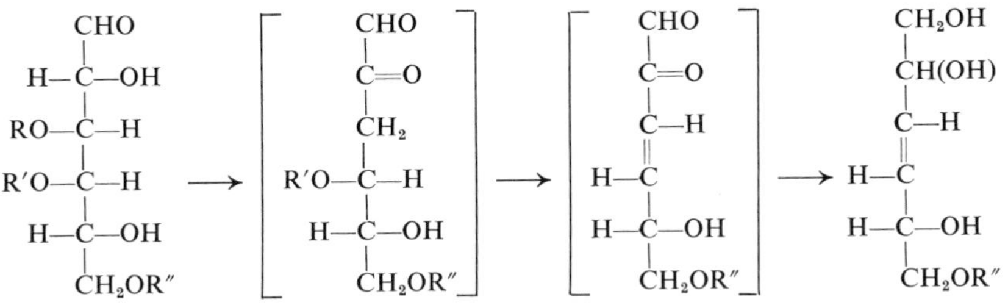

Fig. 15.11   Formation of 3-hexenetetrols during the degradation of 3,4-di-O-substituted hexoses with alkali containing sodium borohydride.

The alkaline degradation of polysaccharides initiated at the reducing end proceeds more rapidly in the presence of calcium ions and is frequently carried out in an excess of oxygen-free lime-water solution at 25–38°, and the reaction is allowed to proceed to completion, often for several weeks or months. Calcium ions are then removed on a cation-exchange resin (H⁺), and the residual polysaccharide may be precipitated from the aqueous solution by the addition of ethanol or acetone as nonsolvent. The degradation products including saccharinic acids may be separated by fractional crystallization of their calcium salts or by suitable chromatographic procedures, for example, on cellulose columns.

The greater susceptibility of pectins (methyl pectates) than of sodium pectate to depolymerization when treated with aqueous alkali led to the proposal that base-catalyzed deesterification is accompanied by β-elimination initiated by the electron-withdrawing carboalkoxy groups and results in rupture of the glycuronan chain [193, 194]. The nature of the reaction has been demonstrated by Neukom and Heim [195] who showed that methyl (methyl 4-deoxy-β-L-*threo*-hex-4-enopyranosid)uronate (**12**) and methyl

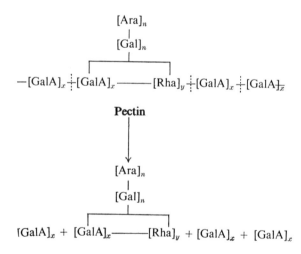

10

11   **12**

D-galacturonate (**11**) are formed when methyl 4-*O*-(methyl α-D-galacto-pyranosyluronate)-(methyl-α-D-galactopyranosid)uronate (**10**) is heated with methanolic sodium methoxide, under which conditions methyl ester groups are re-formed by base-catalyzed ester exchange.

Pectins are also degraded when heated with phosphate buffer at pH 6.8 [196]. Although the extent of reaction is slight, the reaction has been used by Northcote and coworkers [197, 198] who showed that electrophoretically homogeneous pectins, for example, from apple, are degraded with the formation of structurally dissimilar degraded polysaccharides, one a galacturonan of relatively low molecular weight and the other a complex

$$[\text{Ara}]_n$$
$$|$$
$$[\text{Gal}]_n$$

$$-[\text{GalA}]_x \overset{\cdot}{\div} [\text{GalA}]_x \overline{\qquad\qquad} [\text{Rha}]_y \overset{\cdot}{\div} [\text{GalA}]_x \overset{\cdot}{\div} [\text{GalA}]_x$$

Pectin

$$[\text{Ara}]_n$$
$$|$$
$$[\text{Gal}]_n$$

$$[\text{GalA}]_x + [\text{GalA}]_x \overline{\qquad\qquad} [\text{Rha}]_y + [\text{GalA}]_x + [\text{GalA}]_x$$

Fig. 15.12 Formation of two chemically distinct degraded polysaccharides on random degradation of pectin containing an uneven distribution of neutral sugars in the structure.

polysaccharide which contains most of the neutral sugar residues originally present in the parent polysaccharide. Assuming random degradation of the esterified galacturonan chain by the β-elimination mechanism, the results imply that the neutral sugar constituents are unevenly distributed along the galacturonan as shown in Fig. 15.12.

Rees and coworkers [199] have shown that esters of (1→4)-linked glycuronans, for example, 2-hydroxyethyl alginate, undergo degradation by the β-elimination mechanism when treated with methanolic barium methoxide, but no characterizable products were isolated. However, similar treatment of esters of colanic acid and related extracellular polysaccharides from various species of the Enterobacteriaceae results in the liberation of esters of 4,6-O-(1'-carboxyethylidene)-D-galactose, residues of which sugar are linked (1→4) to D-glucuronic acid [199a]. A mechanistically similar reaction has been observed when tetraalkylammonium salts of heparin are treated with diazomethane [200]. The tetrasaccharide (13) containing an unsaturated hexuronic acid residue was isolated.

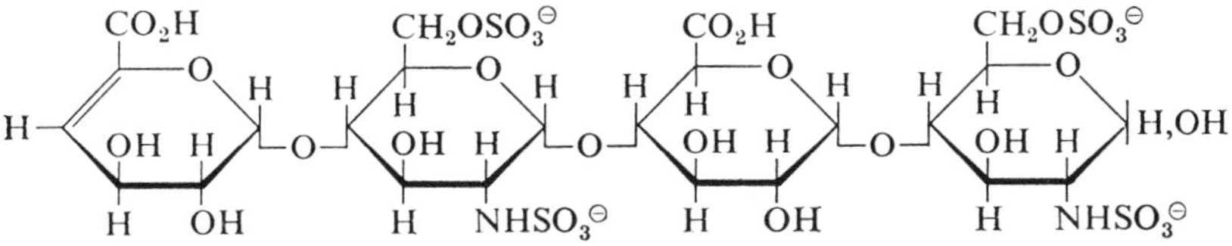

13

Epimerization of a glycosidically linked glucuronic acid residue occurs when the trisaccharide (14) derived from plant glycolipids [201] and 6-O-(α-D-glucopyranosyluronic acid)-myo-inositol (15) are heated with barium hydroxide [202]. Epimerization of D-glucuronic acid to L-iduronic acid takes place without accompanying elimination.

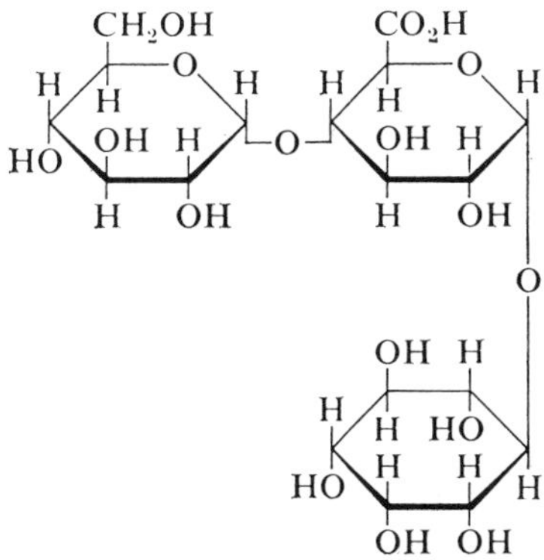

14

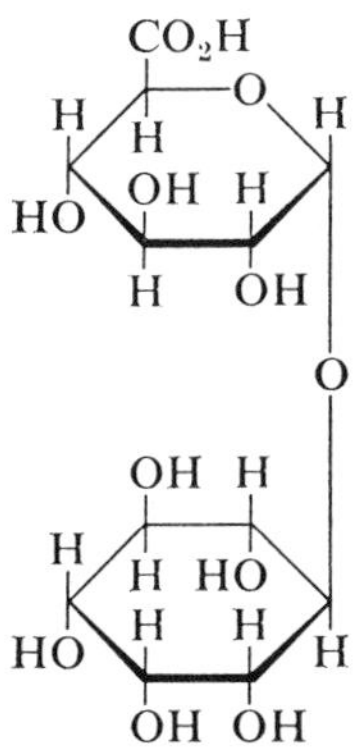

15

Glycoproteins [88] which contain sugar residues glycosidically linked to serine and threonine units in peptide or protein chains are readily degraded by alkali. This type of reaction involves a β-elimination of the glycosyloxy substituent with the formation of α-aminoacrylic acid or α-aminocrotonic acid residues [203, 204]. These amino acids are readily destroyed during hydrolysis, but the formation of unsaturated acids has been established by hydrogenation and subsequent hydrolysis to give alanine (Fig. 15.13) and

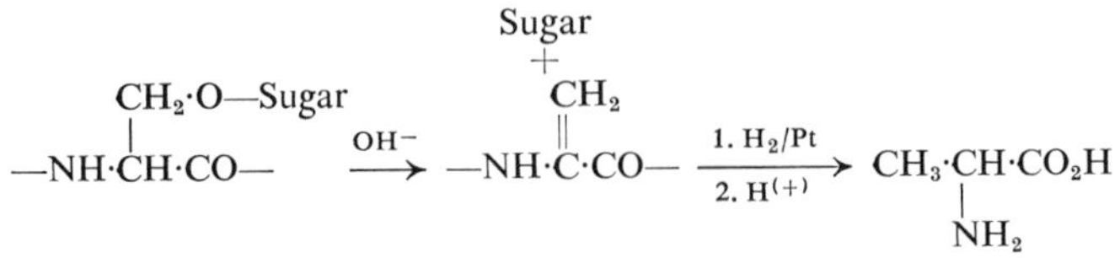

$$
\begin{array}{ccc}
& \overset{\text{Sugar}}{\overset{+}{}} & \\
\underset{|}{\text{CH}_2\cdot\text{O—Sugar}} & \underset{\|}{\text{CH}_2} & \\
\text{—NH·CH·CO—} & \xrightarrow{\text{OH}^-} \text{—NH·C·CO—} & \xrightarrow[\text{2. H}^{(+)}]{\text{1. H}_2/\text{Pt}} \text{CH}_3\cdot\underset{\underset{\text{NH}_2}{|}}{\text{CH}}\cdot\text{CO}_2\text{H}
\end{array}
$$

Fig. 15.13 Base-catalyzed β-elimination of sugar units O-glycosidically linked to serine residues.

α-aminobutyric acid, respectively [203, 204]. The liberated oligosaccharides may undergo further reaction by the "peeling" mechanism with concomitant formation of saccharinic acids or other degradation products. The extent of degradation of oligosaccharides is dependent on both the nature of the alkaline reagent and the types of linkage present.

O-Glycosidic linkages to serine and threonine are now recognized in the submaxillary mucins [205], the polysaccharide-protein complexes involving chondroitin 4-sulfate [206], heparin [207], possibly other aminopoly-saccharides, and the blood group substances [208]. The liberation of the carbohydrate prosthetic group [O-(N-acetylneuraminyl)-(2 → 6)-N-acetyl-D-galactosamine] on mild treatment of ovine submaxillary mucin with alkali [209] or with lithium borohydride [210] led to the postulation of ester linkages (1-O-acyl sugars) to the carboxyl groups of glutamic or aspartic acids. That these reactions involved β-elimination of serine and threonine O-glycosides was recognized by the detection of α,β-unsaturated amino acids by ultraviolet light absorption (241 nm) and after hydrogenation by the increase in the relative proportions of alanine and α-aminobutyric acid [205].

In early work on the aminopolysaccharides, covalent linkages to protein were not detected since isolation frequently involved deliberate treatment with alkali to remove "contaminating" protein. In the blood group substances oligosaccharide chains carry the antigenic determinants [211, 212]. Alkaline degradation of blood group substances has provided a valuable method for isolating oligosaccharides which are not accessible by other methods such as partial acid hydrolysis which results in the ready cleavage of L-fucosyl linkages. Morgan and coworkers [213] first used triethylamine for the alkaline degradation of blood group substances and isolated the tetra-saccharide (**16**) from human group A substance. In a modified procedure

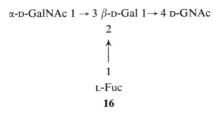

α-D-GalNAc 1 → 3 β-D-Gal 1→ 4 D-GNAc

2

↑

1

L-Fuc

16

Kabat and coworkers have degraded blood group A, B, and H substances with sodium hydroxide containing sodium borohydride, the reducing agent being added with the intention of minimizing the erosion of oligosaccharide chains by the formation of glycitol end groups [214]. In the event degradation still proceeded some way before reduction took place with the formation of a series of oligosaccharides, some of which were terminated by a 6-*O*-substituted 3-hexene-1,2,5,6-tetrol (**17**, designated R as in oligosaccharides **18** and **19** from blood group A substance [192, 215]).

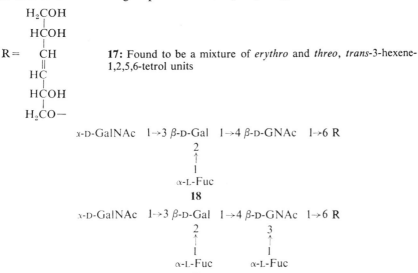

$$R =$$

H₂COH
HCOH
CH
HC
HCOH
H₂CO—

17: Found to be a mixture of *erythro* and *threo*, *trans*-3-hexene-1,2,5,6-tetrol units

x-D-GalNAc 1→3 β-D-Gal 1→4 β-D-GNAc 1→6 R

2

↑

1

α-L-Fuc

18

x-D-GalNAc 1→3 β-D-Gal 1→4 β-D-GNAc 1→6 R

2   3

↑   ↑

1   1

α-L-Fuc   α-L-Fuc

19

Glycans may contain ester groups based on carboxylic, sulfuric, and phosphoric acids. Since these groups are readily hydrolyzed, often with consequent alteration of the glycan structure, the use of alkaline reagents during the extraction and isolation of the glycan from natural sources should be avoided. Carboxylic esters in polysaccharides occur in the form of (*1*) esterified hexuronic acid residues and (*2*) acylated, most commonly acetylated, neutral sugar units. The former type of ester groups are present as methyl α-D-galacturonic residues in pectins. As noted above, since these hexuronic ester residues are 4-*O*-substituted, alkaline hydrolysis may be accompanied by β-elimination resulting in chain cleavages and consequent modification of the polysaccharide structure. Minimum degradation occurs if the saponification of pectins is carried out at pH 12 at 0° [216]. De-O-acylation of esterified neutral sugar residues in glycans likewise takes place very rapidly, but more deep-seated structural changes have not yet been detected. Bouveng [217] has developed a method for the location of the sites of attachment of such ester functions using the sequence of reactions illustrated in Fig. 15.14, in which methyl ether groups replace the original ester groups. In an alternative and simpler procedure [217a] hydroxyl groups are protected as mixed acetals by reaction with methyl vinyl ether, and the method has been used to locate O-acetyl groups in a lipopolysaccharide [217b].

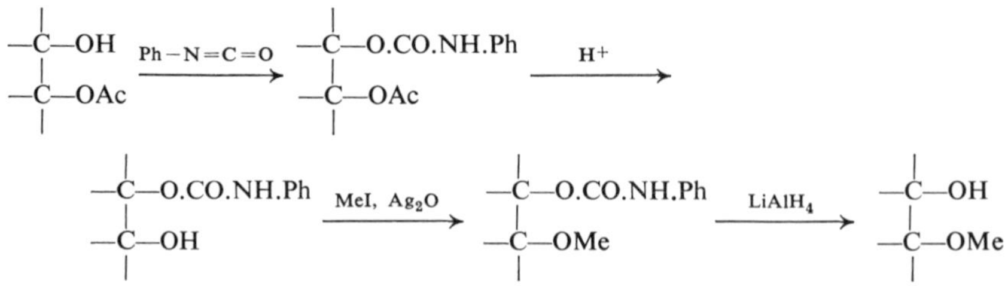

$$
\begin{array}{ccccc}
-\overset{|}{\underset{|}{C}}-OH & & -\overset{|}{\underset{|}{C}}-O.CO.NH.Ph & & \\
& \xrightarrow{Ph-N=C=O} & & \xrightarrow{H^+} & \\
-\overset{|}{\underset{|}{C}}-OAc & & -\overset{|}{\underset{|}{C}}-OAc & & \\
\end{array}
$$

$$
\begin{array}{ccccccc}
-\overset{|}{\underset{|}{C}}-O.CO.NH.Ph & & -\overset{|}{\underset{|}{C}}-O.CO.NH.Ph & & -\overset{|}{\underset{|}{C}}-OH \\
& \xrightarrow{MeI,\ Ag_2O} & & \xrightarrow{LiAlH_4} & \\
-\overset{|}{\underset{|}{C}}-OH & & -\overset{|}{\underset{|}{C}}-OMe & & -\overset{|}{\underset{|}{C}}-OMe \\
\end{array}
$$

Fig. 15.14   Location of O-acyl substituents in glycans.

The basic hydrolysis of sulfate esters in glycans (and other carbohydrates) proceeds extremely slowly unless reaction occurs with the participation of suitably disposed ionizable hydroxyl groups. Where such groups are present, ready intramolecular displacement of sulfate esters takes place with the formation of oxide rings, especially of epoxides and of 3,6-anhydrides. Since epoxides are readily attacked by aqueous base, ring opening occurs with the regeneration of *trans*-diols of both the original and inverted configurations.

Treatment of a glycan sulfate, which carries ester groups suitably disposed for epoxide formation, with aqueous alkali may result in the formation of sugar residues of different configuration from those originally present, the proportion of altered sugar residues depending on the preferred direction of epoxide opening. An example is provided by the complex sulfated polysaccharide from the green seaweed *Ulva lactuca*, which contains *inter alia* sulfated D-xylose residues and on treatment with aqueous alkali gives rise to D-arabinose [218]. These residues could be formed, via different epoxides, from 4-O-substituted D-xylose residues carrying sulfate esters at C-2 or C-3. A distinction between these two possibilities has been made by treatment of the sulfated polysaccharide with methanolic sodium methoxide. Hydrolysis of the resulting modified polysaccharide gave only one methyl ether, namely, 2-O-methyl-D-xylose, which sugar could arise only via residues of 2,3-anhydro-D-lyxose formed from those of D-xylose 2-sulfate (Fig. 15.15).

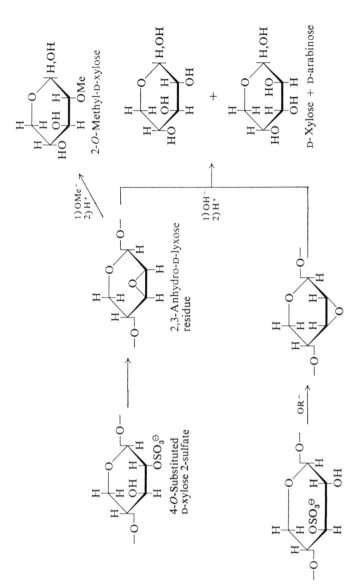

Fig. 15.15 Location of sulfate ester groups via epoxide formation in *Ulva lactuca* polysaccharide.

Sulfate esters attached to C-6 hexose residues are readily displaced intramolecularly by alkoxide ions formed from *cis*-hydroxyl groups at C-3, with the formation of 3,6-anhydrides. Thus the formation of 3,6-anhydro-D-galactose residues when λ-carrageenan is treated with alkali showed that D-galactose 6-sulfate residues are present in the parent glycan [140]. The elegant use of this type of transformation by Anderson and Rees [219] has afforded evidence for a masked repeating structure in porphyran, a sulfated galactan from red seaweeds of the genus *Porphyra*. Porphyran contains an alternating sequence of 3-*O*-substituted D-galactose or 6-*O*-methyl-D-galactose residues, and 4-*O*-substituted L-galactose 6-sulfate or 3,6-anhydro-L-galactose residues. Treatment of porphyran with alkali followed by methylation furnished a methylated polysaccharide (Fig. 15.16) which was indistinguishable from methylated agarose and gave the same derivatives of partially methylated agarobiose on partial methanolysis.

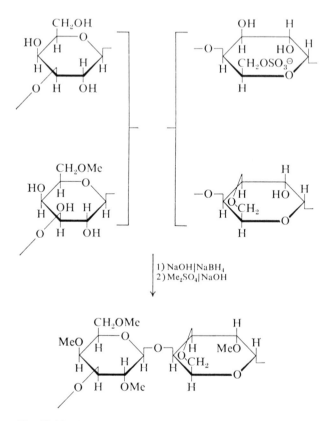

Fig. 15.16 Conversion of porphyran into methylated agarose.

The acid-sensitive 3,6-anhydro sugar units may be created by the action of alkali on residues which carry primary sulfonate esters. An example of the selective formation of such units within a polysaccharide has been provided in recent work on a glucomannan from *Ceratocystis brannea*. This polysaccharide contains α-D-mannopyranose residues joined by (1→6) linkages and α-D-glucopyranose residues unsubstituted at $C_{(6)}$ in side chains. Selective modification of the D-glucopyranose residues by the reaction sequence tritylation, acetylation, detritylation, tosylation, and treatment with alkali was followed by graded acid hydrolysis to remove the 3,6-anhydro-D-glucose residues, leaving the mannan chain virtually intact [219a].

Phosphodiester linkages are present in teichoic acids [220] and in some of the encapsulating substances from type-specific *Pneumococci* [221]. Since these polymeric carbohydrate phosphates carry hydroxyl groups on carbons adjacent to those bearing the ester groups, hydrolysis with alkali takes place very readily (cf., ribonucleic acids 222). The phosphodiester linkages form part of the main chain structure of these materials, and rapid depolymerization takes place with the formation of phosphomonoesters. It should be noted, however, that phosphodiester linkages cannot be located in the parent polymer on the basis of the phosphomonoesters isolated on alkaline hydrolysis since phosphate migration may occur from primary to secondary hydroxyl groups, presumably via 1,2-cyclic phosphates [223, 224].

In the majority of teichoic acids so far examined in which the glycerol or ribitol units are joined by phosphodiester bonds, alkaline treatment gives polyol phosphomonoesters from which the polyol units themselves, which often carry glycosyl substituents, may be obtained on treatment with phosphomonoesterase. Figure 15.17 shows such an alkaline degradation of a ribitol teichoic acid.

Similar degradations are used to isolate the oligosaccharide repeating units from those type-specific capsular substances of *Pneumococci* in which ribitol and glycose residues are joined by phosphodiester linkages. Thus treatment of the specific substance from *Pneumococcus* type VI (**20**) with alkali under mild conditions gave a phosphomonoester from which the crystalline tetrasaccharide (**21**) was isolated in virtually quantitative yield on further treatment with phosphomonoesterase [225].

$$-[-2\ \alpha\text{-D-Gal}p\ 1{\to}3\ \alpha\text{-D-G}p\ 1{\to}3\ \alpha\text{-L-Rha}p\ 1{\to}3\ \text{D-(or L)-Ribitol}\ 1\text{-O}-\overset{\displaystyle \overset{\text{O}}{\|}}{\underset{\displaystyle \underset{\text{OH}}{|}}{\text{P}}}-\text{O-}]_n-$$

20

α-D-Gal*p* 1→3 α-D-G*p* 1→3 α-L-Rha*p* 1→3 D-(or L)-Ribitol

21

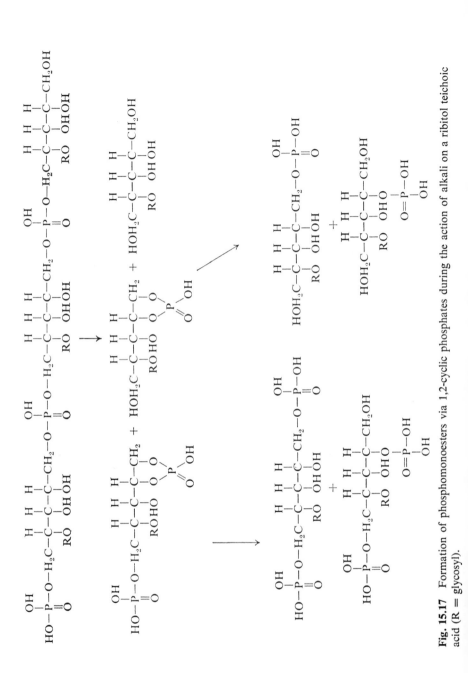

Fig. 15.17 Formation of phosphomonoesters via 1,2-cyclic phosphates during the action of alkali on a ribitol teichoic acid (R = glycosyl).

Nitrous Acid Deamination of Aminopolysaccharides

Nitrous acid reacts with 2-amino-2-deoxy-D-glucose and with its methyl glycosides to give 2,5-anhydro-D-mannose (**22**) [226], which is similarly formed from chitosan (*N*-deacetylated chitin) [227]. A study of the action of nitrous acid on ψ-heparin and chitosan showed that the rate of deamination was much more rapid for the former than for the latter, and that the rates corresponded to those obtained on nitrous acid deamination of methyl 2-amino-2-deoxy-α- and -β-D-glucopyranosides [227], respectively, thus providing evidence for the configurations of the glycosidic linkages in the aminoglycans.

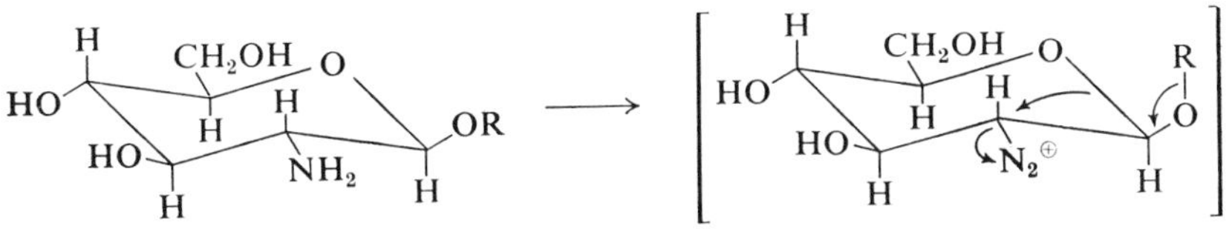

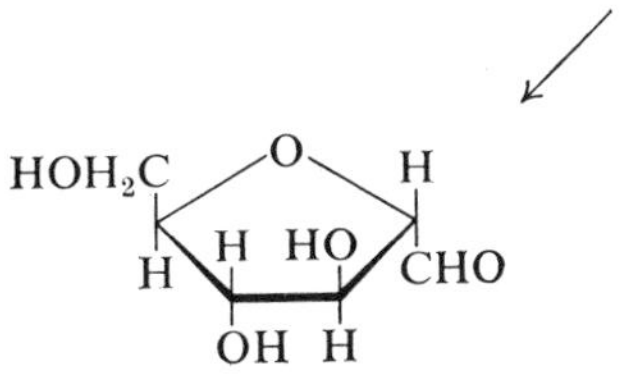

The nitrous acid degradation procedure leads to the selective cleavage of the glycosyl linkage of the glycosamine, which is particularly resistant to cleavage by hydrions. The degradation has been used to obtain oligosaccharides from *Pneumococcus* types 10A and 29 specific substances [228, 229], and thus derive evidence for the sequence of sugar residues (see Appendix).

3   DEGRADATION STUDIES ON CHEMICALLY MODIFIED POLYSACCHARIDES

Degradation Studies on Methylated Polysaccharides

Introduction

The methylation procedure, originated by Purdie and Irvine and extensively used by Haworth, Hirst, and others, is an established method for the analysis of the linkages in polysaccharides [1]. The method consists in the methylation of all the free hydroxyl groups in a polysaccharide followed by cleavage of the glycosidic linkages in the methylated glycan to give *O*-methylglycoses or their derivatives. The nature and proportions of the *O*-methylglycoses provide information on the relative proportions of non-reducing end groups, the degree of branching, the nature of the main linkage types, and the types of interchain linkages at branch points. The method is illustrated by the results expected from the hydrolysis of a segment **23** of a methylated galactan. Hydrolysis of this portion of the methylated glycan affords 2,3,4,6-tetra-*O*-methyl-D-galactose (**24**), 2,4,6-tri-*O*-methyl-D-galactose (**25**), and 2,4-di-*O*-methyl-D-galactose (**26**), which arise from

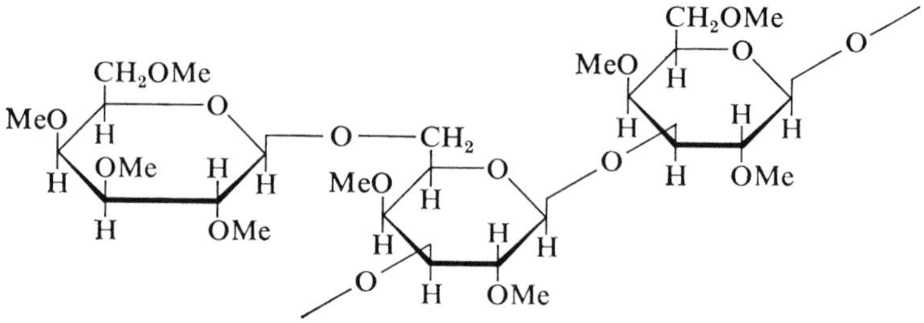

23

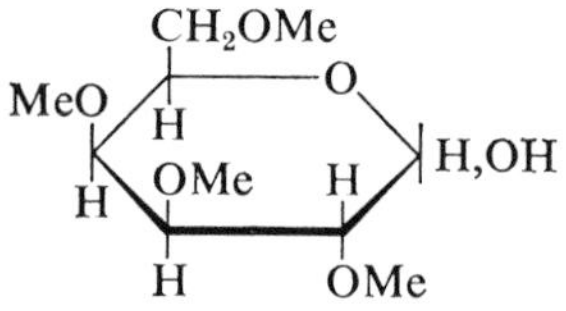

24 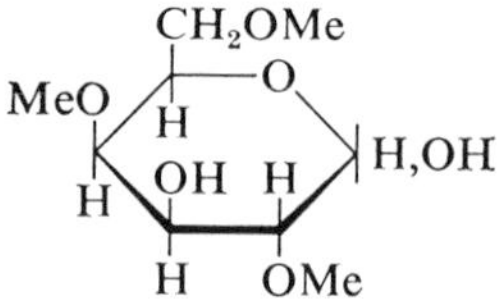   **25** 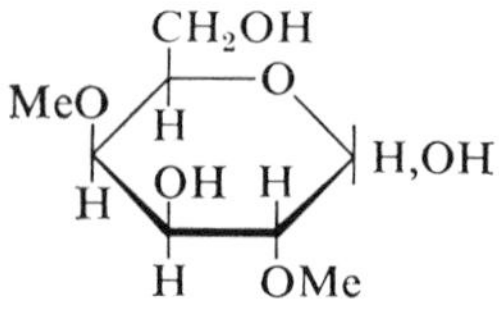   **26**

nonreducing end groups, chain units, and branch points, respectively. The free hydroxyl groups in these hydrolysis products indicate the points through which the individual units are joined to adjacent units in the parent polysaccharide.

The validity of the methylation method requires that the polysaccharide does not undergo chemical modification during the methylation, that the polysaccharide is fully etherified, and that no demethylation takes place during hydrolysis. In practice, however, difficulties have been encountered in achieving complete methylation, and in certain instances specific hydroxyl groups are resistant to etherification. The (1→2)-linked β-D-glucan from *Agrobacteria* species provides a striking example where methylation of hydroxyl groups at C-3 is particularly difficult [230]. The methylation of cellulose in nonswelling solvents has indeed been used to obtain information on the positions involved in hydrogen bonding in the ordered regions of the structure [231]. A small amount of demethylation has been observed during hydrolysis [232]. The structural significance of small amounts of *O*-methylglycoses in hydrolyzates of methylated polysaccharides should therefore be interpreted with caution, particularly if these sugars indicate potential branching, and independent evidence should be sought for the implied structural features.

The methylation method of structural analysis does not reveal the configuration of the glycosidic linkages or the sequence of glycose units in a homopolysaccharide containing more than one type of linkage or in a heteropolysaccharide. Thus, in the example cited previously, several possible partial structures are consistent with the characterization of the three methylated sugars **24, 25,** and **26.** A further problem is encountered when the *O*-methylglycose carries free or hemiacetal-linked hydroxyl groups at C-4 and C-5 since the ring size of the parent sugar residue is not defined. Thus 2,3,6-tri-*O*-methyl-D-galactose (**27**) has been isolated from methylated galactans containing 4-*O*-substituted D-galactopyranose (**28**) and 5-*O*-substituted D-galactofuranose (**29**) residues.

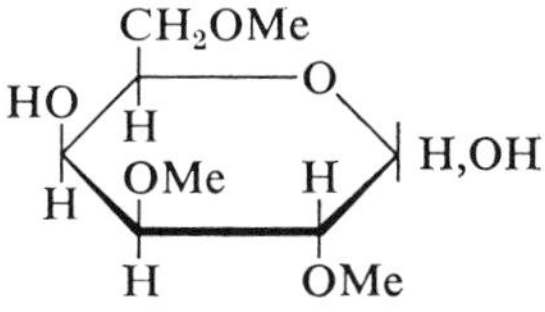

27

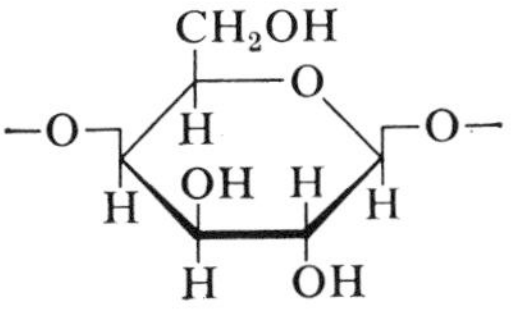

28

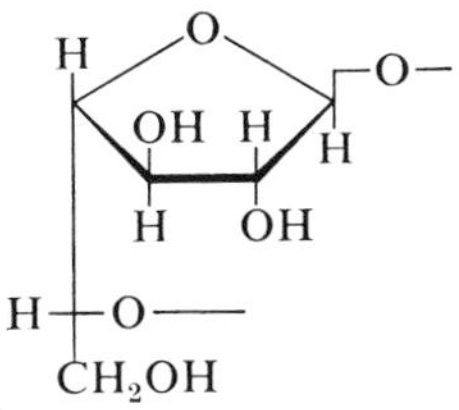

29

Since etherification normally requires the use of basic reagents to effect ionization of hydroxyl groups, the possibility of alkaline degradation, especially of glycans containing 4-*O*-substituted hexuronic acid residues, cannot be ignored. This particular difficulty may be avoided by methylation of the carboxyl-reduced glycan, but new problems of interpretation may be created if the polysaccharide contains residues of a hexuronic acid and the corresponding hexose, for example, D-galacturonic acid and D-galactose. Difficulty has also been encountered in the methylation of sulfated polysaccharides. Although this difficulty may be overcome by prior desulfation [142], it is desirable to be able to compare the cleavage products from the methylated glycan sulfate and the methylated desulfated glycan in order to obtain evidence for the location of the sulfate esters.

Methylation of Polysaccharides

The following procedures have been used for the methylation of polysaccharides [233].

METHYL SULFATE

The methylation of polysaccharides using methyl sulfate and alkali was first developed by Haworth [234] and descriptions are provided in various reviews [6, 233]. Essentially the procedure involves treatment of a vigorously stirred solution of the polysaccharide with approximately equivalent proportions of methyl sulfate and sodium or potassium hydroxide in a nitrogen atmosphere at 10–55°. The first additions of reagents are normally made below 5° in order to minimize alkaline degradation which may occur before the reducing end group has undergone glycosidation to give an alkali-stable unit. Alternatively, the reducing end group may be protected by prior reduction with sodium borohydride [55]. The reaction mixture should be maintained at slight alkalinity throughout the reaction to ionize hydroxyl groups and to neutralize the methyl sulfate. Gross excess of alkali merely leads to nonproductive consumption of methyl sulfate. On the other hand the reaction mixture should not be allowed to become acid since hydrolytic cleavage may occur. Acetone or 1,4-dioxane is often added during a methylation in order to maintain the partially methylated polysaccharide in solution. At the end of the reaction, after removal of organic solvent, the methylated product may sometimes separate out from the reaction mixture, especially on heating, and can be removed mechanically. Alternatively, the methylated glycan is isolated by extraction with chloroform or by dialysis of the aqueous solution to remove inorganic salts, followed by concentration to dryness under reduced pressure.

This methylation procedure suffers from the disadvantage that several treatments with the methylating reagents are usually required. Nevertheless, this method remains the most satisfactory for the nondegradative methylation of acidic glycans. The progress of the methylation is followed by methoxyl determinations on samples of the product or by observation of the disappearance of hydroxyl absorption at ca. 3300 cm^{-1} in the infrared spectrum.

An excellent variation of this method for (1) the further methylation of partially methylated glycans and (2) the simultaneous deacetylation and methylation of acetylated glycans uses tetrahydrofurans as solvent with methyl sulfate and pulverized sodium hydroxide [235]. Other procedures which are satisfactory for the methylation of some polysaccharides involve the use of methyl sulfate with barium oxide and/or barium hydroxide in N,N-dimethylformamide or methylsulfoxide [236], or of methyl sulfate with sodium hydroxide in methylsulfoxide [237].

METHYL IODIDE

The methylation procedure first developed by Purdie [238] using methyl iodide and silver oxide cannot be used for the direct methylation of polysaccharides but is frequently used to complete the etherification of partially methylated glycans and to esterify carboxyl groups in methylated acid glycans. The same procedure may often be carried out more effectively in N,N-dimethylformamide as solvent [239].

An excellent method for the direct methylation of neutral [240] and some acid glycans [241] which are soluble in methyl sulfoxide involves the use of methyl iodide and sodium hydride. Given adequate solubility and the absence of base-sensitive functional groups, this is the method of choice for rapid and complete methylation of glycans. Other methylation procedures with methyl iodide involve (1) barium oxide in methylsulfoxide [242], (2) heating thallium salts (alkoxide and carboxylate) with methyl iodide containing a little anhydrous methanol [243] (this procedure is particularly useful for the direct esterification of acid glycans [244], and (3) sodium amide in liquid ammonia [245, 246].

DIAZOMETHANE

Treatment of polysaccharides in suspension in ethereal diazomethane will slowly methylate hydroxyl groups, but the reaction does not go to completion even on prolonged treatment [247]. Diazomethane-boron trifluoride has been used to methylate carbohydrates containing base-labile substituents [248], but the effect of the Lewis acid on glycosidic linkages has not yet been assessed.

Fission of Methylated Polysaccharides

Fission of methylated polysaccharides is often carried out in an organic or aqueous organic solvent because methylated glycans are seldom soluble in hot water alone. The nature of the depolymerization reaction is dictated partly by the methods to be used for the separation of the glycose derivatives and partly by the necessity of obtaining maximum depolymerization, minimum demethylation of derivatives, and minimum production of artifacts.

The following reagents and conditions have been found to be generally applicable to the cleavage of methylated glycans.

METHANOLIC HYDROGEN CHLORIDE

Methanolysis of a methylated polysaccharide is achieved by heating a 1-2% solution of the material in 1-6% methanolic hydrogen chloride under reflux on a boiling-water bath until the optical rotation of the solution reaches a constant value. Alternatively more resistant methylated glycans may be depolymerized by heating in a sealed tube in a boiling-water bath. The cooled solution is neutralized with silver carbonate or a weak base anion exchange resin, filtered, and the filtrate is concentrated to a syrup under reduced pressure. When the methanolyzate is to be quantitatively analyzed, for example, by gas–liquid chromatography, care must be exercised to avoid inadvertent loss of the more volatile methyl glycosides, especially of fully methylated pentoses and 6-deoxyhexoses, during concentration [249]. If required, hydrolysis of the resulting methyl glycosides of the methylated glycoses is achieved by heating with N sulfuric acid until the solution has a constant optical rotation. The solution is neutralized ($BaCO_3$ or anion exchange resin) and filtered, and the filtrate is then concentrated under reduced pressure. Again care is required to avoid loss of the more volatile fully methylated sugars.

FORMIC ACID [57, 58]

Formic acid (85–98%) at 100° is an excellent solvent and hydrolytic reagent for many methylated polysaccharides, especially the more resistant methylated mannoglycans and glycuronoglycans. The formic acid is removed by distillation under reduced pressure. Since formyl esters are formed, further treatment is necessary with either N sulfuric acid at 100° for 1 hr to give methylated glycoses or methanolic hydrogen chloride to give methyl glycosides, followed by neutralization, as appropriate, and concentration.

HYDROCHLORIC AND SULFURIC ACIDS

Direct hydrolysis of methylated glycans in an aqueous medium is possible but is complicated by their frequent insolubility in hot water. However, methylated glycans are often more soluble at lower temperatures, and if a reasonable degree of depolymerization can be achieved without the methylated glycans coming out of solution, the temperature may then be raised to complete the hydrolysis. Alternatively a water-miscible organic solvent, much as acetic acid, may be added to maintain solution. Alcoholic solvents may result in partial glycoside formation but removal of solvent, for example, by distillation, as depolymerization proceeds ensures that glycoses are formed by hydrolysis. Hydrochloric and sulfuric acids may be neutralized with lead or silver carbonate, or barium carbonate, respectively, or by weak base anion exchange resins.

OXALIC ACID [250]

Methylated glycans, for example, methylated fructans, which are readily hydrolyzed may be depolymerized with oxalic acid in aqueous methanol. Subsequent removal of methanol under reduced pressure is followed by hydrolysis with aqueous oxalic acid, neutralization with calcium carbonate, filtration, concentration of the filtrate to small bulk, and extraction with acetone to give the methylated glycoses.

OXIDATIVE HYDROLYSIS

Complete depolymerization of methylated κ-carrageenan containing 3,6-anhydro-D-galactose residues, which are readily degraded under normal conditions of hydrolysis, has been achieved by oxidative hydrolysis in the presence of bromine [251] (see Fig. 15.18). The glycan contains alternating

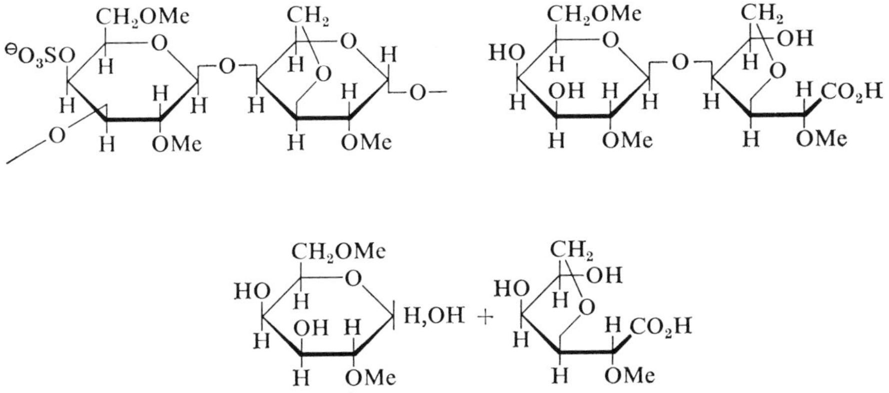

Fig. 15.18 Oxidative hydrolysis of methylated κ-carrageenan.

residues of D-galactose and 3,6-anhydro-D-galactose, and the mild conditions of reaction caused selective cleavage of 3,6-anhydrogalactosyl linkages. After removal of bromine, hydrolysis of the galactosyl linkages was completed without decomposition and the 3,6-anhydrogalactonic acids were isolated in high yields.

Examination of Fission Products of Methylated Polysaccharides

Methylated glycoses and their derivatives from the hydrolyzates of methylated polysaccharides are now almost always separated by chromatographic methods. As in the case of the parent glycoses (see p. 396), identification may be made by paper chromatography, thin-layer chromatography, or gas chromatography of suitable volatile derivatives. Paper ionophoresis in borate buffer is a useful additional technique for the separation of certain mixtures of partially methylated sugars [252].

In the analysis of complex mixtures of O-methyl glycoses acid or basic glycose, derivatives are first removed by absorption on ion exchange resins from which they may be subsequently eluted. Methylated glycans show many of the same differences in rates of hydrolysis of different glycosidic linkages. Thus partially methylated aldobiouronic acids may often be isolated on graded hydrolysis of acid glycans.

Paper chromatographic analysis may be placed on a quantitative basis by elution of the sugars from the paper followed by their determination by colorimetric procedures, for example, with aniline phthalate [253] or the phenol–sulfuric acid reagent [254], or by alkaline hypoiodite oxidation [255].

Gas chromatography [92] now provides the method of choice for the identification of methylated glycoses. Direct methanolysis of methylated glycans leads to complex mixtures of methyl glycosides, since each sugar gives rise to two and sometimes four methyl glycosides. The disadvantage of the increased separation problem may be offset by the greater certainty of identification since the components of the equilibrium mixture of glycosides from any particular sugar are present in characteristic proportions [256]. The separation of the less fully methylated glycoses is aided by conversion into acetates [257] or trimethylsilyl ethers [258]. If methylated aldoses are reduced or oxidized, single derivatives are formed for each component and the gas chromatography of methylated alditol acetates [93] and methylated aldonolactones [259, 260] has been reported. Methylated alditol acetates have proved particularly suitable derivatives for combined gas–liquid chromatography–mass spectrometry [261], which greatly increases the certainty of identification. In the case of methylated alditol acetates from methylated acidic polysaccharides, reduction of carboxyl groups with lithium aluminum deuteride permits those derivatives formed from hexuronic acid residues to be distinguished from those formed from hexose residues [261a]. As noted previously, mass spectrometry alone [262] is somewhat insensitive to stereochemistry, but the technique is very useful in determination of ring size and substitution patterns.

Methylated glycoses may be separated in sufficient quantity for their complete characterization by fractionation by preparative gas chromatography, filter sheet chromatography, or preparative thin-layer chromatography. Column chromatography by partition on cellulose [106] and adsorption on charcoal [107] is commonly used for larger scale separations. The separations achieved by these two procedures are often complementary, and highly complex mixtures of sugars may be completely fractionated with a combination of the two types of column.

The identities of the separated methyl glycoses are established where possible by the preparation of crystalline derivatives [263]. The nature of the parent glycose may be found by chromatographic examination of the free glycose formed on demethylation with boron trichloride [264]. Alternatively, the partially methylated glycose may be fully methylated by treatment with methyl iodide and silver oxide in N,N-dimethylformamide [239], and the product then examined by gas chromatography. The further alkylation of the partially methylated glycose or methyl glycoside with trideuteromethyl iodide followed by mass spectrometry of the fully substituted derivative may be used to determine the substitution pattern in the parent sugar [262].

Hydrolytic Degradation of Reduced Glycuronoglycans

Owing to the difficulty of attaining complete hydrolysis of glycurono-glycans without destruction of the component glycoses, an alternative approach to their analysis is the application of degradation techniques to the glycans produced by reduction of the carboxyl groups to primary alcohol groups. Since this reduction does not involve any configurational changes, deductions from the analysis of the reduced material are applicable to the original glycan.

In addition to the advantage of achieving complete hydrolysis of the reduced glycuronoglycans, graded hydrolysis will often yield oligosaccharides which differ in nature from those obtained from the graded hydrolysis of the original acidic glycan. For example, graded hydrolysis of a glycuronoglycan will yield aldobiouronic acids but, owing to the relative stability to acid hydrolysis of the glycopyranosyluronic acid linkage, the possibility of obtaining oligosaccharides containing neutral glycose residues linked glycosidically to glycuronic acid residues is remote. However, graded hydrolysis of the reduced polymer may give neutral oligosaccharides in which the reducing end group arises from the original acid residue. These differences in "cracking" patterns may be even more marked if acetolysis is used for depolymerization in place of hydrolysis with aqueous mineral acid. For example, the isolation of 4-O-L-rhamnopyranosyl-D-glucose and O-L-rhamnopyranosyl-(1→4)-O-D-glucopyranosyl-(1→6)-D-galactose from carboxyl-reduced gum arabic [131, 132] and carboxyl-reduced *Araucaria bidwillii* gum [265] (see Appendix) would probably not have been achieved using partial hydrolysis with mineral acid. Other examples of differences in fragmentation patterns of acidic glycans and their carboxyl-reduced derivatives are quoted in the Appendix.

The reduction of carboxyl groups in glycuronoglycans may be carried out by the action of sodium borohydride on their methyl esters. Since reduction is accompanied by some ester hydrolysis, two or three cycles of esterification and reduction are normally required to effect almost complete conversion of carboxyl to primary alcohol groups. The formation of methyl esters may be achieved by treatment of the polysaccharide with methanolic hydrogen chloride if no acid sensitive glycosidic linkages are present and if limited depolymerization is of no consequence. Treatment of freeze-dried polysaccharides with ethereal diazomethane has been used as an alternative method of esterification. The possible complication of some accompanying etherification was not observed in the esterification of the *Pneumococcus* Type VIII polysaccharide [55] but has been reported on similar treatment of alginic acid [266]. Treatment of quaternary ammonium salts of heparin with diazomethane results in depolymerization by the β-elimination mechanism with the formation of unsaturated hexuronic acid units [200]. The generality of this type of degradation has not been established. Base-catalyzed degradation by β-elimination has not been reported when glycans containing 4-O-substituted hexuronic esters are treated with sodium borohydride in aqueous solution, but extensive depolymerization has been observed when sunflower pectic esters are reduced in dimethylsulfoxide solution [267].

Variations in the esterification and reduction procedure include (*1*) esterification with ethylene oxide to give 2-hydroxyethyl esters [268] followed by reduction with sodium borohydride [269], and (*2*) esterification with ethylene oxide or similar epoxides, acetylation to render the polysaccharide derivative soluble in tetrahydrofuran, (2-methoxyethyl) ether, or similar ethers, followed by reduction with lithium borohydride [266]. Although reduction is accompanied by de-*O*-acetylation and the glycan becomes insoluble, virtually complete reduction is achieved in one step and no formation of artifacts, for example, ethers, has been detected [266]. Comparison of various methods for the preparation of carboxyl-reduced alginic acid indicates that this latter is the preferred procedure [270]. Some depolymerization, possibly by β-elimination of 4-*O*-substituted hexuronic acid esters during the formation of 2-hydroxyethyl esters, has been observed during the treatment of pectic acid with ethylene oxide [271].

The reduction of carboxyl groups in preference to ester groups by diborane [272] has been used for the preparation of carboxyl-reduced acidic glycans [273]. Acetates or propionates of acidic glycans are treated with diborane in (2-methoxyethyl) ether, tetrahydrofuran, or similar solvent. Although artifact formation is probably less important if diborane is generated externally than *in situ*, the value of the procedure is limited by some reduction of protecting ester groups to ethers [274].

The information obtained from studies on carboxyl-reduced acidic glycans may be limited if a hexuronic acid and the corresponding hexose are constituents of the same polysaccharide, unless independent evidence is available to differentiate hexose residues formed by reduction of hexuronic acid from those originally present. Care is also required in the interpretation of the results of quantitative structural analyses on carboxyl-reduced acidic glycans in view of the experimental difficulties of effecting complete reduction without side reactions.

Hydrolytic Degradation of Oxidized polysaccharides

Oxidations of glycans at primary hydroxyl groups only are considered here, glycol-cleaving reactions having been discussed earlier.

Primary alcohol groups in carbohydrates can be selectively oxidized to carboxylic acid groups by gaseous oxygen in the presence of a platinum catalyst [275]. The selective oxidation may also be carried out on polysaccharides but reaction is very slow and rarely reaches completion. The oxidation of exposed primary alcohol groups in neutral polysaccharides affords glycuronoglycans which undergo graded hydrolysis to yield aldobiouronic acids having structural significance. Thus graded hydrolysis of catalytically oxidized rye flour arabinoxylan gave 3-O-(L-arabinofuranosy-luronic acid)-D-xylose [276] (see Appendix). Since it was known that the L-arabinofuranose residues in the unoxidized polysaccharide could be selectively hydrolyzed to leave a residual degraded xylan, the isolation of the aldobiouronic acid provides direct evidence that the L-arabinofuranose units are linked to the basal xylan chain as single units. Similar structural studies have been reported on European larch ε-galactan [277].

A similar approach has proved of value in the determination of the fine structure of hexopyranoglycans containing a high proportion of (1→6) linkages. Thus graded hydrolysis of the catalytically oxidized polysaccharides obtained from two dextrans from *Leuconostoc mesenteroides* (NRRL B-1416 and NRRL B-1415) containing, respectively, 10 and 14% of (1→3) and (1→4) linkages furnished the aldobiouronic acids 3-O- (**30**) and 4-O-(α-D-glucopyranosyluronic acid)-D-glucose (**31**) [278]. Since linkages other than of the (1→6) type are found only at branch points, it follows that the side chains in both dextrans consist of single D-glucose residues and may be represented in partial structures **32** and **33**. In contrast, catalytic oxidation of another dextran from *Leuconostoc mesenteroides* (NRRL B-512) containing 4% of (1→3) linkages followed by graded acid hydrolysis led to the isolation of the aldobiouronic acid, 6-O-(α-D-glucopyranosyluronic acid)-D-glucose (**34**), showing that in this polysaccharide the side chains contain two or more residues (**35**) [279].

The catalytic oxidation is carried out by passing oxygen through a stirred aqueous solution of the polysaccharide containing sufficient sodium hydrogen carbonate to neutralize the acids formed, for 2–6 weeks at 70°, in the presence of platinum catalyst (Adams platinic oxide after reduction with hydrogen). The catalyst is removed by centrifugation, and the oxidized polysaccharide is precipitated from the aqueous solution by the addition of ethanol and collected.

Alternative methods for the oxidation of primary alcohol groups in polysaccharides include the following. Dinitrogen tetroxide has been used for the oxidation of cellulose and appears to be relatively specific for primary hydroxyl groups [280]. Some degradation probably occurs and the procedure is likely to be unsuitable for glycans containing acid-sensitive linkages. A two-stage oxidation of suitably protected amylose involves treatment with chromium trioxide in acetic acid followed by oxidation with potassium permanganate and results in the formation of a glycan containing 50% of glucuronic acid residues [281]. In a similar oxidation of a partially methylated arabinogalactan oxidation of primary hydroxyl groups of galactopyranose, and possibly arabinofuranose, end groups was achieved using chromium trioxide in acetic acid only [178]. A two-stage procedure specific for the oxidation of primary hydroxyl groups of D-galactose residues employs

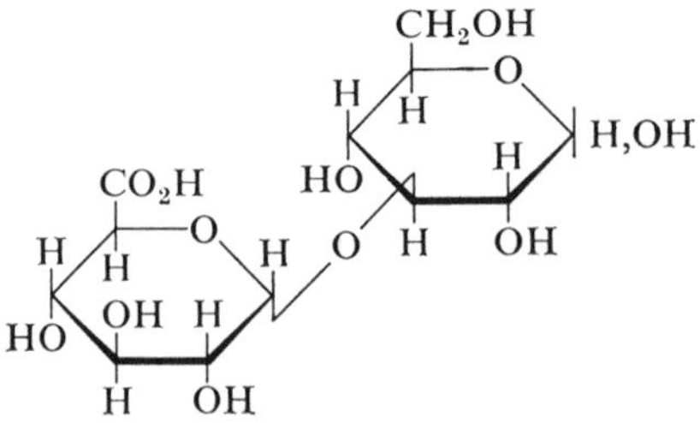

30

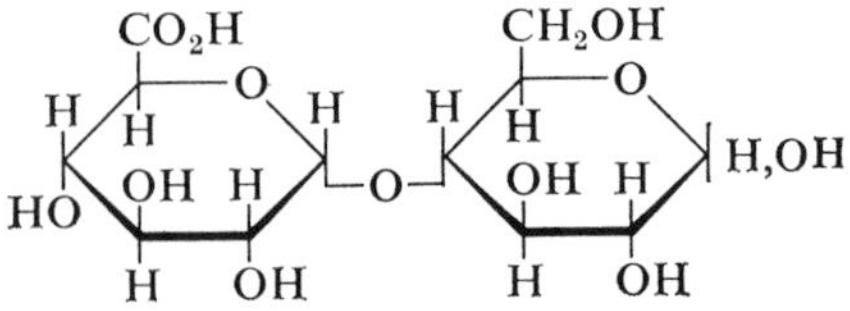

31

α-D-G*p* 1
    ↓
    3
...6 α-D-G*p* 1...

32

α-D-G*p* 1
    ↓
    4
...6 α-D-G*p* 1...

33

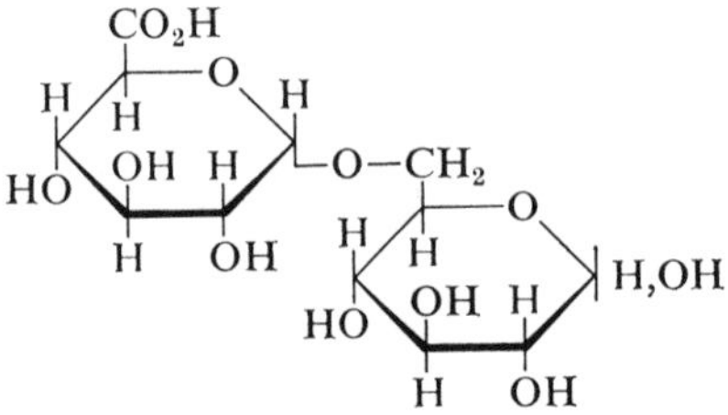

34

α-D-G*p* 1→6 α-D-G*p* 1...

35

D-galactose oxidase giving *aldehydo* sugar residues followed by treatment with sodium hypoiodite to give D-galacturonic acid residues [282]. Approximately 30% of the available D-galactopyranosyl end groups in guaran are oxidized without apparent side reactions. The oxidation, however, of glycans containing nonterminal D-galactose, for example, larch arabinogalactan, appears to be more complex and is accompanied by some degradation.

4 APPENDIX

The following examples illustrate ways in which different fragmentation procedures may be used for the isolation of degradation products with overlapping sequences of sugar residues from polysaccharides and other carbohydrate-containing macromolecules.

$$\longrightarrow -\begin{bmatrix} \beta\text{-D-Gal}p\ 1{\rightarrow}4\ \beta\text{-D-Xyl}p\ 1{\rightarrow}\text{Serine} \\ +\ \beta\text{-D-Xyl}p\ 1{\rightarrow}\text{Serine} \end{bmatrix}$$

(b)

$$\cdots 4\ \beta\text{-D-G}p\text{A}\ 1{\rightarrow}3\ \beta\text{-D-Gal}p\ 1{\rightarrow}3\ \beta\text{-D-Gal}p\ 1{\rightarrow}4\ \beta\text{-D-Xyl}p\ 1{\rightarrow}\text{Serine}$$

$\downarrow$ (a)

(c)

$$\beta\text{-D-G}p\text{A}\ 1{\rightarrow}3\ \text{D-Gal}$$

$$\begin{bmatrix} \beta\text{-D-Gal}p\ 1{\rightarrow}3\ \beta\text{-D-Gal}p\ 1{\rightarrow}4\ \text{D-Xyl}p \\ \beta\text{-D-Gal}p\ 1{\rightarrow}3\ \text{D-Gal} \\ \beta\text{-D-Gal}p\ 1{\rightarrow}4\ \text{D-Xyl}p \end{bmatrix} \longleftarrow$$

Example 15.1 Overlapping segments formed on partial acid hydrolysis of the heparin-protein linkage region (a) with N hydrochloric acid, (b) at pH 1.55, and (c) at pH 3 [17, 207, 283].

(a) $\cdots 4\ \beta\text{-D-G}p\ 1{\rightarrow}3\ \beta\text{-D-Gal}p\text{NAc}\ 1{\rightarrow}4\ \beta\text{-D-G}p\ 1{\rightarrow}3\ \beta\text{-D-Gal}p\text{NAc}\ 1\cdots$

$\downarrow$

$$\beta\text{-D-G}p\ 1{\rightarrow}3\ \text{D-GalNAc}$$

(b) $\cdots 4\ \beta\text{-D-G}p\ 1{\rightarrow}3\ \beta\text{-D-Gal}p\text{N}\ 1{\rightarrow}4\ \beta\text{-D-G}p\ 1{\rightarrow}3\ \beta\text{-D-Gal}p\text{N}\ 1\cdots$

$\downarrow$

$$\beta\text{-D-Gal}p\text{N}\ 1{\rightarrow}4\ \text{D-G}p$$

Example 15.2 Partial acid hydrolysis of (a) carboxyl-reduced chondroitin [284] and (b) N-deacetylated carboxyl-reduced chondroitin [285].

$$\beta\text{-D-Xyl}p\ 1{\rightarrow}4\ \text{D-Xyl}p$$

3

$\uparrow$

1

$\alpha\text{-L-Ara}f$

$\uparrow$ (a)

$$\cdots 4\ \beta\text{-D-Xyl}p\ 1{\rightarrow}4\ \beta\text{-D-Xyl}p\ 1{\rightarrow}4\ \beta\text{-D-Xyl}p\ 1{\rightarrow}4\ \beta\text{-D-Xyl}p\ 1\cdots$$

3

$\uparrow$

1

$\alpha\text{-L-Ara}f$

$\downarrow$ (b)

$$\cdots 4\ \beta\text{-D-Xyl}p\ 1{\rightarrow}4\ \beta\text{-D-Xyl}p\ 1{\rightarrow}4\ \beta\text{-D-Xyl}p\ 1{\rightarrow}4\ \beta\text{-D-Xyl}p\ 1\cdots$$

Degraded xylan

Example 15.3 (a) Enzymic and (b) partial acid hydrolysis of cereal arabinoxylans [60, 286].

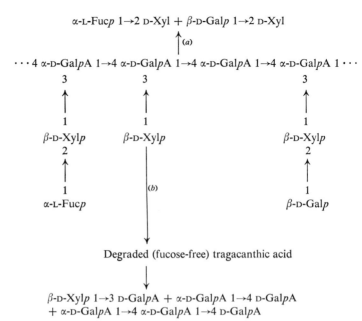

α-L-Fuc*p* 1→2 D-Xyl + β-D-Gal*p* 1→2 D-Xyl

↑(*a*)

··· 4 α-D-Gal*p*A 1→4 α-D-Gal*p*A 1→4 α-D-Gal*p*A 1→4 α-D-Gal*p*A 1 ···

3   3   3

↑   ↑   ↑

1   1   1

β-D-Xyl*p* β-D-Xyl*p* β-D-Xyl*p*

2   2

↑   ↑

1   (*b*)   1

α-L-Fuc*p*   β-D-Gal*p*

Degraded (fucose-free) tragacanthic acid

↓

β-D-Xyl*p* 1→3 D-Gal*p*A + α-D-Gal*p*A 1→4 D-Gal*p*A
+ α-D-Gal*p*A 1→4 α-D-Gal*p*A 1→4 D-Gal*p*A

Example 15.4 (*a*) Acetolysis and (*b*) partial acid hydrolysis followed by enzymic hydrolysis of tragacanthic acid [130].

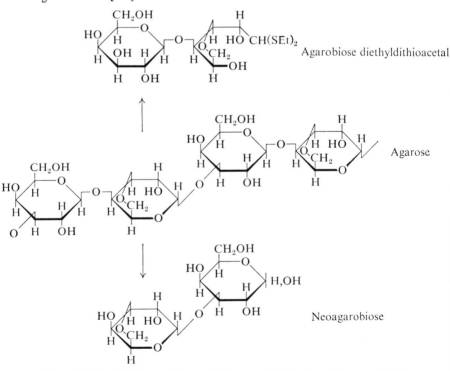

Agarobiose diethyldithioacetal

Agarose

Neoagarobiose

Example 15.5 (*a*) Mercaptolysis and (*b*) enzymic hydrolysis of agarose [287].

436

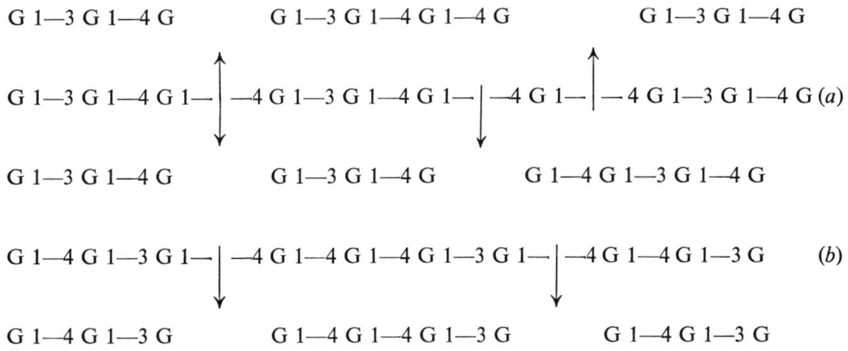

G 1—3 G 1—4 G   G 1—3 G 1—4 G 1—4 G   G 1—3 G 1—4 G

G 1—3 G 1—4 G 1—│ —4 G 1—3 G 1—4 G 1—│ —4 G 1—│ —4 G 1—3 G 1—4 G (*a*)

G 1—3 G 1—4 G   G 1—3 G 1—4 G   G 1—4 G 1—3 G 1—4 G

G 1—4 G 1—3 G 1—│ —4 G 1—4 G 1—4 G 1—3 G 1—│ —4 G 1—4 G 1—3 G   (*b*)

G 1—4 G 1—3 G   G 1—4 G 1—4 G 1—3 G   G 1—4 G 1—4 G 1—3 G

Example 15.6 Enzymic hydrolysis of oat and barley β-D-glucans with (*a*) cellulase from *Streptomyces* sp. QM B814 and (*b*) laminaranase from *Rhizopus arrhizus* QM 1032 [65].

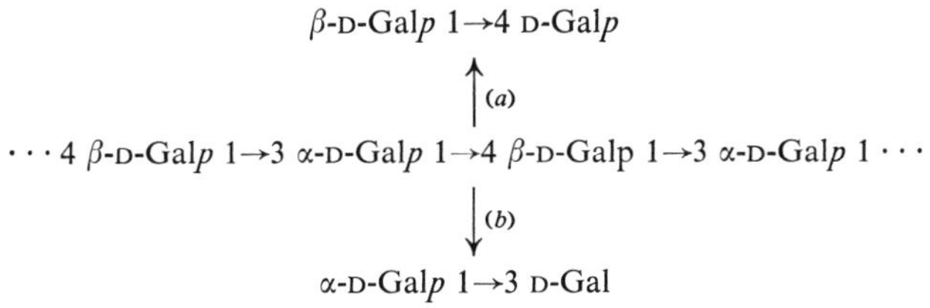

β-D-Gal*p* 1→4 D-Gal*p*

(*a*)

··· 4 β-D-Gal*p* 1→3 α-D-Gal*p* 1→4 β-D-Galp 1→3 α-D-Gal*p* 1 ···

(*b*)

α-D-Gal*p* 1→3 D-Gal

Example 15.7 (*a*) Partial acid hydrolysis and (*b*) acetolysis of desulfated λ-carrageenan [129].

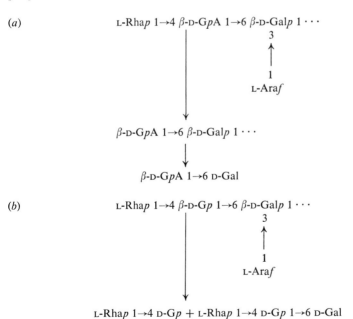

(*a*)   L-Rha*p* 1→4 β-D-G*p*A 1→6 β-D-Gal*p* 1 ···
                                    3
                                    ↑
                                    1
                              L-Ara*f*

β-D-G*p*A 1→6 β-D-Gal*p* 1 ···

β-D-G*p*A 1→6 D-Gal

(*b*)   L-Rha*p* 1→4 β-D-G*p* 1→6 β-D-Gal*p* 1 ···
                                   3
                                   ↑
                                   1
                             L-Ara*f*

L-Rha*p* 1→4 D-G*p* + L-Rha*p* 1→4 D-G*p* 1→6 D-Gal

Example 15.8 (*a*) Partial acid hydrolysis of gum arabic and (*b*) acetolysis of carboxyl-reduced gum arabic [6, 131, 132].

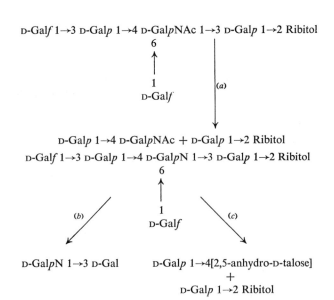

D-Galf 1→3 D-Galp 1→4 D-GalpNAc 1→3 D-Galp 1→2 Ribitol
6
↑
1
D-Galf   (a)

↓

D-Galp 1→4 D-GalpNAc + D-Galp 1→2 Ribitol
D-Galf 1→3 D-Galp 1→4 D-GalpN 1→3 D-Galp 1→2 Ribitol
6
↑
1
D-Galf

(b)   (c)

D-GalpN 1→3 D-Gal   D-Galp 1→4[2,5-anhydro-D-talose]
+
D-Galp 1→2 Ribitol

Example 15.9 (a) Partial acid hydrolysis of neutral (N-acetylated) hexasaccharide, (b) partial acid hydrolysis, and (c) nitrous acid deamination of basic (N-deacetylated) hexasaccharide from Pneumococcus type 10A specific substance [228].

(a)   · · · 4 β-D-GpA 1→2 α-D-Manp 1→4 β-D-GpA 1→2 α-D-Manp 1 · · ·

↓

β-D-GpA 1→2 D-Man
+
β-D-GpA 1→2 α-D-Manp 1→4 β-D-GpA 1→2 D-Man

(b)   · · · 4 β-D-Gp 1→2 α-D-Manp 1→4 β-D-Gp 1→2 α-D-Manp 1→4 β-D-Gp 1→2 α-D-Manp 1 · · ·

↓

β-D-Gp 1→2 D-Man + α-D-Manp 1→4 D-Gp
Gp 1→2 Manp 1→4 Gp + Manp 1→4 Gp 1→2 Man
Gp 1→2 Manp 1→4 Gp 1→2 Man + Manp 1→4 Gp 1→2 Manp 1→4 Gp
Gp 1→2 Manp 1→4 Gp 1→2 Manp 1→4 Gp

Example 15.10 (a) Partial acid hydrolysis of interior chains of leiocarpan A [287] and (b) acetolysis of carboxyl-reduced leiocarpan A [288].

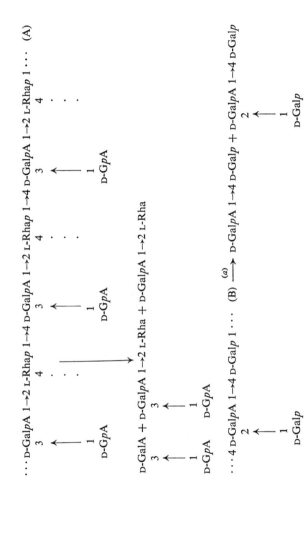

Example 15.11 (*a*) Partial acid hydrolysis of *Sterculia urens* gum, and acetolysis of (*b*) carboxyl-reduced *Sterculia urens* gum and (*c*) carboxyl-reduced *Sterculia urens* gum after degradation by Smith's procedure [289, 290]. (A) and (B) are partial structures of different regions in *Sterculia urens* gum.

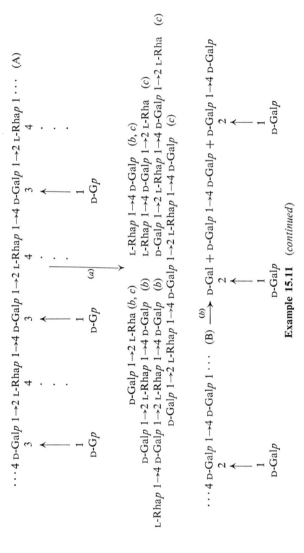

Example 15.11 (*continued*)

(a)

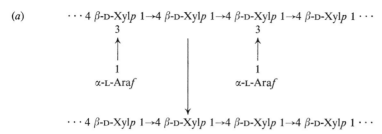

(b)

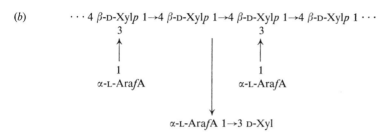

Example 15.12   (*a*) Partial acid hydrolysis of rye flour arabinoxylan [291] and (*b*) partial acid hydrolysis of oxidized rye flour arabinoxylan [276].

References

1. G. O. Aspinall, *Polysaccharides*, Pergamon, Oxford, 1970.
2. P. Karrer and C. Mägeli, *Helv. Chim. Acta*, **4**, 263 (1921).
3. W. J. Whelan, J. M. Bailey, and P. J. P. Roberts, *J. Chem. Soc.*, **1953**, 1293.
4. A. Thompson and M. L. Wolfrom, *J. Amer. Chem. Soc.*, **73**, 5489 (1951).
5. R. L. Whistler and D. F. Durso, *J. Amer. Chem. Soc.*, **73**, 4189 (1951).
6. F. Smith and R. Montgomery, *The Chemistry of Plant Gums and Mucilages*, (ACS Monograph No. 141), Reinhold, New York, 1959.
7. G. O. Aspinall, *Advan. Carbohyd. Chem.*, **24**, 333 (1969).
8. T. J. Painter, *Can. J. Chem.*, **37**, 497 (1959).
9. O. Perila and C. T. Bishop, *Can. J. Chem.*, **39**, 815 (1961).
10. T. J. Painter, in *Methods in Carbohydrate Chem.*, Vol. 5 (R. L. Whistler, ed.), Academic Press, New York, 1965, p. 280.
11. W. Dröge, O. Lüderitz, and O. Westphal, *Eur. J. Biochem.*, **4**, 126 (1968).
12. M. A. Swanson and C. F. Cori, *J. Biol. Chem.*, **172**, 797 (1948).
13. J. N. BeMiller, *Advan. Carbohyd. Chem.*, **22**, 25 (1967).
14. E. L. Hirst, *J. Chem. Soc.*, **1957**, 193.
15. K. K. De and T. E. Timell, *Carbohyd. Res.*, **4**, 177 (1967).
16. B. Capon and B. C. Ghosh, *Chem. Commun.*, **1965**, 586. B. Capon and D. Thacker, *J. Chem. Soc.*, *B*, **1967**, 185.
17. U. Lindahl, *Ark. Kemi*, **26**, 101 (1966).
18. O. Smidsrød, B. Larsen, T. Painter, and A. Haug, *Acta Chem. Scand.*, **23**, 1573 (1969).

19. D. A. Rees, N. G. Richardson, N. J. Wight, and Sir Edmund Hirst, *Carbohyd. Res.*, **9**, 451 (1969).
20. R. C. G. Noggridge and A. Neuberger, *J. Chem. Soc.*, **1938**, 745.
21. E. R. B. Graham and A. Neuberger, *J. Chem. Soc.*, *C*, **1968**, 1638.
22. W. T. Butler and L. W. Cunningham, *J. Biol. Chem.*, **240**, 3449 (1965); **241**, 3882 (1966).
23. L. W. Cunningham, J. D. Ford, and J. P. Segrest, *J. Biol. Chem.*, **242**, 2570 (1967).
24. R. G. Spiro, *J. Biol. Chem.*, **242**, 1923 (1967).
25. A. B. Foster, D. Horton, and M. Stacey, *J. Chem. Soc.*, **1957**, 81.
26. S. J. Pirt and W. J. Whelan, *J. Sci. Food Agr.*, **5**, 224 (1951).
27. V. D. Harwood, *J. Sci. Food Agr.*, **5**, 270 (1954).
28. P. J. Stoffyn and R. W. Jeanloz, *J. Biol. Chem.*, **235**, 2507 (1960).
29. E. L. Hirst and D. A. Rees, *J. Chem. Soc.*, **1965**, 1182.
30. A. S. Perlin, M. Mazurek, L. B. Jacques, and L. W. Kavanagh, *Carbohyd. Res.*, **7**, 369 (1968).
31. M. L. Wolfrom, S. Honda, and P. Y. Wang, *Chem. Commun.*, **1968**, 505.
32. D. H. Ball and J. K. N. Jones, *J. Chem. Soc.*, **1958**, 33.
33. A. B. Foster and D. Horton, *J. Chem. Soc.*, **1958**, 1890.
34. J. K. N. Jones and W. H. Nicholson, *J. Chem. Soc.*, **1958**, 27.
35. A. Thompson, E. Anno, M. L. Wolfrom, and M. Inatome, *J. Amer. Chem. Soc.*, **76**, 1309 (1954).
36. C. N. Turton, A. Bebbington, S. Dixon, and E. Pacsu, *J. Amer. Chem. Soc.*, **77**, 2565 (1955).
37. S. Peat, W. J. Whelan, T. E. Edwards, and O. Owen, *J. Chem. Soc.*, **1958**, 586.
38. M. L. Wolfrom and A. Thompson, *J. Amer. Chem. Soc.*, **77**, 6403; **78**, 4116 (1956).
39. M. L. Wolfrom and A. Thompson, *J. Amer. Chem. Soc.*, **79**, 4212 (1957).
40. S. Peat, J. R. Turvey, and J. M. Evans, *Nature*, **179**, 261 (1957).
41. D. J. Manners, G. A. Mercer, and J. J. M. Rowe, *J. Chem. Soc.*, **1965**, 2150.
42. M. L. Wolfrom, A. Thompson, and R. H. Moore, *Cereal Chem.*, **40**, 182 (1963).
43. K. Freudenberg, W. Kuhn, W. Dürr, F. Bolz, and G. Steinbrunn, *Chem. Ber.*, **63**, 1510 (1930).
44. N. Grassie, *The Chemistry of High Polymer Degradation Processes*, Interscience, New York-London, 1956.
45. W. Kuhn, *Chem. Ber.*, **63**, 1503 (1930).
46. K. Freudenberg, *Chemiker-Ztg.*, **60**, 853, 875 (1936).
46a. S. C. Charms, *Advan. Carbohyd. Chem. Biochem.*, **25**, 13 (1970).
47. G. O. Aspinall, E. L. Hirst, E. G. V. Percival, and R. G. J. Telfer, *J. Chem. Soc.*, **1953**, 337.
48. W. H. Haworth, H. Raistrick, and M. Stacey, *Biochem. J.*, **31**, 640 (1937).
49. C. Galanos, O. Lüderitz, and K. Himmelspach, *Eur. J. Biochem.*, **8**, 332 (1969).
50. P. A. J. Gorin and J. F. T. Spencer, *Can. J. Chem.*, **37**, 499 (1959).
51. W. H. Wadman, *J. Chem. Soc.*, **1952**, 3051.
52. I. H. Bath, *J. Sci. Food Agr.*, **11**, 560 (1960).

53. P. Albersheim, D. J. Nevins, P. D. English, and A. Karr, *Carbohyd. Res.*, **5,** 340 (1967).

54. J. K. N. Jones, *J. Chem. Soc.*, **1950,** 3292.

55. J. K. N. Jones and M. B. Perry, *J. Amer. Chem. Soc.*, **79,** 2787 (1957).

56. H. A. Spoehr, *Arch. Biochem. Biophys.*, **14,** 153 (1947).

57. P. Andrews, L. Hough, and J. K. N. Jones, *J. Chem. Soc.*, **1952,** 3393.

58. P. Andrews, L. Hough, and J. K. N. Jones, *J. Chem. Soc.*, **1953,** 1186. P. Andrews, L. Hough, and D. B. Powell, *Chem. Ind. (London)*, **1956,** 658.

59. J. K. N. Jones and K. C. B. Wilkie, *Can. J. Biochem. Physiol.*, **37,** 377 (1959).

60. G. O. Aspinall, *Advan. Carbohyd. Chem.*, **14,** 429 (1959).

60a. D. A. Hart and P. K. Kindel, *Biochemistry*, **9,** 2190 (1970).

61. P. A. Anastassiadis and R. H. Common, *Can. J. Biochem. Physiol.*, **36,** 413 (1958).

62. P. D. Boyer, H. Lardy, and K. Myrback, eds., *The Enzymes*, 2nd ed., Vol. 4, Academic Press, New York and London, 1960, Chapters 18–27.

63. E. F. Neufeld and V. Ginsburg, eds., *Methods in Enzymology*, Vol. 8, *Complex Carbohydrates*, Academic Press, New York, 1966.

64. G. J. Walker and W. J. Whelan, *Biochem. J.*, **76,** 264 (1960). W. F. H. M. Mommaerts, B. Illingworth, C. M. Pearson, R. J. Guillory, and K. Seraydarian, *Proc. Nat. Acad. Sci. U.S.*, **45,** 791 (1959).

65. F. W. Parrish, A. S. Perlin, and E. T. Reese, *Can. J. Chem.*, **38,** (1960).

66. C. T. Greenwood and E. A. Milne, *Advan. Carbohyd. Chem.*, **23,** 281 (1968).

67. O. Kjølberg and D. J. Manners, *Biochem. J.*, **86,** 258 (1963).

68. W. Banks and C. T. Greenwood, *Arch. Biochem. Biophys.*, **117,** 674 (1966).

69. I. M. Morrison and M. B. Perry, *Can. J. Biochem.*, **44,** 1115 (1966).

70. S. Peat, W. J. Whelan, and G. J. Thomas, *J. Chem. Soc.*, **1956,** 3025.

71. B. I. Brown and D. H. Brown, *Methods Enzymol.* **8,** 395 (1966).

72. H. Bender and K. Wallenfels, *Biochem. Z.*, **334,** 79 (1961).

73. M. Abdullah, B. J. Catley, E. Y. C. Lee, J. Robyt, K. Wallenfels, and W. J. Whelan, *Cereal Chem.*, **43,** 111 (1966).

74. G. K. Adkins, W. Banks, and C. T. Greenwood, *Carbohyd. Res.*, **2,** 502 (1966).

75. E. Y. C. Lee and W. J. Whelan, *Arch. Biochem. Biophys.*, **116,** 162 (1966).

76. J. Larner, B. Illingworth, G. T. Cori, and C. F. Cori, *J. Biol. Chem.*, **199,** 641 (1952).

77. G. N. Bathgate and D. J. Manners, *Biochem. J.*, **101,** 3C (1966).

77a. P. A. J. Gorin, J. F. T. Spencer, and D. E. Eveleigh, *Carbohyd. Res.*, **11,** 387 (1969). G. H. Jones and C. E. Ballou, *J. Biol. Chem.*, **244,** 1043 (1968). G. H. Jones and C. E. Ballou, *J. Biol. Chem.*, **244,** 1052 (1968).

77b. K. Tagawa and A. Kaji, *Carbohyd. Res.*, **11,** 947 (1969).

78. S. A. Barker, G. I. Pardoe, M. Stacey, and J. W. Hopton, *Nature*, **197,** 231 (1963).

79. S. A. Barker, P. J. Somers, and M. Stacey, *Carbohyd. Res.*, **3,** 261 (1966).

80. D. E. Koshland, Jr. and S. S. Stein, *J. Biol. Chem.*, **208,** 139 (1954).

81. A. Linker, K. Meyer, and P. Hoffman, *J. Biol. Chem.*, **219,** 13 (1956).

82. A. Markovitz, J. A. Cifonelli, and A. Dorfman, *J. Biol. Chem.*, **234,** 2343 (1958).

83. J. Ludowieg, B. Vennesland, and A. Dorfman, *J. Biol. Chem.*, **236**, 333 (1961).
84. H. Deuel and E. Stutz, *Advan. Enzymol.*, **20**, 341 (1958).
85. M. Lee and J. D. Macmillan, *Biochemistry*, **7**, 4005 (1968).
86. D. A. Rees, *Biochem. J.*, **81**, 347 (1961).
87. D. J. Tipper, J. L. Strominger, and J. C. Ensign, *Biochemistry*, **6**, 906 (1967)
88. A. Gottschalk, ed., 'Glycoproteins,' Elsevier, Amsterdam, 1966.
89. L. Hough and J. K. N. Jones, in *Methods in Carbohydrate Chem.*, Vol. 1 (R. L. Whistler and M. L. Wolfrom, eds.), Academic Press, New York, 1962, p. 21.
90. G. N. Kowkabany, *Advan. Carbohyd. Chem.*, **9**, 303 (1954).
91. G. W. Hay, B. A. Lewis, and F. Smith, *J. Chromatogr.*, **11**, 479 (1963).
92. C. T. Bishop, *Advan. Carbohyd. Chem.*, **19**, 95 (1964). J. S. Sawardeker, J. H. Sloneker, and A. R. Jeanes, *Anal. Chem.*, **37**, 1602 (1965).
93. S. C. Williams and J. K. N. Jones, *Can. J. Chem.*, **45**, 275 (1967).
94. M. B. Perry, *Can. J. Biochem.*, **42**, 451 (1964).
95. B. Radhakrishnamurthy, E. R. Dalferes, Jr., and G. S. Berenson, *Anal. Biochem.*, **24**, 397 (1968).
96. C. C. Sweeley, R. Bentley, M. Makita, and W. W. Wells, *J. Amer. Chem. Soc.*, **85**, 2497 (1963).
97. M. B. Perry and R. K. Hulyalkar, *Can. J. Biochem.*, **43**, 573 (1965).
98. D. Horton and T. Tsuchiya, *Carbohyd. Res.*, **5**, 426 (1967).
99. W. W. Binkley, *Advan. Carbohyd. Chem.*, **10**, 83 (1955).
100. J. E. Hodge and B. T. Hofreiter, in *Methods in Carbohydrate Chem.*, Vol. 1 (R. L. Whistler and M. L. Wolfrom, eds.), Academic Press, New York, 1962, p. 380.
101. Z. Dische, in *Methods in Carbohydrate Chem.*, Vol. 1 (R. L. Whistler and M. L. Wolfrom, eds.), Academic Press, New York, 1962, p. 477.
102. O. Samuelson, *Anal. Chim. Acta*, **38**, 163 (1967).
103. R. Derungs and H. Deuel, *Helv. Chim. Acta*, **37**, 657 (1954).
104. S. Gardell, *Acta Chem. Scand.*, **7**, 207 (1953).
105. O. Samuelson and H. Strömberg, *Carbohyd. Res.*, **3**, 89, (1966); *Acta Chem. Scand.*, **22**, 1252 (1968).
106. R. L. Whistler and J. N. BeMiller, in *Methods in Carbohydrate Chem.*, Vol. 1 (R. L. Whistler and M. L. Wolfrom, eds.), Academic Press, New York, 1962, p. 47.
107. R. L. Whistler and J. N. BeMiller, in *Methods in Carbohydrate Chem.*, Vol. 1 (R. L. Whistler and M. L. Wolfrom, eds.), Academic Press, New York, 1962, p. 42.
108. J. K. N. Jones and R. A. Wall, *Can. J. Chem.*, **38**, 2290 (1960).
109. Y. C. Lee and R. Montgomery, in *Methods in Carbohydrate Chem.*, Vol. 5 (R. L. Whistler, ed.), Academic Press, New York, 1965, p. 28.
110. B. Weissman, K. Meyer, P. Sampson, and A. Linker, *J. Biol. Chem.*, **208**, 417 (1954).
111. G. O. Aspinall and R. N. Fraser, *J. Chem. Soc.*, **1965**, 4318.
112. G. O. Aspinall, I. W. Cottrell, S. V. Egan, I. M. Morrison, and J. N. C. Whyte, *J. Chem. Soc.*, C, **1967**, 1071.

113. O. Samuelson and L. Wicterin, *Carbohyd. Res.*, **4**, 139 (1967).
114. R. W. Bailey and J. B. Pridham, *Advan. Carbohyd. Chem.*, **17**, 121 (1962).
115. J. M. Bobbitt, *Advan. Carbohyd. Chem.*, **11**, 1 (1956).
116. A. S. Perlin, *Advan. Carbohyd. Chem.*, **14**, 9 (1959).
117. N. K. Kochetkov, O. S. Chizhov, and N. V. Molodtsov, *Tetrahedron*, **24**, 5587 (1968). (a) G. S. Johnson, W. S. Ruliffson, and R. G. Cooks, *J. Chem. Soc. D*, **1970**, 587.
118. J. Kärkkäinen, *Carbohyd. Res.*, **11**, 247 (1969).
118a. J. Kärkkäinen, *Carbohyd. Res.*, **14**, 27 (1970).
119. R. L. Whistler, E. Heyne, and J. Bachrach, *J. Amer. Chem. Soc.*, **71**, 1476 (1951).
120. G. Braun, *Org. Syntheses*, Coll. Vol. **2**, 122 (1943).
121. G. O. Aspinall, R. B. Rashbrook, and G. Kessler, *J. Chem. Soc.*, **1958**, 215.
122. F. Smith and H. C. Srivastava, *J. Amer. Chem. Soc.*, **78**, 1404 (1956).
123. S. Peat, W. J. Whelan, and T. E. Edwards, *J. Chem. Soc.*, **1961**, 29.
124. J. R. Turvey and W. J. Whelan, *Biochem. J.*, **67**, 49 (1957).
125. I. J. Goldstein and W. J. Whelan, *J. Chem. Soc.*, **1962**, 170.
126. Y. C. Lee and C. E. Ballou, *Biochemistry*, **4**, 257 (1965).
127. K. Matsuda, H. Watanabe, K. Fujimoto, and K. Aso, *Nature*, **191**, 278 (1961).
128. S. Peat, J. R. Turvey, and D. Doyle, *J. Chem. Soc.*, **1961**, 3918.
129. C. J. Lawson and D. A. Rees, *J. Chem. Soc., C*, **1968**, 1301.
130. G. O. Aspinall and J. Baillie, *J. Chem. Soc.*, **1963**, 1714.
131. G. O. Aspinall, A. J. Charlson, E. L. Hirst, and R. Young, *J. Chem. Soc.*, **1963**, 1696.
132. G. O. Aspinall and R. Young, *J. Chem. Soc.*, **1965**, 3003.
133. G. O. Aspinall, J. A. Molloy, and C. C. Whitehead, *Carbohyd. Res.*, **12**, 143 (1970).
134. G. O. Aspinall, R. Begbie, and J. E. McKay, *J. Chem. Soc.*, **1962**, 214.
135. G. O. Aspinall and J. N. C. Whyte, *J. Chem. Soc.*, **1964**, 5058.
136. G. Zemplen and D. Kiss, *Chem. Ber.*, **60**, 165 (1927).
137. M. C. Cifonelli, R. Montgomery, and F. Smith, *J. Amer. Chem. Soc.*, **77**, 121 (1955).
138. C. Araki and K. Arai, *Bull. Chem. Soc. Japan*, **30**, 287 (1957), and earlier papers.
139. A. N. O'Neill, *J. Amer. Chem. Soc.*, **77**, 6324 (1955).
140. D. A. Rees, *J. Chem. Soc.*, **1963**, 1821.
141. S. Hirase, *Bull. Chem. Soc. Japan*, **30**, 75 (1957); S. Hirase and C. Araki, *Bull. Chem. Soc. Japan*, **26**, 463 (1953); **27**, 105 (1954).
142. R. G. Kantor and M. Schubert, *J. Amer. Chem. Soc.*, **79**, 152 (1957).
143. T. C. S. Dolan and D. A. Rees, *J. Chem. Soc.*, **1965**, 3534.
144. I. Danishefsky, H. Steiner, A. Bella, Jr., and A. Friedlander, *J. Biol. Chem.*, **244**, 1741 (1969).
145. M. L. Wolfrom, J. R. Vercellotti, and D. Horton, *J. Org. Chem.*, **29**, 547 (1964).
146. L. Malaprade, *C.R. Acad Sci., Paris*, **186**, 382 (1928); *Bull. Soc. Chim. Fr.*, **43**, 683 (1928).

147. R. D. Guthrie, *Advan. Carbohyd. Chem.*, **16**, 105 (1961). I. J. Goldstein, G. W. Hay, B. A. Lewis, and F. Smith, *Methods in Carbohydrate Chem.*, **5**, 361 (1965).
148. J. R. Dixon, J. G. Buchanan, and J. Baddiley, *Biochem. J.*, **100**, 507 (1966).
149. M. Abdel-Akher, J. J. Hamilton, R. Montgomery, and F. Smith, *J. Amer. Chem. Soc.*, **74**, 4970 (1952).
150. F. Smith and R. Montgomery, *Methods Biochem. Anal.*, **3**, 160 (1956).
151. J. F. O'Dea and R. A. Gibbons, *Biochem. J.*, **55**, 580 (1953).
152. C. T. Bishop and F. P. Cooper, *Can. J. Chem.*, **38**, 793 (1960).
153. G. G. S. Dutton and A. M. Unrau, *J. Chromatogr.*, **36**, 283 (1968).
154. C. F. Heubner, R. Lohmar, R. J. Dimler, S. Moore, and K. P. Link, *J. Biol. Chem.*, **159**, 505 (1945).
155. G. Neumüller and E. Vasseur, *Ark. Kemi*, **5**, 235 (1953).
156. L. Hough and M. B. Perry, *Chem. Ind.* (London), **1956**, 769.
157. D. A. Rees and J. W. B. Samuel, *J. Chem. Soc.*, C, **1967**, 2295.
158. B. Larsen and T. J. Painter, *Carbohyd. Res.*, **10**, 186 (1969). T. Painter and B. Larsen, *Acta Chem. Scand.*, **24**, 813 (1970).
159. R. J. Yu and C. T. Bishop, *Can. J. Chem.*, **45**, 2195 (1967).
160. R. Belcher, G. Dryhurst, and A. M. G. Macdonald, *J. Chem. Soc.*, **1965**, 3964, 4543.
161. G. W. Hay, B. A. Lewis, and F. Smith, in *Methods in Carbohydrate Chem.*, Vol. 5 (R. L. Whistler, ed.), Academic Press, New York, 1965, p. 357.
162. P. Fleury and J. Lange, *J. Pharm. Chim.*, [8]**17**, 107 (1933).
163. J. S. Dixon and D. Lipkin, *Anal. Chem.*, **26**, 1092 (1954).
164. G. O. Aspinall and R. J. Ferrier, *Chem. Ind.* (London), **1957**, 1216.
165. W. G. Breck, R. D. Corlett, and G. W. Hay, *Chem. Commun.*, **1967**, 604. R. D. Corlett, W. G. Breck, and G. W. Hay, *Can. J. Chem.*, **48**, 2474 (1970).
166. D. M. W. Anderson, C. T. Greenwood, and E. L. Hirst, *J. Chem. Soc.*, **1955**, 225.
167. A. S. Perlin, *J. Amer. Chem. Soc.*, **76**, 4101 (1954).
168. S. A. Barker and P. J. Somers, *Carbohyd. Res.*, **3**, 220 (1967).
169. L. Hough, D. B. Powell, and M. B. Perry, *J. Chem. Soc.*, **1956**, 4799.
170. T. Nash, *Biochem. J.*, **55**, 416 (1955).
171. H. C. Tun, J. F. Kennedy, M. Stacey, and R. R. Woodbury, *Carbohyd. Res.*, **11**, 225 (1969).
172. V. C. Barry, *Nature*, **152**, 537 (1943).
173. V. C. Barry and P. W. D. Mitchell, *J. Chem. Soc.*, **1954**, 4020.
174. V. C. Barry, J. E. McCormick, and P. W. D. Mitchell, *J. Chem. Soc.*, **1954**, 3692.
175. P. S. O'Colla, *Proc. Roy. Irish Acad.*, **B55**, 165 (1953).
176. T. Dillon, D. F. O'Ceallachain, and P. S. O'Colla, *Proc. Roy. Irish Acad.*, **B55**, 331 (1953); **B57**, 31 (1954).
177. P. A. J. Gorin and J. F. T. Spencer, *Can. J. Chem.*, **43**, 2978 (1965).
178. G. O. Aspinall, R. M. Fairweather, and T. M. Wood, *J. Chem. Soc.* C, **1968**, 2174.
179. V. Zitko and C. T. Bishop, *Can. J. Chem.*, **44**, 1749 (1966).

180. H. S. Isbell, *J. Research Natl. Bur. Standards*, **32**, 45 (1944).
181. R. L. Whistler and W. M. Corbett, *J. Amer. Chem. Soc.*, **77**, 3822, 6328 (1955).
182. B. Lindberg, O. Theander, and M. S. Feather, *Acta Chem. Scand.*, **20**, 206 (1966).
183. W. M. Corbett and J. Kenner, *J. Chem. Soc.*, **1954**, 3281.
184. J. Kenner and G. N. Richards, *J. Chem. Soc.*, **1956**, 2916.
185. R. L. Whistler and J. N. BeMiller, *Advan. Carbohyd. Chem.*, **13**, 289 (1958).
186. G. Machell and G. N. Richards, *J. Chem. Soc.*, **1957**, 4500.
187. G. Machell and G. N. Richards, *J. Chem. Soc.*, **1958**, 1199.
188. R. L. Whistler and J. N. BeMiller, *J. Org. Chem.*, **26**, 2886 (1961).
189. G. O. Aspinall and J. P. McKenna, Abstracts, 152nd National Meeting of the American Chemical Society, Sept. 1966, p. 14E.
190. G. O. Aspinall, C. T. Greenwood, and R. J. Sturgeon, *J. Chem. Soc.*, **1961**, 3667.
191. G. O. Aspinall and K. M. Ross, *J. Chem. Soc.* **1961**, 3774.
192. K. O. Lloyd, E. A. Kabat, and E. J. Layug, and F. Gruezo, *Biochemistry*, **5**, 1489 (1966).
193. B. Vollmert, *Makromol. Chem.*, **5**, 110 (1950).
194. H. Neukom and H. Deuel, *Beih. Z. Schweiz. Forstv.*, **30**, 223 (1960).
195. P. Heim and H. Neukom, *Helv. Chim. Acta*, **45**, 1735, 1737 (1962).
196. P. Albersheim, H. Neukom, and H. Deuel, *Arch. Biochem. Biophys.*, **90**, 46 (1960).
197. A. J. Barrett and D. H. Northcote, *Biochem. J.*, **94**, 617 (1965).
198. R. W. Stoddart, A. J. Barrett, and D. H. Northcote, *Biochem. J.*, **102**, 194 (1967).
199. C. W. McCleary, D. A. Rees, J. W. B. Samuel, and I. W. Steele, *Carbohyd. Res.*, **5**, 492 (1967). C. J. Lawson, C. W. McCleary, H. I. Nakada, D. A. Rees, I. W. Sutherland, and J. F. Wilkinson, *Biochem. J.*, **115**, 947 (1969).
200. J. Kiss, Abstracts of the 5th International Symposium on the Chemistry of Natural Products, London, 1968, E9, p. 249.
201. H. E. Carter, D. R. Strobach, and J. N. Hawthorne, *Biochemistry*, **8**, 383 (1969).
202. H. E. Carter, A. Kisic, J. L. Koob, and J. A. Martin, *Biochemistry*, **8**, 389 (1969).
203. B. Anderson, P. Hoffman, and E. Meyer, *J. Biol. Chem.*, **240**, 156 (1965).
204. E. Tanaka and W. Pigman, *J. Biol. Chem.*, **240**, PC1487 (1965).
205. R. Carubelli, V. P. Bhavanandan, and A. Gottschalk, *Biochim. Biophys. Acta*, **101**, 67 (1965).
206. U. Lindahl and L. Rodén, *J. Biol. Chem.*, **241**, 2113 (1966).
207. U. Lindahl, *Biochim. Biophys. Acta*, **130**, 361 (1966).
208. B. Anderson, N. Seno, P. Sampson, J. G. Riley, P. Hoffman, and K. Meyer, *J. Biol. Chem.*, **239**, PC-2716 (1964).
209. E. R. B. Graham and A. Gottschalk, *Biochim. Biophys. Acta*, **38**, 513 (1960).
210. A. Gottschalk and W. H. Murphy, *Biochim. Biophys. Acta*, **46**, 81 (1961).
211. K. O. Lloyd and E. A. Kabat, *Proc. Nat. Acad. Sci. U.S.*, **61**, 1470 (1968).

212. W. M. Watkins, *Science*, **152**, 172 (1966).
213. T. J. Painter, W. M. Watkins, and W. T. J. Morgan, *Nature*, **206**, 594 (1965).
214. G. Schiffman, E. A. Kabat, and W. Thompson, *Biochemistry*, **3**, 113, 587 (1964).
215. K. O. Lloyd, E. A. Kabat, and E. Licerio, *Biochemistry*, **7**, 2976 (1968).
216. C. Hatanaka and J. Osawa, *Ber. Chara Inst. Landw. Biol., Okayama Univ.*, **13**, 161 (1966).
217. H. O. Bouveng, *Acta Chem. Scand.*, **15**, 96 (1961).
217a. A. N. de Belder and B. Normann, *Carbohyd. Res.*, **8**, 1 (1968).
217b. C. G. Hellerquist, B. Lindberg, S. Svensson, T. Holm, and A. A. Lindberg, *Carbohyd. Res.* **8**, 43 (1968).
218. E. E. Percival and J. K. Wold, *J. Chem. Soc.*, **1963**, 5459.
219. N. S. Anderson and D. A. Rees, *J. Chem. Soc.* **1965**, 5880.
219a. P. A. J. Gorin and J. F. T. Spencer, *Carbohyd. Res.*, **13**, 339 (1970).
220. A. R. Archibald and J. Baddiley, *Advan. Carbohyd. Chem.*, **21**, 323 (1966).
221. M. J. How, J. S. Brimacombe, and M. Stacey, *Advan. Carbohyd. Chem.*, **19**, 303 (1964).
222. D. M. Brown in M. Florkin and E. H. Stotz, eds., *Comprehensive Biochemistry* Vol. 8, Elsevier, Amsterdam, 1963, p. 209.
223. D. M. Brown, *Advan. Org. Chem.*, **3**, 75 (1962).
224. J. R. Cox and O. B. Ramsey, *Chem. Rev.*, **64**, 317 (1964).
225. P. A. Rebers and M. Heidelberger, *J. Amer. Chem. Soc.*, **81**, 2415 (1959); **83**, 3056 (1961).
226. K. H. Meyer and H. Wehrli, *Helv. Chim. Acta*, **20**, 361 (1937).
227. A. B. Foster, E. F. Martlew, and M. Stacey, *Chem. Ind.* (London), **1953**, 825.
228. E. V. Rao, J. G. Buchanan, and J. Baddiley, *Biochem. J.* **100**, 801 (1966).
229. E. V. Rao, M. J. Watson, J. G. Buchanan, and J. Baddiley, *Biochem. J.*, **111**, 547 (1969).
230. P. A. J. Gorin, J. F. T. Spencer, and D. W. S. Westlake, *Can. J. Chem.*, **39**, 1067 (1961).
231. S. Haworth, D. M. Jones, J. G. Roberts, and B. F. Sagar, *Carbohyd. Res.*, **10**, 1 (1969).
232. J. L. Frahn, E. L. Hirst, D. F. Packman, and E. G. V. Percival, *J. Chem. Soc.*, **1951**, 3489.
233. E. L. Hirst and E. Percival, in *Methods in Carbohydrate Chem.*, Vol. 5 (R. L. Whistler, ed.), Academic Press, New York, 1965, p. 287.
234. W. N. Haworth, *J. Chem. Soc.*, **107**, 1 (1915).
235. J. K. Hamilton and H. W. Kircher, *J. Amer. Chem. Soc.*, **80**, 4703 (1958).
236. R. Kuhn and H. Trischmann, *Chem. Ber.*, **96**, 284 (1963).
237. H. C. Srivastava, P. P. Singh, S. N. Garshe, and V. Kirk, *Tetrahedron Lett.* **1964**, 493.
238. T. Purdie and J. C. Irvine, *J. Chem. Soc.*, **83**, 1021 (1903).
239. R. Kuhn, H. Trischmann, and I. Löw, *Angew. Chem.*, **67**, 32 (1955).
240. S. Hakomori, *J. Biochem.* (Japan), **55**, 205 (1964).
241. P. A. Sandford and H. E. Conrad, *Biochemistry*, **5**, 1508 (1966).

242. K. Wallenfels, G. Bechtler, R. Kuhn, H. Trischmann, and H. Egge, *Angew. Chem. Int. Ed.*, **2**, 515 (1963).
243. C. M. Fear and R. C. Menzies, *J. Chem. Soc.*, **1926**, 937.
244. E. L. Hirst and J. K. N. Jones, *J. Chem. Soc.*, **1939**, 1482.
245. I. E. Muskat, *J. Amer. Chem. Soc.*, **56**, 693 (1934).
246. H. S. Isbell, H. L. Frush, B. H. Bruckner, G. N. Kowkabany, and G. Wampler, *Anal. Chem.*, **29**, 1523 (1957).
247. L. Hough and J. K. N. Jones, *Chem. Ind.* (London), **1952**, 380.
248. I. O. Mastronardi, S. M. Flematti, J. C. Deferrari, and E. G. Cros, *Carbohyd. Res.*, **3**, 177 (1966).
249. M. Zinbo and T. E. Timell, *Svensk Papperstidn.*, **68**, 647 (1965).
250. W. H. Haworth, E. L. Hirst, and E. G. V. Percival, *J. Chem. Soc.*, **1932**, 2384.
251. N. S. Anderson, T. C. S. Dolan, and D. A. Rees, *J. Chem. Soc.*, *C*, **1968**, 596.
252. A. B. Foster, in *Methods in Carbohydrate Chem.*, Vol. 1 (R. L. Whistler and M. L. Wolfrom, eds.), Academic Press, New York, 1962, p. 51.
253. J. K. Bartlett, L. Hough, and J. K. N. Jones, *Chem. Ind.* (London), **1951**, 76.
254. M. Dubois, K. A. Gilles, J. K. Hamilton, P. Rebers, and F. Smith, *Anal. Chem.*, **28**, 350 (1956).
255. S. K. Chanda, E. L. Hirst, J. K. N. Jones, and E. G. V. Percival, *J. Chem. Soc.*, **1950**, 1289.
256. G. O. Aspinall, *J. Chem. Soc.*, **1963**, 1676.
257. R. Kuhn and H. Egge, *Chem. Ber.*, **96**, 3338 (1963).
258. S. S. Bhattacharjee and P. A. J. Gorin, *Can. J. Chem.*, **47**, 1207 (1969).
259. G. O. Aspinall and J. A. Molloy, *J. Chem. Soc.*, *C*, **1968**, 2994.
260. G. O. Aspinall, J. A. Molloy, and J. W. T. Craig, *Can. J. Biochem.*, **47**, 1063, (1969).
261. H. Björndal, B. Lindberg, and S. Svensson, *Carbohyd. Res.*, **5**, 433 (1967).
261a. H. Björndal and B. Lindberg, *Carbohyd. Res.*, **12**, 29 (1970).
262. N. K. Kochetkov and O. S. Chizhov, *Advan. Carbohyd. Chem.*, **21**, 39 (1966).
263. J. N. BeMiller, in *Methods Carbohyd. Chem.*, Vol. 5 (R. L. Whistler, ed.), Academic Press, New York, 1965, p. 298.
264. T. G. Bonner, E. J. Bourne, and S. McNally, *J. Chem. Soc.*, **1960**, 2929.
265. G. O. Aspinall and J. P. McKenna, *Carbohyd. Res.*, **7**, 224, (1968).
266. D. A. Rees and J. W. B. Samuel, *Chem. Ind.* (London), **1965**, 2008.
267. V. Zitko and C. T. Bishop, *Can. J. Chem.*, **44**, 1275 (1966).
268. H. Deuel, *Helv. Chim. Acta*, **30**, 1523 (1947).
269. G. O. Aspinall and A. Cañas-Rodriquez, *J. Chem. Soc.*, **1958**, 4020.
270. J. H. Manning and J. W. Green, *J. Chem. Soc.*, *C*, **1967**, 2357.
271. D. A. Rees and I. W. Steele, unpublished results.
272. H. C. Brown and B. C. Subba Rao, *J. Amer. Chem. Soc.*, **82**, 681 (1960).
273. F. Smith and A. M. Stephen, *Tetrahedron Lett.*, **1960**(7), 17, (1960).
274. E. L. Hirst, E. Percival, and J. K. Wold, *J. Chem. Soc.*, **1964**, 1493.
275. C. L. Mehltretter, *Advan. Carbohyd. Chem.*, **8**, 231 (1953).
276. G. O. Aspinall and I. M. Cairncross, *J. Chem. Soc.*, **1960**, 3998.
277. G. O. Aspinall and A. Nicolson, *J. Chem. Soc.*, **1960**, 2503.
278. D. Abbott, E. J. Bourne, and H. Weigel, *J. Chem. Soc.*, *C*, **1966**, 827.

279. B. Lindberg and S. Svensson, *Acta Chem. Scand.*, **22**, 1907 (1968).
280. P. A. McGee, W. F. Fowler, Jr., E. W. Taylor, C. C. Unruh, and W. O. Kenyon, *J. Amer. Chem. Soc.*, **69**, 355 (1947).
281. Y. Hirasaka, *Yakugaku Zasshi*, **83**, 976 (1963); *Chem. Abstr.*, **60**, 4232 (1964).
282. J. K. Rogers and N. S. Thompson, *Carbohyd. Res.*, **7**, 66 (1968).
283. U. Lindahl and L. Rodén, *J. Biol. Chem.*, **240**, 2821 (1965).
284. M. L. Wolfrom and B. O. Juliano, *J. Amer. Chem. Soc.*, **82**, 1673 (1960).
285. K. Onodera, T. Komano, and S. Hirano, *Biochim. Biophys. Acta*, **83**, 20 (1964).
286. C. T. Bishop, *J. Amer. Chem. Soc.*, **78**, 2840 (1956).
287. G. O. Aspinall, J. J. Carlyle, J. M. McNab, and A. Rudowski, *J. Chem. Soc.*, C, **1969**, 840.
288. G. O. Aspinall and J. M. McNab, *J. Chem. Soc.*, C, **1969**, 845.
289. G. O. Aspinall and Nasiruddin, *J. Chem. Soc.*, **1965**, 2710.
290. G. O. Aspinall and G. R. Sanderson, *J. Chem. Soc. C*, **1970**, 2256, 2259.
291. G. O. Aspinall and R. J. Sturgeon, *J. Chem. Soc.*, **1957**, 4469.

Chapter **XVI**

DEGRADATION OF POLYPEPTIDES AND PROTEINS

Angelo Fontana and Ernesto Scoffone

I INTRODUCTION

Protein chemistry has recently undergone a large quantum jump as the reports of the complete structure of γG immunoglobulin [1] (1320 amino acids) and of catalase [2] (505 amino acids) and the *Atlas of Protein Sequence and Structure* [3] illustrate.

Since the primary structure of a protein determines the architecture of the molecule as a whole, including its conformational specificities and in the long run its biological properties, determination of the amino acid sequence is a prerequisite for progress in the physicochemical and biological investigations of proteins.

In this chapter we shall be concerned with methods for the structure elucidation of the protein molecule and in particular degradative techniques.

Although well established paths for this purpose do exist, nevertheless some less generally used techniques will be discussed.

The problem of the primary structure elucidation is so important and the aims are so varied that it is necessary to have available several fundamentally different methods, not only for the sake of mutual control, but also for the solution of particular problems.

Since a complete coverage of all aspects of the degradation of peptides warrants a book rather than a chapter, the reader will find a somewhat personal selection of subject matter, too brief accounts on important methods, and omissions as well. Despite the space limitations, the present chapter was organized in order to provide a relative coherent presentation of the techniques of degradation of peptides and proteins. The availability of recent reviews allowed us to mention briefly used procedures, referring to the pertinent literature.

Recent contributions in the field will be reviewed in detail in order to guide the reader into the current literature. Parts of the field have been reviewed by Witkop [4], who has outlined the principles of selective chemical cleavage. Many aspects of the degradation of polypeptides can be found in the reviews of Thompson [5], of Spencer [6], and in particular in the excellent review of Spande et al. [7]. Reviews by Canfield and Anfinsen [8] and by Smyth and Elliott [9] cover the general problem of sequence work briefly, whereas specific experimental procedures are given in the volume entitled *Enzyme Structure* [10], edited by Hirs, as well as by Bailey [11] and Schroeder [12].

2 END GROUP ANALYSIS

The determination of the N- and C-terminal residues of a protein is a traditional beginning for the elucidation of its sequence. End group analysis gives information if the protein consists of two or more polypeptide chains held together by disulfide bridges (insulin, α-chymotrypsin, immunoglobulin, etc.), and the number of the N-terminal residues will prompt the separation of the chains before starting any sequence work. In addition, the determination of the N-terminal residue is a useful criterion of the homogeneity of the protein sample.

During the chemical or enzymic degradation of a protein, the N-terminal group determination will show the number of newly formed peptides, as well as the amino acids adjacent to the residues where cleavage has occurred.

Principles and techniques of the N-group determination have been extensively discussed elsewhere [10, 11, 13], and only the recent advances in the field will be covered here.

N-Terminal Group Determination

A common feature of all chemical methods for determining the N-terminal amino acid residue is the attachment to the α-amino function at a marker group that will be identified by its color or fluorescence in the presence of all other amino acids after total hydrolysis of the labeled polypeptide chain. Alternatively, N-terminal amino acid may be removed by a suitable reagent without altering the remainder of the chain, thus allowing the stepwise degradation of the polypeptide.

Enzymic methods are also available for the identification of the N-terminal residues. Leucine aminopeptidase releases amino acids sequentially from the amino terminus of a peptide, so that a kinetic study of the amino acids released gives an indication on the N-terminal residue as well as the sequence of the first few amino acids [14].

Arylation

Aromatic halogen compounds with activating substituents, such as nitro groups, react with proteins with the arylation of any or all of the following functional groups: α- and ε-amino, thiol, imidazole, and phenolic hydroxyl [10, 15]. Reaction with thioethers or indolic side chains does not occur. Recently Shaltiel [16] found that dinitrophenyl groups can be displaced selectively from all reactive sites other than amines by 2-mercaptoethanol.

DINITROFLUOROBENZENE

The classical reagent for the N-terminal group determination is 2,4-dinitrofluorobenzene (FDNP) (1) [17–20]. The method involves the formation in alkaline media of an N-dinitrophenyl derivative (2) of a peptide, and the release by hydrolysis with acid of the dinitrophenyl derivative (3) of the amino acid originally present in the N-terminal position. The method has been widely applied and extensive reviews on the subject are available [11, 21–23].

The separation and identification of the DNP-amino acids may be achieved chromatographically on paper [24], thin layer on silica [25, 26] or polyamides [27], or on columns of silicic acid [28, 29]. Liquid–liquid partition has been used in the countercurrent separation of DNP-lysines [30]. The water-soluble DNP-amino acids have been determined quantitatively by elution from IRC-50 column with citrate buffer, pH 5.0. Quantitation has been achieved by recording the yellow color in the 360 mμ region [31].

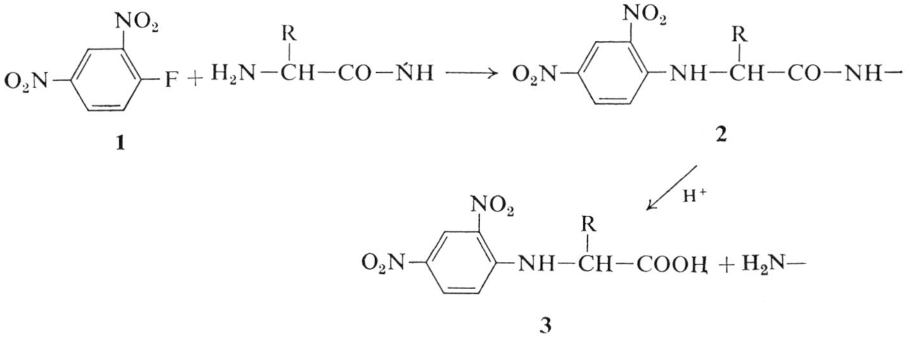

The identification of the DNP-amino acids by mass spectrometry, after conversion to their methyl esters, has been used by Penders [32]. Loss of the methoxycarbonyl group gives an intense peak of the amine component (M minus $COOCH_3$). The advantage of the method lies in the very low quantity of material needed for analysis (micrograms) and in the possibility of analyzing a mixture of DNP derivatives.

Pisano [33] has separated DNP-amino acid methyl esters by gas chromatography. Although unsatisfactory results have been obtained with some DNP-amino acids, the speed and sensitivity of the method suggest that this technique could be of great importance when only very small amounts of material are available.

In order to enhance the sensitivity of the DNP-method, 2,4-dinitro-5-fluoroaniline (DNFA) [34], which has several advantages over the Sanger reagent, has been used. While maxima occur at 350 mμ in the dinitrophenyl derivatives, DNFA shifts the maxima to 400–410 mμ. Moreover these derivatives may be diazotized and coupled to produce azo dyes applicable to detection of very small amounts of amino acids. Dinitroaminophenyl derivatives crystallize easily, and the spread of R_f values in chromatography is quite satisfactory using water–saturated butanol.

The weakest point in the DNP method is the destruction of the N-terminal DNP-amino acid which occurs under the conditions of acid hydrolysis necessary to detach it from the protein. The optimum time of hydrolysis for an end group analysis is the time at which DNP-peptides are no longer present; this is different with every protein, but overnight hydrolysis (16–24 hr) with constant boiling HCl is routine. The most affected DNP-derivatives are those of glycine, cystine, proline, and hydroxyproline [21, 34]. The extent of destruction (25–90%) may also vary from protein to protein and is generally much greater than would be expected for any free DNP-amino acid under the same conditions of hydrolysis in the absence of protein. The acid decomposition of DNP-amino acids is rather complicated: however, among the breakdown products 2,4-dinitrophenol and, to a lesser extent, 2,4-dinitroaniline predominate [21].

Attention should be given to the photodecomposition of DNP-amino acids so that the coupling reaction, hydrolysis, and subsequent chromatographic separation should be preferably carried out in the absence of the light [35, 36]. Thus although the DNP-method has been, and will doubtless continue to be, widely used for end group analysis, it is slowly being superseded by other methods.

NITRO-HALOGENO-PYRIDINES

Alternative procedures to the DNP method involve the use of other arylating agents capable of providing, upon condensation with the α-amino group, anchimeric assistance of the peptide bond cleavage [37, 38].

Nitropyridyl derivatives of peptides (4) are of interest because the heterocyclic nitrogen atom catalyses the hydrolysis of the peptide bond [39–43].

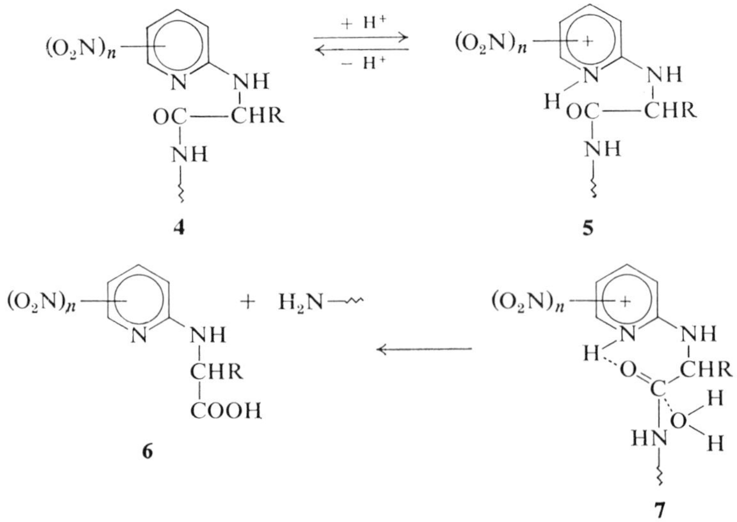

Transition state

A comparison of the rate constants of hydrolysis for 2,4-dinitrophenyl-alanyl-glycine and 3,5-dinitropyridyl-2-alanyl-glycine is significant in considering the behavior of the pyridine derivative; the hydrolysis of the latter is 10^2-10^3 times as fast as that of the DNP-derivative with a comparable concentration of hydrogen ion [43].

Recently Signor et al. [44] have established that the acid hydrolysis of 3- and 5-nitro-2-pyridyl peptides (4) is anchimerically assisted and the reaction proceeds quickly in dilute hydrochloric acid; the pH-rate profiles parallel the ionization curves of the heterocyclic nitrogen atom (5). Because of this neighboring group participation, highly preferential cleavage of the peptide bond at the end of the chain is possible.

In earlier studies [39–43] on the reactivity of some haloheteroaromatic derivatives, it was observed that 2-chloro-3,5-dinitropyridine condenses readily with amino acids and peptides; mononitrochloropyridines, however, are not very reactive and failed to condense under the mild conditions required. Although not as reactive as 2-chloro-3,5-dinitropyridine, mono-nitrofluoropyridines couple with peptides and proteins smoothly in slightly alkaline solution to give practically quantitative yields of the derivatives within 24 hr at 40°.

The nitropyridyl-amino acids (6) are almost quantitatively recovered by hydrolyzing the arylated peptide at 60° in 6N HCl for 2 hr or in 90% formic acid at 100° for 2 hr. They are bright yellow compounds, facilitating chromatographic [45, 46] and photometric determinations.

The possibility of regenerating the parent amino acids from the derivatives so obtained offers further advantages since amino acids can be identified and estimated accurately. Most of the mononitropyridyl derivatives can be converted into the corresponding free amino acids in moderate yields by means of hydrolysis with 2N NaOH at 100°. The derivatives obtained from 2-chloro-3,5-dinitropyridine are even more useful from this point of view, since quantitative recovery of the free amino acids is possible in all cases. Thus the methods involving the mono- and dinitropyridyl derivatives are complementary, the former being more easily hydrolyzed but the latter facilitates identification and estimation.

The nitro-halogeno-pyridines have also been used in a novel approach for the stepwise degradation of peptides.

HOLLEY-HOLLEY METHOD

Holley and Holley [47] investigated the intramolecular participation of the anilino group in splitting the peptide bond. The method involves coupling of peptides with 4-carbomethoxy-2-nitrofluorobenzene (8); then the nitro group can be converted in an amino group by catalytic hydrogenation.

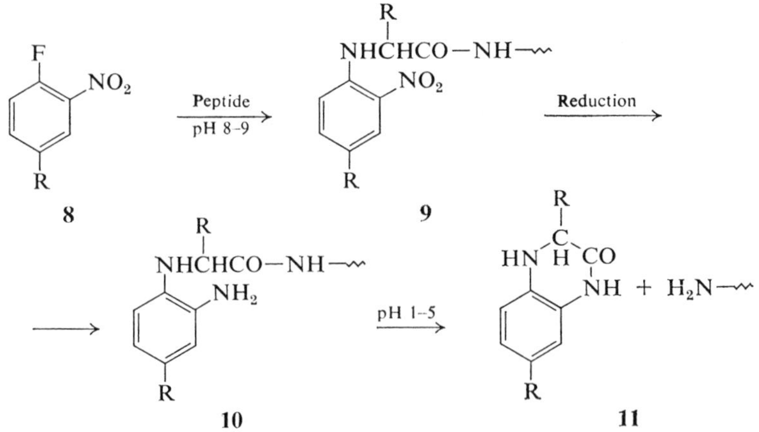

R = -COOCH₃; -NO₂ rendered as $R = -COOCH_3; -NO_2$

Scheme (structures 8, 9, 10, 11):

8 → (Peptide, pH 8–9) → 9 → (Reduction) → 10 → (pH 1–5) → 11 + H_2N—∿

$R = -COOCH_3;\ -NO_2$

An easy and rapid conversion of **10** into the dihydroquinoxalone **11** and into a shorter peptide can then be obtained under mildly acidic conditions. As an extension of the principle, the coupling agent was modified by variation of the 4-substituent [48–52]. By the use of ammonium sulfide the readily accessible 2,4-dinitrophenyl-peptide could be reduced selectively at the 2-nitro group [50]. The product was then cyclized at pH 2–3 by heating at 70° for 15 min to give 7-nitro-3,4-dihydro-2-hydroxyquinoxalone (**11**). Despite the ease and efficiency of the cyclization procedure, the method has not found wide acceptance, principally since it requires a reduction step which can cause complications. The availability of a reagent with a

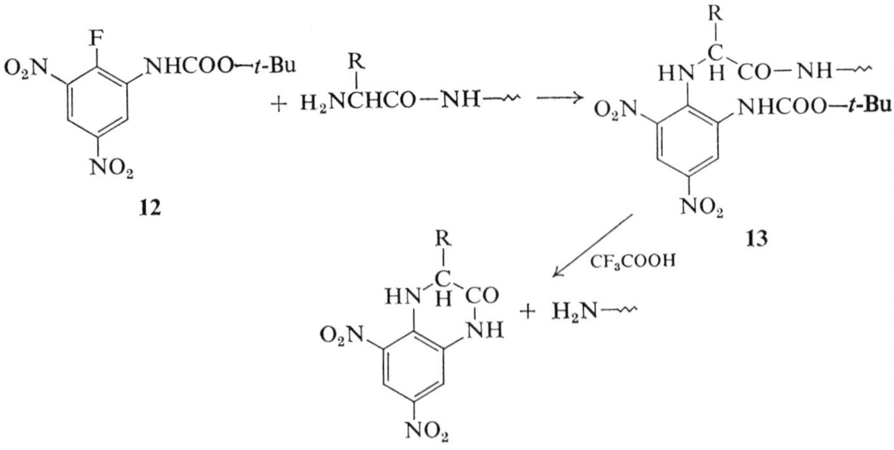

Scheme (structures 12, 13, 14):

12 + H_2NCHCO—NH—∿ → **13** → (CF₃COOH) → **14** + H_2N—∿

preexisting nucleophile at position 2 would simplify the procedure measurably by eliminating the need for a reduction step.

Recently Kirk and Cohen [53] further investigated possible improvements of the method by using butyl-3,5-dinitro-2-fluorocarbamilate (12) as a peptide reagent.

After the coupling of 12 with the peptide, treatment of the arylated compound 13 with trifluoroacetic acid results in the removal of the *t*-butyloxycarbonyl group, and at the same time ring closure to a dihydroquinoxalone derivative (14). Unfortunately a side reaction precluded the application of the method for the sequential degradation of peptides. The derivative 13 undergoes a cyclization to the benzimidazolone 14a, in alkaline media, precluding the subsequent peptide bond cleavage.

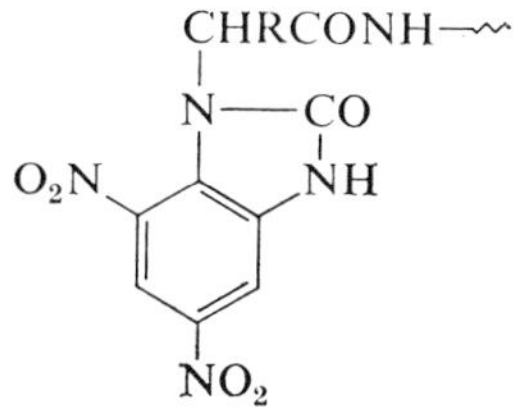

CHRCONH—⌇

N——CO

O₂N   NH

NO₂

14a

Although further modification of the reagent will be necessary to eliminate such a side reaction in the coupling step, the facility and completeness of the cyclization step, as well as the ease of identifying and assaying the resulting amino acid derivative, indicate that further effort is warranted.

Another way to get effective intramolecular participation for peptide bond cleavage involves the use of fluoronitrobenzimidazole [54]. 5,7-Dinitro-4-fluorobenzimidazole (15) is 84 times more reactive than 2,4-dinitrofluoro-

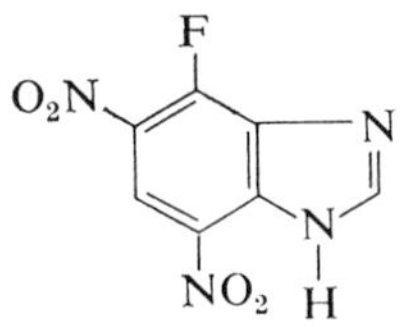

F   F

O₂N   N   O₂N   N

N   N——CH₃

NO₂ H   NO₂ CH₃

15 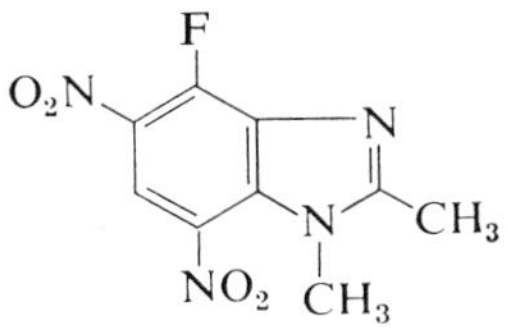   16

benzene toward amino groups. As in the case of arylation with halogeno pyridines it was demonstrated that the rate of hydrolysis is dependent on the degree of protonation of a benzimidazole nitrogen atom. The effectiveness of the intramolecular participation varies with the nature of the peptide side chain and with the degree of alkylation of the imidazole ring. The compound 1,2-dimethyl-5,7-dinitrobenzimidazolyl-4-alanyl-glycine was found to hydrolyze 65,000 times as rapidly as DNP-alanyl-glycine [54].

Cyanate Method

The reaction employed in the cyanate method of Stark and Smyth [55] is in principle similar to the Edman technique except that is not designed to be a degradative one.

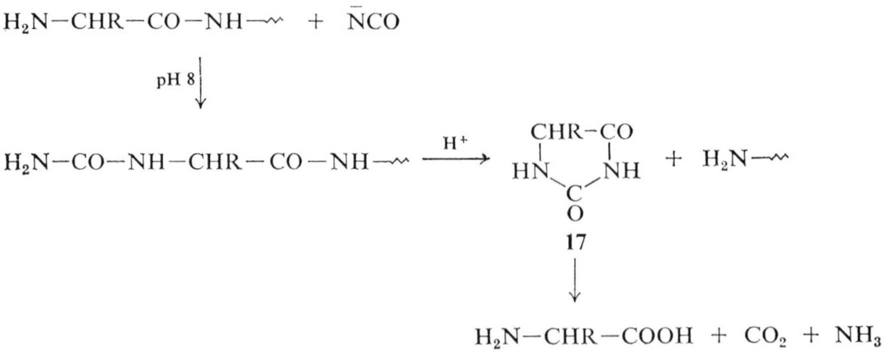

$$H_2N-CHR-CO-NH-\!\!\!\sim + \bar{N}CO$$

$$\downarrow_{\text{pH 8}}$$

$$H_2N-CO-NH-CHR-CO-NH-\!\!\!\sim \xrightarrow{H^+} \underset{O}{\overset{CHR-CO}{\underset{HN\diagdown\diagup NH}{\underset{C}{|}}}} + H_2N-\!\!\!\sim$$

$$\mathbf{17}$$

$$\downarrow$$

$$H_2N-CHR-COOH + CO_2 + NH_3$$

The amino groups of a polypeptide are carbamylated at pH 8 with KNCO, and the N-terminal amino acid is then cleaved through the hydantoin (17) formation by heating with $6N$ HCl at 100° for 1 hr. The hydantoins can be isolated and estimated by hydrolysis to free amino acids [56].

Dansyl Method

Hartley and Massey [57] greatly improved the N-terminal group determination by using 1-dimethylaminonaphthalene-5-sulfonyl (dansyl) chloride (DNS-Cl) (18) as an acylating agent for the amino groups of polypeptides. Upon sulfonamide formation the naphthalene group is strongly fluorescent, allowing easy identification and estimation of DNS-amino acids (19) in nanomolar quantities. In addition these derivatives show increased stability to acid hydrolysis over the corresponding DNP-amino acids [58, 59].

Gros and Labouesse [60] have recently reported a detailed study on the reaction rates between DNS-Cl and water, amino acids, or peptides as a function of pH and temperature. Acid hydrolysis of dansylated proteins releases the dansyl derivative of the N-terminal amino acid with some concomitant decomposition of the DNS-amino acid itself; therefore a hydrolysis time of 4 hr at 110° in $6N$ HCl is suggested [60] except in the case of the DNS-derivatives of valine, leucine, and isoleucine (18 hr hydrolysis)[58].

The separation and identification of DNS-amino acids may be achieved by chromatography on paper [61], or thin-layer chromatography on silica

[60, 62–65]. The electrophoretic separation [58] at different pH values also gives very satisfactory results.

Recently a further enhancement of the sensitivity of the dansyl method for determination of picomole quantities of amino acids has been obtained by using radioactive DNS-Cl [66]. Fingerprints of radioactive dansyl amino acids can be evaluated qualitatively and quantitatively in 10^{-12} mole quantity

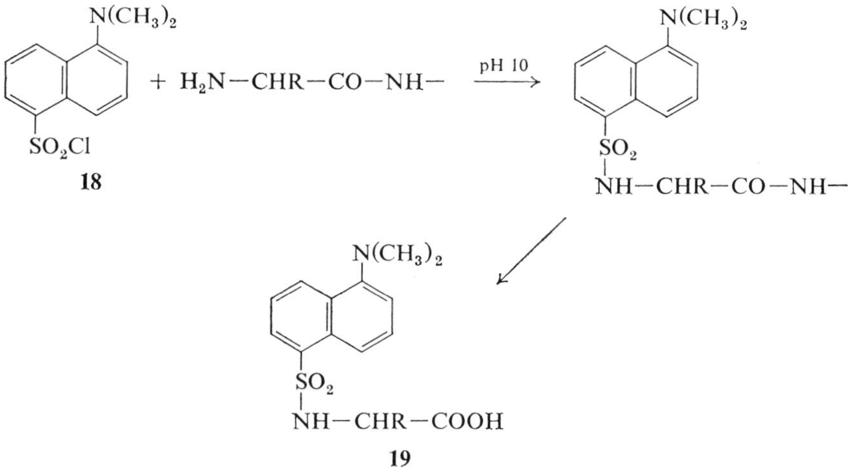

after the fluorescent spots are isolated, brought into solution, and then counted. Alternatively autoradiography may be employed by using high sensitivity X-ray films. Methyl-^{14}C-dansyl chloride has been also used for labeling by fluorescence proteins. The degree of labeling can be determined from the radioactivity content [67].

The high sensitivity of the dansyl group makes this procedure the natural successor to the DNP method and in combination with the Edman degradation (see Section V.A.1), furnishes a powerful tool for the sequence analysis of peptides [58], especially when materials are scarce.

Other Methods

A number of other chemical methods have been proposed for the N-terminal group determination, but most of them are inadequate for general application and seldom used.

Arylsulfonyl groups linked to amino groups are more resistant to acid hydrolysis than peptide bonds, so that arylsulfonyl chlorides have been used as acylating agent of α-amino groups. In addition to dansyl chloride discussed above other halides such as benzenesulfonyl [68], β-naphthalenesulfonyl [69], and p-iodobenzenesulfonyl (pipsyl) chloride [70, 71] have been employed. However, these reagents have not found extensive use since no characteristic marker groups are present; the only exception is pipsyl chloride, which when used containing a radioactive ^{131}I or ^{35}S atom [70–73].

A procedure that holds promise for the rapid assay of N-terminal groups in proteins under very mild conditions has recently been described [74–77]. A tetradentate Co(III) complex, β-[Co(trien)OH(H$_2$O)]$^{2+}$ (trien = triethylene-tetramine), reacts with a peptide (or an amino acid ester) as follows:

$$\beta\text{-[Co(trien)OH(H}_2\text{O)]}^{2+} + \overset{\overset{\text{R}}{|}}{\text{H}_2\text{NCHCO}}\text{—NH}\text{\small\leadsto} \longrightarrow$$

$$\beta\text{-[Co(trien)H}_2\overset{\overset{\text{R}}{|}}{\text{N}}\text{CHCOO]}^{2+} + \text{H}_2\text{N}\text{\small\leadsto}$$

The reaction between the cobalt complex is stoichiometric, occurs in mild conditions, pH 7, and 65°. The amino acid–cobalt complex split off is strongly colored and may be identified by chromatography.

The method has been checked only with simple peptides. Residues with side chains having functions capable of chelation with the cobalt complex, for example aspartic acid, histidine, and cysteine, could obstruct or prevent the cleavage [75].

The interesting and novel topic of selective peptide cleavage mediated by metal ions has been thoroughly reviewed [7].

A rather novel approach for the analysis of N-terminal tyrosine has been suggested [78, 79]. The N-bromosuccinimide (NBS) oxidation of bound

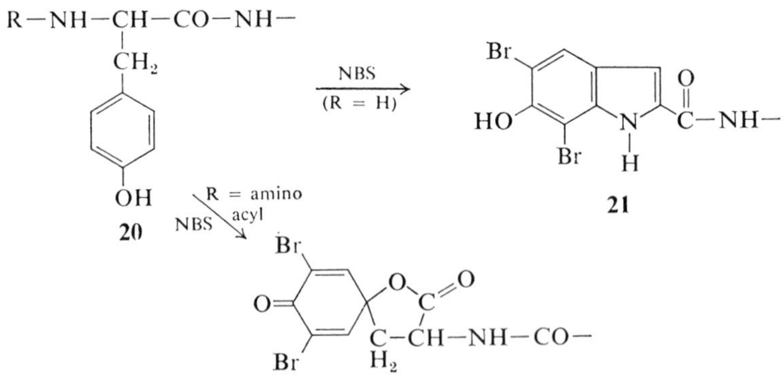

tyrosine (20) occurs in two ways, depending on whether this residue is in the N-terminal position or within the peptide chain. The former gives rise to the chromophore 5,7-dibromo-6-hydroxy-indole-2-carboxamide (21) (λ_{max} 315mμ), while the latter forms the bromodienone (22) (λ_{max} 270 mμ). The indole-chromophore formation permits an easy and rapid spectrophotometric assay for amino-terminal tyrosine, as shown with several model polypeptide sequences.

C-Terminal Group Determination

Determination of the carboxy-terminal amino acid is essential for the sequential analysis of peptides and proteins, but, in contrast to the N-terminal determination, only a few methods, such the hydrazinolysis [80] and the digestion with carboxypeptidases [81], have been successful. Enzymic digestion of large proteins does not always give definite results.

The chemical behavior of the carboxyl group is such that no derivatives stable to acid hydrolysis can be prepared. Recently, the selective labeling of the C-terminal amino acid with tritium [82, 83] has simplified the process.

Hydrazinolysis

One method for the C-terminal analysis involves heating the protein with anhydrous hydrazine [80]. The peptide bonds are cleaved with the formation of amino acid hydrazides, while the C-terminal residue is liberated as the free amino acid. The hydrazinolytic procedure represents the most frequently employed technique, although potential causes of error necessitate verifying the results by other methods [84]. Several amino acids, such as arginine, cysteine, cystine, glutamine, and asparagine, are destroyed. The amino acid hydrazides are not very stable and may undergo decomposition by oxidation or hydrolysis [85], and free amino acids with small side-chains may be slowly converted into their hydrazides. Therefore in the experimental procedures a compromise must be reached between the release of the C-terminal amino acids and the unfavorable side-reactions.

Braun and Schroeder [86] have recently described the use of Amberlite IRC-50 (H+ form) as a catalyst for the hydrazinolysis reaction, proteins being heated at 80° for periods of 10 to 100 hr depending upon the protein under study. The yields of the C-terminal amino acid are 70–100%. Alternatively, increased yields of C-terminal residues have been obtained by including hydrazine sulfate in the reaction mixture [85, 87]. The hydrazinolysis of peptides in molar hydrazine sulfate solution is faster than with hydrazine alone, and can be effected at lower temperatures.

The possibility of undesired hydrolytic cleavage of the hydrazides must be avoided and contact with water must be reduced by rapid separation of the hydrazides. The entire solution can be dinitrophenylated and all di-DNP-hydrazides extracted from the basic solution, while the C-terminal DNP-amino acid anion is not extracted by the organic solvent [35, 87–89]. Alternatively the hydrazides can be removed by reaction with benzaldehyde [80], p-nitrobenzaldehyde [90], isovaleraldehyde [88, 89], or enanthoaldehyde

[91]. The Schiff's bases formed can be separated by filtration if the products are insoluble, or by separation of the phases if the products remain dissolved.

The most elegant separation procedures involve chromatographic separation with the strong cation exchange resin Amberlite IR-120 equilibrated with volatile $0.2M$ pyridine–acetate buffer of pH 3.1. The basic hydrazides are strongly fixed, whereas the free amino acids are rapidly eluted [86]. The C-terminal amino acid eluted is quantitatively estimated with the automatic amino acid analyzer.

Tritium Labeling

In the field of peptide synthesis it is well known that N-acyl-amino acids undergo racemization during the condensation step with the amino components through oxazolone (**23**)–oxazole (**24**) intermediates [92, 93].

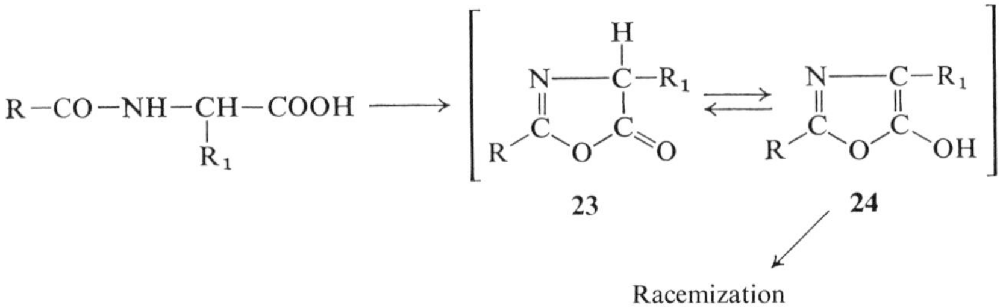

Racemization

Matsuo et al. [82, 83, 94], in an attempt to confirm the above mechanism by using D_2O, found that hydrogen–deuterium exchange occurs on the asymmetric carbon of amino acids under racemizing conditions involving carbanion formation. On this basis selective deuteration of several N-acetyl-amino acids was obtained by action of acetic anhydride or dicyclohexyl-carbodiimide followed by base-catalyzed ring opening in D_2O or T_2O, as shown in the scheme [82].

All C-terminal amino acids may be labeled by this procedure with the exception of aspartic acid, proline, and hydroxyproline. These residues, in the general procedure of the base-catalyzed deuteration or tritiation, are transformed into anhydrides and not into oxazolones. However, labeling of these residues could be achieved by heating in a mixture of deuterated acetic acid and acetic anhydride.

The method seems to be highly promising, since it is highly selective and easily accomplished on a microscale. By using T_2O the isotopically labeled C-terminal amino acid residue can be easily characterized by paper chromatography or similar methods after acid hydrolysis of the tritiated peptide. As a typical example, the radiochromatogram of the hydrolysate of the

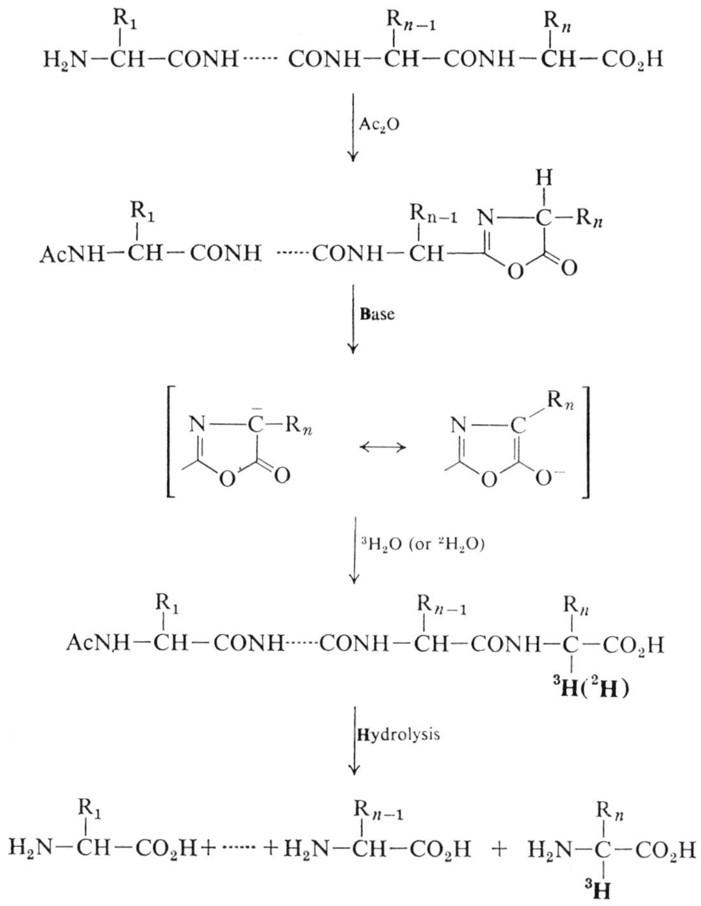

$$\underset{R_1}{\underset{|}{H_2N-CH-CONH}} \cdots\cdots CONH-\underset{R_{n-1}}{\underset{|}{CH}}-CONH-\underset{R_n}{\underset{|}{CH}}-CO_2H$$

$$\downarrow Ac_2O$$

$$\underset{R_1}{\underset{|}{AcNH-CH-CONH}} \cdots\cdots CONH-\underset{R_{n-1}}{\underset{|}{CH}}-$$

(oxazolone ring with R_n)

$$\downarrow \text{Base}$$

$$\left[\text{resonance structures of oxazolone anion with } R_n \right]$$

$$\downarrow {}^3H_2O \text{ (or } {}^2H_2O)$$

$$\underset{R_1}{\underset{|}{AcNH-CH-CONH}}\cdots\cdots CONH-\underset{R_{n-1}}{\underset{|}{CH}}-CONH-\underset{\underset{{}^3H({}^2H)}{|}}{\underset{R_n}{\underset{|}{C}}}-CO_2H$$

$$\downarrow \text{Hydrolysis}$$

$$\underset{R_1}{\underset{|}{H_2N-CH-CO_2H}}+\cdots\cdots+\underset{R_{n-1}}{\underset{|}{H_2N-CH-CO_2H}} + \underset{\underset{{}^3H}{|}}{\underset{R_n}{\underset{|}{H_2N-C}}}-CO_2H$$

tritiated angiotensin II is presented in Fig. 16.1 [83]. The C-terminal phenyl-alanine was unequivocally identified. More recently the C-terminal aspartic acid of neocarzinostatin was determined by tritiation [95]. A critical evaluation of the method and its application to luteinizing hormone has been recently reported [96].

Other Methods

ENZYMIC METHODS

Carboxypeptidases A and B are enzymes which are able sequentially to remove amino acids from the C-terminal residue of peptides so that information can be obtained on the C-terminal sequences of proteins. The method has been extensively applied in protein sequence work in spite of the fact that

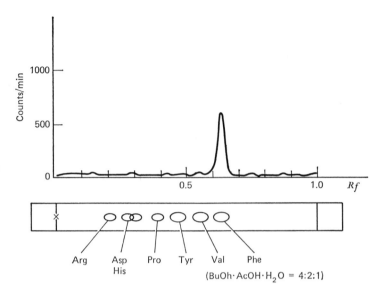

Fig. 16.1   Radiochromatogram of the hydrolyzate of angiotensin II (taken from Matsuo [83]).

the nature of the C-terminal residue plays an important role in determining the rate at which it is released. In addition, since carboxypeptidases are isolated from pancreas, incomplete removal of proteolytic enzymes may introduce further complications due to *endo*-peptidase action. The method is based on the quantitative determination by a suitable method of the amino acids released during the enzymic digestion.

The techniques of digestion with carboxypeptidase A and B have been reviewed [21, 81, 97].

CHEMICAL METHODS

Several attempts have been made chemically to convert the C-terminal carboxyl groups into derivatives amenable to proper identification [98]. The first attempt was made by Fromageot [99] who reduced the carboxyl groups of insulin to the corresponding alcohol groups with $LiAlH_4$ in *N*-ethylmorpholine at 55° for 8 hr. The amino alcohols were separated and identified by paper chromatography. Chibnall and Rees [100] esterified insulin with CH_3OH/HCl and, after reduction of the esters with $LiBH_4$, they were able to identify the amino alcohols derived from the C-terminal asparagine and alanine. However, the procedure was not sufficiently selective since ethanolamine was also obtained from the reduction of a glycyl residue. The process has been further studied [101].

More recently Atassi and Rosenthal [102] have reported the reduction of carboxyl groups in peptides and proteins using diborane in tetrahydrofurane. However, this procedure also is insufficiently selective [103] since amide bonds are affected to some extent. Esters have been reduced to alcohols with more success by LiAlH$_4$ in anhydrous solvents [104]. Although not generally recognized, NaBH$_4$ is also an effective agent for the reduction of ester groups, also in water-containing media [103].

Stark [105] recently reinvestigated the thiohydantoine method of Schlack and Kumpf [106] and applied a modified procedure with success to several peptides.

Masked Terminal Groups

Sometimes the usual N- and C-terminal group determination methods fail to reveal any terminal amino acid. Fractional residues may be only due to impurity of the protein sample. This fact at the beginning of the protein sequence studies was explained in terms of low reactivity of terminal groups due to steric hindrance. Later, when gramicidin S was shown to be a cyclic decapeptide, lacking both N- and C-terminal groups [107], the conclusion was that proteins lacking terminal groups contained cyclic chains or tailed cyclic chain structures. The coat protein of tobacco mosaic virus was the first shown to contain an acetylated amino terminus [108]. From enzymic digests of this protein a peptide with a blocked α-amino group could be isolated. Gramicidin A, B, and C were found to be N-formylated [109]. A large number of peptide and proteins have been found to contain the pyroglutamyl residue [110]. An unusual group has been detected in a hemoglobin component of normal blood, its structure involving the linkage of an aldehyde or ketone to the N-terminal amino group [111]. In all probability, blocking groups of quite different character remain to be detected.

N-Acylated Groups

Peptides and proteins containing *N*-acetyl or formyl group are rather common. In addition to the protein of tobacco mosaic virus, the *N*-acetyl group has been found in several other proteins, such as rabbit muscle myosin, tropocollagen, horse spleen apoferritin, and hyaluronidase [13]. Acetyl blocking groups were also found in calf thymus histones [112, 113]. The N_ε-acetyl-lysine residue occurs in calf thymus histone IV [114–116]. The formyl group has been detected in gramicidin A, B, and C [109], in the bee venom melittin [117] and in the newly synthesized proteins in cells of *Escherichia coli* [118, 119]. Since the formyl group is labile to acidolysis in 1N HCl in methanol [120, 121], formylated peptides can be deacylated by this procedure and then subjected to normal sequence determination. The recent discovery of a deformylase from *E. coli* [122] allows the enzymic removal of this blocking group. By contrast, the N-acetyl group is stable and the sequence determination of acetyl-peptides needs a different approach. They must be degraded by chemical or enzymic methods and the fragments elucidated separately.

The separation of N-acylated-peptides is accomplished by means of chromatographic techniques, taking advantage of the fact that an acidic acetyl-peptide can be obtained by enzymic (pronase, pepsin, chymotrypsin, etc.) digestion of the N-acylated protein. Other peptides in the digest are amphoteric, possessing as they do free α-amino and α-carboxyl groups. If the terminal acetyl-peptide contains basic residues, protection of lysine by reversible acylation and of arginine by condensation with carbonyl compounds is required. In order to obtain an acidic N-acetyl-peptide, trypsin digestion can not be used since the expected acetyl-peptide will contain lysine or arginine as the C-terminal amino acid.

Several procedures are available for the identification of the acetyl or formyl group, depending on the complexity of the acylated peptide and on the quantity available. By hydrazinolysis of the blocked peptide, acetyl hydrazide or formyl hydrazide are formed, which can be separated and identified by chromatography [108, 123].

Acetyl hydrazide was also identified as its dinitrophenyl derivative after dinitrophenylation and extraction with an organic solvent [112]. Analogously, by acylation with dansyl chloride, N-dansyl-N'-acetyl- and N'-formyl hydrazide could be separated and identified by thin-layer chromatography [124]. Methyl acetate is formed from acetyl-peptides by methanolysis in the presence of sulfuric acid for several hours at reflux temperature. The ester may be collected by distillation and identified by gas chromatography or as the hydroxamate by chromatography [125].

Pyroglutamyl Terminus

The presence of a pyroglutamyl-N-terminal residue (**25**) in many peptides

$$O \overset{}{\diagdown} \underset{\underset{H}{|}}{N} \diagup CO{-}NH{-}$$

25

and proteins is common [110]. Such an N-blocked terminal residue arises usually from the cyclization of terminal glutaminyl or glutamyl peptide either enzymically [126] or as an artifact during the separation and isolation of the products after enzymic degradation [127–129]. Glutamine peptides are easily cyclized to lactams (**25**) with deamination [130].

The lack of basicity of the pyrrolidone nitrogen (amide function) prevents the use of general techniques for identifying the N-terminal residues. In addition, degradative techniques (i.e., Edman degradation, are no longer useful for these peptides.

Both difficulties (N-terminal group determination and sequencing) are excellently overcome by mass spectrometry techniques [131]. The heptapeptide of the zymogen of phospholipase A after O,N-permethylation with CD_3I and NaI in dimethylformamide showed an excellent mass spectrum which clearly demonstrated the sequence Pyroglu-Glu-Gly-Ile-Ser-Ser-Arg. The expected N-terminal pyroglutamic acid was indicated by an intense peak at m/e 101, due to the fragment **26**. Analogously, the N-terminal docosapeptide from the λ-chain of pig immunoglobulin, after permethylation,

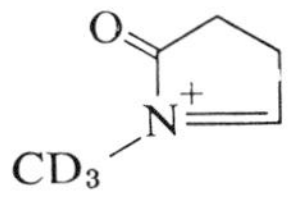

26

was shown to be Pyroglu-Thr-Val-Leu (or Ile) -GlN-Glu- . . ., this sequence accounting for about 75% of the total material.

The pyrrolidone ring (**25**) was converted by Takahashi and Cohen [132] into a pyrrolidine ring (proline) (**27**) by reduction of the peptides with diborane in tetrahydrofuran or tetramethylurea. Yields up to 46% have been

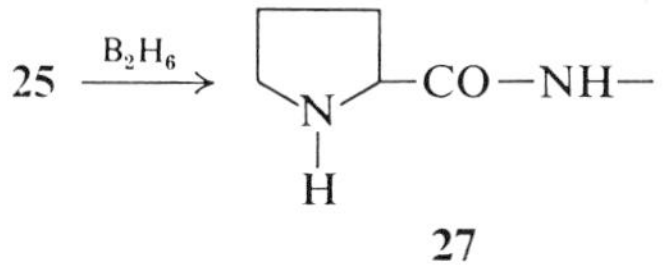

$$25 \xrightarrow{B_2H_6} \quad \text{(pyrrolidine ring)} - CO - NH -$$

27

obtained in the reduction of an octapeptide analog of gastrin. The generation of N-terminal proline by reduction of native and performic acid oxidized bovine γ-globulin was also observed. The method appears promising because of its rapidity and simple execution, although some side reactions were observed (carboxyl and peptide groups partially reduced).

The recent discovery of pyrrolidone carboxylyl peptidase from *Pseudomonas fluorescence*, an enzyme which specifically hydrolyzes amino-terminal pyrrolidonecarboxylyl residues from peptides [133, 134], seems to offer an excellent "laboratory reagent" for overcoming the difficulties presented by these *N*-acylated peptides and proteins. The enzyme was proved to be effective with a number of pyroglutamyl-peptides, including ox fibrinogen B (19 amino acids) [133], and bovine fibrinogen. In this last case phenylalanine was the only amino acid unmasked (identification by the phenylisothiocyanate reaction) during the course of the enzymic digestion, proving a high degree of selectivity of the enzyme [134].

C-Amidated Groups

With the exception of gramicidin A, B, and C, which are amidated with ethanolamine [109], the only blocking group of the C-terminal residue so far found is the amide group. Several polypeptides hormones are known to be amidated, for example, oxytocin [135] and vasopressin contain a C-terminal glycine amide, α-melanocyte stimulating hormone [136] and secretin [137] contain valine amide, gastrin [138] phenylalanine amide, eledoisin [139] and physalaemin [140] methionine amide, and melittin [141] contain glutamine amide. The isolation of the peptide amides was accomplished by chromatographic separation of enzymic digest.

The amidated carboxyl group prevents enzymic digestion with carboxypeptidase as well as chemical methods of degradation from the C-terminus. In this connection the recent finding of C-amidase activity in mammalian tissue extracts is noteworthy [142]. The amidase activity was demonstrated using the C-terminal tetrapeptide amide of gastrin Met-Trp-Asp-Phe-NH_2 as substrate, from which the corresponding peptide with the free carboxyl group was obtained.

3   ENZYMIC CLEAVAGE

The classical approach for the elucidation of the primary structure of a protein involves the selective cleavage of the polypeptide chain in such a fashion that study of the isolated fragments will allow the deduction of the region from which each peptide is derived.

Random splitting by partial acid hydrolysis gives a large number of peptides and is therefore applicable only to relatively short peptide chains. Specific agents which split only at certain sites are of obvious advantage. For instance, an agent which would split only on the right of the residue C of the octapeptide A-B-C-D-D-C-B-A would yield only ABC, DDC, and BA. If another agent capable of selective splitting next to B is available, then this would yield only AB, CDDCB, and A. The isolation and characterization of these six fragments would supply all the information needed to arrive at the original sequence.

Selective peptide bond cleavage by enzymes is the most useful technique for obtaining products, which can be completely characterized with respect to composition and sequence. Of the many enzymes that split internal peptide bonds (*endo*-peptidases), trypsin exhibits the highest degree of substrate specificity; only those bonds involving the carboxyl groups of lysyl and arginyl residues are hydrolyzed.

Chymotrypsin and pepsin are also extensively used and are sufficiently specific in their action for a controlled degradation of proteins. Chymotrypsin approaches trypsin in its discrimination. Usually it catalyzes the hydrolysis of bonds involving the carboxyl groups of tyrosine, phenylalanine, and tryptophan, but it frequently also cleaves linkages between nonaromatic amino acids.

The recent discovery of cathepsin C, a dipeptidyl amino-peptidase, offers a new approach to the use of enzymes in sequence studies [143]. The enzyme sequentially releases dipeptides from the amino end very specifically and does not show any *endo*-peptidase activity. Five dipeptides were cleaved in sequence from the polypeptide hormone β-corticotropin, as clearly indicated in Figure 16.2. Further degradation was prevented by a penultimate prolyl residue. Over 50% of the glucagon molecule was sequentially fragmented to dipeptides. It therefore appears that cathepsin C should yield unambiguous results when applied to oligopeptides of unknown sequence if the release of the dipeptide fragments is followed kinetically by a quantitative method. Keeping in mind that separation and identification of suitable derivatives of dipeptides by gas chromatography–mass spectroscopy is a relatively easy task (see Section 7), the method is promising. The ease with which relatively stable preparations of cathepsin C can be prepared, together with its broad specificity as a dipeptidyl aminopeptidase, suggest that this enzyme might be profitably employed in sequence studies.

When the sequence (see Section 5) of all of the fragments is known, it is possible to deduce the entire amino acid sequence of the protein. The sequential arrangement of the peptides has to be determined by trial-and-error alignment of the tryptic peptides with those chymotryptic and peptic peptides which contains arginine and lysine. The summation of the components of all tryptic peptides must equal the composition of the original protein.

At this point the resultant sequence usually lacks one important feature of the protein, the disulfide bonds, since each cystine is transformed in *S*-carboxymethylcysteine or cysteic acid residues.

A relatively recent development that merits special attention is the description by Brown and Hartley of a procedure, diagonal electrophoresis, for the identification of disulfide bonds in proteins. The method was developed during the study of the disulfide bridges in chymotrypsinogen [144] and other proteins [145].

A mixture of peptides was subjected to paper electrophoresis in one dimension and then oxidized with performic acid directly on the paper. After the oxidation step, the identical electrophoretic operation was repeated with the exception that the paper was turned 90°, so that the electric field was at a right angle to the original line of separated peptides. The paper was then stained to locate the peptides.

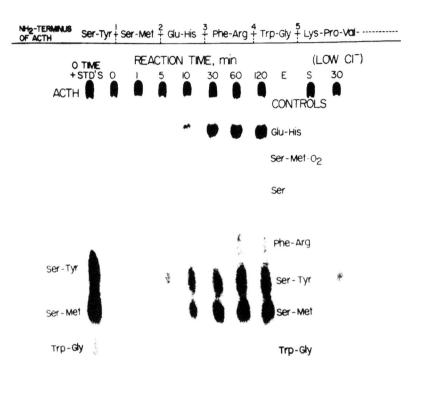

Fig. 16.2 Time course analysis of the degradation of β-corticotropin (ACTH) by rat liver cathepsin C. The origin at the left of the chromatogram contained a zero time aliquot of ACTH and standards. The remaining origin (left to right) represent 0, 1, 5, 10, 30, 60, and 120 min reaction times, followed by an enzyme control (E) and a substrate control (S) (taken from McDonald [143]).

All peptides that had not altered their electrical charge by oxidation were located along the diagonal line. The cystine-containing peptides were converted by oxidation into cysteic acid peptides, that is, to more negatively charged compounds. The peptides not falling on the diagonal line were eluted and identified by their amino acid composition. When several cystine-containing peptides are present in the original mixture, it is possible to match the half-cystine partners as those peptides that migrated in a straight line in the second electrophoretic separation.

The chemistry of the common proteolytic enzymes used for sequence determination has been the object of extensive reviews [8, 146–148].

The enzymic degradation of polypeptides has been reviewed as well, with detailed experimental procedures by Smyth [149]. In the following pages discussion will be confined to the reactions employed for altering the enzymic specificity, a methodological problem which has recently received great attention.

Chemical Modifications Altering the Trypsin Specificity

Trypsin remains the favored enzyme for selective cleavage of polypeptide chains, since it has a well-defined specificity for lysine and arginine residues. This permits the estimation of the number of tryptic peptides obtainable from a protein when the amino acid composition is known. In theory, this number is equal to the sum of lysyl and arginyl residues present in the molecule plus the C-terminal peptide. In order to take further advantage of this enzyme, techniques were devised either to create new sites of tryptic cleavage, that is, aminoethylation of cysteine residues, or to limit cleavage at the lysine or the arginine peptide bond.

The restricted tryptic degradation achieves a twofold purpose. The cleavage of a protein by trypsin usually gives rise to a sizeable number of peptides, separation of which could be rather difficult. The problem of obtaining all of the tryptic peptides in a pure form and in sufficient quantities for the sequence work has many times been the rate limiting step in the elucidation of the primary structure of large proteins. Protecting either residue, lysine or arginine, from tryptic action, will reduce the number of peptides released, so that much easier separation can be achieved. In addition, since the modified amino acid joins two peptides released in the tryptic digest of the untreated protein, the fused peptide will serve to place the previously obtained tryptic peptides in a correct sequence.

In order to obtain the maximum advantage of the use of tryptic cleavage, it is desirable to have *reversible* protecting groups at the amino or guanido functions, so that, after removal of these groups, the isolated peptides can be digested again with trypsin. In this way the separation problem may be simplified and divided into two less complex steps.

Selective and reversible protection of the ε-amino groups of lysine is achieved in a number of ways, but no satisfactory reversible protecting techniques have been devised for the guanido group of arginine.

Reversible Blocking Groups of Lysine Residues

Information available from work dealing with the reversible protection of amino groups [150, 151] has led to the easy employment of the same techniques in protein chemistry for ε-amino group of the lysine residue.

TRIFLUOROACETYL GROUP

The trifluoroacetyl group (TFA), first applied to peptide synthesis for the protection of amino groups by Weygand and Csendes [152], is stable in a neutral or mild acid solutions, but can be readily cleaved in mild alkaline media. Goldberger and Anfinsen [153] used this group for the reversible protection of the lysine residues of ribonuclease. The basic reaction employed involves acylation of lysine residues using S-ethyltrifluoroacetate (**28**) in aqueous solution at pH 10 [154].

$$\text{protein-NH}_2 + \text{CF}_3\!-\!\overset{\overset{\text{O}}{\|}}{\text{C}}\!-\!\text{SC}_2\text{H}_5 \xrightarrow[\substack{1M \text{ piperidine, } 0°}]{\text{pH 10, 25°}} \text{CF}_3\!-\!\overset{\overset{\text{O}}{\|}}{\text{C}}\!-\!\text{NH-protein} + \text{C}_2\text{H}_5\!-\!\text{SH}$$

$$\textbf{28} \qquad\qquad\qquad\qquad\qquad\qquad\qquad \textbf{29}$$

Cleavage of the acyl group from the N_ε-TFA-protein (**29**) or peptide is accomplished by $1M$ piperidine at $0°$. The mercaptan released during the reaction causes disulfide interchange in a protein containing S-S bridges, but this does not represent any drawback in sequence studies since cystine residues are usually reduced and protected by alkylation.

Since alkaline conditions (pH 9.5–10.0) are required for the acylation and since the TFA group is not completely stable under such conditions, careful control of the pH by the use of a pH-stat is recommended. Consequently, to achieve complete blocking of ε-amino groups it is necessary to use an excess of the reagent.

The degree of trifluoroacetylation can be determined by treating the protein derivative with nitrous acid [155, 156]. The deamination occurs only at the free amino groups and the recovery on the analyzer of lysine gives an indirect indication of the degree of acylation. Another way to determine the free amino groups involves the use of the trinitrobenzenesulfonate (TNBS) reaction with amino groups [157, 158].

A disadvantage of the TFA group in some instances [159] is represented by the insolubility of the TFA-protein, rendering the subsequent tryptic digestion impossible or very slow. N_ε-TFA-peptides were also found to form aggregates, so that difficulties were encountered in obtaining pure materials after chromatographic separation [160].

In order to enhance the solubility properties of the modified peptides, tetrafluorosuccinic anhydride (30) has been used as an acylating agent [161]. The presence of a carboxylate anion in the half-amides of tetrafluorosuccinic acid (31) confers good solubility on the acylated peptides. In order to reverse the acylation, exposure of the modified peptide to pH 9.5 and 0° is employed.

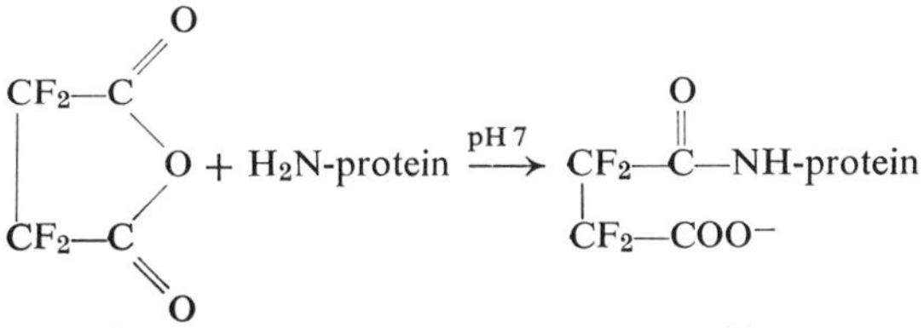

$$CF_2-C \overset{O}{\diagup}_{\diagdown} O \quad CF_2-C \overset{O}{\diagdown}_{\diagup} O \quad + H_2N\text{-protein} \xrightarrow{pH\,7} CF_2-\overset{O}{\overset{\|}{C}}-NH\text{-protein}$$

$$CF_2-COO^-$$

30   **31**

ACETOACETYL GROUP

Scoffone and his coworkers have introduced the acetoacetyl group (ACA) for the reversible protection of amino groups in peptide synthesis [162–164] as well as for the restricted tryptic cleavage at the lysine residues [165, 166]. The reaction involves acylation with diketene (32) at pH 8.5, giving the N_ε-acetoacetyl-protein (33). The removal of the ACA group is accomplished in mild conditions (pH 9.5, $1M$ NH$_2$OH) by hydroxylamine, releasing a

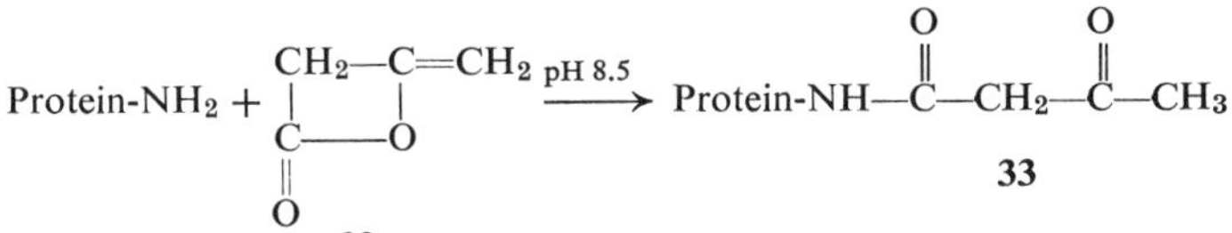

$$\text{Protein-NH}_2 + \underset{32}{\overset{CH_2-C=CH_2}{\underset{C-O}{\overset{\|}{O}}}} \xrightarrow{pH\,8.5} \text{Protein-NH}-\overset{O}{\overset{\|}{C}}-CH_2-\overset{O}{\overset{\|}{C}}-CH_3$$

33

$$\text{Protein-NH}-\overset{O}{\overset{\|}{C}}-CH_2-\overset{O}{\overset{\|}{C}}-CH_3 + NH_2OH \xrightarrow{pH\,7.0}$$

$$\left[\underset{HO-N}{\overset{\text{protein-NH}-\overset{O}{\overset{\|}{C}}-CH_2}{\diagdown}}_{\diagup C-CH_3} \right] \longrightarrow \text{Protein-NH}_2 + \underset{O-N}{\overset{O=C-CH_2}{\diagdown}}_{\diagup C-CH_3}$$

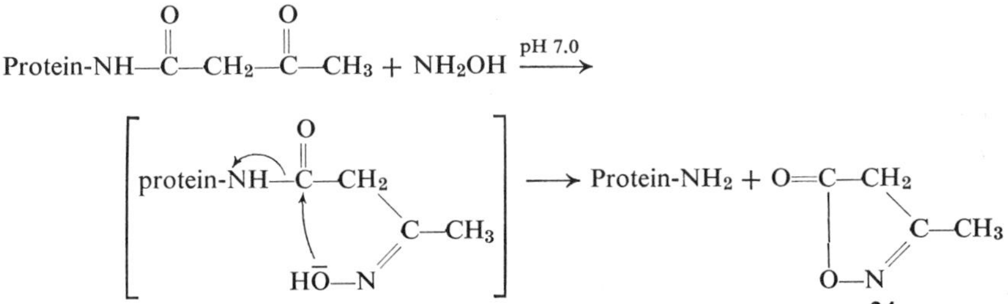

34

5-methyl-oxazolone (34) [163, 164]. Analogous deacylation can also be achieved by means of phenylhydrazine under slightly more drastic conditions (aqueous acetic acid, 30 min, 60°).

The method was successfully applied to ribonuclease [165], in which case tryptic cleavage at the arginyl peptide bonds only was achieved after blocking the N_ε-amino groups of the lysine residues. Acylation of hydroxyl group was also observed, but a simple exposure to alkaline pH reverts the ester formation.

The same chemistry of the β-carbonyl-amides was employed for the cleavage of the serine and threonine peptide bonds after oxidation of the residues to a β-carbonyl moiety [163, 164, 167].

MALEYL GROUP

The reversible blocking of amino groups with maleic anhydride (35) is a very useful technique and has several advantages [168, 169]. The reaction of 35 with proteins is complete in 5 min at pH 9 and 2° with 30 mM reagent, and

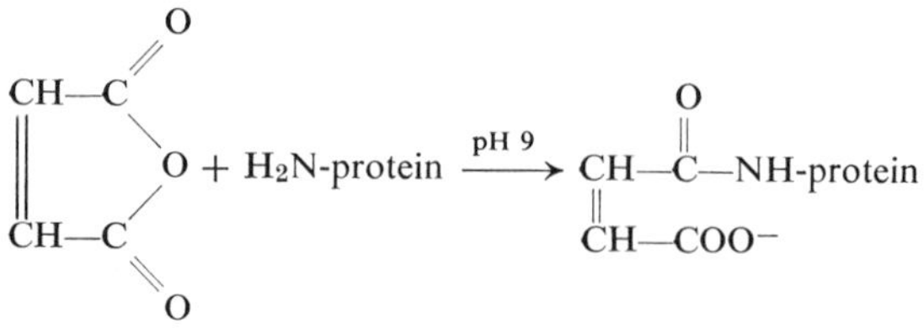

appears to be specific for amino groups, the reaction with other functional groups probably being rapidly reversed. Complete amino-acylation of chymotrypsinogen and β-melanocyte-stimulating hormone was achieved. Another advantage of the method is the possibility of a direct quantitation of the extent of reaction from the spectrum of the maleyl-protein (36) [169]. Maleyl-proteins are generally soluble and disaggregated at neutral pH.

The half-amide of maleic acid is stable at 37° above pH 6, but deacylation occurs readily at pH 3.5, the half-time being 11–12 hr at 37°. A possible catalytic mechanism for this hydrolytic reaction involves an un-ionized β-carboxyl group [170, 171]. These findings may be related to the relatively easy hydrolysis of aspartyl-peptides, in which case an analogous part is played by the β-carboxyl group. The much easier cleavage of maleyl-amides must be related to the presence of the α-β double bond of the maleyl moiety, enhancing the rate of reaction by the proximity effect of the β-carboxyl group.

In a modification of this technique, 2-methylmaleic (citraconic) (37) and 2,3-dimethylmaleic (38) anhydrides were found to react with amino groups of proteins [172]. 2-Methylmaleic amides are more readily cleaved than

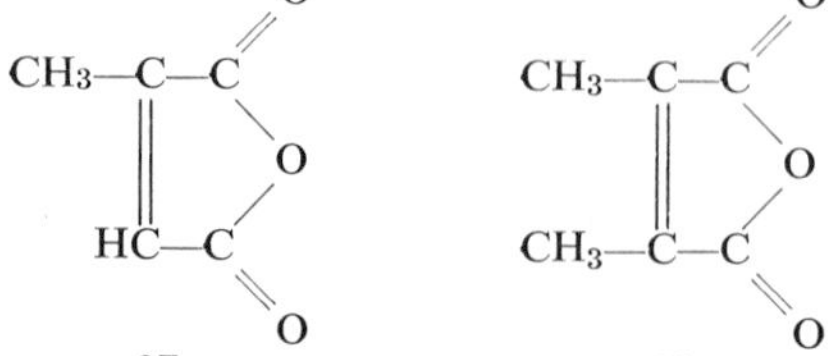

maleic amides but have the disadvantage that two products can be formed, giving multiple forms of the same peptide and thereby complicating the separation. 2,3-Dimethylmaleic amides, which do not give isomeric products, are cleaved too easily during the work-up for the separation of the modified peptides.

The reaction of maleic anhydride with ε-amino groups of lysine residues was usefully employed also for the diagonal electrophoresis of lysine and N-terminal peptides as with β-melanocyte-stimulating hormone [169].

AMIDINATION

Imidoesters (39) react readily with the amino groups of proteins at pH 7–10, at room temperature, giving amidines (40). Methyl acetimidate is the most used reagent. The reaction is carried out with successive addition of

$$\text{Protein-NH}_2 + \begin{matrix} \text{HCl·HN} \\ \diagdown \\ \text{C—R} \\ \diagup \\ \text{CH}_3\text{O} \end{matrix} \longrightarrow \text{Protein-NH—}\overset{\overset{\textstyle \text{NH·HCl}}{\|}}{\text{C}}\text{—R} + \text{CH}_3\text{OH}$$

$$\textbf{39} \qquad\qquad\qquad \textbf{40}$$

several portions of an easily hydrolyzable actimidate over a period of several hours.

Conversion of the ε-amino groups of lysine to the corresponding ε-amidino derivatives renders the lysylpeptide bond resistant to tryptic digestion. The acetimidyl group is easily removed at pH 8.2 in 0.05 M ammonium carbonate–bicarbonate buffer at 37°, or in ammonia–acetic acid buffer at pH 11.3 over a period of 8 hr [173]. Several successful applications of this technique are reported in the literature (insulin, ribonuclease, $E.\ coli$ phosphatase, α_S-casein, bovine serum albumin) [174].

The mild reaction conditions of the amidination permit its wide application. Recently the use has been reported of methylpicolinimidate as a reagent designed to introduce the picolinamidine function, which is able to chelate metals, in specific position of the protein [175].

Since the positive charge at the ε-amino groups of lysine after reaction with amidoesters is preserved, proteins can be modified without gross change in the three-dimensional structure. This property lead to the use of the amidination as a valuable tool for structure-activity studies in enzymes. Bifunctional imidoesters are also employed as reversible cross-linking reagents for amino groups [176–179].

DITHIOCARBAMYLATION

The reaction of carbon disulfide with amino groups was initially employed for sequential degradation purposes [180, 181]. However, the lability of the thiocarbamyl moiety (41) prevented a successful application for this purpose.

$$\text{Protein-NH}_2 + \text{CS}_2 \underset{\text{pH2, aerate}}{\overset{\text{pH 9–10}}{\rightleftharpoons}} \text{Protein-NH—}\overset{\overset{\textstyle \text{S}}{\|}}{\text{C}}\text{—S}^-$$

$$\textbf{41}$$

On the other hand, the easy regeneration in acidic media of the parent free amino groups of the alkali-stable dithiocarbamates prompted the use of the same reaction for the reversible blocking of lysine residues [182]. The *S*-peptide of ribonuclease A and reduced carboxymethylated lysozyme were shown to be cleaved mainly at the arginine residues. However, traces of lysyl cleavage were also detected, indicating incomplete stability of the N_ε-dithiocarbamyl-lysine moieties during the tryptic digestion [182].

OTHER GROUPS

As pointed out above, all reversible N-protecting groups employed in peptide syntheses are in principle useful blocking units for ε-amino groups of lysine residues in proteins.

The carbobenzoxy group (Z) (**42**) was successfully used for cleaving only

$$\text{[benzene ring]}-CH_2O-\overset{\overset{\displaystyle O}{\|}}{C}-NH\text{-protein}$$

42

the arginyl-peptide bonds of oxidized ribonuclease after acylation of the protein with carbobenzoxy chloride in $1M$ $NaHCO_3$, at $-3°$, for ca. 1 hr [183]. A drawback of this protecting group lies in the rather strong deblocking conditions (HBr in CH_3COOH or HCOOH) which cause many side reactions. Hydrogenolysis over Pd catalyst, which is a much milder procedure for removing the carbobenzoxy group, can not be applied generally on account of the possible presence in the protected peptides of sulfur amino acids. The HF treatment of the acylated peptide would appear to be the procedure of choice [184].

The possibility of using the *t*-butyloxycarbonyl group (BOC) (**43**) as a protecting unit for ε-amino groups of lysine residues has been demonstrated

$$CH_3-\overset{\overset{\displaystyle CH_3}{|}}{\underset{\underset{\displaystyle CH_3}{|}}{C}}-O-\overset{\overset{\displaystyle O}{\|}}{C}-NH\text{-protein}$$

43

by the preparation of a BOC-insulin derivative by reaction of insulin with *t*-butyloxycarbonyl azide in anhydrous dimethylformamide [185]. The triacylated derivative was reconverted to crystalline insulin with high biological activity upon removal of the BOC groups with anhydrous trifluoroacetic acid.

Carbonyl reagents deserve special attention for the reversible blocking of lysine residues. Usually the resulting Schiff's bases are readily cleaved in acidic media. However, since these derivatives are not sufficiently stable at neutral or alkaline media, attention has been directed toward the use of reagents which give more stable derivatives. Salicylaldehyde, which gives with amino groups Schiff's bases stabilized by hydrogen bonding, was condensed with ε-amino groups of cytochrome c [186] and with trypsin [187]. β-Dicarbonyl reagents [188] and 5,5,-dimethyl-1,3-cyclohexandione (44) [189], which are stabilized by enamine tatuomerism (45), could also find useful applications.

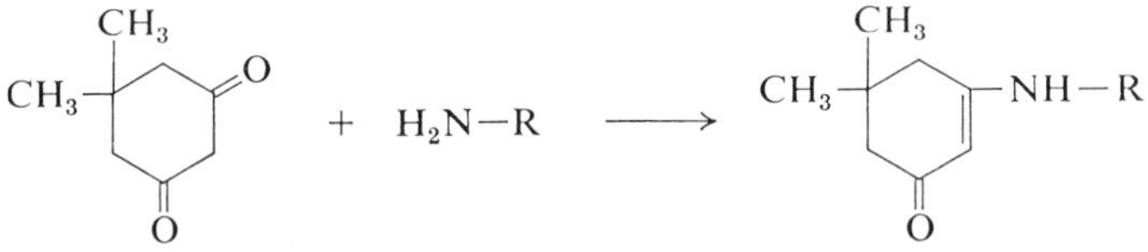

β-Naphthoquinone-4-sulfonic acid reacts readily with primary amino groups at 25° and pH 7.5–9.3, producing colored derivatives with absorption at 480 mμ [190]. 2-Methoxy-5-nitrotropone (46) [191] is a promising reagent which reacts selectively at pH 8.5 with amino groups and is removed by dilute hydrazine.

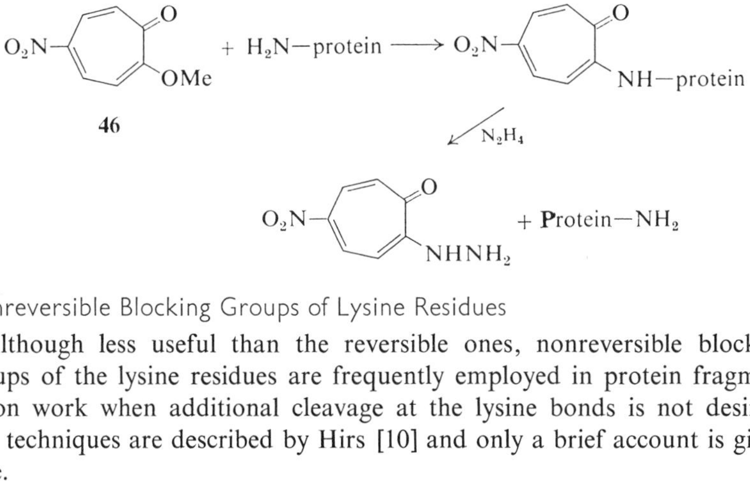

Nonreversible Blocking Groups of Lysine Residues

Although less useful than the reversible ones, nonreversible blocking groups of the lysine residues are frequently employed in protein fragmentation work when additional cleavage at the lysine bonds is not desired. The techniques are described by Hirs [10] and only a brief account is given here.

GUANIDINATION

The reaction of O-methylisourea (47) with proteins in alkaline media (pH 10–11) converts the ε-amino groups of the lysine residues into guanidino functions (48). Thus guanidination converts lysine into homoarginine, and

$$\underset{47}{\text{Protein-NH}_2 + \text{H}_2\text{N}-\overset{\overset{\text{OCH}_3}{|}}{\text{C}}=\text{NH}} \longrightarrow \underset{48}{\text{Protein-NH}-\overset{\overset{\text{NH}}{||}}{\text{C}}-\text{NH}_2 + \text{CH}_3\text{OH}}$$

since this residue is stable to acid hydrolysis, the extent of reaction may be determined by amino acid analysis [192]. On the basis of the results with several proteins, the guanidination appears to occur to a minimal extent at the α-amino group of the polypeptide chain [193–196]. Most probably the selectivity for the ε-amino groups is due to the lower nucleophilicity of the α-amino group and the mild reactivity of the isourea (**47**).

Although the reaction has been mainly used for structure–activity studies [192], its application for blocking the tryptic cleavage at the lysine residues has been described, as in the case of oxidized mercuripapain [197] and α-lactalbumin [198].

In addition to *O*-methylisourea (**47**), which reacts slowly (2–5 days) with amino groups at pH 10–11, *S*-methylisothiourea (**49a**) [199] and 3,5-di-methylguanidopyrazole (**49b**) [200, 201] may be used for more rapid guani-dination, but these compounds do not differentiate between N_α- and N_ε-guanidination. The neutral nitroguanidino group can be introduced at the

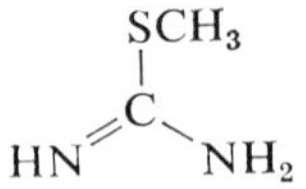

49a

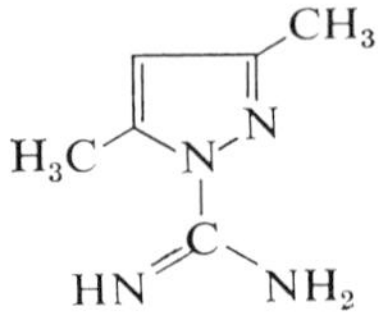

49b

ε-amino groups of lysine by using 2-methyl-1 (or 3) -nitro-2-thiopseudourea (**50**) or 1-nitroguanyl-3,5-dimethylpyrazole (**51**) [202, 203].

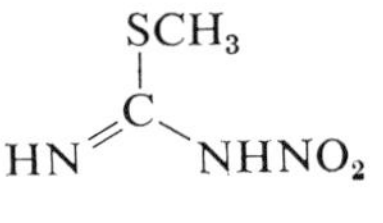

50

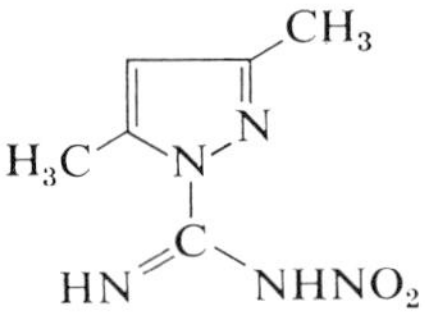

51

CARBAMYLATION

Although the reaction of cyanate with protein side chain groups is not specific and leads to the modification of amino, sulfhydryl, hydroxyl, imi-dazole, and carboxyl groups, its useful application for blocking tryptic cleavage at the lysine residues was verified in the case of carbamylated (**52**) peptide S of ribonuclease [129]. Also leucine aminopeptidase attack was also blocked. However the side reactions mentioned above, and the relatively

$$\text{Protein-NH}_2 + \text{HCNO} \longrightarrow \text{Protein-NH}-\overset{\displaystyle O}{\overset{\displaystyle \|}{C}}-\text{NH}_2$$
52

drastic conditions (pH 8, 50°, 16 hr) required, have limited its use [15, 56]. The reaction can be followed quantitatively by acid hydrolysis and amino acid analysis of the resulting homocitrulline [56].

ACYLATION

Acylation by acid anhydrides or acid halides is extensively employed in protein chemistry, although the reaction often has no satisfactory selectivity for amino groups. Rather than acetic anhydride [204], cyclic anhydrides [205] were found to be milder and more selective reagents. The nonvolatile and stable succinic anhydride has the additional advantage of introducing a negative charge via the carboxylate anion, producing a stable and soluble peptide. Blocking of tryptic cleavage at the lysine residues through acylation has been used in several instances. An instructive example is given by the work of Klippenstein et al, [159] in which succinylation was employed for restricted tryptic cleavage of *Goldfingia gouldii* hemerythrin.

ARYLATION

The easy nucleophilic displacement of halogen from reactive aromatic halogeno compounds by amino groups could be used for irreversible blocking of ε-NH$_2$ groups of lysine residues [15]. Dinitrophenylated-ribonuclease [206] gives the expected five peptides by tryptic digestion of the four arginyl-peptide bonds. Analogous limited cleavage was verified in DNP-α-lactalbumin [198]. Since the DNP-peptides do not show good solubility properties, 4-fluoro-3-nitrobenzenesulfonic acid is preferable as a arylating agent [207, 208].

In the absence of special protein factors, dinitrofluorobenzene reacts with many other side-chain functional groups of the protein, especially sulfhydryl, imidazole, and hydroxyl groups. Treatment of the DNP-protein with β-mercaptoethanol [16] removes all DNP groups except those linked to the amino groups.

Trinitrobenzenesulfonic acid (TNBS) reacts more selectively than dinitro-fluorobenzene with amino groups. providing a useful spectroscopic assay for these groups [157, 209, 210].

OTHER METHODS

Weygand and Tschesche [211] have recently reported the use of 2,2,2-trifluoro-1-(2-ureido-ethansulfonyl)-*N*-benzoyl-ethylamine (**53**) as a regent

$$CF_3-\underset{\underset{OC-}{\overset{|}{NH}}}{\overset{|}{CH}}-SO_2CH_2CH_2NHCONH_2 \quad \xrightarrow[\text{pH 9}]{\text{Protein}-NH_2} \quad CF_3-\underset{\underset{OC-}{\overset{|}{NH}}}{\overset{|}{CH}}-NH-\text{protein}$$

53

for the protection of the lysine residues from tryptic cleavage. Lysozyme was reacted at pH 9 with **53** (1% soluble in aqueous solution) and, after tryptic hydrolysis, the expected peptides from the selective cleavage at the arginyl residues were separated by ion exchange chromatography.

When proteins are treated with aliphatic aldehydes or ketones, the labile Schiff's base enzyme products resulting from the condensation of amino

$$\text{Protein-NH}_2 \xrightarrow{\text{RCHO}} \underset{\textbf{54}}{\text{Protein-N=CHR}} \xrightarrow{\text{NaBH}_4} \underset{\textbf{55}}{\text{Protein-NH—CH}_2\text{R}} \xrightarrow{\text{RCHO}}$$

$$\underset{\underset{\textbf{56}}{\overset{\|}{\underset{\text{CHR}}{}}}}{\text{Protein-}\overset{+}{\text{N}}\text{—CH}_2\text{R}} \xrightarrow{\text{NaBH}_4} \underset{\underset{\textbf{57}}{\overset{|}{\underset{\text{CH}_2\text{R}}{}}}}{\text{Protein-N—CH}_2\text{R}}$$

groups with the carbonyl reagents (**54** and **56**) may be trapped by reduction with sodium borohydride [212, 213]. Amino groups are converted in high yields by this reaction into the corresponding mono- (**55**) or dialkylamino derivatives (**57**). The principal product of the reductive methylation of proteins with formaldehyde has been identified as ε-N,N-dimethyllysine; with acetaldehyde or acetone only the corresponding monoalkylated lysines are formed. The conditions for the reductive alkylation are very mild and the reaction is easily done without special reagents or equipment. The sensitive disulfide bonds were found to survive the reductive action of the borohydride employed.

Since it is known [214] that ε-N,N-dimethyllysine peptide bonds are not cleaved by trypsin, the action of this enzyme on N_ε-alkylated proteins results in cleavage of the arginyl bond only. In addition, since the positive charge at the ε-amino group of lysine is retained, the peptide maps of these tryptic hydrolyzates would be expected to differ only in the number of the peptides. This fact should facilitate the identification and the ordering of the cleaved peptides.

Blocking of Trypsin Cleavage at the Arginine Residues

The masking the ε-amino groups of lysine discussed above represents only a part of the potential use of restricted tryptic hydrolysis of the polypeptide chain. The other aspect involves the blocking of arginyl residues in order to limit cleavage at the lysine groups. The reagents used for the modification of the guanido group usually are dicarbonyl compounds. These reagents may condense also with amino groups, but the formation of the unstable Schiff's base is reversed during the subsequent purification or acidification. In the case of the guanido group, the formation of a heterocyclic ring by intramolecular reaction causes irreversible modification of the side-chain group.

Following the first reports of Itano and his coworkers [215–217], several techniques for the selective modification of the guanido function have been proposed. However, although reversible protection of lysine can be accomplished in a number of ways, the reaction is not applicable to arginine [15].

Because the guanidino group exists in its resonance-stabilized protonated form over the entire pH range of protein stability, it is not surprising that this functional group differs markedly from other basic groups. However, the required features of a modification technique for the guanido function depend upon whether structure–activity studies or sequence work have to be done. Mild reaction conditions (physiological pH, room temperature), so that no change of the protein conformation occurs, are important for studying the role of arginine in biologically active proteins. For protein degradation the reaction must convert the guanido group selectively and quantitatively into a stable derivative.

USE OF DIKETONES

The reaction of cyclohexandione (CHD) (58) with arginine residues in alkaline solution, $0.2N$ NaOH, produces an imidazolinone derivative, which

$$
\underset{\textbf{58}}{
\begin{array}{l}
\text{CO}-\\
|\\
\text{CH}-(\text{CH}_2)_3-\text{NH}-\text{C}\underset{\displaystyle\text{NH}_2}{\overset{\displaystyle\text{NH}}{\diagup}}\\
|\\
\text{NH}-
\end{array}
}
\;+\;
\begin{array}{l}
\text{OC}\\
\text{OC}
\end{array}\!\!\bigcirc
\;\longrightarrow\;
\underset{\textbf{59}}{
\begin{array}{l}
\text{CO}-\\
|\\
\text{CH}-(\text{CH}_2)_3-\text{N}=\text{C}\underset{\displaystyle\overset{\displaystyle}{\text{N}-\text{C}}}{\overset{\displaystyle\text{N}-\text{CO}}{\diagup}}\\
|\\
\text{NH}-
\end{array}
}
$$

was identified, also by synthesis, as N^5-(4-oxo-1,3-diazaspiro[4.4] non-2-ylidene)-ornithine (59) [216–218]. When the reaction is carried out in $0.05N$ NaOH, several products are observed, but the rate of disappearance of arginine is equally high under both conditions. Since arginine is not regenerated by acid hydrolysis of a CHD-treated protein, the extent of modification is easily determined by amino acid analysis. The reaction is selective for the guanido group of arginine. Peptide maps of tryptic digests of CHD-treated chains of hemoglobin showed blocking of hydrolysis at arginyl bonds with no change in position of tryptic peptides not adjacent to arginyl residues [217].

Slobin and Singer [219, 220] have successfully employed this reaction for blocking tryptic cleavage at the arginine residues of the light and heavy chains of immunoglobulins. CHD was also employed to chemically modify the arginyl residues of several inhibitors of proteolytic enzymes [221]. The reaction conditions [217] were modified and a particular buffer (triethylamine, pH 11) was used, but reactivity of the amino groups was detected. Arginyl-peptides were also treated with CHD in order to obtain volatile derivatives for sequence studies by mass spectrometry [222].

The use of benzil (**60**) [215] as a condensing agent for the guanido group has the disadvantage that both reagent and product **61** are insoluble in water. The reaction is carried out in 0.2N NaOH in 60–80% ethanol for 16–18 hr.

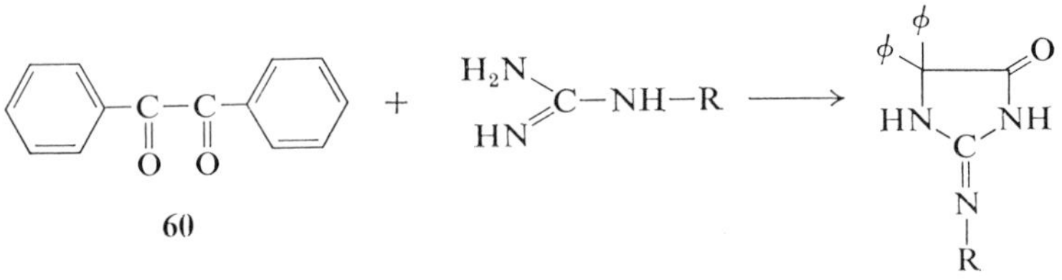

60

61

Salmine was treated with benzil and no cleavage was observed at the arginine residues. Analogously tryptic hydrolysis after Arg^{22} in the performic acid-oxidized β-chain of insulin is blocked, permitting cleavage only at the lysyl bond.

In contrast with the condensation reactions of the guanido group with the diketones observed above, phenanthrenequinone reacts with arginine giving a strongly fluorescent compound, 2-amino-1H-phenanthro[9,10-a]-

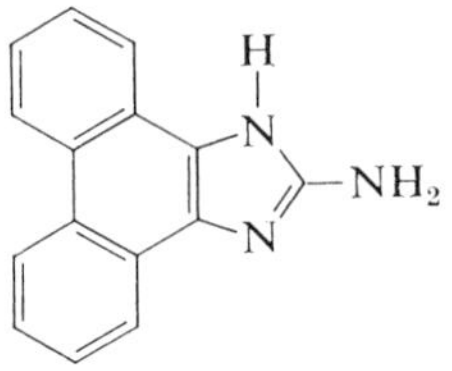

62

imidazole (**62**), which provides an extremely sensitive test for arginine and other monosubstituted guanidines. Because of its sensitivity, the reaction has been used for locating small amounts of arginine-containing peptides in chromatograms [223].

A trimer (**63**) of 2,3-butandione (biacetyl) was found to react with arginine at pH 8.2 [224, 225]. Bovine plasma albumin was treated with the reagent

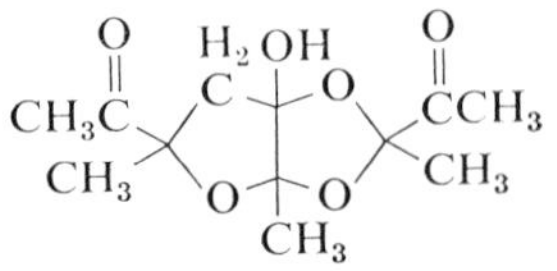

63

(0.4 M) at pH 7, and the loss of arginine after 6 hr, determined by amino acid analysis, was 53%. After 50 hr the loss of lysine and histidine (nearly 30 and 10%) was found [225]. The arginine residues in acetylated rabbit-γ-globulin were modified to the extent of 54%, but it was found that the reagent also condenses with amino groups in a nonspecific manner [226].

USE OF GLYOXAL AND DERIVATIVES

The condensation of glyoxal (**64a**) [227] and related compounds (**64b** and **64c**) [228] with the guanido group occurs under mild conditions (pH 8.6–9.2).

$$H—CO—CO—H \qquad CH_3—CO—CO—H \qquad C_6H_5—CO—CO—H$$

64a   **64b**   **64c**

The nature of the reaction is not yet known. By treatment of arginine with phenylglyoxal a compound containing two phenylglyoxal units was isolated in 13% yield. Two possible structures (**65** and **66**) for this product have been postulated [228].

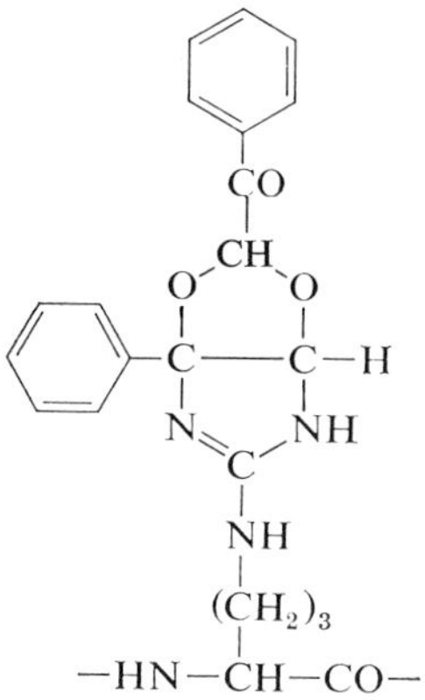

65 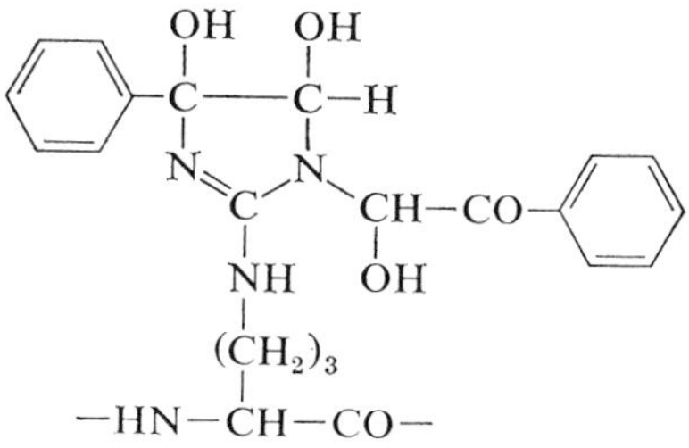   **66**

The mild conditions required for the reaction with these reagents are attractive features, but since quantitative modification of arginyl residues in proteins has not been observed (insulin, lysozyme [227], ribonuclease [228], trypsin and chymotrypsin [229]) the utility of these reagents for protein sequence work is diminished. In addition, the modified guanido group is not stable, being decomposed slowly at neutral and alkaline pH. About 80% of the original arginine can be regenerated by incubation of the phenyl-glyoxal-arginine adduct for 48 hr at 37°. The side reactions of these reagents with other functional groups must also be pointed out, if the reaction is employed for structure–activity studies. With proteins reaction may also occur with the ε-amino groups of lysine residues [212, 228]. In all cases α-amino groups are deaminated to give residues of the corresponding α-ketoacids [228].

PYRIMIDINE RING FORMATION

Ring closure reactions of amidines leading to the formation of heterocyclic compounds [230] are applicable to the modification of the guanido groups of proteins. Numerous possibilities are available, since the principal synthetic methods for 2-amino-pyrimidines involve the use of 1,3-dicarbonyl compounds and guanidines.

Arginine has been condensed [231] with malondialdehyde (67) (commercially available as its precursor 1,1,3,3-tetraethoxypropane) and converted into N_δ-2-pyrimidyl-ornithine (68). The reaction was carried out in

$$
CH_2 \!\!<\!\!{}^{CHO}_{CHO} \;+\; {}^{H_2N}_{H_2N}\!\!\overset{+}{>}\!C\!-\!NH\!-\!protein \;\longrightarrow\; \bigvee\!\!\langle\;\rangle\!-\!NH\!-\!protein
$$

67   68

strong acid conditions (10–12N HCl) for several hours at room temperature. Since the pyrimidyl-ornithine residue absorbs at 315 mμ, the extent of reaction can be conveniently followed by spectrophotometry. Amino acid analysis after acid hydrolysis is not useful since the 2-amino-pyrimidine linkage is not stable in HCl 6N at 100° and ornithine is formed to some extent [231]. The modification of arginine in insulin, lysozyme, and ribonuclease was shown to occur practically quantitatively (83–100%) and, as expected, tryptic cleavage is no longer possible following such modification.

A new useful 1,3-dicarbonyl reagent is the sodium salt of nitromalondialdehyde (69) which condenses readily and quantitatively in alkaline media (0.1–0.5M NaOH, 2–3 hr, 0–5°) with the arginine residues in proteins to form δ-(5-nitro-2-pyrimidyl)-ornithine derivatives (70) [232].

$$
O_2N\!-\!\overset{\displaystyle H}{\underset{\displaystyle H}{\overset{|}{\underset{|}{\overset{\textstyle CO}{\underset{\textstyle CO}{C}}}}}}\!Na^+ \;+\; {}^{H_2N}_{HN}\!\!>\!C\!-\!NH\!-\!R \;\longrightarrow\; O_2N\!-\!\langle\;\rangle\!-\!NH\!-\!R
$$

69   70

The reagent is very soluble in water and the derivatives with N-substituted guanidines have a high molar extinction coefficient (λ_{max} 335 mμ, ε16,000).

The arginine residues of ribonuclease and lysozyme have been completely modified by the reagent. The extent of reaction was determined by spectrophotometry and by amino acid analysis of the modified proteins. Arginine was not present in the acid hydrolysate, the derivative being partly (70–80%) converted by acid hydrolysis into free ornithine [232].

New Sites of Tryptic Cleavage

S-AMINOETHYLATION

Lysine-like residues can be created in a protein by reaction of the sulfhydryl side-chain groups of cysteine with ethyleneimine (71) [233]. The S-(β-aminoethyl)-cysteinyl or thialisine derivatives 72 are now new points of

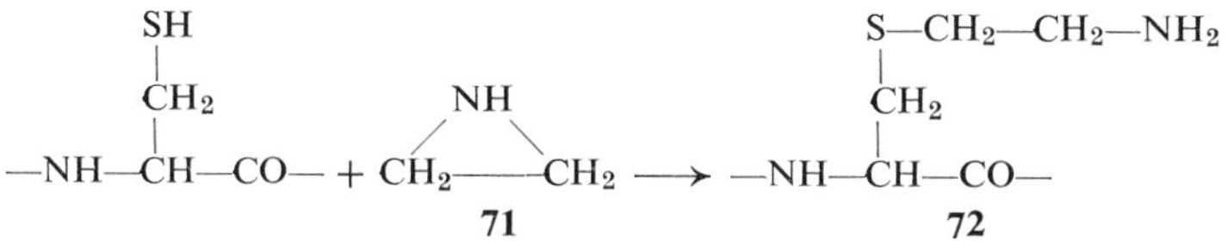

attack for trypsin. The rate of cleavage at this peptide bond is comparatively slow so that a prolonged incubation time is needed. With S-aminoethylated-ribonuclease it was found that cleavage of the thialisine derivatives averaged approximately 56%, while the corresponding hydrolysis of lysyl and arginyl bonds was close to 83% [234].

Instead of ethyleneimine, β-bromoethylamine can be used [235], but intermediate formation of the imine 71 is required before reaction with SH groups. Since the cyclization step was shown to be rate determining, the direct use of 71 is more efficient. The ethyleneimine sulfonium salt (73) of methionine 313 of apocatalase is formed during the ethyleneimine treatment of the protein [236]. Analogously to the cyanogen bromide cleavage of

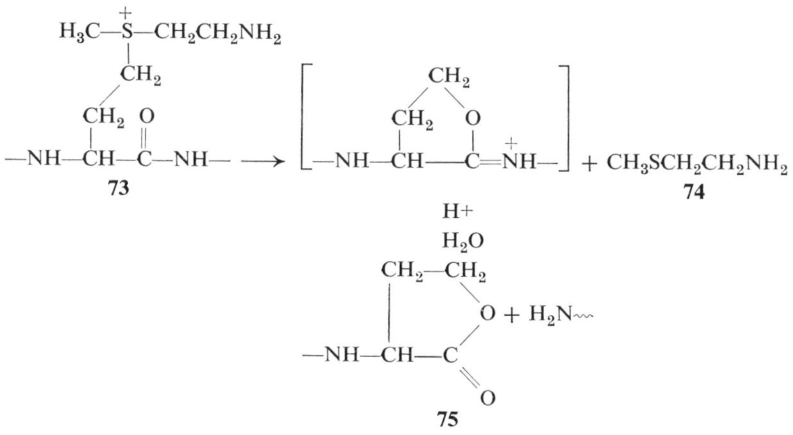

methionine peptides the C-peptide bond was found to be cleaved in yields of approximately 70% with formation of homoserine lactone (75) and methyl-β-aminothioether (74).

Tryptic cleavage at cysteine after its conversion into the thialysine residue has found successful application in many instances, for example, tryptophan synthetase [237], ferredoxin [238, 239], trypsinogen [240–242], tobacco mosaic virus protein [243], and thiroredoxin [244]. The procedure has recently been reviewed by Cole [245].

In addition to cysteine, serine can be converted into a thialysine residue. For this purpose the hydroxyl group of serine is first acylated with tosyl (76) or mesyl (77) chloride. The easy nucleophilic displacement by cysteamine (78)

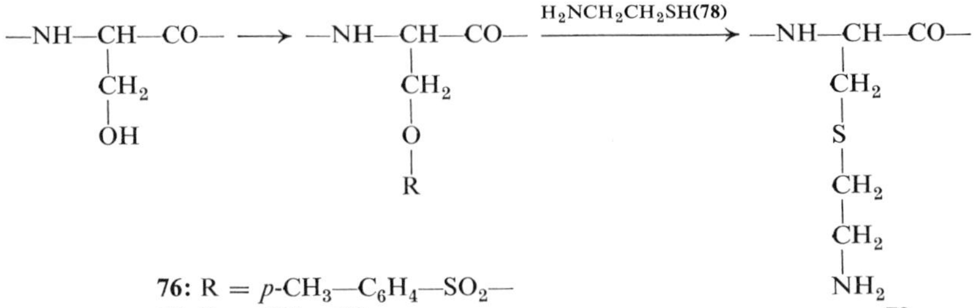

$$
\begin{array}{ccccc}
\text{—NH—CH—CO—} & \longrightarrow & \text{—NH—CH—CO—} & \xrightarrow{\text{H}_2\text{NCH}_2\text{CH}_2\text{SH(78)}} & \text{—NH—CH—CO—} \\
\quad\mid & & \quad\mid & & \quad\mid \\
\quad\text{CH}_2 & & \quad\text{CH}_2 & & \quad\text{CH}_2 \\
\quad\mid & & \quad\mid & & \quad\mid \\
\quad\text{OH} & & \quad\text{O} & & \quad\text{S} \\
& & \quad\mid & & \quad\mid \\
& & \quad\text{R} & & \quad\text{CH}_2 \\
& & & & \quad\mid \\
& & & & \quad\text{CH}_2 \\
& & & & \quad\mid \\
& & & & \quad\text{NH}_2 \\
\end{array}
$$

76: R = p-CH$_3$—C$_6$H$_4$—SO$_2$—
77: R = CH$_3$—SO$_2$—

79

of the tosyl or mesyl group then converts the serine into the thialysine derivative (79) [246, 247]. The procedure was tested on model peptides. The three steps (sulfonylation, displacement with cysteamine, and tryptic digestion) were also carried out without isolation of the intermediates, with over-all yields approaching 70%.

S-CARBOXAMIDOMETHYLATION

Synthetic compounds like N-acetyl-homoglutamine methyl ester (80) are good substrates for trypsin [214]. Likewise, cysteine peptides on treatment

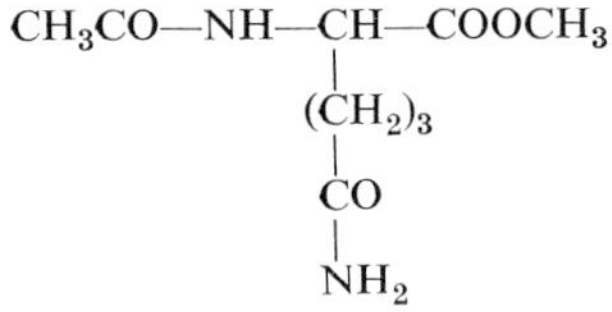

$$
\begin{array}{c}
\text{CH}_3\text{CO—NH—CH—COOCH}_3 \\
\mid \\
\text{(CH}_2)_3 \\
\mid \\
\text{CO} \\
\mid \\
\text{NH}_2 \\
\textbf{80}
\end{array}
$$

with iodoacetamide are converted into S-carboxamidomethylcysteine derivatives (81). Since S-carboxamidomethylation is a procedure routinely

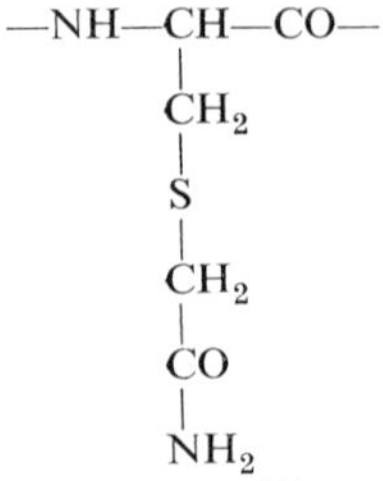

$$
\begin{array}{c}
\text{—NH—CH—CO—} \\
\mid \\
\text{CH}_2 \\
\mid \\
\text{S} \\
\mid \\
\text{CH}_2 \\
\mid \\
\text{CO} \\
\mid \\
\text{NH}_2 \\
\textbf{81}
\end{array}
$$

used for protection of sulfhydryl groups of proteins, the alkylated proteins can be directly employed for tryptic cleavage at the cysteinyl residue.

It is of interest that O-glycyl-serine methyl ester (**82**) is also a good substrate for trypsin [248]; in this case the amino group of glycine occupies the

$$\begin{array}{c} -\text{NH} \\ | \\ \text{CH—CH}_2\text{—O—CO—CH}_2\text{—NH}_2 \\ | \\ -\text{CO} \end{array}$$

82

place of the ε-amino group of lysine.

Blocking of Chymotrypsin Cleavage

Since chymotrypsin does not show the high specificity of peptide bond cleavage of trypsin, no comparative search for methods of blocking the chymotryptic cleavage at the level of the aromatic amino acids has been made. It is conceivable that at least some of the available selective modifications of tyrosyl and tryptophan residues will provide procedures for blocking chymotryptic cleavage at the level of these residues.

Acylation, alkylation, and arylation of the hydroxyl group of tyrosine or modification with cyanuric fluoride [249, 250] are potential reactions to be explored for this purpose. Nitration with tetranitromethane [251–253] does not prevent chymotryptic cleavage of the bond adjacent to the 3-nitro-tyrosyl residue [254], tyrosine-O-sulfate resists cleavage [255, 256], while the diisopropylphosphoryl-tyrosine residue of a modified papain is released slowly by chymotrypsin [257].

The reaction of 2-hydroxy-5-nitrobenzyl bromide (Koshland's reagent) (**87**) with the tryptophan residue [258–260] was shown by Dopheide and Jones [261] to block chymotryptic cleavage at this residue. Porcine pepsin was alkylated with the reagent and the tryptophan peptides were isolated by gel filtration after enzymic digestion. The peptides containing the modified tryptophan residues were retarded by adsorption on the gel, and their elution could be followed by measurement of the color of the hydroxynitrobenzyl group. The amino acid compositions and sequences of four tryptophan peptides have been determined. Unfortunately the reagent reacts with the tryptophan residue giving several compounds (**88**) and others [262–264], so that each tryptophan-containing peptide gave several different peaks in the chromatogram upon gel filtration, even though the amino acid compositions were identical, implying that alternate reaction products of tryptophan and the alkylating reagent were present [261].

The selective reaction of sulfenyl halides, like 2-nitrophenylsulfenyl chloride (**85**) [265–269] could find analogous application to the Koshland process. However, in this case the reaction product is a single 2-thioaryl derivative (**86**).

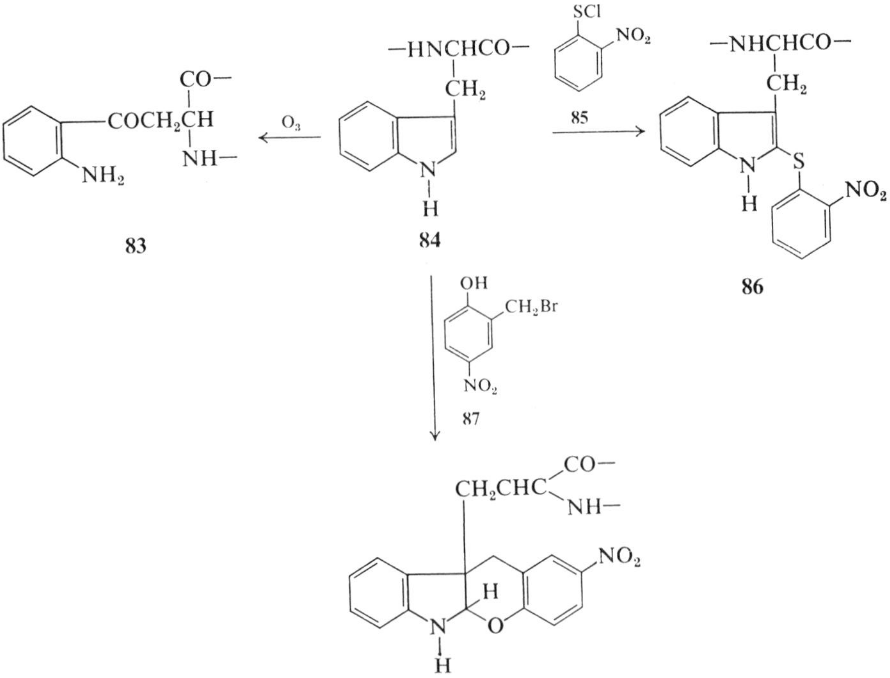

and several other products

Tryptophan-peptides (**84**) after conversion into kynurenyl derivatives (**83**) by ozonization [270–273], were also no longer cleaved by proteolytic enzymes [267].

Blocking of Leucine Aminopeptidase Cleavage

Leucine amino-peptidase (LAP) releases amino acids sequentially from the amino terminus of a peptide. With the exception of the imide bond of proline, which is completely resistant, all amino acids are released at comparable rates [14]. This offers a method for obtaining complete hydrolysis of a protein when analysis for tryptophan, asparagine, and glutamine, which are destroyed in the acid hydrolysis, is wanted. Also, N-acylated peptides are resistant to the enzyme.

The LAP digestion of a polypeptide has also found useful applications in protein sequence work in special cases. The peptide S (**89**) of ribonuclease A has been carbamylated at the N_α- and N_ε-amino groups and then cleaved at the arginine 10 peptide bond, releasing a N-terminal-glutaminyl peptide [129] (N_ε-carbamyl-lysine is resistant to the action of trypsin. The tryptic digest, *without separation*, was treated with LAP, and the amino acids released were those corresponding to the C-terminal 11–20 sequence of the peptide. The acylated N-terminal fragment was resistant to the LAP action.

A chymotryptic peptide derived from papain (**90**) [274] released by LAP

Ser-Ile-Ala-AsN-GlN-Pro-Ser-Val-Val-Leu
90

digestion the four N-terminal amino acids in equal amount. The residual peptide must have an amino terminal residue before a proline residue, in agreement with the specificity of the enzyme [14].

The presence of a β-aspartyl sequence also blocks the LAP digestion [275] so that the comparison between the recovery on the analyzer of the amino acids after acid hydrolysis and after LAP digestion will give information for the amino acids following the β-aspartyl residue.

These examples suggest that LAP could be usefully employed in sequence work. However, at present the major problem is the difficulty of obtaining LAP preparations free of *endo*-peptidase action [147].

A relatively rapid and reliable procedure has been recently described for preparing highly purified LAP from the supernatant fraction of hog kidney. The enzyme did not detectably hydrolyze cytochrome c, rabbit actin, or mercuripapain, proteins whose amino-terminal sequences are refractory to *exo*-peptidase action [276].

4 CHEMICAL CLEAVAGE

The chemical cleavage of polypeptide chains exploits the unique reactivity of the side chains of particular amino acids in the labilization of adjacent peptide bonds by neighboring group participation [4, 37, 38]. The residues investigated so far for this purpose have been methionine, tryptophan, cysteine, and the aromatic, the hydroxy, and the dicarboxylic amino acids. New oxidants, alkylating agents, and nucleophiles are still under study. The methods are an ideal adjunct to the enzymic cleavage and approach enzymes in their selectivity. The subject has been reviewed recently [7, 10, 11, 277–279] and only the commonest and newest procedures are described here.

Cleavage of Disulfide Bonds

Cystine residues may form links between two different chains as in insulin, or they may hold different parts of a single chain together as in ribonuclease. The cleavage of the disulfide bonds of a protein has to be performed before any sequence work in order to separate the peptide chains linked through the SS-bridges. In addition disulfide bonds are reactive groups which can cause complications during structural studies, so that it is necessary to stabilize them as suitable derivatives. Finally, proteins in which the cystine residues are split are more susceptible to attack by proteolytic enzymes.

Three procedures are available for the cleavage of disulfide bonds; reduction, oxidation, and sulfite treatment.

Reduction is commonly accomplished with β-mercaptoethanol or mercaptoethylamine in the presence of $8M$ urea or 4–$6M$ guanidine hydrochloride and EDTA [155, 280, 281]. D,L-1-4-Dithioerythritol (Cleland's reagent, HS–CH_2–$(CHOH)_2$–CH_2–SH) [282], because of its low redox potential and stability to air oxidation, is the reagent of choice for the reduction of disulfide bonds and the maintenance of thiols in their reduced state. The reaction is driven to completion with the favorable formation of a six-membered ring from the oxidation of the dithioerythritol.

The chelating agent EDTA is required to minimize the unfavorable contamination by heavy metal ions which can catalyze reoxidation of the resultant sulfhydryl groups. The SH-groups are usually stabilized by alkylation with iodoacetic acid or its amide [10, 281], ethyleneimine [233, 242] and acrylonitrile [283, 284]. Thioglycolic acid and sodium borohydride are not used as reducing agents, as they cause side reactions [8,10, 11, 285].

Performic acid, prepared by mixing formic acid and hydrogen peroxide, oxidizes cystine to two residues of cysteic acid (91) [10, 20, 286, 287]. Contemporaneously methionine is converted to methionine sulfone (92) and

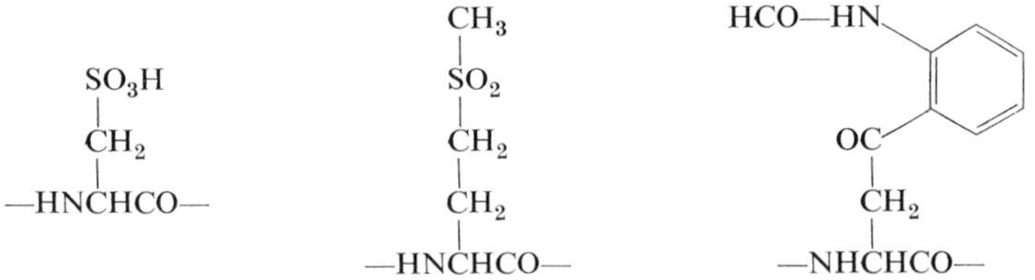

tryptophan to formyl-kynurenin (93) and partly to unknown substances [287–289].

Disulfides are cleaved by sulfites to S-sulfonates (94) and sulfhydryl groups [245, 290]. If the reaction is allowed to proceed in a dissociating

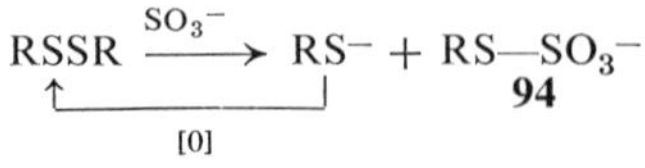

$$\text{RSSR} \xrightarrow{\text{SO}_3^-} \text{RS}^- + \text{RS}\text{—}\text{SO}_3^-$$
$$\overset{\uparrow\underline{\hspace{3cm}}}{\text{[O]}} \qquad\qquad \mathbf{94}$$

medium of high concentrations of urea or guanidine hydrochloride and in the presence of an oxidizing agent, for example, tetrathionate or Cu^{2+} ions [291], all the cystine residues can be converted into the S-sulfonate cysteine derivative. The completely S-sulfonated proteins so obtained are useful in the separation of polypeptides since they are stable in neutral and acidic conditions [291]. An advantage of the S-sulfonate as a blocking group is its ready removal by treatment with excess thiol. Recently a new method has been described for the complete sulfonation of proteins containing cysteine or cystine residues by treating the protein with sodium sulfite and catalytic amounts of cysteine in the presence of oxygen and $8M$ urea [292].

The opening of the disulfide bonds in proteins has been achieved by phosphorothioate (95) [293-296]. The reaction is a nucleophilic heterolytic scission, which proceeds as shown in the equation.

$$RSSR + SPO_3{}^{3-} \longrightarrow RSSPO_3{}^{2-} + RS^-$$
$$\mathbf{95}$$

The Cyanogen Bromide Reaction

The cyanogen bromide reaction for the cleavage of methionyl-peptides remains unequaled in ease, quantitative efficiency, and selectivity [4, 297–299], the proposed mechanism being shown in the scheme.

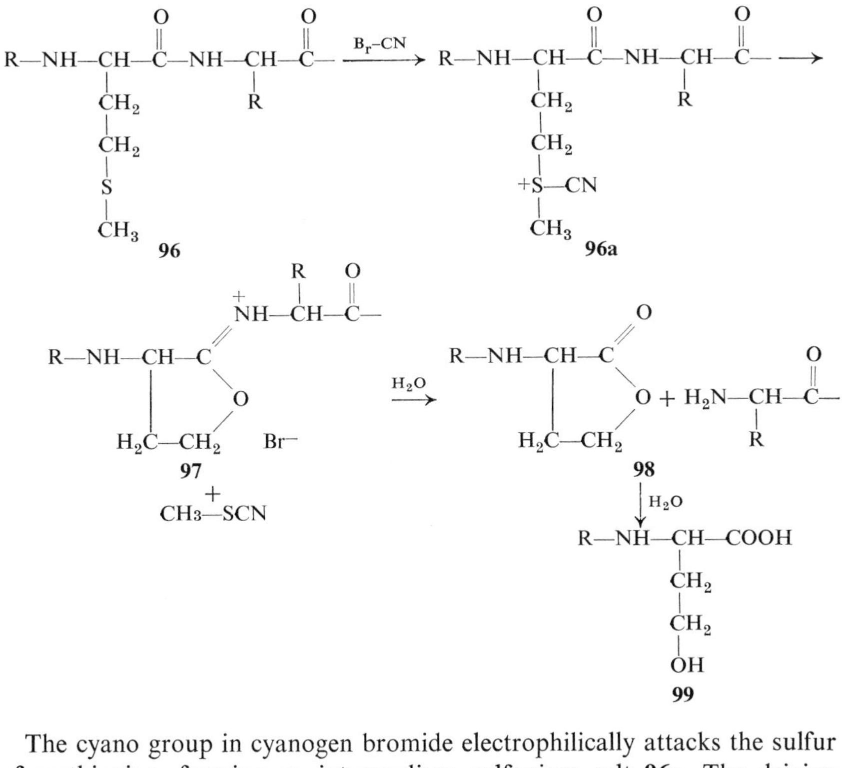

The cyano group in cyanogen bromide electrophilically attacks the sulfur of methionine, forming an intermediary sulfonium salt **96a**. The driving force of the subsequent steps, which lead to cleavage of the methionyl peptide bond, is the participation of the carbonyl group of the methionine carboxyl. This neighboring group effect [37, 38] leads to the formation of an imino lactone **97** which is readily hydrolyzed. In the process, methionine is converted into homoserine lactone (**98**) which is further in part hydrolyzed to homoserine (**99**). Cleavages are often 90–95% complete. The excess of the reagent and the by-products are volatile and are easily removed by lyophilization.

Since the discovery of this reaction and its first application to ribonuclease, it has been applied to structural studies of a wide variety of proteins, ranging from the *S*-peptide of ribonuclease, MW 2000 [300], to large proteins such as thyroglobulin, MW 600,000 [301]. The most complete extensive use of cyanogen bromide was on human immunoglobulin γG [302, 303], in which the light and heavy chains were cleaved at the methionyl residues to release a total of 10 different peptides, which were all separated, identified, and their primary sequence elucidated. The applications of the reaction have been reviewed [7, 299].

The use of 0.1*N* hydrochloric acid as a solvent for the reaction [298] has generally been superseded by 75% trifluoroacetic acid [304] or by 70% formic acid [241].

Recently [2] the reaction has been applied to catalase and the effects of solvent and concentration on the rate of conversion of methionine into homoserine and the extent of cleavage reported [2, 305, 306]. It was found that Met-Thr or Met-Ser bond are cleaved in low yields. The use of trifluoroacetic acid is advantageous when hydroxy amino acids are involved in the cleavage reaction [2]. Intramolecular reaction (**101** and **102**) of the threonyl or seryl residues at the stage of the iminolactone is probable with the reaction course shown.

The final product (**103** or **104**) may be one in which the normal peptide bond has been established, with the difference that the methionine is converted into the homoserine residue. The increased acidity of trifluoroacetic acid may have the effect of preventing the side reaction by protonation of the hydroxyl groups as well as the iminolactone (**100**).

S-Methyl-cysteine, the lower homolog of methionine, does not react with cyanogen bromide. The β-lactone **106**, which would result from a reaction analogous to that of methionine with cyanogen bromide, is not formed. However, if the amino group of *S*-methyl-cysteine is acylated or aminoacylated (**107**), a 1,5-relationship between the oxygen of the carbonyl group of the acyl group (and the β-carbon atom) is introduced and the reaction proceeds with formation of an oxazolinium bromide (**109**) and methylthiocyanate [307, 308].

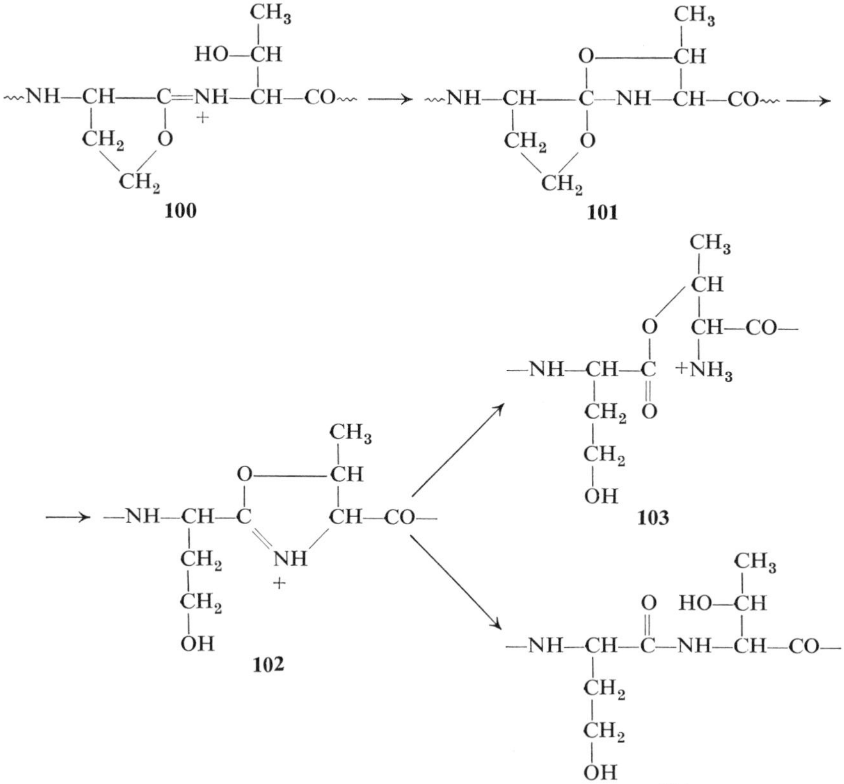

$$
\begin{array}{c}
\text{CH}_3 \\
| \\
\text{HO—CH} \\
| \\
\text{⋯NH—CH——C}=\text{NH—CH—CO⋯} \longrightarrow \\
| \quad\quad + \\
\text{CH}_2 \quad \text{O} \\
\backslash \quad / \\
\text{CH}_2 \\
\textbf{100}
\end{array}
\qquad
\begin{array}{c}
\text{CH}_3 \\
| \\
\text{O———CH} \\
/ \quad | \\
\text{⋯NH—CH——C—NH—CH—CO⋯} \longrightarrow \\
| \quad\quad \\
\text{CH}_2 \quad \text{O} \\
\backslash \quad / \\
\text{CH}_2 \\
\textbf{101}
\end{array}
$$

$$
\longrightarrow \quad
\begin{array}{c}
\text{CH}_3 \\
| \\
\text{O———CH} \\
/ \quad | \\
\text{—NH—CH—C} \quad\quad \text{CH—CO—} \\
| \quad\quad \backslash\backslash\text{NH} \\
\text{CH}_2 \quad\quad + \\
| \\
\text{CH}_2 \\
| \\
\text{OH} \\
\textbf{102}
\end{array}
$$

$$
\begin{array}{c}
\text{CH}_3 \\
| \\
\text{CH} \\
/ \quad | \\
\text{O} \quad \text{CH—CO—} \\
\backslash\backslash \quad | \\
\text{—NH—CH—C} \quad +\text{NH}_3 \\
| \\
\text{CH}_2 \quad \text{O} \\
| \\
\text{CH}_2 \\
| \\
\text{OH} \\
\textbf{103}
\end{array}
$$

$$
\begin{array}{c}
\text{CH}_3 \\
| \\
\text{O} \quad \text{HO—CH} \\
\backslash\backslash \quad | \\
\text{—NH—CH—C—NH—CH—CO—} \\
| \\
\text{CH}_2 \\
| \\
\text{CH}_2 \\
| \\
\text{OH} \\
\textbf{104}
\end{array}
$$

At low temperature the oxazolinium bromide (**109**) is readily hydrolyzed to the O-acyl-derivative (O-seryl-peptide) (**110**) (mechanism A). To complete the cleavage the ester has to be hydrolyzed to produce the two expected fragments. At high temperature, however, the dehydroalanine derivative is formed via a mechanism of β-elimination (mechanism B) [308]. The latter is hydrolyzed to a pyruvic acid derivative (**113**) and N-acetylaminoacylamide (**112**). The reaction was successfully applied with several S-methyl-cysteine peptides [308]. Glutathione has been cleaved by this process [309]. The useful application to proteins remains to be demonstrated.

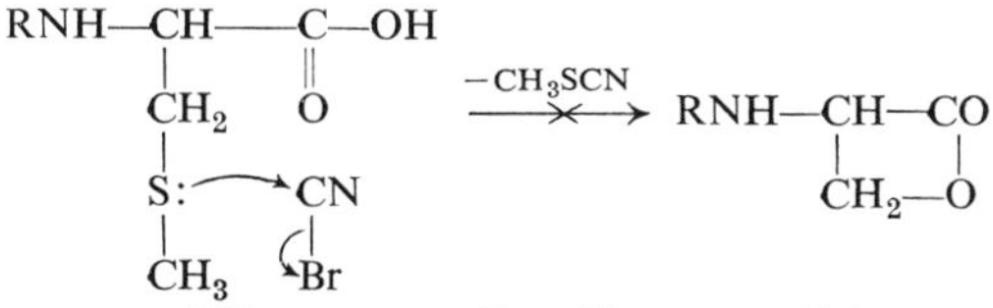

$$
\begin{array}{c}
\text{RNH—CH———C—OH} \\
| \quad\quad \| \\
\text{CH}_2 \quad \text{O} \\
| \\
\text{S:} \quad \curvearrowright \text{CN} \\
| \quad\quad \big\downarrow \\
\text{CH}_3 \quad \text{Br} \\
\textbf{105}
\end{array}
\xrightarrow[\quad]{-\,\text{CH}_3\text{SCN}}
\begin{array}{c}
\text{RNH—CH—CO} \\
| \\
\text{CH}_2\text{—O} \\
\\
\textbf{106}
\end{array}
$$

$$R = H$$

H₃C—C—NH—CH—C——NH   H₃C—C—NH—CH—C——NH
    ‖ | ‖ | ‖ | ‖ |
    O   R   O   CH—CONH—   O   R   O   CH—CONH—
                      /   /
                   CH₂   CH₂
                   | |
                   S: C—Br   ⟶   +S—CN
                   | ‖ |
                   CH₃ N   Br⁻   CH₃
       107   **108**

A | 0-5°
  | 6 days
  ↓

$$H_3C—C—NH—CH—C \quad \overset{+}{N}H_3 \; Br^- \qquad \xleftarrow{H_2O} \qquad H_3C—C—NH—CH—C \overset{+}{=\!=\!=} NH \; Br^-$$

H₃C—C—NH—CH—C   ⁺NH₃ Br⁻   H₂O   H₃C—C—NH—CH—C===NH Br⁻
    ‖ | ‖ | ⟵   ‖ | ‖ |
    O   R   O   C—CONH—   O   R   O   CH—CONH— B
                       /   /
                     CH₂   CH₂
       110   **109** + CH₃SCN

H₃C—C—NH—CH—C—NH₂   ⟵H₂O   H₃C—C—NH—CH—C——NH
    ‖ | ‖   ‖ | ‖ |
    O   R   O   O   O   C
                                           /‖
                                        H₂C   CONH—
       112   **111**

O
‖
+ CH₃—C—CONH—
     113

The analogous derivative *N*-acetyl-*S*-carboxymethylcysteine does not react with BrCN. In proteins this resistance to the reagent can be utilized to

H₃C—CO—NH—CH—COOH
               |
              CH₂
               |
               S
               |
             CH₂COOH
           114

advantage to establish overlapping sequences. The disulfide bonds are reduced and one aliquot is *S*-carboxymethylated, the other *S*-methylated. Only the *S*-methylated protein will be subject to cleavage by cyanogen bromide at the *N*-aminoacyl bond of the alkylated cysteine.

N-Bromosuccinimide Cleavage

The γ-δ-double bond, although part of an aromatic pyrrole (tryptophan) (**115**), phenol (tyrosine) (**116**), or imidazole (histidine) (**117**), will invite 1,5-interaction of the carboxy carbonyl to form spiro-γ-iminolactones when a suitable driving force is provided through a bromonium intermediate, that is, by the action of positive halogen compounds like *N*-bromosuccinimide (NBS) (**121**). The reactive iminolactones (**118–120**) may then be hydrolyzed with

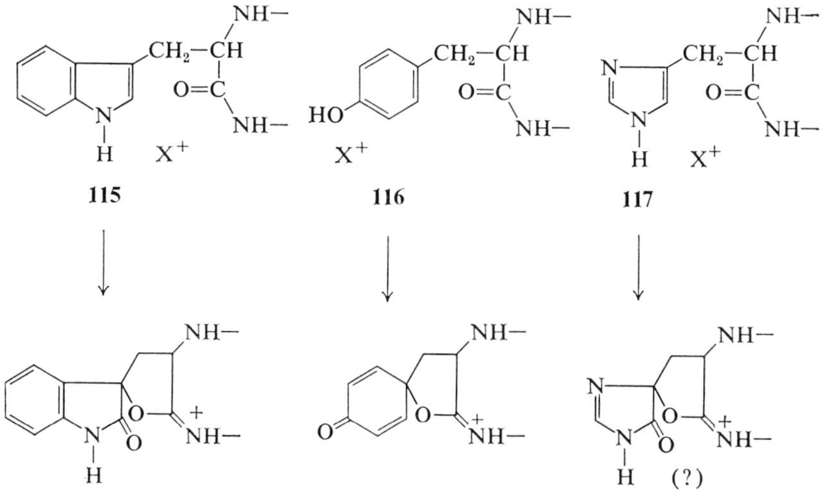

115   116   117

118   119   11 (?)
                                      120

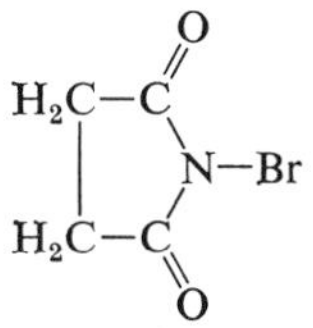

121

the release of the amine component.

Peptides containing unsaturated amino acids with isolated double bonds, as in allylglycine (2-amino-4-pentenoic acid) (**122**), participate in analogous intramolecular displacement reactions.

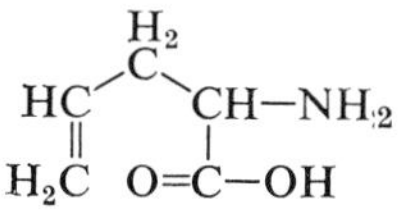

122

In a similar way, by reduction with lithium in methylamine, phenylalanyl peptides (**123**) are converted into tetrahydrophenylalanyl peptides (**124**) which contain the allylglycyl group. The adjacent peptide bond is again cleaved by the action of NBS through analogous iminolactone (125) formation [310].

The NBS cleavage of peptides has been extensively reviewed [4, 7, 311–316].

The NBS oxidation in proteins has been used mainly in connection with structure–activity studies of tryptophan-containing enzymes [317, 318].

With limited amounts of NBS, often at a pH close to neutrality, preferential modification of tryptophan is usually observed. However, *selective* modifications of proteins by NBS must be interpreted with caution. Tryptophan, as well as histidine, and tyrosine residues in lysozyme are all oxidized by NBS at pH 4.0 and 5.5 more or less simultaneously [319].

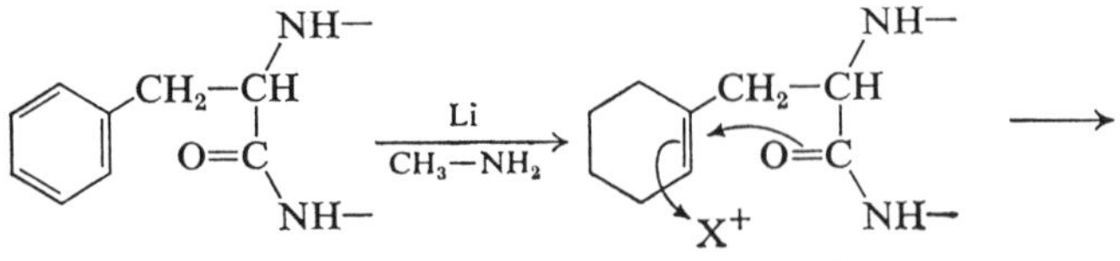

NBS is a reagent extremely reactive, and although modification of tryptophan, tyrosine, and histidine residues is usually accompanied by peptide bond cleavage, several other side reactions are determined by the reagent, such as oxidation of cysteine, cystine, and methionine [315, 319, 320], and decarboxylation of free amino acids [321]. Protective agents, such as formate ion [322] or urea [323] solution, have been used in order to minimize such side reactions.

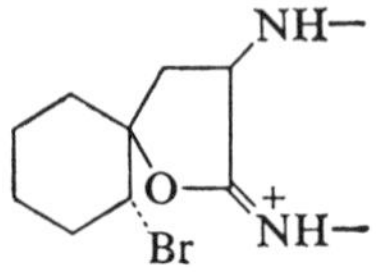

Although the cleavage yields on model peptides are quite satisfactory, cleavages of proteins are rather low (10–50%) [315]. In addition the poor selectivity of the reagent gives rise to multiple forms of the cleaved peptides, so that the isolation of the fragments is a difficult task [324]. In practice, therefore, the rather low yields of cleavage have restricted the use of the NBS cleavage of proteins to a spot-checking technique for sequence studies.

A method has been reported [325, 326] for limiting cleavage next to the tryptophan residue by NBS by converting it to the kynurenyl residue by ozonization [267]. Tryptophan peptides, after reaction with 2-hydroxy-5-nitrobenzylbromide [258], are not susceptible to cleavage [327]. On the other hand, when tryptophan is modified with sulfenyl halides, NBS is still able to cleave the 2-thioaryl-tryptophanyl peptide bond [266–268]. The mechanism of the cleavage most probably involves an iminolactone **126** intermediate, analogous to **118**, formed with tryptophan peptides [328, 329].

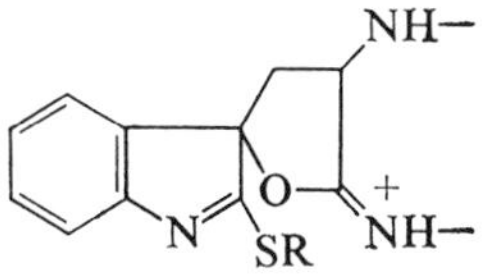

126

A much milder oxidizing agent than NBS is **127** (BNPS-skatole), which was used [320] to selectively modify the single tryptophan residue of staphy-

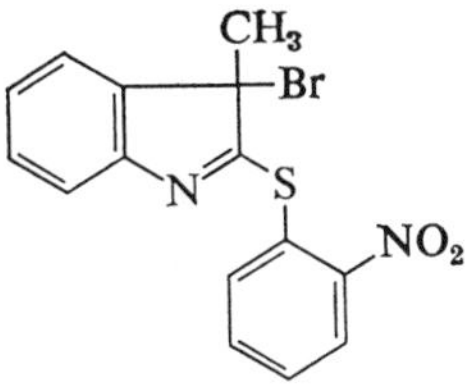

127

lococcal nuclease. The only other amino acid affected is methionine, which is converted by the reagent into its sulfoxide. Reduction with thioglykolic acid reverses the latter process. Cleavage of the tryptophanyl peptide bond also can be achieved using relatively high amounts of reagent and longer reaction times. When the A_1 encephalitogenic protein was treated with 10 equivalents of reagent in 50% acetic acid at 37° for 24 hr the cleavage at the single tryptophanyl-peptide bond occurred selectively to the extent of 50% [330].

The oxidation of the indole nucleus of tryptophan by NBS has been successfully used for the analytical determination of this residue in proteins [318, 331]. The method of assay makes use of the fact that the indole chromophore of tryptophan, absorbing strongly at 280 mμ, is converted on oxidation with NBS into oxindole, a much weaker chromophore at this wavelength.

Cleavage via Dehydroalanine

When cysteine or serine peptides are converted into derivatives in which β-elimination is facilitated, the resulting dehydroalanyl peptide **128** can be induced to cleave at the α-carbon atom [7].

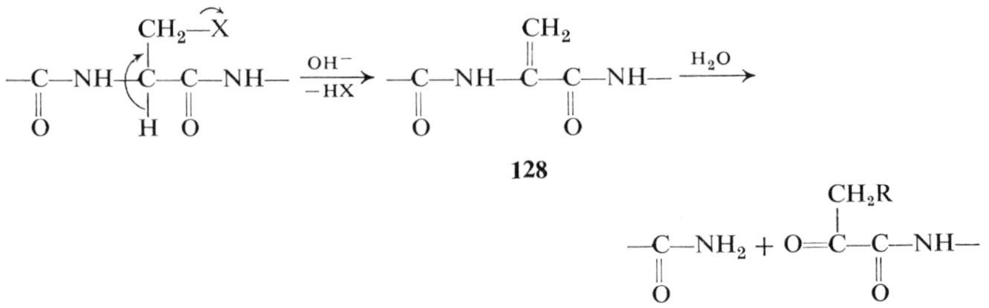

$$\underset{\underset{O}{\parallel}}{C}-NH-\underset{\underset{H}{|}}{\overset{\overset{\frown}{CH_2-X}}{\underset{|}{C}}}-\underset{\underset{O}{\parallel}}{C}-NH- \xrightarrow[-HX]{OH^-} -\underset{\underset{O}{\parallel}}{C}-NH-\underset{\underset{\parallel}{CH_2}}{C}-\underset{\underset{O}{\parallel}}{C}-NH- \xrightarrow{H_2O}$$

128

$$-\underset{\underset{O}{\parallel}}{C}-NH_2 + O=\underset{}{C}-\underset{\underset{O}{\parallel}}{\overset{\overset{CH_2R}{|}}{C}}-C-NH-$$

$$X = OR \text{ or } SR; \ R = H \text{ or } OH$$

Cysteine

The chemical cleavage of cysteinyl peptides [332] involves (*1*) alkylation or arylation of the SH group to dimethyl- [332], cyanomethyl- [307], 2,4-dinitrophenyl- or picryl [333] -sulfonium groups; (*2*) the formation of a dehydroalanine peptide by β-elimination in alkaline media; and (*3*) hydrolysis of the dehydroalanine peptide to a C-terminal peptide amide and a *N*-pyruvoyl (or hydroxy-pyruvoyl)-peptide.

The *N*-pyruvoyl or the *N*-hydroxypyruvoyl group is removed from the peptide fragment by alkaline hydrogen peroxide in 50% yield [334]. The

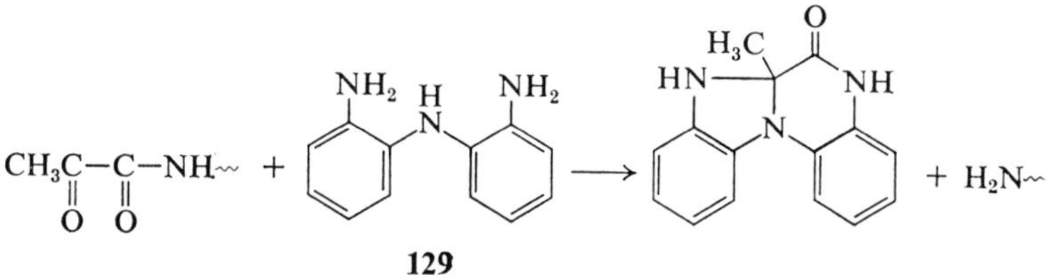

$$CH_3C-\underset{\underset{O}{\parallel}}{C}-NH\sim \ + \quad \text{(diaminodiphenylamine)} \ \longrightarrow \ \text{(product)} \ + \ H_2N\sim$$

129

removal of the blocking group may be effected also by 2,2'-diaminodiphenylamine (**129**) at pH 3–6. The free amino group is liberated in 75–80% yield [333].

The application of these methods to the cleavage of reduced cystine bonds in ribonuclease (eight half-cystines) gave the expected seven new amino terminal residues in 30 to 55% yield [332].

Dehydroalanine peptides are also formed directly from cystine peptides (**130**) by the action of alkali. The β-elimination of the disulfide group occurs easily in the cystine residues of proteins [290]. The breakdown of the disulfide group is initiated by hydrogen abstraction by an attack of a hydroxyl ion followed by release of an *S*-thiocysteine residue **132**. The latter decomposes further into cysteine and free sulfur [335].

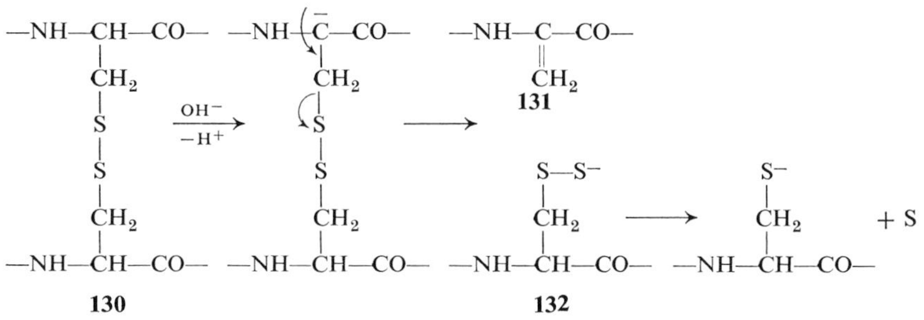

$$\begin{array}{ccc}
-\text{NH}-\text{CH}-\text{CO}- & -\text{NH}-\overset{-}{\text{C}}-\text{CO}- & -\text{NH}-\text{C}-\text{CO}- \\
| & | & \| \\
\text{CH}_2 & \text{CH}_2 & \text{CH}_2 \\
| & | & \mathbf{131} \\
\text{S} \xrightarrow[-\text{H}^+]{\text{OH}^-} & \text{S} & \\
| & | & \\
\text{S} & \text{S} & \text{S}-\text{S}^- \quad\quad \text{S}^- \\
| & | & | \quad\quad\quad\quad | \\
\text{CH}_2 & \text{CH}_2 & \text{CH}_2 \longrightarrow \text{CH}_2 \quad + \text{S} \\
| & | & | \quad\quad\quad\quad | \\
-\text{NH}-\text{CH}-\text{CO}- & -\text{NH}-\text{CH}-\text{CO}- & -\text{NH}-\text{CH}-\text{CO}- \quad -\text{NH}-\text{CH}-\text{CO}- \\
\mathbf{130} & & \mathbf{132}
\end{array}$$

The direction of the initial C–S bond breakage is governed both by steric factors [336] and, in proteins, probably by forces of secondary structure. The dehydroalanine moiety (131) is capable of adding nucleophilic groups, such as the side-chain amino group of lysine to yield (2-amino-2-carboxyethyl)-lysine (lysino-alanine) [337].

In principle, the elimination of disulfide groups from cystine residues could be useful for the cleavage of peptide chains. However, since the C–S bond cleaves on both sides of the unsymmetrical disulfide bridge and since the side reactions involve the dehydroalanine residue formed, the problem has had little investigation.

Serine

Serine peptides may be cleaved through dehydroalanine formation using methods analogous to those described above for serine peptides. By sub-stitution of the serine hydroxyl group of serine by strongly electron-attracting groups, analogous β-elimination process to dehydroalanine peptide can be induced. When O-tosyl-serine peptides (133) are treated with base, the corresponding dehydroalanine peptides are formed in almost quantitative

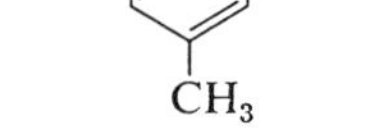

$$\begin{array}{c}
-\text{NH}-\text{CH}-\text{CO}- \\
| \\
\text{CH}_2 \\
| \\
\text{O} \\
| \\
\text{SO}_2 \\
| \\
\bigcirc \\
| \\
\text{CH}_3 \\
\mathbf{133}
\end{array}$$

yields [338, 339]. Analogously the O-mesyl (CH_3SO_2-) [246, 340] and the O-diphenylphosphoryl [$(C_6H_5)_2PO-$] group [338, 341] were used. Chymo-trypsin was selectively O-tosylated at the active serine of the enzyme's catalytic center with tosyl fluoride. Base converted this derivative into anhydrochymotrypsin [342–344]. The essential role of the serine at the catalytic site for the activity of the enzyme was thus demonstrated.

Another potential leaving group for this transformation of serine to dehydroalanine peptides is the O-phosphoryl group present in some phosphoproteins. The kinetics of the β-elimination of phosphate from O-phosphoseryl-residues in proteins has been studied recently [345]. The elimination yields equimolar amounts of phosphate and dehydroalanine in 0.1–1N NaOH at 25–50°. However, some consumption of dehydroalanine occurs by side reactions, for example, by condensation with ε-amino groups of lysine to lysino-alanyl residues [337].

When the O-tosyl group of O-tosyl-serine peptides is displaced by thiolacetate or thiobenzoate, S-acylated derivatives of cysteine are formed (134), from which the S-acyl group can be removed to give the cysteine residue in place of the original serine [246, 340, 346].

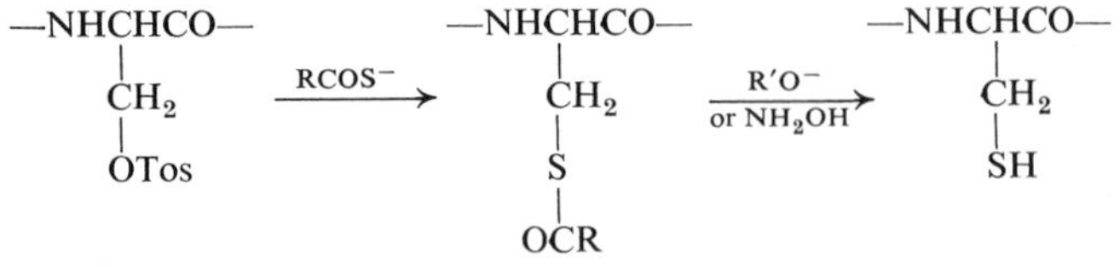

$$
\begin{array}{ccccc}
-\text{NHCHCO}- & & -\text{NHCHCO}- & & -\text{NHCHCO}- \\
| & & | & & | \\
\text{CH}_2 & \xrightarrow{\text{RCOS}^-} & \text{CH}_2 & \xrightarrow[\text{or NH}_2\text{OH}]{\text{R'O}^-} & \text{CH}_2 \\
| & & | & & | \\
\text{OTos} & & \text{S} & & \text{SH} \\
& & | & & \\
& & \text{OCR} & & \\
\end{array}
$$

134

R = —CH$_3$ or C$_6$H$_5$—

When cysteamine (H$_2$NCH$_2$CH$_2$SH) is used as a nucleophilic agent from serine peptides, the corresponding thialysine (S-aminoethyl-cysteine) peptides are obtained, which are subsequently hydrolyzed by trypsin [346, 347].

Cleavage via Carbonylamides

Ring closure reactions of hydrazines or hydroxylamines with ω-carbonyl-amides or esters are well known [230]. The reaction involves condensation with the carbonyl moiety (135) and further cyclization with formation of a heterocycle (136) and release of an amine or ester component (137).

The same basic reaction has been employed for the cleavage of the acetoacetyl (CH$_3$COCH$_2$CO–) group, recently proposed as a reversible N-protective residue in peptide synthesis [162–164] and in protein sequence studies,

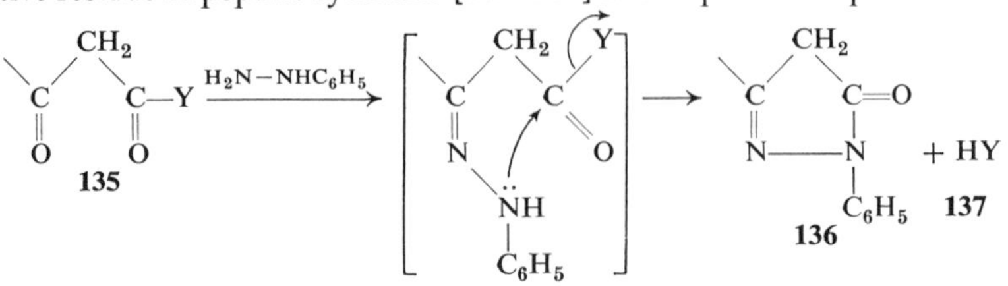

Y = —NHR; —OR

that is, for the reversible protection of the ε-amino groups of lysines [165, 166].

Based on the same rationale, novel procedures have been devised for selective peptide bond cleavage, taking advantage that ω-carbonyl moieties can be generated in the polypeptide chain at the level of tryptophan by its conversion into kynurenine, and of serine and threonine by oxidation of the hydroxyl function.

Tryptophan

Tryptophan peptides (138) are converted into N-formyl-kynurenine peptides (139) by ozonization in anhydrous formic acid [267, 270, 271], preferably containing resorcinol [272]. An analogous procedure was also employed by dissolving simple peptides in ethyl acetate [273]. Ozonolysis converts quantitatively cystine into cysteic acid, methionine to its sulfone, and tyrosine is also slowly oxidized [267, 270]. Likewise, proflavin-sensitized photooxidation of tryptophan-containing peptides [347] or proteins [348] in formic acid converts tryptophan to N-formylkynurenine in higher yields than those achieved through ozonization. The concomitant oxidation of methionine to its sulfoxide is easily reversed by mercaptans.

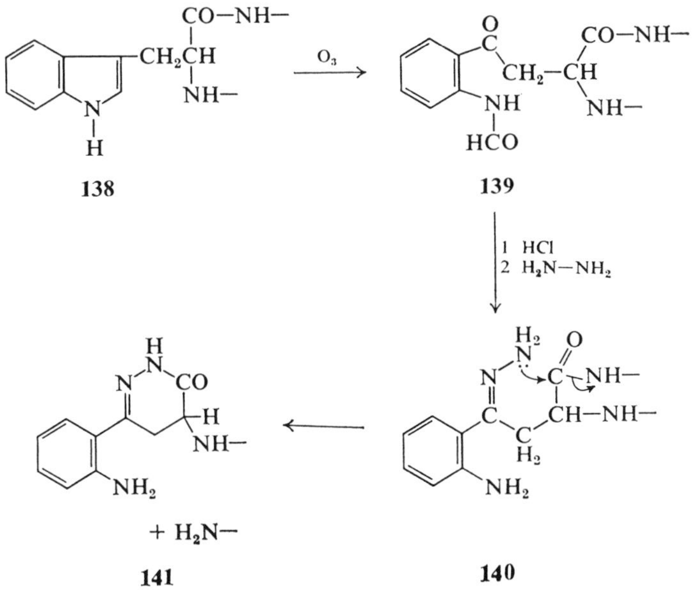

Kynurenine peptides in hydrazine–acetate buffer, pH 3.6 at 100°, form hydrazones 140 which undergo cyclization to tetrahydropyridazones 141

with concomitant cleavage [349, 350]. A tripeptide and three pentapeptides were cleaved by this procedure in nearly 50% yield. However, some non-specific cleavage to an extent of 5–13% also occurred.

Kynurenine peptides are reduced with sodium borohydride [351] or electrolytic reduction at controlled potential in 5 or 50% acetic acid [267, 270, 352] to γ-(o-aminophenyl)-homoserine (**142**). When tryptophan peptides, after ozonization and reduction, are heated for 2 hr at 100° in 0.5M

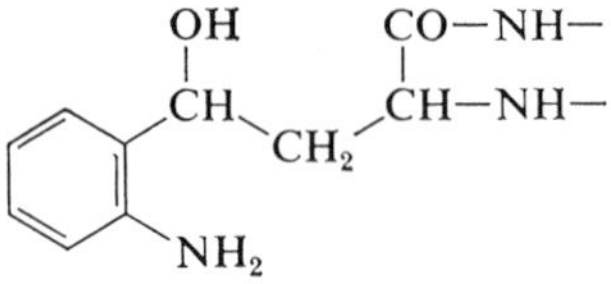

142

NaHCO$_3$, pH 7.0, or in 0.4N H$_2$SO$_4$ for 30 min [267, 270] or 0.2N HCl for several hours [352], both the N- and C-peptide bonds of the modified trypto-phyl residue were cleaved in moderate yields. The catalytic effect of the hydroxyl group on the cleavage of the peptide bond adjacent to the γ-(o-aminophenyl)-homoserine residue was extensively studied [352].

Serine and Threonine

Cleavage of the seryl and threonyl peptides (**143**) at the C-amide [163, 164, 167] is preceded by an oxidation of the hydroxyl group using dicyclohexyl-carbodiimide (DCC) and phosphoric acid in dimethylsulfoxide [353].

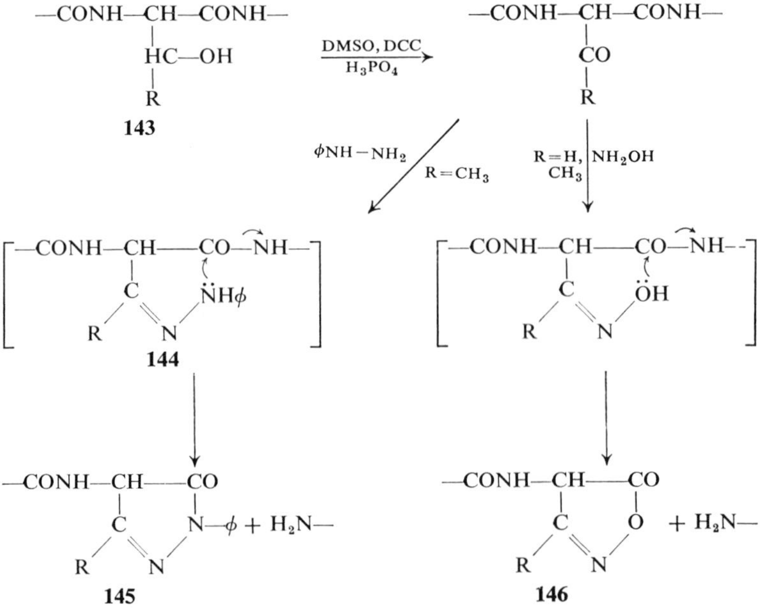

The peptide bond adjacent to the oxidized threonine was cleaved by reaction with phenylhydrazine for a few hours at 25° or 20–40 min at 40–70°. The intermediate phenylhydrazone (**144**) cyclizes to a pyrazolone (**145**) with the release of the adjacent amino terminal. Alternatively, hydroxylamine yielded an iso-oxazolinone (**146**) and the amine moiety [167].

Experiments with Thr-Gly or Thr-Gly-Gly gave yields of 80–90% for the oxidation step and 50–90% for the cleavage step. Preliminary research showed that seryl peptides are also oxidized by the above method to the corresponding β-aldehyde derivatives [167]. Whereas phenylhydrazine failed to cleave these oxidized peptides, hydroxylamine released new N-terminals; reaction products, rates, and yields are still under investigation.

More recent studies showed a satisfactory stability of the most common amino acids during the oxidation of threonine and serine, with the exception of tyrosine which is converted into a methylthiomethyl derivative. Furthermore, it was established that the β-carbonylamide derivative is a racemate, as expected because of the malonic feature of the asymmetric α-carbon [354].

Histidine

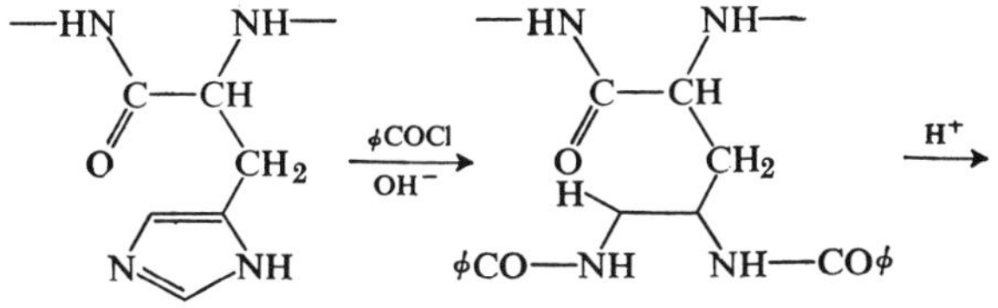

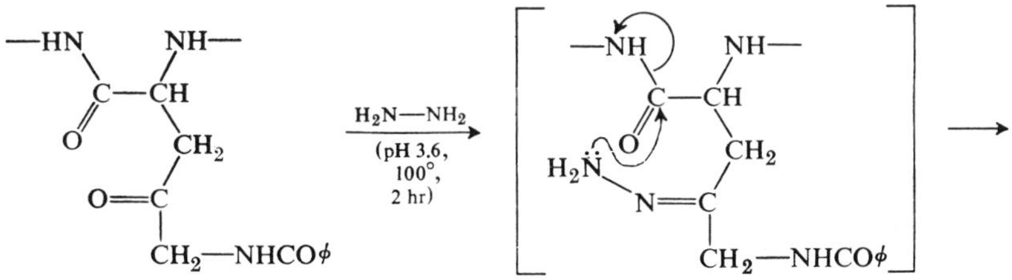

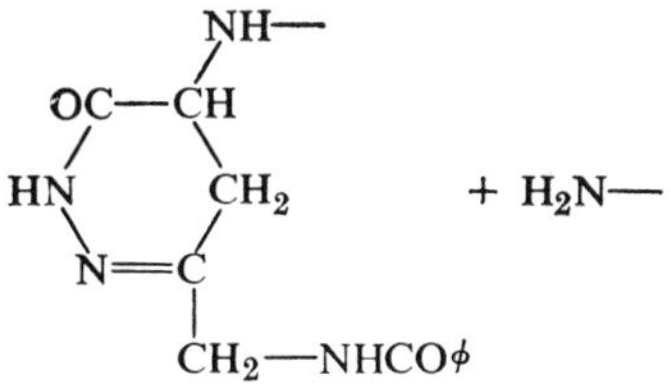

By the action of benzoyl chloride [355] or carbobenzoxy chloride [356] the imidazole ring of histidyl peptides (**147**) is converted into compounds containing the ene-diamine function (148). Mild acid hydrolysis converts 148 into a γ-keto-amide derivative 149.

On treatment with 5–10 equivalents of hydrazine acetate at pH 3.6 at 100° for 2 hr, the keto-amide undergoes cleavage with intramolecular cyclization to a tetrahydropyridazone compound (150) and concomitant release of the amine moiety [350, 357]. Several γ-keto-amide models have been cleaved quantitatively. However, some nonspecific bond cleavage was also found (10%) in peptides lacking the γ-keto group.

Miscellaneous

The oxidative cleavage of tyrosyl peptide bonds recently has been further investigated [358]. Similarly to N-bromosuccinimide, N-iodosuccinimide (NIS) effects the peptide bond cleavage in model compounds like diiodophloretylglycine or N-carbobenzoxy-tyrosyl-glycine (151). The extent of

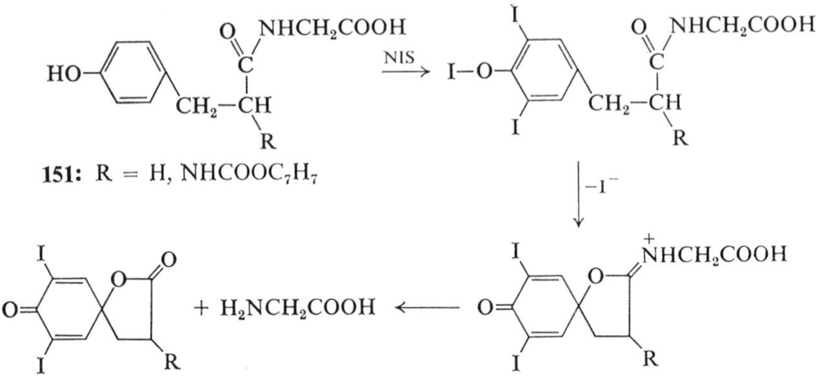

151: R = H, NHCOOC$_7$H$_7$

152

cleavage was 65–95%, as estimated from the dienone lactone (152) formation.

Tyrosyl peptides have been also cleaved by electrolytic oxidation [359–361]. At a platinum anode, analogous dienone lactone (152) conversion is observed with release of an amine moiety.

The cleavage of cystine peptides (153) by cyanide [362–364] appears to be remarkably mild, smooth, and selective.

The thiocyano-derivative 154, arising from the nucleophilic scission of the disulfide bond by cyanide, cyclize to a 2-iminothiazolidine (155) followed by a rapid hydrolysis to 156.

```
-CONH-CH-CONH-                         SH
      |                                 |
      CH2                               CH2
      |              CN-                 |
      S          --------->    -CONH-CH-CONH-
      |           Ph 7-7.4
      S                        -CONH-CH-CONH-
      |                               |
      CH2                             CH2
      |                               |
-CONH-CH-CONH-                       N≡C-S
     153                              154
```

```
-CO-N------CH-CONH-              HN------CH-CONH-
    |      |          H2O             |      |
    C      CH2      -------> -COOH +  ||     CH2
   //  \  /                           C    /
  HN    S                           HN  \ S
      155                              156
```

```
                                      | HCl
                                      | 6N
                                      ↓
                              HN------CH-COOH
                                   |      |
                                   C      CH2
                                  // \   /
                                 HN   S
                                   157
```

The cleavage of oxidized glutathione (158) at pH 7.4 (37°, 72 hr, 100-fold excess cyanide) released reduced glutathione, glutamic acid, and 2-iminothiazolidine-4-formylglycine (159) in quantitative yield [363].

```
γ-Glu-Cys-Gly                                    CONHCH2COOH
         |                                 HN------CH
         S             CN-                      |     \
         |          -------->  γ-Glu-Cys-Gly + Glu +  ||    CH2
         S            pH 7.4           |            C   /
         |                             SH         HN \ S
γ-Glu-Cys-Gly                                       159
     158
```

Bovine pancreatic ribonuclease released nine peptides when treated with 1000-fold excess of cyanide at pH 8, 37°, 48 hr. The peptides were separated and characterized. The composition of the peptides was found to be consistent with the known sequence of ribonuclease and corresponded to peptides resulting from cleavage at the N-acyl peptide bond at each of the eight half-cystines residues of the protein [363].

5 SEQUENTIAL DEGRADATION

After the partial degradation of a protein by enzymic and chemical methods and a clean separation by chromatographic and other techniques of each peptide fragment, the important and difficult task remains of the elucidation of the primary structure (sequence of the amino acids) of the isolated peptides.

It is important in this work that as many sequences as possible be determined by more than one technique, since all procedures can be subject to error. In some cases a danger of misinterpretation is that *trans*-peptidation, for example, peptide bond interchanges, especially occurring during partial acid hydrolysis [20], may lead to the appearance of new peptide sequences in the course of the analytical work. Trypsin, too, was found to initiate *trans*-peptidation reactions [365, 366]. Fortunately this is a rare occurrence, since all peptides isolated in significant yield are usually consistent with a unique sequential arrangement.

It must be stressed that each protein or peptide represents a separate problem. The efficacy of the various tools must be tested at every step, and the most favorable selected. Many sequences have to be confirmed by the use of several methods.

Degradation from the N-Terminus

Edman Degradation

The introduction of a stepwise chemical procedure for the degradation of peptides [367, 368] has provided an important stimulus to amino acid sequence studies. With the exception of results obtained by degradation with *exo*-peptidases, our knowledge of peptide sequence rests almost entirely on results obtained by stepwise degradation with phenylisothiocyanate, the Edman reagent. The series of reactions shown forms the basis of the method [367–371] which has been reviewed [372].

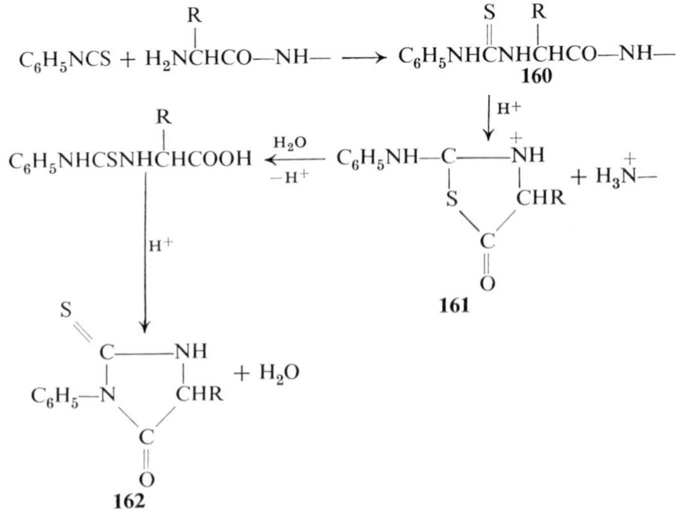

$$\underset{R}{\overset{R}{\big|}}$$
$$C_6H_5NCS + H_2N\overset{R}{\underset{|}{C}}HCO\!-\!NH\!-\!\ \longrightarrow\ C_6H_5NH\overset{S}{\overset{\|}{C}}NH\overset{R}{\underset{|}{C}}HCO\!-\!NH\!-\!$$
160

$$\big\downarrow H^+$$

$$C_6H_5NHCSNH\overset{R}{\underset{|}{C}}HCOOH \ \underset{\underset{-H^+}{\longleftarrow}}{\overset{H_2O}{\longleftarrow}} \ C_6H_5NH\!-\!C\!\overset{+}{-\!}NH + H_3\overset{+}{N}\!-\!$$

with cyclic structure **161** (S, CHR, C, O) and $+ H_3\overset{+}{N}\!-\!$

$$\big\downarrow H^+$$

cyclic structure **162**:
$$S\!=\!C\underset{C_6H_5-N}{\diagup}\!\!\!\overset{C-\!\!-NH}{\diagdown}CHR\ \diagdown C\!\diagup \overset{\|}{O} \ + H_2O$$
162

The first step is the formation of the phenylthiocarbamyl peptide (PTC-peptide) **(160)** by reaction of phenylisothiocyanate with the peptide at pH 8–9 and about 40°. In a second step the terminal residue is cleaved off as a thiazolinone **161** in anhydrous acid [369, 370] and can be converted into the more stable phenylthiohydantoin (PTH) **(162)** in aqueous acid after separation from the peptide.

Edman's original method [367, 368, 371] has been improved through the cumulative experience of many workers, leading to modifications which continue to widen the applicability and sharpen the precision of the method [369, 370, 373–376].

During the early investigations using Edman's method difficulties were encountered on account of the insolubility of large peptides after condensation with phenylisothiocyanate. In the Fraenkel-Conrat modification [377, 378] the protein is adsorbed on paper strips and all reactions are carried out on this support (see below under solid phase degradation methods). After coupling of the peptide with phenylisothiocyanate, extraction of excess of reagents, and degradation of the PTC-peptide, the PTH-amino acid is extracted with acetone and then identified. The procedure, although basically qualitative, may with experience give roughly quantitative data.

By the Fraenkel-Conrat method the amino acid residue which is removed at any step from the paper strip is directly identified, usually by chromatography. Alternatively, the degradation can be followed by the difference in amino acid composition between the original and degraded peptide (subtractive Edman degradation) [379, 380] or by identifying the new amino terminal residue in the peptide after each stage of degradation (sequential degradation plus dansylation [58, 381]).

Since the Edman method is the only widely accepted procedure for the stepwise degradation of a peptide from the amino-terminal end, research has been centered on finding conditions for improving the average yield at each stage so that more degradations can be performed with a single peptide or protein. The repetitive yield of the phenylisothiocyanate degradation is 90–95%. In favorable cases one may expect to produce 15–20 degradation cycles with clearly interpretable results.

However, several side reactions during the chemical manipulations have to be born in mind. Most important of these is the oxidative desulfurization of the phenylthiocarbamyl group [375]. The oxygen dissolved in the medium is sufficient to bring about this reaction. The resulting phenylcarbamyl group acts as a blockage to the degradation since no thiazolinone can be formed. It is therefore advisable to carry out the operations in a nitrogen atmosphere. Metal ions can probably also interfere with the yields of the degradation since thiourea compounds are strong chelating agents [372].

Problems were encountered in the degradation by the Edman method of histidine containing peptides [378, 382]. The reason for this abnormal behavior is not known.

The $\alpha \to \beta$ *trans*-peptidation in aspartyl or asparaginyl peptides [275, 383–385] may occur even under mild conditions, for example, during the isolation of a peptide. The β-aspartyl peptide bond is not cleaved by the Edman reagent [386].

A new type of blocking reaction involves cyclization to **164** of N-terminal S-carboxymethyl-cysteine (**163**) [387]. This complication may be overcome by using other protecting groups for the SH group.

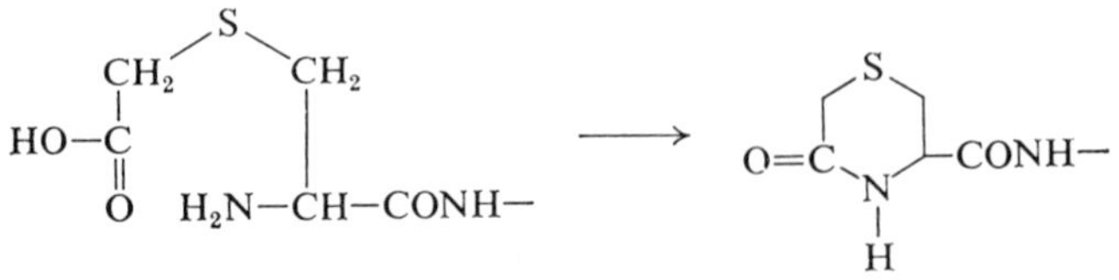

163   164

Analogous cyclization with complete blockage of sequential degradation is observed also with N-terminal glutamine. Release of a glutamine residue after a degradation cycle leads rapidly to pyrrolidone carboxylic acid residue [9].

IDENTIFICATION OF PHENYLTHIOHYDANTOINS

Analysis of the phenylthiohydantoins formed at each successive step of the Edman procedure would seem a more direct and attractive approach to the problem, particularly in view of the automation of the Edman technique (p. 512).

Direct identification of the amino acid removed at each step of the Edman degradation presents several advantages in respect to the subtractive or dansyl procedures (see above), where analysis of an aliquot of the shortened peptide is required in order to determine the missing residue or the new N-terminal amino acid. Loss of sample at each step, accumulation of blocked peptides, and the work and delay involved in repetitive amino acid analyses are disadvantages of the subtractive technique. The "dansyl" technique in conjunction with Edman degradation is widely used and is also very success-ful with short peptides; however, it cannot be readily quantitated and gives poor results with proteins. Such problems do not arise when the cleaved residue is determined by direct analysis.

The amino acid may be regenerated from the PTH-amino acid by hydrolysis in 0.25N barium hydroxide at 140° for 48 hr [388]. In most cases the parent amino acid was recovered, but as expected the PTH derivatives of serine, threonine, arginine, asparagine, and glutamine do not produce the parent amino acids.

Several chromatographic techniques have been devised for the separation and identification of PTH-amino acids. Paper [389, 390] and thin-layer [391, 392] chromatography have been worked out in detail. All PTH-amino acids are well separated by both techniques, permitting unambiguous identification of all amino acids normally encountered in proteins. The PTH derivatives of arginine, histidine, and cysteic acid present some complications, but here alternate spot reactions and paper electrophoretic techniques may be used. For quantitative determinations the PTH-amino acids are separated by paper chromatography [393] and then eluted from the paper and determined spectrophotometrically [393–395]. PTH-amino acids absorb in the region 245–275 mμ and the average molar extinction is 16,000. Partition chromatography on Celite columns with fairly quantitative recoveries has been used [393, 396–398].

Although rapid, and in many instances satisfactory, paper and thin-layer chromatographic methods for direct identification of these compounds are cumbersome, and quantitation at the submicrogram level is difficult.

Gas chromatography has been used recently for the separation and identification of PTH-amino acids [33, 399, 400]. The clean results obtained by this technique indicate that gas chromatography will be the future method of choice in the identification of PTH-amino acids. The difficulties inherent to the seryl, threonyl, asparaginyl, glutaminyl, and lysyl derivatives have been overcome [399] by the use of new thermally stable and more polar silicone liquid phases and a new powerful silylating agent, N,O-bis(trimethylsilyl)-acetamide, which readily converts the less volatile and stable thiohydantoins to trimethylsilyl derivatives having excellent chromatographic properties.

Quantitative analysis of submicrogram amounts of all the PTH-amino acids, except arginine, is possible. Attempts to analyze PTH-arginine have been unsuccessful, presumably because of the high polarity of the guanidinium group and its inability to form a stable silyl derivative. However, since this derivative is always found in the aqueous layer after the conversion step of Edman procedure, quantitative analysis may be achieved by a micro-Sakaguchi test.

The PTH-amino acids are divided into three groups according to their volatility and need for trimethylsilylation. A two-column system, coated with different stationary phases, has been developed for the separation of these groups. In addition, a single-column system was also found suitable, utilizing a blend of three silicone phases, which allows identification of all three groups of PTH-amino acids in two 50-min runs (before and after trimethylsilylation).

When compared to the currently employed paper and thin-layer chromatographic procedures, the gas chromatographic method offers superior resolving power, ease of quantitation, speed, and sensitivity.

Finally, mention should be given to the application of mass spectrometric methods for the identification of PTH-amino acids [401–403].

Since in the mass spectrometric identification of the phenylthiohydantoins no test-substances are needed, the degradation can be performed using different *p*-substituted phenylisothiocyanate; only *one* mass spectrum will be sufficient for the sequence elucidation after several cycles of degradation. Weygand [404] developed this technique using, in succession, *p*-fluoro, *p*-chloro, and *p*-bromo, in addition to the unsubstituted compound. After four cycles the separated four phenylthiohydantoins were combined and the sequence of a pentapeptide unequivocally determined by mass spectrometric analysis of the mixture. *p*-Bromophenylisothiocyanate is a particularly useful reagent since the isotope-ratio ^{79}Br:^{81}Br determines clearly the peaks dealing with the *p*-bromophenylthiohydantoin.

Methylisothiocyanate has been shown to be effective and slightly more reactive than the phenyl derivative [405]. A new approach for sequential degradation, through the introduction of an isotope dilution step and an isotope ratio assay (^{14}N/^{15}N) for the quantitation of the released methylthiohydantoins, has been suggested [405]. The methyl reagent has also been used in the solid state for the sequential degradation of peptides [401, 406].

Methylisothiocyanate-amino acid and -peptide adducts have recently been examined by circular dichroism in order to ascertain if this modified Edman technique could be applied for the determination of the configuration of sequences of amino acid residues in polypeptides. The sign of the 300–320 mμ Cotton effect exhibited by the methylthiohydantoins, obtained either from amino acids or peptides, has been found independent of solvent polarity and related only to the configuration of the asymmetric α-carbon, being positive for the L-series and negative for the D-series. This observation has been tested on all the common amino acid derivatives and found to hold without exception [407].

THE SEQUENATOR

Edman and Begg [376] called the *sequenator* "an instrument which determines the sequence of an ordered linear polymer by repeating a chemical process." As a result of long and intensive investigation, they described a *protein sequenator*, a machine that automatically subjects a protein to sequential degradation. The instrument operates on the principle of the phenylisothiocyanate degradation scheme and the processes performed are essentially the same as those in the manual procedure.

The principle of the sequenator is to immobilize the protein or peptide and allow reagents and solvents to flow over it. The heart of the instrument is its "cup" (see Fig. 16.3), in the sides of which the protein (0.2–1 μmole) has been precipitated in the form of a thin-layer film. Solutions of reactants and extracting solvents are spread inside the cup, which is continuously rotated by an electric motor. The system is ideally suitable for washing away reaction by-products and excess reagents as well as permitting evaporation under reduced pressure of residual solvents because of the large surface of the spinning thin film of the protein.

The sequenator performs only the coupling and cleavage operations. The conversion of the thiazolinone into the phenylthiohydantoin derivative is carried out separately; in this way the design problem of the sequenator is greatly simplified, leaving the conversion reaction and the subsequent identification of the PTH derivative out of consideration.

The steps are as follows:

1. In the spinning cup the protein or peptide is reacted with phenylisothiocyanate in an alkaline buffer and solvent medium to form the phenylthiocarbamyl derivative at the N-terminal amino acid.

2. After the reaction, by-products of the reaction and the excess of reagent are washed out and the residual solvent is evaporated.

3. The phenylthiocarbamyl derivative is then exposed to an anhydrous acid, whereby the thiazolinone derivative of the N-terminal amino acid is formed and split off, separated from the shortened peptide by extraction with an organic solvent, and transferred into a fraction collector. The thiazolinone containing fractions are subsequently separately converted into PTH derivatives and identified by chromatographic methods.

4. The shortened peptide left behind in the spinning cup is then dried in preparation for the next degradation cycle.

The design of the sequenator requires the modification of certain reagents and solvents as compared with the manual procedure. The less volatile N,N,N',N'-tetrakis-(2-hydroxypropyl)-ethylene-diamine (*quadrol*) and heptafluorobutyric acid have replaced N,N-dimethylallylamine and trifluoroacetic acid, respectively. Another deviation from the manual procedure is that the cleavage operation is repeated in each cycle.

The programming of one degradation cycle requires 30 stages, the time required being 93.6 min, which is equivalent to 15.4 cycles in 24 hr.

The application of this procedure to humpback whale myoglobin yielded the sequence of the first 60 amino acid from the N-terminal end by using a sample of only 0.25 μmole of protein. The results are truly excellent.

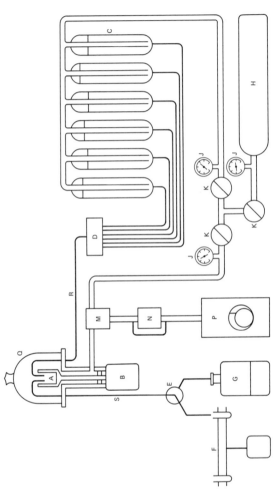

Fig. 16.3 Schematic diagram of the sequenator. The reaction vessel is a cylindrical glass cup (A) mounted on the shaft of an electric motor (B). The cup spins continuously, and solutions and solvents entering the cup are therefore spread as thin films on the walls of the cup. Reagents and solvents enter through the feed line (R) at the bottom of the cup. Extracting solvents climb to the groove where they are scooped off and leave through the effluent line (S). The cup is enclosed in a bell jar (Q), and the system can be evacuated by means of a vacuum pump (P). The system is also thermostated. Reagents and solvents are stored in reservoirs (C), and are admitted to the cup through an assembly of valves (D). The reservoirs are under a constant low pressure of nitrogen supplied by a nitrogen cylinder (H) and pressure regulators (K). The contents of the bell jar are likewise held at a fixed although lower pressure of nitrogen through a similar arrangement. The pressure differential between the reservoirs and the bell jar is constant, and the volume of reagent or solvent admitted to the cup is therefore determined by the time a valve is kept open. The effluent line leads via a 3-way stopcock (E) either to a fraction collector (F) or to a waste container (G). The valves (D) in the assembly and the gas valves (M and N) are operated by solenoids, and the 3-way outlet stopcock (E) and the fraction collector (F) by electric motors. All these functions are governed by an electronic programming unit. The motors driving the cup and the vacuum pump run continuously (taken from Edman and Begg [376]).

The usefulness of the sequenator depends primarily on the yield in the single degradation cycle. An extremely high and specific yield per step is required and is possible only with a combination of the high precision engineering of the instrument and the maximum chemical purity of the solvents and reagents.

Table 16.1 presents the theoretical yields of the derivatives of amino acids after 10, 50, and 100 steps, assuming 99, 95, and 90% yield at each degradation cycle, as well as the working amount of protein needed for performing 100 consecutive degradations if a minimum of 0.1 μmole of the amino acid derivative is needed for its detection and determination.

Table 16.1 Yield of Amino Acid Derivative after Some Steps of Sequential Degradation Assuming Certain Yields at Each Step[a] (taken from Schroeder [12]).

Number of Steps	Yield per Step		
	99%	95%	90%
10	90%	60%	35%
50	60%	8%	0.5%
100	36%	0.6%	0.003%
Amount of starting material[b]	0.3 μmole	17 μmoles	3500 μmoles

[a] The headings of Columns 2, 3, and 4 assume a certain percentage yield at each step. The numbers under these headings give the yield after the designated number of steps in terms of the percentage of the yield on the first step of the sequential degradation

[b] For 100 degradations, if a minimum of 0.1 μmole of amino acid derivative is needed for identification [12].

If it is assumed that a sequence determination ceases to give useful information when the over-all yield has fallen to nearly 30%, a total number of 100 cycles will be feasible only with a 99% yield for each step. However, a yield of 99% is exceptional for an organic reaction, and even with the still excellent yield of 90% it is evident from Table 16.1 that only 10 steps will be workable.

In addition, as with trifluoroacetic acid [408], slow acidolysis of peptide bonds by heptafluorobutyric acid during the course of the degradation work can cause an increasing background of spurious PTH derivatives which eventually makes the identification of the terminal residue impossible.

Special difficulties are likely to be encountered with short peptides. The small difference in solubility between the thiazolinone and the short peptide makes the differential extraction of the former difficult and leads to a rapid drop in yield as the degradation approaches the C-terminal end of the peptide since the peptide is progressively removed from the cup.

As discussed below, attachment of peptides to solid supports will prevent the peptide from being extracted into organic solvents, and in this way make analysis of small peptides amenable to the automated procedure of Edman.

Although still at the experimental stage, the instrument is now commercially available from several companies.

Degradation through Thioketonic Group Participation

In addition to phenylisothiocyanate, other related reagents have been proposed for the stepwise degradation of peptides. It should be noted, however, that none of the reactions involved have been worked out into a generally applicable procedure and therefore only the principles are presented.

Levy [409] demonstrated that carbon disulfide reacts with the amino group of peptides at pH 8 to form substituted dithiocarbonic acids (165). Upon acidification of the dithiocarbamates to pH 1 in aqueous solution at room temperature, they cyclize to 2-thio-5-thiazolidones (166) with simultaneous cleavage of the terminal peptide. The thiothiazolidone derivatives of amino

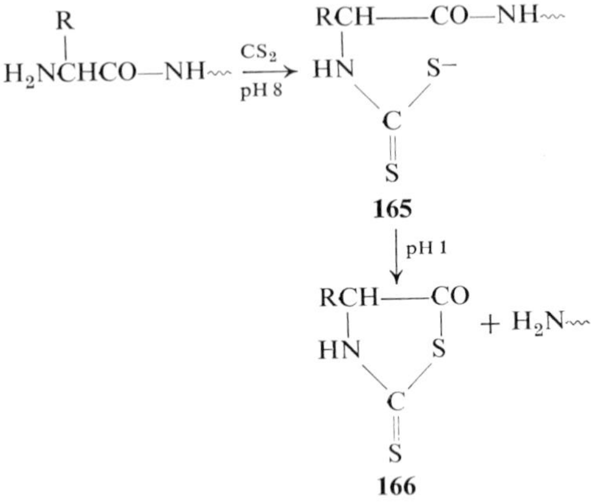

acids absorb at 280 mμ, ε 13,500, and the parent amino acids can be regenerated by heating in 5N HCl. The reaction has been used only with short peptides [72, 181].

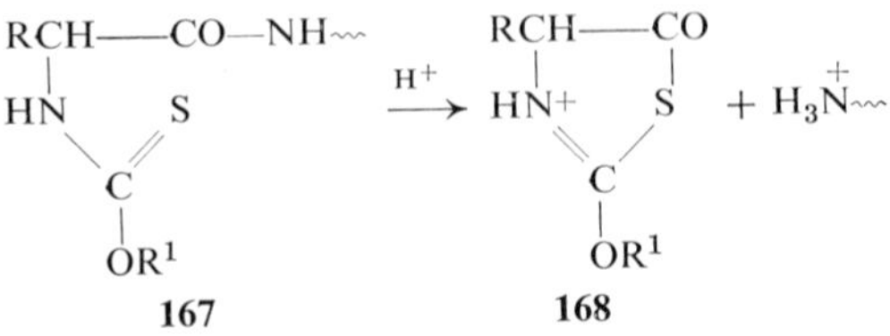

Similarly, an *N*-thioncarbalkoxy peptide (**167**) in acid splits off an 2-alkoxy-5-thiazolinone derivative (**168**) [410].

Elmore and Toseland [411, 412] reported another route to the cleavage of the terminal residue. *N*-Acyldithiocarbamates (**169**) react at pH 8 with the amino group to yield *N*-acylthiocarbamyl derivatives (**170**) which are

$$
\underset{\text{H}_2\text{N}\overset{\text{R}}{\overset{|}{\text{C}}\text{HCO}-\text{NH}}{}\xrightarrow{\text{R}^1\text{CONHCSC}_2\text{H}_5(169)}\quad
\begin{array}{c}
\text{RCH}\!\!-\!\!-\!\!\text{CO}\!-\!\text{NH}\sim \\
| \qquad\qquad | \\
\text{HN} \qquad \text{NH}\!-\!\text{COR}^1 \\
\diagdown \quad \diagup \\
\text{C} \\
\| \\
\text{S} \\
\textbf{170}
\end{array}
$$

↓ Anhydrous HCl

$$
\begin{array}{c}
\text{RCH}\!\!-\!\!-\!\!\text{CO} \\
| \qquad | \\
\text{HN} \quad \text{NH} \\
\diagdown\diagup \\
\text{C} \\
\| \\
\text{S} \\
\textbf{171}
\end{array}
+\ \text{R}^1\text{COOH}\ \xleftarrow{\text{H}_2\text{O}}\
\begin{array}{c}
\text{RCH}\!\!-\!\!-\!\!\text{CO} \\
| \qquad | \\
\text{HN} \quad \text{N}\!-\!\text{COR}^1 \\
\diagdown\diagup \\
\text{C} \\
\| \\
\text{S}
\end{array}
+\ \overset{+}{\text{H}_3}\text{N}\sim
$$

converted into the corresponding thiohydantoin (**171**) with cleavage of the terminal peptide bond, as shown in the scheme.

More recently Barrett [413] described a reaction where *N*-thiobenzoylated peptides (**172**) are easily cleaved in anhydrous acid to form 2-phenyl-5-thiazolinone derivatives (**173**). The cleaved amino acid is best isolated and

$$
\begin{array}{c}
\text{RCH}\!\!-\!\!-\!\!\text{CO}-\text{NH}\sim \\
| \qquad\qquad | \\
\text{HN} \qquad\quad \text{S} \\
\diagdown \quad\diagup\!\!\!= \\
\text{C} \\
| \\
\text{C}_6\text{H}_5 \\
\textbf{172}
\end{array}
\xrightarrow{\text{H}^+}\quad
\begin{array}{c}
\text{RCH}\!\!-\!\!-\!\!\text{CO} \\
| \qquad\qquad | \\
\text{HN}^+ \qquad \text{S} \\
\diagdown \quad\diagup\!\!\!= \\
\text{C} \\
| \\
\text{C}_6\text{H}_5 \\
\textbf{173}
\end{array}
+\ \overset{+}{\text{H}_3}\text{N}-
$$

↓ $\text{C}_6\text{H}_5\text{NH}_2$

$$
\underset{\textbf{174}}{\overset{\text{S}\qquad\text{R}}{\overset{\|\qquad|}{\text{C}_6\text{H}_5\text{CNHCHCONHC}_6\text{H}_5}}}
$$

identified as *N*-thiobenzoylamino acid anilide (**174**), obtained from the corresponding 2-phenylthiazolone (conjugate base of **173**) by brief treatment with aniline in boiling toluene [414]. Alternatively, the thiazolone derivatives of amino acids can be identified by mass spectrometry [145].

Degradation through Nitrogen Participation

In the schemes for sequential degradation discussed so far, the essential feature of the mechanism has been the formation of a thiazolinone. Other procedures do exist, although as yet they have not been worked out in such detail so they can be routinely applied to sequence determinations.

One method [47] exploits the intramolecular participation of the anilino group in the hydrolytic cleavage of the N-terminal peptide bond, the driving force being the formation of a lactam. The procedure has been tried only with short peptides with nonpolar amino acid residues, with an average of 84% yields of cleavage. Later modifications of the procedure [48–52] did not improve significantly the method.

Nitropyridyl-peptides (175) are more easily cleaved in acid media than the corresponding dinitrophenyl derivatives. The hydrolysis is anchimerically assisted by the protonated pyridyl group [43, 44]. The pH region in which the labilizing influence of the neighboring pyridyl species occurs is related to the basicity of the aza function. Since the pH-rate profiles parallel the ionization curves of the heterocyclic nitrogen atom, enhancement of the basicity of the N-atom was expected to displace this catalytic effect to a region of higher pH. In fact, after reduction with sodium borohydride in bicarbonate solution of 3,5-dinitropyridyl-(DNPyr)-peptides (175) to

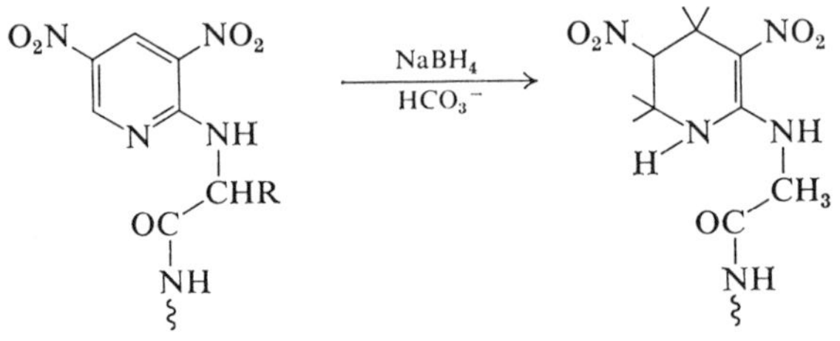

tetrahydro-pyridine derivatives (176) [416], the reduced compounds, after standing at 40°, pH 5–6, for about 12 hr, or by heating at 100° for 5 min, cleaved *quantitatively* at the linkage adjacent to the reduced dinitropyridyl group.

The procedure was tested on several DNPyr-dipeptides; the extent of peptide bond cleavage was estimated by quantitative amino acid analysis of the amino acid set free. The tripeptide DNPyr-Gly-Leu-Tyr quantitatively released the dipeptide Leu-Tyr.

In view of the mild conditions of reaction and the quantitativeness of the cleavage step, the reductive cleavage of the DNPyr-peptides represents a new potentially useful procedure for degradative purposes.

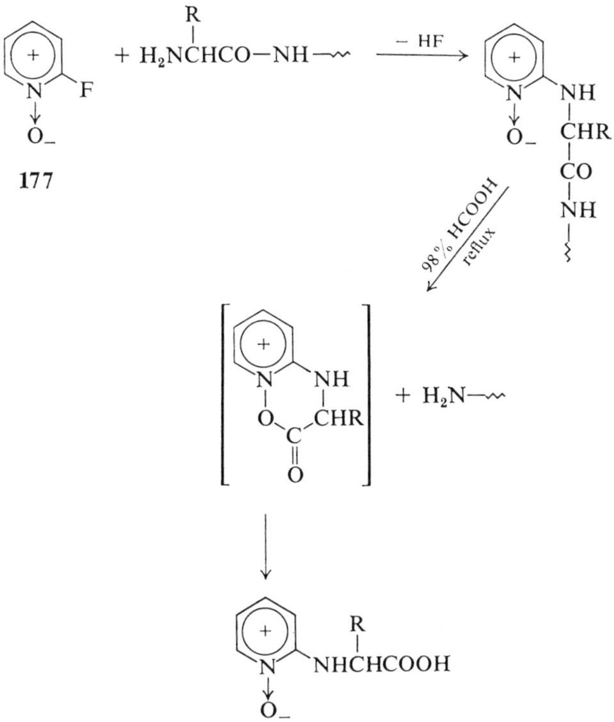

Fluoropyridine N-oxide (**177**) has been used as an arylating agent for peptides [417] to introduce a side chain capable of exhibiting anchimeric assistance for more easy breakage of the first peptide bond. The reaction employed involves release of the first amino acid as a 2-pyridyl-N-oxide compound (**178**) by refluxing in 98% formic acid of the arylated peptide. The method has been applied only to small peptides, and complications are to be expected from reactions with amino acid side chains. In addition some nonspecific cleavage of peptide bonds may occur by the action of hot 98% formic acid.

Solid Phase Degradation

The great success of the synthesis of peptides on a solid support [418] and of the automation of the process suggested the degradation of peptides on a polymeric support [406].

If a peptide is linked to a solid support, the mechanics of performing degradations become greatly simplified since the resin with the linked peptide is very easy to separate by filtration, and tedious extractions procedures are avoided.

In Laursen's procedure [406] the peptide is attached by its C-terminal amino acid, and the degradation carried out using methylisothiocyanate as shown in the scheme.

N-Protection of the peptide is accomplished with methylisothiocyanate and activation with N,N'-carbonyldiimidazole. The activated peptide is then coupled with poly-[(2-aminoethyl)-aminomethyl]-styrene, the cyclization is

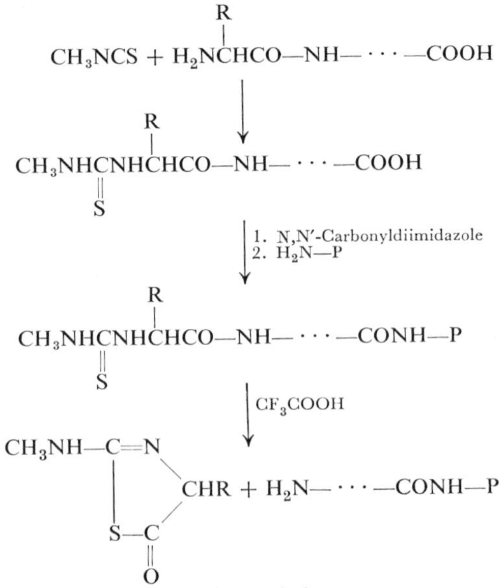

$$
\begin{array}{c}
\text{R} \\
| \\
\text{CH}_3\text{NCS} + \text{H}_2\text{NCHCO—NH—} \cdots \text{—COOH}
\end{array}
$$

$$
\begin{array}{c}
\text{R} \\
| \\
\text{CH}_3\text{NHCNHCHCO—NH—} \cdots \text{—COOH} \\
\| \\
\text{S}
\end{array}
$$

1. N,N'-Carbonyldiimidazole
2. $\text{H}_2\text{N—P}$

$$
\begin{array}{c}
\text{R} \\
| \\
\text{CH}_3\text{NHCNHCHCO—NH—} \cdots \text{—CONH—P} \\
\| \\
\text{S}
\end{array}
$$

CF_3COOH

$$
\begin{array}{c}
\text{CH}_3\text{NH—C}{=}\text{N} \\
| \qquad\qquad \diagdown \\
\qquad\qquad\qquad \text{CHR} + \text{H}_2\text{N—} \cdots \text{—CONH—P} \\
| \qquad\qquad \diagup \\
\text{S—C} \\
\| \\
\text{O}
\end{array}
$$

P = poly-[(2-aminoethyl)-aminomethyl]-styrene

carried out with trifluoroacetic acid, and the residual peptide-resin is separated from the thiazolinone. The polymer is then treated with methylisothiocyanate and the process repeated.

The procedure was tested more successfully using the *t*-butyloxycarbonyl group (cleavable by trifluoroacetic acid), instead of the methylthiocarbamyl group, as a protecting unit of the amino function. The peptide BOC-Gly-Phe-Val-Ala-Pro-Leu-Gly-OH was coupled to the resin and sequentially degraded in approximately 95% yield per cycle.

Another way to get attachment of peptides to solid supports was successfully tried by using a sulfonium resin [419]. When an N-blocked peptide (179) in tetrahydrofuran or dioxane is stirred with a sulfonium resin in the bicarbonate form (180), the peptide is linked by salt formation (181). Heating of the filtered peptide-resin salt *in vacuo* causes esterification (benzyl ester) (182) of the carboxyl group with evolution of the volatile methyl sulfide. The sulfonium resin may be prepared by treating chloromethylated polystyrene-divinylbenzene (1%), 200–400 mesh (183), with methyl sulfide, and

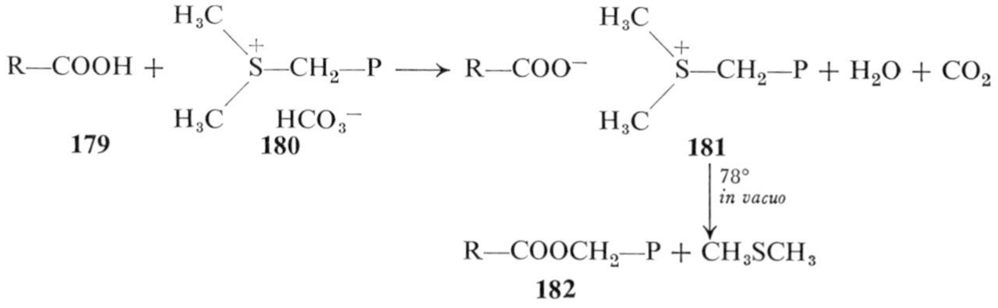

$$R\text{—COOH} + \begin{array}{c} H_3C \\ \diagdown \\ S\text{—}CH_2\text{—}P \\ \diagup \qquad \\ H_3C \qquad HCO_3^- \end{array} \longrightarrow R\text{—COO}^- \begin{array}{c} H_3C \\ \diagdown \\ S\text{—}CH_2\text{—}P + H_2O + CO_2 \\ \diagup \\ H_3C \end{array}$$

$$\qquad\qquad 179 \qquad\qquad 180 \qquad\qquad\qquad\qquad\qquad\qquad 181$$

$$\Big\downarrow {\small \begin{array}{l} 78° \\ \textit{in vacuo} \end{array}}$$

$$R\text{—COOCH}_2\text{—}P + \overset{+}{C}H_3SCH_3$$

$$182$$

P = polystyrene–divinylbenzene (1 %), 200–400 mesh

exchange of the chloride anion with bicarbonate [420]. *t*-Butyloxycarbonyl (BOC) and the 2-nitrophenylsulfenyl (NPS) groups were used as blocking groups of the amino function [421]. While the introduction of the BOC group in peptides requires BOC-azide in anhydrous dimethylformamide with triethylamine [185], the NPS group can be introduced in aqueous tetrahydrofuran or dioxane and triethylamine, the acylation being instan-

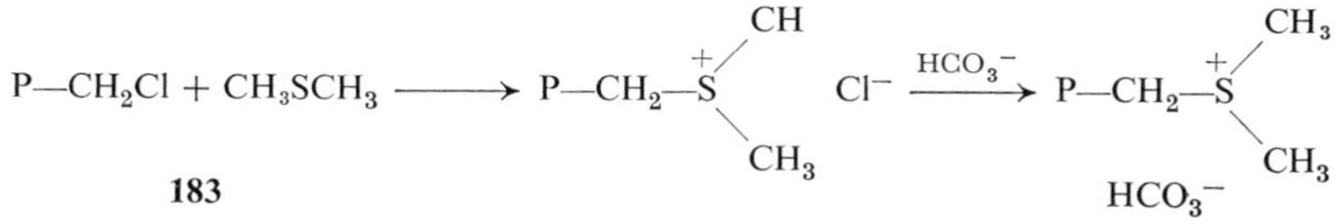

$$P\text{—}CH_2Cl + CH_3SCH_3 \longrightarrow P\text{—}CH_2\overset{+}{\text{—}}S\begin{array}{c}\diagup CH \\ \diagdown CH_3\end{array} Cl^- \xrightarrow{HCO_3^-} P\text{—}CH_2\overset{+}{\text{—}}S\begin{array}{c}\diagup CH_3 \\ \diagdown CH_3\end{array}$$

$$\qquad 183 \qquad\qquad\qquad\qquad\qquad\qquad\qquad\qquad\qquad\qquad HCO_3^-$$

taneous and quantitative. The cleavage of the NPS group from the NPS-peptide-resin is then easily accomplished by washing the resin with gaseous hydrochloric acid in an organic solvents [419].

Dintzis [422] linked an N-protected peptide by means of a water-soluble carbodiimide to a *hydrophilic* amino resin, formed from polyacrylamide beads and ethylene diamine by amide exchange. The peptide was blocked with 2,6-dinitrobenzene-1,4-disulfonate (**184**), which could be removed by treatment with ammonia. Sequential degradation was carried out in an

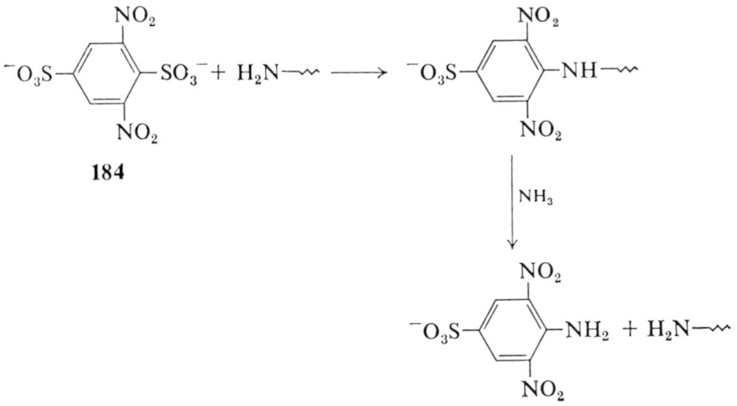

184

aqueous environment using methylisothiocyanate, the cleavage and conversion to a thiohydantoin being effected by cold 6*N* HCl.

Stark [422, 423] used a different approach for the solid phase degradation, linking a peptide by its N-terminus to a support of polystyryl isothiocyanate in pyridine-triethylamine.

During the cyclization step in trifluoroacetic–acetic acid (3:1) the peptide is cleaved from the resin, the thiazolinone or thiohydantoin derivative of the N-terminal amino acid remaining linked to the solid support. The degradation is followed by amino acid analysis of the recovered peptide [380] or by end group analysis [58]. The peptide, after its analysis, is reattached to the isothiocyanate resin and the procedure repeated. Some small peptides [422] from natural sources have been degraded with an average yield of 70% per stage, using 0.5–1 μmole of peptide.

Solid phase degradation will certainly be the approach of choice for automating the process of the sequential degradation of peptides. However, more work is required to find proper reaction conditions, solvents, polymer characteristics, and so on, as well as more elegant methods of attachment of peptides to solid supports.

During the processes of attachment of peptides to the resins discussed above, carboxyl groups of both aspartic and glutamic acid will also react and no thiazolinone or thiohydantoin will be released when the degradation reaches such a residue. If only one aspartic or glutamic acid is present in the peptide, its position in the sequence may be surmised by such a negative result.

As trypsin is the most useful enzyme for the enzymic fragmentation of a protein, and well worked procedures are available for the reversible blocking of the tryptic cleavage at the lysine residues, it is evident that a protein can be cleaved to a number of peptides, all of which, except the C-terminal peptide, bear a C-terminal arginine residue.

Taking advantage of the techniques introduced for the selective blocking of the guanido group, it is expected that by using polymers bearing α- or β-dicarbonyl functional groups, the peptides can be selectively linked to a solid support through the guanido group. Resins of this kind have been prepared, and the scope and limitations of the method are under study [419].

Degradation from the C-Terminus

No fully satisfactory chemical method exists for the sequential degradation of peptides or proteins from the carboxyl terminus.

Early methods for the sequential degradation of peptides by chemical means [98, 106, 424, 425] have been developed by Stark [105] into a method for the sequential degradation from the C-terminus. The reactions involved for a dipeptide are illustrated on page 523.

1. Addition

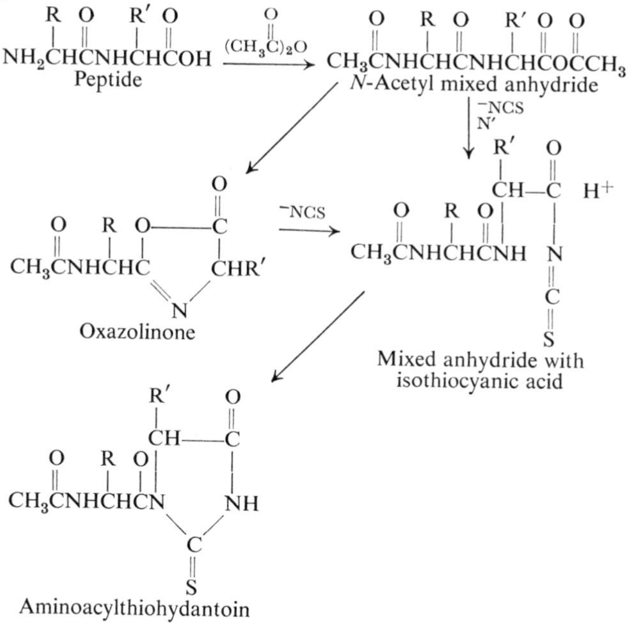

2. Cleavage

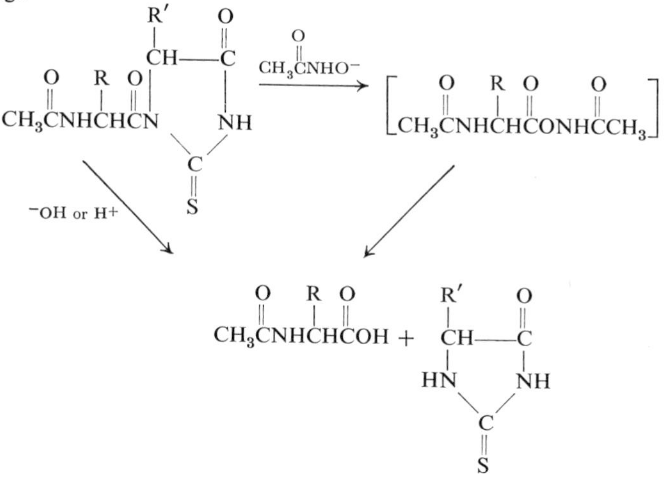

Peptides react with ammonium thiocyanate and acetic anhydride to form peptidylthiohydantoins at their carboxyl termini. After separation from reagents by gel filtration, a peptidylthiohydantoin can be cleaved with acetohydroxamate to a thiohydantoin characteristic of the carboxyl terminus, and an acetylated peptide in which a new residue has become terminal. A number of model peptides derived from ribonuclease and insulin have been degraded with interpretable results, typically for 2 or 3 stages but in the best case for 6 stages.

The method is limited in that carboxyl-terminal aspartic acid and proline are not removed, but all other residues, including asparagine, glutamine, and glutamic acid, do form acylthiohydantoins and are degraded. Removal is particularly poor when a residue precedes an aspartic acid, asparagine, or glutamic acid.

A similar method of preparation of the acylisothiocyanate has been proposed [426]. An acylated peptide is reacted with diphenyl phosphoroisothiocyanatidate (185) to yield the 1-acyl-2-thiohydantoin in good yields.

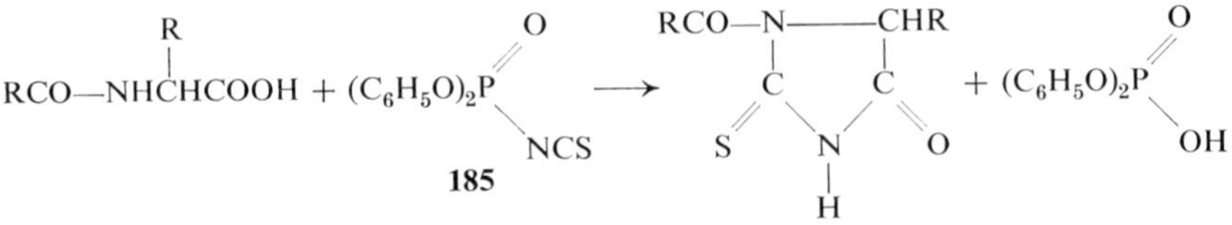

$$RCO-NH\overset{R}{\underset{|}{C}}HCOOH + (C_6H_5O)_2P\overset{O}{\underset{NCS}{\diagup}} \longrightarrow \quad \overset{RCO-N---CHR}{\underset{S}{\overset{|}{C}}\underset{N}{\overset{|}{C}}\underset{H}{\diagdown}O} + (C_6H_5O)_2P\overset{O}{\underset{OH}{\diagup}}$$

185

The method, however, was applied only to very simple peptides.

Khorana [427] proposed a novel, though difficult, general route for the selective removal of the C-terminal residues. He reacted an acyl peptide with p-tolylcarbodiimide ($CH_3-C_6H_4-N=C=N-C_6H_4-CH_3$) to form an acylurea. The latter derivative was then degraded with $0.01N$ NaOH in aqueous ethanol, the C-terminal amino acid being split off as an N-tolylcarbamyl toluidine derivative (186). The product derived from the terminal residue can be converted into the parent amino acid by hydrolysis with acid or alkali.

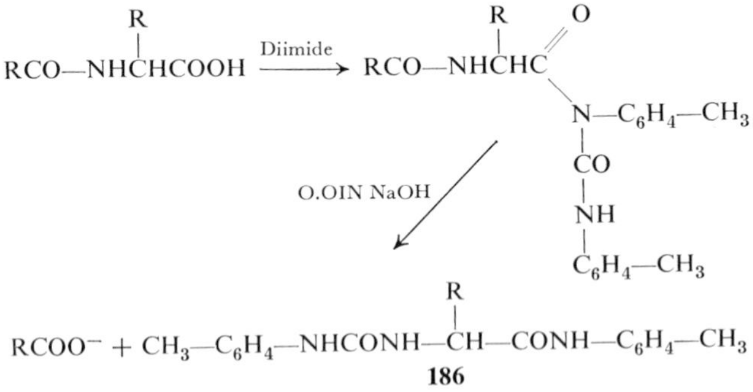

$$RCO-NH\overset{R}{\underset{|}{C}}HCOOH \xrightarrow{\text{Diimide}} RCO-NH\overset{R}{\underset{|}{C}}H\overset{O}{\underset{\diagdown}{C}}$$

$$N-C_6H_4-CH_3$$
$$|$$
$$CO$$
$$|$$
$$NH$$
$$|$$
$$C_6H_4-CH_3$$

$$\xrightarrow{\text{0.01N NaOH}}$$

$$RCOO^- + CH_3-C_6H_4-NHCONH-\overset{R}{\underset{|}{C}}H-CONH-C_6H_4-CH_3$$
186

However, the significant side reaction of the hydrolysis of the acylurea to the original peptide prevents the use of the method for sequential degradation from the C-terminus.

6 MULTIPLE DEGRADATION STRATEGY

The sequential degradation discussed above is intended for single, homogeneous peptides, derived from the enzymic or chemical cleavage of a protein under study. The alignment of the peptides can then be obtained by the overlapping sequences derived from a different cleavage of the protein. Usually the most troublesome and time-consuming step in the sequence work is in obtaining good quantities of pure peptides for subsequent stepwise degradation.

Gray [428] recently proposed a completely different approach and suggested that it is possible to obtain large amounts of sequence data by sequencing unfractioned peptides. Taking advantage of the well-worked methods of cleavage at the methionine residue with cyanogen bromide, at lysine and arginine or cysteine after S-aminoethylation with trypsin, and of the great improvements of the sequential degradation techniques and of the rapid and accurate amino acid analysis, he proposed sequential degradation directly of mixtures of peptides rather than individual peptides isolated in pure form from these mixtures. The identification of the amino acids after each cycle of degradation gives the *nearest neighbor* of the amino acid residue site of cleavage, successively extended to neighbors distal in positions one, two, three, and more from the cleavage point. Cross-correlation between two or more digests enables the deduction of some alignments of the amino acids and finally, if a sufficient number of digests are examined, only one sequence will be found consistent with all the data obtained.

The principle of the technique is delineated in Fig. 16.4 which lists the amino acids which would be removed by 10 steps of degradation on the two digests theoretically obtainable from ribonuclease, after cyanogen bromide cleavage at the methionine residues and after trypsin digestion at the arginine residues after N_ε-trifluoroacetylation of lysines.

As an example, a typical cross-correlation that will give the sequence of amino acids after the arginine (Arg[3]) which is released at the third step of degradation of the cyanogen bromide peptides may be given. Since at the fourth step asparagine, aspartic acid, and threonine are obtained, and since the first cycle of the degradation of the tryptic digest at the arginine residues releases asparagine, cysteine, glutamic acid, and glutamine, asparagine must be the next amino acid to Arg[3]. With similar considerations the sequence Arg-AsN-Leu-Thr-Lys may be uniquely deduced. Whether aspartic acid or serine follows the lysine residue in this sequence can not be decided from the degradation of these two mixtures of peptides, and analysis of different digests is needed in order to decide between these two residues.

Gray [428] was able to show that, if ribonuclease can be split into separate peptide mixtures at arginine, lysine, cysteine, methionine, tyrosine, and histidine residues, and 10 steps of degradation are performed on each set of peptides, then the total sequence of ribonuclease can be determined unequivocally, except for one residue at position 24.

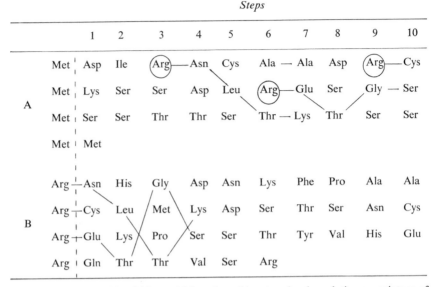

Steps

	1	2	3	4	5	6	7	8	9	10
A	Met Asp Ile Arg—Asn Cys Ala — Ala Asp Arg—Cys									
	Met Lys Ser Ser Asp Leu Arg—Glu Ser Gly — Ser									
	Met Ser Ser Thr Thr Ser Thr — Lys Thr Ser Ser									
	Met Met									
B	Arg—Asn His Gly Asp Asn Lys Phe Pro Ala Ala									
	Arg—Cys Leu Met Lys Asp Ser Thr Ser Asn Cys									
	Arg—Glu Lys Pro Ser Ser Thr Tyr Val His Glu									
	Arg Gln Thr Thr Val Ser Arg									

Fig. 16.4 Amino acids which would be released by stepwise degradation on mixture of peptides derived from ribonuclease by the following cleavage procedures: A, cyanogen bromide cleavage; B, N_ε-acylation of lysines followed by trypsin digestion. At each step the amino acids are listed alphabetically to indicate that their true relationship is unknown. The sequences that can be deduced from these ten degradation steps of the two mixtures of peptides are shown by lines connecting the appropriate residues (taken from Gray [428]).

The method seems suggestive, but its success depends on the assumption that absolutely *clean* results are obtained in each enzymic or chemical degradation.

7 SEQUENCE ANALYSIS BY MASS SPECTROMETRY

The present chemical possibilities for the elucidation of the primary structures of polypeptides and proteins are inadequate for the pressing needs and interests of biochemistry for the solution of biological problems at the molecular level. Determination of the amino acid sequence rapidly and with unequivocal results is not always an easy task, particularly in view of the fact that in the majority of cases only microquantities of substances are available.

A general trend of research in this field is for automated procedures and increases in sensitivity, in order that *microquantities* of material will be amenable for a *rapid* sequence determination.

Analysis by mass spectrometry has recently been found of great interest and seems destined to play an increasingly important part in amino acid sequence determination, and the technique, particularly in view of its possibilities of automation, should compete favorably with other well-established methods.

The basic value of mass spectrometry for biochemical studies lies in the unusual amount, as well as the valuable type, of molecular information that can be obtained from a submicrogram sample, permitting entire research projects to be carried out at the submilligram level.

Spectra with high resolution can be obtained in a few seconds from submicrogram amounts of peptides with molecular weights in the range of 1000 to 2000. No other method gives comparable information in such a direct manner.

It should be pointed out that the fragment ions displayed in the mass spectrum provide the same type of information yielded by the classical technique in which the molecule is chemically degraded and the fragmentation products are studied. It may perhaps also add confidence by providing a completely different approach for checking the results.

Table 16.2 Bibliography on Reviews on Mass Spectrometric Studies of Peptides

Authors	Year	Title	Ref.
Heyns and Grutzmacher	1966	Mass spectrometric analysis of amino acids and peptides	455
Prox and Weygand	1967	Sequence analysis of peptides by gas chromatography and mass spectrometry	456
Lederer and Das	1967	Mass spectrometry of peptides	441
Shemyakin et al.	1967	Mass spectrometric determination of the amino acid sequences in peptides	439
Jones	1967	The mass spectra of amino acid and peptide derivatives	448
Weygand	1968	Trend of development in biochemical analysis	404
Lederer	1968	Mass spectrometry of natural and synthetic peptide derivatives	440
Shemyakin	1968	Primary structure determination of peptides and proteins by mass spectrometry	457
Milne	1969	The application of high resolution mass spectroscopy in organic chemistry	458
Van Lear and McLafferty	1969	Biochemical aspects of high-resolution mass spectrometry	459

Because a wealth of general information is available (see Table 16.2 for a list of the recent reviews on the subject), only a summary of the technique will be given here.

Derivatization of Amino Acids and Peptides

The foremost problem in the analysis of amino acids and peptides by mass spectrometry is the conversion of these rather nonvolatile compounds into derivatives of sufficient vapor pressure to yield useful mass spectra. Free peptides are very nonvolatile, although useful mass spectra were obtained for some simple dipeptides and tripeptides. Dipeptides undergo thermal cyclization to diketopiperazines, and the sequential individuality of the amino acids is thereby lost. The zwitterionic nature of peptides may be changed by converting them into much more volatile compounds by destroying either the acidity of the carboxyl group or the basicity of the amino group, or both. Therefore N-acyl derivatives of oligopeptide methyl, or ethyl, or t-butyl esters have been used. Acetyl, ethoxycarbonyl, benzyloxycarbonyl, formyl, isovaleryl, propionyl, hexanoyl, decanoyl, and stearoyl groups were usefully employed as N-blocking groups.

N-Trifluoroacetyl-peptide esters are suitable derivatives for the separation of the mixtures of peptides, arising from partial acid hydrolyzates, by gas–liquid chromatography [429–431]. A mixture of peptides arising from a partial acid hydrolyzate can be identified by passing the eluant from the chromatographic column directly into the ion source of a fast-scanning mass spectrometer [431]. Therefore, in order to obtain the sequence of a long peptide by combined gas chromatography–mass spectrometry, at first the peptide is partially hydrolyzed and then, after trifluoroacetylation and esterification, the components of the mixture are separated by gas chromatography and identified by mass spectrometry. If a sufficient amount of overlapping peptide fragments is obtained, the total sequence can be derived.

An important factor in the low volatility of peptides is the hydrogen bonding due to the presence of the amide groups. The "permethylation technique" of peptides, introduced by Lederer's group, which eliminated this bonding, was shown to improve greatly the vapor pressure of the peptide [432–436]. The methylation of the peptide bond is accomplished by using silver oxide and methyl iodide in dimethylformamide [432] or with methyl iodide and the methylsulfinyl carbanion [436]. The procedure is also capable of methylating –OH or –SH groups. N-Permethylated peptides have been shown to offer not only the advantage of the volatility, permitting larger oligopeptides to be vaporized in the mass spectrometer, but also a much simpler fragmentation pattern than that of the corresponding nonmethylated derivatives. The simplification of the spectra is certainly due to the decrease of pyrolytic reactions as a result of the lower temperature.

The important N-permethylation technique is sometimes limited by the presence in the peptide backbone by certain functional groups, so that the troublesome side chains, such as the amino, guanido, thioether group, and carboxyl groups, require some additional modification reactions [433, 437].

Methylation of free amino groups produces quaternized methyl iodide salts which lead to derivatives with low vapor pressures. For this purpose amino groups of the peptide are firstly acetylated, before methylation, with methanolic acetic anhydride.

Arginine causes complications due to the specific behavior of its guanidine group. The unfavorable effects of this group may be eliminated by converting the arginine residue into an ornithine residue by refluxing of the arginine-containing peptide with 20% aqueous hydrazine for about 1 hr. The ornithine peptide so obtained is then acylated at the N_δ-amino group. The procedure is limited in its application since some nonspecific cleavage of the peptide bond occurs [438, 439].

A second method of transforming arginine-containing peptides into compounds suitable for mass spectrometry consists of blocking the guanido group by its cyclization with dicarbonyl compounds.

Malonaldehyde and acetylacetone have been used [438, 439] for converting the arginine residue into a pyrimidyl- or dimethylpyrimidyl-ornithine residue. However, since permethylation of the basic pyrimidyl ornithine derivative would give an undesirable quaternized salt, the hydrazinolytic procedure is more suitable [433].

Lenard and Gallop [222] used 1,2-cyclohexandione (see Toi [217]), in modifying arginine residues for mass spectrometry. By combining this modification with acylation and permethylation, volatile compounds were obtained from several peptides, including the nonapeptide bradykinin.

Permethylation of methionine-containing peptides produces a sulfonium iodide at the thioether group [433, 437]. By desulfurization with Raney nickel, methionyl peptides are converted to the corresponding α-aminobutyric acid. Since this amino acid differs in mass from the common aliphatic amino acids, the position of this residue in a peptide, and thus of its methionine precursor, may be easily determined. The same technique could also be used for peptides containing cystine or cysteine, although an unambiguous determination of the sequence position of these residues would be more difficult since their desulfurization gives an alanine residue. The use of deuterated Raney nickel would be an easy way to recognize the newly formed alanine residues [440].

Peptides containing aspartic and glutamic acid are usually esterified by methanol or t-butanol before permethylation. Aspartyl-peptides still gave some difficulties [437].

Fragmentation

In the case of peptides, the fortunate case is that only simple cleavages are possible, the relationship of the fragments to the over-all structure following relatively simple rules. The fragmentation patterns of a variety of peptides show that cleavage of the chain involves two main pathways as shown.

$$
\begin{array}{ccccc}
& R_1 & R_2 & & R_n \\
& | & | & & | \\
A: & YNHCHCO{-}NHCHCO{-}\cdots{-}NHCHCO{-}OR \\
& A_1 & A_2 & & A_n
\end{array}
$$

$$
\begin{array}{ccccc}
& R_1 & R_2 & & R_n \\
& | & | & & | \\
B: & YNHCH{-}CONHCH{-}CO{-}\cdots{-}NHCH{-}COOR \\
& B_1 & B_2 & & B_n
\end{array}
$$

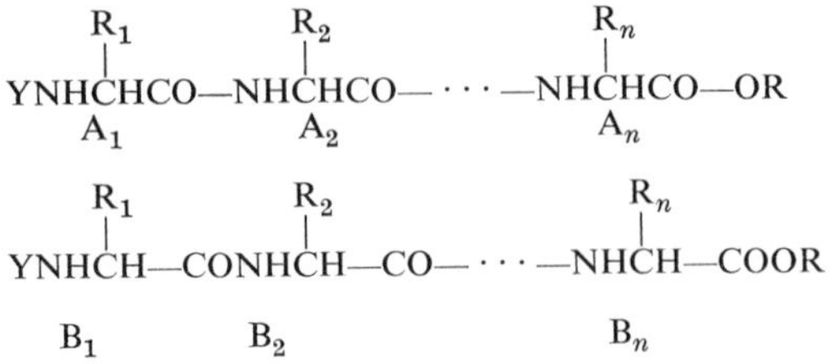

The pathway A, which involves splitting of the peptide bond, is the most useful for sequence determination and is more pronounced in higher peptides than in di- and tri-peptides. The peaks arising from the cleavage of the type A are often accompanied by more or less intense peaks at 28 (CO) mass units lower, due to the fission of type B.

In addition, all amino acid and peptide derivatives show fragmentations characteristic of the side chains. Functional groups in the β-position are eliminated, probably by a thermal reaction; for example, serine derivatives usually show $M{-}H_2O$ peaks. The peaks corresponding to the fragmentation pattern of the aromatic side chains are always prominent features in the mass spectra of peptides containing aromatic amino acids.

For a detailed discussion on the fragmentation patterns of peptides, see Lederer and Das [441].

Sequence Determination

Sequence determination is made possible by the characteristic fragmentations of the polyamide backbone on either side of the carbonyl group, referred to in the above scheme as pathways A and B. Thus the mass spectrum should contain a fragment ion $YNHCHR_1CO^+$ resulting from the cleavage A_1. Since the mass of the N-blocking group Y is known, the exact mass of this ion determines the mass of R_1, which identifies the first amino acid residue. The assignment of R_1 will then be checked by the pathway B cleavage which yields $YNHCHR_1^+$. With R_1 known, the process is continued for the rest of the chain; the second amino acid will be identified from cleavages A_2 and B_2.

Because the mass spectra of peptide derivatives are so complex, data-processing techniques have been used for the automatic sequence determination of peptides.

One computer program is based on recognition of the fact that the structure of the peptide is determined unequivocally by using only the possible fragments which contain one end of the chain [442, 443]. The N-terminal amino acid is identified by checking combinations of the exact mass of the N-protecting group (43.01539 for CH_3CO-), and the exact masses of each of the possible amino acid residues against the observed exact masses. The next amino acid is then identified following the same pattern, starting with the combined theoretical masses of the terminal functional group and the identified residues.

Another program elaborates the sequence by identifying the N-terminal amino acid (and then the others) by checking the combinations of the exact mass of the N-protecting group plus $-NHCH-$ with the masses of each of the possible side chains against with the observed exact masses in the spectrum [444].

A program starting from the C-terminus has been devised, but the need of a molecular ion is a limitation of this method [445].

As a main advantage, the computer techniques allow analysis of mixtures of peptides. As a result of this possibility, with further improvements in methodology and instrumentation, it seems likely that crude mixtures of peptides derived from proteins by enzymic or chemical degradation may be usable for analysis by mass spectrometry. The computer can reconstruct mass spectra, and thus sequences, of the pure components.

Selected Applications

A landmark in the use of mass spectrometry for sequence studies was the elucidation in 1965 of the structure of fortuitine (**187**), an acylnonapeptide ester isolated from *Mycobacterium fortutium* [446]. The mass spectrum, and

$$\overset{\text{Ac}\quad\text{Ac}}{CH_3(CH_2)_n-CO-Val-MeLeu-Val-Val-MeLeu-Thr-Thr-AlA-Pro-OMe}$$
187: $n = 18, 20$

thus the sequence, was obtained from a few micrograms of sample. The compound showed two parent peaks at 1331 and 1359 due to its being a mixture of two homologs containing a C_{20} and C_{22} fatty acid, respectively.

Since this breakthrough the structures of a large number of naturally occurring molecules containing up to 10 amino acids have been determined. A peptidolipid isolated from *Mycobacterium Johnei*, which was supposed to be a tetrapeptidolipid, was shown by mass spectrometry to be a pentapeptide derivative (**188**); amino acid analysis was incorrect due to the difficulty in

$$CH_3(CH_2)_n-CO-Phe-Ile-Ile-Phe-Ala-OCH_3$$
188: $n = 14, 16, 18, 20$

hydrolyzing the hindered Ile–Ile peptide bond [447].

A series of closely related peptidolipid lactones from *Nocardia asteroides*, peptidolipin NA (**179**), Val[6]-peptidolipin NA (**190**), and α-aminobutyryl (Abu)[1]-peptidolipin NA (**191**) were shown by mass spectrometry [440] to have the structures shown.

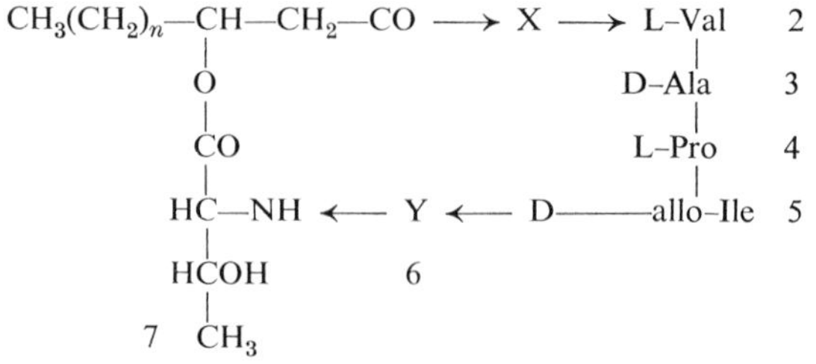

$$
\begin{array}{l}
CH_3(CH_2)_n\!-\!CH\!-\!CH_2\!-\!CO \longrightarrow X \longrightarrow L\text{--}Val \quad 2\\
\qquad\qquad\;\; |\\
\qquad\qquad\;\; O \qquad\qquad\qquad\qquad\quad D\text{--}Ala \quad 3\\
\qquad\qquad\;\; |\\
\qquad\qquad\;\; CO \qquad\qquad\qquad\qquad\quad L\text{--}Pro \quad 4\\
\qquad\qquad\;\; |\\
\qquad\quad HC\!-\!NH \longleftarrow Y \longleftarrow D\text{------}allo\text{--}Ile \quad 5\\
\qquad\qquad\;\; |\\
\qquad\quad HCOH \qquad\qquad 6\\
\qquad\qquad\;\; |\\
\qquad 7 \;\; CH_3
\end{array}
$$

Peptidolipin NA (**189**):	X = L–Thr
	Y = L–Ala
	$n = 16$
Val[6]–peptidolipin NA (**190**):	X = L–Thr
	Y = L–Val
	$n = 16, 17, 18$
α–aminobutyryl (Abu)[1] –peptidolipin NA (**191**):	X = L–α–Abu
	Y = L–Ala
	$n = 16, 17, 18$

The structure of the peptide antibiotic esperin was recently revised [435] using mass spectrometry, after permethylation, and shown to be **192**.

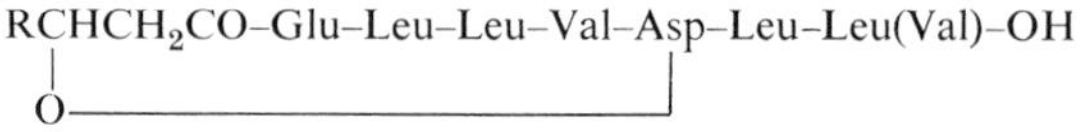

$$
\begin{array}{l}
RCHCH_2CO\text{--}Glu\text{--}Leu\text{--}Leu\text{--}Val\text{--}Asp\text{--}Leu\text{--}Leu(Val)\text{--}OH\\
\;\; |\\
\;\; O\text{-------------------------}|
\end{array}
$$

$$\textbf{192}: R = C_{12}H_{25}\ (45\%)$$
$$R = C_{11}H_2\ (35\%)$$
$$R = C_{10}H_{21}\ (20\%)$$

In addition to several other peptidolipids [440, 448], stendomycin [449], a cyclic tetradecapeptide lactone, and gramicidin A, B, and C [109], *N*-formylpentadecapeptide ethanolamides, were successfully analyzed by mass spectrometry [440].

Cyclic peptides [445] and depsipeptides [450], including enniatin and valinomycin antibiotics [451], have been investigated in a number of laboratories, and a detailed study of the fragmentation of fourteen-membered cyclodepsipeptides, that is, serratamolide and analogs compounds, has been reported [452].

The elucidation of the structure of the cyclic nonapeptide antamanide by classical ways (Edman degradation) was unsuccessful, but partial methanolysis of the cyclopeptide followed by trifluoroacetylation gave a mixture of peptide derivatives which were separated by gas chromatography and analyzed by mass spectrometry. The sequences of the fragments were determined and shown to be consistent with formula **193** for the intact polypeptide [453].

$$
\begin{array}{c}
\text{Leu–Ile–Ile–Leu–Val} \\
\underset{\displaystyle \text{Pro–Pro–Phe–Phe}}{\rule{0pt}{0pt}}
\end{array}
$$

193

The clearest demonstration of the potential of mass spectrometry in sequence analysis of peptides is given by recent work dealing with the complete sequence of feline gastrin [454].

$$SO_3H$$

PyroGlu–Gly–Pro–Trp–Leu(Ala, Glu$_4$)Ala–Tyr–Gly–Trp–Met–Asp–Phe–NH$_2$

194

1. O.lN HCl
2. Chymotrypsin

PyroGlu–Gly–Pro–Trp Leu(Ala, Glu$_4$)Ala–Tyr Gly–Trp Met–Asp–Phe
   a   *b*   *c*   *d*

The heptadecapeptide **194**, isolated in *400 μg quantity* from 200 cat antra, was cleaved with chymotrypsin and the four chymotrypsin peptides were isolated by gel filtration on Sephadex G-25. After esterification and acetylation, and in one case permethylation (peptide *a*), the sequence of all the four peptides was determined and the full sequence of the heptadecapeptide unequivocally determined.

8   CONCLUDING REMARKS

Investigators have available today a variety of refined techniques that can be used for elucidating protein structure. These methods originate mainly from the pioneer work on insulin carried out by Sanger and his colleagues at Cambridge University. Despite the great progress that has been made in a brief period of time, the determination of the amino acid sequence of a protein can not be considered a routine matter. The results of the sequence studies as presented by the *Atlas of Proteins Structure* [3], which lists the sequences of proteins published, are most impressive.

The question of the molecular mechanism of action of many enzymes remains conspicuously unanswered, among the present advances in molecular biology, and one can reasonably assert that some knowledge of structure must logically precede an understanding of function. The amount of structural work to be done is great indeed and the task of determining amino acid sequences is likely to remain with us for quite some time.

For the future it is important that new developments in techniques should permit not only accurate but also *rapid* analysis of protein structures.

References

1. G. M. Edelman, B. A. Cunningham, W. E. Gall, P. D. Gottlieb, U. Rutis-hauser, and M. J. Waxdal, *Proc. Nat. Acad. Sci.*, **63**, 78 (1969).
2. W. A. Schroeder, J. B. Shelton, and J. R. Shelton, *Arch. Biochem. Biophys.*, **130**, 551 (1969).
3. M. O. Dayhoff and R. V. Eck, in *Atlas of Protein Sequence and Structure 1968*, National Biomedical Research Foundation, Silver Spring, Md., 1969.
4. B. Witkop, *Advan. Protein Chem.*, **16**, 221 (1961).
5. E. O. P. Thompson, *Advan. Org. Chem.*, **1**, 149 (1960).
6. E. Y. Spencer, in *Elucidation of Structures by Physical and Chemical Methods*, Vol. 2 (K.W. Bentley, ed.), Wiley-Interscience, New York, 1963, p. 789.
7. T. F. Spande, B. Witkop, Y. Degani, and A. Patchornik, *Adv. Protein Chem.*, **24** (1969).
8. R. E. Canfield and C. B. Anfinsen, in *The Proteins; Composition, Structure and Function*, 2nd ed., Vol. 1 (H. Neurath, ed), Academic Press, New York, 1963, p.311.
9. D. G. Smyth and D.F. Elliott, *Analyst*, **89**, 81 (1964).
10. C. H. W. Hirs, ed., *Methods Enzymo* ., **11** (1967).
11. J. L. Bailey, *Techniques in Protein Chemistry*, 2nd ed., Elsevier, Amsterdam 1967.
12. W. A. Schroeder, *The Primary Structure of Proteins. Principles and Practices for the Determination of Amino Acid Sequence*, Harper-Row, New York, 1968.
13. K. Narita, in *Protein Sequence Determination. Methods and Techniques* (S. B. Needleman, ed.), Springer-Verlag, New York (1970).
14. A. Light, *Methods Enzymol.*, **11**, 426 (1967).
15. L. A. Cohen, *Ann. Rev. Biochem.*, **37**, 695 (1968).
16. S. Shaltiel, *Biochem. Biophys. Res. Commun.*, **29**, 178 (1967).
17. F. Sanger, *Biochem. J.*, **39**, 507 (1945).
18. F. Sanger, *Biochem. J.*, **44**, 126 (1949).
19. F. Sanger and H. Tuppy, *Biochem. J.*, **49**, 463 (1951).
20. F. Sanger and E. O. P. Thompson, *Biochem. Biophys. Acta*, **9**, 225 (1952).
21. H. Fraenkel-Conrat, J. I. Harris, and A. L. Levy, *Methods Biochem. Anal.*, **2**, 359 (1955).
22. R. R. Porter, in *Methods Med. Res.*, **3**, 256 (1950).
23. E. O. P. Thompson and A. R. Thompson, *Progr. Chem. Org. Natural Prod.*, **12**, 270 (1955).
24. G. Biserte, J. W. Holleman, J. Holleman-Dehave, and P. Sautiere, *J. Chromatogr.*, **2**, 225 (1959).
25. J. M. Dellacha and A. V. Fontanive, *Experientia*, **21**, 351 (1965).
26. D. Walz, A. R. Fahmy, G. Pataki, A. Niederwieser, and M. Brenner, *Experientia*, **19**, 213 (1963).

27. K.-T. Wang, J. M. K. Huang, and I. S. Y. Wang, *J. Chromatogr.*, **22**, 362 (1966). K.-T. Wang and I. S. Y. Wang, *J. Chromatogr.*, **24**, 460 (1966).
28. F. C. Green and L. M. Kay, *Anal. Chem.*, **24**, 726 (1952).
29. L. Kesner, E. Muntwyler, G. E. Griffin, and J. Abrams, *Anal. Chem.*, **35**, 83 (1963).
30. D. L. Eaker, T. P. King, and L. C. Craig, *Biochemistry*, **4**, 1473 (1965).
31. M. R. Heinrich and E. Bugna, *Anal. Biochem.*, **28**, 1 (1969).
32. Th. J. Penders, H. Copier, W. Heerma, G. Dijkstra, and J. F. Arens, *Rec. Trav. Chim. Pays-Bas*, **85**, 216 (1966).
33. J. J. Pisano, W. J. A. Vandenheuvel, and E. C. Horning, *Res. Commun.*, **7**, 82 (1962).
34. E. D. Bergmann and M. Bentov, *J. Org. Chem.*, **26**, 1480 (1961).
35. S. Akabori, K. Ohno, T. Ikenaka, A. Nagata, and I. Haruna, *Proc. Japan Acad.*, **29**, 561 (1953). S. Akabori, T. Ikenaka, Y. Okada, and K. Kohno, *Proc. Japan Acad.*, **29**, 509 (1953).
36. D. W. Russell, *Biochem. J.*, **87**, 1 (1963).
37. J. Hine, in *Physical Organic Chemistry*, 2nd ed., McGraw-Hill, New York, 1962, pp. 141–151.
38. B. Capon, *Quart. Rev.*, **19**, 45 (1965).
39. A. Signor, E. Scoffone, L. Biondi, and S. Bezzi, *Gazz. Chim. Ital.*, **93**, 65 (1963).
40. A. Signor and L. Biondi, *Ric. Sci.*, **34**, 165 (1964).
41. A. Signor, L. Biondi, M. Terbojevich, and P. Pajetta, *Gazz. Chim. Ital.*, **94**, 619 (1964).
42. A. Signor, A. Previero, and M. Terbojevich, *Nature*, **205**, 596 (1965).
43. A. Signor and E. Bordignon, *J. Org. Chem.*, **31**, 3447 (1965).
44. A. Signor, E. Bordignon, and G. Vidali, *J. Org. Chem.*, **32**, 1135 (1967).
45. C. Di Bello and A. Signor, *J. Chromatogr.*, **17**, 506 (1965).
46. E. Celon, L. Biondi, and E. Bordignon, *J. Chromatogr.*, **35**, 47 (1968).
47. R. W. Holley and A. D. Holley, *J. Amer. Chem. Soc.*, **21**, 5445 (1952).
48. M. Jutisz and W. Ritschard, *Biochim. Biophys. Acta*, **17**, 548 (1955).
49. V. M. Ingram, *Biochim. Biophys. Acta*, **20**, 577 (1956).
50. E. Scoffone, A. Turco, and M. Scatena, *Ric. Sci.*, **27**, 1193 (1957).
51. E. Scoffone, E. Vianello, and A. Lorenzini, *Gazz. Chim. Ital.*, **87**, 354 (1957).
52. P. De laLlosa, M. Jutisz, and E. Scoffone, *Bull. Soc. Chim. Fr.*, **1960**, 1621.
53. K. L. Kirk and L. A. Cohen, *J. Org. Chem.*, **34**, 395 (1969).
54. K. L. Kirk and L. A. Cohen, *J. Org. Chem.*, **34**, 390 (1969).
55. G. R. Stark and D. G. Smyth, *J. Biol. Chem.*, **238**, 214 (1963).
56. G. R. Stark, *Methods Enzymol.*, **11**, 125, 594 (1967).
57. B. S. Hartley and V. Massey, *Biochim. Biophys. Acta*, **21**, 58 (1956).
58. W. R. Gray, *Methods Enzymol.*, **11**, 469 (1967).
59. H. R. Horton and D. E. Koshland, *Methods Enzymol.*, **11**, 857 (1967).
60. C. Gros and B. Labouesse, *Eur. J. Biochem.*, **7**, 463 (1969).
61. A. A. Boulton and I. E. Bush, *Biochem. J.*, **92**, 11P (1964).
62. N. Seiler and M. Wiechmann, *J. Chromatogr.*, **28**, 351 (1967).

63. Z. Deyl and J. Rosmus, *J. Chromatogr.*, **20**, 514 (1965). D. Morse and B. L. Harecker, *Anal. Biochem.*, **14**, 429 (1966).
64. K. Crowshaw, S. J. Jessup, and P. W. Ramwell, *Biochem. J.*, **103**, 79 (1968).
65. K. R. Woods and K.-T. Wang, *Biochem. Biophys. Acta*, **133**, 369 (1967).
66. V. Neuhoff, F. Von der Haar, E. Schlimme, and M. Weise, *Hoppe-Seyler's Z. Physiol. Chem.*, **350**, 121 (1969).
67. R. F. Chen, *Anal. Biochem.*, **25**, 412 (1968).
68. S. Gurin and H. T. Clarke, *J. Biol. Chem.*, **107**, 395 (1934).
69. R. L. M. Synge, *Chem. Rev.* **32**, 135 (1943).
70. S. Udenfriend and S. F. Velick, *J. Biol. Chem.*, **190**, 733 (1951).
71. J. C. Flecher, *Biochem. J.*, **102**, 815 (1967).
72. M. Levy, *Methods Enzymol.*, **4**, 238 (1957).
73. S. F. Velick and S. Udenfriend, *J. Biol. Chem.*, **191**, 233 (1951).
74. J. P. Collman and D. A. Buckingham, *J. Amer. Chem. Soc.*, **85**, 3039 (1963).
75. D. A. Buckingham, J. P. Collman, D. A. R. Happer, and L. G. Marzilli, *J. Amer. Chem. Soc.*, **89**, 1082 (1967).
76. D. A. Buckingham, L. G. Marzilli, and A. M. Sargeson, *J. Amer. Chem. Soc.*, **89**, 2772, 4539 (1967).
77. D. A. Buckingham, L. G. Marzilli, and A. M. Sargeson, *J. Amer. Chem. Soc.* **89**, 5133 (1967).
78. M. Wilchek, T. Spande, G. Milne, and B. Witkop, *Biochemistry*, **7**, 1777 (1968).
79. M. Wilchek, T. Spande, and B. Witkop, *Biochemistry*, **7**, 1787 (1968).
80. S. Akabori, K. Ohno, and K. Narita, *Bull. Chem. Soc. Japan*, **25**, 214 (1952).
81. R. P. Ambler, *Methods Enzymol.*, **11**, 155 (1967).
82. H. Matsuo, Y. Fujimoto, and T. Tatsuno, *Tetrahedron Lett.*, **39**, 3465 (1965).
83. H. Matsuo, Y. Fujimoto, and T. Tatsuno, *Biochem. Biophys. Res. Commun.*, **22**, 69 (1966).
84. H. Fraenkel-Conrat and C. M. Tsung, *Methods Enzymol.*, **11**, 151 (1967).
85. J. H. Bradbury, *Biochem. J.*, **68**, 482 (1958)
86. V. Braun and W. H. Schroeder, *Arch. Biochem. Biophys.*, **118**, 241 (1967).
87. Y. Kawanishi, K. Iwai, and T. Ando, *J. Biochem.*, **56**, 314 (1964).
88. S. Akabori, K. Ohno, T. Ikenaka, Y. Okada, H. Hanafusa, H. Haruna, A. Tsugita, K. Sugae, and T. Matsushima, *Bull. Chem. Soc. Japan*, **29**, 507 (1956).
89. S. Akabori and T. Ikenaka, *J. Biochem.*, **42**, 603 (1956).
90. G. Braunitzer, *Biochim. Biophys. Acta*, **19**, 574 (1956).
91. K. Kusama, *J. Biochem.*, **44**, 375 (1957).
92. M. Goodman and L. Levine, *J. Amer. Chem. Soc.*, **86**, 2918 (1964)
93. G. T. Young, *J. Chem. Soc.*, **1963**, 1105.
94. H. Matsuo, Y. Kawazoe, M. Sato, M. Ohnishi, and T. Tatsuo, *Chem. Pharm. Bull.* (Tokyo), **15**, 391 (1967).
95. H. Maeda, T. Koyanagi, and N. Ishida, *Biochim. Biophys. Acta*, **160**, 249 (1968).
96. G. N. Holcomb, S. A. James, and D. N. Ward, *Biochemistry*, **7**, 1291 (1968).

97. H. Neurath, in *The Enzymes*, Vol. 4 (P. D. Boyer, H. Lardy, and K. Myrback, eds.), Academic Press, New York, 1960, p. 11.

98. J. P. Greenstein and M. Winitz, *Chemistry of the Amino Acids*, Wiley, New York, 1960.

99. C. Fromageot, M. Jutisz, D. Dreyer, and L. Penasse, *Biochim. Biophys. Acta*, **6**, 283 (1950).

100. A. C. Chibnall and M. W. Rees, *Biochem. J.*, **68**, 105 (1958).

101. J. L. Bailey, *Biochem. J.*, **60**, 173 (1955).

102. M. Z. Atassi and A. F. Rosenthal, *Biochem. J.*, **111**, 593 (1969).

103. O. Yonemitsu, T. Hamada, and Y. Kanaoka, *Tetrahedron Lett.*, **1968**, 3575.

104. P. E. Wilcox, *Methods Enzymol.*, **11**, 605 (1967).

105. G. R. Stark, *Biochemistry*, **7**, 1796 (1968).

106. P. Schlack and W. Kumpf, *Z. Physiol. Chem.*, **154**, 125 (1926).

107. R. Consden, A. H. Gordon, A. J. P. Martin, and R. L. M. Synge, *Biochem. J.*, **41**, 596 (1947).

108. K. Narita, *Biochim. Biophys. Acta*, **28**, 184 (1958); **30**, 352 (1958).

109. R. Sarges and B. Witkop, *J. Amer. Chem. Soc.*, **87**, 2011 (1965).

110. B. Blomback, *Methods Enzymol.*, **11**, 398 (1967).

111. W. R. Holmquist and W. A. Schroeder, *Biochemistry*, **5**, 2489 (1966).

112. D. M. P. Phillips, *Biochem. J.*, **87**, 258 (1963).

113. D. M. P. Phillips, *Biochem. J.*, **107**, 7021 (1968).

114. R. J. De Lange, D. M. Fambrough, E. L. Smith, and J. Bonner, *J. Biol. Chem.*, **244**, 319 (1969).

115. E. L. Gershey, G. Vidali, and V. G. Allfrey, *J. Biol. Chem.*, **243**, 5018 (1968).

116. G. Vidali, E. L. Gershey, and V. G. Allfrey, *J. Biol. Chem.*, **243**, 6361 (1968).

117. G. Kreil and G. Kreil-Kiss, *Biochem. Biophys. Res. Commun.*, **27**, 275 (1967).

118. J. M. Adams and M. R. Copecchi, *Proc. Nat. Acad. Sci. U.S.*, **55**, 147 (1966).

119. R. E. Weber, D. L. Engelhardt, and N. D. Zinder, *Proc. Nat. Acad. Sci. U.S.*, **55**, 155 (1966).

120. J. C. Sheehan and D. H. Yang, *J. Amer. Chem. Soc.*, **80**, 1154 (1958).

121. R. Sarges and B. Witkop, *J. Amer. Chem. Soc.*, **86**, 1862 (1963).

122. G. N. Gussin, M. R. Capecchi, J. M. Adams, J. E. Argetsinger, K. Weber, and J. D. Watson, *Cold Spring Harbor Symp. Quant. Biol.*, **36**, 257 (1966).

123. K. Narita and J. Ishii, *J. Biochem.*, **52**, 367 (1962).

124. G. Schmer and G. Kreil, *Anal. Biochem.*, **29**, 186 (1969).

125. J. Ludwieg and A. Dorfman, *Biochim. Biophys. Acta*, **38**, 212 (1960).

126. M. Messer and M. Ottesen, *Biochim. Biophys. Acta*, **92**, 409 (1964).

127. C. H. W. Hirs, W. H. Stein, and S. Moore, *J. Biol. Chem.*, **221**, 151 (1956).

128. M. A. McDowall and E. L. Smith, *J. Biol. Chem.*, **240**, 281 (1965).

129. D. G. Smyth, W. H. Stein, and S. Moore, *J. Biol. Chem.*, **237**, 1845 (1962).

130. F. Sanger and E. O. P. Thompson, *Biochem. J.*, **53**, 353 (1953).

131. G. H. De Haas, F. Franek, B. Keil, D. W. Thomas, and E. Lederer, *FEBS Lett.*, **4**, 25 (1969).

132. S. Takahashi and L. A. Cohen, *Biochemistry*, **8**, 864 (1969).

133. R. F. Doolittle and R. W. Armentraut, *Biochemistry*, **7**, 1516 (1968).

134. R. W. Armentrout, and R. F. Doolittle, *Arch. Biochem. Biophys.*, **132**, 80 (1969).
135. V. Du Vigneaud, C. Ressler, and S. Trippitt, *J. Biol. Chem.*, **205**, 949 (1953).
136. J. I. Harris, *Biochem. J.*, **71**, 451 (1959).
137. V. Mutt, S. Magnusson, J. S. Jorpes, and E. Dahl, *Biochemistry*, **4**, 2358 (1965).
138. H. Gregory, P. M. Hardy, D. S. Jones, G. W. Kenner, and R. C. Sheppard, *Nature*, **204**, 931 (1964).
139. V. Erspamer and A. Anastasi, *Experientia*, **18**, 58 (1962).
140. V. Erspamer, A. Anastasi, G. Bertaccini, and J. N. Cei, *Experientia*, **20**, 489 (1966).
141. H. Habermann and J. Jentsch, *Hoppe-Seyler's Z. Physiol. Chem.*, **348**, 37 (1967).
142. L. Laster and J. W. Walsh, *Fed. Proc.*, **27**, 1328 (1968).
143. J. K. McDonald, B. B. Zeitman, Th. J. Reilly, and S. Ellis, *J. Biol. Chem.*, **244**, 2693 (1969).
144. J. R. Brown and B. S. Hartley, *Biochem. J.*, **101**, 214 (1966).
145. B. S. Hartley, J. R. Brown, D. L. Kauffman, and L. B. Smillie, *Nature*, **207**, 1157 (1965).
146. R. L. Hill, *Advan. Protein Chem.*, **20**, 37 (1965).
147. J. T. Potts, *Methods Enzymol.*, **11**, 648 (1967).
148. J. A. Rupley, *Methods Enzymology*, **11**, 905 (1967).
149. D. G. Smyth, *Methods Enzymol.*, **11**, 214, 421 (1967).
150. E. Schröder and K. Lübke, *The Peptides. Methods of Peptide Synthesis*, Academic Press, New York, 1965.
151. D. T. Gish, Peptide Synthesis, in *Protein Sequence Determination. Methods and Techniques* (S. B. Needleman, ed.), Springer-Verlag, New York, 1970, p. 276.
152. F. Weygand and E. Csendes, *Angew. Chem.*, **64**, 136 (1952).
153. R. F. Goldberger and C. B. Anfinsen, *Biochemistry*, **1**, 401 (1962).
154. R. F. Goldberger, *Methods Enzymol.*, **11**, 317 (1967).
155. C. B. Anfinsen and E. Haber, *J. Biol. Chem.*, **236**, 1361 (1961).
156. H. Taniuchi, C. B. Anfinsen, and A. Sodja, *J. Biol. Chem.*, **242**, 4736 (1967).
157. A.F.S.A. Habeeb, *Anal. Biochem.*, **14**, 328 (1966).
158. R. Haynes, D. T. Osuga, and R. E. Feeney, *Biochemistry*, **6**, 541 (1967).
159. G. L. Klippenstein, J. W. Holleman, and I. M. Klotz, *Biochemistry*, **7**, 3868 (1968).
160. J. I. Harris and R. N. Perham, *Nature*, **219**, 1025 (1968).
161. G. Braunitzer, K. Beyrenther, H. Fujiki, and B. Schrank, *Hoppe-Seyler's. Z. Physiol. Chem.*, **349**, 265 (1968).
162. F. D'Angeli, F. Filira, and E. Scoffone, *Tetrahedron Lett.*, **10**, 605 (1965).
163. F. D'Angeli, F. Filira, V. Giormani, and C. Di Bello, *Ric. Sci.*, **36**, 11 (1966).
164. F. D'Angeli, E. Scoffone, F. Filira, and V. Giormani, *Tetrahedron Lett.*, **24**, 2745 (1966).
165. A. Marzotto, P. Pajetta, L. Galzigna, and E. Scoffone, *Biochim. Biophys. Acta*, **154**, 450 (1968).

166. A. Marzotto, P. Pajetta, and E. Scoffone, *Biochem. Biophys. Res. Commun.*, **26**, 517 (1967).

167. F. D'Angeli, V. Giormani, F. Filira, and C. Di Bello, *Biochem. Biophys. Res. Commun.*, **28**, 809 (1967).

168. P. J. G. Butler, J. I. Harris, B. S. Hartley, and R. Leberman, *Biochem. J.*, **103**, 78 (1967).

169. P. J. G. Butler, J. I. Harris, B. S. Hartley, and R. Leberman, *Biochem. J.*, **112**, 679 (1969).

170. M. L. Bender, *J. Amer. Chem. Soc.*, **79**, 1258 (1957).

171. M. L. Bender, F. Chloupek, and M. C. Neven, *J. Amer. Chem. Soc.*, **80**, 5384 (1958).

172. H. B. F. Dixon and R. N. Perham, *Biochem. J.*, **109**, 312 (1968).

173. M. L. Ludwig and R. Byrne, *J. Amer. Chem. Soc.*, **84**, 4160 (1962).

174. M. L. Ludwig and M. J. Hunter, *Methods Enzymol.*, **11**, 595 (1967).

175. W. F. Benisek and F. M. Richards, *J. Biol. Chem.*, **243**, 4267 (1968).

176. A. Dutton, M. Adams, and S. J. Singer, *Biochem. Biophys. Res. Commun.*, **23**, 730 (1966).

177. F. Hartman and F. Wold, *J. Amer. Chem. Soc.*, **88**, 3890 (1966).

178. F. Hartman and F. Wold, *Biochemistry*, **6**, 2439 (1967).

179. S. J. Singer, *Adv. Protein Chem.*, **22**, 1 (1967).

180. J. Leonis and A. L. Levy, *Bull. Soc. Chim. Biol.*, **33**, 779 (1951).

181. J. Leonis and A. L. Levy, *C. R. Trav. Lab. Carlsverg*, *Ser. Chim.*, **29**, 57, 87 (1954).

182. T. C. Merigan, W. J. Dreyer, and A. Berger, *Biochim. Biophys. Acta*, **62**, 122 (1962).

183. C. B. Anfinsen, M. Sela, and H. Tritch, *Arch. Biochem. Biophys.*, **65**, 156 (1956).

184. S. Sakakibara and Y. Shimonishi, *Bull. Chem. Soc. Japan*, **38**, 1412 (1965).

185. D. Levy and F. H. Carpenter, *Biochemistry*, **6**, 3559 (1967).

186. J. N. Williams and R. M. Jacobs, *Biochem. Biophys. Res. Commun.*, **22**, 695 (1966).

187. G. E. Means and R. E. Feeney, *Fed. Proc.*, **26**, 831 (1967).

188. E. Dane, F. Drees, P. Konrad, and T. Dockner, *Angew. Chemie*, **74**, 873 (1962).

189. B. Halpern, *Aust. J. Chem.*, **18**, 417 (1965).

190. A. Matsushima, Y. Hachimori, Y. Inada, and K. Shibata, *J. Biochem.* (Tokyo), **61**, 328 (1966).

191. H. Tamaoki, Y. Murase, S. Minoto, and K. Nakanishi, *J. Biochem.* (Tokyo), **62**, 7 (1967).

192. J. R. Kimmel, *Methods Enzymol.*, **11**, 584 (1967).

193. W. A. Klee and F. M. Richards, *J. Biol. Chem.*, **229**, 489 (1957).

194. W. L. Hughes, H. A. Saroff, and A. L. Carney, *J. Amer. Chem. Soc.*, **71**, 2476 (1949).

195. J. Roche, M. Morgue, and R. Baret, *Bull. Soc. Chim. Biol.*, **36**, 85 (1954).

196. C. H. Chervenka and P. E. Wilcox, *J. Biol. Chem.*, **222**, 635 (1956).

197. G. S. Shields, R. L. Hill, and E. L. Smith, *J. Biol. Chem.*, **234**, 1747 (1959).

198. L. Weil and M. Telka, *Arch. Biochem. Biophys.*, **71**, 473 (1957).

199. E. Schutte, *Z. Physiol. Chem.*, **279**, 52, 59 (1943).
200. A. F. S. A. Habeeb, *Biochim. Biophys. Acta*, **34**, 294 (1959).
201. H. Fasold, F. Turba, and W. Wirsching, *Biochem. Z.*, **335**, 86 (1961).
202. A. F. S. A. Habeeb, *Fed. Proc.*, **26**, 703 (1967).
203. H. A. Saroff and R. L. Evans, *Biochim. Biophys. Acta*, **36**, 511 (1959).
204. J. F. Riordan and B. L. Vallee, *Methods Enzymol.*, **11**, 565 (1967).
205. I. M. Klotz, *Methods Enzymol.*, **11**, 576 (1967).
206. R. R. Redfield and C. B. Anfinsen, *J. Biol. Chem.*, **221**, 385 (1956).
207. R. P. Carty and C. H. W. Hirs, *Fed. Proc.*, **24**, 592 (1965).
208. R. P. Carty and C. H. W. Hirs, *J. Biol. Chem.*, **243**, 5244 5254 (1968).
209. T. Okuyama and K. Satake, *J. Biochem.* (Tokyo), **47**, 454 (1960).
210. K. Satake, T. Take, A. Matsuo, K. Tazaki, and Y. Hiraga, *J. Biochem.* (Tokyo), **60**, 12 (1966).
211. F. Weygand and H. Tschesche, *Hoppe-Seyler's Z. Physiol. Chem.*, **350**, 93 (1969).
212. G. E. Means and E. R. Feeney *Biochemistry* **7**, 2192 (1968).
213. G. E. Means *Dissertation Abstr.*, **29B**, 2329 (1969).
214. M. Gorecki and Y. Shalitin, *Biochem. Biophys. Res. Commun.*, **29**, 189 (1967).
215. H. A. Itano and A. J. Gottlieb, *Biochem. Biophys. Res. Commun.*, **12**, 405 (1963).
216. K. Toi, E. Bynum, E. Norris, and H. A. Itano, *J. Biol. Chem.*, **240**, 3455 (1965).
217. K. Toi, E. Bynum, E. Norris, and H. A. Itano, *J. Biol. Chem.*, **242**, 1036 (1967).
218. H. A. Itano and E. Norris, Abstracts, 7th International Congress of Biochemistry, Tokyo, 1967, p. 595.
219. L. I. Slobin and S. J. Singer, *Fed. Proc.*, **26**, 339 (1967).
220. L. I. Slobin and S. J. Singer, *J. Biol. Chem.*, **243**, 1777 (1968).
221. W.-H. Liu, B. Feinstein, D. T. Osuga, R. Haynes, and R. E. Feeney, *Biochemistry*, **7**, 2886 (1968).
222. J. Lenard and P. M. Gallop, *Anal. Biochem.*, **29**, 203 (1969).
223. S. Yamada and H. A. Itano, *Biochim. Biophys. Acta*, **130**, 538 (1966).
224. J. Yankeelov, M. Kochert, J. Page, and A. Westphal, *Fed. Proc.*, **25**, 590 (1966).
225. J. A. Yankeelov, C. D. Mitchell, and T. H. Crowford, *J. Amer. Chem. Soc.* **90**, 1664 (1968).
226. A. L. Grossberg and D. Pressman, *Biochemistry* **7**, 272 (1968)
227. K. Nakaya, H. Horinishi, and K. Shibata, *J. Biochem.* (Tokyo), **61**, 345 (1967).
228. K. Takahashi, *J. Biol. Chem.*, **243**, 6171 (1968).
229. K. Shibata, Abstracts, 7th International Congress of Biochemistry, Tokyo, 1967, p. 413.
230. K. Hofmann, *Imidazole and Its Derivatives*, Part I (*The Chemistry of Heterocyclic Compounds*, Vol. 6), Interscience, New York, 1953, p. 93.
231. T. P. King, *Biochemistry*, **5**, 3454 (1966).

232. A Signor, G. M. Bonora, L. Biondi, D. Nisato, A. Marzotto, and E. Scoffone, *Biochemistry*, **10**, 2748 (1971); A. Signor et al. (1972), manuscript in preparation.

233. M. A. Raftery and R. D. Cole, *Biochem. Biophys. Res. Commun.*, **10**, 467 (1963).

234. B. V. Plapp, M. A. Raftery, and R. D. Cole, *J. Biol. Chem.*, **242**, 265 (1967).

235. H. Lindley, *Nature*, **178**, 647 (1956).

236. W. A. Schroeder, J. R. Shelton, and B. Robberson, *Biochim. Biophys. Acta*, **147**, 590 (1967).

237. J. R. Guest and C. Yanofsky, *J. Biol. Chem.*, **241**, 1 (1966).

238. H. Matsubara, R. M. Sasaki, and R. K. Chain, *Proc. Nat. Acad. Sci.*, **57**, 439 (1967).

239. M. Tanaka, T. Nakashima, A. Benson, H. Mower, and K. T. Yasunobu, *Biochemistry*, **5**, 1666 (1966).

240. T. Hofmann, D. A. Walsh, D. L. Kaufman, and H. Neurath, *Fed. Proc.*, **22**, 528 (1963).

241. E. Steers, G. R. Craven, C. B. Anfinsen, and J. L. Bethune, *J. Biol. Chem.*, **240**, 2478 (1965).

242. M. A. Raftery and R. D. Cole, *J. Biol. Chem.*, **241**, 3457 (1966).

243. C. M. Tsung and H. Fraenkel-Conrat, *Biochemistry*, **5**, 2061 (1966).

244. B. A. Holmgren, R. N. Perham, and A. Baldesten, *Eur. J. Biochem.*, **5**, 352 (1968).

245. R. D. Cole, *Methods Enzymol.*, **11**, 315 (1967).

246. C. Zioudrou, M. Wilchek, and A. Patchornik, *Biochemistry*, **4**, 1811 (1965).

247. L. Polgar, *Acta Biochim. Biophys. Acad. Sci. Hung.*, **3**, 397 (1968).

248. Y. Shalitin, *Bull. Res. Council Israel* **10A**, 34 (1961).

249. K. Kurihara, H. Horinishi, and K. Shibata, *Biochim. Biophys. Acta*, **74**, 678 (1963).

250. M. J. Gorbunoff, *Biochemistry*, **6**, 1606 (1967).

251. J. F. Riordan, M. Sokolovsky, and B. L. Vallee, *J. Amer. Chem. Soc.*, **88**, 4104 (1966).

252. M. Sokolovsky, J. F. Riordan, and B. L. Vallee, *Biochemistry*, **5**, 3582 (1966).

253. M. Sokolovsky, J. F. Riordan, and B. L. Vallee, *Biochem. Biophys. Res. Commun.*, **27**, 20 (1967).

254. R. A. Kenner, K. A. Walsh, and H. Neurath, *Biochem. Biophys. Res. Commun.*, **33**, 353 (1968).

255. J. E. Folk, J. A. Gladner, and K. Laki, *J. Biol. Chem.*, **234**, 67 (1959).

256. A. Anastasi, V. Erspamer, and R. Endean, *Arch. Biochem. Biophys.*, **125**, 57 (1968).

257. I. M. Chaiken and E. L. Smith, *J. Biol. Chem.*, **244**, 4247 (1969).

258. D. E. Koshland, Y. D. Karkhanis, and H. G. Latlam, *J. Amer. Chem. Soc.*, **86**, 1448 (1964).

259. H. R. Horton and D. E. Koshland, *J. Amer. Chem. Soc.*, **87**, 1126 (1965).

260. T. E. Barman and D. E. Koshland, *J. Biol. Chem.*, **242**, 5771 (1967).

261. T. A. A. Dopheide and W. M. Jones, *J. Biol. Chem.*, **243**, 3906 (1968).

262. T. F. Spande, M. Wilchek, and B. Witkop, *J. Amer. Chem. Soc.*, **90**, 3256 (1968).
263. G. M. London, D. Portsmonth, A. Lukton, and D. E. Koshland, *J. Amer. Chem. Soc.*, **91**, 2792 (1969).
264. B. G. McFarland, Y. Inoue, and K. Nakanishi, *Tetrahedron Lett.*, **1969**, 857,
265. A. Fontana, F. Marchiori, L. Moroder, and E. Scoffone, *Tetrahedron Lett.*, **26**, 2985 (1966).
266. A. Fontana, F. Marchiori, R. Rocchi, and P. Pajetta, *Gazz. Chim. Ital.*, **96**, 1301 (1966).
267. E. Scoffone, A. Fontana, and R. Rocchi, *Biochem. Biophys. Res. Commun.*, **25**, 170 (1966).
268. E. Scoffone, A Fontana, and R. Rocchi, *Biochemistry*, **7**, 971 (1968); A. Fontana (1970), unpublished results.
269. F. M. Veronese, E. Boccu', and A. Fontana, *Ann. Chim.* (Rome), **58**, 1309 (1968); A. Fontana, and E. Scoffone, in *Enzyme Structure* (C. H. W. Hirs and S. N. Timasheff, eds.), *Methods in Enzymology*, **25**, 419 (1972), Academic Press, New York.
270. E. Scoffone, A. Previero, C. A. Benassi, and P. Pajetta, in *Peptides, Proceedings of the 6th European Symposium, Athens, 1963* (L. Zervas, ed.), Pergamon, London, 1966, p. 183.
271. A. Previero, E. Scoffone, C. A. Benassi, and P. Pajetta, *Gazz. Chim. Ital.*, **93**, 849 (1963).
272. A. Previero and E. Bordignon, *Gazz. Chim. Ital.*, **94**, 630 (1964).
273. F. M. Veronese, A. Fontana, E. Boccu', and C. A. Benassi, *Gazz. Chim. Ital.*, **97**, 321 (1967).
274. A. Light and J. Greenberg, *J. Biol. Chem.*, **240**, 258 (1965).
275. W. A. Schroeder, J. R. Shelton, J. B. Shelton, J. Cormick, and R. T. Jones, *Biochemistry*, **2**, 992 (1963).
276. S. R. Himmelhoch and E. A. Peterson, *Biochemistry*, **7**, 2085 (1968).
277. G. Bodo, *Fortschr. Chem. Forsch.*, **6**, 1 (1966).
278. B. Witkop, *Science*, **162**, 318 (1968).
279. C. B. Kasper, in *Protein Sequence Determination. Methods and Techniques* (S. B. Needleman, ed.), Springer-Verlag, New York, 1970.
280. F. H. White and C. B. Anfinsen, *Sulfur in Proteins*, Academic Press, New York, 1959, p. 279.
281. A. M. Crestfield, S. Moore, and W. H. Stein, *J. Biol. Chem.*, **238**, 622 (1963).
282. W. W. Cleland, *Biochemistry*, **3**, 480 (1964).
283. L. Weil and T. S. Seibles, *Arch. Biochem. Biophys.*, **95**, 470 (1961).
284. T. S. Seibles and L. Weil, *Methods Enzymol.*, **11**, 204 (1967).
285. C. H. W. Hirs, *Ann. Rev. Biochem.*, **33**, 597 (1964).
286. C. H. W. Hirs, *J. Biol. Chem.*, **219**, 611 (1956).
287. G. Toennies and R. P. Homiller, *J. Amer. Chem. Soc.*, **64**, 3054 (1942).
288. J. Jollès, P. Jollès, and C. Foromageot, *Biochim. Biophys. Acta*, **27**, 298 (1958).
289. C. A. Benassi, E. Scoffone, and F. M. Veronese, *Tetrahedron Lett.*, **1965**, 4389.

290. R. Cecil and J. R. McPhee, *Advan. Protein Chem.*, **16**, 255 (1959).
291. J. M. Swan, *Nature*, **180**, 643 (1957).
292. W. W.-C. Chan, *Biochemistry*, **7**, 4247 (1968).
293. H. Neuman, R. F. Goldberger, and M. Sela, *J. Biol. Chem.*, **239**, 1536 (1964)
294. H. Neuman, J. Z. Steinberg, and E. Katchalski, *J. Amer. Chem. Soc.*, **87**, 3841 (1965).
295. H. Neuman, J. Z. Steinberg, J. R. Brown, R. F. Goldberger, and M. Sela, *Eur. J. Biochem.*, **3**, 171 (1967).
296. H. Neumann and R. A. Smith, *Arch. Biochem. Biophys.*, **122**, 354 (1968).
297. E. Gross and B. Witkop, *J. Amer. Chem. Soc.*, **83**, 1510 (1961).
298. E. Gross and B. Witkop, *J. Biol. Chem.*, **237**, 1856 (1962).
299. E. Gross, *Methods in Enzymol.*, **11**, 238 (1967).
300. E. Gross and B. Witkop, *Biochemistry*, **6**, 745 (1967).
301. P. Nissley, N. Cittanova, and H. Edelhoch, *Biochemistry*, **8**, 443 (1969).
302. M. J. Waxdal, W. H. Konigsberg, W. L. Henley, and G. M. Edelman, *Biochemistry*, **7**, 1959 (1968).
303. M. J. Waxdal, W. H. Konigsberg, and G. M. Edelman, *Biochemistry*, **7**, 1967 (1968).
304. J.-P. Bargetzi, E. O. P. Thompson, K. S. V. Sampath Kumar, K. A. Walsh, and H. Neurath, *J. Biol. Chem.*, **239**, 3767 (1964).
305. B. A. Cunningham, P. D. Gottlieb, W. H. Konigsberg, and G. M. Edelman, *Biochemistry*, **7**, 1983 (1968).
306. K. Narita and K. Titani, *J. Biochem.* (Tokyo), **63**, 226 (1968).
307. E. Gross, C. H. Plato, J. L. Morell, and B. Witkop, Abstracts, 150th National Meeting of the American Chemical Society, Atlantic City, N.J., 1965, p. 60c.
308. E. Gross, J. L. Morell, and P. Q. Lee, Abstracts, International Congress of Biochemistry (Tokyo), 1967, p. XI-535.
309. W. Awad and P. E. Wilcox, *Biochem. Biophys. Res. Commun.*, **17**, 709 (1964).
310. M. Wilchek and A. Patchornik, *J. Amer. Chem. Soc.*, **84**, 4613 (1962).
311. L. K. Ramachandran, *J. Sci. Ind. Res.*, **21C**, 111 (1962).
312. G. C. Stelakatos, *Chim. Chronika*, **27A**, 107 (1962).
313. B. Witkop and L. K. Ramachandran, *Metabolism*, **13**, 1016 (1964).
314. L. K. Ramachandran and B. Witkop, in *Peptides, Proceedings of the 6th European Symposium, Athens, 1963* (L. Zervas, ed.), Pergamon, London, 1966, p. 165.
315. L. K. Ramachandran and B. Witkop, *Methods Enzymol.*, **11**, 283 (1967).
316. K. Han, M. Dautrevaux, and G. Biserte, *Ann. Pharm. Fr.*, **24**, 649 (1966).
317. T. F. Spande, N. M. Green, and B. Witkop, *Biochemistry*, **5**, 1926 (1966).
318. T. F. Spande and B. Witkop, *Methods Enzymol.*, **11**, 506, 528 (1967).
319. M. J. Kronman, F. M. Robbins, and R. E. Andreotti, *Biochem. Biophys. Acta*, **147**, 462 (1967).
320. G. S. Omenn, A. Fontana, and C. B. Anfinsen, *J. Biol. Chem.*, **245**, 1895 (1970); A. Fontana, in *Enzyme Structure* (C. H. W. Hirs and S. N. Timasheff, eds.), *Methods in Enzymology*, **25**, 419 (1972), Academic Press, New York.
321. E. W. Chappelle and J. M. Luck, *J. Biol. Chem.*, **219**, 171 (1957).

322. A. Patchornik, W. B. Lawson, E. Gross, and B. Witkop, *J. Amer. Chem. Soc.*, **82**, 5923 (1960).

323. M. Funatsu, N. M. Green, and B. Witkop, *J. Amer. Chem. Soc.*, **86**, 1846 (1964).

324. E. H. Eylar and G. A. Hashim, *Arch. Biochem. Biophys.*, **131**, 215 (1969).

325. A. Previero and M. A. Coletti-Previero, *Bull. Soc. Chim. Biol.*, **49**, 1059 (1967).

326. A. Previero, M. A. Coletti-Previero, and C. Axelrud-Cavadore, *Arch. Biochem. Biophys.*, **122**, 434 (1968).

327. M. Wilchek and B. Witkop, *Biochem. Biophys. Res. Commun.*, **26**, 296 (1967).

328. A. Fontana and E. Scoffone, in *Mechanisms of Reactions of Sulfur Compounds*, Vol. 4 (N. Kharasch, ed.), Intra-Science Research Foundation, Santa Monica, Calif., 1970.

329. T. F. Spande and A. Fontana, unpublished results.

330. P. R. Burnett and E. H. Eylar, *J. Biol. Chem.*, **246**, 3425 (1971).

331. A. Patchornik, W. B. Lawson, and B. Witkop, *J. Amer. Chem. Soc.*, **80**, 4747 (1958).

332. A. Patchornik and M. Sokolovsky, *J. Amer. Chem. Soc.*, **86**, 1206 (1964).

333. Y. Degani and A. Patchornik, *Abstracts of the VIIth International Congress of Biochemistry, Tokyo, 1967*, p. 11.

334. W. H. McGregor and F. H. Carpenter, *Biochemistry*, **1**, 53 (1962).

335. D. S. Tarbell and D. P. Harnish, *Chem. Rev.*, **49**, 1 (1951).

336. O. Gawron and G. Odstrchel, *J. Amer. Chem. Soc.*, **89**, 3263 (1967).

337. Z. Bohak, *J. Biol. Chem.*, **239**, 2878 (1964).

338. A. Patchornik and M. Sokolovsky, *Proceedings of the Vth European Peptide Symposium*, Pergamon, Oxford, 1963, p. 253.

339. I. Photaki, *J. Amer. Chem. Soc.*, **85**, 1123 (1963).

340. I. Photaki and V. Bardakos, *J. Amer. Chem. Soc.*, **87**, 3489 (1965).

341. G. Riley, J. Turnbull, and W. Wilson, (1957), *J. Chem. Soc.*, **1957**, 1373.

342. D. H. Strumeyer, W. N. White, and D. E. Koshland, *Proc. Nat. Acad. Sci. U.S.*, **50**, 931 (1963).

343. H. Weiner, W. N. White, D. G. Hoare, and D. E. Koshland, *J. Amer. Chem. Soc.*, **88**, 3851 (1966).

344. H. Weiner, C. W. Batt, and D. E. Koshland, *J. Biol. Chem.*, **241**, 2687 (1966).

345. Z. Bohak and E. Katchalski, *International Symposium on Naturally Occurring Phosphoric Esters*, Newcastle upon Tyne, 1967.

346. C. Zioudrou, M. Wilchek, M. Sokolovsky, and A. Patchornik, *Israel J. Chem.*, **2**, 326 (1964).

347. C. A. Benassi, E. Scoffone, G. Galiazzo, and G. Jori, *Photochem. Photobiol.*, **6**, 857 (1967).

348. G. Galiazzo, G. Jori, and E. Scoffone, *Biochem. Biophys. Res. Commun.*, **31**, 158 (1968).

349. M. Morishita, F. Sakiyama, and K. Narita, *Bull. Chem. Soc.* (Japan) **40**, 433 (1967).

350. F. Sakiyama, M. Morishita, T. Sowa, and K. Narita, *Bull. Chem. Soc. Japan*, **39**, 631 (1966).

351. A. Previero, M. A. Coletti-Previero, and P. Jollès, *Biochim. Biophys. Acta*, **124**, 400 (1966).
352. F. M. Veronese, E. Boccu', C. A. Benassi, and E. Scoffone, *Z. Naturforsch.*, **24 b**, 294 (1969).
353. K. E. Pfitzner and J. G. Moffat, *J. Amer. Chem. Soc.*, **87**, 5661 (1965).
354. C. Di Bello, F. Filira, V. Giormani, and F. D'Angeli, private communication.
355. S. Shaltiel and A. Patchornik, *Bull. Res. Council Israel*, **10A**, 48 (1961).
356. K. Narita, F. Sakiyama, and M. Morishita, Abstracts, 7th International Congress of Biochemistry, Tokyo, 1967, p. III-533.
357. M. Morishita, T. Sowa, F. Sakiyama, and K. Narita, *Bull. Chem. Soc. Japan*, **40**, 632 (1967).
358. H. Junek, K. L. Kirk, and L. A. Cohen, *Biochemistry*, **8**, 1844 (1969).
359. H. Iwasaki, L. A. Cohen, and B. Witkop, *J. Amer. Chem. Soc.*, **85**, 3701 (1963).
360. H. Iwasaki and B. Witkop, *J. Amer. Chem. Soc.*, **86**, 4698 (1964).
361. L. A. Cohen and L. Farber, *Methods in Enzymol.*, **11**, 299 (1967).
362. N. Catsimpoolas and J. L. Wood, *J. Biol. Chem.*, **239**, 4132 (1964).
363. N. Catsimpoolas and J. L. Wood, *J. Biol. Chem.*, **241**, 1790 (1966).
364. J. L. Wood and N. Catsimpoolas, *J. Biol. Chem.*, **238**, PC2887 (1963).
365. S. G. Waley and J. Watson, *Biochem. J.*, **55**, 328 (1953).
366. Y. Levin, A. Berger, and F. Katchalski, *Biochem. J.*, **63**, 308 (1956).
367. P. Edman, *Acta Chem. Scand.*, **4**, 277 (1950).
368. P. Edman, *Acta Chem. Scand.*, **4**, 283 (1950).
369. P. Edman, *Nature*, **177**, 667 (1956).
370. P. Edman, *Acta Chem. Scand.*, **10**, 761 (1956).
371. P. Edman, *Arch. Biochem. Biophys.*, **22**, 475 (1949).
372. P. Edman, in *Protein Sequence Determination. Methods and Techniques* (S. B. Needleman, ed.), Springer-Verlag, New York, 1970.
373. P. Edman and K. Lauber, *Acta Chem. Scand.*, **10**, 466 (1956).
374. D. Bethell, G. E. Metcalfe, and R. C. Sheppard, *Chem. Commun.*, **1965**, 189.
375. D. Ilse and P. Edman, *Aust. J. Chem.*, **16**, 411 (1963).
376. P. Edman and G. Begg, *Eur. J. Biochem.*, **1**, 80 (1967).
377. H. Fraenkel-Conrat, *J. Amer. Chem. Soc.*, **76**, 3606 (1954).
378. W. A. Schroeder, *Methods Enzymol.*, **11**, 445 (1967).
379. W. Koningsberg and R. J. Hill, *J. Biol. Chem.*, **237**, 2547 (1962).
380. W. Koningsberg, *Methods Enzymol.*, **11**, XI, 461 (1967).
381. W. R. Gray and B. S. Hartley, *Biochem. J.*, **89**, 379 (1963).
382. G. A. Mross and R. F. Doolittle, *Arch. Biochem. Biophys.*, **122**, 674 (1967).
383. D. G. Smyth, W. H. Stein, and S. Moore, *J. Biol. Chem.*, **238**, 227 (1963).
384. W. D. John and G. T. Young, *J. Chem. Soc.*, **1954**, 2870.
385. D. L. Swallow and E. P. Abraham, *Biochem. J.*, **70**, 364 (1958).
386. K. L. Weber and W. Konigsberg, *J. Biol. Chem.*, **242**, 3563 (1967).
387. D. S. Smyth and S. Utsumi, *Nature*, **216**, 332 (1967).
388. B. Africa and F. H. Carpenter, *Biochem. Biophys. Res. Commun.*, **24**, 113 (1966).

389. W. A. Landmann, M. P. Drake, and J. Dillaha, *J. Amer. Chem. Soc.*, **75**, 3638 (1953).
390. J. Sjöquist, *Acta Chem. Scand.*, **7**, 447 (1953).
391. E. Cherbuliez, B. Baehler, and J. Rabinowitz, *Helv. Chim. Acta*, **43**, 1871 (1969).
392. J.-O. Jeppsson and J. Sjöquist, *Anal. Biochem.*, **18**, 264 (1967).
393. J. Sjöquist, *Biochim. Biophys. Acta*, **41**, 20 (1960).
394. P. Edman and J. Sjöquist, *Acta Chem. Scand.*, **10**, 1507 (1956).
395. H. Niall and P. Edman, *J. Gen. Physiol.*, **45**, *Suppl.*, 185 (1962).
396. J. Sjöquist, *Biochim. Biophys. Acta*, **16**, 283 (1955).
397. J. Sjöquist, *Arch. Kemi*, **11**, 129 (1957).
398. J. Sjöquist, *Arch. Kemi*, **11**, 151 (1957).
399. J. J. Pisano and Th. J. Bronzert, *J. Biol. Chem.*, **244**, 5597 (1969).
400. R. E. Harman, J. L. Patterson, and W. J. A. Vandenheuvel, *Anal. Biochem.* **25**, 452 (1968).
401. A. Dijkstra, H. A. Billiet, A. H. van Doninck, H. van Velthuyzen, L. Maat, and H. C. Beyermann, *Rec. Trav. Chim. Pays-Bas*, **86**, 65 (1967)
402. N. S. Wulfson, V. M. Stepanov, V. A. Puchlov, and A. M. Zyakoon, *Izv. Akad. Nauk USSR, Ser. Khim.*, **1963**, 1524.
403. F. Weygand, Abstracts of The Chemical Society Anniversary Meetings, Exeter, 1967, p. A6.
404. F. Weygand, *Z. Anal. Chem.*, **243**, 2 (1968).
405. T. F. Richards, W. T. Barnes, R. E. Leonis, R. Salomone, and M. D. Waterfield, *Nature*, **221**, 1241 (1969).
406. R. A. Laursen, *J. Amer. Chem. Soc.*, **88**, 5344 (1966).
407. C. Toniolo, *Tetrahedron*, 26, 5479 (1970).
408. K. D. Kopple and E. Bächli, *J. Org. Chem.*, **24**, 2053 (1959).
409. A. L. Levy, *J. Chem. Soc.*, **1950**, 404.
410. G. W. Kenner and H. G. Khorana, *J. Chem. Soc.*, **1952**, 2076.
411. D. T. Elmore and P. A. Toseland, *J. Chem. Soc.*, **1954**, 4533.
412. D. T. Elmore and P. A. Toseland, *J. Chem. Soc.*, **1957**, 2460.
413. G. C. Barrett, *Chem. Commun.*, **1967**, 487.
414. G. C. Barrett and A. R. Khokhar, *J. Chromatogr.*, **39**, 47 (1969).
415. G. C. Barrett and J. R. Chapman, *Chem. Commun.*, **1968**, 335.
416. A. Signor and E. Bordignon, *Tetrahedron*, **24**, 6995 (1968).
417. D. Sarantakis, J. K. Sutherland, C. Tortorella, and V. Tortorella, *J. Chem. Soc., C*, **1968**, 72.
418. R. B. Merrifield, *J. Amer. Chem. Soc.*, **85**, 2149 (1963).
419. A. Fontana, E. Boccii, and F. M. Veronese, *Z. Naturforsch.*, **26b**, 314 (1971).
420. L. C. Dorman and J. Lowe, *J. Org. Chem.*, **34**, 158 (1969).
421. L. Zervas, D. Borovas, and E. Gazis, *J. Amer. Chem. Soc.*, **85**, 3660 (1963).
422. H. Dintzis, unpublished work, cited by G. R. Stark, *Advan. Protein Chem.* **24**, 261 (1969).
423. G. R. Stark, in *Enzyme Structure* (C. H. W. Hirs and S. N. Timasheff, eds.), *Methods in Enzymology*, **25**, 360 (1972), Academic Press, New York.

424. S. W. Fox, T. L. Hurst, J. F. Griffith, and O. Underwood, *J. Amer. Chem. Soc.*, **77**, 3119 (1955).

425. E. Scoffone and A. Turco, *Ric. Sci.*, **26**, 865 (1956).

426. G. W. Kenner, H. G. Khorana, and R. J. Stedman, *J. Chem. Soc.*, **1953**, 673.

427. H. G. Khorana, *J. Chem. Soc.*, **1952**, 2081.

428. W. R. Gray, *Nature*, **220**, 1300 (1968).

429. F. Weygand, B. Kolb, A. Prox, M. A. Tilak, and I. Tomida, *Hoppe-Seyler's Z. Physiol. Chem.*, **322**, 38 (1960).

430. F. Weygand, A. Prox, H. Fessel, and Kun Sun Kwok, *Z. Naturforsch.*, **206**, 1169 (1965).

431. E. Bayer and W. A. Koenig, *J. Chromatogr. Sci.*, **7**, 95 (1969).

432. B. C. Das, S. D. Gero, and E. Lederer, *Biochem. Biophys. Res. Commun.* **29**, 211 (1967).

433. D. W. Thomas, B. C. Das, D. S. Gero, and E. Lederer, *Biochem. Biophys. Res. Commun.*, **32**, 519 (1968).

434. D. W. Thomas, *Biochem. Biophys. Res. Commun.*, **33**, 483 (1968).

435. D. W. Thomas and T. Ito, *Tetrahedron*, **25**, 1985 (1969).

436. E. Vilkas and E. Lederer, *Tetrahedron Lett.*, **1968**, 3089.

437. K. L. Agarwal, R. A. W. Johnstone, G. W. Kenner, D. S. Millington and R. C. Sheppard, *Nature*, **219**, 498 (1968).

438. M. M. Shemyakin, Y. Ovchinnikov, A. Kiryushkin, *Peptides, Proceedings of the 8th European Peptide Symposium, Noordwijk, 1966*, North-Holland, 1967, p. 155.

439. M. M. Shemyakin, Yu. A. Ovchinnikov, E. J. Vinogradova, M. Yu. Feigina, A. A. Kiryushkin, N. A. Aldanova, Yu. B. Alakhov, V. M. Lipkin, and B. V. Rozinov, *Experientia*, **1967**, 423.

440. E. Lederer, "5th IUPAC Symposium on the Chemistry of Natural Compounds, London, July 1968," *Pure Appl. Chem.*, **17**, 489 (1968).

441. E. Lederer and B. C. Das, *Peptides, Proceedings of the 8th European Peptide Symposium, Noordwijk 1966*, North-Holland, 1967, p. 131.

442. M. Senn and F. W. McLafferty, *Biochem. Biophys. Res. Commun.*, **23**, 381 (1966).

443. M. Senn, R. Venkataraghavan, and F. W. McLafferty, (1966), *J. Amer. Chem. Soc.*, **88**, 5593 (1966).

444. K. Biemann, C. Cone, and B. R. Webster, *J. Amer. Chem. Soc.* **88**, 2597 (1966).

445. M. Barber, P. Powers, M. J. Wallington, and W. A. Wolstenholme, *Nature*, **212**, 784 (1966).

446. M. Barber, P. Jollès, E. Vilkas, and E. Lederer, *Biochem. Biophys. Res. Commun.*, **18**, 469 (1965).

447. G. Laneele, J. Asselinean, W. A. Wolstenholme, and E. Lederer, *Bull. Soc. Chim. Fr.*, (1965) **1965**, 2133.

448. J. H. Jones, *Quart. Rev.*, **21**, 302 (1967).

449. I. Muramatsu and M. Bodansky, *J. Antibiot.* (Tokyo), **21**(1), 68 (1968).

450. C. G. McDonald and J. S. Shannon, *Tetrahedron Lett.*, **1964**, 3113.

451. N. S. Wulfson, V. A. Puchkov, B. V. Rozinov, Yu. V. Denisov, V. N. Bochkarev, M. M. Shemyakin, Yu. A. Ovchinnikov, A. A. Kiryushkin, E. I. Vinogradova, and M. Feigina, *Tetrahedron Lett.*, **1965**, 2805.

452. C. H. Hassell and J. O. Thomas, *Tetrahedron Lett.*, **1966**, 4485.

453. Th. Wieland, G. Luben, H. Ottenheym, J. Faesel, J. X. De Vries, W. Konz, A. Prox, and J. Schmid, *Angew. Chem.*, **6**, 209 (1968).

454. K. L. Agarwal, G. W. Kenner, and R. C. Sheppard, *J. Amer. Chem. Soc.*, **91**, 3096 (1969).

455. K. Heyns and H. F. Grutzmacher, *Fortschr. Chem. Forsch.* **6**, 536 (1966); **68**, 587 (1968).

456. A. Prox and F. Weygand, *Peptides, Proceedings of the 8th European Peptide Symposium, Noordwijk, 1966*, North-Holland, 1967, p. 158.

457. M. M. Shemyakin, "5th IUPAC Symposium on the Chemistry of Natural Compounds, London, July 1968," *Pure Appl. Chem.*, **17**, 313 (1968).

458. G. W. A. Milne, *Quart. Rev.*, **23**, 75 (1969).

459. G. E. Van Lear and F. W. McLafferty, *Ann. Rev. Biochem.*, **38**, 289 (1969).

INDEX